BASIC

Mechanical Engineering

M.P. POONIA
Vice-Chairman
All India Council for Technical Education (AICTE)

S.C. SHARMA
Formerly General Manager,
THDC India Limited

KHANNA BOOK PUBLISHING CO. (P) LTD.

Publisher of Engineering and Computer Books

4C/4344, Ansari Road, Darya Ganj, New Delhi-110002
Phone : 011-23244447-48 **Mobile:** +91-99109 09320
E-mail : contact@khannabooks.com
Website : www.khannabooks.com

Price: ₹ 648.00

ISBN: 978-93-86173-33-1

Basic Mechanical Engineering
by M.P. Poonia & S.C. Sharma

First Edition: 2018
Revised Edition: 2022
Reprint: 2023, **2025**

Published by:
Khanna Book Publishing Co. (P) Ltd.
CIN: U22110DL1998PTC095547

Visit us at: **www.khannabooks.com**
Write us at: **contact@khannabooks.com**

Printed in Bharat.

PREFACE

This book, 'Basic Mechanical Engineering' has been written to provide knowledge and insight into various aspects of mechanical engineering. The book is intended as a text book to be used by the students in the technical institutions i.e. Engineering Colleges and Polytechniques. The book covers the syllabii of various Universities on 'Basic mechanical engineering', 'Elements of mechanical engineering', 'Mechanical engineering', 'Introduction to mechanical engineering', and 'Fundamentals of mechanical engineering' for the students of all the disciplines of engineering.

While writing this book, we have kept in mind the requirements of all the students. Adequate attention has been paid to emphasise on basic principles involved in the subject matter. The explanation in the text has been supported with the line diagrams, alongwith numerous solved problems, where required.

The readers will find the book highly useful as a comprehensive text covering the basic principles in simple language and easy to grasp formating. Even though all efforts have been made to avoid mistakes, yet it would be difficult to claim perfection. Comments and suggestions for any improvements in the book shall be gratefully acknowledged.

Authors are thankful to the publisher for their keen interest in bringing out the book in time.

—Authors

CONTENTS

Chapter – 16 : Hydraulic Devices 469–491

Chapter – 17 : Hydro-Electric Power Plants 492–511

Chapter – 18 : Hydraulic Turbines 512–547

CHAPTER – 19 : SOURCES OF ENERGY 548–562

CHAPTER – 20 : ENGINEERING MATERIALS, STRESS AND STRAIN 563–590

CHAPTER – 21 : MANUFACTURING PROCESSES—MECHANICAL WORKING OF METALS 591–607

CHAPTER – 24 : TOOLS, JIGS, AND FIXTURES 683–693

CHAPTER – 25 : MACHINE TOOLS 694–737

CHAPTER – 26 : AUTOMATION AND ROBOTICS 738–765

AUTOMATION

Chapter 1

MECHANICS—AN INTRODUCTION

Learning Objectives

After studying this chapter, the students will learn:

- About the meaning of mechanics
- Definitions, units, fundamentals and measures of force, weight, mass, pressure, volume, dynamics, velocity, acceleration, torque, work, power, energy, system, equilibrium and friction.
- Motion, newtons laws of motion and motion in two dimensions.
- Simple lifting machines.

Mechanics is an area of Science concerned with the behaviour of physical bodies when subjected to force and displacement. In other words it can also be defined as a branch of science which deals with the motion of and forces on objects.

1.1 FORCE

Force is defined as the product of mass and acceleration. The unit of force is Newton (N)

$$1N = 1 \text{ kg} \times 1 \text{ m/sec}^2$$

Thus, we can say, Newton as the force required for producing an acceleration of 1 m/sec^2 on a mass of one kilogram.

A force is the cause which changes or tends to change the state of rest or of uniform motion of a body. The force may be in the form of tension, pull, weight, gravity, friction, stress, pressure, action or reaction. A force always has a magnitude, direction, point of application and line of action.

Force is a vector quantity and hence can be represented by a straight line. The extremity of a line denotes the point of application of the force. The direction of arrow on the line denotes the direction of the force. The arrow head placed on the line indicates the sense of the force, and the length of the line denotes the magnitude of the force.

When two or more forces act upon a body, it may happen that these forces neutralise the effect of one another and the body continues in its state of rest or of uniform motion in a straight line. Such a system of forces is said to be in equilibrium.

S.I. Unit of Force. One newton, is that force, which, when applied to a body having a mass of one kilogram, gives it an acceleration of one metre per second squared—1 m/s^2. The unit of moment is Newton-metre.

Conventional Unit of Force. One kilogram force, is the force which, when applied to a body having a mass of 1 kg, gives it an acceleration of 9.80665 m/s^2 = standard acceleration of free fall. The unit of moment is kgf-metre.

Weight

The weight of a body is that force which, when applied to the body, would give it an acceleration equal to the local acceleration of force fall. Unit of weight is N.

The quantity defined here is commonly called the local "gravitational" force on the body. It should be noted that the "weight" arises not only from the gravitational forces which exist at the place where the body is, but also from the centrifugal force.

The effect of atmospheric buoyancy being excluded, the weight defined is the weight in vacuum.

Mass

Unit of mass is *kilogram.* 1 kilogram is the mass of the platinum-iridium cylinder deposited at the International Bureau of Weights and Measures and declared international proto-type of the kilogram by the First General Conference of Weights and Measures.

$$\text{Force} = \text{Mass} \times \text{Acceleration}$$

$$\boxed{F = m \times a}$$

m = mass, a = acceleration (in m/s^2), F = force (in kgf or N),

Example 1.1 *Mass of a train is 15,000 kilograms. How much tractive force a locomotive should exert so as to accelerate the train starting from rest to reach a velocity of 60 km/h after 1 minute.*

Solution.

$$\text{Acceleration } \alpha = \frac{V}{t} = \frac{60 \text{ km/h}}{1 \text{ min}}$$

$$= \frac{60000}{3600 \times 60} = \frac{5}{18} \text{ m/s}^2$$

$$\text{Tractive force } F = m \times a$$

$$= \frac{15{,}000}{1} \times \frac{5}{18} = \mathbf{4166.7\ N.\ Ans.}$$

Graphical Representation of Forces

Effect of a force depends upon its magnitude, its direction and its point of application. Forces are represented graphically by pointed arrow.

Point of application of force = head of arrow. (A)

Magnitude of force = Length of arrow. (l)

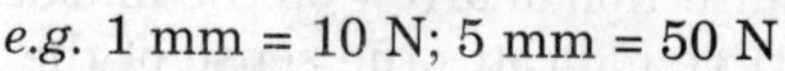

e.g. 1 mm = 10 N; 5 mm = 50 N

Direction of force = Direction of arrow. (OA)

Law of Parallelogram of Forces

When two forces F_1 and F_2 act on a body at an angle, their resultant can be found out by representing the forces graphically and by completing the parallelogram. Length and direction of the diagonal between F_1 and F_2 will represent the magnitude and direction of the resultant.

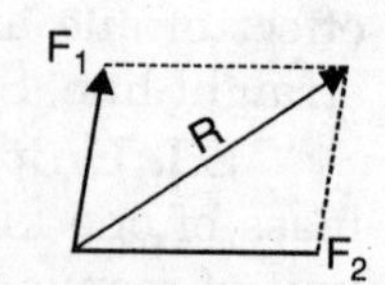

Resolving a Force. By using the parallelogram, a force can be resolved into two (or more) components.

Example (i). If a rope R is to be replaced by two other ropes R_1 and R_2, the forces F_1 and F_2 acting on the individual ropes can be determined by representing graphically the force F in the rope R by the diagonal of the parallelogram and measuring the magnitudes F_1 and F_2 on the adjacent sides of the parallelogram.

Example (ii). If a body weighing 150 N rests on an inclined plane having an inclination of 30°, the component of the weight acting normal to the inclined plane can be found by resolving the force G caused by the weight of 150 N into two components *E* along the plane and *N* normal to the plane by completing the parallelogram having its diagonal *G*.

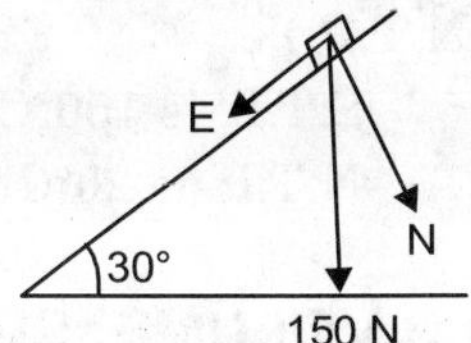

This problem can also be solved by applying trigonometry.

1.2 PRESSURE

Pressure is an effect which occurs when a force is applied on a surface. It is defined as the force per unit area applied to an object in a direction perpendicular to the surface. Mathematically,

$$p = \frac{F}{A}$$

where, p = pressure, F = normal force, and A = Area

Other definitions related to pressure are as follows:

Absolute pressure is the force exerted by a system on a unit area of its boundary.

Gauge pressure is the pressure measured by a pressure gauge above atmospheric pressure.

Vacuum is the pressure measured by pressure gauge below atmospheric pressure.

Atmospheric pressure is the force exerted by air (surrounding the earth) per unit area of surface of earth at sea level. Physical atmosphere is denoted by 'atm' and its value.

= 760 mm of Hg. = 760 Torr

= 1.0332 kgf/cm^2 = 1013223 Pascal $\simeq$ 0.1 MPa

= 1.01325 bar = 14.7 lb/in^2 = 101325 Pascal

Technical atmosphere (used in M.K.S. system) is denoted by 'at' and its value

= 735.5 mm of Hg = 735.5 Torr

= 1 kgf/cm^2 = 14.225 lb/in^2 = 98066.5 Pa $\simeq$ 0.098 MPa

In S.I. unit the pressure is expressed in Newton/m^2 or bar

1 bar = 10^5 Newton/m^2 = 750 Torr = 100000 Pa = 0.1 MPa

= 1.0197 kgf/cm^2 = 14.5038 lb/in^2 = 0.1 MPa

Fig. 1.1 shows the relationship between absolute pressure, gauge pressure, vacuum and atmospheric pressure.

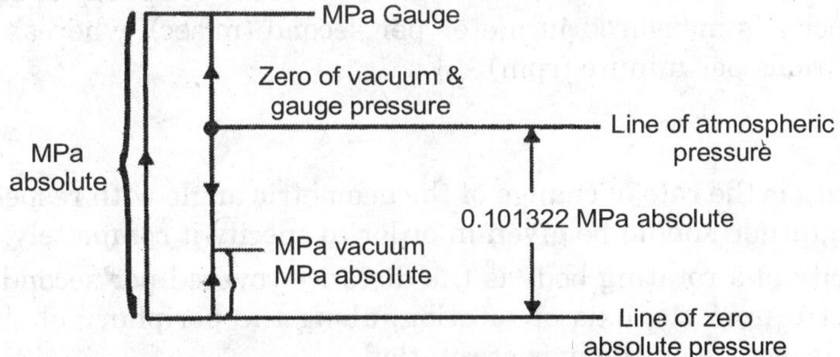

Fig. 1.1 Pressure relationship

Absolute pressure = [(atmospheric pressure) + Gauge pressure – vacuum]

Note. Even when pressure is expressed simply as kgf/cm^2 it implies kgf/cm^2 absolute.

In SI units, 1 Mega Pascal (MPa) = 10^6 Pascal (Pa)

Standard temperature and pressure (S.T.P.) norms temperature and pressure (N.T.P.)

S.T.P. is 150°C temperature and 760 mm of Hg pressure.

N.T.P. is 20°C temperature and 760 mm of Hg pressure.

1.3 VOLUME, SPECIFIC VOLUME AND DENSITY

***(i)* Volume (V)** is the space occupied by a substance. It is measured in cubic metre (m^3).

***(ii)* Specific Volume (*v*)** is the volume occupied by the unit mass of a substance. It is measured in cubic metres per kilogram (m^3/kg).

***(iii)* Density (ρ)** is the mass of unit volume of a substance. It is measured in kilogram per cubic metre (kg/m^3)

The above definitions lead to:

$$v = \frac{V}{m} = \frac{1}{\rho} \quad \text{or} \quad \rho = \frac{m}{V} = \frac{1}{v}.$$

1.4 DYNAMICS

We know that the 'statics' deals with bodies at rest while the dynamics deals with study of bodies in motion. In the study of dynamics, at times the body is treated as a particle and is defined as an object of small dimension in comparison with its range or flight track during motion. Dynamics is generally divided into two main branches as given under :

(i) **Kinematics** is the study of geometry of motion. It is used to relate displacement, velocity, acceleration and time, without reference to the forces causing the motion.

(ii) **Kinetics** is the study of the relation existing between the forces acting on a body, the mass of the body, and the motion of the body. It is used to predict the motion caused by given forces or to determine the forces required to produce a given motion.

1.4.1 Velocity

Velocity is defined as the rate of change of displacement with respect to time. It is a physical vector quantity. Velocity may be either *(i)* linear velocity, or *(ii)* angular velocity. Direction of motion and magnitude should be given in order to specify it completely. In the SI system, linear velocity is measured in meter per second (m/sec); whereas angular velocity is measured in revolutions per minute (rpm).

Angular Velocity

Angular velocity is the rate of change of the geometric angle with respect to time. Direction of rotation and magnitude should be given in order to specify it completely.

Angular velocity of a rotating body is the distance covered per second by a point lying at a distance of 1 metre from the axis of rotation, along the periphery of the circle of 1 metre radius and having its centre on the axis of rotation.

$$\text{Angular velocity } \omega \text{ (rad/s)} = \frac{3.14 \text{ r.p.m.}}{30}$$

$$\boxed{\omega = \frac{\pi \times n}{30} = 0.105 \times n}$$

n = r.p.m., ω = rad/s

ω is approximately = $\frac{1}{10}$ × r.p.m.

$$\text{Angular velocity } \omega \text{ (rad/s)} = \frac{\text{Peripheral speed } v \text{ (m/s)}}{\text{Radius } r\text{(m)}}$$

$$\boxed{\omega = \frac{v}{r}}$$

v = m/s, r = m, ω = rad/s

Uniform Velocity

A body is said to have a uniform velocity if it moves equal distances (s) in equal interval of time. The velocity remains constant. Velocity (v) *is* the distance (s) travelled in unit time (t).

Linear and Curvilinear Motion

$$\text{Velocity} = \frac{\text{Distance covered}}{\text{Time}} \qquad \boxed{v = \frac{s}{t}}$$

$$\text{Distance} = \text{Velocity} \times \text{Time} \qquad \boxed{s = v \times t}$$

$$\text{Time} = \frac{\text{Distance covered}}{\text{Velocity}} \qquad \boxed{t = \frac{s}{v}}$$

Example 1.2 Planing operation

s = Length of workpiece = L = 1500 mm = $\frac{1500}{1000}$ m = 1.5 m

t = Time required per stroke = 5 seconds

then, velocity, $v = \frac{L}{t} = \frac{1.5}{5}$ = **0.3 m/s Ans.**

Rotary Motion

Peripheral speed = Circumference of workpiece multiplied by revolutions per unit time.

$$\boxed{v = \pi d \times n}$$

$$\text{Diameter} = \frac{\text{Peripheral speed}}{\pi \times \text{r.p.m.}}, \text{ i.e. } \boxed{d = \frac{v}{\pi \times n}}$$

$$\text{r.p.m.} = \frac{\text{Peripheral speed}}{\text{Circumference of work}}, \text{ i.e. } \boxed{n = \frac{v}{\pi \times d}}$$

Example 1.3 *n = 750 r.p.m., d = 300 mm = 0.3 m, ω = ?, v = ?*

Solution.

$$\omega = \frac{\pi \times n}{30} = \frac{3.142 \times 750}{30} = 78.5 \text{ rad/s},$$

$$v = \omega \times r = 78.5 \times 0.15 = \mathbf{11.78 \text{ m/s Ans.}}$$

1.4.2 Acceleration

(*i*) *Angular Acceleration* is the rate of change of angular velocity with respect to time.

Units of Angular Acceleration

Radian per second per second (rad/s^2)

Revolution per second per second (revolution/s^2)

Degree per second per second (..... °/s^2)

Conversion factors for these units are same as those for corresponding units of angular velocity.

(*ii*) *Linear Acceleration* is the rate of change of linear velocity vith respect to time.

Units of Linear Velocity

(*a*) In metric system,

Centimetre per second per second (cm/s^2).

(*b*) In British system,

Foot per second per second (ft/s^2).

(*c*) In SI units metre/second/second (m/s^2)

Conversion factors for these units are same as those for corresponding units of linear velocity.

Standard gravity. It is standard acceleration of free fall

$$(g_n) = 9.81 \text{ m/s}^2$$

Uniformly Accelerated Motion

The velocity (v) gradually increases per second at a constant rate of accelevation (a). The acceleration (or retardation) is the gradual rate of increase or decrease of the velocity per unit of time.

Let, a body move from one point to another with a uniform acceleration, whereas,

u = initial velocity,

v = final velocity after time t,

s = distance travelled during time t,

a = acceleration or retardation (also denoted by a).

Then, velocity acquired in t units of time,

$$\boxed{v = u + at} \quad ...(i)$$

Distance travelled in t units of time = Average velocity × Time.

$$S = \left(\frac{v+u}{2}\right) \times t = \left(\frac{u+at+u}{2}\right) \times t$$

$$\boxed{\therefore \quad S = ut + \frac{1}{2}at^2} \quad ...(ii)$$

Now, squaring equation (*i*), we get

$$v^2 = u^2 + 2u\,at + a^2 t^2$$

$$= u^2 + 2a\left(ut + \frac{1}{2}at^2\right)$$

$$\therefore\ v^2 = u^2 + 2as \qquad ...(iii)$$

Vertical Motion Under Gravity

Neglecting resistance of air, if a body dropped from a height will fall freely with a uniform acceleration '*g*'. Then, above equations will be as under :

$$v = u + gt \qquad ...(i)$$

$$s = ut + \frac{1}{2}gt^2 \qquad ...(ii)$$

$$v^2 = u^2 + 2gs \qquad ...(iii)$$

When considering retardation *i.e.*, throwing the body vertically upwards, the values of '*g*' will be negative.

Let a body is projected vertically upward with initial velocity *u*, and let *h* be the height upto which it rises, so that $v = 0$ when the distance travelled is *h*.

Substituting these values, in the formula, $v^2 = u^2 + 2as$:

$$0 = u^2 - 2gh$$

or

$$h = \frac{u^2}{2g} \qquad ...(i)$$

If the body takes '*t*' second to reach the maximum height, substituting the values in the formula, $v = u + at$.

$$0 = u - gt$$

$$t = \frac{u}{g} \qquad ...(ii)$$

Let a body is dropped with initial velocity zero and let *v* be its velocity after falling through distance *h*. Substituting these values in the formula $v^2 = u^2 + 2as$.

$$v^2 = 0 + 2gh$$

or

$$v = \sqrt{2gh} \qquad ...(iii)$$

Example 1.4. *If the velocity increases per second by* 2 *m/s, the acceleration* $a = 2\ m/s^2$. *Find the distance in t seconds.*

If initial velocity $u = 0$,

Velocity after *t* seconds $\boxed{v = at}$ m/s

Acceleration $\boxed{a = \frac{v}{t}}$ m/s²

Distance covered, $s = \frac{v}{2} \times t = \frac{a \times t}{2} \times t = \frac{a}{2} \times t^2 = \frac{2}{2}t^2 = t^2$ metres

Example 1.5 (*i*) *If* $u = 0$, $a = 20\ mm/s^2$, $t = 75\ s$, $v = ?$

$$r = a \times t = 20 \text{ mm/s}^2 \times 75 \text{ s}$$

$$= \frac{20 \times 75\ s}{s^2} = 1500 \text{ mm/s} = 1.5 \text{ m/s}$$

(*ii*) *If* $u = 0$, $v = 8\ m/s$, $t = 4\ s$, $a = ?$, $s = ?$

$$a = \frac{v}{t} = \frac{8 \text{ m/s}}{4\ s} = 2 \text{ m/s}^2 \text{ or } a = 2 \text{ ms}^{-2}$$

$$s = \frac{v}{2} \times t = \frac{4 \text{ m}}{s} \times 4\ s = 16 \text{ m.}$$

1.5 TORQUE (MOMENT OF A FORCE)

A torque is a vector that measures the tendency of a force to rotate an object about some axis. The magnitude of a torque is defined as the product of applied tangential force and the length of the lever arm (radius).

When a force is applied to a body, it has a tendency to turn the body about some point. The turning tendency of a force about a point is called the moment of the force about that point. It is measured by the product of the force and the perpendicular distance of its line of action from the point. The length of the perpendicular from the point on the line of action of the force is called the arm of the moment. The given point is called the "moment centre" or the "Fulcrum."

The moment of a force about a point = Force × Arm of the moment.

Moment of Force (Torque) about a point is defined as the product of the perpendicular distance from that point to the line of action of the force and the force.

Momentum of a particle is defined as the product of its mass and velocity.

Units of Momentum in Metric and British Systems are 'kilogram metre per second (kg m/s)' and 'pound foot per second (Ib ft/s)' respectively. In S.I. units the unit of momentum is N-sec.

1 kilogram metre per second = 7.233 pound foot per second

1 pound foot per second = 0.138 255 kilogram metre per second.

Moment of Momentum (Angular Momentum) of a particle about a point is defined as the product of perpendicular distance from this point to the direction of velocity of the particle and the momentum of the particle.

Units of Moment of Momentum in Metric and British Systems are '(kilogram metre square per second) (kg m^2/s)' and 'Pound foot square per second (Ib ft^2/s)' respectively.

1 kilogram metre square per second = 23.730 4 pound foot square per second.

1 pound foot square per second = 0.042 140 1 kilogram metre squared per second.

Couples

Two equal unlike parallel forces whose lines of action are not the same, are said to form a couple. The perpendicular distance between the lines of action of the forces of the couple is called the "arm of the couple". The product of either force of the couple with the arm of the couple is called the "moment of the couple".

Let each force of the couple be P and the perpendicular distance between their lines of action be p. Then, the moment of the couple = $P \times p$. A couple is therefore denoted by P_p.

1.6 WORK, POWER AND ENERGY

1.6.1 Work

When a body is displaced due to the action of the force, then force is said to do the work. If the force remains constant in magnitude and direction, the work done by it is measured by the product of the force and the displacement in the direction of the force.

In mechanics, work is defined as a product of a force acting on a body and the distance through which the force moves the body along its direction.

$$\text{Work} = \text{Force} \times \text{Distance}$$

Work (W) = Force (F) × Distance covered s (s is measured in the direction of force)

$\boxed{W = F \times s}$ W, the work done is expressed in N-m.

Example 1.6 *When a body weighing 200 N is lifted to a height s = 5 metres or alternatively a force of 200 N applied uniformly to a body displaces it by 5 metres, the work done. Find work done.*

Solution. $W = F \times s = 200\ N \times 5\ m =$ **1000 Nm. Ans.**

The same amount of work is done while lifting a load along an inclined plane, as would have been done in lifting the same load vertically upwards.

Example 1.7 *How much work is done in moving a vehicle weighing 40 kN, through a distance of 0.8 km, along an inclined plane having a gradient 1 : 100 ?*

Solution.

$$F = 40 \text{ kN},\ h = \frac{0.8 \times 1000}{100} = 8 \text{ m}$$

Work done $W = G \times h = 40 \text{ kN} \times 8 \text{ m} =$ **320 kNm. Ans.**

Alternatively. Force F in the upward direction of vehicle along the inclined plane can be found out by using the parallelogram of forces.

G can be resolved into two components G_1 and G_2. $F = G_2$, but acts in opposite direction.

$$\therefore \quad F = G_2 = G \times \sin\alpha = 40 \times 10^3 \times \frac{1}{100} = 400 \text{ N}$$

or $$W = F \times s = 400 \times 800 \text{ m} = \mathbf{32000\ Nm = 320\ kNm.\ Ans.}$$

1.6.2 Power

Power is the rate of doing the work. An agent which performs 550 ft-lb per second is said to be working at the rate of 1 horse power (written as H.P., and termed as British Horse Power).

In S.I. units kilowatt Power P = Work W per unit of time.

If W is in Nm, t in sec, then power is expressed in Nm/s, thus $\boxed{P = \frac{W}{t}}$

If F is in N, s in m, t in sec, then P is expressed in Nm/s, thus $\boxed{P = \frac{F \times s}{t}}$

If F is in N, v in m/s, P is expressed in Nm/sec, thus $\boxed{P = F \times v}$

$$\text{Power } P = \text{Force } F \times \text{Velocity } v.$$

S.I. Unit of Power is Watt (W) or kW = 1000W

$$\text{Watt} = \text{Joule/sec} = 10^7 \text{ erg/sec}$$

Example 1.8 *A load of 200 N lifted vertically upwards to a height h = 5 m in 4 seconds corresponds to a power P = 250 Nm/s.*

Solution.

$$P = \frac{W}{t} = \frac{F \times s}{t} = F \times v$$

$$= \frac{200 \text{ N} \times 5}{4} = \mathbf{250 \text{ Nm/s}} \quad \textbf{Ans.}$$

1 metric horse power = 0.736 kW

$$\boxed{\text{HP} = \frac{F \times v}{75}} \qquad (F \text{ in N})$$

102 N m/s = 1 kW = 1.36 metric H.P.

$$\boxed{\text{kW} \simeq \frac{Fv}{1000}} \qquad (F \text{ in Newtons})$$

Example 1.9 *What should be the rating of a motor for lifting a load of 1200 N to a height of 10 metres in 50 seconds?*

Solution.

$$W = F \times s = 1200 \text{ N} \times 10 \text{ m}$$

$$= 12{,}000 \text{ Nm}$$

$$P = \frac{W}{t} = \frac{12{,}000 \text{ Nm}}{50 \ s} = 240 \text{ Nm/s}$$

$$= \frac{240}{1000} \ k\text{W} = \mathbf{0.24 \text{ kW}}. \quad \textbf{Ans.}$$

1.6.3 Energy

Energy is required for doing any type of work. Energy of a body is the capacity of doing work. The energy is the primary and most universal measure of all kinds of work by human beings and nature. Every thing what happens in the world is the expression of flow of energy is one of the forms.

An energy source may be defined as the one that can provide us with adequate amount of energy in a usable form over a long period of time.

Energy is the physical entity present in a system in different forms, transformable one into the other, of which the most important examples are

(*a*) Mechanical energy

(*b*) Electromagnetic energy

(*c*) Chemical energy

(*d*) Thermal energy

(*e*) Radiant Energy.

Basic Units for Work and Energy are 'Joule' and 'Erg'.

Joule (*J*). It is the amount of work done when the point of application of a force of one newton is displaced through a distance of one metre in the direction of the force.

Erg. It is the amount of work done when the point of application of a force of one dyne is displaced through a distanc of one centimetre in the direction of the force.

$$1 \text{ erg} = 1 \text{ dyne cm} = 10^{-7} \text{ J}$$

Potential and Kinetic Energy

1. Potential Energy. The energy which a system possesses by virtue of its position in a gravity field is called potential energy. It is equal to the work done against the weight in lifting the body from the surface of the earth.

Weight of the body = m.g

where, m = mass of the body, kg

g = acceleration due to standard gravity

= 9.8 m/sec^2.

Therefore, the potential energy (PE) = Work done against gravity

= weight × height raised = m.g.z.

If the system changes height from z_1 to z_2.

$$\Delta PE = m.g.(z_2 - z_1)$$

2. Kinetic Energy (KE). The energy which a system possesses by virtue of its motion is called kinetic energy. It is equal to the work done by a force in moving a body mass through a distance *dx* in the direction of the force, *i.e.*

KE = Work done

Potential energy $\boxed{E_p = F \times h} = m \times g \times h$

where m = Mass. g = Local acceleration of free fall. h = Height (in m),

Then E_p is expressed in N-m.

Kinetic energy $\boxed{E_k = \dfrac{m \times v^2}{2}}$

$$E_k = kg \times \left(\frac{m}{\text{sec}}\right)^2 = \left(kg \times \frac{m}{\text{sec}^2}\right)(m)$$

Then E_k is expressed in Nm.

Example 1.10 *Weight G of drop forge hammer = 200 N, E_k = N-m.*

Height h of fall = 2.5 m, then potential energy

$$E_p = G \times h = 200 \text{ N} \times 2.5 \text{ m} = \mathbf{500 \text{ Nm}}.$$

Kinetic energy attained by the hammer after falling down through the distance of 2.5 m (*i.e.* at the time of striking the piece to be forged) must also be the same as its original potential energy.

Example 1.11 $E_k = \dfrac{mv^2}{2}$ $(\because v^2 = 2gh)$

$= E_k = \dfrac{mgh \times 2}{2}$ (mg = 200 N)

$= \dfrac{200 \times 2.5 \times 2}{2} = 500 \; Nm$

g = Local acceleration of free fall, taken as 9.81 m/s^2.

Efficiency

While transmitting power or changing energy from one form to another, part of the power or energy is lost due to friction, radiation, etc. Hence the output is always less than input.

$$\text{Efficiency } \eta = \frac{\text{Output work}}{\text{Input work}} = \frac{\text{Output power}}{\text{Input power}}$$

$$\text{Percentage } \eta = \frac{\text{Output}}{\text{Input}} \times 100$$

The more η approaches unity, better the machine will be.

Example 1.12 *200 revolutions of the winch handle raise a load of 500 N upto a height of 20 m. Force applied on winch handle is 25 N and radial length of handle is 0.4 m. Find the efficiency.*

Solution. Output work = G × h = 500 N × 20 m = 10,000 Nm

$$\text{Input work} = F \times s = F \times \pi \times d \times 200 \qquad (\text{Distance covered} = \pi dN)$$

$$= 25 \text{ N} \times \pi \times 0.8 \times 200$$

$$= 12{,}560 \text{ Nm}$$

$$\therefore \quad \eta = \frac{\text{Output work}}{\text{Input work}} = \frac{10{,}000 \text{ N}}{12{,}560 \text{ N}} = 0.7961$$

$$= \textbf{79.61\% Ans.}$$

Example 1.13 *Calculate the cutting force required while turning a work piece of 250 mm dia. at 25 r.p.m. using 5 kW drive motor and assuming efficiency of drive as 75%.*

Solution.

$$\eta = \frac{\text{Output power}}{\text{Input power}}$$

$$\text{Output} = \eta \times \text{Input} = 0.75 \times 5 \text{ kW} = 3.75 \text{ kW}$$

$$\text{Output} = F \times v,$$

$$\therefore \quad F = \frac{\text{Output}}{v} = \frac{3.75 \text{ kW}}{\pi dN} \times s$$

$$\therefore \quad F = \frac{3.75 \times 1000 \times \text{m/s}}{\pi \times 0.25 \text{ m} \times \frac{25}{60}} = \textbf{11684 N Ans.}$$

Efficiency of different machines is shown in table 1.1

Table 1.1 Efficiency of Various Machines

S.No.	*Name of the Machine*	*Efficiency*
1.	Electric motors	0.8 to 0.9
2.	I.C. engines	0.25 to 0.30
3.	Steam engines	0.15 to 0.2
4.	Steam turbines	0.18 to 0.22
5.	Water turbines	0.85 to 0.9

1.7 THE SYSTEM

The system is any prescribed and identifiable collection of matter whose behaviour is being investigated. This quantity of mass is enclosed by a boundary and separates it from the *surroundings*. Everything external to the system is the surroundings. Fig. 1.2 shows the usual representation of system, boundary and surroundings.

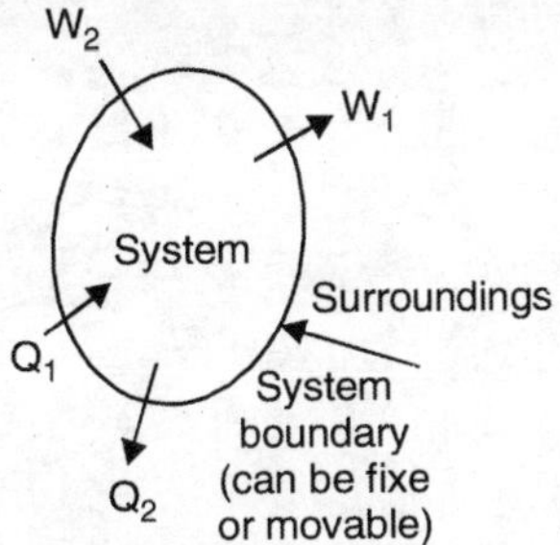

Fig. 1.2 Representation of system, boundary and surroundings

The systems are classified as closed system, open system, adiabatic system and isolated system.

1. Closed System. The boundaries of a system may be flexible and may expand or contract in accordance with the behaviour of the matter keeping the same mass enclosed. If the same mass of matter is used in the investigation the system is called a *closed* system. A closed system permits the passage of energy (heat and work) across it but does not permit the passage of matter; however the boundary itself may move and change the position as and when the system expands or contracts. An example of closed system is a steam power plant consisting of boiler, turbine, condenser and the feed pump (see Fig. 1.3)

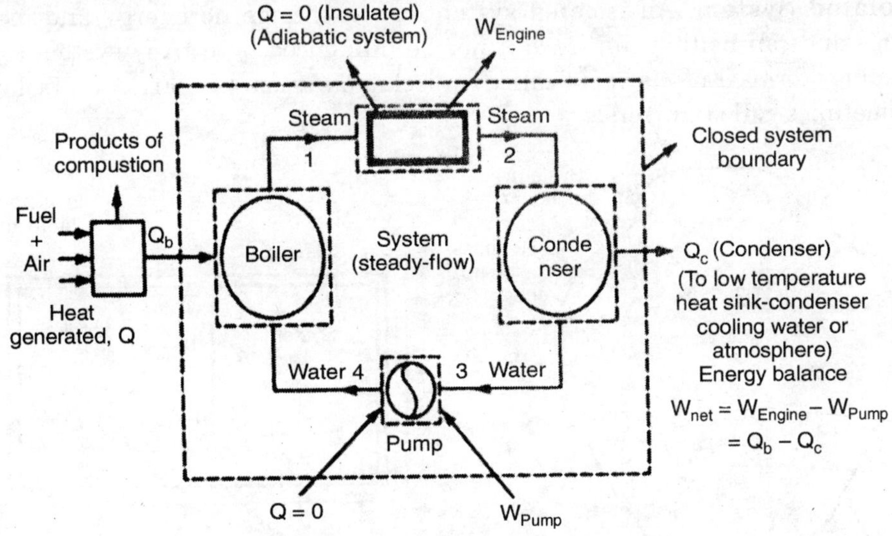

Fig. 1.3 Closed system (steam power plant)

2. Open System. A system which permits the passage of both matter and energy across it, is called an *open system,* (see Fig. 1.4). The mass confined within the boundary may remain constant or may change depending upon the rate at which the matter enters and leaves. The open type of boundary is usually stationary. If the rate of mass and energy transfer with time is constant then the system is known as a *steady flow system.* Other examples of steady flow systems are pumps, compressors, gas turbines and the complete gas turbine plant.

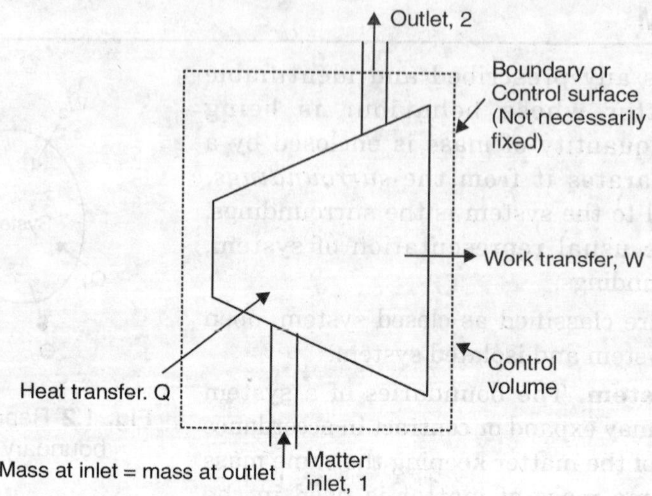

Fig. 1.4 Open system

3. Adiabatic System. An adiabatic system is thermally insulated from its surroundings (*i.e.,* no heat exchange) but exchanges energy in the form of work only.

$$Q = 0\ ;\ W \neq 0$$

Rotary compressors and turbines are the examples of adiabatic systems.

4. Isolated System. An isolated system can exchange no energy and mass with its surroundings and can neither influence it nor be influenced by it. Any system together with its surroundings forms a new system called an isolated system (see Fig. 1.5). Isolated systems are also sometimes called universe.

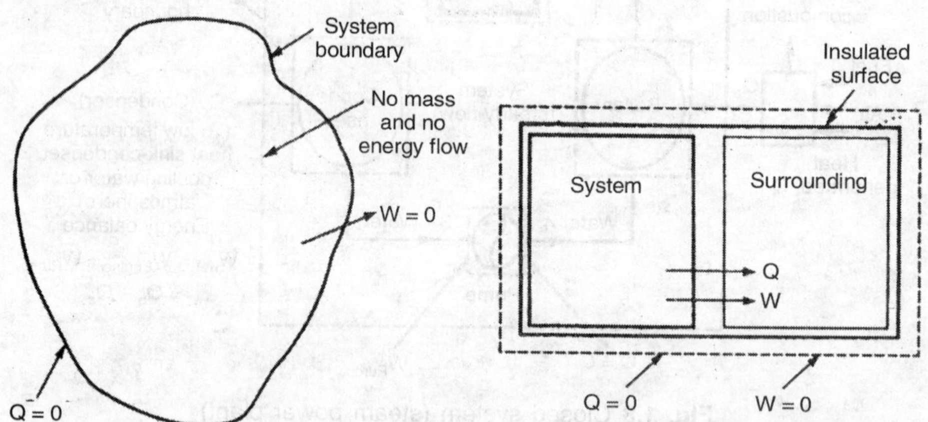

Fig. 1.5 Isolated system

1.8 EQUILIBRIUM

A system is said to be in equilibrium if it does not tend to undergo any further change of its own accord. An isolated system always reaches in course of time a state of equilibrium. Equilibriums are of following types.

1. **Mechanical Equilibrium.** For such systems there should not be any force unbalancing either in the interior of the system or between the system and the surroundings.
2. **Chemical Equilibrium.** For such systems there should be equality of chemical potential *i.e.*, there should not be any chemical reaction.
3. **Thermal Equilibrium.** In this case there should not be any temperature gradient in the system.
4. **Thermodynamic Equilibrium.** When a system satisfies the conditions of mechanical equilibrium, chemical equilibrium and thermal equilibrium, it is said to be in a state of thermodynamic equilibrium.

1.9 FRICTION

The surfaces which appear perfectly smooth to the naked eye, on examination under a microscope are found to consist of raised portion and the depressions. If an attempt is made to move one body over another, a certain resistance to motion is experienced; this resistance which is due to the roughness of the surface is called as "friction" and the force exerted is called the "force of friction".

If a little lubricant is added to the surface, this fills up the microscopic depressions in the apparently smooth surface and thus makes the surface more smooth and reduces the friction between the surfaces in contact.

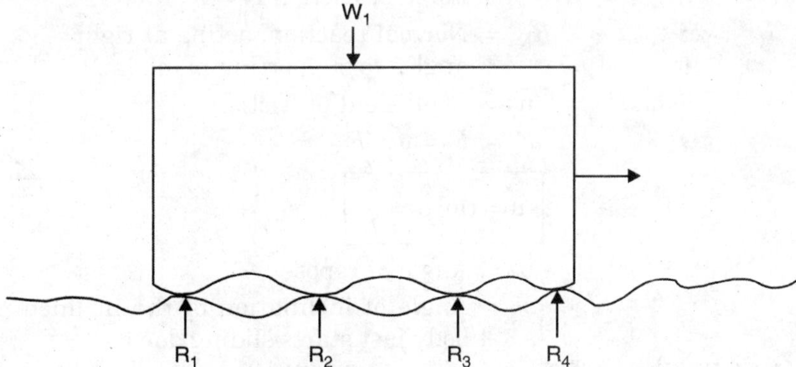

Fig. 1.6 Magnified view of contact surfaces.

Motion of one body relative to another body with which it is in contact is accompanied by a friction force. This force acts tangentially to the surfaces in contact in a direction opposing motion. (see Fig. 1.6).

Value of force, $F = \mu R$,

where R = normal reaction between the surfaces and μ = coefficient of friction.

Value of μ depends on the nature of surfaces which are in contact.

Laws of Dry Friction

Dry friction is the force of friction between unlubricated surfaces. It is governed by the following laws:

1. The direction of the force of friction is opposite to that in which it is moving or has the tendency to move relative to the other.

2. So long as there is no relative motion between the bodies, the force of friction between them is a self-adjusting force, only that much comes into play as is just sufficient to prevent motion.
3. There is a limit to the magnitude of static friction. The maximum value F_s is called limiting static friction.
4. When maximum friction is acting, the angle between the resultant reaction and the normal to the surface is called the angle of friction.
5. The coefficient of friction depends on nature of surfaces and does not depend on area of contact.

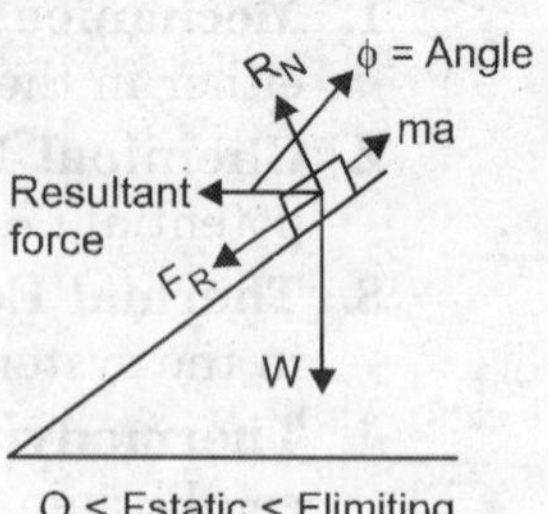

Sliding Friction

The frictional resistance F_f depends upon material of the body and sliding surface, its surface finish and the weight of the sliding body. For a body sliding uniformly on a horizontal plane, the driving force F equals the frictional resistance F_f. The ratio of the frictional force F_f to the normal reaction R_N (*i.e.*, the reaction caused due to the weight of the body at right angles to the supporting plane XX). This ratio is known as "Coefficient of friction (μ)".

$$\mu = \frac{F_f}{R_N}$$

where

F_f = Force of friction N = F

R_N = Normal reaction, acting at right angles to supporting plane in N.

μ = Coefficient of friction.

$F = F_f = \mu \times R_N$

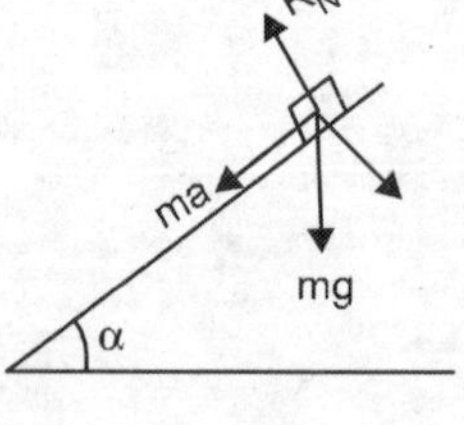

$$\mu = \tan\alpha = \frac{h}{l}$$

where

α = Angle of repose
= Angle of inclination of the inclined plane when the body just starts sliding down.

Example 1.14 *Find the driving force F, which will just make the body weighing 500 N to slide on a horizontal plane, assuming μ = 0.1.*

Solution.

$$F = F_f = \mu \times R_N = 0.1 \times 500 \text{ N}$$
$$= \mathbf{50\ N.}$$ **Ans.**

Rolling Friction

In rolling, the contact of roller with the surface on which the roller rolls is theoretically a point or line contact. However in practice, due to deformation of roller or surface or both, the contact is not a line or point contact.

The weight G will act through the centre of the roller at right angles to contact surface, and will have a lever arm = f, where f is the coefficient of rolling friction (for steel wheel upon steel rail, f = 5 mm).

By Law of Moments,

$$F_f = F = \frac{f}{r} \times R_N = \frac{f}{r} \times G$$

where r = Radius of roller in cm,
F_f = Frictional resistance = Driving force, F
G = Weight = Normal reaction, R_N.

1.10 NEWTON'S LAWS OF MOTION

1. A particle remains at rest or continues to move in a straight line with uniform velocity until acted upon by an external force.
2. Sum of components of external forces ; which act on a particle in a given direction, is equal to the rate of change of momentum in that direction.
3. When particles interact, the forces of action and reaction are equal in magnitude and line of action but opposite in direction.

Centre of Gravity (G)

It is the point at which the distributed weight of a system acts as a single force. Position of centre of gravity is found by analysing or experiment. In the former case, moments are taken about three mutually perpendicular axes giving coordinates of G,

$$\bar{x} = \frac{\Sigma\,\delta_{mg}\,.\,x}{\Sigma\,\delta_{mg}},\ \bar{y} = \frac{\Sigma_{mg}\,.\,y}{\Sigma\,\delta_{mg}} \text{ and } \bar{z} = \frac{\Sigma\,\delta_{mg}\,.\,z}{\Sigma\,\delta_{mg}}$$

where δ_m = an elementary mass at distance x, y or z from the respective axes. Table 1.2 gives some centre of gravity data.

Table 1.2. Centres of Gravity and Moments of Inertia

Figure	G	I
Plane circular sector	$\bar{x} = \frac{2r\sin\theta}{3\theta}$ (Centroid)	$I_{xx} = \frac{r^4}{4}\left(\theta - \frac{\sin 2\theta}{2}\right), I_{yy} = \frac{r^4}{4}\left(\theta - \frac{\sin 2\theta}{2}\right)$ (Second moment of area)
Cylinder, radius, r	—	$I_{xx} = \frac{mr^2}{4} + \frac{ml^2}{12}, I_{zz} = \frac{mr^2}{2}$
Half cylinder, radius r	$\bar{x} = \frac{4r}{3\pi}$	$I_{xx} = I_{yy} = \frac{mr^2}{4} + \frac{ml^2}{12}, I_{zz} = \frac{mr^2}{2}$
Rectangular parallelopiped	—	$I_{xx} = \frac{m}{12}(a^2 + l^2)$
Sphere	—	$I = \frac{2mr^2}{5}$
Hemisphere, radius r	$\bar{x} = \frac{3r}{8}$	$I_{xx} = \frac{2mr^2}{5}$
Circular cone, basic radius r	$\bar{x} = \frac{\ell}{4}$	$I_{xx} = \frac{3mr^2}{10}, I_{yy} = \frac{3mr^2}{20} + \frac{ml^2}{10}$
Slender rod	—	$I_{yy} = \frac{ml^2}{12}$

Moment of Inertia (I) about an axis xx, *i.e.*,

$$I_{xx} = \Sigma\, \delta^2_{mx} = mk^2_{xx},$$

where x = perpendicular distance of the elementary mass m from axis xx and k_{xx} = radius of gyration about axis xx.

Moment of Momentum or Angular Momentum of a Body. It is the moment of linear momentum of body about axis of rotation, 0.

In Fig. 1.7., Angular momentum = $\Sigma\delta_m\, \omega_x \,.\, x = \Sigma\delta\, \omega\, x^2 = \omega\, I_0$. And by parallel axes theorem,

$I_0 = I_G + m \times OG^2$

so that angular momentum = $\omega\,(I_G + M \times OG^2)$

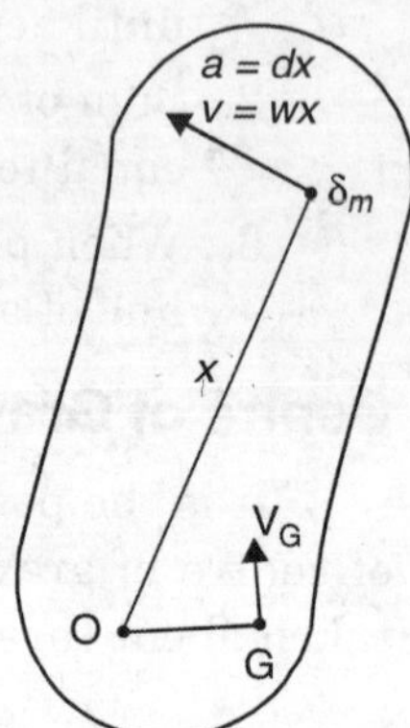

Fig. 1.7

Angular momentum of a system remains constant unless the system is acted on by an external force.

Angular impulse. It is the product of torque by time, *i.e.*

Angular impulse = $Tt = I\alpha\,.t = I(\omega_2 - \omega_1)$.

Angular kinetic energy about an axis

$$= \Sigma\frac{1}{2}\delta m(\omega x^2) = \frac{1}{2}I_0\omega^2$$

But, $I_0 = I_G + m \times OG^2$ so that angular kinetic energy

$$= \frac{1}{2}\omega^2(I_G + m \times OG^2) = \frac{1}{2}I_G\,\omega^2 + \frac{1}{2}\frac{1}{2}mv_a^2$$

Work due to torque. It is the product of torque by angular distance.w = T.θ.

Power due to Torque. It is the rate of work ; $P = \dfrac{T\theta}{t} = T\omega$

1.11 MOTION IN TWO DIMENSIONS

Basic equations of motion contained in above concepts where $v = \dot{x}, a = \ddot{x} = \dot{v}$, $\omega = \theta$, $\alpha = \theta = \omega$.

If formal equations to displacement, velocity or acceleration are available, the above relations may be, integrated or differentiated or graphically if experimental data evaluated. Examination of several cases :

(1) ***Constant Acceleration.***

Linear Motion	Angular Motion
Eqns. (2) below are obtained.	$\theta = \omega_1\, t + \frac{1}{2}\alpha t^2$
Suffix 1 refers to values when $t = 0$.	$\omega_2 = \omega_1 + \alpha t$ ...(1)
Suffix 2 refers to values when $t = t$.	$\omega_2^2 = \omega_1^2 + 2\alpha\theta$

$$\left\{\begin{aligned} x &= v_1 t + \frac{1}{2}at^2 \\ v_2 &= v_1 + at \\ v_2^2 &= v_1^2 + 2ax \end{aligned}\right\} \qquad ...(2)$$

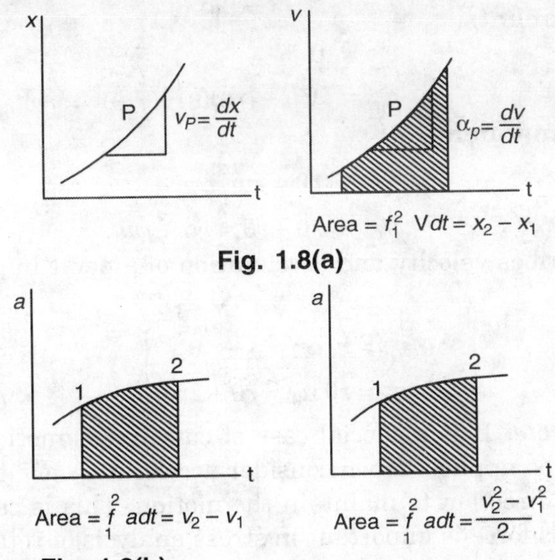

Fig. 1.8(a)

Fig. 1.8(b)

2. *Linear or Angular Motion with Variable Acceleration.* fig 1.8(a) & 1.8(b) indicate relations which apply to linear motion.

3. *Curvilinear Motion.* In this, there are linear as well as angular components, (see Fig. 1.9) may be described by different methods :

V, P, dv, dx, v_p, A, B, x

$AB = V_p \frac{d_v}{d_x} = a_p$

Fig. 1.9

(*i*) Method 1 describes velocity and acceleration of a point by x and y components along two axes at right angles Fig. 1.10 (*a*);

$$v_x = v \cos \theta, \ v_y = v \sin \theta,$$
$$a_x = a \cos \phi, \ a_y = a \sin \phi,$$

(*ii*) Method II describes velocity and acceleration of a point by normal and tangential components at the point Fig. 1.10 (*b*);

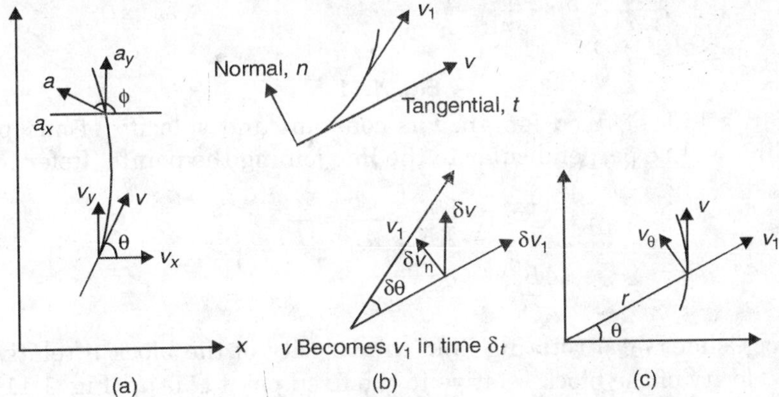

Fig. 1.10

Normal Components

$$v_n = 0$$

$$a_n = V\dot{\theta} = (r\dot{\theta})(\dot{\theta}) = r(\dot{\theta})^2 = rw^2 \qquad ...(3)$$

Tangential Components

$$v_t = r\dot{\theta} = rw$$

$$a_t = r\ddot{\theta} + r\dot{\theta} = r\alpha + \dot{r}w$$

(*iii*) Method III describes velocity and acceleration of a point in polar coordinates r and θ, Fig. 1.10 (*c*) ;

$$\left.\begin{aligned} v_r &= \dot{r},\ ar = r - r\,\dot{\theta}^2 \\ v_\theta &= r\dot{\theta}\ a_\theta = r\ddot{\theta} + 2\dot{r}\,\dot{\theta} \end{aligned}\right\} \qquad ...(4)$$

4. *Motion in a Circle.* It is a special case of curvilinear motion in which r is constant and above methods may be applied. If we consider acceleration $\omega^2 r$ towards the centre, there should be a force in this direction to maintain the motion. This is centripetal force. Outward centrifugal reaction to this force is important in stress analysis of rotating bodies.

5. *Velocity and Acceleration in Mechanisms.* For determining the motion of points in mechanisms, it is usually convenient to draw velocity and acceleration vector diagrams using the principles indicated above.

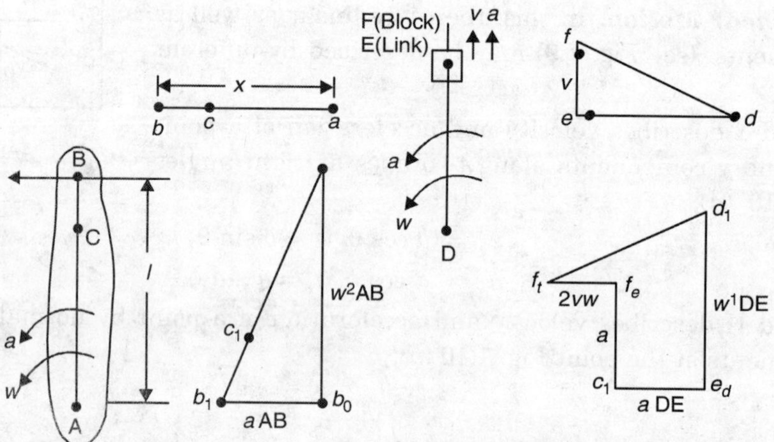

Fig. 1.11

Velocities. In a link of fixed length, r is constant and velocity of one point relative to another on the link will be perpendicular to the line joining the points. Refer to Fig. 1.11 (*a*),

$$\omega = \frac{v_{ba}}{AB} = \frac{ab}{AB} = \frac{x}{l}.$$

Also $$\frac{AB}{AC} = \frac{ac}{ab}.$$

When a block slides on a rotating link, the velocity of the block v relative to the link is along the link. Velocity of the block relative to the fixed point D is df (Fig. 1.11).

Accelerations. In a link of fixed length Fig. 1.11, B will have a centripetal acceleration $\omega^2 AB = a_1 b_a$ towards A and a tangential acceleration $\alpha . AB$. Acceleration of B relative to A is $a_1 b_1$.

When a block slides along a rotating link, it may have a linear acceleration along the link relative to the link. Whether a is + ve, – ve or zero, there will also be a tangential acceleration 2 $v\omega$. This is called "Coriolis acceleration" and its direction is determined by rotating the sliding velocity vector through 90° in the direction of the link angular velocity ω. Fig. 1.11 (*b*) indicates four acceleration components of sliding block *F*.

6. *Gyroscopic Effects.* Fig. 1.12 indicates a rotor with moment of inertia I_{xx} about axis *OX*. Rotor has angular velocity ω_x about *OX* which itself rotates about axis *OY* at right angles to *OX* with angular velocity ω_y.

Due to change (in direction) of angular momentum vector *OP* to *OQ* in time δt, there should be a torque which may be represented in direction by the vector PQ. Magnitude of the torque is *PQ* divided by δ_0. Gyroscopic reaction torque *T* is opposite in sign to this torque and is supported by the bearings of the rotor shaft,

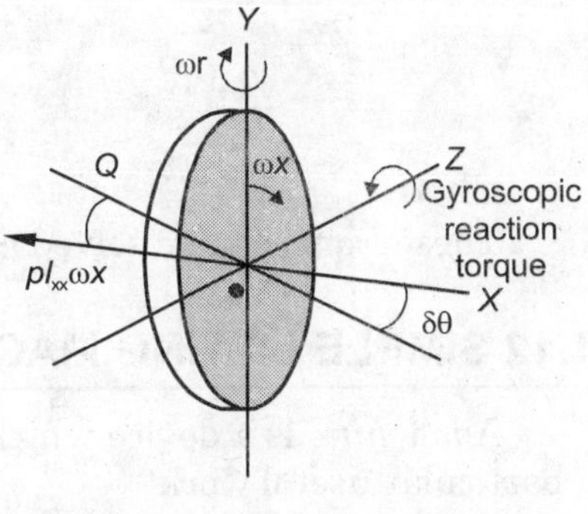

Fig. 1.12 A rotor with moment of inertia I_{xx} about axis *OX*.

$$T = I_{xz}\,\omega_x\,\frac{\delta\theta}{\delta_0}$$

$$= I_{xx}\,\omega_x\,\omega_y \qquad \text{...(5)}$$

Gyroscopic reaction torque occurs in all situations where a rotor is precessed *viz.*, automobile wheels on corners, aircraft turbine rotors in turns or loops, ship turbine rotors in pitch or turns.

Where precession is not *convenient* in *OZX* plane, angular momentum vectors are resolved. Gyroscopic effects are useful in design of missile and spacecrafts and in navigation.

Balancing

Rotating and reciprocating parts of machines at high rotational speeds must be balanced so as to avoid excessive stresses and vibration.

Rotating Masses

Forces required to be balanced are the centrifugal reactions to the circular motion. If all the rotating masses are in same plane (Fig. 1.11), balancing is achieved by placing a single extra mass in the plane. If this is inconvenient, more than one mass is used. These masses are placed at the greatest possible radius so as to reduce their volumes.

Conditions for balancing such two dimensional problems are :

$$\Sigma F_x = \Sigma\, mw^2 r \sin\theta = 0,$$

$$\Sigma F_y = \Sigma\, mw^2 r \cos\theta = 0,$$

where, w = angular velocity of the shaft,

r = radius of the mass *m*,

θ = its angular position relative to the axis *y*.

Axes *x* and *y* are perpendicular in the plane in question. Equations may be solved graphically with a single force polygon.

If all the masses are not in the same plane, the centrifugal reactions will have moments about any plane which should also be balanced. Two masses in different planes are to achieve complete balancing. Convenient planes are chosen for these masses and moments taken to give conditions for balancing ;

$$F_x = \Sigma\, mw^2\, r \sin \theta = 0$$
$$F_y = \Sigma\, mw^2\, r \cos \theta = 0$$
$$M_x = \Sigma\, mw^2\, r \sin \theta\, .\, a = 0$$
$$M_y = \Sigma\, mw^2\, r \cos \theta\, .\, a = 0$$

where a = distance of mass m from the chosen plane.

These Eqns. can be solved graphically by a moment or couple polygon and a force polygon.

1.12 SIMPLE LIFTING MACHINES

A *machine* is a device which receives energy in some available form and uses it for doing a particular useful work.

A *lifting machine* may be defined as a device to overcome a force or load applied to one point by means of another force called effort, applied at another point. In most cases, the effort is smaller than load.

Therefore a machine has following three features:

(*i*) The point of application of effort.

(*ii*) Machanism, which is a working part that makes up a machine.

(*iii*) The point where useful work is done, or load is lifted.

Mechanical Advantage. The ratio of load lifted to the effort applied is known as Mechanical Advantage (MA),

Thus, $$\text{MA} = \frac{\text{Load lifted by a machine (W)}}{\text{Effort applied (P)}}$$

Velocity Ratio. The ratio of the distance moved by the effort to the distance moved by the load is known as Velocity Ratio (V.R.),

Thus, $$\text{V.R.} = \frac{\text{Distance moved by the effort (D)}}{\text{Distance moved by the load (d)}}$$

Work Output. It is the work done by the machine on the load. Therefore, output of a machine = $W \times d$.

Work Input. It is the work done by the effort during the process. Therefore, input of the machine = $P \times D$.

Efficiency. The ratio of the output of a machine to its input is known as its efficiency.

Thus, efficiency $$(\eta) = \frac{W \times d}{P \times D} = \frac{W/P}{D/d} = \frac{M.A.}{V.R.}$$

When the output and the input of a machine are equal, it is said to be the ideal (or perfect) machine. But in practice output is always less than the input because some of the energy is lost in overcoming friction.

QUESTIONS

1.1 Define Force. What is moment of a force?

1.2 Under what conditions forces of different magnitudes produce equal moment with respect to one and the same centre.

1.3 Define Torque and Couples.

1.4 Distinguish between mass and weight.

1.5 What is graphical representation of forces.

1.6 Define dynamics. Explain briefly its main branches.

1.7 Distinguish between linear velocity and angular velocity.

1.8 Define work and friction.

1.9 Distinguish between sliding friction and rolling friction.

1.10 Explain vector and scalar quantities and give two examples of each.

1.11 (*a*) What are the laws of friction?

(*b*) Explain briefly coefficient of friction, angle of friction and angle of repose.

1.12 Distinguish between traction and friction.

1.13 What is a friction? How it affects the performance of a machine?

1.14 (*a*) Lever is the simplest machine. Justify.

(*b*) What are different kinds of levers?

1.15 What do you understand by machine, velocity ratio, mechanical advantage and efficiency of machine ? What is the relationship between them?

1.16 Define the following.

Linear velocity and linear acceleration. Angular velocity and angular acceleration.

1.17 Explain the following.

(*i*) Friction, (*ii*) Centrifugal force, and (*iii*) Coefficient of friction.

1.18 Define the following.

Speed, velocity, uniform velocity, acceleration and retardation.

1.19 Distinguish between angular and linear acceleration.

1.20 Define Work, Power, and energy.

1.21 Explain friction, laws of dry friction.

1.22 Write a short note on equilibrium situation.

1.23 What are the Newton's laws of motion.

1.24 Write short notes on centre of gravity, moment of Inertia.

1.25 Define work, force, energy and power. Write their units in S.I. system of units.

1.26 Define mass, weight, volume, density and specific volume.

1.27 Define and differentiate potential and kinetic energy.

1.28 What is a system? Discuss open and closed systems.

1.29 What is equilibrium? Explain.

1.30 Define and write the unit of pressure. Differetiate between absolute pressure, gauge pressure and vaccum pressure.

Chapter 2

TRANSMISSION OF POWER

Learning Objectives

After studying this chapter, the students will learn:

- Transmission of power and components, like coupling, types of drives and pulleys.
- Transmission of power using belt-drive, rope drives, chain drives and gear drives.
- Types of gears and gear trains.

2.1 INTRODUCTION

The power generated by the Prime Mover is transmitted to the desired machine through a bar capable of turning about its axis is called "Shaft". The portion of a shaft inside bearing is called a "Journal". A. small shaft which is used for light duty is called "Spindle".

A source of power is required to drive machine tools and for manufacturing processes. Now a days electric power is the main source which is converted into rotational energy by means of an electric motor. It is further transmitted to the machine tools which is converted into the required motion by the internal mechanism of the machine.

Power Transmission is the movement of energy from its place of generation to a location where it is applied to perform useful work. Power is defined as unit of energy per unit time. In S.I. units;

$$\text{Watt} = \frac{\text{Joule}}{\text{Second}} = \frac{\text{Newton} \times \text{meter}}{\text{second}}$$

Power transmission has two main functions, (*i*) transmission of power, and (*ii*) changing the speed. It must also serve to bridge the physical gap between the location of the motor shaft and the driven shaft.

The motion or power from a prime mover or electric supply to machine or from one shaft (driver shaft) to the other shaft (driven shaft) may be transmitted by belts, ropes, chains or gears. Belts and ropes are used when the drive is flexible and the two shafts are at a long distance. Belts and ropes are mounted on the pulleys and transmit motion due to friction. Chains and gears are used when the drive is rigid, without any slip and when the two shafts are at a small distance.

2.2 METHODS OF POWER DRIVE

1. Direct Drive. In actual practice, it is not possible to get the length of the shaft more than 10 meters. In such cases shafts are coupled in order to get required lengths for the transmission of power. This is achieved with the help of flanges, sleeves, cotters, keys, pins etc.

A shaft coupling is used for transmitting rotary motion directly from one shaft to another. This method is used where one speed is economical for both the prime mover and the machine e.g., motor-pump set, turbo-generator set etc.

Following two types of couplings are commonly used:

(*a*) Rigid couplings (*b*) Flexible couplings

(*a*) Rigid Couplings. Rigid couplings are those which connect shafts in rigid and in axial alignment. These are further sub-divided as: (*i*) Muff or box coupling, and (*ii*) Flange coupling.

(*b*) Flexible Couplings. These are used to connect two shafts which are not in exact alignment. These couplings absorb sudden shocks of driving shafts and also prevent reversed stresses when one or both shaft deflect under load. When alignment can not be absolutely ensured, or where either the drive or the machine is liable to sudden fluctuations of speed or vibrations, which must not be communicated suddenly to the other machine, a flexible coupling should be used.

2. Non-positive Drive. In this method the drive is liable to slip and the angular velocity ratio of driving and driven member is not constant. These drives depend upon the friction between the driver and driven member on the one hand and the connector between them on the other. Examples of this method are:

1. Belt drives,
2. Rope drives, and
3. Friction drives.

3. Positive Drive. In this method, the transmission of power is not dependent upon friction but transmitted at constant speed without any slippage. Examples of this method are:

1. Gear drives,
2. Chain drives etc.

2.3 JOINTS AND COUPLINGS

As discussed above, we come across a number of situations where we want to increase the length of shafts/rods. The length of rods can be increased by providing certain joints at proper places by means of cotters, jibs, threads, pins etc.

Similarly to overcome difficulties for obtaining suitable lengths of shafts, these are coupled together by certain arrangements, called couplings.

2.3.1 Joints

Some common types of joints are being discussed here as under:

1. Cotter Joints

A cotter is a flat wedge piece of metal usually of rectangular in section. It is uniform in thickness by tapering in width generally in one side. These are usually made of mild steel or wrought iron. The important types of cotter joints are:

1. Sleeve and cotter joint.
2. Spigot and socket joint.

1. Sleeve and Cotter Joint. A sleeve and cotter joint consists of one sleeve, two rods and two cotters. The end of the two rods are inserted from opposite sides in a steel box or sleeve. Cotters are inserted through the slots. Sleeve and rods are subjected to axial forces whereas cotters are subjected to shearing forces. Fig. 2.1 shows details and assembly of a sleeve and cotter joint.

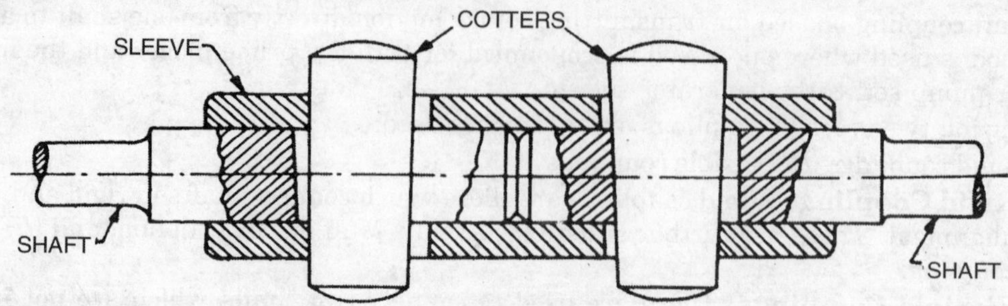

Fig. 2.1 Sleeve and cotter joint.

2. Spigot and Socket Joint. In this joint, a socket is formed by enlarging the end of one rod and in this socket the spigot end of the other rod is fitted. A cotter is driven tightly through the slots (in the rods). (see Fig. 2.2).

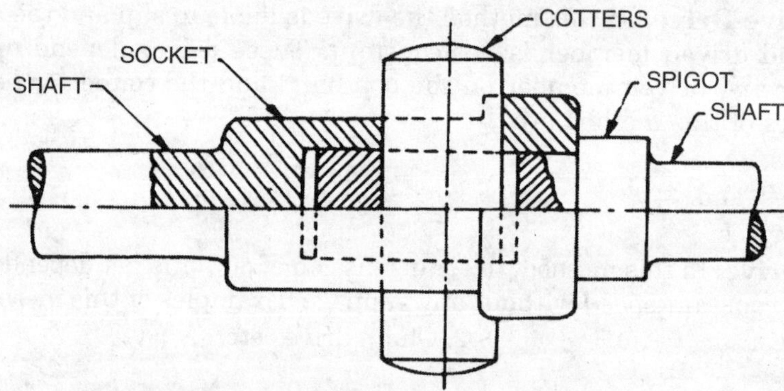

Fig. 2.2 Spigot and socket joint.

2. Adjustable Joints

This type of joint is used to fasten two cylindrical rods axially. It consists of a threaded coupler and two round or cylindrical threaded rods. The threaded rods are screwed to a coupler which has a threaded hole. (see Fig. 2.3).

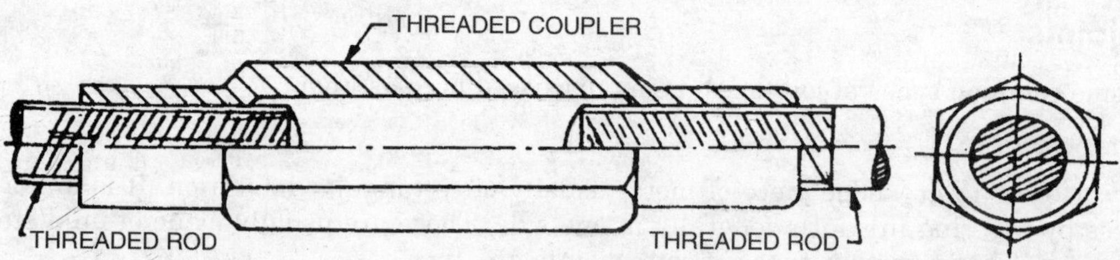

Fig. 2.3 Adjustable joint.

The coupler is hexagonal in shape in the centre and round at both the ends in order to facilitate the cylindrical bars or rods to tighten or loosen with the help of spanner when required.

3. Knuckle or Pin Joint

This joint is used for rods which may not be in alignment but their axes intersect each other. It allows angular movement between the rods and it is not a rigid joint. Two rods are connected by means of a pin and hence the name is given to it as a *pin joint*. It is commonly used when a reciprocating motion is to be converted into a rotary motion or vice versa. The principal characteristics of a pin joint are as follows:

1. It permits considerable angular movement between the rods.
2. It may readily be connected or disconnected.

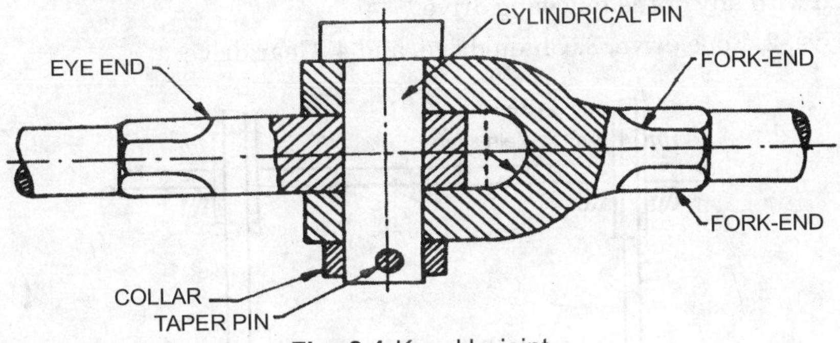

Fig. 2.4 Kunckle joint.

2.3.2 Couplings

In actual practice, it is not possible to get the length of the shaft more than 10 metres. In such cases, shafts are coupled in order to get required lengths for the transmission of power. This is achieved with the help of flanges, sleeves, keys pins etc.

Common types of couplings are:

1. Box or muff coupling.
2. Flange coupling.

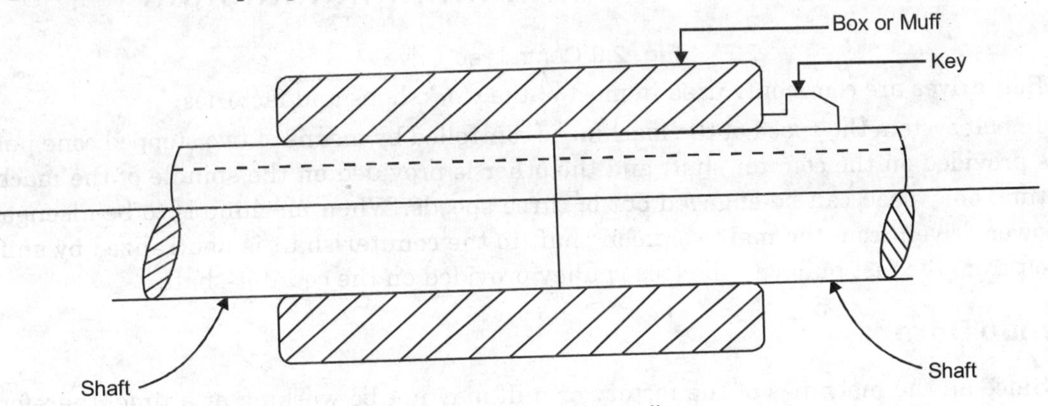

Fig. 2.5 Box or muff coupling.

2.4 TYPES OF DRIVES

Generally power is transmitted by following three kinds of drives:

1. Centralized Drive

This system consists of a single large prime mover which is employed in running all the machines of the factory or workshop. Referring Fig. 2.6, the power from prime mover is transmitted to the main common shaft from where to the counter-shaft and then to the machine spindle. The different Machines in their turn are driven by the main common shaft. Each machine is provided with its own counter-shaft.

The shaft which drives the power is called the Driver or Driving-shaft. The shaft which receives the power from driver is called the Driven-shaft or the Follower shaft. The system can be employed with any of the following drive:

1. Belt drive, 2. Rope drive, 3. Chain drive, and 4. Gear drive.

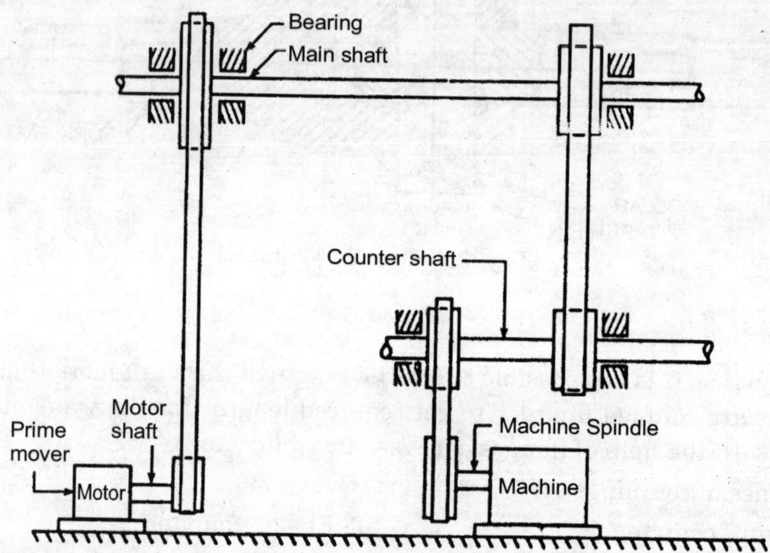

Fig. 2.6 Centralized Drive.

Belt drives are commonly used in most of the workshops and factories.

In belt system the speed of the machine is controlled by means of two stepped cone pulleys one is provided on the counter shaft and the other is provided on the spindle of the machine. At a time one speed can be engaged out of three speeds. When machine is to be disengaged, the power driven from the main common shaft to the counter-shaft is neutralized by shifting the belt from the fast pulley to the loose pulley provided on the counter-shaft.

2. Group Drive

Since all the machines of the factory or mill may not be working at a time, therefore to employ a single large prime mover is simply wastage of power. Under such circumstances the machines of similar category are grouped together and each group is provided with separate prime mover of desired capacity.

In this drive there are many prime movers as there are group of machines. Each group is provided with a common shaft and acts like a small centralised drive. In a group drive the power cost may be high when all the machines are networking. It is however not as high as that in a centralized drive in identical conditions. (see Fig. 2.7).

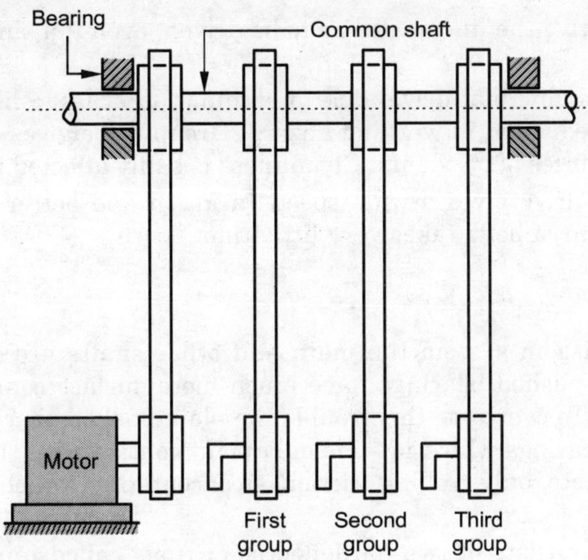

Fig. 2.7 Group drive.

3. Individual Drive

In job order industry generally there are irregular works for which variety of machines are employed for a limited period. Under such circumstances both the centralized drive and the group drive are inconvenient and have more operative cost, therefore another system is employed in which each machine is driven by its own prime mover or motor which is called individual drive. This system makes the best utilization of power and machine hours. Each motor should be of sufficient capacity to carry the full load of the machine. The individual drive omits overhead shafting and belting therefore the system is installed very quickly and economically. This system is very suitable for mobile equipment and for re-arrangement of plant layout. (see Fig. 2.8).

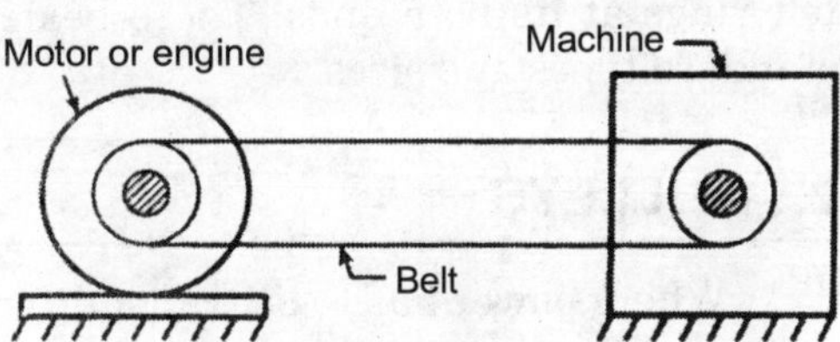

Fig. 2.8 Individual drive.

Group Versus Individual Drive

(*a*) ***Advantages in favour of Group Drive***

(*i*) Cost of motors and control gears required is less in case of group drive. A number of small motors are relatively more costly than those of higher capacity motor e.g. ten 1 kW motors may cost about three times as much as a single 10 KW motor.

(*ii*) The supervision and maintenance of a number of small motors cost more than those of a few large machines.

(*b*) ***Disadvantages of Group Drive***

(*i*) For group drive, in case of motor failure, all the machines become idle, where as in case of individual drive, a motor failure of one machine will cause only that particular machine remain idle and this does not affect the working of other machines.

(*ii*) In group drive the line shaft has to be driven, even if a single machine is required to run.

(*iii*) In case of individual drive, due to elimination of large line shaft and long cross belts etc. reduction in working hazard. Group drive gives clumsy appearance and greater chances of accidents. Cleanliness is badly affected in group drive.

(*iv*) Individual drive gives a wide speed variation and better control of speed range. Replacement of belts takes very little time.

Ball Bearings

In power transmission system the main and other shafts are supported by means of bearings. As the plain-bushed bearings have much more surface contact thereby developing more "sliding friction." To overcome this trouble the plain-bushed bearings have been replaced by the Ball or Roller bearings which provide only point contact with the shaft. Since the balls or rollers can freely rotate, only "rolling friction" is encountered which is small in comparison with sliding friction.

The ball bearing consists of two hardened steel rings called inner and outer race. The inner race is fastened on the rotating shaft and outer race is made rigid with the housing. Hardened steel balls or rollers are located between the ways of the two races. A cage or separator may be used which keeps the rolling members at the proper distance from each other. To prevent the entry of the dust and grit into the bearing, the housing is provided with two oiled felt rings at its both ends. These bearings are manufactured as complete units in a range of standard types and sizes.

2.5 PULLEYS

Where power has to be transmitted between two shafts which are at considerable distance apart, a belt or rope drive is frequently used. In such drives the power transmitted depends upon the friction between the belt or rope and the pulley rim. The cross-section of the pulley rim either may be flat with wedged section or circular section.

Pulleys with flat cross-section of rim are most commonly used. In practice the rim surface is made slightly crowned, it means that the surfaces below the belt is designed slightly convex which prevents the belt running off the pulley. In case of grooved rims the grooves are designed so that the belt or rope does not rest on the bottom of the groove, but is wedged between the sides of the groove.

Pulleys are fixed on the shaft by means of keys. For efficient drive, the ratio between the drive and the driven pulleys should not exceed 6 to 1, otherwise the belt will not grip the smaller pulley sufficiently. Pulleys may be made of cast iron, steel or wood. Hubs of the pulleys may be either solid, split or divided type. In case of split or divided type the system is provided with clamping bolts so that the shaft may be gripped tightly.

For power drives the wooden pulleys are much lighter and superior but they are not suitable in the moistured climate. The steel pulleys made from sheets are light and strong; and having less slippage of power than a cast iron pulley. Its smaller weight reduces the weight imposed on the line shaft thus reduces the frictional losses. With the use of steel pulley we can use maximum safe speed of the transmission system.

2.6 BELT-DRIVES

In power transmission flat and V-belts are normally used. The flat belts are most extensively used in practice and are made of leather, textile, rubber and balata. A high grade belt is made of leather. Textile belts are made of cotton built up of three to ten layers of duck stitched together. Rubber belts, consisting of layers of fabric impregnated with a rubber composition and having a thin layers of rubber on the faces are used where endless is necessary.

Sometimes the belts are also made out of some proprietory brands consisting of layers of camel woven hair or cotton cemented together and impregnated with balata and other gums. The two ends of the belt may be fastened together by cementing, wire lacing, rawhide lacing, metal hinges or metal clamps etc.

The V-belts are moulded of rubberised cotton fabric and cord in trapezoidal section and are used with grooved pulleys. The V-belt drive is more compact, positive, quiet and shock absorbing, and can transmit larger velocity ratios, but is comparatively costlier.

Types of Belts

Belts may be classified in following ways:

1. According to Shape

Belts may be (*i*) Flat belts, (*ii*) V-belt, or (*iii*) Circular belt or rope.

(*i*) *Flats belts* are used mainly where a moderate amount of power to be transmitted and two pulleys are not more than 10 meters apart (*ii*) *V-belts* are used where a great amount of power is to be transmitted. These belts are made of fabrics impregnated and covered with rubber and moulded to a trapezoidal shape. (*iii*) Circular ropes made of manila or hamper cotton are used for transmitting power. These are used for moderately long distance usually more than 10 meters and generally upto 24 to 30 meters. For transmitting large powers, a number of ropes of small diameters are preferred over single rope of large diameter. Steel or wire rope are used for transmission of power in cases where the distance between the shafts is large and extra strength is required.

2. According to Materials

Belts used for power transmission are made of: (*i*) Leather, (*ii*) Cotton or Fabric, (*iii*) Rubber, (*iv*) Balata (a gummy substance like rubber)

Arrangements of Belts

(*i*) Open Belt Drive. This system is employed when both the driving and driven shafts are parallel; and the power is transmitted in the same direction. Referring Fig. 2.9, one half of the belt engages the top part of the driver and driven pulleys; and the other half engages the bottom parts. Driver pulley moves in the clockwise direction, the driven pulley has also the same direction. The direction of the pulleys and the belt is so arranged that the slack side is above and the tight side is below the pulleys by which the natural sag aids belt-wrapping round the pulleys. This reduces slippage and increase the efficiency of the power transmission system.

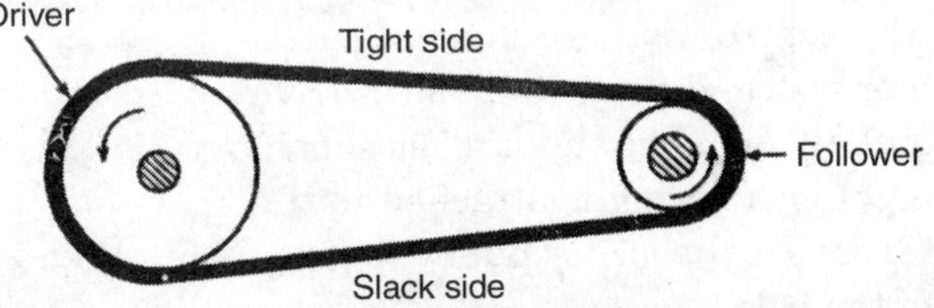

Fig. 2.9 Open belt drive.

***(ii)* Crossed Belt Drive.** This system is also employed when the shafts are in parallel positions and the power is driven in the reversed direction. Referring Fig. 2.10, the driver pulley moves in the clockwise direction and the driven pulley moves in the anti-clock-wise direction or *vice versa*. Belt is engaged on the pulleys as shown in Fig. 2.10, so that the natural sag of the belt helps it in wrapping round the pulleys.

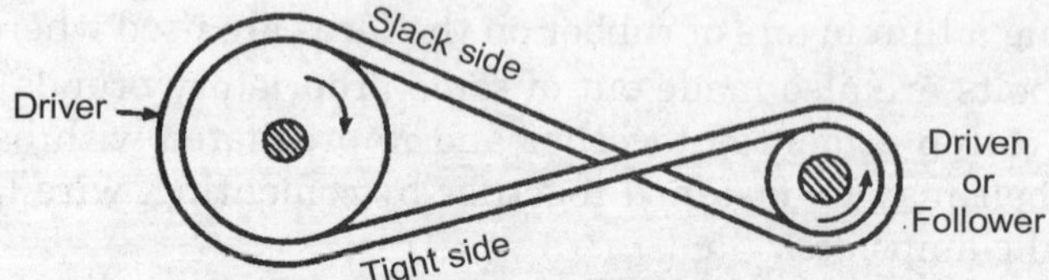

Fig. 2.10 Crossed belt drive.

***(iii)* Angular Belt Drive.** Sometimes power is transmitted when the shafts are arranged at an angle instead of parallel position. Drive under this condition is called Angular belt drive. In an angular drive the planes of the pulleys do not lie in the same plane but meet at an angle. Therefore the belt as it leaves, each pulley should lie in the plane of the other pulley to which the belt is about to move. (see Fig. 2.11).

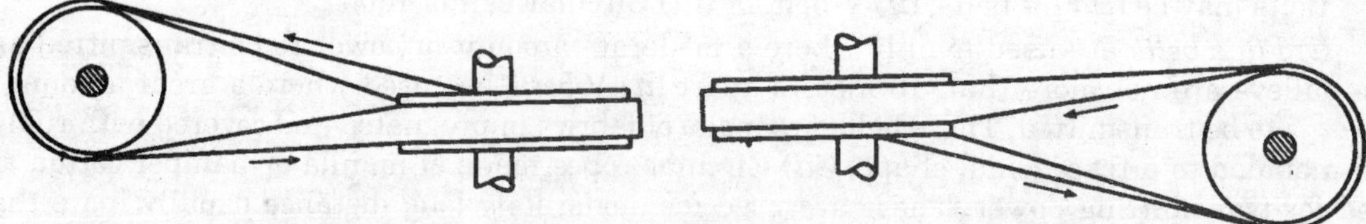

Fig. 2.11 Angular belt drive.

***(iv)* Guide and Jockey Pulleys.** When direct drive from a shaft is not possible, and power is to be transmitted in altered direction, then small pulleys are used to serve the purpose as shown in Fig. 2.12, such pulleys are called "Guide pulleys". Many similar arrangements according to the positions of the shaft may be made to transmit the power.

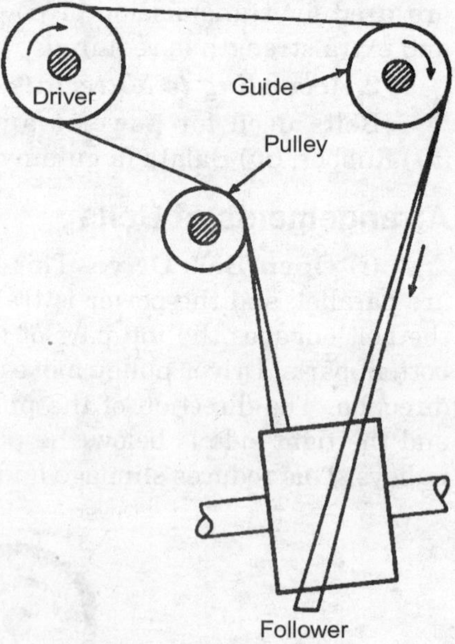

Fig. 2.12 Use of guide pulleys.

In short belt drives, where the velocity ratio between the pulleys is high, the maximum power which can be transmitted by the belt is much smaller than for a long drive operating under otherwise identical conditions. The reasons are that the arc of contact on the smaller pulley is reduced, the tightening effect of the catenary path followed by the free positions of belt between the pulleys is less marked, when the centre distance is small. It is however, possible to improve conditions in a short drive by providing a "jockey pulley" or "idler support" as shown in Fig. 2.13. A dead weight or a spring is attached to the arm in such a way that it causes the jockey pulley to press against the slack side of the belt.

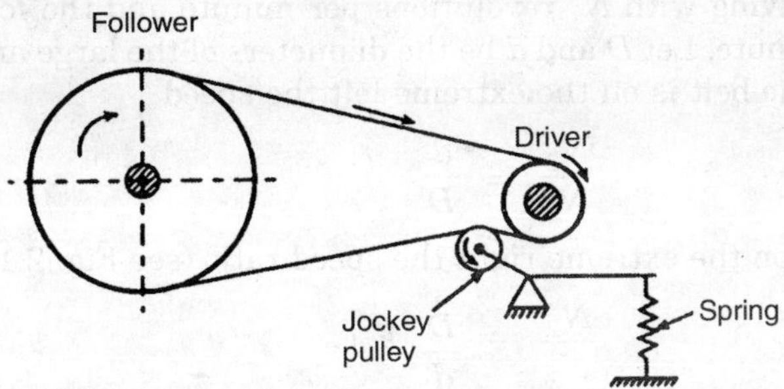

Fig. 2.13 Use of jockey pulley.

***(v)* Drive through Step Pulleys.** Step pulleys are cast iron pulleys having several steps of different diameters on which a belt may run. They are used for varying the velocity ratio between a pair of parallel shafts by simply shifting the belt from one step of the pulley to the other. They work in pairs, one on the counter-shaft and the other on the machine spindle as shown in Fig. 2.14. The diameters of the corresponding steps must be such that the same belt can operate the different pairs of steps.

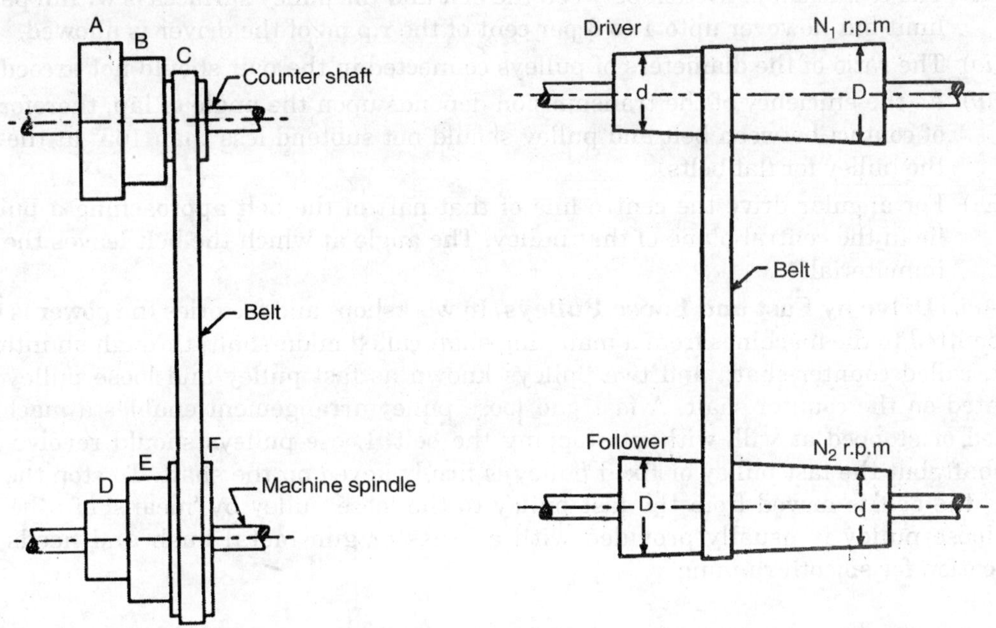

Fig. 2.14 Step pulleys. **Fig. 2.15** Speed cones.

The relation between the speeds and diameters of the each pair of steps is in the usual manner.

$$\frac{\text{Speed of driver in r.p.m.}}{\text{Speed of follower in r.p.m.}} = \frac{\text{Diameter of follower}}{\text{Diameter of driver}}$$

***(vi)* Speed Cones.** The speed cones resemble truncated cones which are used like step pulleys. In case of step pulleys, the variation of speeds are regular and not gradual. Whenever it is desired to have gradual variation of speed, speed cones can be used as shown in Fig. 2.15.

If driver is revolving with N_1 revolutions per minute and the follower is to revolve with N_2 revolutions per minute. Let D and d be the diameters of the large and small end respectively of the cones. When the belt is on the extreme left the speed

$$\frac{N_2}{N_1} = \frac{d}{D} \qquad ...(i)$$

and when the belt is on the extreme right the speed ratio (see Fig. 2.15).

$$\frac{N_2}{N_1} = \frac{D}{d}$$

By adjustment of the position of the belt any speed ratio between (*i*) and (*ii*) can be obtained.

Conditions of Belt Drive

(*i*) When the distance between shafts is within permissible limits. Belt can transmit however, upto a distance of about 10 m with a maximum surface speed of 25 m per second.

(*ii*) The drive depends upon the friction between the belt and pulleys. Therefore it is used where exact velocity ratio between the driver and the follower shaft is not required.

(*iii*) The coefficient of friction between the belt and the pulley surfaces is within permissible limit but however upto 1 to 3 per cent of the r.p.m. of the driver is allowed.

(*iv*) The ratio of the diameters of pulleys connected in the pair should not exceed 6 : 1.

(*v*) As the efficiency of the transmission depends upon the angle of lap, therefore the arc of contact between belt and pulley should not subtend less than 165° at the centre of the pulley for flat belts.

(*vi*) For angular drive the centre line of that part of the belt approaching a pulley must lie in the central plane of that pulley. The angle at which the belt leaves the pulley is immaterial.

(*vii*) **Drive by Fast and Loose Pulleys.** In workshops and factories the power is generally transmitted to the machines from a main line-shaft called main-shaft, through an intermediate shaft, called counter-shaft, and two pulleys known as fast pulley and loose pulley that are mounted on the counter-shaft. A fast and loose pulley arrangement enables a machine to be started or stopped at will, without stopping the belt. Loose pulleys should revolve freely on the shaft, but the fast pulley or fixed pulley is firmly keyed on the shaft. To stop the machine shaft, the belt is moved from the fast pulley to the loose pulley by 'means' of a belt shifter. The loose pulley is usually provided with a brass or gun metal bush and needs efficient lubrication for smooth running.

2.7 POWER TRANSMITTED BY A BELT

Let T_1 be the tension in the tight side of the belt and let T_2 be the tension in the slack side. The effective turning force at the circumference of the follower is the difference of the tension in the tight and slack sides of the belt. T_1 being greater than T_2, the difference of tensions is $(T_1 - T_2)$. Therefore effective pull of the belt is equal to T.

or

$$T_1 - T_2 = T$$

This force T is called the driving force or driving tension. If V is the linear velocity in metre per second, and if T *is* the driving force in N. Then

Work transmitted per second = $T \times V$ Nm.

$$\therefore \quad \text{Power in kW} = \frac{T\omega}{1000}$$

If the belt passes over a pulley which makes N r.p.m., and which has a diameter D is metre, then

$$\text{Speed of-the belt } w = \frac{\pi DN}{60} \text{ metre per second}$$

$$\therefore \quad \text{Power P} = \frac{\pi DNT}{60 \times 1000}$$

or

$$\text{P} = \frac{(T_1 - T_2)\,V}{1000} \quad \text{(where } V \text{ is in metre per sec.)}$$

2.8 VELOCITY RATIO OF A BELT DRIVE

The velocity ratio of a belt drive is defined as the ratio of the number of turns of the driving pulley to the number of turns of the driven pulley in a unit of time.

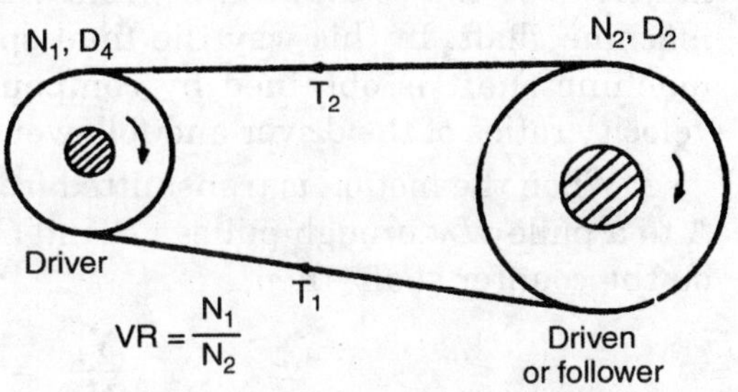

Referring Fig. 2.16, let D_1 and D_2 be diameters of the driver and follower respectively, and let N_1 and N_2 denote their speeds in r.p.m.

Therefore surface of the driver = $\pi D_1 N_1$,

and surface speed of the follower = $\pi D_2 N_2$..

If there is no slip between the belt and the pulley then both the speed must be equal, *i.e.*,

$$\pi D_1 N_1 = \pi D_2 N_2 \quad ...(i)$$

or

$$D_1 N_1 = D_2 N_2 \text{ or } \frac{N_1}{N_2} = \frac{D_2}{D_1}$$

$$\therefore \quad \text{Velocity ratio} = \frac{N_1}{N_2} = \frac{D_2}{D_1}. \quad ...(ii)$$

N₁, D₄
N₂, D₂
T₂
Driver
T₁
VR = N₁/N₂
Driven
or follower

Fig. 2.16 Velocity ratio.

The equation (*ii*) is true whether the belt is open or crossed. With an open belt the sense of rotation of the two pulleys is same, while with a crossed belt this is opposite.

If t is the thickness of the belt, then

$$\text{Velocity ratio} = = \frac{D_2 + t}{D_1 + t} \quad ...(iii)$$

Belt Slip

The driver carries the belt along its periphery because of friction between two. The same makes the belt to drive the follower as well. Thus as soon as the force to be transmitted becomes greater than the frictional force, the belt starts slipping over the pulley rim so that the power transmitted is less. In practice the slip occurs under all conditions. It can be reduced but cannot be avoided completely. This decreases the efficiency of the drive.

$$\frac{N_2}{N_1} = \frac{D_1}{D_2}\left(\frac{100 - S_1}{100}\right)\left(\frac{100 - S_2}{100}\right)$$

or $$\frac{N_2}{N_1} = \frac{D_1}{D_2}\left(\frac{100 - S}{100}\right)$$

where, S_1 = Percentage slip between the driver pulley and the belt,

S_2 = Percentage slip between the belt and the driver pulley.

S = Total percentage slip between the driver and the follower.

Taking into consideration the belt thickness also, then equation becomes

$$\frac{N_2}{N_1} = \frac{(D_1 + t)}{(D_2 + t)} \frac{(100 - S)}{100}$$

Counter-shaft Drives

Referring Fig. 2.17 the power from main shaft is driven to the counter shaft from which to the machine shaft. In this way the final speed in the machine shaft is obtained by compounding the velocity ratios of the driver and follower pairs.

When the motion is transmitted from a pulley A to a pulley D through pulleys B and C, mounted on the counter shaft, then,

$$\frac{N_4}{N_1} = \frac{D_1}{D_2} \times \frac{D_3}{D_4}$$

Here D_1, D_2, D_3 and D_4 are the diameters of the pulley A, B, C and D respectively, and N_1 and N_4 is the speed of A and D respectively.

Therefore,

$$\frac{\text{Velocity of the last shaft}}{\text{Velocity of the first shaft}} = \frac{\text{Dia. of the drivers multiplied}}{\text{Dia. of the followers multiplied}}$$

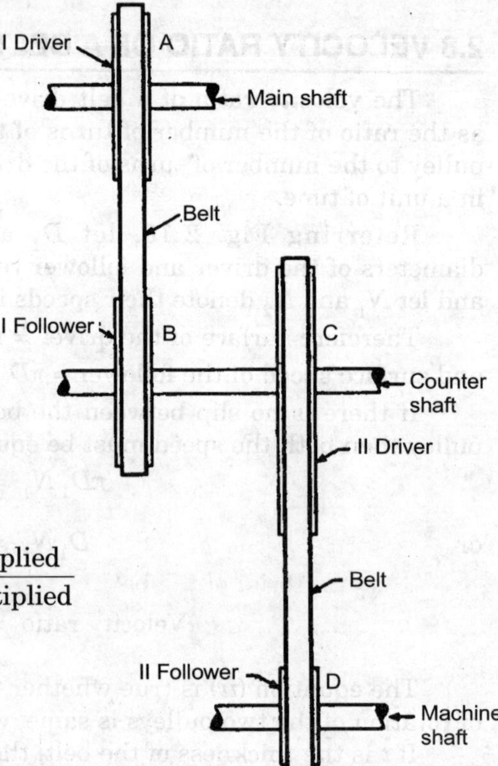

Fig. 2.17 Counter-shaft drive.

Ratio of Driving Tensions

The ratio of driving tensions in a belt just on the point of the slipping is given by

$$\frac{T_1}{T_2} = e^{\mu\theta}$$

This relation is also known as Euler's Formula for Belt Drive, where e is the base of the Napierian log = 2.718, μ = co-efficient of friction for the belt on the pulley, and θ = angle of lap or arc of circumference embraced by the belt in radians.

Taking log on both sides, the above equation becomes

$$\log \frac{T_1}{T_2} = \mu\theta \log e$$

$$= 0.434\ \mu\theta, \text{ (in circular measure)}$$

$$= 0.007578\ \mu\theta, \text{ (if } \theta \text{ is in degrees).}$$

Here the **"angle of lap"** is the angle subtended at the centre of a pulley by that arc of its circumference which is in actual contact with the belt as shown in Fig. 2.18.

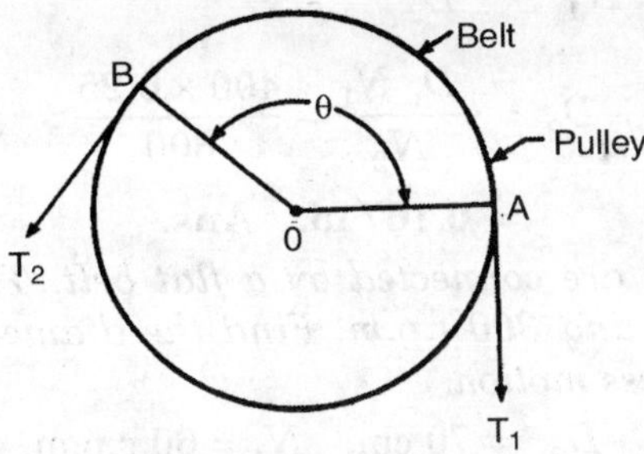

Fig. 2.18 Angle of lap.

2.9 CARE OF BELTS

Driving surface of the belt should be kept clean and free from any dirt and other foreign matter. Such dirt, if allowed to get into the surface, forms into lumps, tearing or distorting the driving face of the belt and preventing it from forming proper contact with the face of the pulley.

Care should be taken to prevent oil or grease from getting at the belt. It will cause loss of power by slip; and is particularly harmful to cotton belts, tending to separate the plies. An application of ground chalk will absorb the oil on a belt and make it workable for a time. Sometimes occasional washing and brushing in warm water also serve the purpose. In case of leather belts, the grain side is the correct driving side which drives nearly twice the power conveyed by the flesh side.

Creep of the Belt

As the material of the belt is elastic, the belt length elongates more on tight side than on the slack side, resulting in unequal stretching. Since the peripheral speed of a pulley depends upon the length of belt received by it, therefore, the driven pulley shall run at a smaller peripheral speed than the driving pully. Thus a partial slip occurs. This phenomenon is called a *creep*.

Example 2.1 *The circumferential speed of the driven pulley is 5 metres per second and the belt slip is 3 per cent. What is the circumferential speed of the follower?*

Solution. Circumferential speed of the follower

= 97 per cent of the circumferential speed of the driver

$$= 5 \times \frac{97}{100} \text{ metre per second}$$

= **4.95 metre per second. Ans.**

Example 2.2 *A pulley 0.25 m diameter and running at 400 r.p.m. drives a pulley on a machine. If the machine is to run at 600 r.p.m., what should be the diameter of the pulley on it?*

Solution. D_1 = 25 m. N_2 = 600 r.p.m. N_1 = 400 r.p.m. D_2 = ?

$$\therefore \quad \frac{N_2}{N_1} = \frac{D_1}{D_2}$$

$$\therefore \quad D_2 = \frac{D_1 N_1}{N_2} = \frac{400 \times 0.25}{600}$$

$$= \mathbf{0.167\ m.} \quad \textbf{Ans.}$$

Example 2.3 *Two pulleys are connected by a flat belt. The sum of their diameters is 0.7 m and speeds are 60 r.p.m. and 300 r.p.m. Find the diameters of the pulleys, if the belt thickness is 4 mm. Assume slipless motion.*

Solution. $D_1 + D_2 = 70$ cm. $N_1 = 60$ r.p.m. and $N_2 = 300$ r.p.m.

$t = 4 \text{ mm} = 0.004$ m.

$$\because \quad \frac{N_2}{N_1} = \frac{D_1 + t}{D_2 + t} \quad \text{or} \quad \frac{300}{60} = \frac{D_1 + 0.004}{D_2 + 0.004}$$

$$\therefore \quad \frac{D_1 + 0.004}{D_2 + 0.004} = 5 \quad \text{or} \quad D_1 + 0.004 = 5\,D_2 + 0.02$$

$$\therefore \quad D_1 = 5\,D_2 + 0.02 - 0.004 = 5\,D_2 + 0.016$$

But $D_1 + D_2 = 0.7$ (given)

$$\therefore \quad 5D_2 + 0.016 + D_2 = 0.7 \quad \text{or} \quad 6D_2 = 0.7 - 0.016$$

$$\therefore \quad D_2 = \frac{0.6984}{6} = \mathbf{0.1164\ m.} \quad \textbf{Ans.}$$

and $D_1 = 5 \times 0.1164 + 0.016 = \mathbf{0.5836\ m.}$ **Ans.**

Example 2.4 *It is required to drive a lathe of diameter 0.3 m. at 600 r.p.m. from a shaft rotating at 400 r.p.m. Calculate the diameter of the pulley required on the driving shaft. Assuming 5% slip.*

Solution. $D_2 = 0.3$ m $D_1 = ?$ $N_1 = 600$ r.p.m. $N_2 = 400$ r.p.m. $S = 5\%$

$$\because \quad \frac{N_2}{N_1} = \frac{D_1}{D_2} \times \frac{(100 - S)}{100}$$

$$\therefore \quad D_1 = \frac{N_2 \times D_2 \times 100}{N_1\,(100 - S)}$$

or

$$D_1 = \frac{600 \times 0.3 \times 100}{400\,(100 - S)} = \frac{600 \times 0.3 \times 100}{400 \times 95}$$

$$= \mathbf{0.4737\ m.} \quad \textbf{Ans.}$$

Example 2.5 *A pulley driven by a belt has a diameter of 0.6 m and a speed of 300 r.p.m. The tension on the tight or slack sides are respectively 1200 N and 400 N. Find the power transmitted if, (i) there is no slip and (ii) there is a slip of 4 per cent.*

Solution. (*i*) There is no slip,

$D = 0.6$ m. $T_1 = 1200$ N $N = 300$ r.p.m. $T_2 = 400$ N

Effective tension on the belt,

$$T = T_1 - T_2 = 1200 - 400 = 800 \text{ N}$$

$$\text{Speed of the belt } V = \frac{\pi DN}{60} = \frac{22 \times 0.6 \times 300}{7 \times 60}$$

$$= 9.43 \text{ metre per second.}$$

$$\text{Power} = \frac{TV}{1000} = \frac{800 \times 9.43}{1000} = \mathbf{7.544\ kW}$$

(*ii*) **When there is slip 4 per cent**

$$\text{Power} = \frac{TV}{1000} \times \frac{(100 - S)}{100} = \frac{800 \times 9.43 \times 96}{1000 \times 100} = \mathbf{7.242\ kW}$$

Example 2.6 *In a belt drive the speed and diameter of the follower are respectively 150 r.p.m., and 1.2 m. and the power transmitted is 20 kW. Find the effective pull in the belt, neglecting belt slip.*

Solution. $N = 150$ r.p.m. $D = 1.2$ metre $P = 20$ kW

Let T be the effective pull, then we have

$$\therefore \quad P. = \frac{\pi DNT}{60 \times 1000}$$

$$\therefore \quad T = \frac{P. \times 60 \times 1000}{\pi \times D \times N} = \frac{20 \times 60 \times 1000}{\pi \times 1.2 \times 150}$$

$$= \mathbf{2123\ N} \quad \textbf{Ans.}$$

Example 2.7 *A belt drives a pulley of 0.9 m. diameter at 200 r.p.m. The effective tension in the belt is 300 N. Find the horse-power that the belt can transmit.*

Solution. $D = 0.9$ metre $N = 200$ r.p.m. $T = 300$ N

$$P. = \frac{\pi DNT}{60 \times 1000} = \frac{\pi \times 0.90 \times 200 \times 300}{60 \times 1000}$$

or
$$P. = \mathbf{2.826\ kW} = \frac{2.826 \times 10^3}{746}$$

$$= 3.7901 \text{ Horse Power (H.P.)} \qquad \text{(Since 1 HP = 746 W)}$$

Example 2.8 *A driver pulley 1.2 m. in diameter has a speed of 250 r.p.m. and drives a line shafting at 500 r.p.m. If the power transmitted is 10 kW; find the diameter of the follower and the effective pull in the belt, if*

(*i*) *There is no belt slip,*

(*ii*) *There is a belt slip of 4 per cent.*

If the tight side tension is 3 times that on the slack side, find the tension.

Solution. (*i*) There is no belt slip,

$D_1 = 1.2$ metre $D_2 = ?$ $N_1 = 250$ r.p.m. $N_2 = 500$ r.p.m. $P. = 10$ kW $T = ?$

$$\because \quad \frac{N_2}{N_1} = \frac{D_1}{D_2}$$

$$\therefore \quad D_2 = \frac{N_1 \times D_1}{N_2} = \frac{250 \times 1.2}{500} = 0.6 \text{ metre} \quad \textbf{Ans.}$$

Linear velocity of the belt, $V = \dfrac{\pi D_1 N_1}{60}$ metre per second

$$= \frac{22 \times 1.2 \times 250}{7 \times 60}$$

$= 15.71$ metre per second.

$\because \quad P = \dfrac{TV}{1000}$

$\therefore \quad T = \dfrac{P \times 1000}{V} = \dfrac{10 \times 1000}{15.71}$

$=$ **636.54 N Ans.**

Let, tension on slack side, $T_2 = F$ N

Therefore tension on tight side,

$T_1 = 3F$ N $\qquad (\because \; T_1 = 3\,T_2$ given)

$\therefore$ Effective tension on the belt,

$$T = T_1 - T_2 = 3\,F - F$$

But, $3F - F = 636.54$ or $2\,F = 636.54$

$\therefore \quad F = 318.3$ N

$\therefore$ Slack side tension = **318.3 N Ans.**

Tight side tension = 318.3 × 3 = **954.9 N Ans.**

***(ii)* There is a belt slip of 4 per cent,**

$$\because \quad \frac{N_2}{N_1} = \frac{D_1}{D_2} \times \frac{(100 - S)}{100}$$

$$\therefore \quad D_2 = \frac{N_1 \times D_1 (100 - S)}{N_2 \times 100}$$

or $$= \frac{250 \times 1.2}{500} \times \frac{(100 - 4)}{100} = \frac{250 \times 1.2 \times 96}{500 \times 100}$$

$= 0.576$ metre **Ans.**

Circumferential speed of the follower,

$$V = \frac{\pi\, D_2\, N_2}{600} \text{ m/sec}$$

$$= \frac{\pi \times 0.576 \times 500}{60} \text{ m/sec}$$

$$\therefore \quad P = \frac{TV}{1000}$$

$$\therefore \quad T = \frac{P \times 1000}{V} = \frac{10 \times 1000 \times 60}{\pi \times 0.576 \times 500}$$

$=$ **663.5 N Ans.**

Let F be the tension on the slack side. Then 3 F is the tension on the tight side.

Thus $3F - F = 663.5$ or $2F = 663.5$

$\therefore$ F = **331.75 N Ans.**

$\therefore$ Tension on tight side = 3 × 333.25 = **995.25 N Ans.**

Example 2.9 *For a belt drive calculate the horse-power transmitted by a belt running at 30 m/sec. Assume μ = 0.3, θ = 175° and the maximum tension in the belt to be 500 N.*

Solution. $T_1 = 500 \text{ N} \quad \theta = 175° = \dfrac{175 \times \pi}{180} = 3.05$

$\mu = 0.3 \quad V = 30 \text{ m/sec.}$

$\therefore \quad \dfrac{T_1}{T_2} = e^{\mu\theta}$ or $\dfrac{50}{T_2} = e^{0.3 \times 3.05}$

Taking log both the sides

$\log 500 - \log T_2 = 0.3 \times 3.05 \times \log e$

or $\log 500 - \log T_2 = 0.915 \log 2.718$ [$\because e = 2.718$]

Now values taken from log table we have

$2.6990 - \log T_2 = 0.915 \times 0.4343$

or $2.6990 - \log T_2 = 0.3971$

or $-\log T_2 = -2.6990 - 0.3971$ or $\log T_2 = 2.3019$

Taking antilog we have $T_2 = 200 \text{ N}$

$\therefore$ Effective tension transmitted by the belt, $T = (T_1 - T_2) = 500 - 200 = 300$ N.

$$P = \frac{T \times V}{1000} = \frac{30 \times 300}{1000} = \mathbf{9\ kW} \quad \textbf{Ans.}$$

Example 2.10 *Calculate the diameter of a lathe pulley when the speed of the lathe is 600 r.p.m. and the mainshaft pulley has a diameter of 0.5 m. and a speed of 240 r.p.m., while the first follower and second driver pulley on the countershaft have respectively diameters of 0.3 m. and 0.4 m. when*

(i) there is no belt slip and

(ii) there is a belt slip of 5 per cent.

Solution. Referring Fig. 15.10, we have,

(*i*) **There is no belt slip.** Let N_a, N_b, N_c and N_d be the r.p.m. and D_a, D_b, D_c and D_d be the diameters respectively of the pulleys A, B, C and D.

D_d = ? D_b = 0.3 m. N_d = 600 r.p.m. D_c = 0.4 m. D_a = 0.5 m. N_a = 240 r.p.m.

Consider drive from pulley A to pulley B.

$$\therefore \quad \frac{N_b}{N_a} = \frac{D_a}{D_b}$$

$$\therefore \quad N_b = \frac{D_a \times N_a}{D_b} = \frac{0.5 \times 240}{0.3}$$

$$= 400 \text{ r.p.m.}$$

$\because$ Pulleys B and C are rigid on the same counter-shaft.

Therefore their r.p.m. will be same.

$$\therefore \quad N_b - N_c = 400 \text{ r.p.m.}$$

Now consider drive from pulley *C* to pulley *D*, we have

$$\therefore \quad \frac{N_d}{N_c} = \frac{D_c}{D_d}$$

$$\therefore \quad D_d = \frac{N_c \times D_c}{N_d} = \frac{400 \times 0.4}{600}$$

$$= \mathbf{0.266 \text{ m.}} \quad \textbf{Ans.}$$

(*ii*) **There is belt slip of 5 per cent.** Consider drive from pulley *A* to pulley *B*,

$$\because \quad \frac{N_b}{N_a} = \frac{D_a}{D_b} \times \frac{(100 - S)}{100}$$

$$\therefore \quad N_b = \frac{D_a \times N_a}{D_b} \times \frac{95}{100} = \frac{0.5 \times 240}{0.3} \times \frac{95}{100}$$

$$= 380 \text{ r.p.m.}$$

$$N_b = N_c = 380 \text{ r.p.m.}$$

Now consider drive from pulley *C* to pulley *D*, we have

$$\because \quad \frac{N_d}{N_c} = \frac{D_c}{D_d} \times \frac{(100 - S)}{100}$$

$$\therefore \quad D_d = \frac{N_c \times D_c (100 - S)}{N_d \times 100} = \frac{380 \times 0.4 \times 95}{60 \times 100}$$

$$= \mathbf{0.24 \text{ m. (approx)}} \quad \textbf{Ans.}$$

2.10 ROPE DRIVES

Rope drive is used to transmit power over a moderately long distance. The horizontal distance apart of the shaft centre in a rope drive should not be less than 10 m, but may be made as much as 30 m. One big advantage of rope drive is that a number of separate drives may be taken from one of the driving pulleys. Ropes are made of cotton, manila or hemp. Cotton ropes being usually considered the best on account of their greater driving power and flexibility.

In rope drive, diameter of the pulley is greater than thirty times the diameter of the cotton rope, and about forty five to sixty times the diameter of the manila or hemp rope. In transmitting large powers, it is better to use a number of ropes of small diameter rather than a single rope of large diameter. The pulleys are grooved and the groove angle varies from 40° and 60° but is generally 45°.

The rope comes only in contact with the faces and not with the bottom of the groove as shown in Fig. 2.19 (*a*), so that a better wedging action is obtained. But it is not the case with a steel rope, as it does not come in contact with the sides of the groove, because it is liable to injure the sides very badly. It, therefore, rests on a leather lining provided at the bottom of the groove as shown in Fig. 2.19 (*b*).

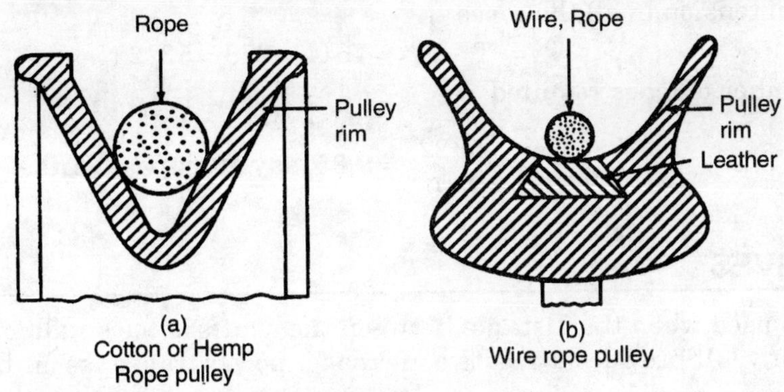

Fig. 2.19 Rope Drives.

$$\text{Powre transmitted by ropes} = \frac{T \times V}{1000} = \frac{T \times \pi DN}{1000 \times 60}$$

where T = Effective tension transmitted through rope,

V = Linear velocity of rope in metre/second

D = Diameter of the pulley in metre.

N = r.p.m. of the pulley.

The relation between two tensions T_1 and T_2 is given by

$$\frac{T_1}{T_2} = e^{\frac{\mu\theta}{\sin\alpha}}$$

where, e = The base of the napierian log = 2.718

μ = Coefficient of friction for the rope on the pulley.

θ = Angle of lap in radians.

2α = Angle of the groove of the pulley.

Example 2.11 *A rope drive transmits 50 kW at 150 r.p.m. from a grooved pulley with an included groove angle of 45°. The mean diameter of the pulley measured between the rope centres is 1.37 m. The angle of embrace of the ropes is 180° and the coefficient of friction 0.25. If the permissible tension for each rope is 540 N. determine the number of ropes required.*

Solution. We have $\frac{(T_1 - T_2)\,V}{1000} = 50$

where, $V = \pi \times 1.37 \times 150 = 646 \text{ m/min.} = 10.767 \text{ m/s}$

so that, $T_1 - T_2 = \frac{50 \times 1000}{10.767} = 4644 \text{ N}$

Since, $\mu = 0.25, \quad \theta = 180° = \pi \text{ radian}, \quad \alpha = 22.5°$

we get, $\frac{\mu\theta}{\sin\alpha} = \frac{0.25 \times \pi}{0.382} = 2.06$

Thus, $\frac{T_1}{T_2} = e^{\frac{\mu\theta}{\sin\alpha}} = e^{2.06} = 7.85$

$\therefore \quad 7.85T_2 - T_2 = 4644 \quad \text{or} \quad 6.85T_2 = 4644$

$$T_2 = \frac{4644}{6.85} = 678 \text{ N}$$

Therefore total tension in all the ropes

$$T = T_1 + T_2 = 4644 + 678 = 5322 \text{ N.}$$

Hence the number of ropes required

$$= \frac{5322}{540} = \mathbf{9.85, say\ 10\ ropes.}\quad \textbf{Ans.}$$

2.11 CHAIN DRIVES

Chain drive is used when the distance between the shafts is too small for belt drive and too large for gear drive. It provides a positive means for power transmission because there is no power slip. The maximum distance for which chain drive is employed is about 5 metre.

A chain is a sort of continuous band made of rigid links hinged together. The chain wraps round the wheels called the sprockets. A sprocket is a toothed wheel having projections and recesses, the former fitting into the cavities of chains links and *vice versa.* The transmission efficiency of this system is as high as 99 per cent. Heat, cold and moisture have no effect on chains and so this forms of drive may be used when belt drive is not suitable on account of high or low temperature and moisture.

$$\text{Power (in kW) transmitted by chain} = \frac{FV}{1000}$$

where, F = Force (in N) on the sprocket or teeth of a chain wheel.

V = The speed of the chain in metre per second.

Types of Chains.

There are two practical considerations for an efficient chain drive:

1. The chain links should enter and leave the sprocket with a minimum of noise and loss of power.
2. The wear of the joints should be a minimum.

In order to meet these considerations, the chains are divided in two groups.

1. Roller Chain. A roller chain is shown in Fig. 2.20 (*a*) which is most commonly used in general purpose drives bicycles and the rear drive of a motor cycle. Here *D* are the outer links, held together with the inner links *A*, separated by the rollers *R,* through which pass the pins *C* which are riveted at their ends. In between the rollers and the pins are provided the bushes, *B,* such that the rollers can turn freely on the bushes while the bushes can turn freely on the pins. The contact surfaces are well lubricated to reduce friction to a minimum.

2. Silent Chains. A silent chain is shown in Fig. 2.20 (*b*) and is also sometimes called the "inverted tooth chain" because of its shape. This is limit up of a series of thin strips, each having two projections or teeth with a recess in between. The working faces of the teeth have an included angle of 60°, the recess being sufficiently deep to keep clear of the tooth tips. The thin strips are held together by means of hardened steel pins, riveted at their ends. The pins have bushes surrounding them so that the chain is rendered flexible for the wrapping and unwrapping action. It should be noted that the inverted tooth chain merely rides on the teeth and is absolutely free from the sliding action. Consequently, the operation of such a chain is noiseless.

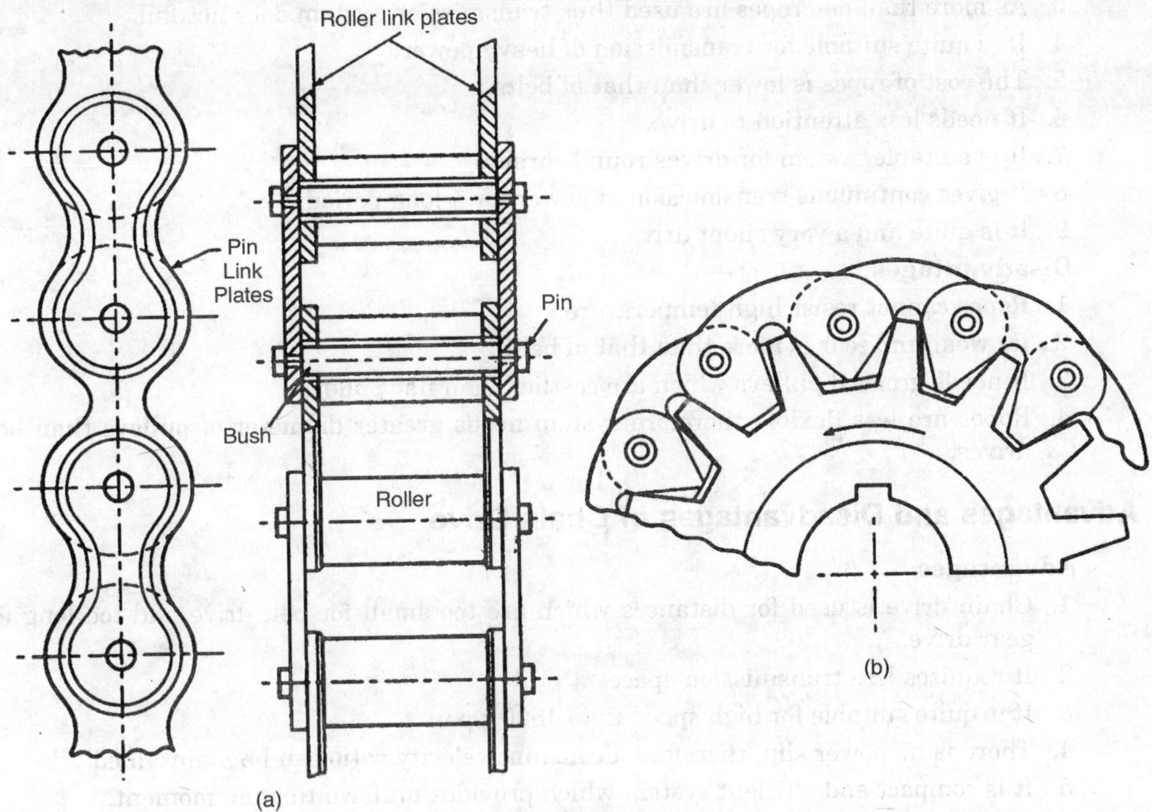

Fig. 2.20 Types of chain.

Advantages and Disadvantages of Belt Drive

Advantages

1. Surface speed of the belt can be obtained up to 1500 metres per minute.
2. Power can be transmitted through the shafts having span distance from 5 to 10 metres.
3. It is lighter and cheaper method of power transmission.
4. Its cost of installation is lower than the other systems.

Disadvantages

1. Due to belt slip a constant velocity ratio cannot be maintained.
2. It is heavier and costlier than rope drive.
3. It is convenient method of drive for short and too large span of shafts.
4. Belt drive is a noisy method of power transmission.

Advantages and Disadvantages of Rope Drives

Advantages

1. Rope drive is suitable for long distances *i.e.*, more than 10 metres, where the other methods are not suitable.
2. There is less power slip than in belt drive.

3. As more than one ropes are used thus transmission system does not fail.
4. It is quite suitable for transmission of heavy power.
5. The cost of ropes is lower than that of belts.
6. It needs less attention to drive.
7. It is suitable system for drives round corners.
8. It gives continuous transmission of power for a long period.
9. It is quite and a very silent drive.

Disadvantages

1. Ropes cannot resist high temperature.
2. Its wear and tear is more than that of belts.
3. It needs grooved pulleys which are costlier than flat pulleys.
4. Ropes are less flexible therefore system needs greater diameter of pulleys than belt drives.

Advantages and Disadvantages of Chain Drive

Advantages

1. Chain drive is used for distances which are too small for belt drive and too long for gear drive.
2. It requires less transmission space.
3. It is quite suitable for high speed upto 1500 r.p.m.
4. There is no power slip, therefore a constant velocity ratio can be maintained.
5. It is compact and efficient system which provides uniform turning moment.
6. It does not require an initial tension as required in belt drive system.
7. It provides positive drives with high efficiency of transmission.
8. A wide range of power transmission can be obtained very easily.
9. Its life is more than that of belt or rope.
10. Temperature and moisture do not effect on its working.

Disadvantages

1. Chain drive is not suitable for too large distances *i.e.,* exceeding 5 metres.
2. Its initial cost is more than that of belt or rope drives.
3. After certain period, this system creates a noise in the drives.

Selection of Transmission System

The belt drive is very common for transmitting the power from prime mover or motor to main shaft of the system. Because it can attain a surface speed upto 1500 metres per minute and its installation cost is also low.

Sometimes gear drive is also used for this purpose but it is not suitable for high speeds. Comparatively its installation cost is more than that of belt drive.

Chain drive is suitable for positive drives in the places where too low or too high temperature and moisture are created. It can attain an angular speed of 1500 r.p.m. Its installation is costlier than that of belt drive.

Rope drive is not suitable for main shaft drives because its uses are limited to light loads. Its wear and tear is also more than that of belts.

2.12 GEAR DRIVES

Gears or Toothed Wheels

Gears are suited for the transmission of any amount of power, and the slip will be completely avoided. Thus the toothed gears are used where an exact velocity ratio is to be transmitted. The only limitation upon the gear drives is that the distance between the shaft centres should be small, as otherwise the size and cost of the wheels make gearing prohibitive.

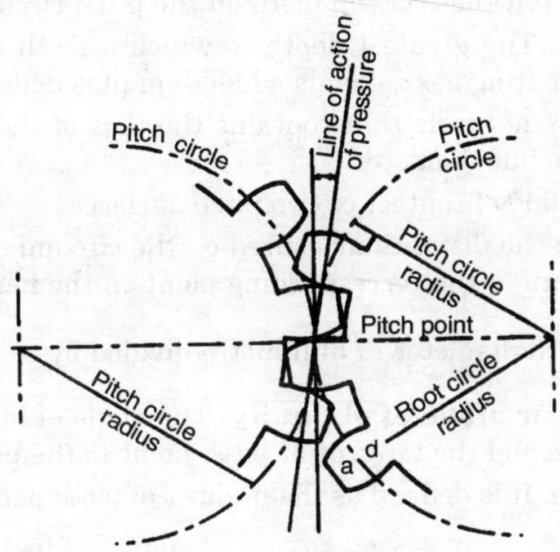

Fig. 2.21 Gear Drive.

Terms used in Gears

Since the spur gearing is used over a wide range of articles from small watches, precision measuring instruments, machine tools to gear boxes fitted in motor cars and aero-engines etc. The common terms in general use in connection with the spur gear are as shown in Fig. 2.22, and are as given under:

[1. Thickness, 2. Circular pitch, 3,5. Face, 4. Tip or crest, 6. Flank, 7. Bottom or root, 8. Dedendum circle or root circle, 9. Addendum circle or Tip circle, 10. Pitch circle, 11. Addendum, 12. Dedendum.]

(*i*) **Pitch circle.** An imaginery circle through the pitch point having its centre at the axis of the gear. The diameter of the pitch circle is called pitch diameter.

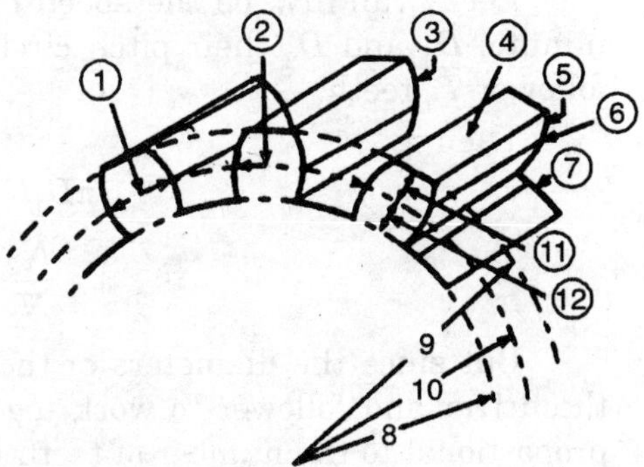

Fig. 2.22 Spur gear elements.

(*ii*) **Pitch point.** The point of contact of two pitch circles.

(*iii*) **Root circle.** The circle that contains the bottoms of the teeth. The diameter of this circle is called root diameter.

(*iv*) **Addendum.** The radial height from the pitch circle to the tip of the tooth.

(*v*) **Dedendum.** The radial depth from the pitch circle to the bottom of the tooth space in the mating gear.

(*vi*) **Clearance.** The radial distance from the top of the tooth to the bottom of the tooth space in the mating gear.

(*vii*) **Face of tooth.** The surface of the tooth above the pitch circle.

(*viii*) **Flank of tooth.** The surface of the tooth below the pitch circle.

(*ix*) **Tooth surface.** Includes both face and flank.

(*x*) **Thickness.** The thickness of the tooth on the pitch circle.

(*xi*) **Working depth.** The greatest depth to which a tooth of one gear extends into the tooth space of a mating gear, equals addendum plus dedendum minus the clearance.

(*xii*) **Outside circle.** The circle that contains the tops of the teeth. The diameter of this circle is called outside diameter.

(*xiii*) **Pitch line.** The line of contact of two pitch surfaces.

(*xiv*) **Circular pitch.** The distance measured on the circumference of the pitch circle from a point of one tooth to the corresponding point on the next tooth.

(*xv*) **Module.** The pitch diameter in millimetre divided by the number of teeth, *i.e.*, $\frac{d}{N}$.

(*xvi*) **Pressure angle or angle of obliquity.** The angle of inclination between the line of action of pressure and the tangent at pitch point to the pitch circle.

(*xvii*) **Diametral pitch.** It is defined as the number of teeth per inch of pitch circle diameter.

Thus,
$$\text{D.P.} = \frac{\text{Number of teeth}}{\text{Pitch circle diameter in inches}} = \frac{N}{d}$$

Velocity Ratio of Gears

Let N_1 and N_2 be the speed of the driver and follower respectively in revolutions per minute, D_1 and D_2 their pitch circle diameter, and let the driver contain T_1 teeth and the follower T_3 teeth.

Then

$$\therefore \quad \pi D_1 N_1 = \pi D_2 N_2$$

$$\therefore \quad \frac{N_2}{N_1} = \frac{D_1}{D_2}$$

But since the diameters of the pitch circles are proportional to the circumference, and the driver and follower to work together, must have the same pitch, the circumferences are proportional to the number of teeth in the wheels.

$$\therefore \quad \frac{N_2}{N_1} = \frac{D_1}{D_2} = \frac{T_1}{T_2}$$

Idle Wheel

Referring Fig. 2.23, that the direction of rotation of two gears directly in contact will be opposite to each other. If instead of direct gearing the driver D with the follower F, a wheel I is introduced between the driver D and the follower F, the driver D will drive the wheel I, and I will drive F, the effect of I being to make the wheel F turn in the same direction as D instead of in the opposite direction if it is connected directly with D. The wheel I is called the "idle wheel" as it has no effect on the velocity ratio. The function of an idle wheel is, therefore, to change the direction of rotation of the train.

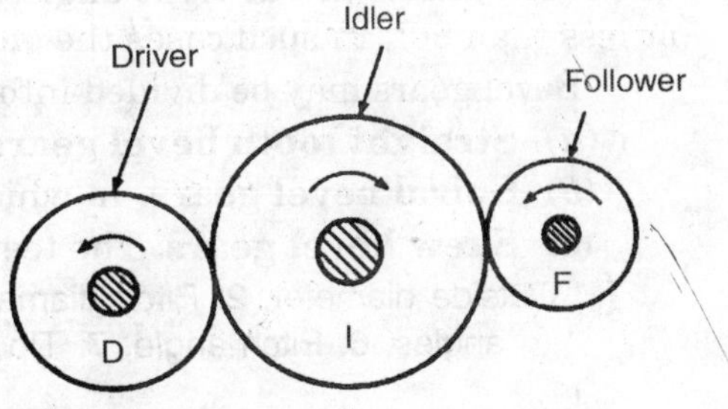

Fig. 2.23 Function of an idle wheel.

Power Transmitted by Gearing

In a train of gear wheels the work put into the train by the first driver must be equal to the work obtained from the last follower, and assuming there is no loss due to friction, this must be equal to the work transmitted.

Let, F = Driving force at pitch line in N.

D = Diameter of pitch circle in metres.

V = Velocity in metres per second at pitch circle.

N = Number of revolutions of wheel per minute.

Then work transmitted = $P \times V$ Nm

$$\therefore \quad \text{Power kW} = \frac{P \times V}{60000} \quad \text{or} \quad \text{P.} = \frac{F \times \pi DN}{60000}$$

or

$$\text{kW} = \frac{\pi DNP}{60000}.$$

2.13 TYPES OF GEARS

1. Spur Gears

These are used to transmit power between two shafts which are parallel and co planer, as shown in Fig. 2.24. The gears are cylindrical, the teeth being cut straight and parallel to their axes, so that the contact takes place along a straight line. The larger wheel is called the spur wheel and the smaller one the pinion. The change gear wheels for screw-cutting in a lathe are of this type.

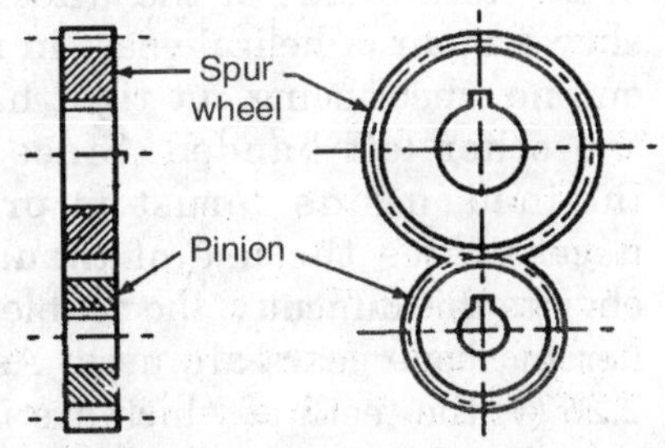

Fig. 2.24 Spur Gearing.

2. Bevel Gears

When two shafts, the axes of which intersect, are to be connected by gearing, the wheels are known as bevel gears shown in Fig. 2.25. In a bevel gear the teeth are cut on a conical surface, such as would be represented by a truncated cone. In the great majority of bevel gear drives the shafts are at right angles, but the angle between the shafts may be either greater or less than 90°. In such cases the gears are called "angular bevel gears".

Bevel gears may be divided into four classes:

(*a*) Straight tooth bevel gears. In which the elements of the tooth surfaces are straight.

(*b*) Spiral bevel gears. In which the teeth are curved on a spiral.

(*c*) Skew bevel gears. The teeth surfaces are hyperboloids and the teeth are straight.

[1. Outside diameter, 2. Pitch diameter, 3. Dedendum, 4. Addendum, 5. Root angle or cutting angles, 6. Pitch angle, 7. Tip angle of face angle, 8. Face width, 9. Back angle].

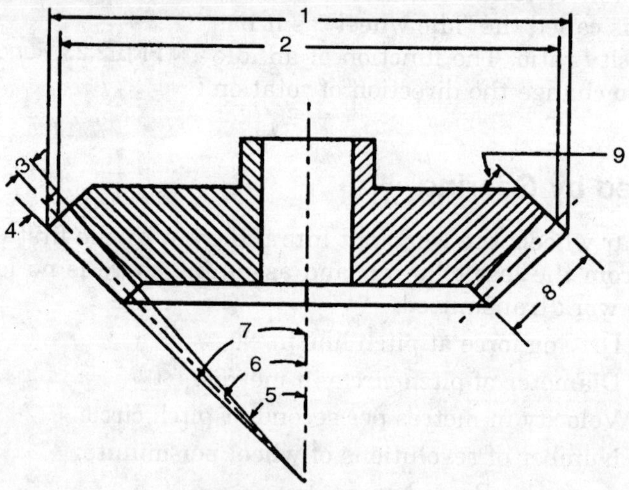

Fig. 2.25 Bevel gear.

(*d*) Hypoid gears. They possess the main characteristics of skew bevel gears, but they have curved teeth.

3. Helical Gears

Like the spur gears, these are also used to transmit power between two shafts whose axes are co-planer and parallel. Here the teeth cut on the pitch surfaces are inclined at an angle of 20° to the axis of the wheel. Fig. 2.26. (*a*) shows a pair of helical gears in mesh, the teeth on one wheel being cut right handed while on the other left handed. Since the teeth are inclined, a side thrust is produced which necessitates the use of thrust bearing. To obviate this difficulty, the double helical gears or herring bone gears are used. As shown in Fig. 2.26 (*b*) the teeth of which are inclined in both the directions on each wheel so that the side thrusts are counter balanced.

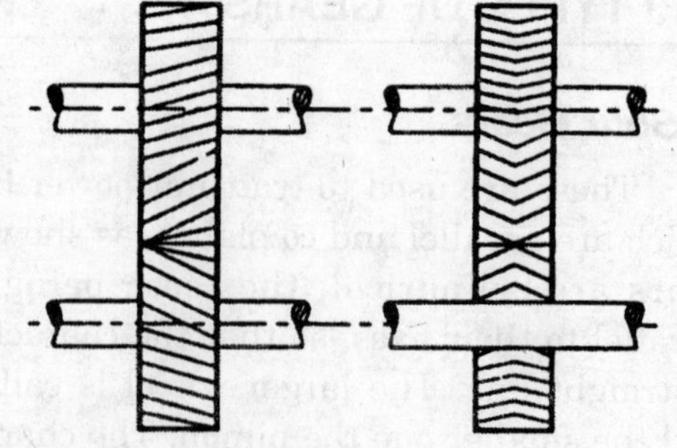

(*a*) Single helical gear (*b*) Double helical gear.

Fig. 2.26 Helical gears.

4. Spiral Gears

These are used when the axes of the shafts are non-parallel and non-intersecting as shown in Fig. 2.27. The pitch surface of both the wheels are cylindrical. The teeth on the spiral gears may be either the same hand or opposite hand. Since the teeth of the mating gears have point contact, spiral gears are suited for the transmission of small power only.

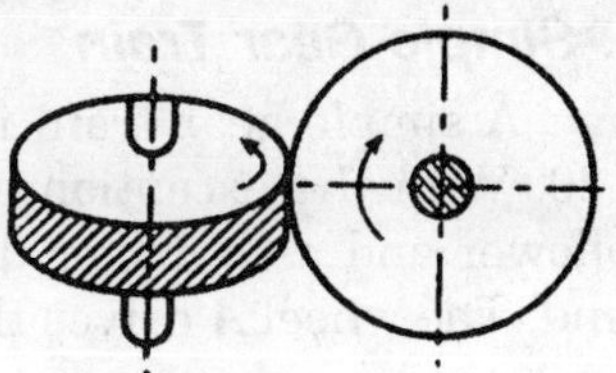

Fig. 2.27 Spiral gears.

5. Worm Gears

Referring Fig. 2.28, that the worm gearing is a form of spiral gearing where the axes of the shafts are in different planes and at right angles to each other. Here the diameter of the driving wheel which is called the "worm" is very small. The worm is actually in the form of a screw which rotates about its axis in fixed bearings. The wheel which meshes with the worm is called the "worm wheel". It has its teeth cut on a curved surface fitting the curvature of the worm. The teeth on worm wheel are inclined to mesh with the teeth of the worm properly. Worm gearing is capable to transmit very high velocity ratios.

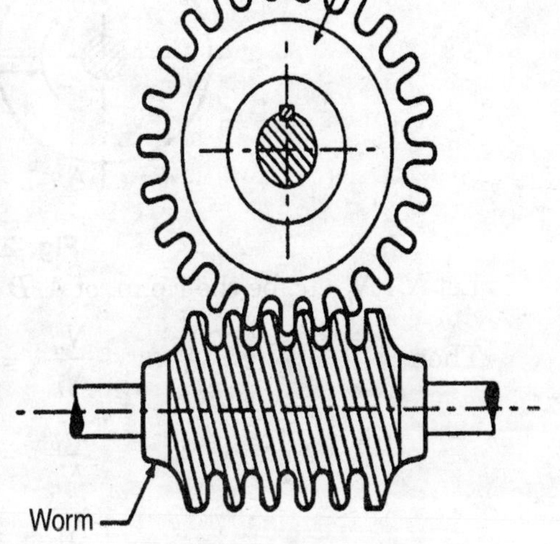

Fig. 2.28 Worm gearing.

6. Rack and Pinion

Referring Fig. 2.29, here the pinion is in the form of a small spur gear and the rack is a spur gear of infinite diameter. Thus the rack in fact assumes the shape of a straight gear. Such an arrangement may be used where the rotary motion is to be converted into a rectilinear motion and vice versa.

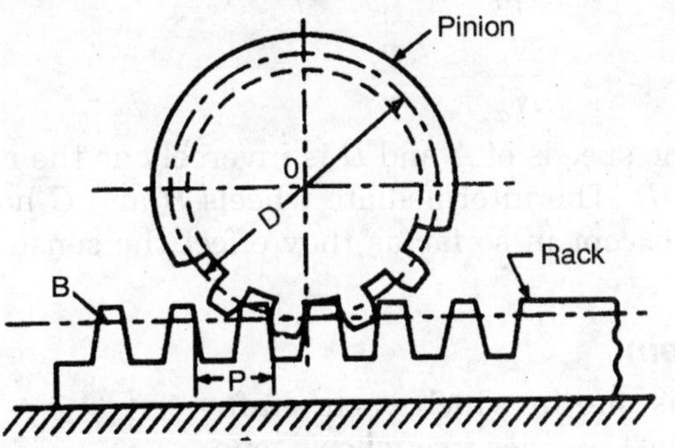

Fig. 2.29 Rack and Pinion.

2.14 GEAR TRAINS

Any combination of gear wheels by means of which motion is transmitted from one shaft to another shaft is called a gear train. A single gear train may include any or all of the different kinds of gears-spur, bevel, spiral etc. The most common types of gear trains are as given under:

1. Simple Gear Train

A simple gear train is one in which each shaft carries one wheel only as shown in Fig. 2.30. With the exception of the first and last wheels in the train, each wheel acts both as a follower and as a driver. It is seen that in this example the gear wheels are all of the same kind. The wheel A drives the wheel B, while B drives C and C drives D.

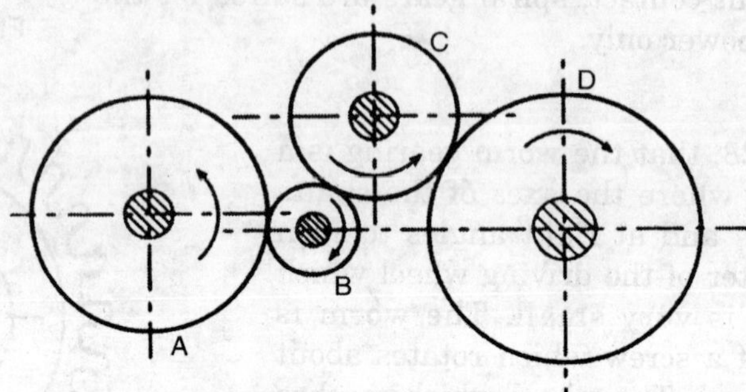

Fig. 2.30 Simple gear train.

Let N_a, N_b etc. be the r.p.m. of A, B etc. and T_a, T_b etc. be the numbers of teeth on A, B etc.

Then
$$\frac{N_a}{N_b} = \frac{T_b}{T_a} \quad \text{...}(i)$$

$$\frac{N_b}{N_c} = \frac{T_c}{T_b} \quad \text{...}(ii)$$

$$\frac{N_c}{N_d} = \frac{T_d}{T_c} \quad \text{...}(iii)$$

Multiplying (*i*), (*ii*) and (*iii*), we have

$$\frac{N_a}{N_b} \times \frac{N_b}{N_c} \times \frac{N_c}{N_d} = \frac{T_b}{T_a} \times \frac{T_c}{T_b} \times \frac{T_d}{T_c}$$

or
$$\frac{N_a}{N_d} = \frac{T_d}{T_a}.$$

Hence the ratio of the speeds of A and D is inversely as the ratio of the numbers of teeth on the two wheels A and D. The intermediate wheels B and C have no effect on the velocity ratio given by the train, except in so far as they effect the sense of rotation of D for a given sense of rotation of A.

2. Compound Gear Train

A compound train is one in which each shaft, except, the first and the last, carries two wheels, one of which acts as the follower and receives its motion from a wheel on a second shaft, while the other acts as a driver and transmits motion to a wheel on a third shaft. An example of compound train is shown in Fig. 2.31, which consists entirely of spur gears. It is clear that the two gears C and B which are keyed to the same shaft must revolve at the same speed.

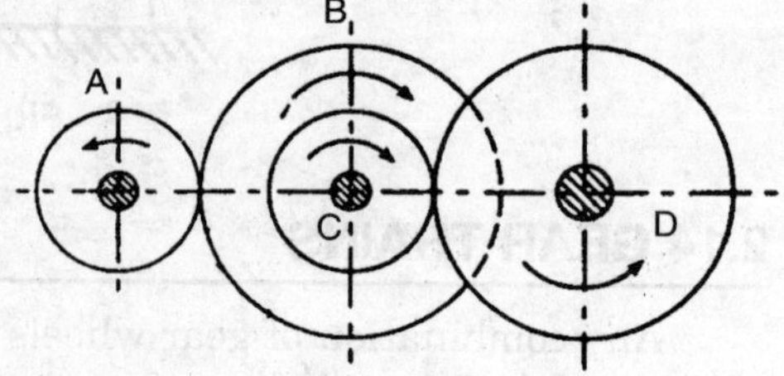

Fig. 2.31 Compound gear train.

Let N_a, N_b etc. be the r.p.m. of *A*, *B* etc. and T_a, T_b etc. be the numbers of teeth on *A*, *B* etc.

Then
$$\frac{N_a}{N_b} = \frac{T_b}{T_a} \quad \text{...(i)}$$

$$\frac{N_c}{N_d} = \frac{T_d}{T_c} \quad \text{...(ii)}$$

Multiplying (*i*) and (*ii*), we have

$$\frac{N_a}{N_b} \times \frac{N_c}{N_d} = \frac{T_b}{T_a} \times \frac{T_d}{T_c}$$

or
$$\frac{N_a}{N_b} \times \frac{N_b}{N_d} = \frac{T_b}{T_a} \times \frac{T_d}{T_c} \quad (\because \; N_c = N_b)$$

$$\therefore \quad \frac{N_a}{N_d} = \frac{T_b}{T_a} \times \frac{T_d}{T_c}$$

The direction of rotation of each of the wheels is shown by an arrow. The advantages of a compound train are that a large speed ratio may be obtained with wheels of small diameter and that the driver may be transmitted round bends or corners, and even, if necessary, returned upon itself so that the driving and driven shafts are co-axial.

Types of Gear Drives

1. Spur Gear Train

(*i*) *Single gear train:* No. of teeth × rpm of driving gear = No. of teeth × rpm of driven gear

i.e.,
$$\boxed{T_1 \times n_1 = T_2 \times n_2}$$

$$\therefore \quad T_1 = \frac{T_2 \times n_2}{n_1}, \; T_2 = \frac{T_1 \times n_1}{n_2}, \; n_1 = \frac{T_2 \times n_2}{T_1} \text{ and } n_2 = \frac{T_1 \times n_1}{T_2}$$

Gear Ratio = Inverse ratio of rpm of gears.

$$\boxed{\frac{T_1}{T_2} = \frac{n_2}{n_1} = \frac{d_1}{d_2}}$$

Gear ratio, *i*
$$\boxed{i = \frac{n_1}{n_2} = \frac{d_2}{d_1} = \frac{T_2}{T_1}}$$

Distance a of the axes,
$$a = \frac{d_1 + d_2}{2}$$

where d = pitch circle dia.

(*ii*) *Double gear train:*

T_1, T_3 = Driving gears

T_2, T_4 = Driven gears

Speeds
$$n_2 = n_3 = \frac{T_1 \times n_1}{T_2}, \quad n_4 = \frac{T_3 \times n_3}{T_4}$$

$$n_4 = n_1 \times \frac{T_1}{T_2} \times \frac{T_3}{T_4}$$

Total Transmission Ratio, $i = i_1 \times i_2$

Speed, $$n_4 = \frac{n_1}{i}, i = \frac{n_1}{n_4} = \frac{T_2}{T_1} \times \frac{T_4}{T_2}$$

2. Worm Gear Train

R.p.m. of worm × Thread number of worm = r.p.m. of worm wheel × No. of teeth of worm wheel.

$$n_1 \times T_1 = n_2 \times T_2$$

$$\text{Transmission ratio, } i = \frac{\text{Speed of worm}}{\text{Speed of worm wheel}}$$

$$i = \frac{n_1}{n_2}$$

$$i = \frac{\text{No. of teeth of worm wheel}}{\text{Thread no. of worm}}$$

$$i = \frac{T_2}{T_1}$$

Note: For single-start thread worm, $T_1 = 1$
For double-start thread worm, $T_1 = 2$.

Gear Dimensions

D_p = Diametral pitch = $\dfrac{\text{No. of teeth}}{\text{No. of inches in pitch dia.}}$

C_p = Circular pitch = Distance from centre of teeth measured on pitch circle.

Example 2.12 *Calculate the tooth-pitch for the module of 10 mm.*

Solution. ∴ $$\text{Module} = \frac{\text{Diameter of pitch circle}}{\text{Number of teeth}}$$

or $$10 = \frac{D}{N}$$

∴ $$D = 10N$$

$$\text{Tooth pitch} = \frac{\text{Circumference of pitch circle}}{\text{Number of teeth}} \quad \text{...(i)}$$

$$= \frac{\pi D}{N} = \frac{\pi \times 10 \times N}{N} = 10\pi$$

$$= \mathbf{31.43\ mm.}$$ **Ans.**

Example 2.13 *Show that in a wheel-train angular speed of the wheels is inversely proportional to their respective diameters.*

Solution. Let ω be the angular speed and u be the linear speed of a wheel.

$$\therefore \quad u = \omega r$$

$$\therefore \quad \omega = \frac{u}{r}$$

When two wheels are in mesh their linear speeds remain same, *i.e.* constant = c.

$$\text{Thus} \quad \omega = \frac{u}{r} = \frac{c}{r} = \frac{2c}{d}$$

$$\text{or} \quad \omega = \frac{2c}{d}$$

Here ω and d both are variable and 2c is constant.

$$\therefore \quad \omega \propto \frac{1}{d}. \textbf{ Hence proved.}$$

Example 2.14 *A larger gear has 80 teeth and small mating gear has 10 teeth. If the module is 4 mm, calculate the distance between the centres of small gears.*

Solution. Let d_1 be the pitch circle diameter of larger gear and d_2 that of small gear.

$$\therefore \quad \text{Module } m = \frac{d_1}{N_1} \quad \text{or} \quad d_1 = N_1 \times m$$

$$= 80 \times 4 = 320 \text{ mm.} \quad \text{...(i)}$$

Similarly for smaller gear,

$$d_2 = N_2 \times m = 10 \times 4$$

$$= 40 \text{ mm} \quad \text{...(ii)}$$

∴ Centre distance between the gears

$$= \frac{d_1 + d_2}{2} = \frac{320 + 40}{2}$$

$$= \textbf{180 mm. Ans.}$$

Example 2.15 *A wheel has 50 teeth of module 5 mm. Find the pitch diameter and the circular pitch.*

Solution.

$$\therefore \quad m = \frac{d}{N} \quad \text{(where } d \text{ be the pitch circle dia.)}$$

$$\therefore \quad d = N \times m = 50 \times 5 = \textbf{250 mm. Ans.}$$

$$\therefore \quad p \times N = \pi d \quad \text{(where } p \text{ be the circular pitch).}$$

$$\therefore \quad p = \frac{\pi d}{N} = \frac{\pi \times 250}{50}$$

$$= \textbf{15.71 mm. Ans.}$$

Example 2.16 *Find out the r.p.m. of a 30 teeth worm wheel when it is being driven by a worm shaft whose lead is equal to 3 pitch revolving at a speed of 300 r.p.m.*

Solution. $\therefore \quad \frac{N_2}{N_1} = \frac{T_1}{T_2}$ or $\frac{N_2}{300} = \frac{3}{30}$

$$\therefore \quad N_2 = \frac{300 \times 3}{30} = 30 \textbf{ r.p.m. Ans.}$$

Example 2.17 *The shafts with their centres approximately 800 mm apart are to be connected by a pair of toothed gears so as to transmit a velocity ratio of 2.8. If the module pitch of the gears is 12 mm, find the number of teeth in each wheel and the exact distance between the shaft centres.*

Solution. We have,

$$L = r_1 + r_2 = 800 \text{ mm. (app. given)} \quad ...(i)$$

and

$$\frac{N_2}{N_1} = \frac{r_1}{r_2} = 2.8 \quad ...(ii)$$

From (*i*) and (*ii*) we have,

$$r_1 = 589.5 \text{ mm}$$

and

$$r_2 = 210.5 \text{ mm.}$$

Since the pitch for both the gears is same,

$$m = \frac{D_1}{T_1} = \frac{D_2}{T_2} = 12$$

$$\therefore \quad T_1 = \frac{2r_1}{12} = \frac{2 \times 589.5}{12} = 98.25$$

$$T_2 = \frac{2 \times r_2}{12} = \frac{2 \times 210.5}{12} = 35.10$$

As T_1 and T_2 must be the exact numbers, assume $T_1 = 98$ and $T_2 = 35$; and their V.R. also must be 2.8.

Thus we have,

$$D_1 = 12 \times 98 = 1176 \text{ mm.}$$

$$D_2 = 12 \times 35 = 420 \text{ mm.}$$

$\therefore$ Exact distance between the shaft centres,

$$L = \frac{D_1 + D_2}{2} = \frac{1176 + 420}{2}$$

$$= \textbf{798 mm.} \quad \textbf{Ans.}$$

Example 2.18 *In a worm and worm-wheel, the worm-wheel has 40 teeth, the worm is single threaded, the diameter of the load drum or barrel is 18 cm., the length of the handle is 39 cm., and the thickness of the rope wound on the barrel is 3/2 cm. Determine the effort required to raise a load of 1000 kg., assuming efficiency of the system to be 25%.*

Solution. Velocity Ratio

$$= \frac{2LT_2}{T_1(D + d)} = \frac{2 \times 39 \times 40}{1 \times \left(18 + \frac{3}{2}\right)} = 160$$

Effort,

$$P = \frac{W}{\eta \times V.R.} = \frac{1000}{0.25 \times 160}$$

$$= \textbf{25 kg.} \quad \textbf{Ans.}$$

Example 2.19 *A compound gearing is used to transmit power from one shaft to another. If the number of teeth in the first driving wheel is 20 and the driver wheel is 80 and the second driver wheel is 25 and the driven wheel is 75, and the r.p.m. of the driving shaft is 1000, determine the r.p.m. of the driven shaft.*

Solution. Referring Fig. 2.31, we have

$T_a = 20 \quad N_a = 1000$ r.p.m. $\quad T_b = 80 \quad N_d = ? \quad T_c = 25 \quad T_d = 75$

Considering drive from gear *A* to *B*, we have

$$\frac{N_b}{N_a} = \frac{T_a}{T_b}$$

$$\therefore \quad N_b = \frac{T_a \times N_a}{T_b} = \frac{20 \times 1000}{80}$$

$$= 250 \text{ r.p.m.}$$

$$\because \quad N_b = N_c \quad \text{(As mounted on the same shaft)}$$

$$\therefore \quad N_c = 250 \text{ r.p.m.}$$

Considering drive from gear *C* to *D*, we have

$$\frac{N_d}{N_c} = \frac{T_c}{T_d}$$

$$\therefore \quad N_d = \frac{T_c \times N_c}{T_d} = \frac{25 \times 250}{75}$$

$$= \mathbf{83.33 \text{ r.p.m.}} \quad \textbf{Ans.}$$

Example 2.20 *Arrange a compound train of wheels for reducing the speed in two stages from 1,000 r.p.m. to 50 r.p.m. in the same direction. The wheels available range from one of 15 teeth to that of 120 teeth; rising by steps of 5. There is only one wheel having a particular number of teeth in a given stage. Sketch the arrangement you adopt.*

Solution. Here velocity ratio,

$$\frac{N_d}{N_a} = \frac{50}{1000}$$

$$= \frac{1}{20} = \frac{1}{4} \times \frac{1}{5} = \frac{15}{60} \times \frac{20}{100}$$

$$= \frac{T_a}{T_b} \times \frac{T_c}{T_d}$$

There may be several other arrangements of this ratio but here we adopt the arrangement with the gears having teeth as given under:

$$T_a = 15, \quad T_b = 60$$
$$T_c = 20 \quad T_d = 100$$

The arrangement is shown in Fig. 2.32.

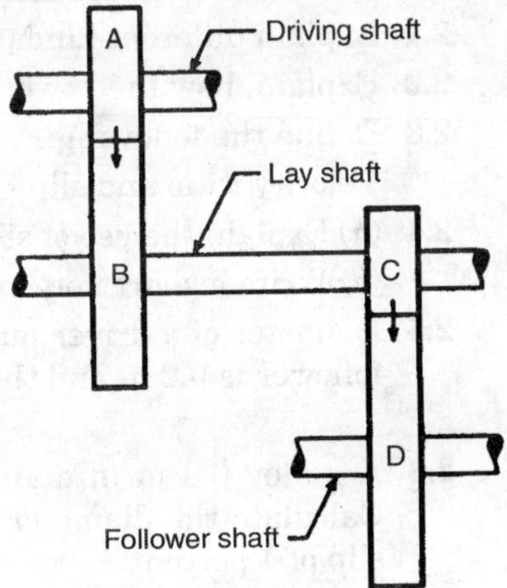

Fig. 2.32 Compound gear Train.

2.15 CLUTCHES AND BRAKES

1. Clutches

A clutch is used to permit connections or disconnections, when required, of the couplings of two working parts. The clutch is located between the prime mover and the transmission, and disconnects the prime mover from the power transmission system at the will of the operator (or driver in automobiles), resulting the prime mover to run without driving the machine tool (or the automobile).

The clutch can be defined as a device used to connect a driving shaft to driven shaft so that the driven shaft may be started or stopped when required without stopping the driving shaft.

Different types of clutches have been discussed in the chapter on ***"Automobile engineering."***

2. Brakes and Braking System

Brake is a device by means of which artificial frictional resistance is applied to a moving machine member in order to retard or stop the motion of a machine. The braking operation is reverse of the acceleration. In the braking operation, the kinetic energy is converted into heat, which is dissipated into the atmosphere.

Different types of brakes and braking systems have been discussed in detail in chapter-12 of this book. The major functional difference between a clutch and a brake is that a clutch is used to keep the driving and driven member moving together whereas brakes are used to stop a moving member or to control its speed.

QUESTIONS

2.1 Explain different kinds of power drives.

2.2 Explain, how the power is transmitted through belt drive?

2.3 Define the following:
Velocity ratio and slip in belt drive.

2.4 (*a*) Explain the use of step pulleys
(*b*) Write a short note on "care of belts used in power transmission."

2.5 Diameter of a driver pulley is 0.5 m and its speed is 250 r.p.m. If the diameter of the follower is 0.2 m and the slip is 3 percent, what is the speed of the follower?
(**Ans.** 606.25 rpm)

2.6 A pulley 0.3 m in diameter is driven at 450 rpm by a shaft revolving at 225 rpm. Calculate the diameter of the shaft pulley, if (*i*) There is no slip, and (*ii*) There is a slip of 4 percent.
(**Ans.** 0.6 m, 0.625 m)

2.7 A, B, and C are three steps of a pulley keyed to the counter-shaft, D, E and F are the corresponding steps of the pulley keyed to the machine shaft. If A = F = 0.4 m, B = E = 0.3 m and C = D = 0.2 m in diameter, find the range of speeds available on the machine shaft, the speed of the counter shaft being 400 rpm.
(**Ans.** 800 rpm, 400 rpm, and 200 rpm.)

2.8 Explain the use of chain drive in Transmitting the power.

2.9 Explain different types of chains used in power transmission.

2.10 Explain following terms used in Gear drive:
Crest, root, pitch circle, addendum, dedendum.

2.11 Derive the formula used to calculate velocity ratio of gears, and the power transmitted by Gearing.

2.12 Explain different type of gears.

2.13 Explain briefly: Idle gear, Gear Trains.

2.14 Explain and illustrate the functions of "rack and pinion" and "worm and worm-wheel."

2.15 Distinguish between a simple wheel train and a compound wheel-train.

2.16 (*a*) What are the functions of an idle-wheel in a wheel train?

(*b*) What are the advantages of gear-drive compared to flat belt drive.

2.17 Distinguish between spur gears, bevel gears, and spiral gears.

2.18 Discuss the advantages and disadvantages of group drive as against individual drive.

2.19 Illustrate with examples various methods used for transmitting power in a machine shop. Give reasons for adopting the particular drive in each case.

2.20 Describe the open belt and cross belt systems of drives used in power transmission.

2.21 Describe the various materials used for manufacturing belts.

2.22 Discuss merits and demerits of V-belt drive over flat belt drive.

2.23 Discuss various types of pulleys commonly used for the transmission of power.

2.24 Drive an expression for calculating power transmitted in a belt drive.

2.25 Suggest a method to transmit power from the driver shaft to driven shaft in the following circumstances: (*a*) when the two shafts have the same axis. (*b*) when the shafts are parallel to each other, (*c*) when the shafts are perpendicular to each other are intersecting and large velocity ratio is required, (*d*) when 1 meter apart two parallel shafts are required to transmit large amount of power.

2.26 What are belt and rope drives? Which of them is preferred and why?

2.27 Explain the phenomenon of (*i*) Creep, and (*ii*) Slip in belts.

2.28 How the chain - drives are classified? Discuss different common types of chains.

2.29 What are stepped or cone pulleys? Where are they used?

2.30 Name the different types of gear trains and state where each of them is used.

2.31 What is a clutch? What is its function? Where is it installed? Name the types of clutches in common use.

2.32 What is a friction clutch? Where is it used? Name the types of friction clutches.

2.33 What is a brake? Name the materials used for brake lining. State the difference between a clutch and a brake.

2.34 Name different types of brakes. Which type of brake is mostly used in automobiles and why?

FUNDAMENTALS OF THERMODYNAMICS AND HEAT TRANSFER

Chapter

3

Learning Objectives

After studying this chapter, the students will learn:

- Fundamentals of thermodynamics including important units, like temperature, heat, work, energy enthalpy and power.
- Zeroth, first and second law of thermodynamics.
- Thermodynamic processes.
- Heat engines and heat pumps.
- Carnot cycle and carnot theorem.
- Real gas, vapour and perfect gas.
- Heat transfer; conduction, connection and radiation
- Heat exchangers.

3.1 INTRODUCTION

Engineering thermodynamics is concerned with evaluation of energy transfers in the working substance of a machine. Particular part of working substance is called the 'system' and is separated from its surroundings. Two types of systems required are:

(*i*) Closed or non-flow system enclosed by boundary enveloping constant mass, and

(*ii*) Open or flow system within a control surface through which there is a mass flow rate.

The state of a system is defined by properties *e.g* pressure, temperature, volume and energy. Properties are usually expressed specifically *i.e* per unit mass. Two independent properties determine the state of a simple or pure substance.

Two recognized modes of energy transfer: (*i*) Work (transfer) and (*ii*) Heat (transfer). Work and heat (*i*) are not properties, (*ii*) are not contained in a system and (*iii*) can not be stored. They are transient phenomena, which may appear when a system changes its state. Work and heat may also be specifically expressed.

Thermodynamics is the field of science which deals with (*i*) energies possesed by gases and vapors (*ii*) their transformation in terms of heat and work and (*iii*) their relationship with the properties of the system. Thermodynamics also includes kinetic theory of gases and its application to the gas laws; the variation of the specific heats of gases are also considered in thermodynamic calculations. Other extensions include the design of rotary compressors, industrial gas turbines and turbo-jet units for air craft propulsion. Refrigeration has been extended to include chemical refrigerators and liquefaction of oxygen.

Applied thermodynamics deals with such special applications of energy transfers as steam generation, power generation, gas commpression and refrigeration.

In simple words thermodynamics can be defined as a science which deals with the heat and work. Heat is a basic source of energy available naturally whereas work is generally derived from heat. Heat is released by the combustion of fuels and it is converted into mechanical work with the help of an engine. A part of the heat energy is converted into work while the remaining is rejected at lower temperature to the surroundings. A machine which is commonly used for converting heat into mechanical work is known as **heat engine.** The ***working substances*** which are commonly used in the heat engines are gases and vapours. In heat engines, a fluid is always used to receive heat, expand and produce work output.

In **refrigerators** and **heat pumps** also a fluid is used to receive heat at low temperature and after an increase of pressure, rejects heat at a higher temperature. These working substances should have convenient properties for the purpose. The common working substances are used in the form of either gas or vapour. Air, steam and refrigerants are commonly used in work developing (engines) and work absorbing devices (compressor and refrigerators). It is essential that the properties of the working fluids should be known in order to understand their behaviour under different pressure and temperature conditions.

A mixture of air and fuel is used as working substance in internal combustion engines which are used to drive automobile vehicles, locomotives and ships. Steam is used as working substance in steam power plants. Ammonia and Freon-12 are used for ice plants and refrigerators respectively.

Thermodynamic System

A thermodynamic *system* is any space or any prescribed and identifiable group of matter enclosed by a *boundary* which may be real or imaginary. Everything outside the boundary which would affect the behaviour of the system is called the *surrounding.* Fig. 3.1 shows the usual representation of a thermodynamic system, system boundary and surroundings.

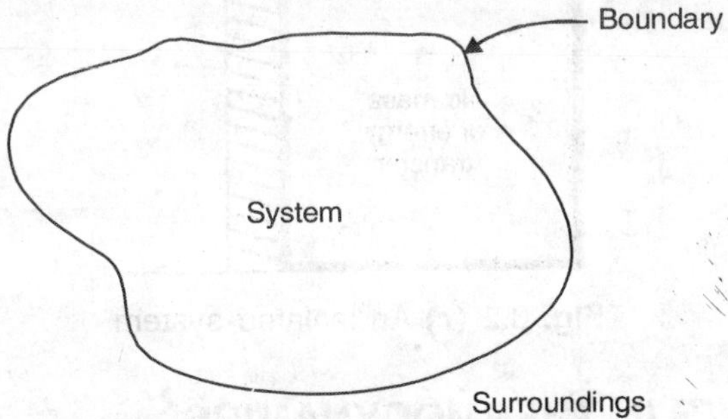

Fig. 3.1 A Thermodynamic System.

For example, compressed air bounded by the inner surface of the vessel may be considered as a thermodynamic system, the inner surface of the vessel will be considered as the system boundary while everything beyond the inner surface will be the surrounding.

A system and its environment together are said to constitute universe.

The system, as discussed in Chapter 1, can be classified as closed system, open system, and isolated system.

1. In closed system (also called batch or non-flow system) no matter may enter into or leave the system but energy may however, cross the system boundary (see Fig. 3.2(a)). It may however be noted here that the boundary of a closed system is not necessarily be fixed. It may move or change its position as and when the system expands or contracts, due to the transfer of energy from or to the surrounding.

2. In open system (also called flow system) the boundaries are not closed but have one or more openings through which mass transfer may also take place in addition to energy transfer (see Fig. 3.2(b)). The mass confined within the system boundary may remain constant or may change depending upon the rate of which matter enters into or leaves the system.

If the rate of mass transfer and energy transfer is constant with time, the system is called ***steady flow system.***

The boundary of an open system is usually stationary. Boilers, turbines nozzles, air compressors, etc., are the examples of an open system.

3. In isolated system neither mass nor energy is allowed to cross the system boundary (see Fig. 3.2(c)). Further such systems are not affected by the surroundings.

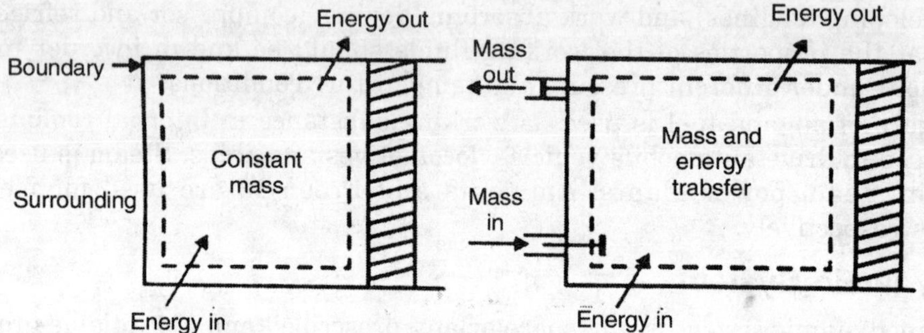

Fig. 3.2 (*a*) A closed system. **Fig. 3.2 (*b*)** An open system.

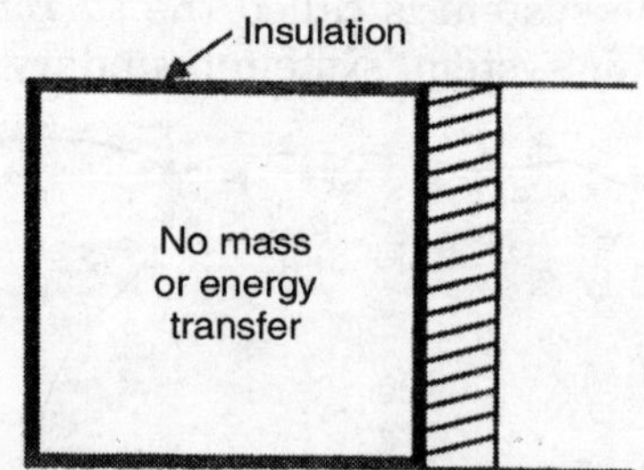

Fig. 3.2 (*c*) An isolated system.

3.2 IMPORTANT UNITS IN THERMODYNAMICS

3.2.1 Temperature

It is an intensive thermodynamic property and sensed directly. A body which feels hotter than another is said to have higher temperature. Similarly, a body having lower temperature is called a cold body. The temperature of a body is measured by a thermometer. There are two well known scales for the measurement of temperature.

***(i)* Centigrade Scale.** (Used by Celsius in 1742). This scale is obtained by dividing the range of temperature between the freezing and boiling points of water under atmospheric pressure into 100 equal divisions; the zero reading is at the freezing point of water. Then, freezing point of water = 0°C boiling point of water = 100°C

***(ii)* Fahrenheit Scale (first used in 1665).** This scale divides the range of temperature between the freezing and boiling points of water under atmospheric pressure into 180 equal divisions. Its zero is 32 divisions below the freezing points of water; hence, in Fahrenheit units,

Freezing point of water = 32°F

Boiling point of water = 32 + 180 = 212°F

To convert Fahrenheit thermometer readings to centigrade readings,

$$°C = \frac{100}{180} [°F - 32] \qquad ...(3.1)$$

Absolute Units of Temperature

The zero reading on centigrade and Fahrenheit scales are arbitrarily chosen for simplicity and will entirely suffice if it is required to know only the changes of temperature during a process or the temperature of a substance with respect to some known reference. But, when the readings are to be used in equations of fundamental laws, we must use temperature readings whose reference point is the true or absolute zero of temperature. According to Charles' law, it has been calculated that a perfect gas will occupy no volume at absolute zero temperature. According to kinetic theory of gases, the velocity of molecules of gases reduces to zero at zero degree absolute temperature.

Absolute temperature may be stated in °C or in °F. The point of absolute zero temperature is calculated to occur at 273°C, or 492°F, below the freezing point water. Then absolute temperature in °C = thermometer reading in °C + 273 ...(3.2)

absolute temperature in °F = thermometer reading in °F + 460 ...(3.3)

Absolute temperatures in degrees C are known as degrees Kelvin, or °K.

Absolute temperatures in degrees F are known as degrees Rankine, or °R.

Fig. 3.3 shows the relationship between different scales of temperature

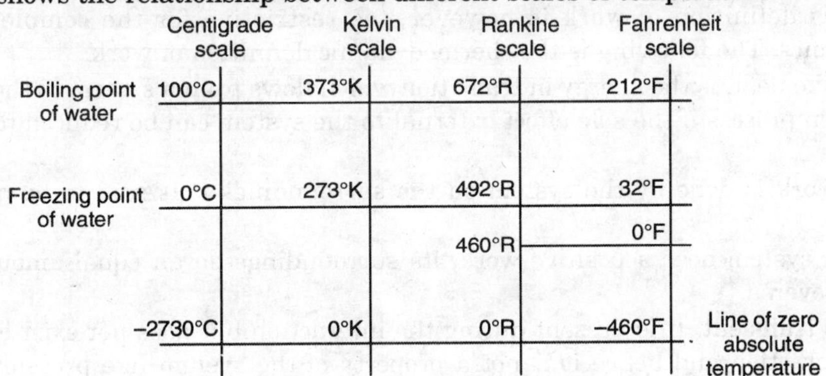

Fig. 3.3 Temperature relationship

3.2.2 Heat

Heat is energy in transition which flows from one system at higher temperature to another at lower temperature unaided when they are brought in contact. Heat, like work, is transitory in nature; it is never contained in the system. If heat is transferred from surrounding to the system, it is said to be positive for the system.

One CHU (Centigrade Heat Unit) is the amount of heat required to raise the temperature of one pound of water through 1°C.

One BTU (British Thermal Unit) is the amount of heat required to raise the temperature of the pound of water through 1°F.

One kcal is the amount of heat required to raise the temperature of one kg of water through 1°C.

S.I. unit of heat is Joule. It is the work done when a force of one Newton is applied over a displacement of one meter. It is also unit of energy or work. 1 cal = 4.19 Joules

Specific Heat

Specific heat of a substance is the heat required to raise the temperature of unit mass through unit degree of temperature. Solids and liquids have one specific heat, but gases have two specific heats.

Specific heat at constant pressure (C_p) is the amount of heat required to raise the temperature of unit mass of gas through unit degree temperature when the pressure is kept constant during heat transfer.

Specific heat at constant volume (C_v) is the amount of heat required to raise the temperature of unit mass of gas through unit degree temperature when the volume of gas is kept constant during heat transfer.

The specific heat at constant pressure is always greater than that at constant volume by the amount of work done in constant pressure expansion of unit mass of gas when change in temperature is one degree.

The ratio of C_p and C_v of any gas is constant and is denoted by γ

For air C_p = 1 kJ/kg K, C_v = 0.716 kJ/kg K and $\gamma = 1.4$. ...(3.4)

3.2.3 Work

In mechanics, work done is given by the product of force and displacement in the direction of force. This definition of work is however, too restrictive for the complex problems of thermodynamics. The following is the thermodynamic definition of work.

Work, like heat, is the energy in transition which flows from a system to the surroundings during a given process if the sole effect external to the system can be reduced to the raising of a weight.

When work is done by the system on the surroundings it is said to be positive for the system.

When a system does a positive work, its surroundings do an equal amount of negative work and viceversa.

Work is transient. It is present during the interaction but does not exist either before or after the interaction and hence it is not a property of the system like pressure, volume and temperature having finite values in a system at equilibrium state.

Work is also transferred in flow of electricity. The mechanical work, however, is virtually the only kind of work which can be conveniently delivered in any appreciable quantity to gaseous systems.

3.2.4 Energy

A system is said to possess energy when it is capable of doing some work.

Potential energy is a stored energy of a system by virtue of its height above some reference surface.

If m kg of a fluid is allowed to fall through a distance of h meters, then

$$\text{P.E. lost} = m.h. \text{ kg-m} \quad \text{...(3.5)}$$

$$= m.g.h. \text{ Newton-m or Joules} \quad \text{...[3.5(a)]}$$

Kinetic energy is a stored energy of a system by virtue of its motion. If 'm' kg of fluid is moving with a velocity of v m/sec., then,

$$\text{K.E.} = \frac{1}{2}\, m.v^2 \text{ Newton-m or Joules} \quad ...(3.6)$$

Flow energy is the energy due to transfer of mass across the boundaries of a system. To understand clearly the meaning of flow energy, let us consider a pipe through which a fluid is flowing under pressure (Refer Fig. 3.4).

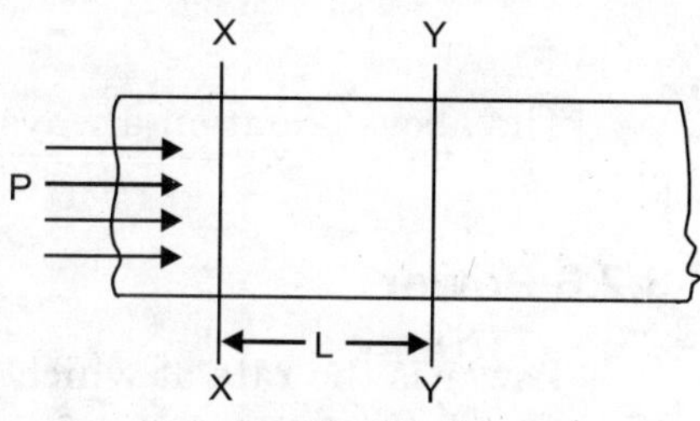

Fig. 3.4

Let A = area of flow in pipe in m^2)

L = distance between planes $X-X$ and $Y-Y$ (in m)

p = pressure of the fluid (in N/m^2)

Assuming the pipe to contain an imaginary piston, the face of which is located to X-X plane, then force exerted on the imaginary piston by the upstream fluid = $p.\,A.$ N

The work done by the upstream in pushing fluid (imaginary piston) from X-X plane to $Y-Y$ plane

$$= \text{force} \times \text{displacement}$$

$$= (p.A).L \text{ Nm or J}$$

or flow energy $= p.v$ Nm ...(3.7)

where v = volume swept by the imaginary pistion.

Since this work has been done upon the fluid, then it possesses energy by virtue of this work.

Internal energy is a stored energy by a system by virtue of its raised temperature above a reference temperature (absolute zero temperature). It is denoted by U. There is no definite condition at which U is zero.

Internal energy of a substance may also be defined as the stored energy by virtue of its molecular arrangement and activity.

According to Joule's law of gases the internal energy of a gas is a function of its temperature. It follows from this that the change of internal energy is proportional to the change of temperature.

or ΔU = constant $(T_2 - T_1)$ per kg of gas.

It has been found that this constant is C_v; hence

$$\Delta U = m.C_v(T_2 - T_1) \text{ kJ} \quad ...(3.8)$$

where C_v = specific heat at constant volume $\frac{\text{kJ}}{\text{kg.K}}$

Above equation always gives the change in internal energy, no matter how the gas is heated, cooled, expanded or compressed.

3.2.5 Enthalpy

Enthalpy or total heat energy of a system is the sum of its internal energy (U) and external energy Pv due to its pressure (P) and volume (v).

or Enthalpy $H = U + Pv$ kJ ...(3.9)

Like internal energy, there is no definite condition at which the enthalpy is zero. This makes hardly any difference since in dealing with thermodynamic process we are only concerned with the change of enthalpy. Thus

Change of enthalpy $\Delta H = m.C_v (T_2 - T_1) + P_2v_2 - P_1v_1$

$$\Delta H = m.C_p (T_2 - T_1) \text{ kcal for m kg of gas} \quad ...(3.10)$$

The above equation always gives the change in enthalpy irrespective of thermodynamic process.

3.2.6 Power

Power is the rate at which work is done. In S.I. system, the unit of power is watt, which *is* equivalent to 1 Nm/s. But for convenience kW is often used as the unit of power.

$$1\text{W} = 1 \text{ J/sec} = 1 \text{ Nm/sec.}$$

The unit of energy in S.I. system is kW hr which is equivalent to

$$= 1000 \times 3600 \text{ J} = 36 \times 10^5 \text{ Nm} = 3600 \text{ kJ}$$

3.3 PROPERTIES, STATE AND THERMODYNAMIC PROCESS

The state of a system at any equilibrium condition can be identified by a statement of any two independent properties, for example, pressure, volume and temperature of the system. If a property, or properties are changed, then the state is changed.

A series of infinitesimal changes in state of a system between any two states is called the path of change of state.

The state of a system can be represented by a diagram whose coordinates are properties. When the system goes from one equilibrium state to another equilibrium state due to energy transfer, the new state is represented by another point on the property diagram as shown in Fig. 3.5.

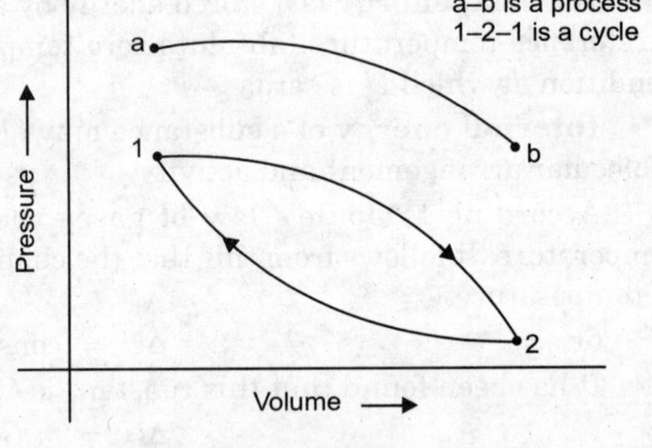

Fig. 3.5

The continuous series of states followed by a working medium as it liberates, transfers, transforms or receives energy is called a thermodynamic process.

Typical thermodynamic processes are the expansion and compression of a gas, conversion of water into steam and vice versa refrigeration and heating. A thermodynamic process can be analyzed by investigating into the changes in the properties of a system and the energy transfers which may have taken place.

The properties like internal energy, volume are proportional to the mass in the system are called **extensive properties.** The properties like temperature and pressure are not dependent on the mass of the system and are called **intensive properties.**

Thermodynamic Equilibrium

The properties of the working substance (temperature, pressure, volume) in the system are required to be measured to fix the state of the system. As the system goes through a process (change of state), the state of the system at each instant should be known in order to define the process. The state of the system can be fixed only if the system is in equilibrium. A system is said to be in thermodynamic equilibrium if no spontaneous change in the properties of the system can occur even after a catalytic or other such small disturbance. Such a system is in complete balance with the restraints offered by the surroundings for a system to be in thermodynamic equilibrium it must be in physical, chemical & thermal equilibrium simultaneously.

Process

A process is said to occur when the system undergoes a change in state. If the process occurs at a very fast rate, the intermediate equilibrium conditions are not definable. For convenience, the process is assumed to occur at such a rate that intermediate equilibrium conditions can be described.

Cyclic Process

If the number of processes in sequence bring the system back to its initial state, the system is said to execute a cycle or cyclic process. The net effect of a cyclic process may be conversion of the heat received into work (engine) or maintaining a system at a lower temperature than the surroundings (refrigerator) by means of work input.

As discussed earlier the cyclic processes are classified as closed cycles and open cycles. In a **closed cyclic** process, the same working substance is used again and again and only heat and work transfers take place between system and surrounding. The steam power plant and refrigeration system work on closed cycle.

In an **open cycle** system, the working fluid once used during that cycle process is thrown out and new mass of working fluid is taken in during the next cyclic process. I.C. engines and gas turbines work on open cycle.

3.4 ZEROTH LAW OF THERMODYNAMICS

Equality of temperature of two or more bodies is defined as that condition in which no change in observable properties occur when they are brought in communication with each other. *Two bodies which are equal in temperature to a third body are equal in temperature to each other.* This statement is called the *Zeroth Law of Thermodynamics.* Fig. 3.6 shows three bodies *A, B* and *C*. Suppose *A* and *B* are at equal temperature, so that no change occurs in physical properties when they are brought into contact with each other. Similarly, let *A* and *C* be at equal temperature so that no change occurs when they are brought into contact. If now *B* and *C* are placed in contact it is an observed fact that there occurs no change in their physical properties. Therefore *B* and *C* must be equal in temperature.

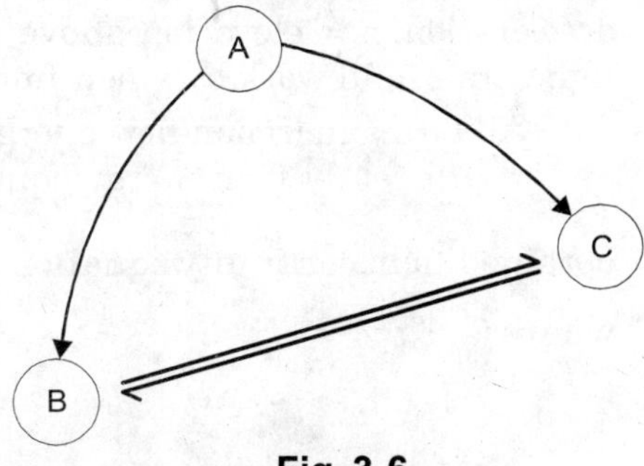

Fig. 3.6

This, zeroth law deals with thermal equilibrium and must precede the three laws (first, second and third laws) of thermodyamics. Therefore this law was called the Zeroth Law of thermodynamics.

3.5 FIRST LAW OF THERMODYNAMICS

The law of conservation of energy states that the energy can neither be created nor destroyed. However energy can be converted from one form into another form.

The first law of the thermodynamics is a more rigorous statement of this general principle with reference to energy in the form of heat and Mechanical work. The first law of thermodynamics states that heat and work are mutually convertible. However it does not say anything about the possibility and method of conversion of heat into work or work into heat.

Both open and closed systems obey law of conservation of energy. First law states that "the final energy of a system is equal to the initial energy plus that added or subtracted by heat and work transfers". It embraces a sign convention *viz.* the heat added to the system will be considered as +ve and the heat rejected from the system is –ve. Work done by the system will be +ve and the work done on the system as –ve. This is expressed as:

$$Q - W = \Delta E \qquad ...(1)$$

where Q = Heat transfer, W = Work transfer, and ΔE = change in energy in the sense final energy minus initial energy.

When applied to non-flow system, only internal energy U is relevant and non-flow energy equation obtained is:

$$Q - W = \Delta U \qquad ...(2)$$

This may be specifically expressed as

$$q - w = \Delta u \qquad ...(3)$$

where q = specific heat transfer, w = specific work transfer, and Δu = specific internal energy change.

When applied to flow system, kinetic energy and potential energy are also considered and steady flow energy equation obtained as:

$$Q - W_x = m\,\Delta\left(h + \frac{V^2}{2} + gz\right)$$

or specifically as

$$q - w_x = \Delta\left(h + \frac{V^2}{2} + gz\right) \qquad ...(4)$$

where Q = heat transfer rate. W_x = Work transfer rate or power, m = mass flow rate, h = specific enthalpy given by $h = u + pv$, p = pressure, v = specific volume, g = gravitational acceleration, z = elevation above some datum and V = velocity. Suffix 'x' attached to work indicates useful work obtained from the system.

All terms in (*i*) non-flow energy equation will be expressed in joules (*J*) or if specific, J/kg and in (*ii*) steady flow energy equation in watts (W) or if specific, J/kg. Mass flow rate m is obtained from continuity equation $m = \dfrac{AV}{v}$ ...(5)

where
A = cross sectional area of flow
V = mean velocity
v = volume

Idealized processes are frictionless and have heat transfers over infinitely small temperature differences. They are termed 'Reversible' and may be represented by mathematical relation between properties as:

$$pv^n = \text{constant}$$

It is called "Polytropic process law".

According to first law of thermodynamics, heat and work are mutually convertible.

The first law of thermodynamics is also known as the law of conservation of energy *i.e.* energy is neither created nor destroyed but one form of energy can be transformed to another form. Hence, the total energy associated with an energy conversion remains constant.

Joule, a scientist, carried out experimental tests in the year 1840-1849, and found that the ratio 'J' between the amount of work and the quantity of heat that would produce identical effects in the system had the same value.

$$J = \frac{\text{Work}}{\text{Heat}} = \text{Constant} \qquad \text{...(6)}$$

J is also known as "mechanical equivalent of heat", and has following values:

$$1\,J = 426.7 \text{ N/kcal}$$
$$= 4.18 \text{ J/cal.}$$

Limitation of the First Law of Thermodynamics

Assuming no changes in chemical composition of the system and taking it to be homogeneous and in thermodynamic equilibrium at all stages of process, the first law of thermodynamics for a closed system informs about the change of internal energy of the system along with transfer of heat and work across the system boundaries. Even for steady flow system, we are concerned with change of enthalpy of system along with transfer of heat and work neglecting changes of potential and kinetic energies. In other words first law of thermodynamics gives us relation between heat, work and properties of the system.

When heat is supplied to a system, without doing any external work, all the heat energy supplied is stored in the system in the form of increased internal energy (property). Now question arises, whether we can take work out of the system equivalent to the heat energy supplied and stored by the system in the form of increased internal energy. The answer is no. In other words, it is impossible to convert all the heat energy supplied to a system into work.

The above observation is valid in general for all polytropic processes, where transfer of heat and work and changes in the properties of the system occur simultaneously.

The following observations too are evident of the fact that processes proceed in certain direction and not in opposite direction although reversed processes do not violate the first law:

1. Consider a thermally insulated vessel containing water, let the work be done by a paddle wheel when the weights are being lowered. The temperature and consequently the internal energy of water would increase. But it is not possible to perform the process in reverse direction i.e. with the cooling of water to original state, the paddle wheel mechanism will not turn and raise the weights.
2. Consider a car wheel which is being stopped by applying brake. The brake and drum will get heated and the increase in internal energy will be equal to the decrease in kinetic energy of the wheel. The first law of thermodynamics will be still valid if on cooling, the brake and drum lose internal energy and rotate the wheel again. But, the reversed process however does not actually occur.

3. In free expansion of a perfect gas from high pressure chamber to an evacuated chamber through a valve, there is no change in temperature of the gas. There is however increase in volume and fall in pressure. But it is not possible for the gas at low pressure to rush back through the valve and compress itself into its original chamber.

From the above observations it is concluded that the first law of thermodynamics does not help to predict whether the system under consideration will or will not undergo a process it does not impose any restriction on the amount of conversation of energy from one form into another form. It also does not indicate whether all types of energies are equally amenable to transformation in other forms. Therefore we cannot determine as to what portion of heat energy may be converted into useful work. In actual practice it is known from experience that additional restrictions do exist and all form of energy are not equally amenable to transformation. Thus, it becomes necessary to formulate other laws to give information about the restrictions to be followed in designing various devices involving transformation of energy.

Thus it is concluded that the first law of thermodynamics is necessary but not sufficient condition for a process to occur.

3.6 SECOND LAW OF THERMODYNAMICS

From the observations made in the preceding article it is clear that the first law of thermodynamics only fixes the exchange rate between heat energy and mechanical work. It neither specifies the conditions under which the conversion of heat energy into mechanical work is possible nor the direction in which heat energy transfer can take place.

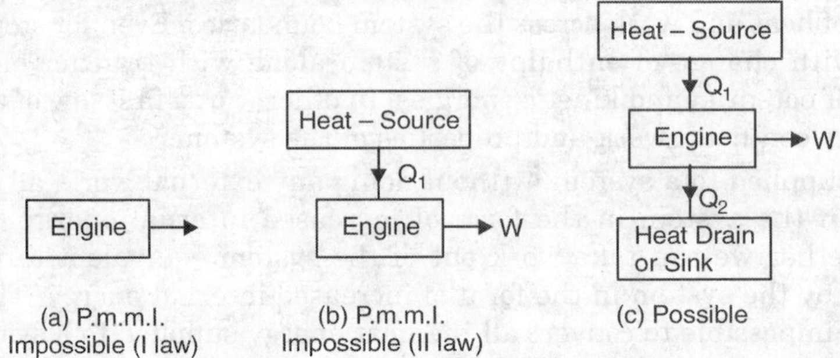

Fig. 3.7

According to the first law of thermodynamics, a perpetual motion machine of the first kind (a machine which produces power without consuming any energy) shown in Fig. 3.7 (*a*) is impossible.

The **second law of thermodynamics can be stated by two well known statements** as follows:

1. Kelvin Planck Statement. It states: "It is impossible to construct an engine working on cycle process, whose sole purpose is to convert all the heat supplied to it into equivalent amount of work".

Thus according to the second law a perpetual motion machine of the second kind (100%, η engine violating second law) shown in Fig. 3.7(*b*) is impossible. Fig. 3.7(*c*) shows the possible engine which receives Q_1 amount of heat from the heat source at higher temperature T_1, rejects Q_2 amount of heat to the heat drain or sink at lower temperature T_2 and delivers $[W = (Q_1 - Q_2)]$ amount of work to the surrounding per cycle continuously. Thus

$$\text{Thermal } \eta \text{ of the engine} = \frac{W}{Q_1} = \frac{Q_1 - Q_2}{Q_1} \qquad ...(3.11)$$

2. Clausius Statement. "It is impossible for the heat to flow from a body at lower temperature to another body at higher temperature without the expenditure of mechanical work". See Fig. 3.8(*a*).

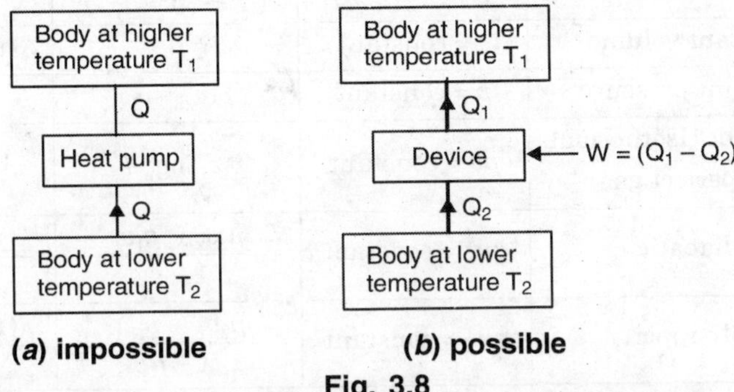

(*a*) **impossible** (*b*) **possible**

Fig. 3.8

If it is required to transfer heat energy Q_2 from a body at lower temperature T_2 to another body at higher temperature T_1, mechanical work $W = (Q_1 - Q_2)$ will have to be spent to do so. See Fig. 3.8(*b*).

Thus, Coefficient of performance (C.O.P.) of device as refrigerator

$$= \frac{Q_2}{W} = \frac{Q_2}{Q_1 - Q_2} \qquad ...[3.12(a)]$$

and C.O.P. of device as heat pump

$$= \frac{Q_1}{W} = \frac{Q_1}{Q_1 - Q_2} \qquad ...[3.12(b)]$$

Although two statements are different, yet the two are inter-linked *i.e.*, it is impossible to have a system satisfying one statement and violating the other.

There is no direct proof for the second law of thermodynamics. Only indirect proof that is known that no engine or heat pump has been constructed violating the second law of thermodynamics.

When a series of processes brings a working substance back to its initial state, a cycle has been performed. In a cycle, there is no resultant change of state so that there can not be any change in the value of the property energy and the first law becomes

$$\sum_{cycle} Q = \sum_{cycle} W$$

where $\sum_{cycle}$ means "the sum around the cycle."

Primary object is to have a continuous work transfer from a continuous heat transfer. Second law of thermodynamics states "this can not be achieved without wasting a proportion of initial heat transfer by a second heat transfer to a lower temperature sink". In other words "it is impossible to build a machine, which will give continuous positive work transfer exchanging heat with a single reservoir".

Table 3.1. Work Transfer in Processes

S.No.	*Name of process*	*Law*	*Non – flow process ; specific work* $w = \int_1^2 pdv$	*Flow process ; specific work* $w_x = -\int_1^2 vdp$
1.	Constant volume	v = constant	0	$v(p_1 - p_2)$
2.	Constant pressure	p = constant	$p(v_2 - v_1)$	0
3.	Hyperbolic (Isothermal for a perfect gas)	pv = constant	$p_1v_1 \log_e \frac{v_2}{v_1}$	$p_1v_1 \log_e \frac{p_1}{p_2}$
4.	Adiabatic	pv^k = constant	$\frac{p_2v_2 - p_1v_1}{1-k}$	$\frac{k(p_2v_2 - p_1v_1)}{1-k}$
5.	Polytropic	pv^n = constant	$\frac{p_2v_2 - p_1v_1}{1-n}$	$\frac{n(p_2v_2 - p_1v_1)}{1-n}$

Note: $k = C_p / C_v$

In a heat engine with heat transfer Q_1 from a hot reservoir and heat transfer Q_2 to a cold reservoir giving work transfer W ; by first law: $W = Q_1 - Q_2$. Measure of success of any heat engine is known as "Thermal efficiency, η",

where
$$\eta = \frac{\text{Net work transfer from engine}}{\text{Heat transfer to engine}} = \frac{W}{Q_1}$$

In an initial heat engine, all processes and all energy transfers will be reversible. Engine is termed 'reversible' and it will have maximum possible efficiency. It is known as 'Carnot efficiency'.

Considering second law mathematically, leads to two fundamental properties: (*i*) Thermodynamic temperature T and (*ii*) Entropy S related in such a way that for reversible process from a state 1 to state 2, specific heat transfer

$$q = \int_1^2 T\, ds$$

Also, carnot efficiency for reversible heat engine may be expressed by

$$\eta = 1 - \frac{T_2}{T_1}$$

where T_1 = maximum temperature in the cycle (hot reservoir), T_2 = minimum temperature (cold reservoir). Above equation is an important relation as it tells max. possible efficiency attainable from given max. and min. temps. from any power producing plant, using thermal energy.

In other words, second law of thermodynamics also explains the following:

(*i*) No process is possible whose sole effect is the removal of heat from a single reservoir and the performance of an equivalent amount of work in a perpetual motion machine of the second kind.

(*ii*) Heat, by itself, can not flow from a lower temperature system (heat sink) to a system at a higher temperature (heat source).

Note. A fuel cell, utilizing chemical energy converted directly to work, is not restricted by Carnot efficiency.

Considering the process further in which q is zero (called adiabatic), if process is reversible, then

$\int T\, ds = 0$ and process is one of constant entropy.

This is a very useful concept as many real processes are approximately adiabatic and may be idealized for analysis as 'reversible adiabatic processes'. In real process, friction causes temp. of working substance to reach a higher value than in ideal process. Thus, "in real adiabatic processes, the entropy always increases".

The process efficiency for an adiabatic process is known as "Isentropic efficiency" and is used for determining real work from ideal work obtained by analysis. For an expansion, isentropic efficiency,

$$\eta_{isen.} = \frac{\text{Actual work transfer}}{\text{Ideal work transfer}}$$

and for a compression,
$$\eta_{isen.} = \frac{\text{Ideal work transfer}}{\text{Actual work transfer}}$$

Machines usually considered adiabatic include steam and gas turbines, gas compressors, and nozzles.

In practice two distinct groups of machine are used:

(*i*) Those in which working substance experiences liquid vapour phase change, and

(*ii*) Those in which working substance remains a gas at all times.

For common working substances, empirical data tables and charts are available. Values of enthalpy, entropy and energy changes in any process may be determined from these tables. Normal entry to a table is by pressure and temperature, but in case of wet vapour, these properties are not independent, and dryness fraction x is used with pressure and temperature:

Dryness fraction,
$$x = \frac{\text{Mass of dry vapour}}{\text{Mass of mixture}}$$

3.7 HEAT ENGINES AND HEAT PUMPS

A **heat engine** is a device that, operating in a cycle, produces work and exchanges with heat sources and sinks. The heat exchanges in the cycle usually takes place involving a fluid (called the working fluid). Fig. 3.9 (*a*) shows a heat engine receiving heat Q_1 from source which is a high temperature thermal energy reservoir at constant temperature T_1. The heat engine, working in a cycle converts part of this heat to work W, and rejects the remaining waste heat Q_2 to a low temperature thermal energy reservoir (sink) at a constant temperature T_2.

By law of conservation of energy, $W = Q_1 - Q_2$.

The thermal efficiency of the heat engine is defined as ;

$$\text{Thermal efficiency} = \frac{\text{Net work output}}{\text{Total heat output}}$$

$$\eta_{\text{thermal}} = \frac{Q_1 - Q_2}{Q_1} = 1 - \frac{Q_2}{Q_1}$$

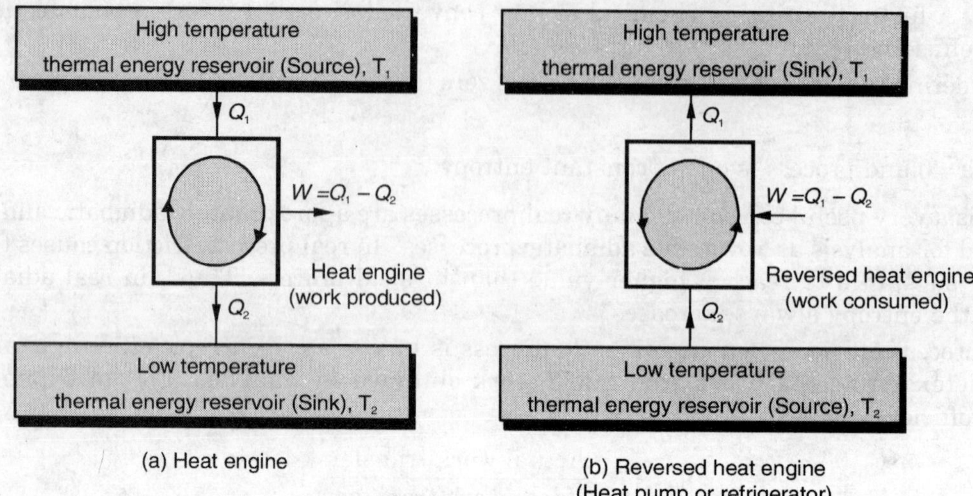

(a) Heat engine

(b) Reversed heat engine (Heat pump or refrigerator)

Fig. 3.9 Heat engine and reversed heat engine.

Example of heat engine is a steam power plant.

Reversed Heat Engine

If a heat engine is thermodynamically reversed, it is called a reversed heat engine as shown in Fig. 3.9 (*b*). A reversed heat engine abstracts heat Q_2 from a low temperature thermal energy reservoir at constant temperature T_2 (source), is supplied external work W, and working in a cycle delivers total of $(Q_2 + W) = Q_1$ in the form of heat, to a high temperature energy reservoir at constant temperature T_1 (sink).

$$W = Q_1 - Q_2$$

Reversed heat engines are of two types as under:

1. Heat Pump. If the objective of reversed heat engine is the supply of heat to high temperature energy reservoir, it is called a heat-pump, and its performance is given by:

$$\text{Performance (C.O.P.) of heat pump} = \frac{\text{Desired Output}}{\text{Required Output}} = \frac{Q_1}{W} = \frac{Q_1}{Q_1 - Q_2}$$

2. Refrigerator. If the objective of the reversed heat engine is the abstraction of heat from low temperature reservoir, it is called a refrigerator, and its performance is given by:

$$\text{Performance (C.O.P.) of refrigerator} = \frac{\text{Desired Output}}{\text{Required Output}} = \frac{Q_2}{W} = \frac{Q_2}{Q_1 - Q_2}.$$

Example 3.1 *A heat engine is used to drive a heat pump. The heat transfers from the heat engine and from the heat pump are used to heat the water circulating through the radiators of a building. The η of the heat engine is 27% and the COP of heat pump is 4. Show that the ratio of the heat transfer to the circulating water to the heat transfer to the engine is 1:81.*

Solution.

$$\eta \text{ of heat engine} = \frac{\text{Work done}}{\text{Heat supplied}}$$

or

$$0.27 = \frac{Q_1 - Q_2}{Q_1}$$

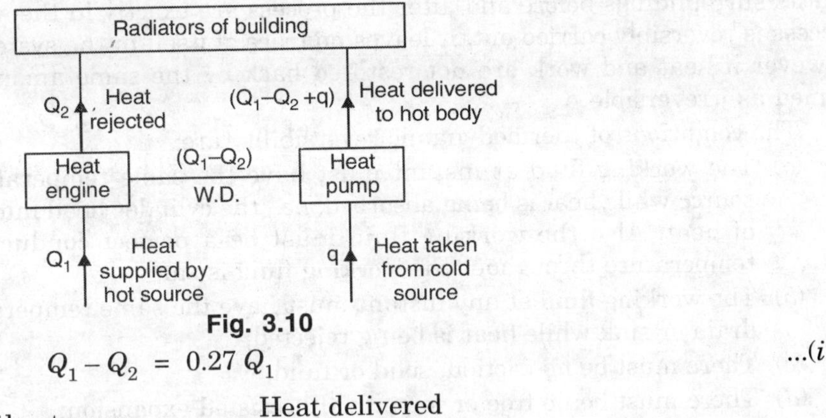

Fig. 3.10

or $$Q_1 - Q_2 = 0.27\, Q_1 \quad ...(i)$$

$$\text{C.O.P. of heat pump} = \frac{\text{Heat delivered}}{\text{Work done to drive pump}}$$

or $$4 = \frac{(Q_1 - Q_2) + q}{(Q_1 - Q_2)} \quad ...(ii)$$

Substituting value of $(Q_1 - Q_2)$ from Eq. (*i*) in Eq. (*ii*) we get

$$4 = \frac{0.27Q_1 + q}{0.27Q_1} = 1 + \frac{1}{0.27} \cdot \frac{q}{Q_1}$$

or $$\frac{q}{Q_1} = 0.81 \quad ...[ii(a)]$$

Now heat transfer to the circulating water of radiators = (heat rejected by engine + (heat delivered by pump)

$$= Q_2 + Q_1 - Q_2 + q$$

$$= (Q_1 + q)$$

and heat transfer to the engine = Q_1

$\therefore$ $$\text{Ratio} = \frac{Q_1 + q}{Q_1} = 1 + \frac{q}{Q_1}$$

$$= 1 + 0.81 \quad [\because \text{From Eq. } (ii)a]$$

or ratio = 1 : 81 hence proved.

3.8 REVERSIBLE AND IRREVERSIBLE PROCESSES

Let us consider a process in which system moves from state point (1) to state point (2), causing a change in its internal energy *(dU)*, transforming work (– δ*W*) *(i.e.*, work is done on the system by the surroundings) and rejecting heat (δ*Q*) to the surroundings such that $-\delta Q = dU - \delta W$.

If now the system desires to move back from state point (2) to state point (1), it may or may not be possible for the system to absorb heat (δ*Q*) from the surroundings and deliver work (δ*W*) to the surroundings. A process is thermodynamically reversible if it can be carried out in reverse direction in exactly the same manner as in the forward direction and if the system and its surroundings before and after the process are exactly in the same state. Thus when a process is reversibly carried out, it leaves no trace of itself in the system and its surroundings. However if heat and work are not restored back by the same amounts, the process will be termed as irreversible.

The conditions of thermodynamic reversibility are:

(*a*) The working fluid at instant must have the same temperature as that of the heat source while heat is being absorbed, *i.e.*, the cylinder head must be a perfect conductor of heat. Also the working fluid must be a perfect conductor of heat so that the temperature throughout the working fluid is same;

(*b*) The working fluid at any instant must have the same temperature as that of the heat drain or sink while heat is being rejected;

(*c*) There must be no friction, solid or fluid;

(*d*) There must be no free or imperfectly resisted expansion;

(*e*) The process must proceed in series of equilibrium state *i.e.*, the process must proceed at infinitely slow pace.

Isothermal and adiabatic processes are the only processes which are reversible. But even adiabatic process becomes irreversible when internal friction (i.e. friction between the molecules of working fluid and cylinder surface or friction among the molecules) is generated. Examples of this are in turbines and rotary compressors where there is friction due to high speed.

The concept of reversibility, though hypothetical, is very important in thermodynamics because a reversible process is the most efficient process. Reversible process alone can be truly represented on P-V diagram. In practice thermodynamic reversibility can be approached but cannot be achieved.

3.9 REVERSIBLE CYCLE (CARNOT CYCLE)

In an isothermal expansion the whole of the heat supplied is converted into mechanical work, so obviously this process should be employed in any engine which is to be efficient. For the continuous development of power, however, a cycle is essential and this involves a change of temperature. An isentropic expansion will produce a change of temperature without transference of heat and its associated loss.

Using these processes alone, Sadi Carnot, in 1824, discovered a heat engine cycle which for a given temperature range would give the highest possible thermal efficiency; other cycles give as high, but none higher.

Carnot specified a non-conducting cylinder and piston which is a perfectly free but gas tight and a cylinder head so arranged that it could be a perfect heat conductor as well as a perfect heat insulator. Between the piston and cylinder head was entrapped the working fluid (see Fig. 3.11).

For the analysis of Carnot cycle applied to an ideal heat engine, consider the cylinder to contain air at temperature T_1, pressure p_1 and volume v_1 (state point (1) on *p-v* diagram) at the beginning of stroke. Heat is added isothermally from an infinite external source such that transfer of heat from it does not affect its temperature T_1 through a perfectly conducting cylinder head into a perfectly conducting gas, resulting in isothermal expansion to state point (2). The pressure and the volume become p_2 and v_2 respectively. During this process (1 – 2) external work is done by the air on the piston ($W_1 = RT_1 \log_e \frac{v_2}{v_1}$ per kg of air) with corresponding heat absorbed by the air given by Q_1.

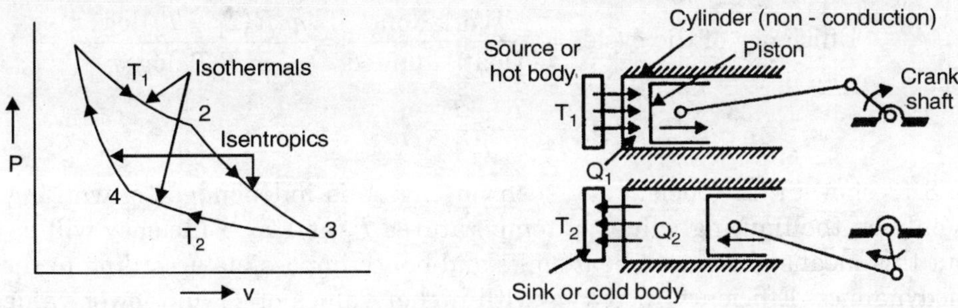

Fig. 3.11

At (2) the source of heat is removed and the cylinder head which was a perfect conductor is now insulated perfectly. The air in the cylinder is now allowed to expand further to state point (3) isentropically. The temperature falls from T_1 to T_2. No heat is exchanged during the process (2 – 3) but external work is done at the expense of internal energy.

Now at (3) air is brought into contact with the sink of infinite capacity (such that transfer of heat to it does not increase its temperature) at temperature T_2 the insulation is removed from the cylinder head to make it again a perfect conductor. The air is now compressed isothermally from state point (3) to state point (4). During this process, external work is done on the air $\left(W_2 = RT_2 \log_e \frac{v_3}{v_4} \right)$ and a corresponding heat Q_2 is rejected by the air to the sink.

At (4) sink is removed and the cylinder head which was a perfect conductor is again insulated perfectly. The air in the cylinder is compressed further to state point (1) isentropically. The temperature increases from T_2 to T_1. No heat is exchanged during the process falls from T_1 to T_2. No heat is exchanged during the process (4 – 1) also but external work is done on the air with the increase in internal energy. In this way cycle is completed.

Since the air returns back to original state point (1), hence according to first law of thermodynamics, net work done during the cycle must equal the net heat transfer during the cycle. Hence work done per cycle

$$= \left[m.RT_1 \log_e \frac{v_2}{v_1} - m.RT_2 \log_e \frac{v_3}{v_4} \right]$$

$$= (T_1 - T_2)\, m.R \log_e r$$

where $$r = \frac{v_2}{v_1} = \frac{v_3}{v_4}.$$

It is obvious that since the cycle of operation is limited by two pairs of parallel curves (two isothermal and two isentropics), air should be compressed isothermally as it had expanded isothermally.

Hence $$r = \frac{v_2}{v_1} = \frac{v_3}{v_4}.$$

or in other words, the value of expansion ratio and compression ratio must be equal; otherwise the cycle will not close.

$$\text{Efficiency of the cycle} = \frac{\text{Work done}}{\text{Heat supplied}} = \frac{m.R(T_1 - T_2)\log_e r}{m.R\, T_1 \log_e r}$$

or
$$\eta = \frac{T_1 - T_2}{T_1} \quad ...(3.13)$$

Eq. 3.13 shows that efficiency of such an engine is independent of working fluid but it depends only on the limiting values of temperatures T_1 and T_2. Efficiency will be 100% when $T_2 = 0$ but this means absence of heat sinks and hence impossible according to the second law of thermodynamics. Efficiency increases with higher values of T_1 and lower values of T_2. But temperature T_2 is generally the atmospheric temperature and T_1 is controlled by the properties of the materials available for the construction of the engine, and with the ability to lubricate piston.

Implication of Carnot Cycle. Carnot cycle gives the maximum possible thermal efficiency which can be obtained from a given temperature limits. Therefore it becomes the aim of an engineer to construct an engine working on Carnot cycle. The cycle is however impracticable due to the following reasons.

(*i*) Isothermal and adiabatics of the gas have a very insignificant difference in their slopes and as such the cycle is very narrow as compared to its length and therefore, a very small amount of useful work is obtained corresponding to a very long stroke length;

(*ii*) During isothermal operation the piston should move very slowly, so as to allow the transfer of heat to maintain constant temperature whereas for adiabatic operations the piston movement should be very fast so that gas does not have time to transfer heat.

The variation of speed during a stroke of piston of a practical engine is not possible;

(*iii*) In actual working of the engine the friction exists between the moving parts which makes the cycle irreversible.

Example 3.2 *By using perfect gas show that for Carnot cycle the cyclic integration of* $\frac{\delta Q}{T} = 0$.

Solution. *Refer Fig. 3.11 Let* $p_1\, v_1\, T_1$; $p_2, v_2,$ *and* $T_1\, p_3, v_3,$ *and,* $T_2\; p_4, v_4, T_2$ *denote the state of the working fluid at points 1, 2, 3 and 4 respectively.*

For isothermal process 1 – 2

$$\int \frac{\delta Q}{T} = \frac{mRT_1}{J.T_1} \log_e \frac{v_2}{v_1}$$

For isentropic process 2 – 3 $\int \frac{\delta Q}{T} = 0$

For isothermal process 3 – 4

$$\int \frac{\delta Q}{T} = \frac{mR}{J} \log_e \frac{v_4}{v_3} = -\frac{mR}{J} \log_e \frac{v_3}{v_4}$$

And finally for isentropic process 4 – 1

$$\int \frac{\delta Q}{T} = 0$$

Therefore, cycle integration of $\frac{\delta Q}{T}$

$$= mR \log_e \frac{v_2}{v_1} - mR \log_e \frac{v_3}{v_4}$$

But
$$\frac{T_1}{T_2} = \left(\frac{v_3}{v_2}\right)^{\gamma - 1} = \left(\frac{v_4}{v_1}\right)^{\gamma - 1}$$

$$\therefore \quad \frac{v_3}{v_2} = \frac{v_4}{v_1} \text{ or } \frac{v_3}{v_4} = \frac{v_2}{v_1}$$

$\therefore$ Cyclic integration of $= \frac{\delta Q}{T} = 0$.

3.10 CARNOT'S THEOREM

From the second law of thermodynamics two important statements about the reversible cycles can be proved. These statements which are known as Carnot's theorem are:

(*i*) No heat engine, operating in cycles between two reservoirs (heat source and heat drain or sink) at constant temperatures (T_1 and T_2) can have a greater efficiency than a reversible engine (e.g. Carnot's engine) operating between the same two reservoirs.

(*ii*) All reversible engines operating between two reservoirs at constant temperature have equal efficiencies.

The proof is as follows:

Let there be a heat source at constant temperature T_1 and a heat sink at lower constant temperature T_2. Imagine a reversible engine absorbing 100 kJ of heat from the heat source and rejecting 40 kJ of heat to the heat sink, the thermal efficiency of the engine being 60%. If this engine is reversed to work as heat pump, it will take 60 kJ of work to drive it, another 40 kJ of heat will be taken from the heat sink and will deliver 100 kJ of heat to the heat source as shown in Fig. 3.12.

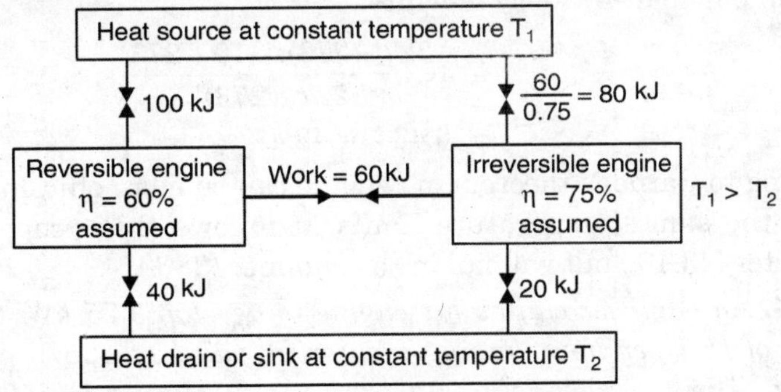

Fig 3.12

For the sake of discussion let us assume an irreversible engine having efficiency 75% (greater than that of the reversible engine), and driving the reversible engine without any losses. Now for the 60 kJ of work (to be delivered to the reversed engine), the irreversible engine will absorb 80 kJ of heat from the heat source and reject 20 kJ of heat to the sink.

Now taking reversed engine and irreversible engine whole as a system, (100 – 80 = 20 kcal) of heat is being pumped to the heat source at higher temperature T_1 from the heat sink at lower temperature T_2 without consuming any external work, which is contrary to the second law of thermodynamics. Hence the assumption that irreversible engine is more efficient than the reversible is wrong.

By assuming another reversible engine in place of the irreversible engine and applying the same reasoning as above, it will be seen that one reversible engine cannot be more efficient than another reversible engine working between the same heat reservoirs and between the same temperature limits. Hence, the conclusion is derived that the reversible engines are the most efficient engines and all reversible engines have same efficiency working between the same temperature limits.

Example 3.3. *It is proposed to utilize 9000 kJ of heat per hour from a heat source at constant temperature 227°C. The engine is designed to produce 1 kW operating continuously between the above heat source and atmosphere as heat sink. The atmospheric temperaturic averages 37°C in Summer and 10°C in Winter. Comment on the performance of the proposed engine.*

Sol. 1 kW = 3600 kJ/hr

$$\therefore \text{ Efficiency of the proposed engine} = \frac{\text{Work output}}{\text{Heat absorbed}}$$

$$= \frac{3600}{9000} = 40\%$$

Maximum possible efficiency i.e. Carnot engine efficiency in summer.

$$= \frac{T_1 - T_2}{T_1} = \frac{(227 + 273) - (37 + 273)}{(227 + 273)}$$

$$= 0.38 \text{ or } 38\%$$

and maximum possible efficiency in winter

$$= \frac{(227 + 273) - (10 + 273)}{(227 + 273)}$$

$$= 0.434 \text{ or } 43.4\%$$

Since according to Carnet's theorem, no engine can be more efficient that Carnot's engine working between the same temperature limits, it follows that proposed engine can work successfully in winter (43.4%) but will fail in the summer (38%).

Example 3.4. *An engineer claims his engine to develop 3.75 kW. On testing, the engine consumes 0.44 kg of fuel per hour having a calorific value of 42670 kJ/kg. The maximum temperature recorded in the cycle is 1400°C and the minimum is 350°C. Find whether the engineer is justified in his claim.*

Sol. Heat supplied to the engine = .44 × 42670 kJ

Work output of the engine = 3.75 × 3600

= 13500 kJ/hr

$\therefore$ Efficiency of the engine $= \dfrac{\text{Work output}}{\text{Heat supplied}}$

$$= \frac{13500}{0.44 \times 42670} = 0.719 \quad \text{or} \quad 71.9\%$$

Maximum possible efficiency of a reversible engine (*i.e.*, Carnot's engine), working between above temperature limits

$$= \frac{(1400 + 273) - (350 + 273)}{(1400 + 273)}$$

$$= 0.6276 \text{ or } 62.76\%$$

As no engine can be more efficient than Carnot's engine working between the same temperature limits, it follows, that the claim of the engineer is wrong.

Example 3.5 *A Carnot's engine absorbs heat from an infinite heat source A and rejects heat to another infinite heat sink B. Half of the work developed by this engine is used to drive an electric generator (100% efficient) and remaining half is used to drive a reversed Carnot's engine receiving heat from the infinite heat reservoir B and rejecting heat to an infinite heat reservoir C. Express the heat rejected to C by the reversed engine as a percentage of heat supplied from A to the first Carnot's engine and calculate the heat rejected per hour to C if 300 kW is supplied to the generator.*

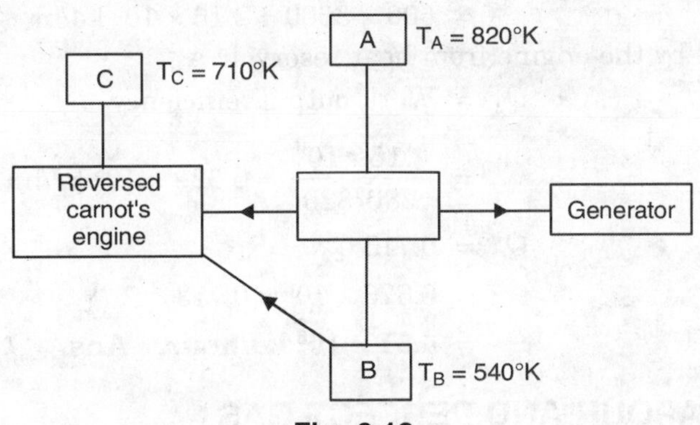

Fig. 3.13

Sol. Efficiency of Carnot's engine $= \dfrac{T_A - T_B}{T_A}$

$$= \frac{820 - 540}{820} = \frac{280}{820} = 0.3415 \quad \text{or} \quad 34.15\%$$

Let Q_A be the amount of heat abstracted from the heat source A, then work output of Carnot's heat engine

$$= \eta \times \text{heat absorbed} = \frac{280}{820} \times Q_A$$

Work input to reversed Carnot's engine

$$= \frac{1}{2} \times \frac{280}{820} Q_A$$

Coefficient of performance of the reversed Carnot's engine

$$= \frac{T_C}{T_C - T_B} = \frac{710}{710 - 540}$$

$$= \frac{710}{170}$$

But, by definition, C.O.P. of heat pump (reversed heat engine)

$$= \frac{\text{Heat rejected to heat reservoir C}}{\text{Work input to reversed heat engine}}$$

∴ Heat rejected to heat reservoir *C*.

$$Q_C = \text{C.O.P} \times \text{Work input}$$

$$= \frac{710}{170} \times \frac{1}{2} \times \frac{280}{820} \cdot Q_A$$

$$\frac{Q_C}{Q_A} = 0.713 \quad \text{or} \quad \mathbf{71.3\%} \quad \textbf{Ans.}$$

Power output of heat engine = 2 × power input to generator

$$= 2 \times 300 \text{ kW} = 600 \text{ kW}$$

$$= 600 \times 3600 = 2.16 \times 10^6 \text{ kJ/hr}$$

∴ Heat absorbed by the engine from heat reservoir A

$$Q_A = \text{Work output /efficiency}$$

$$= \frac{2.16 \times 10^6}{280/820} = 6.326 \times 10^6 \text{ kJ/hr}$$

$$Q_C = 0.713\, Q_A$$

$$= 6.326 \times 10^6 \times 0.713$$

$$= \mathbf{4.51 \times 10^6 \textbf{ kJ/hour.}} \quad \textbf{Ans.}$$

3.11 REAL GAS, VAPOUR AND PERFECT GAS

A **real gas** is a substance which cannot be liquefied by the application of pressure, however, large it may be, so long as the temperature remains above the critical temperature of the substance. From the engineering points of view, gas is a substance whose evaporation from its liquid state is complete. Substance which are thought of as gases (for example oxygen, nitrogen, hydrogen, etc.,) are highly superheated at normal atmospheric conditions. Critical temperatures of oxygen, nitrogen, and hydrogen are – 119°C, – 147°C and – 240°C respectively.

A **vapour** is a partially, evaporated liquid, and consists of the pure gases state together with liquid particles in suspension. Its behavior is different from that of a gas, as the substance is liable to further evaporation or condensation no changes of temperature and pressure. Thus unlike gases, vapors can be liquefied by applying pressure at constant temperature. The laws of gases do not apply to vapors. The vapors commonly used in engineering practice are steam, ammonia, carbon dioxide, sulphur dioxide. The respective critical temperatures of NH_3, SO_2 and H_2O are 130°C, 157°C and 374.15°C. It means their temperature is below their critical temperatures at normal atmospheric conditions.

A **perfect gas or an ideal gas** is one which strictly obeys all the gas laws (e.g. Boyle's law, Charles's law, Regnault's law, Joule's law and Avogadro's law) under all conditions of temperature and pressure.

The characteristic gas equation $pv = RT$, which is true for ideal gases only, is derived with the following assumptions:

(*i*) There is no intermolecular force of attraction.

(*ii*) The space occupied by the molecules themselves is extremely small when compared with the volume of the space in which gas is confirmed.

No gas known can fulfil the above assumptions, because there exists a force of attraction between the molecules and also the volume occupied by the molecules cannot be completely neglected. Van der waals suggested an equation of state of real gases modifying the characteristic equation of a gas

$$p.v = RT$$

to the form
$$\left(p + \frac{a}{v^2}\right)(v - b) = RT \quad \text{...(3.14)}$$

where 'a' = a constant which takes into account the intermolecular force of attraction

and b = volume of the gas when liquefied, at 273°C temperature that is when its molecules have ceased to vibrate and are packed together as closely as possible.

$\frac{A}{v_2}$ = increase of pressure due to intermolecular force of attraction.

The deviation of real gases from the ideal gases is more at high pressures and low temperatures.

3.11.1 Resisted and Unresisted Expansion

In resisted expansion the force exerted by the fluid on a piston is opposed by an equal force exerted by the piston on the fluid and the pressure of the fluid at any instant is almost uniform throughout the cylinder. All the effort of the fluid is used up in moving the piston, *i.e.* the boundary, and not in moving itself. Reciprocating engines are examples of fully resisted expansion. Fully resisted expansion is a reversible process. The work done in a resisted expansion is called displacement work. It is also called *P.dv* work because the area under the process on *P-V* diagram is equal to work done.

Unresisted or free expansion occurs when a fluid is allowed to expand suddenly into vacuum chamber through an orifice of large dimensions. Throttling or wire drawing (when applied to vapours) is also an unresisted expansion which occurs when a gas, or vapour, is expanded through an aperture of minute dimensions, such as a narrow throat or a slightly opened value; it is illustrated by the leak of a fluid though a crack in a vessel. Unresisted expansion is an irreversible process. No work is done. Such expansion cannot be truly represented on p–v diagram and, therefore, may be indicated by a dotted line.

A perfect gas is defined as one having constant specific heat capacity, which obeys simple equation of state $pv = RT$, where R is specific gas constant, which may be obtained from equation $R = \frac{R_0}{M}$, where M = molecular weight of gas, and R_0 = molar or universal gas constant, 8.3143 kJ/kg-mol-K

Considering specific heat capacities, enthalpy and internal energy is given by

$$\Delta h = c_p \Delta T \quad \text{and} \quad \Delta u = c_v \Delta T$$

Specific entropy change may be obtained from $\Delta s = c_p \log_e \frac{T_2}{T_1} - R \log_e \frac{p_2}{p_1}$

For reversible adiabatic perfect gas process $\frac{T_2}{T_1} = \left(\frac{p_2}{p_1}\right)^{R/c_p}$

$$\frac{R}{c_p} = \frac{\gamma - 1}{\gamma}; \quad c_p - c_v = R$$

Perfect Gas Relations

1. Equation of state $pv = RT$
2. Specific enthalpy change $\Delta h = c_p \Delta T$
3. Specific internal energy change $\Delta u - c_v \Delta T$
4. Specific entropy change

$$\Delta s = c_p \log_e \frac{T_2}{T_1} - R \log_e \frac{p_2}{p_1}$$

5. Specific heat capacities

$$c_p - c_v = R$$

6. Polytropic process law

(*i*) $p_1 v_1^n = p_2 v_2^n$, (*ii*) $T_1 v_1^{n-1} = T_2 v_2^{n-1}$, (*iii*) $\frac{T_2}{T_1} = \left(\frac{p_2}{p_1}\right)^{(n-1)/n}$

7. Adiabatic process law:

(*i*) $p_1 v_1^{\gamma} = p_2 v_2^{\gamma}$ (*ii*) $T_1 v_1^{\gamma-1} = T_2 v_2^{\gamma-1}$ (*iii*) $\frac{T_2}{T_1} = \left(\frac{p_2}{p_1}\right)^{(\gamma-1)/\gamma}$

8. Isothermal process law $p_1 v_1 = p_2 v_2$

Mixtures of gases (*e.g.*, air) may be treated as a single perfect gas by using Dalton's law, which states that:

(*i*) Total pressure exerted by a mixture of gases = sum of partial pressures, which each component would exert, if it alone occupied the volume of mixture at the temp. of the mixture, and

(*ii*) Enthalpy, energy and entropy of a mixture of gases = sum of respective enthalpies, energies and entropies, which the components would have if each alone occupied the volume of the mixture at the temp. of the mixture. This gives:

$$p = \Sigma p_t, \quad m = \Sigma m_t, \quad u = \frac{1}{m} \Sigma m_t u_t,$$

$$h = \frac{1}{m} \Sigma m_t h_t \text{ and } \quad s = \frac{1}{m} \Sigma m_t s_t,$$

where suffix t represents t^{th} component.

By further analysing: $R = \frac{1}{m} \Sigma m_t R_t, \quad M = \frac{m}{\Sigma \frac{m_t}{M_t}}$

$$c_p = \frac{1}{m} \Sigma m_t \, C_p, \quad c_v = \frac{1}{m} \Sigma m_t \, c_v$$

and $$\gamma = \frac{c_p \text{ mixture}}{c_v \text{ mixture}}$$

3.12 HEAT TRANSFER

As we know that heat is a form of energy, and heat transfer takes place between any two systems due to temperature difference.

Energy transfer occurs from a system of higher temperature to a system of lower temperature according to the second law of thermodynamics without any work transfer. But heat transfer can take place from a system of lower temperature to a system of higher temperature by adding external work. The subject dealing with the rate of heat transfer is called the "heat transfer".

The subject of heat transfer is applied in the design of boilers, heat engines, refrigerators, cold storages, air conditioners etc.

Heat transfer occurs through Conduction, Convection and Radiation.

1. Conduction

In conduction heat transfer takes place due to transfer of thermal energy from one molecule to another molecule without any appreciable motion of molecules within a solid matter. For example, when a metallic bar is heated at one end, the other end is also heated up gradually due to heat being transferred due to conduction along the bar.

Dimensional conduction is governed by Fourier's law:

$$Q' = -k\frac{dT}{dx},$$

where, Q'' = heat transfer rate per unit area, k = thermal conductivity of material, and dT/dx = temperature gradient in the direction of heat transfer.

Thermal conductivity is not constant. It is usually averaged over temperature range considered unless range is very large. Table 3.2 gives some values of k. Fourier's equation is now applied to four common situations and solutions obtained by integration.

Table 3.2. Thermal Conductivity of Materials at 20°C in watt per meter-kelvin (W/mk)

Material	*Conductivity*	*Material*	*Conductivity*
Cast iron	51.9	Bronze (75—25)	26.0
Aluminium	204	Lead	34.6
Duralumin	164	Mercury	8.7
Magnesium	171	Nickel	90
Copper	386	Silver	407
Brass (70 — 30)	111	Zinc	112
Constantan	22.7	Tin	64
Tungsten	163	Glasswool	0.04

Glass	0.762	Loose cork	0.036
Bakelite	0.232	Loose earth	0.521
Concrete	0.8—1.4	Silk	0.036
Gypsum plaster	0.485	Engine oil	0.145
Common brick	0.692	Freon-12	0.073
Facing brick	1.32	Water	0.597
Cotton	0.059	Methyl chloride	0.163

Thermal Conductivity of Steel at 20°C (W/mk)

Carbon steel	35—55
Nickel steel	12—50
Chrome steel	30—60
Ni— Cr steel	14
Cr— Ni steel	16
Silicon steel	31
Manganese steel	38
Tungsten steel	62

Thermal Conductivity at given Temperature

Fire clay	500°C	1.28
Fire clay	1100°C	1.4
Asbestos	0°C	0.151
Asbestos	100°C	0.192
Wool felt	30°C	0.0519
Magnesite	204°C	3.81
Magnesite	1204°C	1.9
Ice	0°C	2.22
Diatomaceous earth	204°C	0.242
Diatomaceous earth	872°C	0.312

Electrical Analogy to Conduction. Expressions may be compared to an Ohm's law type relation in which Q'' or Q' is analogous to current I, $(T_1 - T_2)$ to potential difference V, and $\Sigma\left(\frac{\Delta x}{k}\right)$ *or* $\Sigma\frac{\log_e(r_{outer}/r_{inner})}{2\pi\, k_{layer}}$ to resistance R for a cylindrical body.

Accordingly, these latter terms are sometimes called "thermal resistance" of the situation. Thermal resistances in series (as in multi layer wall) may be added as electrical resistances in series. Above method often gives a quick solution and also dominating resistance to be identified easily. Problems in unsteady conduction or three-dimensional conduction are also solved by Fourier's law.

2. Convection

Heat convection occurs when bulk flow of a fluid (gas or liquid) carries heat along with the flow of matter in the fluid. The flow of fluid may be forced by external processes, or sometimes (in gravitational fields) by buoyancy forces caused when thermal energy expands the fluid (for example in a fire plume), thus influencing its own transfer. The latter process is often called "natural convection". All convective processes also move heat partly by diffusion, as well. Another form of convection is forced convection. In this case the fluid is forced to flow by use of a pump, fan or other mechanical means.

Molecules of liquid and gases have freedom of motion and in moving, they carry over their energy. Hence heat transfer, from one region to another in fluids is due to the motion of molecules, and such a mode of heat transfer is called "heat transfer by convection". When the motion of the molecules of fluid is caused by variation of density due to difference of temperature, then the heat transfer is by "free convection". When the motion of the molecules of fluid is caused by an external energy such as a pump or a blower, the heat transfer is by "forced convection".

Heat transfer rate per unit area from a surface due to convection, $Q'' = \alpha\theta$, where θ = temperature difference between surface and bulk of fluid above it.

An electrical analogy be made and it may be seen that for a cylindrical surface, $Q = \dfrac{\theta}{\left(\dfrac{1}{2\pi r\alpha}\right)}$, and for a plane surface/wall, $Q'' = \dfrac{\theta}{(r\alpha)}$, and thermal resistances are $\dfrac{1}{2\pi r\alpha}$ and $\dfrac{1}{\alpha}$ respectively.

In convection heat transfer determination of $\propto$ (surface heat transfer coefficient) is main problem. Several methods are available to make reasonable estimates.

Modes of Convection:

(*i*) *Natural or Free Convection.* In this fluid motion is by buoyancy due to density gradients.

(*ii*) *Forced Convection.* In this fluid motion is induced by a pump, and

(*iii*) Both free and forced convection may be associated with laminar or turbulent flow.

In turbulent flow, heat transfer rates are greater than those in laminar flow.

In forced convection, heat transfer rates are greater than in free convection.

Hence, forced turbulent convection is used when high rates are required. Determining whether flow is laminar or turbulent in any situation, is determined by the value of Rayleigh Number (*Ra*) in free convection and by the value of Reynolds Number (*Re*) in forced convection.

Dimensional analysis and experiment is one of the methods used to form eqnations, in which $\propto$ is involved in the non-dimensional Nusselt Number (Nu). Fig. 3.6 gives results, which may be applied to plane and cylindrical surfaces and also indicates correct method of application.

3. Radiation

All substances (solid, liquid and gases) at all temperature levels emit thermal energy in the form of electromagnetic waves and absorb similar energy emerging from the neighbouring substances. This kind of heat transfer is by "thermal radiation". This mode of heat transfer does not need any material medium. When the electromagnetic waves strike the surface of another substance, part of the energy is reflected, a part transmitted through the substance and the remainder absorbed by the substance.

Heat transfer by radiations becomes increasingly important as temperature rises. It is because Stefan-Boltzman law determining the amount of radiation contains the fourth power of thermodynamic temperature.

Radiation theory is based on an ideal black body, which absorbs all radiation incident upon it. A grey body is also considered, which absorbs the same constant proportion of incident radiation at all wavelengths. The amount absorbed is determined by the absorptivity ∝ of the body ($\alpha < 1$).

Kirchoff's law indicates that "the emissivity E of a grey body is equal to its absorptivity α". Real surfaces have emissivities (and absorptivities), which vary with wavelength and temperature. Table 3.3 gives radiation laws.

For solving radiation problem, it is decided as to how much of the radiation emitted by radiator is intercepted by the receiver. It is a geometrical problem and is accounted for by a geometrical or configuration factor F or any situation.

Table 3.3. Laws of Radiation

Law	*Equation*	*Remarks*
Stefan-Boltzmann	$E''_b = \sigma T^4$	E'' = Emissive power in W/m^2, σ = Stefan-Boltzmann constant $= 5.66 \times 10^{-8}$ W/m^2 K^4 Suffix b refers to black body radiation.
Planck	$E_b\lambda = \dfrac{C_1\lambda^{-5}}{e^{C_2/\lambda_2 T}}$ where $C_1 = 3.74 \times 10^{-16}$ Wm^2, and $C_2 = 0.0144$ mK	This gives emissive power at wavelength λ and here $\int_\theta^\infty E_{b\lambda}\, d\lambda = \sigma T^4$
Wein	$\lambda_{max} T$ = constant $= 2.9 \times 10^{-3}$ mK	
Kirchoff	$\alpha = \varepsilon$ for a grey body	$\varepsilon = \left[\dfrac{E''}{E_b''}\right]$
Lambert	$i_\phi = i_n \cos\phi$, where $i = dE''/d\,W$, ω = solid angle.	This law determines special distribution of radiation from a point. It is expressed in terms of intensity of radiation in a direction normal to the surface i_n, where $i_n = \varepsilon\sigma T^4/\pi$.

Notes

(*i*) For real bodies, monochromatic emissivity at wavelength λ and temp. $T\epsilon_\lambda = \left(\dfrac{E''_\lambda}{E''_{b\lambda}}\right)$. As this varies with both wavelength and temp. it is convenient to eliminate wavelength effect by integrating radiation overall λ. Total emissivity obtained can then be used for determining total radiation from a surface at a given temp. T. Thus, $\in = \left[\dfrac{\int_0^{\infty} \varepsilon_\lambda \; E_{b\lambda} \; d\lambda}{\sigma \, T^4}\right]_T$.

(*ii*) In a radiation problem, application of approximation may require careful consideration. Absorptivity at wavelength of incident radiation is required coupled with emissivity of emitted radiation. When radiation is in narrow wavelength bands, error may be caused by use of total emissivities or total absorptivities.

It is not easy to calculate configuration factors except in simple cases. These may be obtained by experiment or from graphs drawn for all common situations.

If configuration factor is defined for a situation by F_{12} = fraction of energy emitted by 1 *i.e.*, intercepted by 2, then energy intercepted = $A_1\sigma \in_1 T_1^4 F_{12}$ and of this α_2 $(A_1\sigma\in_1 T_1^4 \; F_{12})$ will be absorbed by 2. The remaining is reflected.

A similar expression for the energy emitted by 2 *i.e.*, intercepted and absorbed by 1 = α_1 $(A_2 \, \sigma \, \varepsilon_2 \, T^4 \, F_{21})$. It may be shown that: $A_1 \, F_{12} = A_2 \, F_{21}$.

Radiation, which is not absorbed, is reflected, and some will be incident on the emitter. This effect results in a series of terms to be added to evaluate the net heat transfer rate.

3.13 HEAT EXCHANGERS

A heat exchanger is a tool built for efficient heat transfer from one fluid to another, whether the fluids are separated by a solid wall so that they never mix, or the fluids are in direct contact. Heat exchangers are widely used in refrigeration, air conditioning, space heating, power generation, and chemical processing. One common example of a heat exchanger is a car's radiator, in which the hot coolant fluids cooled by the flow of air over the radiator's surface.

Common types of heat exchanger Clows include parallel flow, counter flow and cross flow. In parallel flow, both fluids move in the same direction while transferring heat; in counter flow, the fluids move in opposite directions; and in cross flow, the fluids move at right angles to each other. Common constructions for heat exchanger include shell and tube, double pipe, extruded finned pipe, spiral fin pipe, u-tube, and stacked plate.

When engineers calculate the theoretical heat transfer in a heat exchanger, they must contend with the fact that the driving temperature difference between the two fluids varies with position.

Heat Dissipation

A heat sink is a component that transfers heat generated within a solid material to a fluid medium, such as air or a liquid. Examples of heat sinks are the heat exchangers used in refrigeration and air conditioning systems, and the radiator in a car (which is also a heat exchanger).

Heat exchangers involve convection at the two surfaces of and conduction through a solid boundary separating two layers of fluid. For solving problems, these resistances are considered in series:

If U = Overall heat transfer coefficient per unit area for a plane heat exchanger,
and U' = Overall heat transfer coefficient per unit length for a cylindrical heat exchanger, then

$$Q = UA\theta \text{ or } Q = U'L\theta$$

From earlier considerations, $\frac{1}{U} = \frac{1}{\alpha_1} + \left(\frac{\Delta x}{k}\right)_{12} + \frac{1}{\alpha_2}$ (1), and

$$\frac{1}{U'} = \frac{1}{2\pi r_1 \alpha_1} + \frac{\log_e \frac{r_2}{r_1}}{2\pi k_{12}} + \frac{1}{2\alpha r_2 \alpha_2}$$

Values of α_1 and α_2 are determined for mean conditions and therefore θ in Eq. (1) must be for mean conditions.

It may be shown that θ must be "log mean temperature difference" given by

$$\theta = \frac{\theta_1 - \theta_2}{\log_e \frac{\theta_1}{\theta_2}},$$

where θ_1 and θ_2 are differences in two stream temperatures at the ends of the heat exchanger.

QUESTIONS

3.1 What is thermodynamics? Explain laws of thermodynamics.

3.2 (*a*) What is a perfect gas? Explain.
(*b*) What is combustion? Explain.

3.3 Explain the cycle used in steam plants.

3.4 Explain the working of a gas turbine.

3.5 (*a*) What is refrigeration?
(*b*) What is the unit of refrigeration.

3.6 What is Heat transfer? Explain.

3.7 Define conduction, convection, and radiation.

3.8 Explain the working of heat-exchangers.

3.9 Define the first law of thermodynamics and discuss why it is known as the law of conservation of energy.

3.10 Define first law of thermodynamics and prove the interconvertibility of heat and work.

3.11 Narrate the second law of thermodynamics.

3.12 Define heat engine and heat pump.

3.13 Discuss the factors due to which a carnot heat engine cycle is not practicable in practice.

3.14 (*a*) Define thermal conductance and resistance.
(*b*) Define convective and radiative conductance.

3.15 Define the terms: system, system boundary and surroundings.

3.16 What is (*i*) open system; (*ii*) closed system; and (*iii*) isolated system? Give practical examples of each.

3.17 Define thermodynamic properties. How these properties are classified?

3.18 State whether the following properties of a system are intensive and extensive: Mass, specific weight, volume, specific volume, density, pressure, temperature, surface area, velocity, strain, elevation, electric charge, magnetization, potential energy, kinetic energy and flow energy.

3.19 What is a system? What are surroundings? What are the different forms of energy transfer that occur between the system and the surroundings?

3.20 What are the differences between ideal and real gases? Give examples.

3.21 What do you understand by an ideal gas?

3.22 What are the properties of an ideal gas? Under what conditions does a real gas behave as an ideal gas?

3.23 Define the terms: State of a system, path of change of state and thermodynamic process.

3.24 Define specific heat at constant pressure and specific heat at constant volume, which is greater and why? Give the relationship between these two.

3.25 What are different units of temperatures? Give relationship between them.

3.26 Define: absolute pressure, vacuum, gauge pressure.

3.27 Give the thermodynamic definition of work. What are positive work and negative work?

3.28 What are resisted expansion and un-resisted expansion? In which case the work done is $\int p.dv$?

3.29 Show that the work done in the steady flow process is given by $\int v.dp$.

3.30 Explain what do you understand by "heat transfer" and "work transfer". Are these properties of a system?

3.31 Prove the relationship between the two principal specific heats and the characteristic gas constant for a perfect gas.

3.32 What is thermodynamic cycle?

3.33 Explain the difference between heat and internal energy. What is enthalpy?

3.34 Define internal energy and prove that it is a property of a system.

3.35 State the zeroth law of thermodynamics. What is its importance?

3.36 State and explain the first law of thermodynamics.

3.37 Discuss and derive the equation of conservation of energy as applied to an open system under steady state conditions.

3.38 How the first law of thermodynamics is applied to a closed system undergoing a non-cyclic process?

3.39 With the help of the first law of thermodynamics, prove that the internal energy is a property?

3.40 What are the limitations of the first law of thermodynamics?

3.41 State and prove the second law of thermodynamics.

Chapter

PROPERTIES OF GASES

4

Learning Objectives

After studying this chapter, the students will learn:

- Laws of perfect gases.
- Thermodynamic processes constant volume, constant pressure, constant temperature gases, constant internal energy processes adiabatic process and throttling process.
- Steady flow process and its applications such as steam boilers, steam nozzles, steam turbines steam condensers, air compressors, throttle valve etc.
- Entropy.

4.1 INTRODUCTION

As already discussed in previous chapter, a gas which follows the gas laws at all ranges of pressure and temperature is considered as an ideal or perfect gas. Real gases tend to follow the gas laws at low pressures or high temperatures or at both. The range of low pressure is different for different gases. Oxygen, nitrogen, hydrogen, air etc., within certain temperature limits are regarded as perfect gases. A gas is a state of a substance for which evaporation from its liquid state is complete.

A vapour is a state of a substance when evaporation from its liquid state is partial. Hence the vapour always contains liquid particles in suspension. Thus steam is considered as vapour. A vapour becomes dry when it is completely evaporated. Further heating of dry vapour is known as superheating and the state of such vapour is known as superheated vapour. The gas laws do not apply to vapours.

Dry and superheated vapour approach the behaviour of that of a perfect gas.

The present chapter deals with the behaviour of the gases and the laws controlling their behaviour.

4.2 LAWS OF PERFECT GASES

The pressure exerted by a gas (p), the volume occupied by it (v) and its temperature (T) control the physical properties of the gas. The effect of changes of temperature and pressure on the behavior of gases can be known by applying gas laws which were deduced from experimental results. *These laws do not apply to vapours.*

4.2.1 Boyle's Law

It state that *"the volume of a given mass of a gas is inversely proportional to its absolute pressure, if the temperature is kept constant"*..

Mathematically $v \propto \frac{1}{p}$, if temperature (T) is kept constant

or $$p.v = \text{constant}$$

If a graph is plotted between p and v, the curve so obtained will be similar to the curve depicted in Fig. 4.1. The curve is a rectangular hyperbola and therefore the process following the law $p.v = C$ is called *hyperbolic expansion or compression.*

Also, the expansion or compression of a gas at constant temperature is known as an isothermal. Hence, hyperbolic process is also isothermal process for gases. However, it is not true for vapours.

If (p_1, v_1) and (p_2, v_2) apply to initial and final states of a gas respectively, then Eq. 4.1 can be written as

$$p_1 . v_1 = p_2 . v_2 = \text{constant} \quad ...(4.1)$$

when $T_1 = T_2$

Boyle's law is not absolutely accurate. It holds correctly for small ranges of pressure; with the large pressure ranges obtained in modern engineering practice the application of this law causes errors up to 3%. This discrepancy is not taken into account in case of gases like air, hydrogen and nitrogen, etc., except at very low temperatures or at very high pressures.

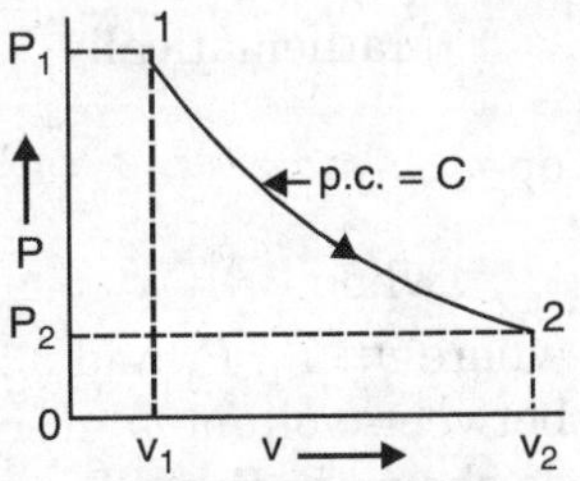

Fig. 4.1

Example 4.1 *A gas occupies 2.5 m³ at a pressure of 0.3 MPa. The gas is compressed isothermally to a final pressure of 0.65 MPa. the volume of the gas at the end of compression.*

Sol.

$$p_1 = 0.3 \text{ MPa}, v_1 = 2.5 \text{ m}^3$$
$$p_2 = 0.65 \text{ MPa}, v_2 = ?$$

Applying Eq. 4.1(*a*), *i.e.*,

$$p_1.v_1 = p_2.v_2$$
$$0.3 \times 2.5 = 0.65 \times v_2$$

or

$$v_2 = \mathbf{1.154 \text{ m}^3} \quad \textbf{Ans.}$$

4.2.2 Charles Law

Charles found that if a given mass of any gas is heated at constant pressure the change in volume varied directly with the temperature range. This changes in volume is the same for all perfect gases.

He found that with perfect gases the volume changed by $\frac{1}{273}$ of its initial volume, at 0°C, for each centigrade degree change in temperature provided, the pressure exerted by the gas is maintained constant.

Let v_0 = volume of a given mass of a gas at 0°C.

v_t = volume of the same mass of the gas at $\pm t$ °C.

Then according to Charles

$$v_t = v_0 \pm \frac{t}{273} v_0 \quad ...(4.2)$$

It follows from Eq. 4.2 that if temperature 't' falls below (–) 273°C the volume of the gas will become negative which is not possible. Hence (–) 273°C is considered to be the lowest possible temperature at which volume of a gas would be zero. This is considered to be the point of absolute zero temperature. The corresponding temperature for the Fahrenheit scale is (–) 460°F.

Eq. 4.2 can be written as

$$v_t = v_0\left(\frac{273 \pm t}{0 + 273}\right)$$

or
$$\frac{v_t}{T} = \frac{v_0}{T_0} \qquad \text{...(4.2}a\text{)}$$

where T and T_0 are absolute temperatures in degrees Kelvin.

According to Eq. 4.2(a) Charles's law can be stated as *"The volume of a given mass of a gas varies directly as its absolute temperature, provided the pressure exerted by the gas is maintained constant."*

Mathematically,
$$v \propto T$$

or
$$\frac{v}{T} = \text{constant} \qquad \text{...[4.2.(}b\text{)]}$$

when
$$p_1 = p_2$$

where v_1, T_1, P_1 and v_2, T_2, P_2 relate to initial and final states of the gas. If a graph is plotted between volume v absolute temperature T, then the curve so obtained will be a straight line as shown in Fig. 4.2

By careful measurements it is found that Charles' law is not absolutely accurate for large ranges of temperature; it will cause errors of about 5% within the extreme range of modern practice; but this discrepancy is ignored in engineering calculations.

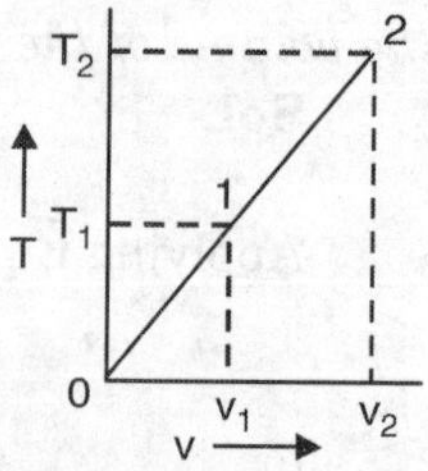

Fig. 4.2

Example 4.2. *0.15 m³ of gas at 47°C is heated at constant pressure to a final temperature of 227°C. Find the final volume, change in volume. Also find the work done by the gas if the constant pressure is 0.5 MPa.*

Sol.

$$v_1 = 0.15 \text{ m}^3,$$
$$T_1 = 47 + 273 = 320°\text{K} \quad v_2 = ?$$
$$T_2 = 227 + 273 = 500°\text{K}$$

Since heat is supplied at constant pressure, hence Charles' law can be applied. Thus

$$\frac{v_2}{T_2} = \frac{v_1}{T_1}$$

or
$$v_2 = \frac{500}{320} \times 0.15 = 0.2344 \text{ m}^3 \quad \textbf{Ans.}$$

$$\text{Change in volume} = v_2 - v_1 = 0.2344 - 0.15$$
$$= \mathbf{0.0844\ m^3\ (Increase)} \quad \textbf{Ans.}$$
$$\text{Work done by the gas} = p.dv = 0.5 \times 10^6 \times .0844$$
$$= \mathbf{42200\ Nm} \quad \textbf{Ans.}$$

Combination of Boyle's law and Charles law (Characteristic Gas Equation)

Let p_1, v_1 and T_1 be the absolute pressure, volume and absolute temperature of a perfect gas respectively.

If this gas is compressed or expanded to a volume v_2 and pressure p_2 at constant temperature T_1, then according to Boyle's law.

$$p_1v_1 = p_2v_2 \qquad \text{...(}i\text{)}$$

Now, let this gas be further heated or cooled from temperature T_1, volume v_2, to temperature T_2, volume v'_2 keeping pressure P_2 constant.

Then according to Charles law,

$$\frac{v'_2}{T_2} = \frac{v_2}{T_1} \qquad ...(ii)$$

Eliminating v_2 with the help of equations (*i*) and (*ii*).

We have $$p_1 v_1 = \frac{p_2 v'_2 T_1}{T_2} \Rightarrow \frac{p_1 v_1}{T_1} = \frac{p_2 v'_2}{T_2}$$

i.e., $$\frac{P.v}{T} = \text{constant} \qquad ...(4.3)$$

If v is the volume of unit mass of the gas, then the constant in Eq. 4.3 is usually denoted by R and is called the *characteristics gas constant.* Then

$$P.v = RT \qquad ...[4.3(a)]$$

Eq. 4.3(*a*) is called the equation of a *perfect gas or characteristic equation of a gas.* For '*m*' kg of gas, Eq. 4.3(*a*) can be written as

$$P.v = mRT \qquad ...(4.3(b))$$

When v is the volume of '*m*' kg of gas.

Units of R. If p is in N/m^2, v in m^3 . m in kg and T in degrees Kelvin then R will be in N.m/kg °K.

If p is in N/m^2, v in m^2, m in kg and T in °K then R will be in Joules per kg per degree Kelvin.

For atmospheric air R = 96 ft-lb centigrade units.

= 53.3 ft-lb Fahrenheit units.

= 29.27 N-m per kg per degree Kelvin.

= 287.14 Joules per kg per °K. = 8.314 J/mol °K

The Mole, Universal Gas Constant, Molar Heats:

kg-mole is defined as a mass of one mol (22.41 m^3 at N.T.P.) of a gas equal to M kg where M is the molecular weight of the gas.

If characteristic gas constants of gases are multiplied by their corresponding molecular weights, then it is interesting to note that the product is a constant quantity for all gases. The constant is called *Universal gas constant or Molar constant* and may be denoted by *Ru*.

The values of *Ru* in different systems of units are as follows:

Ru = 848 N-m per kg mol per degree Kelvin.

= 8314 Joules per kg mol per degree Kelvin.

= 1544 ft-lb per lb mol per degree Rankine.

= 2780 ft-lb per lb mol per degree Kelvin.

Molar heat is defined as the quantity of heat required to raise the temperature of one mol (22.41 m^3 at N.T.P.) of a gas at constant pressure (K_p) or at constant volume (K_v). Thus

$$K_p = M.C_p \text{ and } K_v = M.C_v$$

where M is molecular weight.

Example 4.3 *Find the volume of a gas at 0°C temperature and 0.10332 MPa pressure when its mass is equal to the molecular weight expressed in kg.*

Sol. We know that

$$p.v = 848\,T.$$

or

$$1.0332 \times 10^4 \times v = 848 \times (273)$$

$$v = \mathbf{22.4065\ m^3}\ \textbf{Ans.}$$

22.4065 m³ of a gas at N.T.P. (one mol.) weighs equal to the molecular weight of the gas in kg.

Example 4.4 *A quantity of a gas initially at pressure 0.1 MPa, temperature 87°C and volume 0.05 m³ is compressed to final pressure 0.7 MPa and volume 0.015 m³. Find the final temperature of the gas.*

Sol.

$$p_1 = 0.1 \times 10^6 \text{ Pa},\ v_1 = 0.05 \text{ m}^3$$

And

$$T_1 = (87 + 273)°\text{K}$$

$$p_2 = 0.7 \times 10^6 \text{ Pa},\ v_2 = 0.015 \text{ m}^3 \text{ and } T_2 = ?$$

Applying characteristic equation of gas (Eq. 4.3)

$$\frac{p_1 v_1}{T_1} = \frac{p_2 v_2}{T_2}$$

$$\therefore \quad T_2 = \frac{p_2 v_2}{p_1 v_1} \times T_1 = \frac{0.7 \times 10^6 \times 0.015}{0.1 \times 10^6 \times 0.05} \times (273 + 87)$$

$$= \textbf{756°K or 483°C}\quad \textbf{Ans.}$$

Example 4.5 *A quantity of a gas occupies 0.15 m³ at 1.2 MPa and 107°C. Find the quantity of the gas. Take $C_p = 1.0$ and $C_v = 0.71$.*

Sol. $p = 1.2$ MPa, $v = 0.15$ m³, $T = (107 + 273)$°K

$$C_p = 1.0 \text{ kJ/kg °K},\ C_v = 0.71 \text{ kJ/kg °K}$$

Using the equation

$$C_p - C_v = R$$

We have

$$R = (1.0 - 0.71) = 0.29 \text{ kJ/kg °K} = 290 \text{ kJ/kg °K}$$

Applying characteristics equation of gas

$$P.v = mRT \quad \text{or} \quad m = \frac{1.2 \times 10^6 \times 0.15}{.290 \times 380}$$

$$= \textbf{1.63 kg}\quad \textbf{Ans.}$$

Example 4.6 *Find the density of a gas at S.T.P. if the specific heat at constant pressure and at constant volume are 1.0 and 0.71 kJ/kg °K respectively.*

Sol. S.T.P. means temperature = 15°C.

And pressure = 0.1 MPa

Using the characteristic equation of gas

$$p.v = mRT \quad \text{or} \quad p.v. = m\,.\,(C_p - C_v)T$$

or

$$0.1 \times 10^6 \times v = 1 \times (1.0 - 0.71) \times (15 + 273)$$

$$v = 0.835 \text{ m}^3 \text{ per kg of gas.}$$

$$\therefore \quad \text{Density} = \frac{1}{v} = \mathbf{1.197\ kg/m^3}\ \textbf{Ans.}$$

4.2.3 Regnault's Law

Regnault states that the *specific heats of a gas at constant pressure* (c_p) *and that at constant volume* (c_v) *remain constant at all temperatures and pressures.*

For many calculations in thermodynamics this law is assumed to hold good; but actually, there is a variation in the values of c_p and c_v for any gas. Modern accurate experiments have proved that the value of the specific heats increases as the temperature of the gas increases. The value of the specific heats may increase by about 20% with the extreme temperature ranges used in modern engineering practice.

4.2.4 Joule's Law

Joule's law states that the internal energy of a given quantity of a gas is a function of the temperature only from this it follows that the change of internal energy is directly proportional to the change of temperature.

$$U_2 - U_1 = \text{a constant } (T_2 - T_1)$$

It will be proved later (constant volume heating) that this constant is c_v per kg of gas; hence

$$U_2 - U_1 = m.c_v\,(T_2 - T_1) \quad ...(4.4)$$

Eq. 4.4 gives the change in internal energy always, no matter how the gas is heated or expanded.

Joule proved this law by subjecting a quantity of a gas to a free expansion and measuring the initial and final temperatures. He found that the temperatures remained constant. He argued that as no heat was supplied and as no work was done, there could be no change in internal energy. Or applying law of conservation of energy.

$$dU = dQ - dW$$

i.e.,

$$dU = 0 - 0 \text{ hence } dU = 0$$

But, the experiment proved that the change of temperature = 0 hence change of internal energy is proportional to change of temperature.

or

$$U_2 - U_1 = \text{a constant } (T_2 - T_1)$$

Example 4.7 *A quantity of a gas occupies 0.1 m³ at 1.5 MPa pressure and 100°C. Find the change of internal energy if the temperature is raised to 250°C. Take* c_p *= 1 KJ/Kg °K and* c_v *= 0.71 kJ/kg °K.*

Sol.

$$P_1 = 1.5 \times 10^6 \text{ Pa}, \; v_1 = 0.1 \text{ m}^3$$

And

$$T_1 = (100 + 273)°\text{K} \quad T_2 = (250 + 273)°\text{K}$$

$$c_p = 1.00, \; c_v = 0.71 \text{ kJ/kg °K}$$

$$R = (c_p - c_v)$$

$$R = (1.00 - 0.71) \times 1000 = 290 \text{ J/kg/°K}$$

$$p_1 v_1 = mRT_1 \quad \text{or} \quad m = \frac{p_1 v_1}{RT_1}$$

$$= \frac{1.5 \times 10^4 \times 0.1}{290 \times 373} = 1.387 \text{ kg}$$

$$U_2 - U_1 = m \,.\, c_v\,(T_2 - T_1)$$

$$= 1.387 \times 0.71\,(250 - 100)$$

$$= \textbf{147.7 KJ increases} \quad \textbf{Ans.}$$

Example 4.8 *A quantity of gas occupying 0.1 m³ at 0.8 MPa is heated at constant pressure until volume becomes 0.2 m³. The initial temperature is 197°C and c_p = 1.00 kj/kg-K and c_v = 0.71 kj/kg-K. Find the change in internal energy.*

Sol.

$$p_1 = 0.8 \times 10^6 \text{ Pa}, \ v_1 = 0.1 \text{ m}^3,$$
$$T_1 = (197 + 273)°\text{K} = 470°\text{K}$$
$$p_2 = p_1, v_2 = 0.2 \text{ m}^3, T_2 = ?$$

Applying Charles law
$$\frac{v_2}{v_1} = \frac{T_2}{T_1}$$

or
$$T_2 = (197 + 273) \times \frac{0.2}{0.1} = 940°\text{K}$$
$$R = (c_p - c_v) = (1.00 - 0.71) \times 1000$$
$$= 290 \text{ J/kg °K}$$

Applying characteristic gas equation
$$p_1 v_1 = mRT_1$$
$$0.8 \times 10^6 \times 0.1 = m \times 290 \times (197 + 273)$$
$$m = 1.7 \text{ kg}$$
$$\text{Change in internal energy} = U_2 - U_1 = mc_v (T_2 - T_1)$$
$$= 1.7 \times 0.71 (940 - 470)$$
$$= \textbf{567 increase} \quad \textbf{Ans.}$$

4,2.5 Avogadro's Law (or Hypothesis)

In 1811, the Italian scientist Avogadro discovered that equal volumes of all perfect gases under similar conditions of temperature and pressure contain equal number of molecules.

Thus if two ideal gases are contained in two vessels of same volume at the same temperature and pressure, each gas will contain equal number of molecules, say n. Denoting the molecular weights of the gases by M_1 and M_2 and their respective masses by m_1 and m_2; then

$$m_1 = K \,.\, n \,.\, M_1$$

and
$$m_2 = K \,.\, n \,.\, M_2$$

where K is constant of proportionality. Therefore, we get,

$$\frac{m_1}{m_2} = \frac{M_1}{M_2} \quad \text{or} \quad \frac{\rho_1 \, v}{\rho_2 \, v} = \frac{M_1}{M_2} \qquad (\because \ m = \text{density} \times \text{volume})$$

or
$$\frac{vs_2}{vs_1} = \frac{M_1}{M_2}\left(\because \rho = \frac{1}{vs} = \frac{1}{\text{specific volume}}\right)$$

or
$$M_1 \, vs_1 = M_2 \, vs_2 = M \,.\, vs = \text{constant} \qquad ...(4.5)$$

Thus it can be seen that the product of molecular weight and specific volume (i.e. volume of one kg) is constant for all ideal gases at same temperature and pressure. The product of mole and specific volume will be the volume of one mole and obviously the volume of one mole (called molar volume) is same for all ideal gases at the same temperature and pressure.

The molar volume at N.T.P. is given by

$$v_{mol} = \frac{M.R.T}{p} = \frac{R_u \cdot T}{p} \qquad (R_u = \text{Universal gas constant})$$

$$= \frac{8314 \times (0+273)}{1.0332 \times 10^5} = 22 \text{ m}^3 \text{ per kg mol.}$$

This molar volume is same for all ideal gases at N.T.P.

Thus 22.41 m^3 of H_2 at N.T.P. has a mass of 2 kg

Similarly 22.41 m^3 of N_2 at N.T.P. weighs 28 kg

and 22.41 m^3 of O_2 at N.T.P. weighs 32 kg

and so on.

4.3 THERMODYNAMIC PROCESSES (METHODS OF HEATING AND EXPANDING)

There are several ways in which a gas may be heated or expanded and it is very important that the students should be able to distinguish clearly between these methods, which are known as thermodynamic processes.

These methods also apply to the cooling or compression of gases; cooling is a negative heating and compression, or contraction as a negative expansion.

Thermodynamic process is a change in the state of the working fluid as a result of energy. The properties of the working fluid like volume, pressure and temperature change during the process. The internal energy of the fluid also changes as the temperature changes. The heat or work is either applied or taken out of the system during the process.

Work Done During the Process

Consider that the state of a system changes from state 1 to state 2 as shown in the Fig. 4.3 during the process. The work during the process 1 – 2 is given by

$$W = \int_1^2 (p \cdot dv) \text{ Nm.}$$

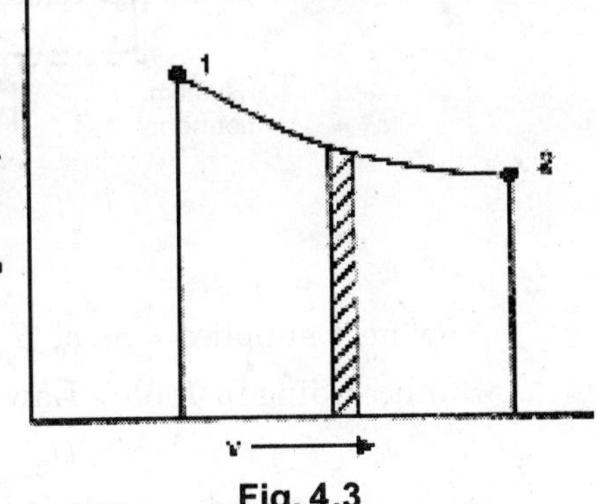

Fig. 4.3

Enthalpy. The enthalpy is the total energy of the substance due to the molecular kinetic energy as well as pressurisation and is defined as:

$$h = u + p.v \quad \text{(for unit mass)}$$

or

$$H = U + P.V$$

where u = Internal energy

$p.v$ = Energy due to the pressurisation.

where enthalpy and internal energy are in *Joule*, P is in N/m^2 and V is in m^3.

If the system changes from state 1 to state 2 the change in enthalpy is given by ;

$$\Delta H = H_2 - H_1 = U_2 - U_1 + p_2V_2 - p_1V_1$$
$$= mC_v(T_2 - T_1) + mR(T_2 - T_1)$$

Substituting the value of $R = c_p - c_v$

$$\Delta H = mc_v (T_2 - T_1) + m (c_p - c_v)(T_2 - T_1)$$
$$= m\, c_p (T_2 - T_1)$$

While executing the different processes, the change in the properties (volume, pressure, temperature, internal energy and enthalpy) are to be calculated in addition to work and heat transfers.

4.3.1. Constant Volume Process

When a gas is heated in a completely closed vessel, the process is called a constant volume process. This results in an increase in pressure and temperature $\left(\frac{P_2}{P_1} = \frac{T_2}{T_1}\right)$ of the system and this is represented by process 1 – 2 on $p - v$ diagram as shown in Fig. 4.4(*b*). As the boundary of the system is rigid (i.e. no change in volume), there is no displacement work (i.e. p.dv = 0). This also becomes clear from the fact that area under the constant volume process on p.v diagram is zero. Thus applying law of conservation of energy

$$dQ = dW + dU$$

Since $$dW = 0$$

$\therefore$ $$dQ = dU \quad ...(4.6)$$

i.e. all the heat supplied is stored in the system as internal energy.

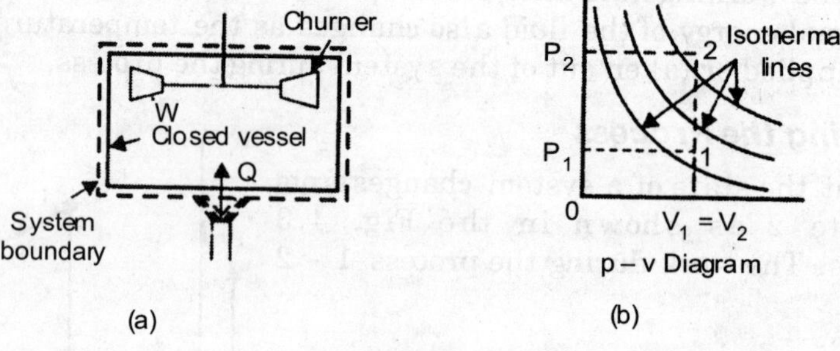

Fig. 4.4

Now heat supplied = $m.\, c_v (T_2 - T_1)$ kcal

But according to Joule's Law the change in internal energy

$$U_2 - U_1 = m.\ \text{constant}\ (T_2 - T_1).$$

As heat supplied = increase in internal energy

or $$m\,.\, c_v (T_2 - T_1) = m\,.\ \text{a Constant}\ (T_2 - T_1)$$

$\therefore$ $$\text{constant} = c_v$$

Thus change in internal energy = $m\,.\, c_v (T_2 - T_1)$ ...[4.6(*a*)]

Thus ratio of the internal energy to the corresponding increase in temperature is expressed as derivative and is called the specific heat at constant volume c_v

$$c_v = \left[\frac{\partial U}{\partial t}\right] \text{ constant volume} \qquad \text{...[4.6(b)]}$$

The change of system from state point 1 to state point 2 at constant volume can also be achieved by supplying external work to the system by an action of electric churner as shown in Fig 4.4(*a*). The system is insulated against transfer of heat energy ($\partial Q = 0$).

In such a case, $\partial Q = 0 = \partial U + \partial W$

or
$$\partial U = -\partial W \qquad \text{...[4.6(c)]}$$

The negative sign indicates that work is supplied to the system for increase of its internal energy.

Since however the internal energy is a point function, it is immaterial how the process is carried out, its value for given points will remain the same and hence its value will be same as for constant volume heating process.

A cooling process from state point 2 to state point 1 for the system can be easily carried out by rejection of heat to atmosphere.

$$\begin{aligned}
\text{Change in enthalpy} &= (H_2 - H_1)\\
&= (U_2 + p_2v_2) - (U_1 + p_1v_1)\\
&= (U_2 - U_1) + (p_2 \cdot v_2 - p_1 \cdot v_1)\\
&= mc_v\,(T_2 - T_1) + (mRT_2 - mRT_1)\\
&= m\,(T_2 - T_1)\,(c_v + R)\\
H_2 - H_1 &= m.c_p\,(T_2 - T_1) \qquad \text{...(4.7)}
\end{aligned}$$

4.3.2 Constant Pressure Process (Charles Law)

Let a cylinder contain a volume of perfect gas ($v_1 = abcd$) Fig. 4.5(a), and let there be a sliding piston of area 'A' floating on the gas and supported by it. Let '*W*' be the weight of (piston + dead weights resting on piston). As the weight '*W*' is entirely supported by the pressure of the gas, it follows that

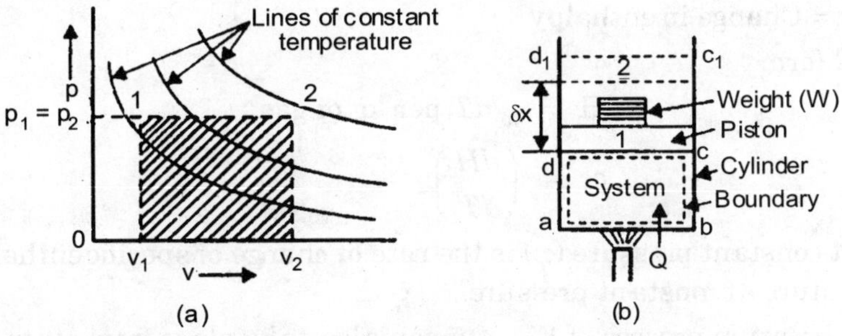

Fig. 4.5

$$\text{Intensity of pressure} = \frac{W}{A}.$$

Now let heat be supplied to the gas. As the heat is supplied the volume and temperature will increase following Charles law ($v \propto T$). Thus causing the piston to move upwards until the piston occupies the position d_1c_1, but the gas will remain at the same pressure $\left(\frac{W}{A}\right)$. It is assumed that the piston moves upwards slowly and the system (gas) passes through a number or equilibrium positions keeping pressure constant. This is termed heating at constant pressure.

Fig. 4.5(a) shows the above process on p-v diagram from equilibrium state point (1) to state point (2) (a horizontal line)

The movement of piston changes the system boundary and displacement work is done by the system against a constant force (W).

Or displacement work $= W \times dd_1$

$$= p.A.dd_1 \qquad \left[\because p = \frac{W}{A}\right]$$

$$= p \text{ (change in volume)}$$

$$= p(v_2 - v_1) \text{ Nm} \qquad \text{...(4.8)}$$

$$\because p.v. = m.R.T.$$

$$= mR(T_2 - T_1) \text{ Nm} \qquad \text{...(4.8(}a\text{))}$$

$$= mR\,(T_2 - T_1) \text{ kJ} \qquad \text{...(4.8(}b\text{))}$$

$$\text{Heat supplied} = m.c_v\,(T_2 - T_1) \text{ kJ} \qquad [\because \text{ constant pressure heating}]$$

$$\text{Change in internal energy} = m.c_v(T_2 - T_1)$$

Applying first law of thermodynamics

$$m.c_p\,(T_2 - T_1) = (T_2 - T_1) + m.cv\,(T_2 - T_1)$$

or $$c_p = R + c_v$$

or $$c_p - c_v = R \qquad \text{...(4.9)}$$

$$\text{Change in enthalpy} = \text{H}_2 - \text{H}_1$$

$$= U_2 + P_2v_2 - U_1 - P_1v_1$$

or $$(H_2 - H_1) = m.c_p(T_2 - T_1) \qquad \text{...(4.10)}$$

Thus for constant pressure process

Heat transfer = Change in enthalpy

In differential form

$$\text{dH} = c_v\,dT \text{ per kg of gas.}$$

or $$c_p = \left(\frac{dH}{dT}\right) \qquad \text{...(4.11)}$$

i.e., specific heat at constant pressure (c_p) is the rate of change of specific enthalpy with respect of absolute temperature at constant pressure.

The constant pressure process of Fig. 4.5 can also take place from state point 2 to state point 1 by allowing the system to cool by discharging heat to atmosphere.

4.3.3 Constant Temperature or Isothermal Process (Process Following Boyle's law)

Fig. 4.6(a) shows an arrangement for performing isothermal process under ideal conditions. It consists of a cylinder piston mechanism. The cylinder head is in contact with a constant temperature large reservoir (source/sink). The system (perfect gas) is also at the same temperature (T).

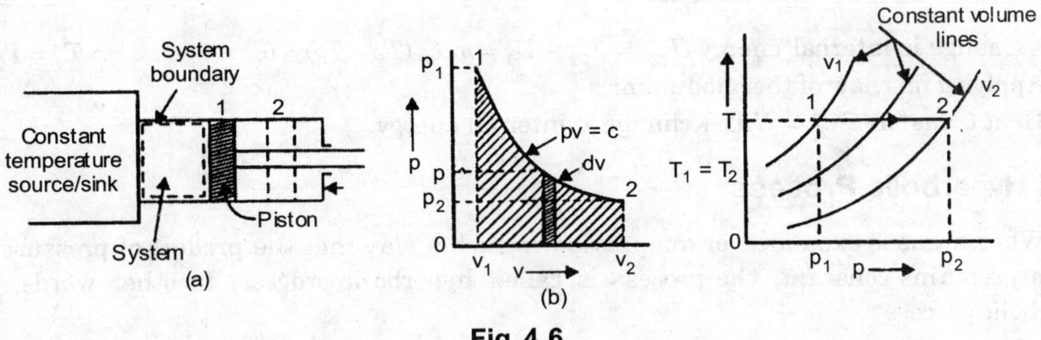

Fig. 4.6

Now imagine the piston to move slightly to the right, so that volume of the gas changes by dv and consequently pressure and temperature of the gas drop by dp and dT respectively. As the system boundary moves, therefore, displacement work ($P.d.v.$) is present. Due to drop in temperature dT, heat will now flow from the source to the system till the system reaches the temperature T again. As the source is of large capacity, there will be no change in temperature of the source due to transfer of heat. There will be no change of internal energy, because the system is restored to the original temperature. The piston may move again by a small distance to the right and similar chain of events would follow.

It should be noticed that during an isothermal expansion the piston would have to move very slowly so as to give time for the heat to flow through the cylinder head. For this reason isothermal expansions are not used in modern heat engines. The speed of internal combustion engines is too high to allow for the necessary amount of heat to be absorbed. Thus isothermal process will be possible when the process is carried out in quasi-static (thermal equilibrium) manner and when surroundings are maintained at constant temperature. A similar explanation can be given for isothermal compression.

According to Boyle's law, isothermal process must also be hyperbolic process in case of perfect gases.

The equation of the process is

$$p.v = C \quad ...(4.12)$$

or

$$p_1 v_1 = p_2 v_2 = \text{constant} \quad ...(4.12(a))$$

$$\text{Displacement work} = \int_1^2 p.dv = \int_1^2 C.\frac{dv}{v} \qquad [\because p.v = C]$$

$$= C \log_e \frac{v_2}{v_1}$$

or

$$W_{12} = (p_1.v_1 \text{ or } p_2.v_2) \log_e \frac{v_2}{v_1} \quad ...(4.13)$$

$$= mRT \log_e \frac{v_2}{v_1} \text{ Nm} \quad \text{...(4.13(a))}$$

or

$$W_{12} = m.R.T \log_e \frac{v_2}{v_1} \text{ kJ} \quad \text{...(4.13(b))}$$

Change in internal energy $U_{12} = U_2 - U_1 = m.c_v (T_2 - T_1) = 0 \quad \because T_1 = T_2 = T$

Applying first law of thermodynamics

Heat transfer (Q_{12}) = W.D + change in internal energy.

4.3.4 Hyperbolic Process

When a gas is expanded or compressed in such a way that the product of pressure and volume remains constant, the process is called hyperbolic process. In other words, in a hyperbolic process

$$p_1v_1 = p_2v_2 = \text{constant}$$

or

$$p.v = C$$

The displacement work, heat exchange etc., are the same as in case of isothermal process. It must be noted that

(*a*) Isothermal and hyperbolic process are same only in case of ideal or perfect gases and not in case of real gases and vapours.

(*b*) hyperbolic process means $p.v$ = constant and isothermal means T = constant.

(*c*) For wet vapour, isothermal and isobaric process are same, due to latent heating.

4.3.5 Constant Internal Energy Process

When a gas is expanded in such a way that the heat absorbed by the gas is equal to the work done by the gas, then internal energy of the gas will remain constant and the process will be called a constant internal energy process. The internal energy of the gas will also remain constant if it is compressed in such a way that the heat rejected by the gas is equal to the work done on the gas.

An isothermal process may also be treated as a constant internal energy process, if specific heat of the gas at constant volume (c_v) remains absolutely constant during the process.

An example of a constant internal energy process is that of an irreversible process occuring when a perfectly insulated chamber has a partition and contains a gas at some pressure on one side of the partition and vacuum on the other side of the partition, and the partition gets punctured [Fig. 4.7(*a*)]. The specific volume of gas will increase, the pressure of the gas will decrease and temperature may also decrease. But, as

Heat transfer = 0 [$\because$ insulated]

Work transfer = 0 [$\because$ unresisted expansion]

$\therefore$ Change in internal energy = 0

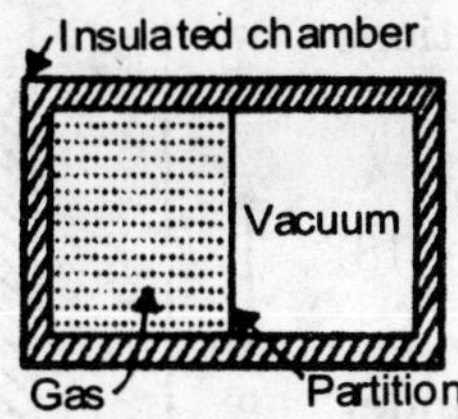

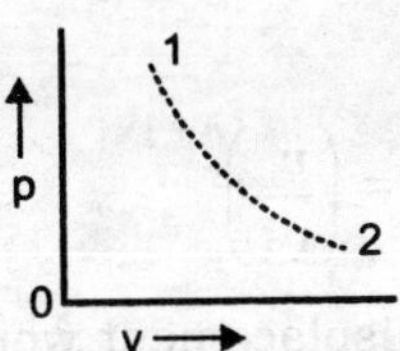

Fig. 4.7

The above process is irreversible and is shown on $p.v$ diagram in Fig. 4.7(b). The process is shown dotted because no predictable changes in pressure and volume occur during this process.

or $$Q_{12} = Q_2 - Q_1 = (p_1 v_1 \text{ or } p_2 v_2) \log_e \frac{v_2}{v_1} + 0$$

or $$Q_2 - Q_1 = (p_1 v_1 \text{ or } p_2 v_2) \log_e \frac{v_2}{v_1} \text{ kJ} \quad ...(4.14)$$

or $$Q_2 - Q_1 = mRT \log_e \frac{v_2}{v_1} \text{ kJ} \quad ...4.14(a)$$

Change in enthalpy $(H_{12}) = V_2 - V_1 = mc_v (T_2 - T_1) = 0$ $\quad \because T_1 = T_2 = T$

The ratio $\frac{v_2}{v_1}$ *is termed as expansion ratio or compression ratio depending upon whether the process is expansion or compression.*

4.3.6 Expansion following a General Law PV^n = Constant (or Polytropic Process)

A gas may be expanded or compressed in such a way that the curve of the process follows the law pv^n = constant. Where the index n may have any value depending upon the method of carrying out the process. For example:

When $n = 0$, the expression $pv^n = c$ becomes $p = c$ that is constant pressure process

When $n = 1$, the expression $pv^n = c$ becomes $p.v = c$ that is hyperbolic process

When $n = \infty$, the expression $pv^n = c$ becomes $v = c$ that is constant volume process

And it will be proved later in case of adiabatic process that

$$n = \frac{c_p}{c_v} = \gamma$$

for adiabatic process. Thus for polytropic process

$$p.v^n = c \quad ...(4.15)$$

or $$p_1 v_1^n = p_2 v_2^n = c \quad ...[4.15(a)]$$

also applying characteristic Eq. of gas

$$\frac{p_1 v_1}{T_1} = \frac{p_2 v_2}{T_2} \quad ...(4.16)$$

Eliminating v_1 and v_2 with the help of Equations 4.15(a) and 4.16, we get

$$\frac{T_2}{T_1} = \left(\frac{p_2}{p_1}\right)^{\frac{n-1}{n}} \quad ...(4.17)$$

and now eliminating p_1 and p_2 with the help of equations 4.15(*a*) and 4.16, we get

$$\frac{T_2}{T_1} = \left(\frac{v_1}{v_2}\right)^{n-1} \quad ...(4.18)$$

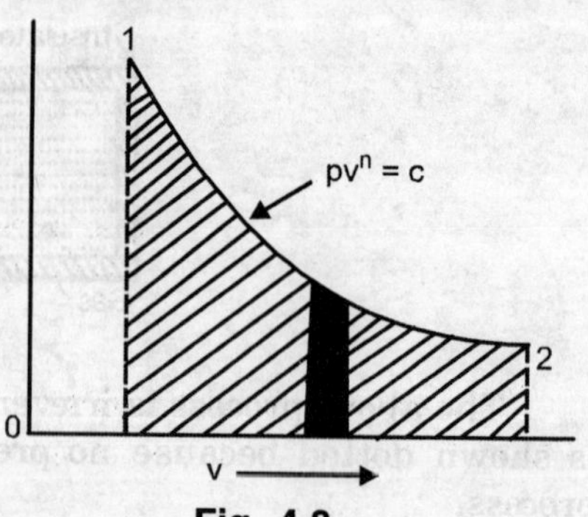

Fig. 4.8

Refering to Fig. 4.8 displacement work = area under the process on $p.v$ diagram or p-v plane

$$W_{12} = \int_1^2 p.dv = \int_1^2 \frac{c}{v^{-n}}.dv \qquad \because PV^n = C$$

$$= C\left[\frac{v^{-n+1}}{-n+1}\right]_{v_1}^{v_2}$$

or $$W_{12} = (p_2 v_1^n \text{ or } p_2 v_2^n)\left[\frac{v_2^{1-n} - v_1^{1-n}}{1-n}\right]$$

or $$W_{12} = \frac{p_1 v_1 - p_2 v_2}{n-1}\text{Nm} \quad ...(4.19))$$

also $$W_{12} = \frac{m.R(T_1 - T_2)}{(n-1)}\text{Nm} \quad ...(4.19(a))$$

and $$W_{12} = \frac{m.R(T_1 - T_2)}{J(n-1)}\text{kJ} \quad ...(4.19(b)$$

Change in internal energy $U_{12} = mc_v(T_2 - T_1)$ kJ ...(4.20)

Heat exchange $Q_{12} = W_{12} + U_{12}$

$$= \frac{m\,.\,R(T_1 - T_2)}{(n-1)} + m\,.\,c_v(T_2 - T_1)$$

$$= \frac{m\,.\,R(T_1 - T_2)}{(n-1)}\left[1 - \frac{c_v\,.\,J}{R}(n-1)\right]$$

$$= (\text{W.D. in kcal})\left[1 - \frac{(n-1)}{(\gamma-1)}\right] \qquad \because c_p\, c_v = R$$

or $$Q_{12} = (\text{W.D. in kcal})\left(\frac{\gamma - n}{\gamma - 1}\right) \quad ...(4.21)$$

Eq. 4.21 can also be modified as follows

$$Q_{12} = \left(\frac{\gamma - n}{\gamma - 1}\right).\frac{mR(T_1 - T_2)}{(n-1)}$$

$$= \left(\frac{\gamma - n}{\gamma - 1}\right).\frac{mRT_1}{(n-1)}\left(1 - \frac{T_2}{T_1}\right)$$

$$= \left(\frac{\gamma - n}{\gamma - 1}\right)\frac{p_1 v_1}{J(n-1)}\left[1 - \left(\frac{v_1}{v_2}\right)^{n-1}\right]$$

$$\because \quad \frac{T_2}{T_1} = \left(\frac{v_1}{v_2}\right)^{n-1} \quad \text{[Ref. Eq. 4.18]}$$

$$= \left(\frac{\gamma - n}{\gamma - 1}\right)\frac{p_1 v_1}{(n-1)}\left[1 - \frac{1}{r^{(n-1)}}\right] \quad \text{where } r = \left(\frac{v_2}{v_1}\right)$$

or

$$Q_{12} = \left(\frac{\gamma - n}{\gamma - 1}\right)\frac{p_1 v_1}{(n-1)}\left[1 - \frac{1}{r^{(n-1)}}\right] \text{ kJ} \quad \text{...(4.21)}$$

$$\text{Change in enthalpy} = H_{12} = m.c_v(T_2 - T_1) \quad \text{...(4.22)}$$

Index of the process "n" $\quad p_1 v_1^{\,n} = p_2.v_2^{\,n}$

Taking log we have

$$\log p_1 + n \log v_1 = \log p_2 + n \log . v_2$$

or

$$n\,[\log v_1 - \log v_2] = \log p_2 - \log p_1$$

$$n = \frac{\log \dfrac{p_2}{p_1}}{\log \dfrac{v_1}{v_2}} \quad \text{...(4.23)}$$

4.3.7 Reversible Adiabatic Process or Isentropic Process

In a reversible adiabatic process there is no heat exchange between the surroundings and the system but the boundary of the system moves giving displacement work at the expense of its own internal energy. Therefore the temperature of the system will fall during expansion and will rise during compression. Such a process is possible if the system is thermally insulated from the surroundings and there is no friction. In the absence of friction no frictional heat is generated and therefore entropy (to be dealt later) of the system remains constant. Thus a reversible adiabatic process is also called an isentropic process.

In case of gas turbines, centrifugal and axial flow air compressors although the system is isolated from the surroundings, *i.e.,* there is no heat flow between the system and the surroundings, but the velocities of air are so high that heat will be generated in the system due to friction which increases the entropy of the system. The adiabatic process under such conditions is called an **irreversible adiabatic process** and cannot be regarded as an isentropic process. For simplicity the word adiabatic is assumed to mean a reversible adiabatic or isentropic process.

Law of an Adiabatic Process. Up to the present, nothing has been said concerning the equation or law for an adiabatic process; let it be assumed to follow the general law $pv^n = a$ constant, and then find the particular value of the index 'n' which will satisfy the conditions for an adiabatic process.

From Eq. 4.19 (b) work done W_{12}

$$= \frac{mR(T_1 - T_2)}{(n-1)} \text{ kJ}$$

and also from Eq. 4.20

Change in internal energy $U_{12} = mc_v (T_2 - T_1)$ kcal

Applying law of conservation of energy,

heat exchange $Q_{12} = W_{12} + U_{12}$

i.e. $$0 = \frac{mR(T_1 - T_2)}{(n-1)} + m.c_v(T_2 - T_1)$$

($\because$ $Q_{12} = 0$ for adiabatic procers)

or $$\frac{mR(T_1 - T_2)}{(n-1)} = m.c_v\,(T_1 - T_2)$$

or $$\frac{R}{(n-1)} = c_v \qquad ...(i)$$

But $$c_p - c_v = R \text{ or } \gamma - 1 = \frac{R}{c_v}$$

or $$c_v = \frac{R}{(\gamma - 1)} \qquad ...(ii)$$

Comparing Eq. (*i*) and (*ii*) we get

$$\frac{R}{(n-1)} = \frac{R}{(\gamma-1)}$$

or $$n = \gamma$$

Thus the equation for adiabatic process becomes pv^{γ} = constant.

This is known as **poisson's law.**

In proving $n = \gamma$, it was assumed that the process follows the $pv^n = C$, but this involves empirical assumption and hence the proof cannot be called a rigorous one. The rigorous proof is as follows.

Consider *m* kg of a perfect gas following adiabatic process when the gas has a pressure *p*, temperature *T* and volume *v*. Let *dp*, *dv* and *dT* be very small increments of pressure, volume and temperature respectively. Then according to the law of conservation of energy.

$$O = p\,.\,dv + m.C_v.dT \qquad ...(iii)$$

[$\because$ $dQ = 0$]

Now differentiating the characteristic equation of perfect gases,

$$p.v = mRT. \text{ We get}$$

$$p.dv + v.dp = mR.dT \qquad ...(iv)$$

From Eq. (*iii*) and (*iv*)

$$O = p\,.\,dv + c_v\left(\frac{p.dv + v.dp}{R}\right)$$

or $$O = p.dv(R) + c_v\,(p.dv + v.dp) \qquad \because \text{ On multiplying by } R$$

or $$O = p.dv(R + c_v) + c_v.v.dp$$

or $$O = p.dv.cp + c_v \,.\, v.dp \qquad (\because\ c_p - c_v = R)$$

Dividing both sides of above equation by p, v, c_v, we get

$$O = \gamma \frac{dv}{v} + \frac{dp}{p}$$

Integrating both sides $$c_1 = \gamma \,.\, \log_e v + \log_e p$$

or $$c_1 = \log_e pv^\gamma$$ where c_1 constant of integration

Taking antilog $$e^{c1} = pv^\gamma$$

or $$pv^\gamma = C \qquad ...(4.24)$$

where C is another constant

for two end states reached during process:

or $$p_1 v_1^\gamma = p_2 v_2^\gamma = C \qquad ...[4.24(a)]$$

4.3.8 Free Expansion

Consider a system shown in Fig. 4.9 consisting of a perfect gas in an insulated container. The container has two compartments separated by a partition. One compartment contains perfect gas and the other compartment is in the state of perfect vacuum. Now if partition is suddenly removed the gas will expand and fill up the whole container as shown in Fig. 4.9(*b*). Since the container is insulated against heat, the system does not exchange any heat during this process and hence the process (expansion) of the system can be called as adiabatic process. However the difference between this adiabatic expansion and adiabatic expansion dealt under previous article lies in the fact that this expansion has no resistance to overcome (unresisted expansion) which in previous case the expansion was against the resistance offered by piston (resisted expansion). Hence this expansion is called unresisted adiabatic expansion or free expansion.

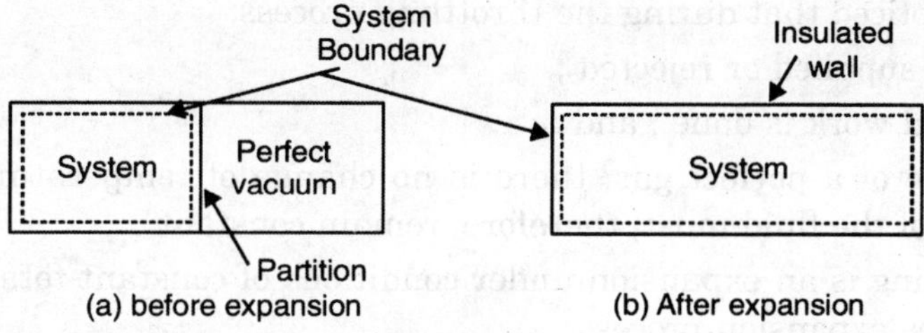

Fig. 4.9

Since it is unresisted expansion, no displacement work is done by the system.

Because heat transfer and work transfer during the process are zero. Therefore according to law of conservation of energy $(dQ = p \,.\, dv + cv \cdot dT)$. There is no change of internal energy of the system and consequently there is no change of temperature of the system ($c_v\ dT = 0$). The process may therefore be called constant internal energy process.

Since the temperature of the system during the process remains constant, hence the enthalpy of the system must also remain constant ($dH = c_p \,.\, dT$) or the process may also be called as constant enthalpy process.

Since internal energy during the process is constant, the flow work ($p.v$) must also remain constant. Thus it may be concluded that during a free expansion process the product of absolute pressure and specific volume remains constant.

It may also be noted that adiabatic and isothermal process described under previous articles can be carried out in reversed direction (reversible process), but it is not possible to retrace back the free expansion process without the help of external work (irreversible process).

Further since the process takes place suddenly the process is not quasistatic and therefore cannot be truly represented on *p-v* diagram. For this process, though, the integral $\int p.dv$ has a value and yet the work done is zero.

4.3.9 Throttling Process

This type of expansion occurs when a gas is expanded through an aperature of minute dimensions such as a narrow throat or a slightly opened valve; it is illustrated by the leak of a fluid through a crack in a vessel.

During the throttling process the expanding fluid is forced through the aperature by its pressure, but the aperature is so narrow that the frictional resistance between the fluid and the walls reduces the fluid velocity to a negligible amount ; consequently, the fluid oozes through the orifice with only a small amount of kinetic energy. Its kinetic energy has been converted into heat by friction which in the case of a perfect gas, warms up the fluid to its original temperature (actually, there is a very slight change of temperature which is too small to affect engineering problems). When applied to vapours the process is sometimes called *wire drawing.*

A throtlling process should not be confused with an expansion through a large orifice or nozzle. In the latter cases the frictional resistance is relatively small and the fluid leaves with a large velocity and consequently, with a large kinetic energy. A fluid expanding through a large orifice or nozzles approximates to an adiabatic process.

It should be noticed that during the throttling process:

(*a*) No heat is supplied or rejected ;

(*b*) No external work is done ; and

(*c*) In the case of a perfect gas, there is no change of temperature. The total heat (or enthalpy) of the fluid must, therefore, remain constant.

Hence, throttling is an expansion under conditions of constant total heat; in this respect it resembles the free expansion process.

Throttling process is an irreversible, unresisted process and cannot be truly represented on p-v diagram.

Solved Examples on Thermodynamic Processes

Example 4.9. *2 kg of gas at 27°C is heated in a reversible, non-flow, constant volume process, till the pressure is doubled. Determine (i) Final temperature (ii) work done (iii) the change in internal energy (iv) the change in enthalpy and (v) the heat exchange.*

Take $R = 287.17\ J/kg/°k$

$c_p = 1\ kJ/kg\text{-}°C.$

Sol. For constant volume process from characteristic equation of gas

(*i*) $\frac{P_1}{T_1} = \frac{P_2}{T_2}$ or $T_2 - T_1\left(\frac{P_2}{P_1}\right)$

$= (27 + 273)\,(2)$

$=$ **600°K Ans.**

(*ii*) Displacement work = *p.dv*

$=$ **zero Ans.** $\because dv = 0$

(*iii*) Change in internal energy

$= U_2 - U_1 = m.c_v(T_2 - T_1)$

$= m\left(c_p - R\right)(T_2 - T_1) \qquad \left(\because\ c_p - c_v = R\right)$

$= 2(1000 - 287)\,(600 - 300)$ J/kg °K = 427800 J/kg °K

$=$ **427.8 kJ increase. Ans.**

(*iv*) Change in enthalpy $= m \,.\, c_p(T_2 - T_1)$

$= 2 \times 1.0\,(600 - 300)$ kJ

$=$ **600 kJ increase.**

(*v*) Heat transfer = W.D. + Change in internal energy

$=$ **0 + 427.8 = 427.8 kJ Ans.**

(+) sign for heat transfer means heat is supplied to the gas.

Example 4.10 *A quality of gas occupying 0.07m³ at 0.9 MPa pressure and 27°C temperature is heated at constant pressure until the volume increases to 0.14 m³. The specific heats of the gas at constant pressure and at constant volume are respectively, 1.0 and 0.717 kJ/kg/°C. Find (i) characteristic gas constant ; (ii) mass of the gas ; (iii) final temperature; (iv) change in internal energy ; (v) work transfer ; (vi) heat transfer ; (vii) change in enthalpy.*

Sol. $p_1 = 0.9$ MPa, $v_1 = 0.07$ m³, $T_1 = (27 + 273) = 300$°K

$p_2 = p_1, v_2 = 0.14 \text{ m}^3, c_p = 1.0, c_v = 0.717$

(*i*) $R = (c_p - c_v) = 1.0\,(0.717)$

or $R =$ **0.293** kJ/kg/°K = 293 J/kg/°K **Ans.**

(*ii*) $p_1v_1 = mRT_1$

$0.9 \times 10^6 \times 0.07 = m \times 293 \times 300$

$m =$ **0.716 kg Ans.**

(*iii*) $\frac{v_1}{T_1} = \frac{v_2}{T_2}$

or $T_2 = T_1\frac{v_2}{v_1} = \frac{300 \times 0.14}{0.07}$

or $T_2 =$ **600°K or 327°C Ans.**

(*iv*) Change in internal energy $= m.c_v(T_2 - T_1) = 0.716 \times 0.717 \times (600 - 300)$

$=$ **154 kJ. Ans.**

(*v*) Work transfer $= p_1 (v_2 - v_1)$

$= 0.9 \times 10^6 (0.14 - 0.07) = 63000$ Nm

$=$ **14.8 Ans.**

(*vi*) Heat transfer = (work transfer = change in internal energy)

$= 14.8 + 71.16$

$=$ **85.96** (absorbed) **Ans.**

(*vii*) Change in enthalpy = heat transfer for constant pressure process. **Ans.**

Example 4.11 *A stationary mass of gas is compressed without friction from the initial state of 0.3 m³ and 0.105 MPa absolute to final state of 0.15 m³ and 0.105 MPa absolute there is a transfer of 36.67 kJ of heat from the gas during the process. How much does the internal energy of the gas change?*

Sol. $p_1 = p_2 = 0.105$ MPa (absolute)

$= 0.105 \times 10^6$ MPa (absolute)

$v_1 = 0.3\ m^3,\ v_2 = 0.15\ m^3$

Now, work done during compression

$${}_1W_2 = (p_1 \text{ or } p_2)(v_1 - v_2)$$

$$= 0.105 \times 10^6 (0.15 - 0.3)$$

$$= -15750 \text{ Nm.}$$

$$= -\frac{15750}{1000}$$

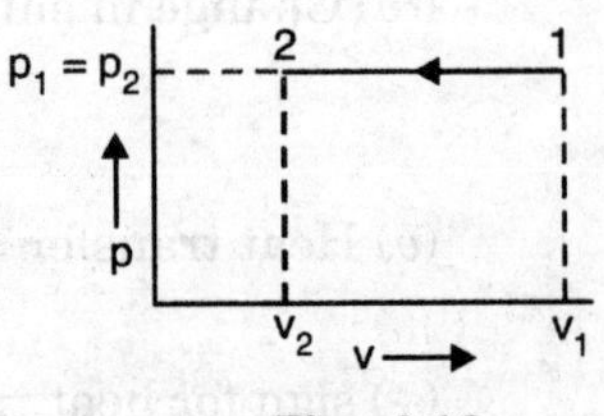

Fig. 4.10

or $${}_1W_2 = -15.75 \text{ kJ}$$

Here negative sign for work done indicated that work is done on the gas.

Applying law of conservation of energy to the closed system

$${}_1Q_2 - {}_1W_2 = {}_1U_2$$

∴ Change in internal energy during compression

$${}_1U_2 = {}_1Q_2 - {}_1W_2$$

$$= -36.67 - (-15.75)$$

[∵ heat is rejected to the surrounding during compression, ∴ negative sign for ${}_1Q_2$]

or $${}_1U_2 = \mathbf{-20.92\ kJ} \quad \textbf{Ans.}$$

Here negative sign for ${}_1U_2$ indicates decrease in internal energy.

Example 4.12 *2 kg of air at a pressure of 0.7 MPa occupies a volume of 0.3 m³ ; this air is expanded to a volume of 1.5 m³ isothermally. Taking $c_p = 1$, $c_v = 0.718$. Find (i) Final pressure (ii) work transfer ; (iii) heat transfer; (iv) change in internal energy ; (v) change in enthalpy.*

Sol. $m = 2$ kg., $p_1 = 0.7$ MPa, $v_1 = 0.3\ m^3$, $v_2 = 1.5\ m^3$, $p_2 = ?$

(*i*) For isothermal or hyperbolic process

$$p_1v_1 = p_2v_2$$

or $$p_2 = \frac{0.7 \times 0.3}{1.5} = p_2 = \mathbf{0.14\ MPa} \quad \textbf{Ans.}$$

(*ii*) Work transfer = $p_1 v_1 \log_e \dfrac{v_2}{v_1}$

$$= 0.9 \times 10^6 \times 0.07 \log_e \frac{0.14}{0.07} = 63000 \times 0.693$$

$= 43659$ J = **43.659 kJ** work is done by the gas **Ans.**

(*iii*) Heat transfer

$(Q_{12}) = W_{12} + U_{12} = 43.659 + 0$ $\quad [\because U_{12} = 0 \text{ as } T_2 = T_1]$

$=$ **43.659 kJ** (absorbed by gas) **Ans.**

(*iv*) Change in internal energy

$U_{12} = 0$ as above.

(*v*) Change in enthalpy $\quad H_{12} = mc_v(T_2 - T_1)$

$= 0$ **Ans.** $\quad \because [T_2 - T_1]$

Example 4.13 *0.2 kg of air at a pressure of 0.1 MPa and a temperature of 27°C is compressed to a pressure of 1 MPa according to the law $pv^{1.25}$ = C. Determine:*

(*i*) Initial and final parameters of the air;

(*ii*) Work done on or by the air;

(*iii*) Heat exchange;

(*iv*) Change in internal energy;

(*v*) Change in enthalpy.

Take $c_p = 0.1$, $R = 293$

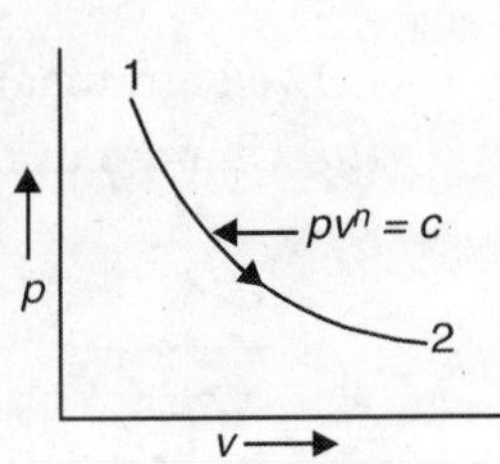

Fig. 4.11

Sol. $m = 0.2$ kg; $P_1 = 0.1$ MPa ; T_1 (27 + 273) = 300°K

$p_2 = 1$ MPa; $n = 1.25$

(*i*) $p_1 v_1 = mRT_1$

or $$v_1 = \frac{0.2 \times 293 \times 300}{0.1 \times 10^6}$$

$= \mathbf{0.1756\ m^3}$ **Ans.**

$$\frac{T_2}{T_1} = \left(\frac{p_2}{p_1}\right)^{\frac{n-1}{n}}$$

or $$T_2 = 300(10)^{\frac{1.25-1}{1.25}} = 475.5°\text{K}$$

$=$ **202.5°C** **Ans.**

$$p_1 v_1^n = p_2 v_2^n$$

$$v_2 = \left(\frac{p_1}{p_2}\right)^{\frac{1}{n}} \times v_1$$

$$= 0.1756\left(\frac{0.1}{1}\right)^{\frac{1}{1.25}} = \mathbf{0.0278\ m^3} \quad \textbf{Ans.}$$

(*ii*)
$$W.D. = \frac{p_1v_1 - p_2v_2}{(n-1)} = \frac{mR(T_1 - T_2)}{(n-1)}$$

$$= \frac{0.2 \times 287\,(300 - 475.5)}{(1.25 - 1)}$$

$$= -40295 \text{ Nm} = \mathbf{-40295\ J} \qquad [\because 1\text{Nm} = 1\text{J}]$$

Negative sign for W.D. means work is done on the air. **Ans.**

(*iii*) Heat exchange $Q_{12} = W.D. + U_{12}$

$$= \text{W.D.} + m\left(c_p - R\right)(T_2 - T_1)$$

$$= -40295 + 0.2 \times \left(1000 - 287\right)(475.5 - 300)$$

$$= -40295 + (0.2 \times 713 \times 175.5)$$

$$= -40295 + 25026 = 15269 \text{ J}$$

$$= \mathbf{15.269\ kJ} \quad \textbf{Ans.}$$

(–) ve sign means rejection of heat

(*iv*) Change in internal energy

$$= m\left(c_p - \frac{R}{J}\right)(T_2 - T_1)$$

$$= \mathbf{32.56\ kJ} \text{ Increase} \quad \textbf{Ans.}$$

(*v*) Change in enthalpy $= m \,.\, c_p\,(T_2 - T_1)$

$$= 0.2 \times 0.1 \times (475.5 - 300)$$

$$= \mathbf{35.1\ kJ.} \quad \textbf{Ans.}$$

Example. 4.14 *A cylinder contains 0.18 m³ of gas at a pressure of 0.1 MPa abs and temperature 47°C. If this gas is compressed polytropically to 1/12th of its volume and the pressure is then 2 MPa absolute. find; (i) the mass of the gas; (ii) the temperature at the end of compression; (iii) the index of compression ; (iv) the change in internal energy; and (v) the heat rejected during compression.*

Take $C_v = 0.83$ *kj/kg-°C ;* $c_p = 1.08$ *kj/kg-°C.*

Sol. Refer Fig. 4.11

$p_1 = 0.1$ MPa abs. $\quad p_2 = 2$ MPa abs. $\quad v_1 = 0.18 \text{ m}^3$

$v_2 = \dfrac{0.18}{12} = 0.015 \text{ m}^3 \quad T_1 = (47 + 273)°\text{K} = 320°\text{K}$

(*i*) Now, $P_1\, v_1 = mRT_1$ $\qquad \left[\because c_p - c_v = \dfrac{R}{J}\right]$

$$P_1\, v_1 = m\,(c_p - c_v)\, T_1$$

$$1 \times 10^5 \times 0.18 = m\,(c_p - c_v)\, T_1$$

$$\therefore \quad m = \frac{10^5 \times 0.18}{1000 \times (1.08 - 0.83) \times 320} = 0.225 \text{ kg.}$$

(*ii*) $$\frac{P_1 v_1}{T_1} = \frac{P_2 v_2}{T_2}$$

∴ $$T_2 = T_1 . \frac{p_2}{P_1} . \frac{v_2}{v_1}$$

$$= 320 \times \frac{2}{0.1} \times \frac{1}{12} = 533.33°\text{K}$$

or $$t_2 = \textbf{260.33°C} \quad \textbf{Ans.}$$

(*iii*) $$P_1 v_1^n = P_2 v_2^n$$

and ∴ $$n = \frac{\log \frac{P_2}{P_1}}{\log \frac{v_1}{v_2}} = \frac{1.301}{1.0792}$$

or $$n = \textbf{1.2055} \quad \textbf{Ans.}$$

(*iv*) Change in internal energy $\Delta U = m.c_v (T_2 - T_1)$

$$= 0.225 \times 0.833 (533.33 - 320)$$

$$= \textbf{40 kJ} \quad \textbf{Ans.}$$

(*v*) W.D. $$= \frac{P_1 v_1 - P_2 v_2}{(n-1)}$$

$$= \frac{(0.1 \times 0.18 - 2 \times 0.015) \times 10^6}{0.2055} = \frac{(0.18 - .03) \times 10^6}{0.2055} \text{Nm}$$

$$= - 58390 \text{ J} = \textbf{–58.39 kJ} \quad \textbf{Ans.}$$

Now, Q = W.D. + ΔU = – 58.39 + 40 = **– 18.39 kJ** (Rejected)

Example. 4.15 *1 kg of air at a pressure of 1 MPa absolute and 20°C expands in a cylinder to a pressure of 0.1 MPa absolute according to the law* $pv^{1.3} = c$. *If for air* $\gamma = 1.4$ *; characteristic gas constant is 287 J/kg-°K ; estimate (a) work done (b) interchange of heat between cylinder walls and air, and (c) change in internal energy,*

Sol. Ref. Fig. 4.8

m = 1 kg ; P_1 = 1 MPa ; T_1 = (20 + 273)°K

p_2 = 0.1 MPa ; γ = 1.4, R = 287 J/kg-°K

$$T_2 = T_1 \left(\frac{p_2}{p_1}\right)^{\frac{n-1}{n}} \qquad \text{[Ref. eq. 4.17]}$$

$$= 293 \left(\frac{0.1}{1}\right)^{\frac{1.3-1}{1.3}} = 172.15°\text{K}$$

$$\text{W.D.} = \frac{p_1 v_1 - p_2 v_2}{(n-1)} = \frac{mR(T_1 - T_2)}{(n-1)}$$

$$= \frac{1 \times 287(293 - 172.15)}{(1.3 - 1)} = 115610\,\text{J}$$

$$= \textbf{11.561 kJ. Ans.}$$

Change in internal energy $= m.c_v.(T_2 - T_1)$

or $$U_{12} = m\left[\frac{R}{(\gamma - 1)}\right](T_2 - T_1)$$

$$= 1\left[\frac{287}{426.7\,(1.4 - 1)}\right](172.15 - 293) = 86710\,\text{J}$$

$$= \textbf{– 86.71 kJ} \text{ (decrease)} \quad \textbf{Ans.}$$

Heat interchange

$$Q_{12} = W_{12} + U_{12} = 115.6 - 86.71$$

$$= \textbf{28.9 kJ} \text{ absorbed by air.} \quad \textbf{Ans.}$$

Example 4.16 *Gas at 0.15 MPa abs. and 20°C in a closed vessel is compressed to 1 MPa abs. Its temperature then becomes 180 °C. If the compression follows the law $p.v^n = c$, find the value of n.*

Sol.Ref. Fig. 4.9 and Eq. 4.17

$$\frac{T_2}{T_1} = \left(\frac{p_2}{p_1}\right)^{\frac{n-1}{n}} \quad \text{or} \quad \frac{n-1}{n} = \frac{\log \dfrac{T_2}{T_1}}{\log \dfrac{p_2}{p_1}}$$

or $$1 - \frac{1}{n} = \frac{\log \dfrac{(180 + 273)}{(20 + 273)}}{\log \dfrac{1}{0.15}} = \frac{\log \dfrac{453}{293}}{\log 6.67} = \frac{0.4357}{1.897} = 0.23$$

$$\therefore \quad n = \textbf{1.298} \quad \textbf{Ans.}$$

Example 4.17 *0.5 kg of air at 180°C expands adiabatically to three times its original volume and during the process, the temperature falls to 16°C. The work done during the process is 58500 Nm, calculate c_p and c_v.*

Sol. Referring Fig. 4.12

$$T_1 = (180 + 273) = 453°\text{K}$$

$$T_2 = (16 + 273) = 289°\text{K}$$

For adiabatic process,

$$\frac{T_1}{T_2} = \left(\frac{v_2}{v_1}\right)^{\gamma - 1}$$

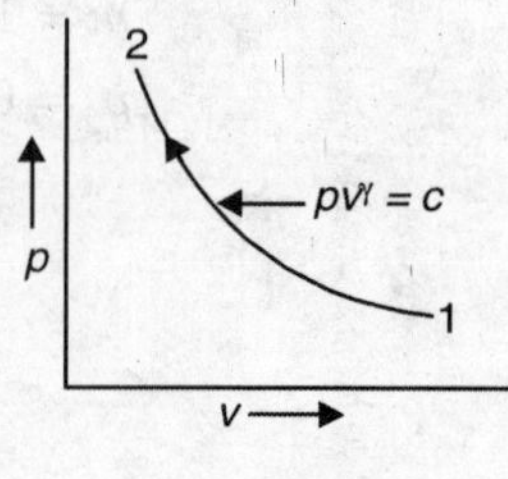

Fig. 4.12

or $$\gamma - 1 = \frac{\log \dfrac{T_1}{T_2}}{\log \dfrac{v_2}{v_1}} = \frac{\log \dfrac{453}{289}}{\log 3} = \frac{0.449}{1.098}$$

or $$\gamma = 1.4087$$

or $$\frac{c_p}{c_v} = 1.4087 \qquad \ldots(i)$$

For work done during the adiabatic process

$$W_{12} = \frac{mR(T_1 - T_2)}{(\gamma - 1)}$$

or $$58500 = \frac{0.5 \times R(453 - 289)}{(1.4087 - 1)} \text{ or } R = \frac{58500 - 0.4087}{0.5 \times 164} = 291.6$$

or $$(c_p - c_v) = 291.6 \text{ J/kg-°k} \qquad ...(ii)$$

or $$c_p - c_v = \frac{291.6}{1000} = 0.291.6 \text{ kJ/kg-°k}$$

From Eq. (*i*) and (*ii*)

$$\mathbf{c_v = 0.713, \quad c_p = 1.005} \quad \textbf{Ans.}$$

Example. 4.18 *Volume swept by the piston in a gas engine cylinder is 0.0294 m³ and clearance volume is 0.00508 m³. Assuming adiabatic compression ($pv^{1.4}$ = constant), determine the temperature at the end of compression. Take temperature at the beginning of compression = 100°C.*

Sol. Using Fig. 4.13

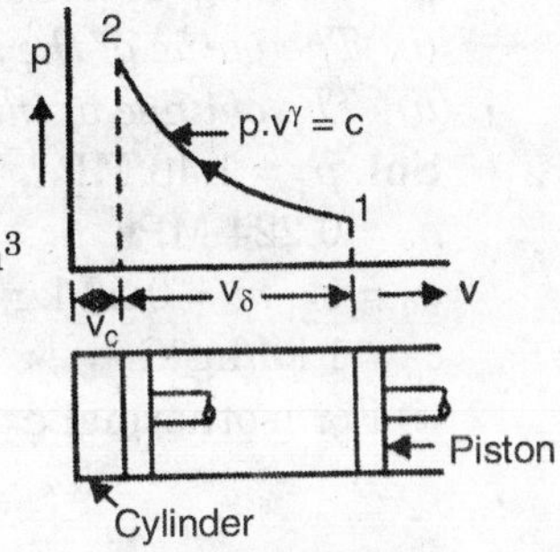

$$v_1 = v_c + v_s,$$

$$T_1 = (100 + 273) = 373°\text{K}$$

$$v_1 = 0.00508 + 0.0294 = 0.3448 \text{ m}^3$$

$$v_2 = v_c = 00508 \text{ m}^3$$

Now, $$\frac{T_2}{T_1} = \left(\frac{v_1}{v_2}\right)^{\gamma - 1}$$

Fig. 4.13

∴ $$T_2 = 373\left(\frac{0.03448}{0.00508}\right)^{(1.4-1)}$$

$$= \mathbf{802.3°K} \qquad \textbf{Ans.}$$

Example. 4.19 *6 kg of air at a pressure of 0.6 MPa abs. occupies a volume of 2 m³; the air is then expanded adiabatically to a volume of 5 m³. Find the temperature and pressure at the end of expansion, the work done and heat exchange. Take R = 287 J/kg-°K, C_p = 0.996, c_v = 0.707 kJ/kg-°k.*

Sol. $m = 6$ kg, $p_1 = 6 \times 10^5$ MPa abs. $v_1 = 2$ m³

$V_2 = 5$ m³, $p_2 = ?$, $T_2 = ?$, $W_{12} = ?$, $Q_{12} = ?$

$$p_1 v_1 = mRT_1$$

or $$T_1 = \frac{6 \times 10^5 \times 2}{6 \times 287} = 683.3°\text{K}$$

$$\gamma = \frac{c_p}{c_v} = \frac{0.238}{0.169} = 1.4083$$

$$\frac{T_1}{T_2} = \left(\frac{v_2}{v_1}\right)^{\gamma - 1}$$

or $$\frac{683.3}{T_2} = \left(\frac{5}{2}\right)^{(1.4083-1)}$$

$$\therefore \quad T_2 = \mathbf{454°K} \quad \textbf{Ans.}$$

$$p_1 v_1^{\gamma} = p_2 v_2^{\gamma}$$

or

$$p_2 = p_1\left(\frac{v_1}{v_2}\right)^{\gamma} = 0.6\left(\frac{2}{5}\right)^{1.4083}$$

$$= 0.165 \text{ MPa}$$

Work done

$$W_{12} = \frac{mR(T_1 - T_2)}{(\gamma - 1)}$$

$$= \frac{6 \times 287(686.3 - 454)}{(1.4083 - 1)} = 979722 \text{ J} = \mathbf{979.722\ kJ} \quad \textbf{Ans.}$$

Heat exchanged = 0 because adiabatic process

Example. 4.20 *A gas initially at 1.43 MPa and 360°C expanded isothermally to a pressure of 0.224 MPa. It is then cooled at constant volume till the pressure falls to 0.102 MPa. Finally an adiabatic compression brings the gas back to the initial state. The mass of the gas is 0.23 kg and c_p = 1 kJ/kg°C. Draw the P-V diagram and determine,*

(i) The value of the adiabatic index of compression, and

(ii) The change of internal energy of the gas during the adiabatic process.

Sol. p_1 = 1.43 MPa, $T_2 = T_1$ = (360 + 273)°K

p_2 = 0.224 MPa, p_3 = 0.102 MPa,

$v_2 = v_3$, m = 0.23 kg,

c_p = 1 kJ/kg°C

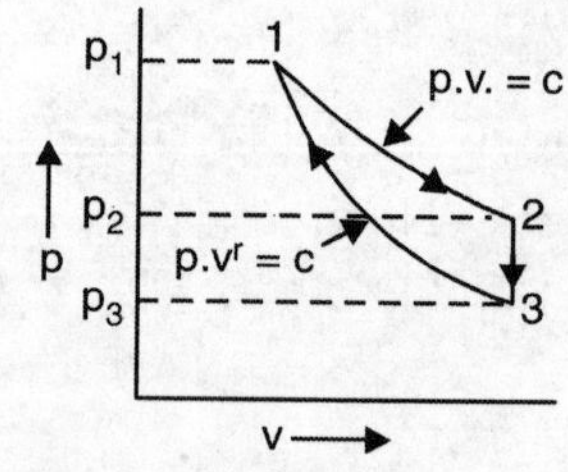

Fig. 4.14

(*i*) For isothermal expansion 1-2

$$P_1 v_1 = P_2 v_2$$

$$\therefore \quad v_2 = v_1 \cdot \frac{P_1}{P_2}$$

$$= v_1 \cdot \frac{1.43}{0.224} = 6.384\ v_1$$

$v_2 = v_3 = 6.384\ v_1$ for constant volume cooling (2–3). For adiabatic compression (3 – 1)

$$P_1 v_1^{\gamma} = P_3 v_3^{\gamma}$$

$$\therefore \quad \gamma = \frac{\log \frac{P_1}{P_3}}{\log \frac{v_2}{v_1}} = \frac{\log \frac{14.3}{1.02}}{\log \frac{6.384 \cdot v_1}{v_1}} = \frac{1.1467}{0.8051}$$

$$= \mathbf{1.4243} \quad \textbf{Ans.}$$

(*ii*) For constant volume cooling (2 – 3)

$$\frac{P_2}{T_2} = \frac{P_3}{T_3} \quad \text{or} \quad T_3 = T_2 \cdot \frac{P_3}{P_2}$$

$$= T_1 \cdot \frac{P_3}{P_2} = (360 + 273)\frac{1.02}{2.24}$$

$$= 288.241°\text{K}$$

Now change of internal energy during adiabatic compression

$$= m.\, c_v.(T_1 - T_2)$$

$$= 0.23 \times \frac{1}{1.4243}(633 - 288.241) \qquad (\because\ c_v = \frac{c_p}{\gamma})$$

$$= \mathbf{55.6\ kJ} \quad \textbf{Ans.}$$

Example. 4.21 *0.75 m³ of H_2 gas is initially at a pressure of 0.1 MPa and temperature 290°K. It is compressed isentropically to 1.5 MPa. Next it is expanded at constant temperature to original volume. Finally, heat rejection takes place at constant volume and the gas pressure is restored to the original condition, find:*

(i) Pressure, volume and temperature at the end of each operation.

(ii) Heat added during isothermal expansion;

(iii) Change of internal energy during each process.

For H_2 $R = 4200\ J/kg°K$; $c_p = 14.25\ kJ/kg°K$.

Sol.

$$v_1 = 0.75\ \text{m}^3,\quad P_1 = 0.1\ \text{MPa},$$

$$T_1 = 290°\text{K}$$

$$P_2 = 1.5\ \text{MPa}$$

$$T_2 = T_3 \quad v_3 = v_1$$

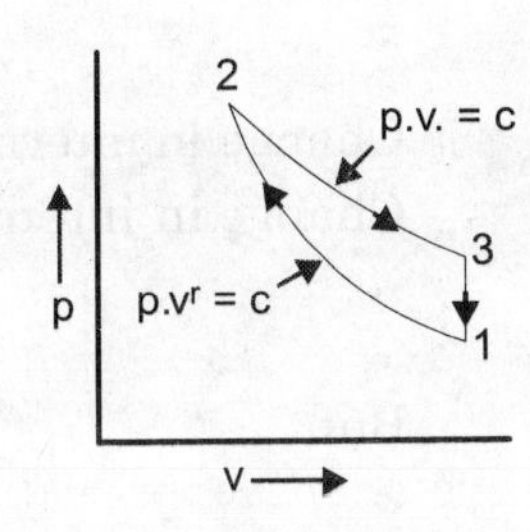

Fig. 4.15

(i)

$$P_1 v_1^{\gamma} = P_2 v_2^{\gamma}$$

$$0.1\,(0.75)^{\gamma} = 1.5\,(v_2)^{\gamma}$$

But

$$\gamma = \frac{1}{1 - \dfrac{R}{c_p}} = \frac{1}{1 - \dfrac{4200}{14.25 \times 1000}}$$

$$= 1.4075$$

$$\therefore \quad v_2 = 0.75\left(\frac{0.1}{1.5}\right)^{\frac{1}{1.4075}}$$

$$= \mathbf{0.1095\ m^3} \quad \textbf{Ans.}$$

$$\frac{v - 1}{v} = \frac{0.4075}{1.4025} = 0.2895$$

$$T_2 = T_1\left(\frac{P_2}{P_1}\right)^{\frac{\gamma - 1}{\gamma}} = 290\ (15)^{0.2895}$$

$$= \mathbf{635.1°K} \quad \textbf{Ans.}$$

$$P_2 v_2 = P_3 v_3$$

or

$$1.5 \times 0.1095 = P_3 \times \times 0.75$$

$$\therefore \quad P_3 = \frac{1.5 \times 0.1095}{0.75} = \mathbf{0.219\ MPa} \quad \textbf{Ans}$$

(ii) During isothermal expansion, change in internal energy is zero

$\therefore$ Heat supplied = W.D.

$$= P_2 . v_2 \log_e \frac{v_3}{v_2}$$

$$= 1.5 \times 10^6 \times 0.1095. \log_e \frac{0.75}{0.1095}$$

$$= 0.164 \times 1.924 \times 10^6 \text{ Nm} = 315 \text{ kN-m} = \mathbf{315\ kJ} \quad \textbf{Ans.}$$

(*iii*) Change in internal energy during isentropic compression

$$= -\text{(W.D)} \qquad [\because \text{ no heat exchange}]$$

$$= -\left[\frac{P_1 v_1 - P_2 v_2}{(\gamma - 1)}\right]$$

$$= -\left[\frac{0.1\times 10^6 \times 0.75 - 1.5\times 10^6 \times 0.1095}{0.4075}\right]$$

$$= \frac{10^6(0.075 - 0.164)}{0.4075 \times 1000} = \mathbf{219\ kJ} \text{ increase} \quad \textbf{Ans.}$$

Change in internal energy during isothermal expansion = zero

Change in internal energy during heat rejection

$$= m.c_v(T_1 - T_3)$$

But $m = \dfrac{P_1 v_1}{RT_1} = \dfrac{10^5 \times 0.75}{4200 \times 290}$

$$= 0.0614885 \text{ kg}$$

and $c_p - c_v = R$

$\therefore$ $c_v = c_p - R = 14.25 \times 10^3 - 4200$

$$= 10050 \text{ J/kg-°K} = 10.05 \text{ kJ/kg-°K}$$

$\therefore$ Change in I.E. $= 0.0614885 \times 10.05\ (290 - 635.1)$

$$= \mathbf{-\ 213.2\ kcal} \text{ (decrease)} \quad \textbf{Ans.}$$

4.4 STEADY FLOW PROCESS

In the previous sections we have discussed applications of first law of thermodynamics to a closed system, where the substance remains in the system and there is no flow in or out of the system. Such systems are known as 'non-flow process'. But in actual practice, there is a continuous flow of working fluid in and out from the system such as in air compressors, I.C. engines, steam engines, pumps etc. Such a process is called **'flow process'** and the system is known as **'open system'.**

Flow processes may be classified into "steady flow process" and "unsteady flow process". In most practical problems the rate at which the fluid flows through the system is constant. with respect to time. This type of flow is known as "steady flow".

Steady Flow Energy Equation.

In steady flow processes:

(*i*) The mass flow rate of the working fluid within the control volume remains constant with respect to time.

(*ii*) The energy possessed by the working fluid at any point in the control volume remains constant and the rate at which heat and work cross the control volume boundary is constant, and,

(*iii*) The state of the working fluid at any point in the control volume is constant.

In unsteady flow process, the rate of mass of working fluid into the control volume is not constant; and there is change in the energy possessed by the working fluid. For example; filling or evacuation of vessels.

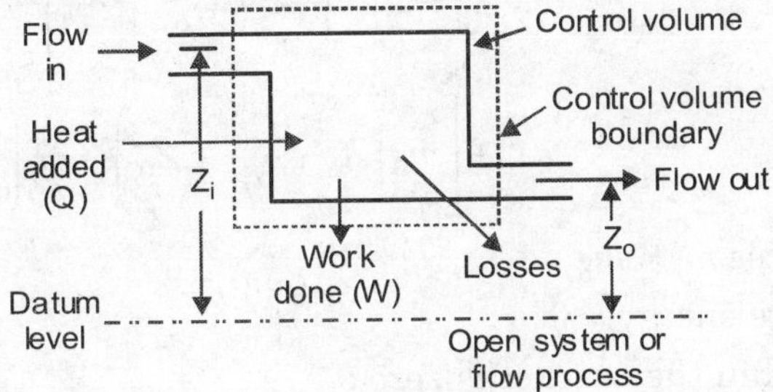

Fig. 4.16

Referring to Fig. 4.16, the energy balance required by the law of conservation of energy (an alternative statement of the first law of thermodynamics) for unsteady flow process may be written as follows:

$_1Q_2 - {}_1W_2$ = change in storage energy + change in flow energy

$$= m\left[\left(U_2 + \frac{c_2^2}{2g_e} + \frac{g}{g_e}\cdot Z_2\right) - \left(U_1 + \frac{C_1^2}{2g_e} + \frac{g}{g_e}\, Z_1\right)\right]$$

$$+ \left[m_0\left(U_0 + \frac{C_0^2}{2g_e} + \frac{g}{g_e}\cdot Z_0 + P_0 v_0\right) - m_i\left(U_i + \frac{C_i^2}{2g_e} + \frac{g}{g_e}\, Z_i + P_i v_i\right)\right] \quad ...(4.25)$$

where suffixes 1 and 2 denote the initial and final conditions during a given time within the system (or control volume) and suffixes '*i*' and '*o*' denote the inlet and outlet conditions of the working fluid.

and U = random molecular energy (or internal energy);

$\frac{C^2}{2g_e}$ = the kinetic energy;

$\frac{g}{g_e}\cdot Z$ = the potential energy;

$P.v$ = Flow work;

$_1Q_2$ = heat supplied during a given time;

$_1W_2$ = work done during the same time in heat units;

g = local acceleration due to gravity;

g_c = standard acceleration due to gravity;

v = specific volume of the working fluid;

m = mass of the working fluid within the control volume;

c = velocity

m_i = mass of inflow of working fluid;

m_0 = mass outflow of the working fluid;

Substituting $h = U + Pv$ in Eq. 4.25

where h is the enthalpy

And assuming $g = g_e$ for all practical purposes then

$$_1Q_2 - {}_1W_2 = m_i\left[\left(U_2 + \frac{C_2^2}{2g} + Z_2\right) - \left(U_1 + \frac{C_1^2}{2g} + Z_1\right)\right]$$

$$+\left[m_0\left(h_e + \frac{C_0^2}{2g} + Z_0\right) - m_i\left(h_1 + \frac{C_i^2}{2g} + Z_i\right)\right] \quad ...(4.25(a))$$

For steady flow process $m_i = m_o$

And change in storage energy = 0

i.e. the first term within the bracket is zero ;

Then Eq. 4.25(a) reduces to

$$_1Q_2 - {}_1W_2 = m_i\left[\left(h_0 + \frac{C_0^2}{2g} + Z_o\right) - \left(h_i + \frac{C_i^2}{2g} + Z_i\right)\right] \quad ...(4.25(b))$$

Eq. 4.25(b) is called the steady flow energy equation and is abbreviated as S.F.E.E.

If the effect of gravity is neglected, (i.e. $Z_i = Z_o$), which is very common in thermodynamics; Eq. 4.25 (b) reduces to

$$_1Q_2 - {}_1W_2 = m_i\left[(h_o - h_i) + \frac{1}{2g}\left(C_0^2 - C_i^2\right)\right] \quad ...(4.25(c))$$

If $C_i = C_o$, the Eq. 4.25(c) further reduces to

$$_1Q_2 - {}_1W_2 = m_i\,(h_o - h_i) \quad ...(4.25(d))$$

In a non-flow process the flow work is zero, (*i.e.*, $P.v = 0$)

Then, $$_1Q_2 - {}_1W_2 = (u_0 - u_i)\,m_i \quad ...(4.25)(e))$$

The conditions used in the derivation of S.F.E.E. are

(*i*) The composition, state and velocity of the streams crossing the boundary of the control volume do not change with time at entrance and exit.

(*ii*) The state of the fluid at any point within the control volume is same at all times.

(*iii*) The rate of flow of mass into the control volume is constant and is equal to the mass rate of flow out of it.

(*iv*) The rates of heat and work transfers across the control volume boundary are constant.

The term control volume used is the space bounded by imaginary boundaries around the components under consideration.

The difference between system and control volume lies in the fact that the boundary of the system may change, but the boundary of the control volume is assumed to be rigid.

4.4.1 Applications of Steady Flow Energy Equation

1. Steam Boiler

In case of steam boiler (refer Fig. 4.17).

(*a*) There is no appreciable change in velocity of working fluid and consequently change in kinetic energy may be neglected *i.e.*,

$$\frac{C_2^2 - C_1^2}{2g} = 0$$

(*b*) There is no shaft work (i.e., no work transfer).

(*c*) There is no appreciable change in elevation between inlet point (1) and exit point (2) and may be neglected *i.e.*,

$$Z_2 - Z_1 = 0.$$

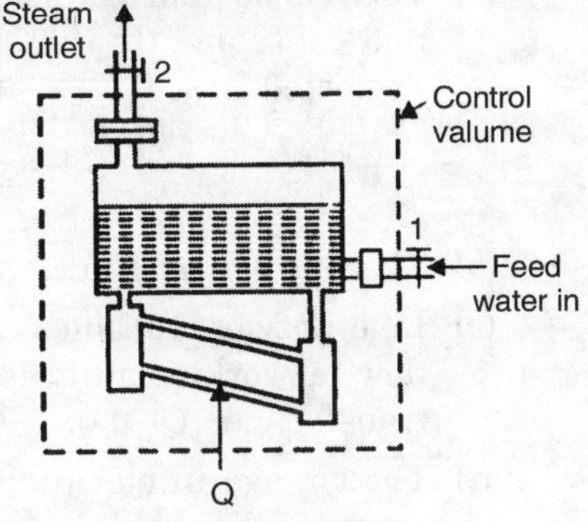

Fig. 4.17 Boiler

By selecting a control volume as shown in Fig. 4.17, boiler can be treated as steady flow system. The steady flow energy equation reduces to

$$_1Q_2 = m_i\ [h_2 - h_1]$$
$$= \{m_i\left[(U_2 + p_2v_2) - (U_1 + p_1v_1)\right] \quad \text{...(4.26)}$$

2. Steam Nozzle

A steam nozzle is characterised by following features. (Refer Fig. 4.18)

(*a*) Velocity of the working fluid increases and consequently kinetic energy at the expense of pressure (or enthalpy).

(*b*) There is no shaft work i.e. $_1W_2 = 0$.

(*c*) The nozzles are thermally insulated and the working fluid flow at very high rate and therefore, heat transfer is negligible i.e. $_1Q_2 = 0$.

(*d*) The change in elevation between the entrance and exit points of the nozzle is negligible i.e. $Z_2 - Z_1 = 0$.

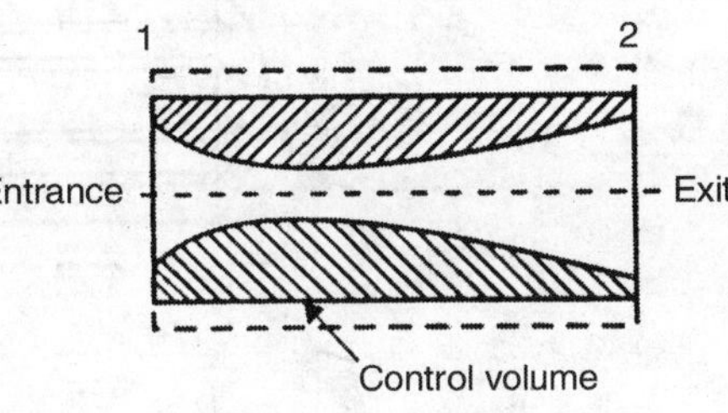

Fig. 4.18 Nozzle

Therefore the steady flow energy equation reduces to

$$0 = \left[(h_2 - h_1) + \frac{C_2^2 - C_1^2}{2g}\right]$$

or

$$\frac{C_2^2 - C_1^2}{2g} = (h_1 - h_2) \quad \text{...(4.27)}$$

3. Steam Turbines

A steam turbine is characterised by the following features. (Refer fig. 4.19)

(a) There is no appreciable change of velocity of working fluid and consequently the change in kinetic energy may be neglected i.e.,

$$\frac{C_2^2 - C_1^2}{2g} = 0$$

(b) There is work output.

(c) Like nozzles, turbines are also thermally insulated and flow of working fluid is fast and thus there is no heat transfer, i.e. ${}_1Q_2 = 0$.

(d) The change in elevation between the entrance and exit points of the turbine is negligible i.e.,

$$Z_2 - Z_1 = 0.$$

Fig. 4.19 Turbine

Therefore the steady flow energy equation reduces to

$$O - {}_1W_2 = m_i\,[h_2 - h_1]$$

or

$${}_1W_2 = m_i\,[h_1 - h_2] \qquad ...(4.28)$$

4. Steam Condensers

Refer Fig. 4.20. A steam condenser is characterised by the following features:

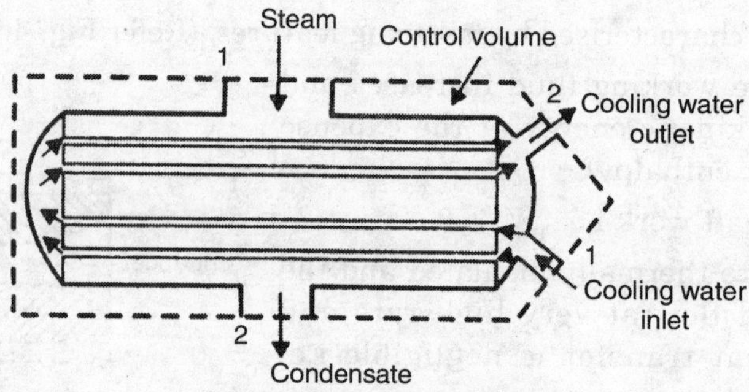

Fig. 4.20 Condenser

(a) There is heat transfer from steam to cooling water.

(b) There are two steady flow mass steams (steam and cooling water).

(c) There is no shaft work i.e. ${}_1W_2 = 0$.

(d) No change of velocity of either steam and therefore

$$\frac{C_2^2 - C_1^2}{2g} = 0.$$

(e) No appreciable change in elevation between the entrance and exit points of the condenser and change in potential energy may be neglected i.e.

$$Z_2 - Z_1 = 0.$$

Let m_c = mass of cooling water flowing per hour

m_v = mass of steam water flowing per hour

h_1c = enthalpy of cooling water at entrance in kJ/kg

h_2c = enthalpy of cooling water at exit in kJ/kg

h_1v = enthalpy of steam at entrance in kJ/kg

h_2v = enthalpy of steam at exit in kJ/kg

Then $${}_1Q_2 = (h_2c - h_1c) \times m \quad ...(4.29)$$

And $${}_1Q_2 = (h_2v - h_1c) \times m \quad ...4.29(a)$$

Heat exchange (${}_1Q_2$) in Eq. 4.29 is positive whereas in Eq. 4.29(*a*) it is negative.

5. Air Compressors

(a) Reciprocating Air Compressor. Refer Fig. 4.21. Following points should be borne in mind while dealing with reciprocating air compressors;

(*a*) The receiver is of sufficiently large capacity so that fluctuation of flow is reduced to a minimum and thus reciprocating compressor may be regarded as a steady flow machine.

(*b*) Due to low flow rate and large surface areas (fins) in contact with the surroundings, besides a deliberate heat removal to reduce work of compression, there is appreciable heat transfer.

(*c*) Work is done on the working fluid.

(*d*) There is no appreciable change in kinetic energy and potential energy of the working fluid and therefore may be neglected ; i.e.,

$$\frac{C_2^2 - C_1^2}{2g} = 0 \text{ and } Z_2 - Z_1 = 0$$

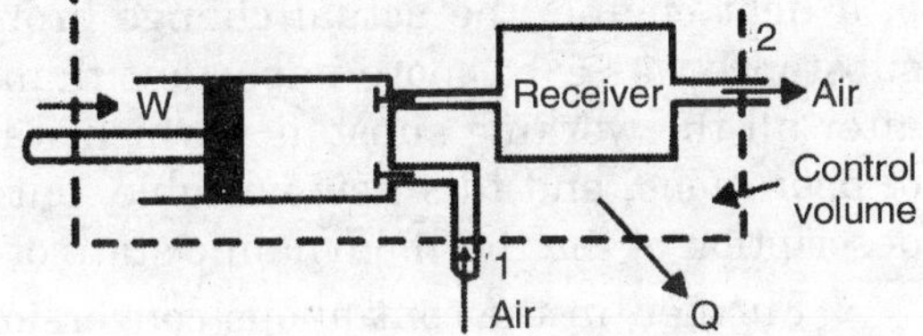

Fig. 4.21 Reciprocating Compressor

Thus S.F.E.E. for unit mass may be written as

$$({}_1Q_2) - ({}_1W_2) = m(h_2 - h_1) \quad ...(4.30)$$

(b) Rotary Compressor.

Refer Fig. 4.22.

These compressors are characterised by the following features.

(*a*) Work is done on the working fluid.

(*b*) Heat is rejected by the working fluid intercoolers to reduce work of compression.

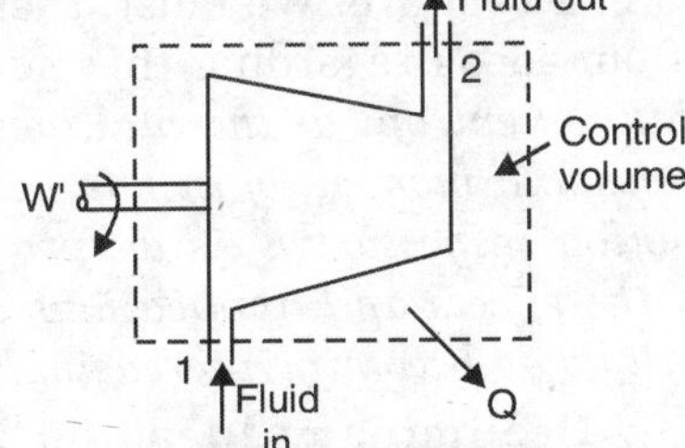

Fig. 4.22 Rotary Compressor

(*c*) Changes in kinetic energy and potential energy are negligible.

(*d*) It may be regarded as a steady flow. Thus S.F.E.E. reduces to

$${}_1Q_{2} - {}_1W_2 = m(h_2 - h_1) \quad ...(4.31)$$

6. Throttle Valve

Refer Fig. 4.23.

In throttling process, following points should be kept in mind.

(a) No work is done, *i.e.*, ${}_1W_2 = 0$.

(b) No changes in kinetic energy and potential energy

i.e., $\frac{C_2^2 - C_1^2}{2g} = 0$

Fig. 4.23 Throttle Valve

(c) No exchange of heat *i.e.*, ${}_1Q_2 = 0$.

(d) Flow may be regarded as steady flow.

Thus S.F.E.E. may be written as

$$h_1 - h_1 = 0 \quad \text{...(4.32)}$$

4.5 ENTROPY

During the study of thermodynamic changes, Clausius felt need of another variable of state to describe the condition of a substance, along with the usual ones, like temperature, pressure, volume and internal energy. In 1854, he introduced the term entropy.

The necessity for a new variable arose for many reasons. Thus, for example, in the case of a heat engine, the actual change brought about in the thermal condition of the working substance was even more important than the convertibility of heat into work, because it was after all the working substance which was subjected to a thermodynamic change in the cycle of operations, and this new variable, entropy, was found to be eminently suitable for a true description of the thermodynamic state of working substance.

Further, in the continuous conversion of heat into work, dealt with by the second law of thermodynamics, the direction of change is of major importance than the mere amount of energy change of the system. This too is easily handled by entropy.

Then, again we identify an isothermal by the magnitude of its temperature, which remains constant all along it. So, also, we must have a physical quantity to identify an adiabatic, and, entropy proved to be just that quantity, remaining constant all along the adiabatic.

Entropy is just as real a physical quantity as temperature or pressure, but there is nothing physical to represent it, and therefore, it becomes difficult to actually visualise it and understand its exact nature. We must, therefore, look for such analogy which may help us in the forming of our ideas regarding this new variable. *We may liken entropy to mass or inertia and may think of entropy as thermal inertia which bears to heat motion a relation similar to that which mass and moment of inertia bear to linear and rotational motion respectively. As we proceed along with our study of the effects and applications of entropy, its significance will go on becoming increasingly clearer to us.*

Preliminary Ideas on Entropy and Adiabatic Curves

Let us draw a number of isothermals, I_1, I_2, I_3, and I_4 on a *P-V* diagram, at temperatures T_1, T_2, and T_3 etc., as shown in Fig. 4.24. If we also draw a series of adiabatics A_1, A_2, A_3, and A_4, etc., on the same *P–V* diagram, so as to intersect these isothermals in points like *a, b, c, d...h*, then clearly, each four sided figure like abcd, dcef, etc. constitutes a Carnot cycle. Thus, considering the cycle

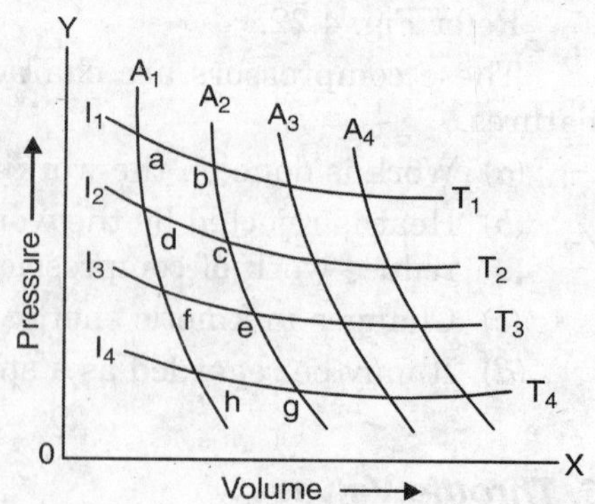

Fig. 4.24 Entropy and Adiabatics.

abcd, if be the heat absorbed at constant higher temperature T_1 along the upper isothermal and Q_2, the heat rejected at constant lower temperature T_2 along the succeeding lower isothermal, then from the theory of a Carnot engine, we have

efficiency $$\eta = \frac{Q_1 - Q_2}{Q_1} = \frac{T_1 - T_2}{T_1}$$

or $$1 - \frac{Q_2}{Q_1} = 1 - \frac{T_2}{T_1}$$

$$\therefore \quad \frac{Q_2}{T_2} = \frac{Q_1}{T_1}$$

Similarly, considering the cycle, 'dcef', heat Q_2 is absorbed along isothermal 'dc' at T_2 and heat Q_3 rejected along the isothermal 'ef' at the T_3. So that,

$$\frac{Q_2}{T_2} = \frac{Q_3}{T_3}$$

and, for the cycle 'fegh', we have $\dfrac{Q_3}{T_3} = \dfrac{Q_4}{T_4}$.

Exactly similar relations are also true for the cycles between adiabatics A_1, A_2, A_3, and A_4, except that the values of Q_1, Q_2, Q_3, etc. may now be different. We thus find that

$\dfrac{Q_1}{T_1} = \dfrac{Q_2}{T_2} = \dfrac{Q_3}{T_3} = \dfrac{Q_4}{T_4} =$ constant. Or, in general, $\dfrac{Q}{T}$ remains constant during transition; between two given adiabatics, say A_1 and A_2, whatever the isothermal chosen, where Q is the heat absorbed or rejected at temperature (absolute) $T°$ of the isothermal. Since this ratio is determined solely by the two adiabatics in question, it may be taken as a measure of the difference between them. This constant ratio $\dfrac{Q}{T}$ is spoken of as the change of entropy between a given pair of adiabatics considered.

In the case considered, the change from the state represented by A_1, to that represented by A_2 is an isothermal one but, it is by no means necessary. All that is required is that the change must be reversible.

If the adiabatics in question be closed together, so that the heat absorbed or rejected at an absolute temperature $T°$, is a small quantity dQ, the small change in entropy ds is given by

$$ds = \frac{dQ}{T}.$$

It follows, therefore, that the change of entropy $(s_2 - s_1)$ in passing from adiabatic A_1 to A_2 is given by

$$s_2 - s_1 = \int_{A_2}^{A_1} \frac{dQ}{T} \qquad \text{...(4.33)}$$

Further, it can be easily seen that since during an adiabatic change, no heat is allowed to enter or leave the system, $dQ = 0$ and, therefore, the change in entropy, $\dfrac{dQ}{T}$, is also zero. In other words, the entropy of a system remains constant during an adiabatic change. That is why adiabatics are referred to as isentropics, or curves of constant entropy.

The change of entropy in passing from the adiabatic A_1 to A_2 is constant, irrespective of the path taken between them. In other words, the change of entropy of a system, when it passes from one state to another, depends solely upon its initial and final states, and is quite independent of the manner in which the change is brought about.

This is a point of considerable importance. Because, it at once distinguishes entropy from heat, the quantity of which absorbed or rejected as it is known, does depend upon the path along which, or the manner in which, the change in the system, takes place.

It is thus clear that the entropy of a system is a function of its actual state, like pressure and volume, etc., and does not depend upon its previous history.

4.5.1 Change of Entropy in a Carnot Cycle (Reversible Cycle)

A Carnot cycle consists, as we know, of two isothermals and two adiabatics, all of which are reversible. During the two adiabatics, there is no change in entropy (since heat is neither taken in nor given out), so that the only changes in entropy that occur do so during the two isothermals. Of these, one is an isothermal expansion, during which heat Q_1 is absorbed at constant absolute temperature T_1 and the gain of entropy is, therefore, $\frac{Q_1}{T_1}$. The other is an isothermal compression, during which heat Q_2 is given out at a constant absolute temperature T_2 and so the loss in entropy is $\frac{Q_2}{T_2}$. Now, since in a Carnot cycle. $\frac{Q_1}{T_1} = \frac{Q_2}{T_2}$, it follows that the gain in entropy is equal to the loss in entropy, or the change in entropy through the cycle is zero. Mathematically, given the loss of entropy a negative sign, we have

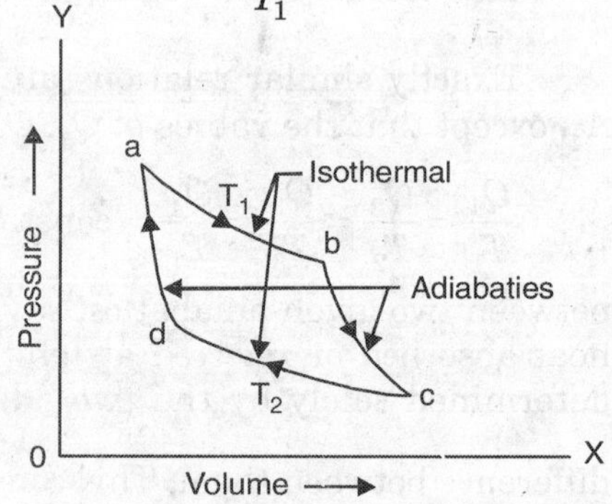

Fig. 4.25 Carnot cycle

$$\frac{Q_1}{T_1} = \frac{Q_2}{T_2} \text{ Or } \frac{Q_1}{T_1} + \frac{Q_2}{T_2} = 0 \text{ for the whole cycle}$$

or
$$\oint \frac{dQ}{T} = 0 \qquad \text{...(4.34)}$$

the integral sign with the circle, referring to the complete cycle.

Thus, during a reversible cycle, the entropy of a system remains unchanged.

4.5.2 Concept of Entropy from Availability of Heat

Considering a Carnot cycle (most efficient cycle), it is known that, of the quantity Q_1 of heat, taken in from the source at temperature T_1, and a quantity Q_2 is rejected to the sink at temperature T_2, so that the quantity of heat availabile for useful work is only $(Q_1 - Q_2)$. Now the efficiency of the engine working on Carnot cycle is given by

$$\eta = \frac{Q_1 - Q_2}{Q_1} = \frac{W}{Q_1} = \frac{T_1 - T_2}{T_1}$$

hence
$$W = \left(\frac{T_1 - T_2}{T_1}\right) Q_1$$

or quantity of heat available for useful work = $\left(\frac{T_1 - T_2}{T_1}\right)$ × heat absorbed by the system.

If the lowest available temperature of the sink be T_0, we shall have,

Quantity of heat available for useful work $W = \left(\frac{T_1 - T_2}{T_1}\right). Q_1$ which is certainly greater than before, since $T_0 < T_2$.

This quantity Kelvin called "available energy"; because obviously, this is the maximum amount of heat energy which, with the coldest available sink, can be converted into useful work.

It thus follows at once that the amount of available energy depends upon the difference of temperature $(T_1 - T_0)$ between the source and sink. So that, with a given source at T_1^0 the lower the temperature of sink the greater the amount of available energy. And, if we could have the sink at the absolute zero, i.e., when $T_0 = 0$ the available energy will be the maximum, meaning thereby that the entire quantity of heat, taken in from the source, will be available for useful work.

Since there are many processes at work (*e.g.*, conduction, radiation, etc,) pending to equalise temperature large differences of temperature between the source and the sink are not easily available. Because, suppose, we have Q units of heat stored up in the steam in a boiler at temperature T_1 ; then if the lowest available temperature for sink (condenser) is T_0, the maximum work that can be obtained.

$$W = \left(\frac{T_1 - T_0}{T_1}\right) Q = \left(1 - \frac{T_0}{T_1}\right) Q.$$

But, if this heat is first to be transferred into cylinder before it can be converted into work, and its temperature in the cylinder becomes T_2, (due to conduction, radiation, etc) then obviously, the energy now available is only.

$$W_1 = \left(\frac{T_2 - T_0}{T_2}\right) Q = \left(1 - \frac{T_0}{T_2}\right) Q.$$

So that, loss of available energy due to conduction, radiation, etc., is

$$W - W_1 = Q \,.\, T_0 \left(\frac{1}{T_2} - \frac{1}{T_1}\right).$$

Which is clearly, a positive quantity, since $T_1 > T_2$.

Since, as mentioned above temperature differences, in nature, always tend to be minimised, the available energy from a system tends to decrease. Kelvin, therefore, enunciated the principle that the available energy of the universe tends to decrease.

Thus this is, in effect, the same result as the one enunciated by Clausius may be seen from the fact that $Q\left(\frac{1}{T_2} - \frac{1}{T_1}\right)$, in the expression for $(W - W_1)$ above, is clearly the increase in the entropy of the system of the two bodies during conduction or radiation.

So that, loss of available energy from a system for a given maximum temperature T_0 of the sink = T_0 (gain in entropy). Thus as the entropy of a system increases, the available energy decreases. This is equivalent to saying that as the entropy of a system increases, the unavailability of energy from it also increases. We may, therefore, think of entropy of a system as a measure of the unavailability of entropy from it.

4.5.3 Technical Importance of Entropy

There are at least two good reasons which make entropy as important for technical purposes: (*i*) a knowledge of entropy of a substance is essential to determine its "work value" for thermal processes, and (*ii*) temperature - entropy diagrams of the cycle of actual heat engines corresponding to their $P - V$ diagrams can be drawn, which enable not only their efficiencies to be calculated much more easily but also any deviations from their proper normal working to be detected.

Calculation of Entropy

Let dQ be the quantity of heat absorbed during a small reversible process, working between two slightly differing temperatures T and $(T + dT)$. Then

$$dS = \frac{dQ}{T}$$

Considering 1 kg, mass of the substance, if c be its specific heat, we have $dQ = C.\ dT$, hence

$$dS = c\ .\ \frac{dT}{T}$$

Therefore, change of entropy $(S_2 - S_1)$ for the process between temperatures T_1 and T_2 is given by the expression

$$S_2 - S_1 = c\int_{T_1}^{T_2} \frac{dT}{T} = c\ (\log_e T_2 - \log_e T_1)$$

or

$$S_2 - S_1 = 2.3026.\ c.\ (\log_{10} T_2 - \log_{10} T_1) \qquad ...(4.35)$$

Let us apply Eq. (4.35) to calculate the entropies of 1 kg. of ice, water and steam at various temperatures.

(1) Entropy of Ice at 0°C and 1 kg of ice at 0°C is obtained by removing L_i kcal of heat of 1 kg. of water at 0°C. So that, here, heat rejected by water = L and the temperature at which this occurs = 0°C = 273°K. And, therefore, if S_i be the entropy of ice at 0°C assuming entropy of water at 0°C to be zero, we have

$$S_i = \frac{L_i}{273}$$

where L_i = Latent heat of ice

(2) Entropy of Water at Saturation Temperature t_s°C. As in the case of ice above, so also here, entropy of water at t_s°C or $(t_s + 273)$ or T_s°K

= entropy of water at 0°C + 2.3026°C [$\log_{10} T_s - \log_{10} 273$]

where C is the specific heat of water (equal to 1). Denoting this entropy water at t_s°C by S_t, we have

$$S_t = 2.3026\ [\log_{10} T_s - \log_{10} 273] \qquad ...(4.36)$$

(3) Entropy of Steam at Saturation Temperature t_s°C. Since 1 kg of water at its saturation temperature (or boiling point) absorbs L_s kcal of heat (*i.e.,* latent heat of steam) to be converted into steam at same temperature, the entropy of steam S_g is given by the equation

$$S_g = S_f + \frac{L_s}{T_s} \qquad \text{(For dry saturated steam)}$$

or

$$S_g = S_f + (S_{fg}) \qquad ...(4.37)$$

where S_{fg} = entropy of evaporation

However, in case of wet steam with dryness fraction x, the evaporation is partial (*i.e.*, only x kg of water being converted into steam out of 1 kg of water) and the heat absorbed will be $x\ L_s$ per kg of wet steam. Therefore,

$$\text{entropy of wet steam} = S_f + \frac{x\,.\,L_s}{T_s}$$

$$S_{wet} = S_f + xS_{fg} \quad \text{...(4.37(a))}$$

or

$$S_{wet} = S_f + x(S_g - S_f) \quad \text{...(4.37(b))}$$

(4) Entropy of Superheated Steam at Temperature T_{sup} °K.

Using the Eq. (4.35), entropy of superheated steam is given by

S_{sup} = entropy of dry-saturated steam + change of entropy during super heating

or

$$S_{sup} = S_g + 2.3026\, c_p\, (\log_{10} T_{sup} - \log_{10} T_s) \quad \text{...(4.38)}$$

where c_p = specific heat of superheated steam.

4.5.4 Thermodynamic Process on (*T* – *S*) Diagrams

(1) Constant Pressure Process. For this process,

$$(S_2 - S_1) = c_p \log_e \frac{T_2}{T_1} - \frac{R}{J} \log_e \frac{p_2}{p_1} \quad \text{...(4.39)}$$

Since $p_2 = p_1\ \because$ $\frac{p_2}{p_1} = 0$ $[\because \log_e 1 = 0]$

$\therefore$ Change of entropy of a constant pressure process is by the expression.

$$(S_2 - S_1) = c_p \log_e \frac{T_2}{T_1} \quad \text{...(4.40)}$$

Constant pressure processes on $T - S$ diagram are shown in Fig. (4.26) by logarithmic curves.

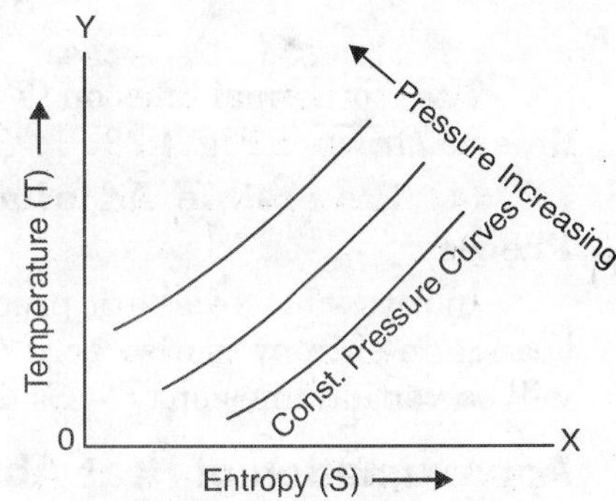

Fig. 4.26 Isobric Process on T-S-graph

(2) Constant Volume Process. For this process,

$$(S_2 - S_1) = R \log_e \frac{v_2}{v_1} + c_v \log_e \frac{T_2}{T_1} \quad \text{...(4.41)}$$

Since, $v_2 = v_1$ $\therefore \log_e \frac{v_2}{v_1} = 0$

$\therefore$ Change of entropy for a constant volume process is given by the expression:

$$S_2 - S_1 = c_v \log_e \frac{T_2}{T_1} \quad \text{...(4.42)}$$

It will be noticed that when Eq. (4.42) plotted with absolute temperature as ordinate and entropy as abscissa the curve is also a logarithmic curve (Fig. 4.27).

Since $c_p > c_v$, the constant volume curves will be steeper than the constant pressure curves. In other words, for the same range of temperature the change in entropy in constant volume process will be less than the change in entropy in constant pressure process.

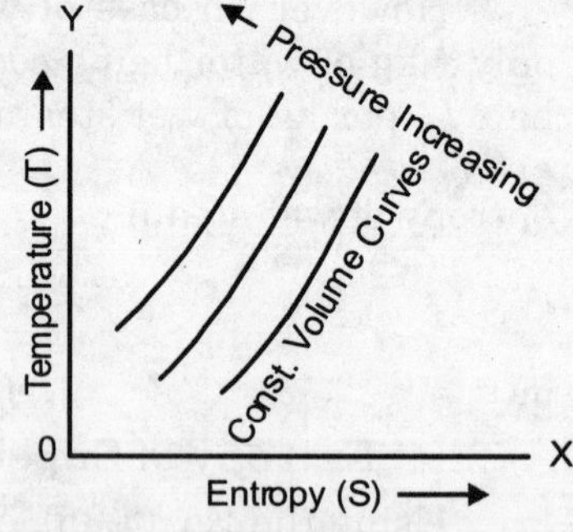

Fig. 4.27 Isochoric process on T-S diagram

(3) Isothermal Process. Since $T_1 = T_2$ and $\log_e \dfrac{T_2}{T_1} = 0$

Change of entropy in an isothermal process is given by the expression

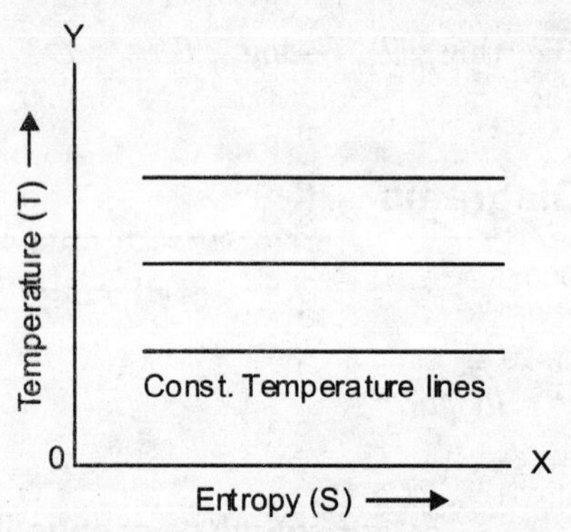

Fig. 4.28 Isothermal process on T-S-Graph

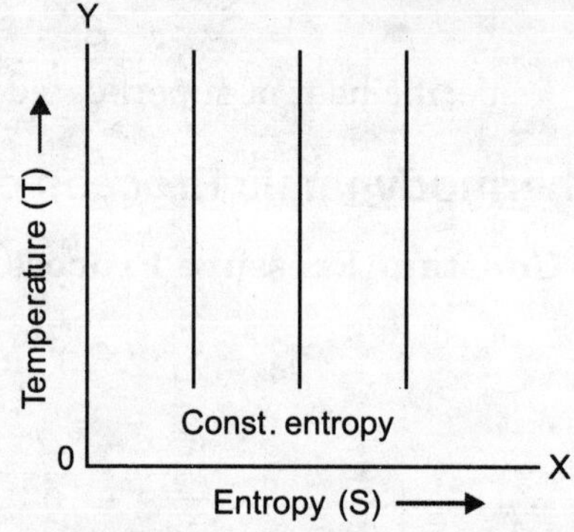

Fig. 4.29 Reversible Adiabatic process

$$S_2 - S_1 = R \log_e \frac{p_1}{p_2} = R \log_e \frac{v_2}{v_1} \qquad \text{...(4.43)}$$

The isothermal lines on $(T - S)$ diagram will be horizontal lines as shown in Fig. 4.28.

(4) Reversible Adiabatic Process or Isentropic Process.

In reversible adiabatic process, since $dQ = 0$, therefore, the change in entropy is also zero. Thus reversible adiabatic lines will be vertical lines on $(T - S)$ diagram as shown in Fig. 4.29.

Approximation of Heat Absorbed

Let the curve (1 – 2) represent the heating of a perfect gas on $(T - S)$ diagram from absolute temperature T_1 to T_2 as shown in Fig. 4.30. The entropy changes from S_1 to S_2. The curvature of curve (1 – 2) is usually small and the curve may, therefore, be assumed to be a straight line.

Fig. 4.30 Heat obsorption on T-S graph

Heat absorbed = area under the curve (1 – 2) shown shaded

$$\text{or Heat absorbed} = \left(\frac{T_2 + T_1}{2}\right) \times (S_2 - S_1) \qquad \text{...(4.44)}$$

Example 4.22 *An ideal gas of molecular weight 30 and specific heat ratio 1:38 is compressed according to the law $pv^{1.25}$ const, from a pressure of 0.1 MPa abs and 15°C to a presssure of*

1.6 MPa abs. Calculate the temperature at the end of compression, the heat received or rejected and work done by the gas during the process, and the change in entropy. Assume 1 kg mass of the gas. Use only calculated values of c_p and c_v.

Solution. $$\frac{T_2}{T_1} = \left(\frac{p_2}{p_1}\right)^{\frac{n-1}{n}}$$

or $$T_2 = (15 + 273)\left(\frac{1.6}{0.1}\right)^{\frac{1.25-1}{1.25}} = 1.741 \times 288$$

$$= \mathbf{501.48\ °K.}\quad \textbf{Ans.}$$

Molecular weight × characteristic gas constant = R_v

Or characteristic gas constant $R = \dfrac{8.314}{30} = 0.277$ kJ/kg°K.

Now $$c_p - c_v = R \text{ and } \frac{c_p}{c_v} = \gamma$$

or $$c_p - c_v = 0.277 \text{ and } \frac{c_p}{c_v} = 1.38$$

∴ $$c_v = \frac{0.277}{1.38 - 1} = 0.729 \text{ kJ/kg°K.}$$

Change in internal energy $dU = c_v\,(T_2 - T_1)$

$$= 0.729\,(501.48 - 288)$$

$$= \mathbf{155.6\ kJ/kg.}\quad \textbf{Ans.}$$

$$\text{Work done} = \frac{p_1v_1 - p_2v_2}{(n-1)} = \frac{R(T_1 - T_2)}{(n-1)}$$

$$= \frac{0.277(288 - 501.48)}{(1.25 - 1)}$$

$$= \mathbf{-236.5\ kJ/kg.}\quad \textbf{Ans.}$$

By law of conservation of energy

$$dQ = dW + dU$$

or heat transferred $= -236.5 + 155.6$

(Rejected) $= -80.9$ kJ/kg

Using Eq. 4.39 $$S_2 - S_1 = c_p \log_e \frac{T_2}{T_1} - \log_e \frac{p_2}{p_1}$$

$$= 1.38 \times 0.729 \log_e \frac{501.41}{288} - 0.277 \log_e \frac{16}{1}$$

$$= 0.558 - 0.768 = \mathbf{0.21\ kJ/kg\ °K.}\quad \textbf{Ans.}$$

Example 4.23 *A Vessel of capacity 3 m³ contains air at a pressure of 0.15 MPa abs. and a temperature of 25°C. Additional air is now pumped into the system until the pressure rises to 3 MPa abs. and the temperature rises to 60°C. Determine the mass of air pumped in and express* *the quantity as a volume at a pressure of 0.1 MPa abs. and a temperature of 20°C. If the vessel is allowed to cool until the temperature is again 25°C calculate the pressure in the vessel. Determine the quantity of heat transferred and change of entropy of the gas during the cooling process only. Neglect the effect of heat capacity of the vessel. Assume air as an ideal gas.*

Solution. Pumping process

Let m_1 be the initial mass of air and m_2 the final mass

Then
$$m_1 = \frac{p_1 v_1}{RT_1} = \frac{1.5 \times 10^5 \times 3}{287 \times (273 + 25)} = 5.26 \text{ kg}$$

And,
$$m_2 = \frac{p_2 v_2}{RT_2} = \frac{30 \times 10^5 \times 3}{287 \times (273 + 60)} = 94.17 \text{ kg}$$

Therefore, mass of air pumped in $= m_2 - m_1$

$$= 94.17 - 5.26$$

$$= \mathbf{88.9\ kg.}$$ **Ans.**

Volume of mass of air pumped in at a pressure of 1.02 N/cm² abs. and 20°C is given by

$$V = \frac{mRT}{p} = \frac{88.9 \times 287 \times (20 + 273)}{10^5}$$

or $V = \mathbf{74.76\ m^2}$ **Ans.**

Pressure at the end of cooling at constant volume

$$\frac{p_1 v_1}{T_1} = \frac{p_2 v_2}{T_2} \text{ or } p_2 = \frac{p_1}{T_1} \times T_2 \qquad \because v_1 = v_2$$

or
$$p_2 = \frac{0.15 \times 10^6}{(25 + 273)} \times (60 + 273) = 0.15 \times 10^6 \times \frac{333}{298}$$

$$= \mathbf{0.167\ MPa}$$ **Ans.**

Change of entropy **during** constant volume process is given by the Eq. 4.42

$$S_2 - S_1 = \left[C_v \log_e \frac{T_2}{T_1} \right] \times m_2$$

$$= 0.71 \log_e \frac{(60 + 273)}{(25 + 273)} = 0.71 \log_e \frac{333}{298}$$

$$= \mathbf{0.079\ kJ/°K}$$ **Ans.**

Example 4.24 *A vessel of 2.5 m³ capacity contains 1 kg mole of N_2 at 100°C:*

(i) Evaluate the specific volume and pressure ;

(ii) If the gas is cooled to 30°C, calculate final pressure, changes in specific internal energy, specific enthalpy and specific entropy.

Ratio of specific heats is 1.4. One kg mole nitrogen is 28 kg.

Solution. (*i*) Specific volume $= \dfrac{\text{Total volume}}{\text{Mass of the gas}}$

$$= \frac{2.5}{28} = \mathbf{0.08928\ m^3/kg}$$ **Ans.**

Pressure of the gas is obtained from the Eq.

$$p_1v_1 = mRT_1 \text{ or } p_1 = \frac{1 \times 8.314 \times 10^{-3} \times (100 + 273)}{2.5}$$

$$= \textbf{1.24 MPa.} \quad \textbf{Ans.}$$

(*ii*) Again $\quad \dfrac{p_1v_1}{T_1} = \dfrac{p_2v_2}{T_2}$ or $p_2 = \dfrac{p_1}{T_1} \times T_2 \qquad \because v_1 = v_2$

$$= \frac{1.24}{(100 + 273)} \times (30 + 273) = \frac{1.24 \times 303}{373}$$

$$= 1.007 \text{ MPa}$$

Change in specific internal energy $= c_v (T_2 - T_1)$

$$= \frac{R}{(\gamma - 1)}(T_2 - T_1) \quad \left[\because c_p - c_v = R \text{ or } \gamma - 1 = \frac{R}{c_v}\right]$$

$$= \frac{(30 - 100) \times 8.314}{28 \times (1.4 - 1)} = -\frac{8.314 \times 70}{0.4 \times 28}$$

$$= \textbf{– 51.9 kJ} \text{ (decrease).} \quad \textbf{Ans.}$$

Change in specific enthalpy $= c_p (T_2 - T_1)$

$$= \frac{R(T_2 - T_1)}{\left(1 - \dfrac{1}{\gamma}\right)} = \frac{8.314(30 - 100)}{28 \times \left(1 - \dfrac{1}{1.4}\right)} = \frac{-582}{28 \times 0.286}$$

$$= \textbf{– 72.75 kJ/kg} \text{ (decreases)} \quad \textbf{Ans.}$$

Change in specific entropy at constant volume cooling using Eq.

$$S_2 - S_1 = c_v \log_e \frac{T_2}{T_1} = \frac{8.314}{28 \times (1.4 - 1)} \log_e \frac{30 + 273}{(100 + 273)}$$

$$= 0.74 \log \frac{303}{373} = \textbf{– 0.154 kJ/kg°K} \text{ decrease.} \quad \textbf{Ans.}$$

Example 4.25 *2.5 kg of air 27°C and 1 MPa is expanded to 5 times its initial volume, the law of expansion being $Pv^{1.3} = c$. Find the change in entropy. Take R = 287 kJ/kg°K and c_p = 1.00 kJ/kg°K.*

Solution. As $\quad c_p - c_v = R$ or $c_v = 1.0 - 0.287$

$$= 0.713$$

From polytropic process $p_1v_1^n = p_2v_2^n$

And $\quad \dfrac{p_1v_1}{T_1} = \dfrac{p_2v_2}{T_2}$

We have $\quad \dfrac{T_1}{T_2} = \left(\dfrac{v_2}{v_1}\right)^{n-1} = (5)^{(1.3 - 1)} = (1.62)$

Change in entropy $\quad S_2 - S_1 = mc_v \left(\dfrac{\gamma - n}{n - 1}\right) \log_e \dfrac{T_1}{T_2}.$

$$= 2.5 \times 0.713 \left(\frac{1.4 - 1.3}{1.3 - 1}\right) \log_e 1.62 = 1.78 \times 0.333 \times 0.482$$

$$= \mathbf{0.286\ kg\ °K.}\quad \textbf{Ans.}$$

Example 4.26 *0.284 m^3 of air of initial condition of 0.15 MPa abs. and 20°C is compressed to a pressure of 0.45 MPa abs., and its volume reaches to 0.11 m^3. Calculate the change in entropy.*

$$R = 287\ J/kg°K,\ c_v = 0.71,\ c_p = 1.0\ kJ/kg°K$$

Solution. Using characteristic equation of gas

$$p_1 v_1 = mRT_1$$

or
$$m = \frac{1.5 \times 10^5 \times 0.284}{287 \times (20 + 273)} = 0.506 \text{ kg}$$

Using Eq. 4.39(*b*)

Change in entropy
$$S_2 - S_1 = \left[c_p \log_e \frac{v_2}{v_1} + c_v \log_e \frac{p_2}{p_1}\right]$$

or Total change in entropy
$$S_2 - S_1 = m\left[c_p \log_e \frac{v_2}{v_1} + c_v \log_e \frac{p_2}{p_1}\right]$$

$$= 0.506\left[\log \frac{0.11}{0.284} + 0.7 \log \frac{4.5}{1.5}\right]$$

$$= 0.506 \times 1.0 \times (-0.05) + 0.78$$

$$= \mathbf{-0.487 + 0.78 = 0.3}\quad \textbf{Ans.}$$

QUESTIONS

4.1 State the four laws of a perfect gas.

4.2 Prove that the heat supplied in constant volume heating is equal to $mc_v(T_2 - T_1)$.

4.3 What is adiabatic process? Derive an expression for the work done during the adiabatic expansion for an ideal gas.

4.4 Differentiate between adiabatic and isentropic processes.

4.5 Prove that for a polytropic process

$$\frac{T_1}{T_2} = \left(\frac{v_2}{v_1}\right)^{n-1} = \left(\frac{p_1}{p_2}\right)^{\frac{n-1}{n}}$$

where suffix 1 and 2 refer to initial and final states.

4.6 Derive the following expressions for a polytropic process;

(*a*) $W = \frac{p_1 v_1}{n-1}\left[1 - \frac{1}{r^{n-1} - 1}\right]$; (*b*) $\Delta Q = \left(\frac{\gamma - n}{\gamma - 1}\right) \times$ Work done;

(*c*) $\Delta H = \frac{\gamma}{\gamma - 1}(p_2 v_2 - p_1 v_1)$; (*d*) $\Delta U = \frac{1}{\gamma - 1}(p_2 v_2 - p_1 v_1)$;

4.7 What are the conditions of steady flow process?

4.8 What is a control volume?

4.9 Explain the term "work of introduction" in a flow process.

4.10 What is a throttling process? What is the difference between free expansion and throttling process?

4.11 Give two famous statements of second law of thermodynamics and prove their equivalence.

4.12 Give the Kelvin-Plank statement of second law and briefly explain its importance in energy technology.

4.13 Write short note on the second law of thermodynamics.

4.14 What are perpetual motion machine of the first kind and second kind?

4.15 Discuss reversibility. Name some reversible process.

4.16 Briefly explain a reversible process.

4.17 State the conditions of thermodynamic reversibility.

4.18 Explain the Carnot cycle and derive an expression for its efficiency as applied to a heat engine.

4.19 Why Carnot cycle cannot be realised in practice?

4.20 Write short note on reversible and irreversible process.

4.21 Distinguish between a refrigerator and a heat pump. If both are reversible, how do their C.O.P. values differ?

4.22 Prove that of all heat engines operating between a given constant temperature source and a given constant temperature sink, none has a higher efficiency than a reversible engine.

4.23 What is thermodynamic temperature scale?

4.24 Show that two reversible engines operating between two reservoirs at constant temperatures have the same efficiency.

4.25 Define the entropy and explain how it is useful in the study of thermodynamics.

4.26 Explain how change in entropy is a measure of irreversibility?

4.27 Show that entropy is a property of a system.

4.28 Explain the physical significance of entropy.

4.29 Explain the difference between isentropic process and adiabatic process.

4.30 What is Clausius inequality?

4.31 What is meant by "available energy"?

4.32 Sketch the following thermodynamic processes on a T–S diagram;

(*i*) constant volume; (*ii*) constant pressure;

(*iii*) isothermal; (*iv*) reversible adiabatic;

(*v*) polytropic.

4.33 Discuss briefly the concept of entropy.

4.34 During the working stroke of an engine, the internal energy of the working fluid is reduced from 2.824 kJ to 198.7 kJ and 104.6 kJ of heat is transferred to the cylinder walls. What is the work done by the working fluid in N-m. **(418500 Nm)**

4.35 1.5 kg of air is heated at constant volume from a pressure of 1 kgf/cm^2 abs and temperature 27°C to a pressure of 0.25 MPa abs. Calculate the change in entropy. Take R = 287 Nm/kg°K and c_p = 1.0 for air. **[0.985 kJ/kg-°K]**

4.36 0.35 m^3 of a perfect gas at constant pressure of 0.25 MPa abs. is heated from 77°C to 250°C and is then cooled at constant volume to its initial temperature. Calculate the overall change in entropy. Take c_p = 1.0 and c_v = 0.72.

[m = 0.837 kg, $S_2 - S_1$ = + 0.08065, $S_3 - S_2$ = – 0.05713, $S_2 - S_1$ = 0.02352 kJ/kg-°K]

4.37 0.04 m^3 of a perfect gas at a pressure of 2.8 MPa abs and temperature 2200°K expands in an engine cylinder to a volume of 0.17 m^3 and a pressure of 0.42 MPa abs. Taking R = 304 Nm/kg°K and c_v = 0.795 kJ/kg°c.

[T_2 = 1402.5°K, m = 0.1669 kg, $S_2 - S_1$ = 0.00301 kJ/kg-°K]

4.38 0.25 m^3 of a gas at a pressure of 14 kgf/cm^2 abs and temperature 60°C expands to a pressure of 0.35 MPa abs polytropically following the law $PV^{1.3}$ = constant. Find the change in entropy. Take c_p = 2.17 and c_v = 1.54 kJ/kg°K.

[R = 64, m = 1.6423 kg. T_2 = 241.6°K, $S_2 - S_1$ = 0.0673 kJ/kg-°K]

4.39 A petrol engine has a bore of 57 mm and a stroke of 60 mm compression ratio 7 : 1. The temperature and pressure at the beginning of compression stroke are 70°C and 1.033 N/cm^2 abs respectively. The index of compression is 1.3. Find the change in entropy. Take R = 287 N/kg°K and c_v = 0.72.

[CyL, Volume = 279.3 cm^3, m = 0.0002873 kg,
T_2 = 614.66°K, $S_2 - S_1$ = (–) 0.0000096 kJ/kg-°K]

4.40 State Boyle's law, Charles 'law and Avogadro's hypothesis. Prove that universal gas equation is given by $Pv = RT$, where R is a gas constant.

4.41 Explain why gases have two specific heats? Prove that;

$$c_p - c_v = R/J.$$

Chapter

PROPERTIES OF STEAM 5

Learning Objectives

After studying this chapter, the students will learn:

- Properties of steam.
- Formation of steam under constant pressure.
- Critical point of steam.
- How to use steam tables.
- Specific volume of wet steam, dry steam and superheated steam.
- External work during evaporation and internal energy of steam.
- Pressure-volume diagram and temperature-entropy diagram of steam.

5.1 INTRODUCTION

Water, which is very important from engineering point of view, exist in three phases *i.e.*, solid (ice), liquid (water) and gas (steam), the phase depending upon the prevailing temperature and pressure. It has been noticed that whenever a change of phase occurs, all the mass of the substances does not change its phase simultaneously. For example, during the process of melting of ice, the substance in the container remains in mixed phase, *i.e.*, ice and water at constant temperature at a given pressure. Similarly, when water is, heated to a boiling point (saturation temperature) it starts evaporating. During the process of evaporation of water, it remains in the mixed phase of liquid and gas (water and steam) in the container at constant saturation temperature at a given pressure. The temperature of the substance at a given pressure does not vary so long as the substance remains in mixed phase. A gaseous phase near the liquid phase is known as vapour.

Steam is used for heating and as a working fluid in the steam engine and steam turbine plants. The laws of perfect gases do not apply to vapours, hence variations between their properties are depicted on charts and tables. The zero of properties of steam is taken as 0°C; because then complication due to a change in phase from ice to water are avoided. For convenience, also, our calculations are based on 1 kg of water. The following subscripts are used in this book to specify the various phases of water.

f - refers to a property of the saturated liquid ;

g - refers to a property of the saturated vapour

f_g - refers to the change of phase from liquid to vapour.

5.2 FORMATION OF STEAM UNDER CONSTANT PRESSURE

In practice, steam is generated from water at certain constant pressure. This may be simulated by considering an upright cylinder fitted with a frictionless piston loaded to produce the desired pressure. Commencing therefore with 1 kg of water at 0°C, and having a specific volume v_{f_0} m^3, the first operation is to force water into the cylinder against a constant pressure p MPa abs. Treating water as incompressible, and assuming no heat flow from an external source, mechanical work, to the extent, $pv_{f_0} \times 10^6$ J is done by the feed pump in raising the piston. This is said to be external energy.

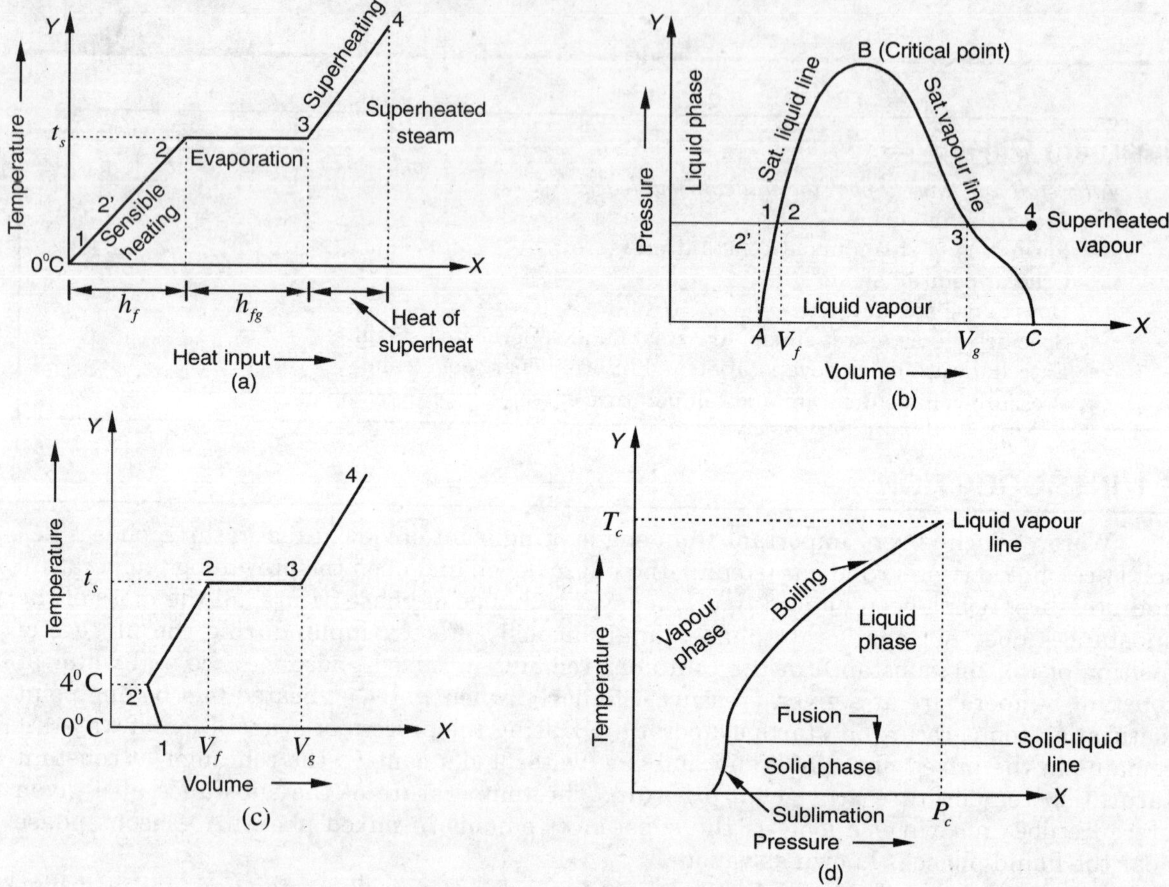

Fig. 5.1 Formation of steam

Next, the application of heat to water causes an increase in temperature, until, at a certain saturation temperature t_s, water becomes heat saturated and vapour appears. The saturation temperature is different for different pressures, it is 100°C at atmospheric pressure (0.101 MPa). Saturation temperature increases with increase of pressure. In this process there is a reduction in the specific volume until the temperature is about 4°C and subsequently there is an increase in the specific volume until the boiling point. This process of heating is called sensible heating and is shown by the line 1-2′-2 [Fig. 5.1.(*a*)]. The heat required to raise 1 kg of water from 0°C to saturation temperature is represented by h_f and termed as sensible heat.

The continued addition of heat now merely produces more steam without any increase in temperature. This constant temperature is called the saturation temperature or temperature of formation of steam or boiling point corresponding to the given pressure. The amount of steam formed is proportional to the amount of heat added. The process of change of phase from liquid to gas at constant temperature and pressure is called evaporation or vaporisation and is shown by the line 2-3 [Fig. 5.1(*a*)]. The amount of heat required for complete evaporation of 1 kg water at constant saturation temperature is represented by h_{f_g} and is called latent heat of vaporisation. During this process of evaporation, there is a considerable increase in the specific volume. The specific volume of dry-saturated steam is represented by v_g. The latent heat of vaporisation consists of;

(*i*) The heat expended in overcoming the external resistance to change in volume from v_f to v_g,

$$p(v_g - v_f) \times 10^6 \text{ J};$$

and (*ii*) The heat expended in overcoming the internal molecular resistance to change of phase from saturated water to dry-saturated vapour (steam), i.e., change in internal energy, $(u_g - u_f)$ or u_{fg}

or
$$h_{fg} = p(v_g - v_f) \times 10^6 + u_{fg} \quad ...(5.1)$$

The quantity $p(v_g - v_f) \times 10^6$ is often referred to as the external heat of vaporisation, while u_{fg} is known as the Internal latent heat and is stored as internal energy or molecular energy. The amount of this heat is independent of the way in which this state had been reached.

In practice the volume of water v_f is very small compared to the volume of dry-saturated vapour v_g; v_f is, therefore, neglected and Eq. 4.44 becomes

$$h_{fg} = p.v_g \times 10^6 + u_{fg} \quad ...[5.1(a)]$$

When the last drop of suspended water has been evaporated, the steam is known as dry-saturated steam—saturation here refers to the heat content of the steam, since any further addition of heat produces a superheated steam.

Commencing with water at 0°C, the total heat (h_g) which must be supplied to produce dry saturated steam is given by

$$h_g = h_f + h_{fg} \quad ...(5.2)$$

The Eq. (5.2) is of great importance in thermodynamics, and is built up from the terms

(*i*) Work done by the feed pump;

(*ii*) External work due to the expansion of water;

(*iii*) Internal energy acquired during sensible heating;

(*iv*) External work due to the formation of steam; and

(*v*) Internal energy due to a change of phase.

As long as there is still some water left unevaporated in the container, the steam formed will not be pure steam, but will have small particles of water in suspension ; such steam is known as wet steam and may be classified as a vapour. Wet steam, being vapour, has no specific heats.

The dryness fraction of wet steam is the ratio of the weight of actual dry-steam to the weight of the wet steam containing it,

Let, x = dryness fraction

W_g = weight of dry steam contained in wet steam.

W_f = weight of water in suspension in wet steam.

Then
$$x = \frac{W_g}{W_g + W_f} \quad ...(5.3)$$

If weight of wet steam be 1 kg then $x = W_g$, *i.e.,* dryness fraction of steam is equal to the weight of dry steam in 1 kg wet steam. The dryness fraction when expressed in percentage is called the quality of steam.

No steam can be perfectly dry whilst it is in contact with the water from it is being formed.

It will be obvious that the latent heat required to produce steam of dryness fraction x is $x.h_{fg}$, since only the fraction x kg is evaporated. Hence the total heat of a wet steam is

$$H_{wst} = h_f + xh_{fg}. \quad ...(5.4)$$

Wetness fraction is another term associated with wet steam. It is the ratio of the weight of water in suspension to the total weight of wet steam containing it. Thus, wetness fraction

$$y = \frac{W_f}{W_f + W_g} = 1 - \frac{W_g}{W_f + W_g} \quad ...(5.5)$$

or

$$y = 1 - x \quad ...[5.5(a)]$$

The wetness fraction when expressed in percentage is called priming.

Superheated steam is produced from a dry saturated steam by the continued application of heat, provided there is a moderate amount of thermal insulation between water and its steam. If this insulation is not provided, the heat supplied will go to evaporate more water. For this reason superheating is usually carried out in superheater separate from the boiler. The process of superheating is shown by the line 3-4 Fig. 5.1(*a*). Heat absorbed during superheating is termed as heat of superheat and is given by

$$c_p(t_{sup} - t_s)$$

Where 't_{sup}' is the temperature of superheated steam and c_p is the average specific heat of superheated steam of the given constant pressure. In general C_p is a function of pressure and temperature and may be obtained from steam tables by taking the ratio

$$c_p = \frac{\text{Difference in total heat of superheated and dry saturated steam}}{(t_{\text{sup}} - t_s)}$$

The difference $(t_{sup} - t_s)$ is known as the degree of superheat.

The total amount of heat for formation of superheated steam,

$$h_{sup} = h_g + c_p\,(t_{sup} - t_s) \quad ...(5.6)$$

5.3 TRIPLE POINT AND CRITICAL POINT

If the heating of deep frozen ice (ice at a temperature lower than freezing point) is conducted at pressures lower than atmospheric pressure, the following points will be observed:

(*i*) Melting point of ice rises;

(*ii*) Boiling point of water decreases; and

(*iii*) Change of volume during evaporation increases.

When pressure is reduced to 0.00061 MPa absolute, the melting point and boiling point coincide and sensible heating and melting of ice, and sensible heating and evaporation of water lie in a single horizontal line. The saturation temperature corresponding to the above pressure is termed as triple point temperature and the horizontal line through triple point temperature is called triple line. The value of triple point temperature is 0.01°C. The point is an internationally agreed point for the absolute temperature scale. Ice, water and steam consist in a thermodynamic equilibrium in a closed container only at this pressure. At pressure lower than triple point pressure, the ice directly becomes steam without melting.

Now, if the heating is conducted at pressure above atmospheric pressure, the following points will be observed:

(*i*) Melting point of ice decreases;

(*ii*) Boiling point of water increases;

(*iii*) Change of volume during evaporation decreases; and

(*iv*) Latent heat decreases.

When pressure is raised to 22.5 MPa absolute, the change of volume during evaporation and also latent heat falls to zero and at this point water directly gets converted into steam without going through the process of evaporation. This point is referred to as the critical point and is denoted by the subscript *C*. The corresponding temperature, pressure and specific volume are T_c = 647.3°K, p_c = 22.5 MPa abs. and v_c = 0.00317 m³/kg. Fig. 5.1 (*d*) shows the variation of saturation temperature with pressure.

5.4 STEAM TABLES

It is almost impossible to have a single equation like characteristic gas equation which can satisfy the various states of pure substance like water in all its phases. In fact it is difficult to have single equation for various states of same phase because it does not behave like a perfect gas except in superheated region.

The properties of steam which are likely to be used for engineering calculations are saturation temperature, specific volume, specific enthalpy, specific internal energy and specific entropy, etc., at any given pressure. These properties for saturated liquid and dry saturated vapour, obtained by experiments and calculations, are tabulated against pressure or corresponding saturation temperature. The sample of such tables is given below:

Table 5.1 Saturated Water and Steam (Pressure in MPa)

p	t_s	*Specific volume m³/kg*		*Specific internal energy kJ/kg*		*Specific enthalpy, kJ/kg*			*Specific entropy kJ/kg/k*		
MPa	°C	v_f	v_g	u_f	u_g	h_f	h_{fg}	h_g	s_f	s_{fg}	s_g
0.0035	26.7	0.001003	39.50	111.9	2412.1	111.9	2438.4	2550.3	0.301	8.130	8.523

Table 5.2 Saturated Water and Steam (Temperature °C)

t_s	*p*	*Specific volume m³/kg*		*Specific internal energy kJ/kg*		*Specific enthalpy*			*Specific entropy kJ/kg/k*		
°C	*Pascal*	v_f	v_g	u_f	u_g	h_f	h_{fg}	h_g	s_f	s_{fg}	s_g
5	872.1	0.0010001	147.1	21.0	2382.3	21.0	2489.6	2510.6	0.076	8.950	9.026

The properties of superheated steam are tabulated separately. These properties depend not only on pressure but also on temperature of superheated steam (t_{sup}) or degree of superheat ($t_{sup} - t_s$). These properties could have been easily calculated, had c_p remained constant for whole range of temperature rise. In fact c_p varies. The experiments have shown that:

(*i*) At first, c_p decreases rapidly with increase of temperature and after reaching a certain minimum value, it then begins to rise slowly.

(*ii*) At constant temperature, c_p is higher for higher pressures. However, its dependence on pressure decreases at higher temperatures.

Sometimes, another variable called Gibb's function (G) is also given in steam tables. It is defined as

$$G = T \,.\, S - H.$$

where H is the enthalpy of wet steam.

S is the entropy of wet steam.

T is the absolute saturation temperature of wet steam.

Important conversion factors for steam tables.

1 bar = 10^5 N/m^2 = 10^5 Pa = 0.1 MPa, 750.062 mm H_g = 1.01972 kgf/cm^2

1 KJ = 10^3 N/m = 101.792 kgf-m = 0.277778 *Wh* = 238.846 cal

1 KJ/kg = 0.239 kcal/kg.

5.5 SPECIFIC VOLUME OF STEAM

1. Wet Steam

The specific volume of wet steam is equal to the sum of the volumes of x kg of dry-saturated steam with a specific volume v_g m^3/kg and $(1 - x)$ kg of water with a specific volume V_f m^3/kg, *i.e.*

Specific volume of wet steam = $x \,.\, v_g + (1 - x)\, v_f$ ...(5.7)

The values of v_g and v_f can be directly read from steam tables at any specified pressure or saturation temperature.

The dryness fraction of wet steam (x) can vary between zero to one.

The volume of water in wet steam is very small in comparison to the volume of dry-saturated steam at lower pressure and hence the volume of water $(1 - x)\, v_f$ is neglected ; therefore,

Specific volume of wet steam = $x,\, v_g$ ...[5.7(*a*)]

However, at high pressures, the volume of water in wet steam is significant and cannot be neglected.

2. Dry-Saturated Steam

Specific volume of dry-saturated steam can either be directly read from steam tables or calculated with the help of Clapeyron's equation which is given below:

Clapeyron's equation, $(v_g - v_f) = \dfrac{J(h_g - h_f)}{T_s} \cdot \dfrac{dT_s}{dp}$ ...(5.8)

The value of $\dfrac{dT_s}{dp}$ is obtained from the slope of temperature—pressure curve of steam which is plotted from experimental data Clapeyron's equation is derived from the Carnot theorem.

3. Superheated Steam

The specific volume of superheated steam at any specified pressure (p) and temperature (t_{sup}) is directly read from steam tables of superheated steam.

The approximate specific volume of superheated steam may be calculated by assuming that steam behaves as a perfect gas from the commencement of the superheating stage. This assumption gives accurate results for a large superheated temperature range at a low pressure; but will give discrepancies for a small range of superheating at high pressure.

Let p = pressure under which steam is superheated;

v_g = specific volume of dry-saturated steam at pressure p (from steam tables);

T_s = temperature of formation of steam at pressure p in degrees Kelvin (from steam tables)

T_{sup} = final temperature to which steam is superheated in degrees Kelvin;

v_{sup} = specific volume of superheated steam at pressure p.

Then, applying Charles law,

$$\frac{v_g}{T_s} = \frac{v_{sup}}{T_{sup}}$$

From which
$$v_{sup} = v_g \cdot \frac{T_{sup}}{T_s} \qquad ...(5.9)$$

5.6 EXTERNAL WORK DURING EVAPORATION

The latent heat absorbed by the steam during evaporation includes an amount of external work due to the large increase in volume when changing from water to steam.

Consider the complete evaporation of 1 kg of water.

Let v_f = specific volume of water in cub. Metre/kg.

p = absolute pressure at which evaporation of water takes place in MPa.

v_g = specific volume of dry-saturated steam at pressure p, in cub metre/kg.

During evaporation the volume increases from v_f to v_g under a constant pressure p ; external work is thus done by the steam in increasing its volume, and the energy for performing this work is obtained during the absorption of the latent heat. This work is known as the external work of evaporation.

External work done per kg of steam = $(v_g - v_f) \cdot p \times 10^6$ J (dry-saturated)

$$= (v_g - v_f) \cdot p \times 10^6 \text{ J} \qquad ...(5.12)$$

At low pressures, v_f is very small as compared with the volume of the steam (v_g) it forms, hence, the term v_f in Eq. 5.12 may be neglected as small (The results are of sufficient accuracy for engineers). Then Eq. (5.12) may be written,

External work done per kg of steam (dry-saturated)

$$= v_g . p \times 10^6 \text{ J} \qquad ...[5.12(a)]$$

If the steam raised is wet, with a dryness fraction x, then,

External work done per kg of wet steam = $x \cdot v_g \cdot p \times 10^6$ J ...[5.12(b)]

If the steam is superheated the volume of 1 kg of superheated steam must be substituted for v_g in Eq. [5.12(a)].

As this amount of external work is included in the latent heat of the steam, which consists of the true latent heat plus the external work of evaporation, this true latent heat is called the internal latent heat, and may be found by:

Internal latent heat = $h_{fg} - v_g \cdot p \times 10^6$ ...(5.13)

5.7 INTERNAL ENERGY OF STEAM

Internal energy of steam is the actual heat energy above the freezing point of water stored in the steam; and it may be found by subtracting the external work of evaporation from the total heat (or enthalpy),

Let u_g = internal energy of 1 kg of dry-saturated steam at an absolute pressure p.

h_g = total heat of 1 kg of dry-saturated steam at the absolute pressure p.

$$\text{hence } u_g = h_g - v_g.p \times 10^6 \quad ...(5.14)$$

If the steam is wet, with a dryness fraction of x, internal energy of 1 kg of wet steam

$$u_{wet} = (h_f + xh_{fg}) - x \cdot v_g \cdot p \times 10^6 \text{ J} \quad ...[5.14(a)]$$

If the steam is superheated to a temperature of T_{sup} degrees Kelvin,

Internal energy of 1 kg. of superheated $u_{sup} = [h_g + c_p (T_{sup} - T_s)]$

$$- \frac{v_g \cdot T_{sup} \cdot p \times 10^6}{Ts} \quad ...[5.14(b)]$$

Let 1 kg. of steam *of pressure* p_1, dryness fraction x_1 and specific volume v_{g1}, be expanded in some manner to a pressure p_2, dryness fraction x_2, and specific volume v_{g2}.

Then, $$u_1 = (h_{f1} + x_1 h_{fg1}) - x_1 \cdot v_{g1} \cdot p \times 10^6$$

And $$u_2 = (h_{f2} + x_2 \alpha_2 h_{fg2}) - x_2 \cdot v_{g2} \cdot p \times 10^6$$

Then applying the law of conservation of energy to this expansion,

Heat absorbed during expansion (Q) = external work (W) done on piston + increase of $(U_2 - U_1)$ during internal energy

The equations for the work done on piston will be the same as these evolved for perfect gases.

The above results are for 1 kg of steam only. If the heat absorbed required for a given weight of steam, those results must be multiplied by that given weight.

5.8 PRESSURE—VOLUME DIAGRAM FOR STEAM

In the *p-v* diagram of water, Fig. 5.2, the curve *AHB* represents the saturated liquid line and gives the volume of 1 kg of water at saturation temperature. The curve *BC* represents the saturated vapour line. The part of the diagram to the left of the curve *AHB* is known as the liquid region and the part enclosed by the curve *AHBC* is known as the wet region (mixture of liquid and vapour). To the right of the curve *BC* and bounded by the critical isothermal T_4 is the superheated region.

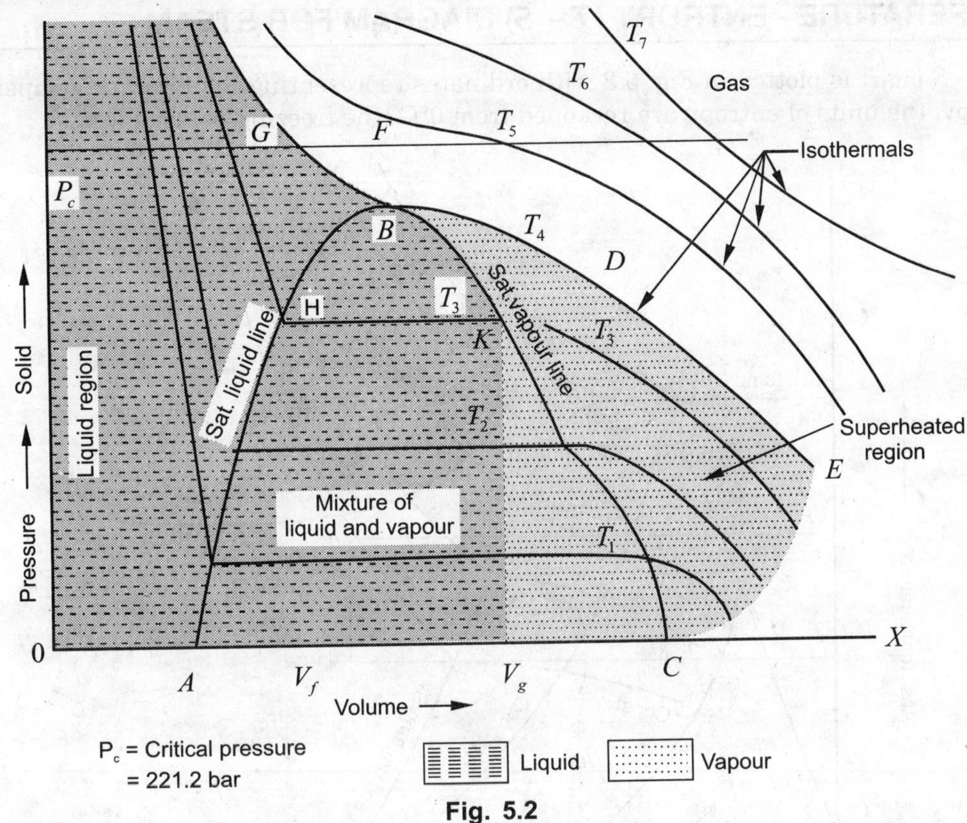

Fig. 5.2

Considering 1 kg. of water at any pressure, point H on saturated liquid line represents the volume of water v_f at saturation temperature T_3. The horizontal line HK represents evaporation at constant pressure and constant temperature. Hence, it follows, that for wet steam an isothermal or constant pressure (Isobaric) expansion actually represents the absorption of the latent heat. Point K on saturated vapour line represents the volume v_g of dry-saturated steam at saturation temperature T_3. Beyond the curve EC, *a* constant expansion is not an isothermal expansion.

It is seen from the diagram that the specific volume of steam decreases as the pressure increases and at critical pressure, the volume of saturated liquid is equal to the volume of dry-saturated vapour.

An adiabatic expansion of wet steam follows approximately the law $p.v^{1.33}$ = constant.

If the steam is superheated the expansion lines on the p–v diagram will be the same as for a gas.

The curve in Fig. 5.2 is plotted from experimental data on the particular vapour, properties beyond the experimental range being predicted from experiments on a vapour of similar atomic structure, but which is more amenable to test.

A typical system of isothermals is shown, and it will be seen that a jog occurs as each isothermal passes through a change of state in all cases except at and above B. Any increase in pressure above B removes the phenomena of evaporation and condensation, the change from the liquid to the gaseous field takes place continuously and homogeneously across the critical isothermal. Paradoxical as it may seem, no separation of liquid and vapour occurs during a constant pressure compression FG; yet subsequent isothermal expansion GH shows that a liquid has existed.

5.9 TEMPERATURE - ENTROPY ($T-S$) DIAGRAM FOR STEAM

A $T-S$ chart is plotted in Fig. 5.3 with ordinates representing the absolute temperature and entropy; the units of entropy are reckoned from 0°C (the freezing point of water).

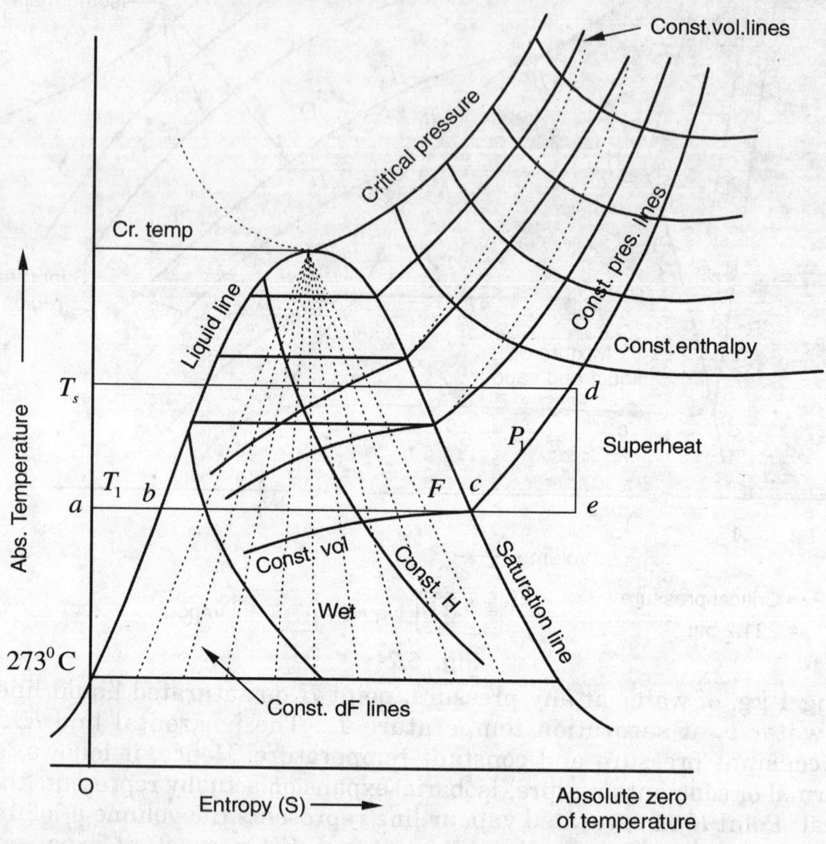

Fig. 5.3

Consider 1 kg of steam at absolute pressure p_1; the absolute temperature of formation at this pressure will be T_1, the value of T_1 can be obtained from the steam tables.

Then, $$S_{f1} = C_{pw} \,.\, log_e \frac{T_1}{273} = ab$$

Where, S_{f1} = entropy of water at saturation temperature T_1 and pressure p_1

C_{pw} = heat capacity of water at constant pressure p_1.

Let this value of S_{f1} be represented by ab to a convenient scale, and plot this distance on the $T-S$ diagram of Fig. 5.3. Repeat this for other pressures of steam and draw a curve through the points so obtained. This curve is known as the saturated liquid line, and will give the value S_f at all pressures and temperatures. Below 200°C, the saturated liquid line is concave on the left because of the reduction of specific heat of water with an increase in temperature. Above 200°C the rapid increase in C_{pw} reverses the concavity of the curve and ultimately turns it over at the critical temperature.

Again, considered the steam at absolute pressure p_1, during evaporation the change of entropy of this steam will be

$$S_{fg_1} = \frac{h_{fg1}}{T_1}$$

Let this be represented by the distance b_c. The point C will represent dry-saturated steam at pressure p_1. Repeat this for other pressure, and draw a curve through the points so obtained. This curve is known as dry-saturated steam line (or saturation line), and its horizontal distance from the vertical ordinate represents the total entropy of dry-saturated steam at any pressure.

If the steam at pressure p_1, had been wet, with a dryness fraction x, the entropy of evaporation would be

$$\frac{x.h_{fg1}}{T_1}$$

Let this be represented by bF and be plotted to the same scale. Then, aF is the total entropy of wet steam at pressure $p_{1,}$ and dryness fraction is the ratio of bF to bc

or $$x = \frac{bF}{bc}$$

Let this steam at pressure p_1 be now superheated by increasing its absolute temperature T_1 to T_s. Then increase in entropy due to superheating

$$C_{ps} \,.\, \log_e \frac{T_3}{T_1}$$

Let this increase of entropy be represented by the horizontal distance. Draw a horizontal line through T_s and vertical line through point *e;* let these lines intersect at *d.* Then point *d* represents the condition of the superheated steam at pressure p_1. Repeat this for other values of T_s the temperature T_1 remaining the same as p_1 is constant. Draw the curve *cd* through the points so obtained; then this curve represents the entropy of superheated steam at pressure p_1, for all temperatures of superheat. Actually the curve *cd* is a constant pressure line; similar constant pressure lines can be drawn for superheated steam at other pressures.

It should be noticed that the area between the liquid line and the saturation line represents wet steam. The area to the right of the saturation line represents superheated steam, and is known as the region of superheat. Finally, the area to the left of saturated liquid line is known as liquid region.

In Fig. 5.3 the evaporation lines at all pressures have been divided into ten equal parts. Curves have been drawn through each corresponding tenth part of these lines. These curves represent the dryness fraction of the steam at any temperature and pressure; they are known as dryness fraction lines or constant quality lines. It should be observed that when $x = 1/2$ the constant quality line is almost vertical, and therefore a change in temperature at constant S does not affect the dryness. With $x < 1/2$ the steam becomes drier for a corresponding expansion, whilst for $x > 1/2$ the steam becomes wetter, for a fall in temperature.

T – S charts for steam available for use have all the lines plotted on them taking into account the variation in specific heat. Thus the equation for change in entropy are only explanatory here.

1. Isothermal Lines on T – S Diagram

It is seen from the *T-S* diagram of Fig. 5.3 that an isothermal expansion line for steam is a horizontal line. During the wet steam period bc, the horizontal represents the evaporation of the water to steam. The horizontal line in the superheat region (*ce*) represent the isothermal expansion of super heated steam which behaves in a similar way to a gas.

2. Adiabatic Lines on T – S Diagram

A reversible adiabatic expansion is an expansion at constant entropy because there is no heat transfer; it is, therefore, isentropic. An adiabatic expansion will be a vertical line on a *T–S* chart.

3. Constant Total Heat (or Enthalpy) Lines on T–S Diagram

Constant total heat *H* lines are of value in solving problems where a change of pressure occurs without any external work being done, i.e. throttling process. They may be plotted by selecting a certain total heat, and by reference to steam tables to obtain the corresponding pressures and degree of superheat that will give this total heat. Knowing these pressures and superheats, the appropriate steam table will give the corresponding value of *S*.

In the wet region the dryness fraction, for a given total heat *H*, must first be computed from the equation $H = Hf + x.\ H_{fg}$, hence

$$S = C_{pw} \log_e \frac{T_1}{273} + \frac{x\ .\ Hfg}{T_1}$$

4. Lines of Constant Volume on T–S Diagram

Constant volume lines on the *T–S* diagram may be found by plotting the dryness fraction of the wet steam as the pressure falls. Consider 1 kg. of dry-saturated, steam of absolute pressure p_1 and absolute temperature T_1 being represented by the point *A* on the *T – S* diagram of Fig. 5.4.

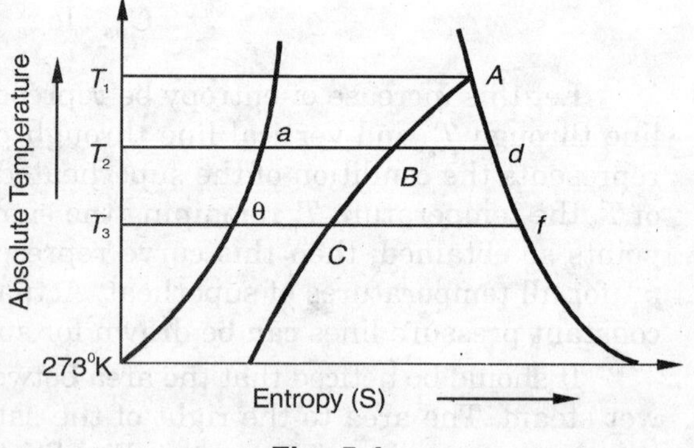

Fig. 5.4

From steam tables, volume of 1 kg of steam at abs. pressure $p_1 = v_{g1}$.

Let this steam cool at constant volume until its abs, pressure is p_2, corresponding saturation temperature T_2, and specific volume v_{g2}; the last two values are obtained from steam tables for pressure p_2. This steam after it has cooled will be wet with an unknown dryness fraction x_2. As the cooling is at constant volume,

i.e. $$v_{g1} = x_2 v_{g2} \quad \text{or} \quad x_2 = \frac{v_{g1}}{v_{g2}}$$

Then, on the horizontal line through T_2 mark off the point *B* such that

$$\frac{aB}{ad} = x_2$$

Then volume at *A* = volume at *B*.

Next consider a pressure p_2 of this constant volume cooling, let the corresponding saturation temperature be T_3, the specific volume v_{g3}, and the unknown dryness fraction x_3,

Then volume $$= x_2 v_{g2} = x_3 v_{g3}$$

From which $$x_3 = \frac{x_2\ v_{g2}}{v_{g3}}.$$

On the horizontal line through T_3 mark off the point C such that

$$\frac{ec}{e_f} = x_3.$$

Then the point C will represent the condition of the steam at pressure p_3 for this constant volume cooling. Other similar points can be obtained in the same way for other pressure. The curve drawn through the points A, B, C, etc., will be a constant volume line (volume being v_{g1}) for this steam. In the same way, constant volume lines for volumes v_{g2}, v_{g3}, etc. can be drawn and should be observed that they converge towards the origin but never actually reach the liquid line since steam is always present even at extremely low temperature, whilst the dryness lines converge to the critical temperature.

5. Total Heat on T – S Diagram

Let the point A (Fig. 5.5) represents the condition of one kg of wet steam on a $T - S$ diagram. From A draw a horizontal to the liquid line (AB), and a vertical to the absolute zero of temperature line (AG). Then total heat at A above absolute zero of temperature

$$H = \int T.ds = \text{area of T.S diagram up to point } A = \text{area } ABFG$$

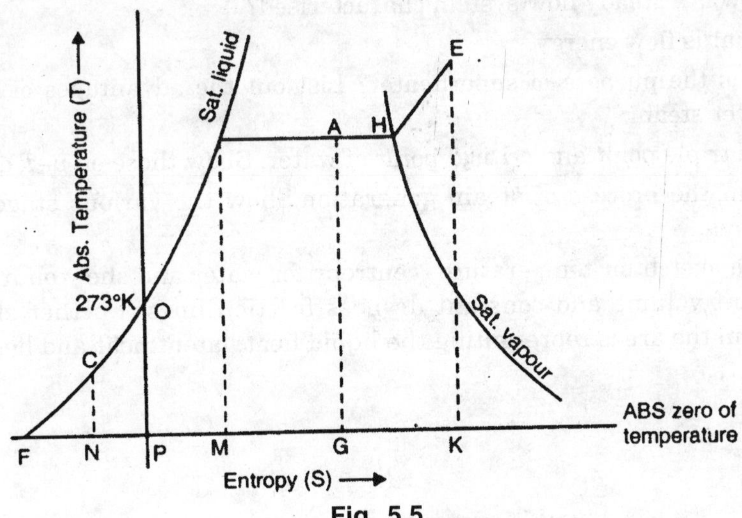

Fig. 5.5

The point O on the saturated liquid line at a temperature of 273°K represents the freezing point of water.

Then, total heat at point O, above abs. zero of temperature = area OFP.

Hence, total heat at point-A, above freezing point of water = area $ABFG$ – area OFP = area $ABOPG$.

Hence it will be seen that the total heat at any point above abs. zero of temperature is the enclosed area up to F, but the total heat above the freezing point of water is the area up to the line OP.

Consider a point C on the liquid line below 273°K.

Total heat at C above abs. zero of temperature = area CFN and total heat at C above 273°K = area CFN – area OFP = – area $OCNP$.

In this case the liquid heat is negative.

Next consider a point *E* in the superheat area.

Total heat at *E* above abs. zero of temperature. = area *EHBFK* and total heat at *E* above 273°K = area *EHBOPK*.

QUESTIONS

5.1 Explain the following terms:
(*a*) Dry saturated steam, (*b*) Superheated steam,
(*c*) Wet steam, (*d*) Sensible heat,
(*e*) Latent heat, and (*f*) Superheat of steam.

5.2 Explain the terms: (*a*) specific volume, (*b*) density and internal energy in respect of steam.

5.3 What is adiabatic process? Explain with examples.

5.4 Define state, process and a cycle.

5.5 (*a*) How is a steady flow system characterised?
(*b*) What is flow energy?

5.6 What is the purpose of superheater? List out the advantages of superheated steam over wet steam.

5.7 Define triple point and critical point of water. Show these on *p–T* diagram.

5.8 Explain the process of steam generation.Show the various stages on *p–v* and *T–S* diagrams.

5.9 Draw a sketch for temperature – entropy for water and show on it constant pressure, constant volume and constant dryness fraction lines. Further show on a separate diagram the areas representing the liquid heat, latent heat and heat of superheat.

FUELS AND COMBUSTION OF FUELS

Chapter

6

Learning Objectives

After studying this chapter, the students will learn:

- Different types of fuels, such as coal, liquid fuels and gaseous fuels.
- Composition and analysis of fuels.
- Chemical reactions in combustion of fuels.
- Minimum air required for complete combustion.
- Calorific value of fuels.

6.1 FUELS AND THEIR CLASSIFICATION

Fuel is defined as any material that combines chemically with oxygen and liberates heat. The primary fuels which are burned to release heat and generate steam in boilers are fossil fuels in the form of coal, fuel oil and natural gas, which represent the remains of plant and animal life that are preserved in sedimentary rocks. Besides these, industrial wastes like blast furnace gas, coke oven gas, refinery gas, sugar factory refuse, saw mill wood dust, rice husk, etc., are also used.

Carbon and hydrogen are the main constituents of the fuel which produce the heat. Most of the solid and liquid fuels contain a small percentage of sulphur also (1 to 3%). Fuels are commonly used for power generation, industrial applications and for transportation. These are also used for heating and cooling purposes.

Fuels are generally classified as natural fuels and prepared fuels. These are further classified as solid, liquid and gaseous fuels.

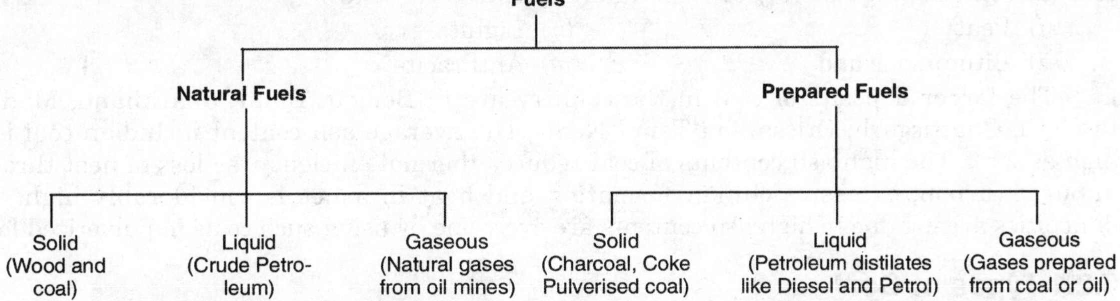

Natural solid fuels include wood and different types of coals. Wood is used as domestic fuel for cooking in rural areas, but it is not considered as a commercial fuel. Different types of coals are discussed later in this chaper.

Prepared solid fuels used for different purposes are:

1. **Wood Charcoal.** It is prepared by heating wood with a limited supply of air below 280°C, and contains 80 to 90% carbon. It is used for various metallurgical processes and also used for cooking and heating.

2. **Coke.** It is made from bituminous coal by heating continuously for 42 to 48 hrs in the absence of air in a closed chamber at 500 to 700°C for soft coke. It is hard in structure, dull black in colour, porous, and smokeless. These are generally used in blast furnaces.
3. **Briquetted Coal.** It is produced from finely ground coal by moulding under pressure with or without binding material. The commonly used binding materials are: Coal-tar, crude oil, or clay. It reduces the loss of fuel through the grate openings of the boilers and thus increases the efficiency.
4. **Pulverised Coal.** The low grade and high ash content coal is powdered to produce pulverised coal. It can be mixed with air more uniformly and burns most efficiently. Most power plants now use such pulverised fuel.

6.2 COAL

Coal is a very important fuel used in homes, as well as furnaces in industries, power plants and locomotives. It forms the backbone of all schemes aimed at the industrial progress of the country and hence it is of utmost economic value. Coal is formed by the transformation of buried vegetation matter into the carbonaceous product, over a long period of time. In composition, therefore, it consists of carbon with moisture, volatile matter and some mineral residuals.

The relative proportion of fixed (or pure) carbon, in relation to volatile matter and moisture, decides the quality of coal. The greater the percentage of fixed carbon present in a coal sample, the better will be its quality.

The initial decomposition of the buried accumulated vegetation matter into carbonaceous product, takes place through several biochemical processes. With this initial decomposition, starts the process of transformation into coal. The effect of heat and pressure due to further consolidation results in the expulsion of moisture and volatile constituents, giving rise to coal of higher grades. As this process of dynamo-chemical transformation continues, improved varieties of coal get produced.

Soil pressure, heat and movement of the earth's crust distilled off some of the bog's gaseous matter to form brown coal, or lignite. Continuing subterranean activity reduces the coal's gaseous content progressively to form different ranks. These are:

(*i*) Peat
(*ii*) Lignite
(*iii*) Bituminous and
(*iv*) Anthracite.

The larger deposits of coal in the country are in Bengal, Bihar, Jharkhand, Madhya Pradesh, Chattisgarh, Orissa and Tamil Nadu. The average ash content in Indian coal is as high as 20%. The high ash contents of coal reduces thermal efficiency, as loss of heat through unburnt carbon, excessive clinker formation and heat in ashes is considerably high. The difficulties arising due to high ash contents are overcome by using such coals in pulverized form.

6.2.1 Types of Coal

Following are the main types of coal:

1. Peat. This is the first stage in the transformation of vegetation matter into the coal. It has only partly carbonised vegetation matter. Chemically, it contains a large percentage of moisture, and other ingradients including carbon, sulphur, nitrogen and hydrogen. Peat is generally used as a fertiliser, or as a fuel in dried forms. The high moisture content greatly reduces its calorific value, to about 730 kcal/kg as against 3000 to 5600 kcal/kg when it is dried, and therefore is not suitable for power plants and mostly used for domestic and other purposes.

It is sun dried to remove the greater part of moisture and converted into briquettes before being transported.

2. Lignite. Lignite is the next stage in the development of coal. It also contains high percentage of moisture (30 to 45%) but can be dried by exposing to air, and can bring down the moisture upto 6%. It is brown in colour and exhibits a woody structure. It can be used as fuel in pulverised form. Its heat value range from 7000 to 11500 kJ/kg.

3. Bituminous Coal. This is the real coal formed from lignite, variety and is most popular form of coal used for all purposes. It has low moisture content and has high caking power (caking is the process in which the volatile matter is driven off from the coal when heated leaving behind particularly pure carbon and formed product is known as coke). Bituminous coal is a black coloured, brittle substance, in which plant residuals are not visible. In chemical composition, bituminous coals show considerable variation, with carbon from 40 to 90%, moisture from 5 to 20%, ash from 6 to 12%, and volatile constituents from 10 to 45%. The bituminous coal shows a typically banded appearance.

Low volatile matter and high caking coals are prepared for metallurgical industries, while high volatile matter and low caking bituminous coals have great utilities for gas making purposes. Its heat value varies from 9,500 to 16,000 kJ/kg.

Inferior quality of bituminous coal, also known as *Sub-Bituminous coals* are similar to lignite but have lost the woody structure and contains 50% or less moisture. This type of coal is also known as black lignite and is also used in the form of briquettes or pulverised state.

Highest grade of bituminous coals are also known as *Semi-Bituminous coal.* These, when burnt generate little smoke and are of low volatile variety, which is intermediate between bituminous and anthracite coal in properties. It is most widely used for steam raising in thermal power plant. It has low percentage of moisture, ash and sulphur and has high available hydrogen. It has carbon content ranging between 90-93%, volatile matter between 10-20% and oxygen between 2 to 4%.

4. Anthracite Coal. Anthracite is the last stage in the process of transformation of buried vegetation into coal and it contains highest percentage of carbon 92-98% and the percentage of volatile matter is below 8%. It burns only at high temperature with little or no flame, and pulverisation of anthracite is very difficult, therefore, it is used only on the grate with forced draft. The anthracite is clean, dense, hard, non-caking. It burns freely uniformly and smokelessly. Its heating value is between 14,500 to 16,000 kJ/kg.

6.3 ANALYSIS (COMPOSITION) OF COAL

There are two types of coal analysis used to find the commercial value of the coal. Both the analysis are done on a mass percent basis.

6.3.1 Proximate Analysis

The proximate analysis gives the percentage, by weight, of moisture, volatile matter, fixed carbon and ash, determined in that order. The proximate analysis of most coals indicates the following ranges of various constituents:

Constituents	Percentage
Moisture	3 to 30%
Volatile matter	3 to 50%
Ash	2 to 30%
Fixed (pure) carbon	16 to 62%

The constituents given by the proximate analysis generally decide the adoption of coal and design of power plant for handling, combustion and ash removal, and dust removal system.

1. Moisture. All coal contains some percentage of moisture either as inherent or free moisture. The inherent moisture is the combined moisture and is held in the pores of the coal. The inherent moisture is never removed from the coal used for power plants as it is a costly procedure. The percentage of inhernet mcisture is determined by heating the 1 gm sample of coal is subjected to a temperature of about 105°C for a period of 1 hour. The loss in weight of the sample gives the moisture content of the coal. The free moisture is defined as moisture present in the coal which can be removed just by exposing the coal to the natural air flow or by drying with the help of air at 50%.

Determining the free moisture in the coal is important as it affects the working of the power plant, as its presence in the coal adds to the transportation and storing cost of coal. It also decreases the heating value of the fuel and part of heat generated is also carried away as vapours with the exhaust gases. In addition, it quenches the fire in the furnace of boiler. Therefore, before using in the power plant, the coal is dried and free moisture is removed from it.

2. Volatile Matter. The volatile matter present in the coal may be combustible gases such as methane, hydrogen, carbon monoxide and other hydrocarbons, or incombustible gases like carbon dioxide and nitrogen. The volatile matters (combustible or non-combustile) affect the furnace volume, while the presence of incombustible gases are always undesirable as they do not add in heat value but increase the volume of furnace required.

When 1 gm sample of coal is placed in a covered platinum crucible and heated to 950°C and maintained at that temperature for about 7 min, there is a loss in weight due to the elimination of moisture and volatile matter. The latter may now be determined since moisture has been calculated from the previous test.

3. Ash. Ash is also an undesirable constituent of coal. The ash present in the coal is of two forms as fixed ash and free ash. The fixed ash comes from the original vegetable matter and it can not be removed before burning the coal. The free ash comes with coal in the form of clay, shale and pyrites. The free ash can be reduced or removed by mechanical processing of coal such as washing and screening.

The presence of ash also increases transportation, handling and storage costs, and decreases the heat value of fuel. It also involves additional cost in ash disposal. The presence of ash with coal is also responsible for early wear of the furnace walls, burning apparatus and feeding mechanism. Ash is also responsible for formation of clinkers which may choke air passages through the grate of the furnace and remains attached with unburned fuel resulting into loss of fuel. This factor is considered while deciding the type of the combustion chamber, burners and feeding system used in the power plant.

Ash can be determined, by subjecting 1 gm sample of coal in an uncovered crucible to a temperature of about 720°C until the coal is completely burned, a constant weight is reached, which indicates that there is only ash remaining in the crucible.

Fixed carbon is the difference between 100% and the sum of the percentages of moisture, volatile matter and ash.

The proximate analysis of coal gives: Fixed carbon (FC) + Volatile matter (VM) + Moisture (M) + Ash (A) = 100% by mass.

6.3.2 Ultimate Analysis

The coal substance consists of organic compounds of carbon, hydrogen, and oxygen derived from the original vegetable matter. The ultimate analysis shows the following components on mass basis: carbon (C), hydrogen (H), oxygen (O), nitorgen (N), sulphur (S), moisture (M), and ash (A). Therefore,

$$C + H + O + N + S + M + A = 100\% \text{ of mass.}$$

Since the proximate analysis of coal does not give any idea about the suitability of coal for the purpose of heating, and also do not help in finding out the calorific value of coal, therefore ultimate analysis is used to find out the chemical analysis of coal like carbon, hydrogen, oxygen, nitrogen, sulphur and ash.

Ultimate analysis of most coals indicates the following ranges of various constituents:

Constituents	Percentages by mass
Carbon	50 - 95%
Hydrogen	1 - 5.5%
Oxygen	2 - 40%
Sulphur	0.5 - 3%
Nitrogen	0.5 - 7%
Ash	2 - 30%

The role played by each constituent in adopting coal for power plant is discussed below:

***(i)* Carbon.** Higher percentage of carbon gives high heat value and reduces the size of combustion chamber required and vice-versa.

***(ii)* Hydrogen.** A part of the hydrogen existing in the coal in combined form with oxygen is known as inherent moisture. This inherent moisture is objectionable as it carries heat with flue gases without playing any role in combustion. High percentages of free hydrogen is always desirable as it increases the heating value of the coal.

***(iii)* Oxygen.** The oxygen which is present in coal is always in combined form, with hydrogen. Therefore, low percentages of oxygen is always desirable as it reduces the percentages of hydrogen available for free heating.

***(iv)* Nitrogen.** The nitrogen in coal does not play any role in combustion.

***(v)* Sulphur.** It adds a little heating value, but has many undesirable characteristics. The high percentage of sulphur is highly objectionable because the sulphur is responsible for slagging, clinkering, corrosion of air-preheaters and economisers, spontaneous combustion during storage and air pollution.

***(vi)* Ash.** Ash is a residue from combustion, while clinker is caused by the melting of this ash. It contains silica, alumina, ferric oxide, calcium oxide, magnesium oxide, alkalies and 1 to 2% sulphur. The formation of clinker on the grate depends upon the percentages of constituents in ash and the temperature over the grate.

6.4 LIQUID FUELS

Liquid fuels include petroleum and its derivatives. Petroleum is generally believed to have been formed during past geological ages from decayed marine life, both vegetable and animal. Dead marine animals and vegetable matter accumulated for millions of years ultimately got transformed into oil, mainly in sedimentary rocks, by pressure and heat. Crude petroleum oil deposits accumulate in voids in the rocks and sands below the earth's surface. Fairly dense earth strata cover most deposits. Oil has an underlying body of water and pockets of pressurised natural gas above. Oil well drilled through the layer earth strata penetrates the deposits. The pressure forces the gases and oil to the surface. After the pressure has diminished, the oil should be pumped.

There is also possibility that some oils were formed through the decomposition of vegetable matter mainly, particularly where coal has been formed in adjacent territory. Liquid fuels are an excellent energy source, as they are clean, easy to handle, easy to store and easy to burn. These are primarily a mixture of hydrocarbon compounds, which may also contain nitrogen, oxygen and sulphur. The most of the hydrocarborns belong to the paraffin series, like methane (CH_4), ethane (C_2H_6), propane (C_3H_8) and butane (C_4H_{10}) which are gaseous, and pentane (C_5H_{12}), hexane (C_6H_{14}) and octane (C_8H_{18}) which are liquid at STP.

The crude petroleum, after it is extracted from the oil fields is taken to refinery where it is heated, vaporised and cooled at different heights. Each separated oil is further refined into various products, such as bitumen, diesel, kerosene, petrol, etc. The wastes of the crude oil is also taken to other parts of a refinery, where it is used to manufacture other materials like plastics, fertilizers, paints, rubbers, wax, etc.

As mentioned earlier, India is quite deficient in production of oil and natural gas, and a large quantum of crude is being imported to fulfil our domestic requirements. The *on-shore* (under the land) deposits of petroleum and natural gas (which usually occur together) are found in Assam, Gujarat, Krishna-Godavari basin (K.G. basin) and western Rajasthan. Similarly *off-shore* (from below the sea) oil is being produced from Bombay high and Gandhar.

Classification of Oil

Liquid fuels are hydrocarbons and are classified as light oils and heavy oil. Some commonly used liquid fuels for power engineering are:

(*i*) Gasoline or petrol, (*ii*) Paraffins,
(*iii*) Diesel oil, (*iv*) Heavy fuel oil

Liquid fuels are commonly employed in I.C. engines and for steam generation.

During distillation of petroleum, gasoline is obtained between the temperature range of 30-200°C, whereas kerosene oil is obtained between the range of 150-300°C and diesel oil is obtained between 200–320°C. The calorific value of the gasoline is 46,881 kJ/kg and that of kerosene oil is 46,442 kJ/kg. The calorific values of diesel oil and heavy oil are 46,442 kJ/kg and 43,933 kJ/kg respectively.

The fuel oil used for industrial or domestic purposes are obtained by refining the crude oil. A refining process as stated earlier produces gasoline, Kerosene, lubricating distillate and a residual oil, tar or petroleum coke. This in turn processed to produce the wide array of petroleum products found in the market, like motor fuels, kerosene fuel oils, lubricating oils, paraffin, asphalt and coke oils.

The required physical properties of fuel oil are specific gravity, viscosity, specific heat, volatility, pour point, flash point, carbon residue, heating value, ash, moisture, sediment and sulphur. The pour point is the lowest temperature at which the oil will flow under standard pressure conditions. The flash point is the minimum temperature at which the oil may be ignited. It decides the safety of fuel and ease of ignition. Viscosity is the measure of resistance to the oil flow through the pipes and nozzles. This affects the cost of pumping the fuel. The percentage of sulphur should be as minimum as possible, as it results in the corrosion of different parts in the plants. For better combustion fuel should have moisture and sediment as small as possible. The higher percentage of hydrogen will result into moisture carrying the heat more, thereby reducing the overall combustion efficiency of the plant.

Main advantages of liquid fuels over coal are:

(*i*) Higher calorific value.
(*ii*) Economy in space.
(*iii*) Cleanliness.
(*iv*) Easy control of consumption.
(*v*) No loss of heat during storage.
(*vi*) Less operational labour requirement.
(*vii*) The storage and handling is easy and economical.
(*viii*) Elimination of wear and tear of grate bars.
(*ix*) Combustion losses are less.
(*x*) Liquid fuels leave very little ash after burning hence it saves from ash handling problems.

6.5 GASEOUS FUELS

The gaseous fuels are of two types, namely, natural gas and manufactured gas. Since manufactured gas is costly, only natural gas is used for steam generation. **Manufacture gases** are: blast furnace gas, coal gas, producer gas, water gas etc. As discussed earlier, **natural gas** may be considered as an associate product of petroleum which exists naturally under the surface of the earth. Natural gas comes out of the gas wells and petroleum wells. About one-third of the natural gas produced is found with petroleum.

Natural gas contains 60 to 95% of methane (CH_4), with small amounts of other hydrocarbons such as ethane (C_2H_6) upto 16% napthene, CO_2, and N_2.

The natural gas is carried through pipes to distances which are hundreds of kilometers from the source. The natural gas is a colourless, odourless, and non-poisonous gas. These gases have calorific value between 33,500 to 50,200 kJ/m^3.

Commonly used manufactured gases are:

1. Blast Furnace Gas. This gas is obtained as a by-product from blast furnace used for producing pig iron. Its approximate composition is CO = 30%, N_2 = 52%, H_2 = 3%, CO_2 = 2%, CH_4 = 3%.

The calorific value of this gas is 4060 kJ/m^3

2. Coal Gas. Coal gas is a by-product obtained during the destructive distillation of coal. Its calorific value is 31800 kJ/m^3. Its approximate composition is CH_4 = 35%, H = 45%, CO = 8%, N = 6%, CO_2 = 2% and other hydrocarbons = 4%. This gas is also known as **Town gas or city gas.**

3. Producer Gas. Producer gas is produced during incomplete combustion of coke in current of air. Its main constituents are nitrogen and carbon monoxide. Its approximate composition is N = 62%, CO = 23%, H_2 = 6%, CO_2 = 5%, CH_4 = 4%. Its calorific value is 5025 kJ/m^3.

$$\underset{\text{Coke}}{2C} + \underset{\text{Air}}{(O_2 + N_2)} \rightarrow \underset{\uparrow\ \text{Producer gas}}{2CO + N_2}$$

4. Water Gas. Water gas is obtained by passing a blast of steam through a deep bed of red hot coke. Its main constituents are CO_2, CO and H_2.

$$\underset{\text{Coke}}{C} + \underset{\text{Steam}}{H_2O} \rightarrow \underset{\text{Water gas}}{CO + H_2} \uparrow$$

5. Compressed Natural Gas (CNG) is now being considered as the alternative fuel for automobiles.

6. Liquid Petroleum Gas (LPG) refers to hydrocarbons, such as propane, propylene, butane, butylene, etc., which are liquefied under moderate pressure and at normal temperatures. It is used widely as domestic fuel and also to supplement natural gas flow.

Advantages of Gaseous Fuels

1. The length and spread of the flame, and temperature in the combustion chamber can be easily controlled.
2. These are free from solid and liquid impurities and therefore burns completely.
3. These do not produce ash or smoke.
4. Minimum excess air is required for complete combustion resulting in better efficiency.
5. Transport through pipe is easy.
6. A clean atmosphere around the plants using the gaseous fuel.

The main problem with gaseous fuels is that these are readily inflammable and require large storage volume capacity.

6.6 COMBUSTION OF FUELS

As discussed earlier, fuel is a substance which produces heat when burned in the presence of air (oxygen). The amount of heat generated per kg of fuel during complete combustion is known as **calorific value** of the fuel. Carbon and hydrogen are the main constituents of the fuel which produce heat. Fuels are used mainly for power generation, industrial applications and for transportation. Fuels are also required for heating and cooking purposes. Fuels are generally classified as (*i*) natural fuel (also known as **primary fuels),** such as solid fuels (wood, peat and coal), liquid fuels (crude petroleum), and gaseous fuels (natural gases from oil mines); and (*ii*) Prepared fuels (also known as **secondary fuels)** such as solid fuels (charcoal, coke, pulverised coal), liquid fuels (petroleum distillation like diesel, petrol, alcohols), and gaseous fuels (prepared gases from coal or oil).

Fuel Combustion deals with various reactions taking place concerning different elements which constitute the fuel. The combustion of fuels may be defined as a chemical combination of oxygen in the atmospheric air and hydro-carbons. It is usually expressed both.

(*i*) qualitatively, and (*ii*) quantitatively

by equations known as chemical equations which indicate the nature of chemical reactions taking place.

Adequate supply of oxygen is very essential for the complete combustion of fuel in order to obtain maximum amount of heat from a fuel.

Combustion of fuels is accomplished by mixing fuel and air at elevated temperture. The combustion process may be simply expressed as follows:

Fuel + Air = Products of combustion + Heat liberated

The oxygen contained in the air unites chemically with carbon, hydrogen and other elements in fuel to produce heat. The amount of heat liberated during the combustion process depends on the amount of oxidation of the constituents of fuel and the nature of fuel (chemical composition of fuel).

In order that the combustion of fuel may take place with high efficiency, the following conditions must be fulfilled:

1. The amount of air supplied should be such that it is sufficient to burn the fuel completely. Complete combustion of fuel means complete oxidation of all the combustible material in the fuel. A deficiency of air causes incomplete combustion of fuel which result in considerable unburnt fuel being discharged from the furnace whereas too much supply of air simply dilutes the gases and cools the furnace.
2. The air and fuel should be thoroughly mixed so that each combustible particle comes in intimate contact with the oxygen contained in the air.
3. The fuel should remain in the furnace for sufficient time till it gets burnt completely.
4. The temperature in the furnace should be high enough to ignite the incoming air fuel mixture.

Products of Combustion

The complete combustion of fuel produces various gases such as carbon dioxide (CO_2), sulphur dioxide (SO_2), water vapour, nitrogen (N_2) and oxygen (O_2). Nitrogen comes from air supply and oxygen from excess air. Water vapour is produced from the following three sources:

(*i*) Moisture originally contained in the coal

(*ii*) Vapour produced by combustion of hydrogen

(*iii*) The water vapour of atmospheric humidity.

If all the carbon present in the fuel does not get burn completely then carbon monoxide (CO) is also produced. The flue gases will have considerable amount of carbon monoxide in them if the oxygen supply is less. However, large excess of air would mean that a large amount of sensible heat would be lost in flue gases. Analysis of flue gases give a correct idea of how the fuel is burning.

6.7 CHEMICAL REACTIONS

6.7.1 Combustion of Solid and Liquid Fuels

The combustion process involves chemical reactions. The combustible elements in fuels consist of carbon, hydrogen and sulphur.

The chemical equations representing the combustion of C, H_2, S, CH_4. etc., are described as follows:

(*i*) Combustion of Carbon

$$C + O_2 \rightarrow CO_2$$

Substituting the values of molecular weight in equation.

$$12 + 16 \times 2 = 12 + 16 \times 2 \quad \Rightarrow \quad 12 + 32 = 12 + 32$$

$$1 + \frac{8}{3} = 1 + \frac{8}{3} \quad \Rightarrow \quad 1 + \frac{8}{3} = \frac{11}{3}$$

This means that 1 kg of carbon requires 8/3 kg of oxygen for its complete combustion and produces 11/3 kg of carbon dioxide.

If the amount of oxygen supplied is not sufficient the combustion of carbon is incomplete and the product of combustion will be carbon monoxide.

$$2C + O_2 \rightarrow 2CO \Rightarrow 2 \times 12 + 16 \times 2 = 2\,(12 + 16)$$

$$24 + 32 = 56 \Rightarrow 1 + \frac{4}{3} = \frac{7}{3}$$

which means that 1 kg of carbon needs 4/3 kg of oxygen to produce 7/3 kg of carbon monoxide. Further burning of CO produces CO_2.

$$2CO + O_2 \rightarrow 2CO_2$$

$$2(12 + 16) + 16 \times 2 \rightarrow 2(12 + 32) \Rightarrow 56 + 32 \rightarrow 88$$

$$1 + \frac{4}{7} \rightarrow \frac{11}{7}$$

This means that 1 kg of CO needs 4/7 kg of oxygen and produces 11/7 kg of CO_2.

***(ii)* Combustion of Hydrogen.** Burning of hydrogen with oxygen produces water vapours,

$$2H_2 + O_2 \rightarrow 2H_2O$$

$$2(1 + 1) + 16 \times 2 \rightarrow 2(2 + 16)$$

$$4 + 32 \rightarrow 36 \Rightarrow 1 + 8 \rightarrow 9$$

This means that 1 kg of hydrogen combines with 8 kg of oxygen to produce 9 kg of water.

***(iii)* Combustion of Sulphur.** When sulphur burns with oxygen it produces sulphur dioxide.

$$S + O_2 \rightarrow SO_2 \Rightarrow 32 + 16 \times 2 \rightarrow 32 + 16 \times 2$$

$$32 + 32 \rightarrow 32 + 32 \rightarrow 64$$

1 kg of sulphur + 1 kg of oxygen $\rightarrow$ 2 kg of sulphur dioxide

***(iv)* Combustion of Methane CH_4**

$$CH_4 + 2O_2 \rightarrow CO_2 + 2H_2O$$

$$(12 + 4) + 2 \times 32 \rightarrow (12 + 32) + 2 \times 18$$

$$16 + 64 \rightarrow 44 + 36$$

$$1 + 4 \rightarrow \frac{11}{4} + \frac{9}{4}$$

1 kg methane + 4 kg oxygen $\rightarrow$ 11/4 kg. Carbon dioxide + 9/4 kg of water.

The various values are summarised in Table 6.1.

Table 6.1 Oxygen Required for Combustion.

Substance	*(Oxygen*	*Production of combustion (kg)*			
(1 kg)	*reqd. kg)*	CO	CO_2	H_2O	SO_2
C	8/3	–	11/3	–	–
CO	4/7	–	11/7	–	–
C	4/3	7/3	–	–	–
H_2	8	–	–	9	–
S	1	–	–	–	2
CH_4	4	–	11/4	9/4	–

Chemically, the hydrogen of the fuel and oxygen of the air form water, and the carbon and the oxygen form carbon dioxide and sometimes carbon monoxide. These oxidation processes, when accompanied by the rapid liberation of heat, are known as combustion. To start combustion, the hydrocarbons and the oxygen must be brought to certain minimum ignition temperature.

The Mole. The kilogram mole called simply the *'mole'* is widely used in combustion calculations. The *mole* is that quantity of a substance which is equal in mass to the molecular weight of the substance. The one kilogram mole of hydrogen has a mass of 2 kilograms, one kilogram mole of nitrogen has a mass of 28 kilograms and one kilogram mole of CO_2 has a mass of 44 kilograms.

At standard temperature and pressure (0°C and 760 mm of Hg), the volume occupied by 1 mole of any ideal gas is the same and is given by;

$$V = 22.48 \text{ m}^3/\text{ kg mole}$$

The atomic and molecular weights of different elements and compounds are listed below:

Table 6.2: Atomic and Molecular Weights

Substance	*Chemical system*	*Atomic weight*	*Molecular weight*
Hydrogen	H_2	1	2
Oxygen	O_2	16	32
Nitrogen	N_2	14	28
Carbon	C	12	12
Sulphur	S	32	32
Carbon monoxide	CO	–	28
Methane	CH_4	–	16
Ethane	C_2H_6	–	30
Ethylene	C_2H_4	–	28
Acetylene	C_2H_2	–	26
Carbon dioxide	CO_2	–	44
Sulphur dioxide	SO_2	–	64
Steam or water	H_2O	–	18

6.7.2 Combustion of Gaseous Fuels

Gaseous fuels are usually measured by volume (in cubic metres). The various chemical equations are described as follows:

***(i)* Combustion of Hydrogen**

$$2H_2 + O_2 \rightarrow 2H_2O$$

$$2 \text{ vol.} + 1 \text{ vol.} \rightarrow 2 \text{ vol.}$$

$$1 \text{ cu-metre} + \frac{1}{2} \text{ cu-metre} \rightarrow 1 \text{ cu-metre.}$$

This means that one cu-metre of hydrogen requires ½ cu-metre of oxygen to produce one cu-metre of water.

***(ii)* Combustion of Methane.** When methane (Marsh Gas) burns with oxygen it gives CO_2.

$$CH_4 + 2O_2 \rightarrow 2H_2O + CO_2$$

$$1 \text{ vol.} + 2 \text{ vol.} \rightarrow 2 \text{ vol} + 1 \text{ vol.}$$

1 cu-metre + 2 cu-metre → 2 cu-metre + 1 cubic-metre.

Thus 1 cu-metre of methane needs 2 cu-metre of oxygen for combustion and produces 2 cu-metre of water and 1 cu-metre of CO_2.

***(iii)* Combustion of CO.** When carbon monoxide burns in oxygen, it gives CO_2.

$$2CO + O_2 \rightarrow 2CO_2$$

$$2 \text{ vol.} + 1 \text{ vol.} \rightarrow 2 \text{ vol.}$$

1 cu-metre + 1/2cu-metre → 1 cubic metre.

Thus one cu-metre of CO needs $\frac{1}{2}$ cu-metre of oxygen to produce one cu-metre of CO_2.

***(iv)* Combustion of Ethylene (C_2H_4).** When C_2H_4 burns in oxygen it gives CO_2 and H_2O.

$$C_2H_4 + 3O_2 \rightarrow 2CO_2 + 2H_2O$$

1 vol. + 3 vol. → 2 vol. + 2 vol.

1 cu-metre + 3 cu-metre → 2 cu-metre + 2 cu-metre.

Thus 1 cu-metre of ethylene combines with 3 cu-metre of oxygen to produce 2 cu-metre of CO_2 and 2 cu-metre of water vapours.

The combustion products of various gaseous fuels are summarised in Table 6.3.

Table. 6.3 Combustion Products of Gaseous Fuels.

Gas (1 cu-metre)	*Oxygen reqd. (cu-metre)*	*Products of combustion (cu-metre)*	
		CO_2	H_2O
H_2	$\frac{1}{2}$	—	1
CH_4	2	1	2
CO	$\frac{1}{2}$	1	—
C_2H_4	3	2	2

6.8 MINIMUM AIR REQUIRED

The weight of air required for the complete combustion of fuel is calculated from the analysis of fuel. To calculate the amount of air required for complete combustion of fuel, firstly the oxygen required for burning each of the constituent fuel is calculated and then the air required is found out. The atmospheric air consists of oxygen, nitrogen is found out. The atmospheric air consists of oxygen, nitrogen and small amount of carbon dioxide and other gases such as neon, argon, krypton, etc. For calculating the air required for burning a fuel, the following composition of air can be taken:

By weight:	Oxygen = 23%	Nitrogen = 77%
By volume:	Oxygen = 21%	Nitrogen = 79%

If it is found that the fuel already contains some amount of oxygen, then it should be deducted from the calculated value of oxygen.

The combustion of all combustible constituents requires oxygen but in practice, only air can be supplied. For calculation purposes, it is assumed that air is a mixture of only oxygen and nitrogen. This means that one kg of O_2 is associated with 77/23 = 3.35 kg of N_2 in (1 + 3.35) = 4.35 kg of air. Similarly, one m^3 of O_2 is associated with 79/21 = 3.76 m^3 of N_2 in (1 + 3.76) = 4.76 m^3 of air.

The nitrogen being inert gas does not take part in chemical reactions. Its pressure in air, on the other way, retards the mixing process of fuel and oxygen, lowers the temperature of combustion and carries away heat when goes with the exhaust gases.

Adequate supply of oxygen is essential for complete combustion of fuel and hence for obtaining the maximum amount of heat. If the contents of fuel are C% of carbon, H_2% of hydrogen, O_2% of oxygen, and S% of sulphur respectively, then the mass of oxygen required for complete combustion is determined as under:

$$\text{Oxygen required per kg of fuel} = \frac{1}{100}\left[\frac{8}{3}\text{C} + 8\left(\text{H} - \frac{0}{8}\right) + \text{S}\right]$$

As 1 kg of carbon requires 8/3 kg of O_2, 1 kg of hydrogen requires 8 kg of O_2 and 1 kg of sulphur requires 1 kg of O_2.

As air contains 23% of oxygen by mass, the **amount of air required per kg of fuel** for complete combustion is given by:

$$= \frac{1}{100}\left[\frac{8}{3}\text{C} + 8\left(\text{H} - \frac{0}{8}\right) + \text{S}\right] \times \frac{100}{23}$$

$$= \frac{1}{23}\left[\frac{8}{3}C + 8\left(H - \frac{0}{8}\right) + S\right]$$

Similarly, as air contains 21% of oxygen by volume, the **volume of air required per m^3 of gaseous fuel** for complete combustion is given by

$$\frac{1}{21}\left[\frac{H_2}{2} + \frac{CO}{2} + 2CH_4 + 3C_2H_4\right]$$

The quantity of air so calculated is known as *'theoretical'* or *'stoichiometric'* air required for 'perfect combustion' that based on perfect mixing of fuel and air with neither substance left over after completion of the process. It is the minimum quantity of air required.

6.9 EXCESS AIR

As a deficiency of air would cause the formation of smoke and carbon monoxide and as the theoretical amount of air would require perfect mixing of fuel and air molecules, an excess of air is needed for good clean combustion. Too much excess air also causes wastage of fuel.

Excess air is the difference between the amount of air actually supplied and that theoretically required. The percentage of excess air is based on the theoretical amount, as follows

$$\text{Excess air} = \frac{\text{Actually air supplied} - \text{theoretical amount}}{\text{Theoretical amount}}$$

Weight of Excess Air Supplied

The weight of excess air required during combustion of coal is calculated from the weight of unused oxygen in flue gases after CO present in flue gases is burnt to CO_2.

Let W_1 = Weight of flue gases per kg of fuel

W_2 = Percentage weight of oxygen present in flue gases

W_3 = Percentage weight of CO present in flue gases

As 1 kg of CO need 4/7 kg of oxygen to be burnt to CO_2.

∴ Oxygen required to burn W_3 kg of CO to CO_2.

$$\frac{W_3}{100} \times \frac{4}{7} = W_4 \text{ (say)}$$

$$\therefore \quad \text{Excess Oxygen } W_5 = \frac{W_2}{100} - W_4$$

Weight of excess oxygen per kg of fuel (W_6) = Weight of excess oxygen per kg flue gas × Weight of flue gas per kg of fuel

$$W_6 = W_5 \times W_1$$

Therefore, the weight of excess air supplied

$$= W_6 \times \frac{100}{23}$$

Proper control of the right amount of excess air maintains optimum combustion efficiency. CO_2 and O_2 in combustion gases are index of excess air. Air feed should be controlled so that optimum amount of CO_2 or O_2 is produced. Smoke formation and slagging of boiler surfaces also play an important part in determining the optimum excess air. Practical conditions of fuel type furnace arrangements and heat transfer arrangement determine the total amount of air needed for complete combustion.

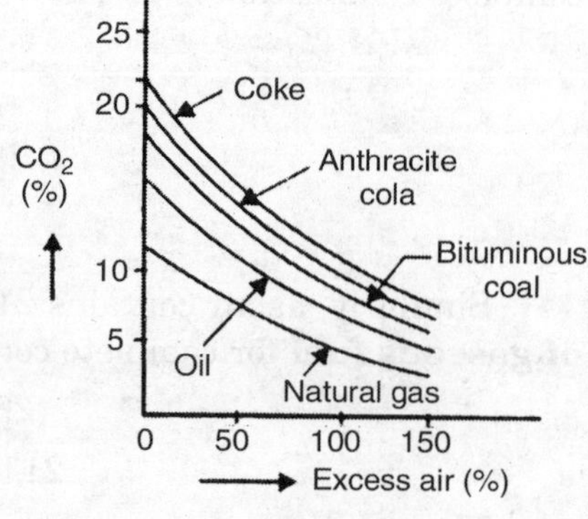

Fig. 6.1

The amount of CO_2 in flue gases depends on type of fuel and excess air supplied to the furnace. Fig. 6.1 shows typical variation of CO_2 in flue gases (per cent by volume) and excess air (per cent) for complete combustion of various types of fuels.

The total amount of air needed for complete combustion of a fuel depends on following factors:

(*i*) Type of fuel

(*ii*) Furnace arrangements.

(*iii*) Heat transfer surface arrangement.

Typical values of excess air supplied, expressed as percentage of the quantity theoretically required are as follows:

Hand fired boiler furnace	:	100
Mechanically stokered furnace	:	40
Petrol engine	:	20
Oil engine	:	20

6.10 PRINCIPAL STAGES OF COMBUSTION

The combustion of fuel is a complicated physical and chemical process in which the combustion elements of the fuel combine with the oxygen of air with the evolution of heat attended by a sharp rise in temperature and formation of flame. During the burning of any fuel two stages are observed:

(*i*) Ignition (*ii*) Combustion

Ignition is the period during which the fuel is gradually raised in temperature. On attaining a definite temperature the fuel is ignited and stable combustion sets in. When solid fuel is introduced into the furnace, moisture is first removed and the volatiles begin to be liberated. The resultant gaseous products of the fuel decomposition are gradually heated to the ignition point and burn in a flame over the solid part of the fuel. Combustion of the gaseous substances, heats the coke which begins to burn stably when the ignition points is reached. At this stage maximum temperature is generated. Burning down is the final stage in the combustion solid fuel. Gasification and the combustion of solid elements are completed in this stage and enough heat is generated to maintain combustion at a sufficiently high temperature.

Liquid fuel should be first atomised to increase its area of contact with the air. An additional phase in the firing of liquid fuel is evaporation. A drop of fuel oil entering the spray of burning fuel is heated and gradually reaches the temperature at which the fuel components begin to evaporate. The gases formed burn in the oxygen of air and increase the temperaure of drop. At a definite temperature the molecules of the drop begin to disintegrate. The drop begins to burn at its surface. The oxygen penetrates inside where partial combustion takes place and the gases formed inside the drop further intensify the combustion.

During different stages of combustion of fuel the required quantity of air should be supplied. Complete combustion of fuel takes place depending upon the quantity of air supplied.

6.10.1 Complete Combustion

It is the process in which the combustible elements of fuel combine chemically with the oxygen of air at a definite temperature.

The flue gas produced consist of CO_2, SO_2, water vapour (H_2O), oxygen (O_2) and nitrogen (N_2).

6.10.2 Incomplete Combustion

A deficiency in air supplied causes incomplete combustion of fuel which results in considerable unburnt fuel burning discharged from the furnace along with ash and slag. The presence of carbon monoxide gas (CO) in the combustion products indicates incomplete combustion. When liquid or gaseous fuel is fired incomplete combustion is accompained by soot formation.

6.10.3 Weight of Carbon in Flue Gases

The weight of carbon contained in 1 kg of flue gases can be calculated from the amounts of CO_2 and CO present in it.

During the complete combustion of C to CO_2.

$$C + O_2 = CO_2 \Rightarrow 12 + 16 \times 2 = 12 + 16 \times 2$$

$$12 + 32 = 44$$

$$1 + \frac{32}{12} = \frac{44}{12} \Rightarrow 1 + \frac{8}{3} = \frac{11}{3}$$

Thus 1 kg of carbon on combustion produces 11/3 kg of CO_2. Hence, 1 kg of CO_2 will contain 3/11 kg of carbon.

Now when carbon burns in insufficient supply of oxygen then the combustion of carbon is incomplete and products of combustion will be carbon monoxide.

$$2C + O_2 = 2CO$$

$$2 \times 12 + 16 \times 2 = 2(12 + 16)$$

$$24 + 32 = 56 \quad \text{or} \quad 1 + \frac{32}{24} = \frac{56}{24}$$

$$1 + \frac{4}{3} = \frac{7}{3}$$

Thus 1 kg of carbon produces 7/3 kg of CO. Hence, 1 kg of CO contains 3/7 kg of carbon.

$\therefore$ W_1 = weight of carbon in 1 kg of flue gases

$$= \frac{3}{11} CO_2 + \frac{3}{7} CO$$

Weight of Flue Gas Per kg of Fuel Burnt. The actual weight of dry flue gases can be obtained by comparing the weight of carbon present in flue gases with the weight of carbon in the fuel.

Let, W_2 = weight of carbon in 1 kg of fuel

W = weight of flue gas per kg of fuel burnt W_2/W_1

where W_1 = Weight of carbon in 1 kg of flue gases.

6.10.4 Conditions for Proper Burning (Combustion) of Fuel

The various conditions that should be established for proper burning of fuel are as follows:

(*i*) Correct mixing and ratio of fuel and air

(*ii*) Enough time to burn the fuel completely

(*iii*) High temperature flame

(*iv*) Turbulent mixing of fuel and air

(*v*) Proper proportioning of furnace dimensions.

6.10.5 Temperature of Fuel Combustion

The combustion of fuel is always accompained by heat losses. Therefore, the real temperature or actual temperature of combustion is always lower than the theoretical temperature of combustion which is obtained in ideal cases without heat losses. Table 6.4 shows the theoretical temperature of combustion for various fuels in degree centigrade.

Table 6.4: Temperature of Combustion for Fuels.

Type of fuel	*Excess AIR coefficient*			
	1.0	1.3	1.5	2.0
Anthracite	2270	1854	1665	1300
Lignite	1875	1590	1425	1150
Peat	1700	1510	1370	1110
Fuel oil	2125	1740	1580	1265
Gas	2000	1749	1478	1167

6.11 HEAT OF COMBUSTION (CALORIFIC VALUE OF FUELS)

As we know that, the *Calorific value* (CV) of a solid or liquid fuel is defined as the amount of heat given out by the complete combustion of 1 kg of fuel in kJ. However, the calorific value of gaseous fuel is expressed in kJ per m^3 of gas at specify temp & pressure. The calorific value of basic fuel constituents are as under:

Table 6.5: H.C.V. and L.C.V. for Fuel Constituents

Fuel constituent burning of products	*HCV (kJ / kg)*	*LCV(kJ / kg)*
$C \rightarrow CO_2$	33,780	–
$C \rightarrow CO$	10,120	–
$CO \rightarrow CO_2$	10,080	–
$S \rightarrow SO_2$	9,280	–
$H \rightarrow H_2O$	1,44,200	121,200

The calorific values of the fuels are of two types, higher calorific and lower calorific values.

Higher calorific values (HCV) of fuel is the sum of the heat liberated by the combustible constituents like C, H_2, S.

$$CV = 33780\ C + 144200\ H_2 + 9280\ S$$

where, C, H_2 and S are the weight of carbons, Hydrogen and sulphur per kg of fuel burnt.

If the fuel contains O_2 kg of oxygen then it is assumed that the oxygen is always present in the combined form of water (H_2O). Then the amount of H_2 available for burning is reduced and it is equal to $\left[H_2 - \frac{O_2}{8}\right]$, as 8 kg of O_2 is required per kg of H_2 for complete combustion.

Therefore, the CV of the fuel, when it contains O_2 is given by

$$HCV = 33780C + 144200\left(H_2 + \frac{O_2}{8}\right) + 9280S.$$

This value of heat evolved is known as HCV of fuel and the formula is known as *Dulong's formula.*

Lower Calorific Value (LCV) of Fuel. Since the latent heat of the water vapour formed, from the hydrogen of the fuel, is carried away by the vapour with the exhaust gases, therefore, the actual amount of heat evolved is reduced by the amount of heat carried away by the water vapour. This value of heat is known as LCV of fuel.

$$LCV = HCV - (9\ H_2) \times 2{,}460.$$

where, H_2 is the amount of hydrogen in kg per kg of fuel and 2,460 kJ is the latent heat of one kg of steam formed.

The heat evolved by the complete combustion of one cubic meter of gaseous fuel is known as calorific value of fuel. Gaseous fuels also have HCV and LCV like liquid and solid fuels.

General practice is to use LCV of fuel for all practical purposes, as this is the net heat available for use.

SOLVED PROBLEMS

Example 6.1 *A steam power station of 200 MW capacity uses coal of calorific value of 25,600 kJ/kg. The thermal efficiency of the station is 30% and electrical generation efficiency is 92%. Determine the coal required per hour when the plant is working on full load.*

Solution. Assume W is the weight of coal supplied per hour.

∴ Mechanical energy available

$$= W \times 25600 \times 0.3 \text{ kJ/hr}$$

Electrical energy available = W × 25600 × 0.3 × 0.92 kJ/hr

200 MW plant if operated for 1 hr, then its heat equivalent is

$$= 200 \times 1000 \times 3600 \text{ kJ/hr}$$

$$\therefore\ W \times 25600 \times 0.3 \times 0.92 = 200 \times 1000 \times 3600$$

$$W = \frac{200 \times 1000 \times 3600}{25600 \times 0.3 \times 0.92}$$

= 101,900 kg/hr = **101.95 tons/hr. Ans.**

Example 6.2 *A coal having a calorific value of 28,900 kJ/kg is supplied to a power station. The boiler, turbine and generator efficiency are 83%, 32% and 97% respectively. If the coal consumption of the station is 10^5 kg/hr, determine the capacity of the power plant in MW.*

Solution. As,

$$P = \frac{W \times C.V. \times \eta_b \times \eta_t \times \eta_g}{3600 \times 1000}$$

where, P is the power in MW,

W is the weight of coal in kg per hour, and

η_b, η_t and η_g are the boiler, turbine and generator efficiencies.

$$\therefore\quad P = \frac{100000 \times 28900 \times 0.83 \times 0.32 \times 0.97}{3600 \times 1000}$$

= **207 MW Ans.**

Example 6.3 *A small generating plant of 1 MW capacity uses gas of a calorific value of 4000 kJ/m³. The overall efficiency of the plant is 30%, determine the volume of gas required per hour when the plant is running at full load condition.*

Solution. As,

$$P = \frac{(V \times CV) \times \eta}{3600}$$

where, P = power in kW,

V = volume of the gas supplied in m^3 per hour

η = overall efficiency of the plant.

$$1000 = \frac{V \times 4000 \times 0.3}{3600}$$

$$\therefore \quad V = \frac{1000 \times 3600}{4000 \times 0.3} = \textbf{3000 m}^3\textbf{/hr.} \quad \textbf{Ans.}$$

Example 6.4 *A sample of coal has the following analysis by weight: C = 80%, H_2 = 10%, O_2 = 4%, S = 1% and remaining is ash. Determine the higher and lower calorific values of the coal.*

Solution. As,

$$\text{HCV} = \frac{1}{100}\left[33780C + 144200\left(H - \frac{0}{8}\right) + 9280S\right]$$

By putting the values,

$$HCV = \frac{1}{100}\left[33780 \times 80 + 144200\left(10 - \frac{4}{8}\right) + 9280 \times 1\right]$$

$$= 33780 \times 0.8 + 144200 \times 0.095 + 9280 \times 0.01$$

$$= 27024 + 13699 + 92.80$$

$$= 40816 \text{ kJ/kg} \quad \textbf{Ans.}$$

As

$$LCV = HCV - \frac{9H}{100} \times 2460$$

$$= 40816 - \frac{9 \times 10}{100} \times 2460 = 40816 - 2214$$

$$= \textbf{38602 kJ/kg} \quad \textbf{Ans.}$$

Example 6.5 *The percentage composition of coals from two different mines are given below:*

	C	H_2	O_2	*S*	*Ash*
Coal I	90	3	2	2	3
Coal II	65.5	6.65	17.5	1.8	8.55

Determine which coal is more suitable considering the heat evolved by each type of coal.

Solution. As we know that, a coal having more LCV is better.

Coal I

$$HCV = \frac{1}{100}\left[33780C + 144200\left(H - \frac{0}{8}\right) + 9280S\right]$$

$$= \frac{1}{100}\left[33780 \times 90 + 144200\left(3 - \frac{2}{8}\right) + 9280 \times 2\right]$$

$$= 30402 + 3965.5 + 185.6 = 34553 \text{ kJ/kg.}$$

$$LCV = HCV - \frac{9H}{100} \times 2460 = 34553 - \frac{9 \times 3}{100} \times 2460$$

$$= 34553 - 664 = \textbf{33889 kJ/kg}$$

Coal II

$$\therefore \quad HCV = \frac{1}{100}\left[33780C + 144200\left(H - \frac{0}{8}\right) + 9280S\right]$$

$$= \frac{1}{100}\left[33780 \times 65.5 + 144200\left(6.65 - \frac{17.5}{8}\right) + 9280 \times 1.8\right]$$

$$- 22060 + 6435 + 165 = 28660 \text{ kJ/kg}$$

$$LCV = HCV - \frac{9H}{100} \times 2460 = 28660 - \frac{9 \times 6.65}{100} \times 2460$$

$$= 28660 - 1472 = \mathbf{27188\ kJ/kg}$$

Therefore, coal I is better as it gives higher LCV **Ans.**

Example 6.6 *The percentage composition of a fuel by weight is C = 87%, H_2 = 4.5%, O_2 = 1.5% and remaining is ash. Determine the: (a) minimum weight of air required for complete combustion of one kg of fuel, and (b) volume of air per kg of fuel burnt.*

Solution. (*a*) As, minimum weight of O_2 required per kg of fuel

$$= \frac{1}{100}\left[\frac{8C}{3} + 8\left(H - \frac{0}{8}\right) + S\right]$$

$$= \frac{1}{100}\left[\frac{8 \times 87}{3} + 8\left(4.5 - \frac{1.5}{8}\right) + 0\right]$$

$$= \frac{8 \times 0.87}{3} + \frac{0.345}{1}$$

$$= 2.32 + 0.345 = \mathbf{2.665\ kJ/kg\ of\ fuel}$$

$$\therefore \quad \text{Air required} = \frac{100}{23} \times 2.665 = \mathbf{11.5\ kJ/kg\ of\ fuel.} \quad \textbf{Ans.}$$

(*b*) As, density of O_2 at NTP

$$= \frac{\text{Molecular weight}}{\text{Molar volume}} = \frac{32}{22.4} = 1.43 \text{ kg/m}^3$$

$$\therefore \text{ Volume of } O_2 \text{ required,} = \frac{\text{Weight of } O_2}{\text{Density of } O_2} = \frac{2.665}{1.43}$$

$$= 1.84 \text{ m}^3 \text{ Kg of fuel}$$

As air contains 21% O_2 by volume.

∴ Volume of air supplied/kg of coal burnt

$$= \frac{100}{21} \times 1.84 = \mathbf{8.78\ m^3} \quad \textbf{Ans.}$$

Example 6.7 *A sample of flue gases contain CO_2 = 15%, CO = 2.5%, O_2 = 1.5% and remaining is nitrogen by volume, (a) Convert this volumetric analysis into weight analysis by percentage and (b) Also find the weight of carbon in one kg of flue gas.*

Solution. (*a*) The process of converting volumetric analysis to weigth analysis is given in the table below:

Constituent	*Volume per m*^{3} *of gas (a)*	*Molecular weight (b)*	*Proportional weight in kg* $C = a \times b$	*Percentage weight* $d = \frac{C}{\Sigma C} \times 100$
CO_2	0.150	44	$0.150 \times 44 = 6.60$	$\frac{6.6}{30.46} \times 100 = 21.6\%$
CO	0.025	28	$0.025 \times 28 = 0.70$	$\frac{0.70}{30.46} \times 100 = 2.3\%$
O_2	0.015	32	$0.015 \times 32 = 0.48$	$\frac{0.48}{30.46} \times 100 = 1.57\%$
N_2	0.810	28	$0.810 \times 28 = 22.68$	$\frac{22.68}{30.46} \times 100 = 74.53\%$
Total	1.000	–	$\Sigma C = 30.46$	100%

Thus the sample of flue gases contains $CO_2 = 21.6\%$, $CO = 2.3\%$, $O_2 = 1.57\%$, $N_2 = 74.53\%$ by weight.

(*b*) As,
$$C + O_2 = CO_2 \text{ and } 2C + O_2 = 2CO$$
$$12 + 32 = 44 \text{ and } 24 + 32 = 56$$

From the above equations, we can say that one kg of CO_2 contains $\frac{12}{44} = \left(\frac{3}{11}\right)$ kg of carbon,

and one kg of CO contains $\frac{24}{56} = \left(\frac{3}{7}\right)$ kg of carbon.

∴ Amount of carbon carried by one kg of the gas

$$= 0.216 \times \frac{3}{11} + 0.023 \times \frac{3}{7} = 0.059 + 0.00985$$

$$= \mathbf{0.06885 \text{ kg of carbon/kg of flue gas.}}$$ **Ans.**

Example 6.8 *A sample of gas mixture contains* $H_2 = 14\%$, $CH_4 = 2\%$, $CO = 24\%$, $CO_2 = 5\%$, $O_2 = 2\%$ *and* $N_2 = 53\%$ *by volume. If 50% excess air is supplied, determine the volume of air supplied per* m^3 *of gas burnt.*

Solution. As, volume of oxygen required for complete combustion of 1 m^3 of gas,

$$= \frac{1}{100}\left[\frac{H}{2} + \frac{CO}{2} + 2CH_4\right] = \frac{1}{100}\left[\frac{14}{2} + \frac{24}{2} + 2 \times 2\right]$$

$$0.07 + 0.12 + 0.04 = 0.23 \text{ m}^3$$

∴ Volume of O_2 to be supplied

$$= 0.23 - O_2 \text{ (already available)}$$
$$= 0.23 - 0.02 = 0.21 \text{ m}^3/\text{m}^3 \text{ of gas}$$

∴ Volume of air to be supplied for complete combustion

$$= \frac{100}{21} \times 0.21 = 1.00 \text{ m}^3/\text{m}^3 \text{ of gas.}$$

∴ Actual air supplied 1.00×1.5 = **1.5 m³/m³ of gas Ans.**

Example 6.9 *Percentage composition by weight of a sample of coal was found to be as follows:*

$$C = 24\%; \quad H_2 = 5\%; \quad O_2 = 8\%; \quad Ash = 63\%$$

It was also observed that the dry flue gas had the following composition by volume:

$$CO_2 = 10\%; \quad CO = 2\%; \quad O_2 = 13\%; \quad N_2 = 75\%$$

Determine the following:

(a) Minimum weight of air required for complete combustion of 1 kg of coal:

(b) Weight of excess air required per kg of coal.

Solution. (*a*)

Substance	*Weight per kg of coal*	*Oxygen required per kg of substance*	*O_2 requied per kg of coal*
C	0.24	8/3	0.64
H_2	0.05	8	0.40
O_2	0.08	—	0.08
Ash	0.63	—	—

Total oxygen required = 0.64 + 0.40 – 0.08 = 0.96 kg.

∴ Minimum weight of air required per kg of coal

$$= \frac{0.96 \times 100}{23} = \mathbf{4.18\ kg} \quad \textbf{Ans.}$$

(*b*) The composition of dry flue gas is given by volume. It can be converted into composition by weight as follows:

Gas	*Volume per m^2 of flue gas (V)*	*Molecular weight (M)*	*Proportional weight (W)* $W = V \times M$	*Weight per kg of flue gas (W)* $w = \frac{W}{\Sigma W}$
CO_2	0.10	44	4.40	0.146
CO	0.02	28	0.56	0.018
O_2	0.13	32	4.16	0.140
N_2	0.75	28	21.00	0.696
			$\Sigma W = 30.12$	

Weight of carbon per kg of coal = 0.24 kg

Weight of carbon per kg of the flue gas = Weight of carbon in 0.146 kg of CO_2 + Weight of carbon in 0.018 kg of CO

$$= \frac{3}{11} \times 0.146 + \frac{3}{7} \times 0.018 \text{ kg}$$

$$= 0.04 + 0.008 = 0.048 \text{ kg}$$

∴ Weight of flue gas per kg of coal $= \dfrac{0.24}{0.048} = 5$ kg

Weight of excess oxygen per kg of flue gas = Amount of oxygen in flue gas – Oxygen required to burn CO.

$$= 0.140 - \frac{4}{7} \times 0.018 = 0.13 \text{ kg.}$$

∴ Weight of excess oxygen per kg of coal = 5 × 0.13 = 0.65 kg.

∴ Weight of excess air per kg of coal = $\dfrac{0.65 \times 100}{23}$ = **2.83 kg Ans.**

QUESTIONS

6.1 What are the various types of fossil fuels? Explain.

6.2 What is meant by composition of fuel? Give percentage composition of some fuels.

6.3 Explain ultimate analysis to find various constituents of a solid fuel.

6.4 What are the advantages of liquid fuels and gaseous fuels?

6.5 What are the basic constituents of fuels? Mention the Higher Colorific Value of main constituents.

6.6 State the difference between ultimate and proximate analysis of coal.

6.7 Discuss different types of coals giving their properties and fields of applications.

6.8 List out the merits and demerits of liquid fuels over solid fuels.

6.9 Define calorific value of fuel. Differentiate HCV and LCV of a fuel and state which is used in practical calculations and why?

6.10 What are the basic constituents of fuels? Which of the constituent supplies a large portion of heat?

6.11 Define calorific value of heat. Differentiate between HCV and LCV of a fuel and state which is used in practical calculations and why?

6.12 What is an emulsion? How does an emulsion of water in heavy oil promote combustion?

6.13 What is combustion? What are the basic conditions to be fulfilled to burn a fuel efficiently?

6.14 Why is excess air required to be supplied for combustion? What is dilution coefficient?

6.15 Write short notes on:

(*a*) Products of combustion.

(*b*) Chemical reactions in combustion process of fuels.

6.16 Discuss the minimum requirement of air for the complete combustion of fuel.

THERMAL POWER PLANTS: 1—STEAM BOILERS (GENERATORS)

Chapter 7

Learning Objectives

After studying this chapter, the students will learn:

- General layout of steam power station.
- Construction and working of coal fired plants.
- Construction and working of different types of steam boilers including high pressure boilers and their mountings and accessories.
- Draught system for seam boilers and calculation of chimney height.
- Boiler performance.
- Inspection and maintenance of boiler.

7.1 INTRODUCTION TO THERMAL POWER PLANTS

The development of thermal, nuclear and hydel plants becomes a neccessity for assuming sound supply of power in the country. The mix of three depends upon the availability of resources in the particular area, but it is always economical and desirable to develop all the sources of power for the progress of the country.

The thermal power plant generates the power in two major phases, the first convert the heat energy of the coal or oil into steam and the second to convert the energy in the steam into electricity. The first process is carried out by the boiler unit and the second by the turbine-generator unit. Thus, power plant is an assembly of equipment that produces and delivers mechanical and electrical energy. Mechanical equipment includes turbines, condensers, heat exchangers, feed water pumps etc. Electrical equipment of a power station includes generators, transformers, switch-gears, exciters, circuit-breakers etc.

Steam is an important medium of producing mechanical energy. Steam has the advantage that it is raised from water which is available in abundance, and it does not react much with the materials of the equipment of power plant and is stable at the temperature required in the plant. Steam is used to drive steam engines, steam turbines etc. Steam power station is most suitable where fossil fuel is available in abundance. Out of the total power developed in India, about 60% is thermal power.

No two power plants are similar, therefore different problems are faced in each new power station. The differences in the problems are responsible for the lack of uniformity of treatment of design and layout. Power station design, constrution and operation are firmly bound together that each should go hand in hand in all problems. Various items of plant may be arranged in relation to one another in many ways, and it is necessary to know the behaviour of such plant under normal and abnormal condition of operation.

7.2 GENERAL LAYOUT OF STEAM POWER STATION

Power plants may be arranged in many ways, but the general scheme of their layout has centered in the arrangement of the major two equipment, *viz.*, boilers and primemovers. But the general layout of all the thermal power plants consist of mainly 4 circuits (paths) as mentioned hereunder.

7.2.1 Coal and Ash Circuit (Fuel Path)

In this circuit, the coal from storage is fed to the boiler through coal handling equipment for the generation of steam. Ash produced due to the combustion of coal is removed to ash storage through ash-handling system. The fuel path is a combination of elements in which solid fuel is transported, crushed, ground, and delivered to the boiler furnace for combustion. The fuel path comprises the crushing equipment, conveyers, crushed-fuel bunker, grinding mill, and pulverized fuel duct leading to the furnace. Fuel is transferred by conveyors upto the crushed fuel bunkers.

7.2.2 Air and Gas Circuit

In this circuit, air is supplied to the combustion chamber of the boiler either through F.D. (Forced Draft) or I.D. (Induced Draft) fan or by using both. Thus dust from the air is removed before supplying to the combustion chamber. The exhaust gas carrying sufficient quantity of heat and ash are passed through the air-heater where the exhaust heat of the gases is given to the air and then it is passed through the dust collectors where most of the dust is removed before exhausting the gases to the atmosphere through chimney.

In this circuit *air path* includes a combination of elements for suction of atmospheric (cold) air, its preheating, transport and supply into the furnace. The air path comprises a cold air duct, air heater, hot-air duct, and burners.

This circuit also has a *gas path,* in which the combustion products flow from the furnace into the atmosphere. The gas path begins in the boiler furnace and passes through the superheaters, economizers, air-heaters, ash collector and chimney.

Since air and gas paths are connected in the series, it is also called *gas-air path.* The transition from one to the other takes place in the boiler furnace space. In this circuit, for proper flow, air is transported by blowers into the furnace at a pressure higher than the atmospheric, while all gas ducts are at a pressure below the atmospheric. This system is known as balanced draft.

7.2.3 Feed Water and Steam Circuit

The steam generated in the boiler is fed to the steam prime-mover (steam engine or steam turbine) to develop the power. The steam coming out of the prime-mover is condensed in the condenser and then fed to the boiler again, with the help of pump. The condensate is heated in the feed heaters using the steam tapped from different points of the turbine. Since some of the steam and water is lost passing through different components, feed water is supplied from external source to compensate this loss. The feed water supplied from external source is passed through the purifying plant to reduce the dissolved salts so as to avoid scaling of the boiler tubes.

7.2.4 Cooling Water Circuit

Cooling water is taken from the upper side of the river, it is passed through the condenser and heated water is discharged to the lower side of the river. Such system of cooling water supply (known as open system) is possible if adequate cooling water is available through out the year. When the adequate water is not available, then the water coming out from the condenser is cooled either in cooling pond or cooling tower. In this case quantity of water required is affected by partly evaporating the water. To compensate the evaporation loss, the water from the river is continuously supplied. When the cooling water coming out of the condenser is cooled again and supplied to the condenser, the system is known as closed system.

Main Components (Elements) of a Thermal Power Plant (TPP)

Different systems and components used in thermal power plants are:

1. Coal handling system
2. Drought
3. Ash and dust handling system
4. High pressure boilers
5. Prime-movers
6. Condensers and cooling towers
7. Feed water purification plant
8. Water pumping system
9. Thermal efficiency enhancing components like, economiser, superheater, feed heater etc.

Planning for Layout of TPP

Following factors are considered while planning the layout of steam-electric power station:

1. Arrangements for taking the coal, oil or gas to the boilers via bunkers (while using coal).
2. Disposal of combustion products from boilers, also, the disposal of ash, Flue dust and gases.
3. Carrying steam generated in the boilers, to the turbines through shortest possible route.
4. Delivery of cooling water to and discharge from the turbine to condensing plant to recool. Where recooling is necessary, spray ponds and/or cooling towers will be required.
5. Arrangements for delivering the electrical energy to the switch boards.
6. Location of control room.
7. All similar items such as bunker bays, boilers, turbines, generators, transformers and other mechanical and electrical components are arranged in a systematic manner e.g., in parallel or in rows.
8. Chimneys are erected independently of the station building with one chimney serving two or more boilers.
9. Main flue and draft fans should be located behind the boiler house.
10. Selection, use and location of fuel handling equipment.
11. Transformers should be located in front of the turbine house.

Flow-sheet of a TPP

Flow sheet of a thermal power station is shown in Fig. 7.1. As already discussed, a thermal power plant using steam as working medium works on the Rankine cycle and has two major items namely boiler and turbine. Steam is generated in the boiler, expanded in the turbine, condensed in the condenser and then fed into the boiler. A number of auxiliary units are provided with the boiler and turbine.

A boiler uses coal or oil as the fuel, for which a storage is required in the plant. A coal or fuel handling unit is necessary to maintain regular supply of fuel to the boiler. A huge quantity of coal is required for large TPPs and the cool fired boilers produce about 10 to 15% of the total weight of the coal fired. A TPP of 400 MW capacity requires about 5×10^6 to 6×10^6 kg of coal per day and produces 0.5×10^6 to 0.6×10^6 kg of ash per day. For this an effective and reliable ash handling system is required to remove ash from furnace and transfer it to ash storage.

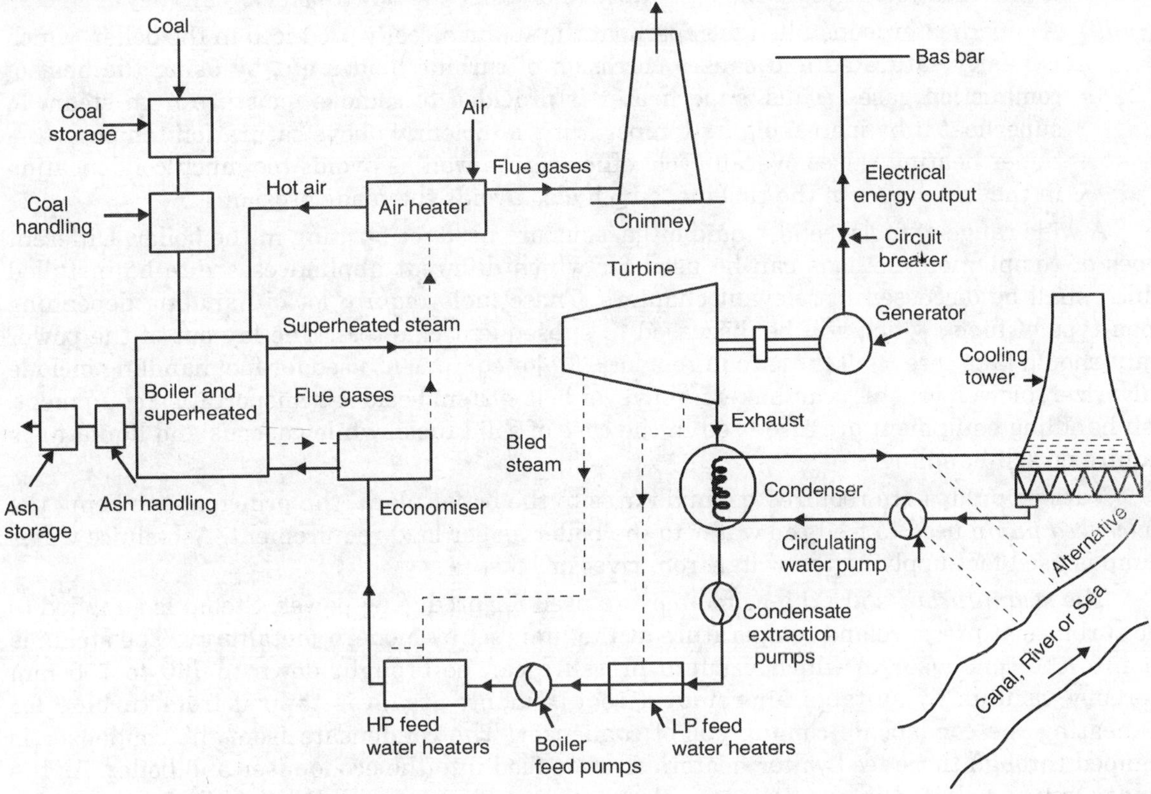

Fig. 7.1 Thermal power station – flow sheet.

Air is supplied to the furnace through the forced/induced draft system. Fresh air is heated in the air heaters to make use of the heat of flue gases. The steam generated in the boiler is expanded in turbine to produce power. Exhaust steam is passed to the condenser, where it is condensed and then fed to the boiler with the help of a pump, by passing through high and low pressure heaters, where condensate is heated using the steam tapped from different points of the turbine (Bled steam).

When steam and water pass through different components of the system, some amount is lost, which is compensated by supplying the feed water from external source. This feed water is purified so as to avoid scaling of boiler tubes due to dissolved salts in water.

Water is used in the condenser for condensing the steam. While the water is circulated through the condenser, it gets heated, which is then cooled in the cooling tower and recirculated. This system of cooling is known as *closed system.* In *open system,* which is used when plenty of water is available in the river, the cooling water is taken from the upper side of the river and discharged to the lower side of the river.

For heat economy and to increase thermal efficiency of the plant, various modifications are made in the Rankine cycle. Some of the important steps taken are:

(*i*) Putting an *economizer,* having a horizontal sets of tubes, in the gas path after gases leave the boiler. Thus, economizer transfers the heat of waste furnace gases to water being fed to the boiler.

(*ii*) Tubular or regenerative type **air-heater** is added to further cool the furnace gases by transferring some of the heat of the gases to the combustion air. Since economizer and air heater add resistance to the gas flow through the boiler, an *induced draft fan* is provided to overcome the resistance and assist the flow to stack (chimney).

(*iii*) A superheater is installed to superheat the steam already produced in the boiler, which is nearly saturated and causes corrosion of turbine blades etc, by using the heat of combustion gases. Thus, superheater is provided to remove moisture from steam to superheat it by increasing its temperature sufficiently above saturation temperature. Super heating raises overall cycle efficiency as well as avoids too much condensation in the last stages of the turbine, which also avoids the blade erosion.

A wide range of fuels, solid, liquid or gaseous are used for burning in the boiler. Different types of combustion methods can be used for which different appliances are to be installed which shall be discussed in relevant chapters. These fuels require lot of handling depending upon type of fuels, which will be discussed in subsequent chapters. The lay out of the power plant should take care of all these requirements. Major equipment used for fuel handling include pulverizer, blower, crusher, carshakers, conveyer belt system besides transportation equipment. Ash handling equipment are also used in the case of solid fuels, while gaseous and liquid fuels require no ash-handling.

Various pumps are required as auxiliaries to the boiler plant, the principal one being the *boiler feed pump* used to feed the water to the boiler as per load requirement. Ash-sluice water pump is used for supplying the water from river or lake.

The *steam-turbines* are the prime-movers used for producing power. Steam is supplied to the turbines at pressure and temperature at the limit set by modern metallurgy. The steam is let into the condenser in which vacuum pressure can be brought down to 730 to 750 mm mercury vacuum by suitable air removal requirement. Steam is tapped from turbine for preheating the condensate coming out of condenser. The condensate from the condenser is pumped through these feed water heaters so as to feed into the economiser and boiler. As the condensate is collected in the bottom of the condenser, it is removed, generally by the steam jet air pump. One of the feed water heaters is usually of the de-aerating type so as to eliminate oxygen and dissolved gases from the feed water.

The turbine drives a generator which produces electrical energy at a suitable voltage. This energy is stepped up in the voltage by a transformer and sent out over an electrical transmission system. Auxiliary transformers step down transmission or generator voltage to auxiliary voltage for use in driving auxiliary motors in the power plant. A substation receives power from the generators on one or more buses, steps up the voltage and sends out power. The substation also houses protective equipment, like lighting arresters. Considerable station equipments are also required for special purposes, some of them are: dust collectors to remove the flyash from the flue gases; special soot blowers to clean the steam generating surfaces of the boilers, air compressors; machine shop equipment to carry out repair and maintenance of the plant, ventilation system, control room etc.

7.3 WORKING OF COAL-FIRED THERMAL POWER PLANT

The coal is delivered to the facility by railway wagons, barges or trucks/dumpers. The coal handling system unloads the coal, stocks, crushes and conveys it to storage silos. Coal from the silos is then pulverized to a fine powder and blown into the steam generator (boiler), where it is mixed with air and combusted to release energy for the generation of steam.

The steam generator produces steam at the desired pressure and temperature, superheats and reheats as it proceeds through the cycle.

The steam generated is passed through steam turbine where part of its thermal energy is converted into mechanical energy which is further used for generating electrical power. Thus steam turbine generator converts the thermal energy (enthalpy rise) of the superheated and reheated steam to electrical energy. Steam exhausted from the turbine is condensed to liquid in the condenser. The condensate pumps feed the water through the low pressure regenerative feed water heaters to de-aerator. Boiler feed pumps move the de-aerated liquid through the high-pressure feed water heaters back to the boiler.

The forced draft fans supply combustion air to the boiler and the primary air fans transport pulverized coal into the burners. Induced draft fans remove the flue gases from the furnace and exhaust them through the stack (chimney) into the atmosphere.

The cooling water for the condenser is supplied by the circulating water system, which takes the heat removed from the condensers and rejects it to the cooling tower or some other heat sink such as cooling lake, river or sea.

Combustion gases coming out of the boiler are treated for removal of flyash particulates, sulphur oxides, nitrogen oxides before release through the exhaust stack (chimney).

7.4 INTRODUCTION TO STEAM BOILERS

Steam boilers (also known as steam generators) are used to convert water into steam at constant pressure. In these the energy liberated by the combustion of fuel is used to increase the enthalpy of the fluid passing through the boiler. Some of the energy liberated tends to leave with the flue gases through chimney, an attempt is made to regain this energy to maximum possible extent by installing economiser, and air preheater. The steam generated by the boilers, is mostly used for steam engines or steam turbines in power plants. The steam is also used in process industries.

Constructional features of boilers vary from manufacturer to manufacturer, however these can be classified as Fire Tube Boiler, and Water Tube Boiler. In fire tube boiler, hot gases pass through the tubes surrounded by water, while in water tube boilers, water is made to pass through large number of tubes over which hot gases flow. Capacity of boiler is indicated by the maximum amount of steam generated in kilogram per hour at a particular pressure.

For selecting a boiler following factors are considered:

(*i*) The quantity of steam required and at what pressure.
(*ii*) Floor area available.
(*iii*) Probable load factor.
(*iv*) Initial cost, operating cost and maintenance cost.
(*v*) Type of Fuel and water available.

A good boiler is that boiler which has following qualities:

(*i*) It should be reliable and should be able to work satisfactorily under variable load conditions.
(*ii*) The tubes should not choke and be strong enough to meet corrosion.
(*iii*) It should be equipped with all safety devices and accessories.
(*iv*) It should be economical from fuel consumption point of view low initial cost, and low operation and maintenance cost.
(*v*) It should be light and occupy less space.
(*vi*) It should be easy to maintain, repair or inspect.

Fire Tube v/s Water Tube Boilers

	Fire tube boilers	*Water tube boilers*
1.	Slow evaporation due to large amount of water is heated of a time.	Quicker evaporation.
2.	Capacity is small.	High capacity, hence used by power plants.
3.	Suitable for low pressures.	Suitable for high pressure.
4.	Occupies larger floor area for a given power.	Occupies less floor area for same power.
5.	Bursting may cause serious damages, but has less chances of bursting.	Bursting may cause less damages, but has more chances of bursting.
6.	Water treatment is is not very necessary.	Water treatment is necessary, since small deposit in the tube may cause overheating resulting in bursting.

7.5 FIRE TUBE BOILERS

Commonly used fire tube boilers are explained here under:

1. Cochran Boiler

It is a fire tube vertical multi-tubular boiler (see Fig 7.2). The fuel (coal) is fed through a fire gate in a fire box, over a circular fire grate. Fire box has a hemispherical dome to help in full combustion. The flue gases enter the combustion chamber from where they pass through the number of horizontal tube to the chimney *via* a smoke box. The water is thus heated while coming into contact with outer surface of the fire tubes, and forms into steam, which is then collected into the hemispherical dome of the boiler.

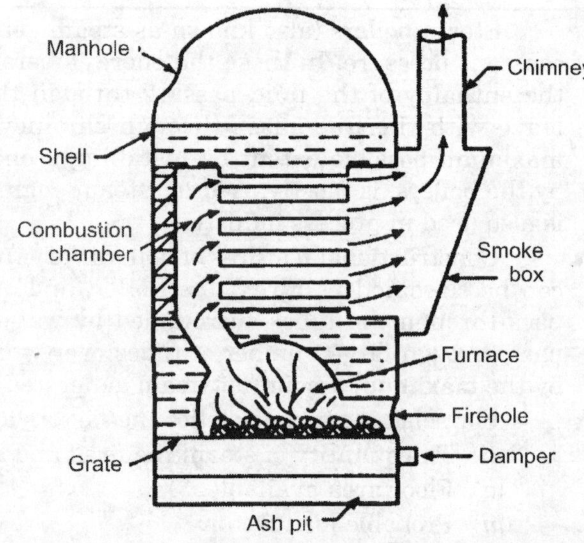

Fig. 7.2 Cochran Boiler.

The boiler is mounted with (*i*) water level indicator (*ii*) steam pressure gauge (*iii*) safety valve (*iv*) stop valve (*v*) Feed check valve. Fuse plug is provided at the top of the fire box, to extinguish the fire if temperature rises beyond the permissible limit. A man hole is provided in the top steam dome to enable a man to enter the boiler for cleaning and maintenance.

Working pressure of this type of boiler is generally upto 7 kg/cm^2 and hence are not common, except where small quantity of steam is required.

2. Lancashire Boiler

This is a fire tube boiler having internally fired two horizontal fire tubes. Each of the flue tube has a fire grate where fuel is fed through fire holes. The water shell is having water and is placed in the brickwork setting the hollow brickwork works as the flue path, which ensures complete combustion of fuel.

Flue gases traverse the flue tubes from front to the back, where it dips down and pass under the boiler shell back to the front. In the front these gases are divided into paths and pass by the sides of the shell and traverse from front to the back and finally escape through the chimney. This path of flue gases enables the utilisation of heat available in the flue gases to the maximum extent.

These boilers have a normal working pressures upto 1.5 MPa, and are used for process steam requirement like sugar or textile industries.

3. Locomotive Boiler

These are horizontal, multi-tubular internal fired boiler and can meet sudden fluctuations in demands of steam (see Fig. 7.3). It has an inclined fire grate over which coal is fed through the fire hole. An ashpan is provided below the grate from where air supply is controlled by means of a damper. In these boilers, an artificial draught is produced by means of exhaust steam passing through a blast pipe provided under the chimney.

The barrel of boiler contains numerous horizontal tubes through which flue gases pass from the fire box to smoke box. One steam dome called 'hump' is provided on the top of the barrel, in which a stop valve is placed and is operated from the boiler cabin. This steam passes from the stop valve to the superheater.

Water level indicator and pressure gauges are provided in the driver's cabin. A fuse plug is provided at the crown of the fire box, which melts when water falls beyond danger level, and water rushes on to the fire grate to extinguish the fire.

Locomotives are mostly used in railways.

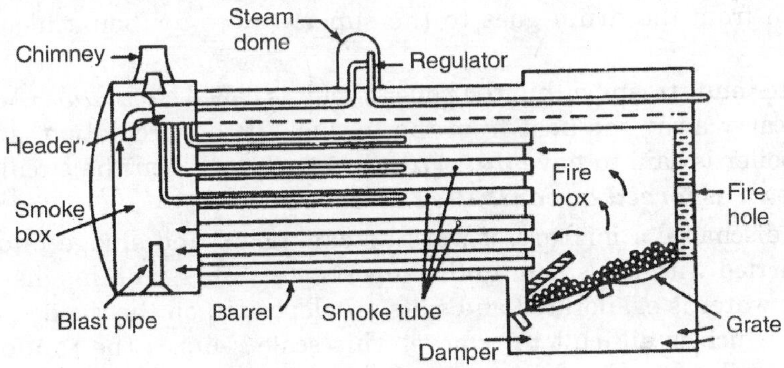

Fig. 7.3 Locomotive Boiler.

7.6 WATER TUBE BOILERS

Water tube boilers are of several types, out of which Babcock and wilcox boiler are most common and are being described hereunder.

Babcock and Wilcox Boiler

It is horizontal, multitubular (inclined) boiler and is used in power houses (see Fig. 7.4). It has large number of water tubes, which are inclined to the horizontal at an angle of 15° and connected to the shell by means of uptake and down take headers.

The water is heated in the inclined tubes while it is traversing in them, and steam is generated which is collected in the shell, known as steam and water drum. The furnace is located below the tubes and hot gases move upwards, then downwards, and finally upwards before escaping through the chimney. Coal is fed at one end by hoppers over the moving chain, and by the time it reaches the other end the coal is completely burnt.

Steam thus generated and collected in the shell is led to the superheater.

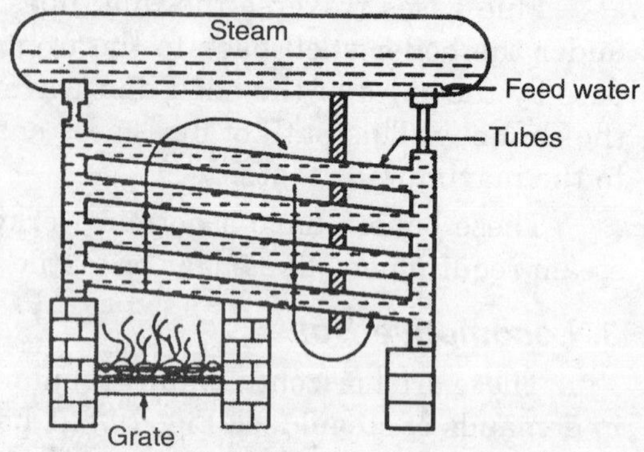

Fig. 7.4 Principle of Babcock and Wilcox water tube boiler.

The water tube boilers are used in the large capacities (usually pressure above 1.7 MPa and steam generation more than 120 kg/sec). Positive fluid circulation assures good heat transfer and enables building large capacity units. Tubes and drums are smaller in size as compared to fire-tube boilers. This enables high steam pressures feasible. Tubes, external to the drums, can be placed for best heat absorption.

In a water-tube boiler, feed water is heated in three main kinds of heat exchangers *viz,* **economizer, evaporator and super-heater.** Feedwater enters the economiser where it is heated by the outgoing flue gases till it is saturated liquid at that pressure and then it is fed to the steam drum of the boiler. Saturated water falls through the downcomer into the bottom header and moves up through the riser where water is partially boiled back into the drum. Saturated steam from the drum goes to the superheaters for being heated to the desired temperature.

Flow of water and stream within the boiler circuit is called *circulation.* Adequate circulation is necessary to carry away the heat from the furnace. If the circulation is caused by density difference, the boiler is said to have ***natural circulation.*** When the circulation is caused by a pump, it is known as ***forced or controlled circulation.***

The risers discharge a mixture of water, steam, foam and sludge into the steam drum. Any moisture carried with the steam to the superheater tubes contains dissolved salts. In the superheater, the water is evaporated and salts are deposited on the inside surface of the tubes to form a scale, which is difficult to remove. This scale reduces the rate of heat absorption, and ultimately leading to the failure of superheater tubes by overheating and rupture. Therefore, utmost care should be taken so that no damage is done to it by the excessive carryover of moisture with steam. The water should therefore be adequately treated prior to feeding it to the boiler as make up.

Mechanical separators, like baffles, screens and cyclones are housed inside the drum for seperation of steam-water mixture. Baffle plates change or reverse the steam flow direction to assist gravity separation, act as primary separators. Screens made of wire mesh act as secondary separators where the individual wires attract and intercept the fine droplets. The accumulating water drops then fall by gravity back to the main water body.

A boiler drum is said to be "priming", if there is too much moisture carry over because of a high water level or a high steaming rate. Erratic feedwater control and rapid changes in steaming rate can indude priming. Therefore, a certain water level should be maintained to give adequate steam space and reduce priming.

Foaming is a condition resulting from the formation of bubbles on the water surface, and is responsible for the delivery of impure steam. Concentrated solid precipitates at drum bottom are removed through blowdown, which may be either intermittent or continuous. Blowing down of boiler water is the removal of some of the concentrated saline water and its replacement with relatively dilute feedwater. The amount of blow down is governed by the allowable solids concentration and the purity of feed water.

7.7 HIGH PRESSURE BOILERS

Special features of these boilers are: generating tubes, water walls, steam generating drums, forced circulation, small tubes, lack of drums, pressurised combustion, multiple tube circuits, extreme compactness and low weight. It is a common practice to use high pressure and temperature steam in power plants to increase the efficiency of the plant and to reduce the cost of electricity production. The modern high pressure boilers used for power generation are of steam capacities 8 to 180 kg/s and above with a pressure upto 16 MPa and maximum steam temperature of about 550°C.

By using high pressure boilers, low grade fuels can be burnt easily, a saving in costly construction material can be achieved and a higher efficiency is possible. High pressure boilers are water-tube boilers, use pulverised coal firing and in such boilers furnace walls are water cooled and water tubes form the walls of the furnace. They use slag type furnace.

7.7.1 Unique Features of High Pressure Boilers

The high pressure boilers possess special features which make them more advantageous. The unique features are as follows:

1. Method of Water Circulation. In the boilers water circulation is maintained with the help of pump which forces the water through the boiler plant. However, sub-critical boilers may use natural circulation.

Circulation by a pump ;

(*i*) ensure positive circulation

(*ii*) increases evaporative capacity of the boiler, and

(*iii*) requires less number of drums.

2. Type of Tubing. These boilers generally use several sets of tubings through which water passes. This arrangement helps:

(*i*) to reduce the pressure loss

(*ii*) to achieve better control over the quality of the steam.

3. Improved Method of Heating. In high pressure boilers following improved methods of heating may be used to increase the heat transfer rate.

(*i*) To save heat by evaporation of water above critical pressure of the steam.

(*ii*) To heat water by mixing the superheated steam. It gives high heat transfer coefficient.

(*iii*) To use high water velocity inside the tubes and to increase gas velocity above sonic velocity. This will increase overall heat transfer coefficient.

Advantages

The various features (advantages) of such boilers which make them useful are as follows:

(*i*) In high pressure boiler pumps are used to maintain forced circulation of water through the tubes of the boiler. This ensures positive circulation of water and increases evaporative capacity of the boiler and less number of steam drums will be required.

(*iii*) The heat of combustion is utilised more efficiently by the use of small diameter tube in large number and in multiple circuit.

(*iii*) Pressurised combustion is used which increases rate of firing of fuel thus increasing the rate of heat release.

(*iv*) Due to compactness less floor space is required.

(*v*) It is economical to use high pressure and high temperature steam produced by such boilers.

(*vi*) All the parts are uniformly heated, therefore, the danger of over-heating is reduced.

(*vii*) The efficiency of the plant is increased to about 40%.

(*viii*) The tendency of scale formation is eliminated due to high velocity of water through the tubes.

7.7.2 Types of High Pressure Boilers

The various types of high pressure boilers commonly used are as follows:

(*i*) La Mont Boiler (*ii*) Loeffler Boiler

(*iii*) Benson Boiler (*iv*) Velox Boiler

1. La Mont Boiler. Fig. 7.5 shows the schematic arrangement of the boiler. Feed water from hot well is supplied to the separating drum (boiler) through the economiser. Water circulating pump maintains the forced circulation of water through the tubes in evaporation section. The steam produced flows to the drum. Dry steam from the top of drum is then passed through a convection superheater and the superheated steam flows to the turbine. This type of boiler has a working pressure of 17 MPa.

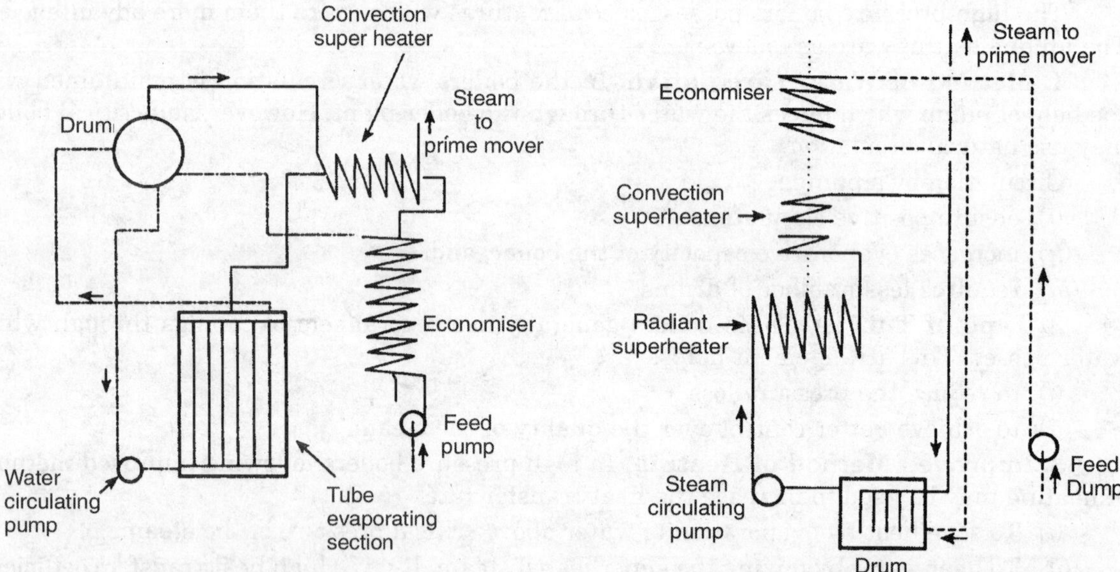

Fig. 7.5 La Mont Boiler **Fig. 7.6** Loeffler boiler

2. Loeffler Boiler. In this boiler (Fig. 7.6) the drum is placed away from the furnace. Feed water passing through the economiser moves to the drum through which superheated steam is also passing. The superheated steam converts the water into saturated steam. A steam circulating pump is used to circulate steam through radiant and convective superheaters, where steam is heated to the required temperature. A part of this steam flows to the prime mover and the remaining steam flows to the drum. Such boilers have a working pressure of 13.5 MPa.

3. Benson Boiler. The schematic arrangement of the boiler is shown in Fig. 7.7. This boiler does not use any drum. The feed water after circulating through the economiser tubes flows through the radiant parallel tube section (*T*) to evaporate partly. The steam water mixture produced then moves to the transit section (*TS*) where this mixture is converted into steam. The steam is now passed through the convection superheater (*CS*). Such boilers have a working pressure of 23 MPa.

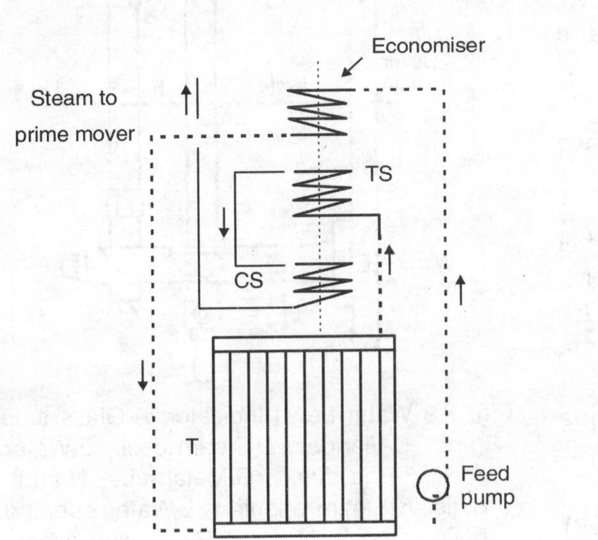

Fig. 7.7 Benson boiler.

Steam to
prime mover
Convection
super heater
Steam
separating
section
WC
Feed
pump
AC
GT
Economiser

Fig. 7.8 Velox boiler.

4. Velox Boiler. In this boiler (Fig. 7.8) pressurised combustion is used. The axial flow compressor (*AC*) driven by gas turbine (*GT*) raises the incoming air from atmospheric pressure to furnace pressure. The combustion gases after heating the water and steam flow through the gas turbine to the atmosphere. The feed water after passing through the economiser is pumped by a water circulating pump (*WC*) to the tube evaporating section (*T*). Steam separated in steam separating section flows to the super-heater.

This boiler can generate a pressure of about 8.4 MPa.

7.7.3 Modern Trends in Generating Steam

Some of the special features of generating steam using high pressure boilers are as follows:

(*i*) Use of forced circulation of water inside the boiler.

(*ii*) Use of small tubes in large number in order to increase the surface area of heat receiving surface.

(*iii*) Treating the feed water thoroughly before use.

(*iv*) Use of multiple tubes circuits.

(*v*) Use of pressurised combustion.

(*v*) Use of superheaters.

The process of steam generation can be divided into three stages namely:

(*i*) heating of water upto boiling point

(*ii*) evaporation of boiling water and its conversion into dry saturated steam,

(*iii*) transformation of dry saturated steam into superheated steam.

7.8 BOILER MOUNTINGS

Boilers are mounted with various equipments and instruments for safe and satisfactory working of a boiler. Commonly used boiler mountings are: water level indicators ; steam pressure gauge ; safety valve ; stop valve, feed check valve ; fusible plug, low water and high pressure alarm etc.

1. Water Level Indicator.

It is a water gauge, the function of which is to indicate correctly the level of water in the boiler.

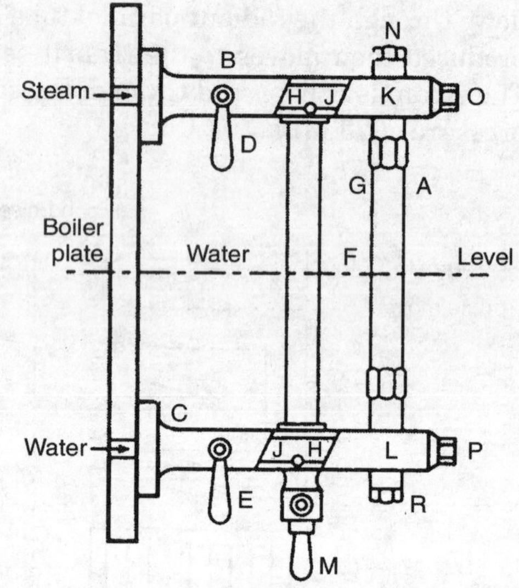

Fig. 7.9 Water Level Indicator. *A*-Glass tube, *B* and *C*-Flanges, *D*-Steam cock, *E*-Water cock, *F*-Red mark, *G*-Metal tube, *H* and *J*-Balls, *K*-Steam opening, *L*-Water opening, *M*-Drain cock; *N, O, P, R*-Screwed caps.

Referring Fig. 7.9 to check the water level in the boiler follow the procedure given under:

1. Close steam and water cocks (*D* and *E*), thereby cutting off the supply of steam and water to the indicator tube.
2. Open drain cock *M*, thereby draining out the steam and water available inside the tubes.
3. Blow through a number of times by opening and closing the steam cock *D*.
4. Close the cock *D* and blow through by opening and closing the water cock *E*.
5. Close the cock *E,* and then the drain cock *M.*
6. Now open the water cock *E* first and the steam cock *D* next. Thereby rising the water level in the glass tube equal to the level of water level inside the boiler.
7. Compare the water level with the safe water level indicated by red mark *F* on the glass tube.

The balls *H* and *J* remain in their normal position as shown in the Fig. 7.9. But when the glass tube breaks the rush of steam in the upper casting throws the ball *H* to fit in the opening *K,* thereby preventing the leakage of steam into the boiler room. Similarly the rush of water in the lower casting throws the ball *J* to fit in the opening *L,* thus preventing the leakage of water.

2. Pressure Gauge.

It is an instrument by means of which the pressure exerted inside a boiler can be measured.

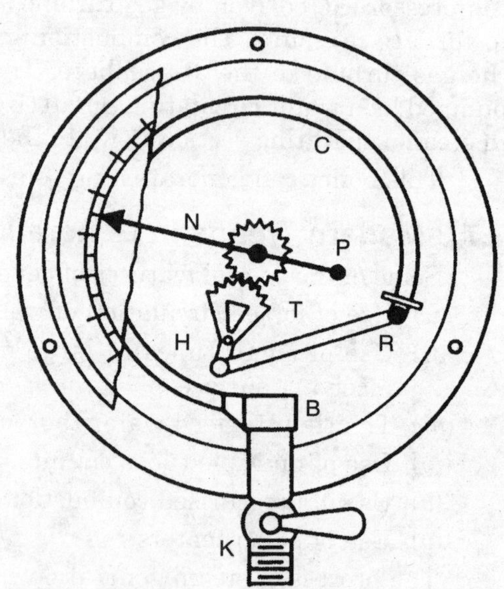

Fig. 7.10 Bourdon Pressure Gauge. *C*-Sprint tube, *B*-Block, *R*-Rod, *H*-Hinge point, *T*-Toothed sector, *P*-Pinion, *N*-Pointer, *K*-Cock.

Referring Fig. 7.10 the gauge consists of a spring tube which is of elliptical cross-section and its shape is like the letter *C*. One end of the spring tube is connected with the blow *B* and the other end is closed. The block *B* is connected with a siphon tube which is full of water. When the cock *K* is turned to the opened position then the steam pressure forces the water from the siphon into the spring tube. Thereby changing the tube in circular section and straightening its length.

The movement of the closed end of the tube is transmitted and magnified by the rod *R*, which is then further transmitted to a toothed sector *T*. This toothed sector is hinged at the point *H* and it is meshed with a pinion *P* fixed to the spindle of the pointer *N*. The pointer may move over the dial indicating the pressure.

3. Safety Valves.

The function of a safety valve is to discharge the steam from the boiler automatically when the steam pressure exceeds the normal limit. These are generally mounted on the top of the boiler shell. They are of the following types.

(a) *Dead Weight Safety Valve.* For working, referring Fig. 7.11 when the steam pressure inside the boiler is greater than the working pressure then the steam passes through the pipe *P* and lifts the valve *V* with its weight *W*. Thus passing out the excess steam and decreasing the boiler pressure. When the pressure becomes normal then the valve *V* again rests on its seat '*S*' thereby preventing the flow of steam through this unit.

Fig. 7.11 Dead weight safety valve. *A*-valve, *S*-valve seat, *W*-weight, *D*-carrier, *P*-pipe.

(b) *Lever Safety Valve.* For working, referring Fig. 7.12. When the steam pressure inside the boiler becomes greater than the normal working pressure, the valve *V* is lifted from its seat *S*, with the sturt *R*, lever *L* and the weight *W*. Thus the excess steam passes out through the passage between the valve and the valve seat, thereby decreasing the boiler pressure. The weight *W* can slide and adjusted on the level *L*, the position depends upon the safe pressure desired.

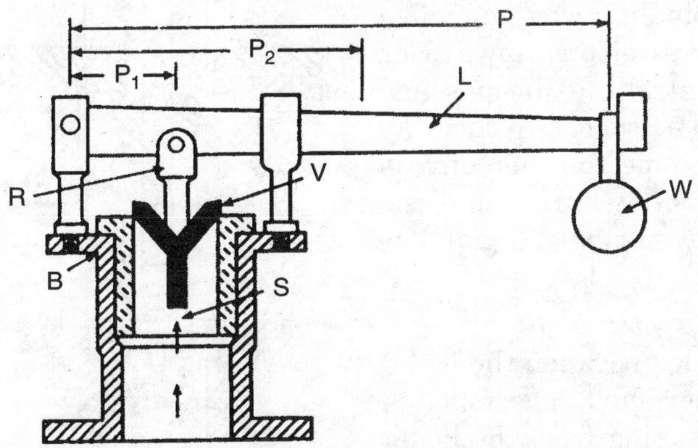

Fig. 7.12 Lever safety valve.
L–Lever, *V*–Valve, *S*–Valve seat, *W*–Weight, *R*–Sturt, *B*–Block.

(c) Spring Loaded Safety Valve. Referring Fig. 7.13, the unit consists of two valves *V*, each of which is placed over a valve seat *S*, fixed over a branch pipe are further connected to the boiler shell by means of a common block. The lever *L* has two pivots *P*, each of which is placed over each valve. The valves *V* are pressed on their seats *S*, against the steam pressure by means of a helical spring *R* provided between the lever *L* and the block.

When the boiler, pressure exceeds its normal working limit then the valves are lifted against the spring thereby passing out the excessive steam and decreasing the boiler pressure. The link *K is* provided to prevent the lever *L* from being blown away due to the breaking of the spring.

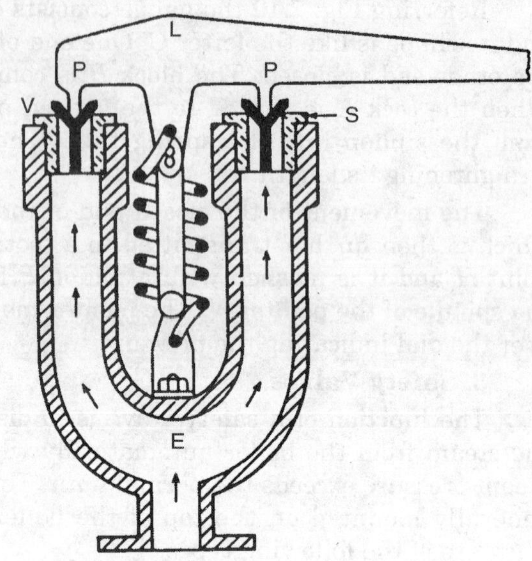

Fig. 7.13 Spring loaded safety valve.
R–Spring, *E*–Steam pipe, *L*–Lever, *V–Value,* S–*Valve seat,* P–*Pivot, K*–Link.

(d) High Steam and Low Water Safety Valve. A common cause of boiler failure is too low a water lever in the boiler. This results in the overheating and softening of the combustion chamber walls, furnace walls and boiler tubes etc. To safeguard against this low water level danger, a low water safety valve is provided. This valve, in practice, is normally combined with another safety valve which blows off when the pressure of steam exceeds the working limit. Sometimes an alarm whistle is also fitted to give an indication of a dangerously low water level.

4. Stop Valve.

The function of the stop valve or junction valve is to regulate the flow of steam from one steam pipe to the other or from the boiler to the steam pipe.

Referring Fig. 7.14, when the hand wheel *H* is rotated thus the spindle *P* carrying valve *V* at its lower end moves either upward or downward. Thereby controlling the passage between the valve *V* and its seat *S*. In this way flow of steam from one pipe to other pipe is controlled. The leakage of steam around the spindle *P* is prevented by means of a gland or stuffing box *G*.

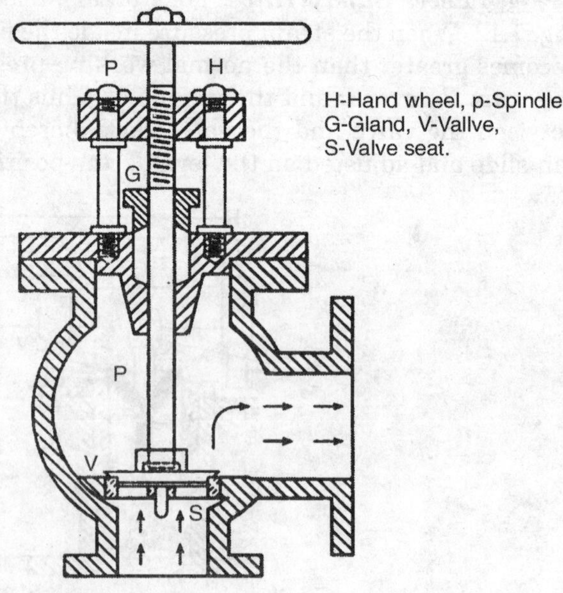

Fig. 7.14 Stop valve.

5. Feed Valve.

This valve is provided between the boiler feed pump and the boiler shell. The function of it is to stop water escaping from the boiler when the pump is not working.

Referring Fig. 7.15 when the pipe pressure is more than the boiler pressure then the valve *V* is lifted upwards from its seat thereby feeding the water, into the boiler. On the other hand when the boiler pressure is more than the pump pressure, then the valve may be closed to its seat automatically due to the pressure of water from the boiler, thereby preventing the escaping of water from the boiler. The lift of the valve V is regulated by the end of the spindle *E* which is attached with the hand wheel *H*.

6. Blow-off Cock.

The blow-off cock is fitted to the lowest part of the boiler. The functions of it are:

(*a*) To empty the boiler when necessary for cleaning, inspection and repair.

(*b*) To discharge a portion of water when the boiler is in operation to blow out mud, scale or sediments periodically.

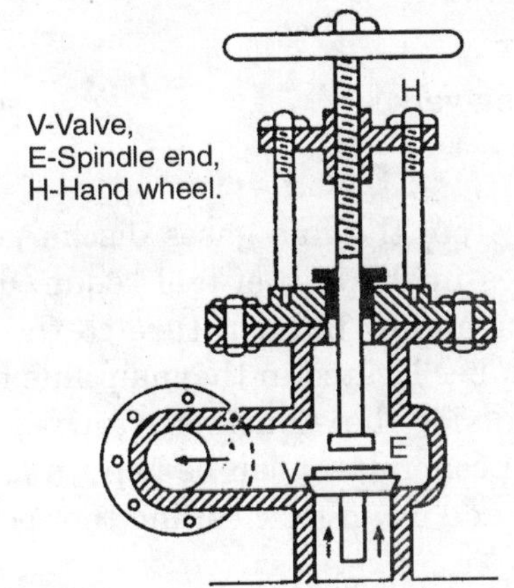

Fig. 7.15 Feed valve.

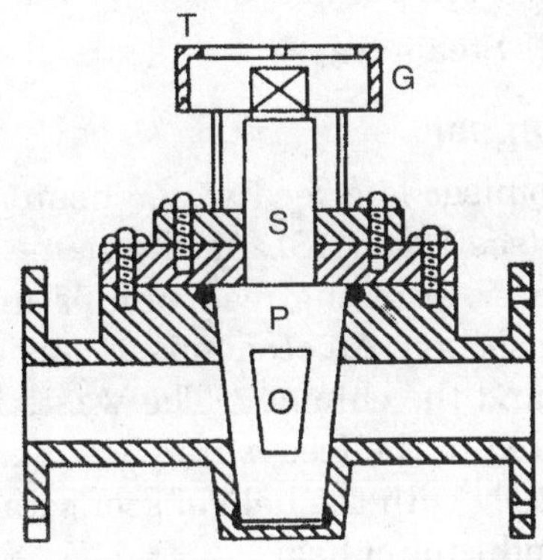

Fig. 7.16 Blow-off cock. *C*-Casing, *P*-plug, *S*-Spindle, *O*-Guard, *T*-slot

Referring Fig. 7.16, it is closed position of the cock. The conical plug *P* is fitted accurately into a Similar casing *C*. The plug *P* consists of a rectangular opening *O*. When the spindle *S* attached with the plug is rotated either in the direction by 90°, the opening *O* of the plug is coincided with the line of the passage of the casing *C*, thereby discharging the boiler water containing mud scale, sediments etc.

7. Fusible Plug.

The function of a fusible plug is to protect the boiler against the damage due to overheating for low water level. It is generally placed over the crown of the furnace or over the combustion chamber.

Referring Fig. 7.17, the outermost plug *A* is screwed to crown plate *B*, the plug *D* is screwed to plug *A* and the innermost plug *F is* locked to *D* by means of a fusible metal annulus *G*.

Under normal conditions, it remains at constant temperature as it is covered with water on one side and the hot gases on the other side. When the water level

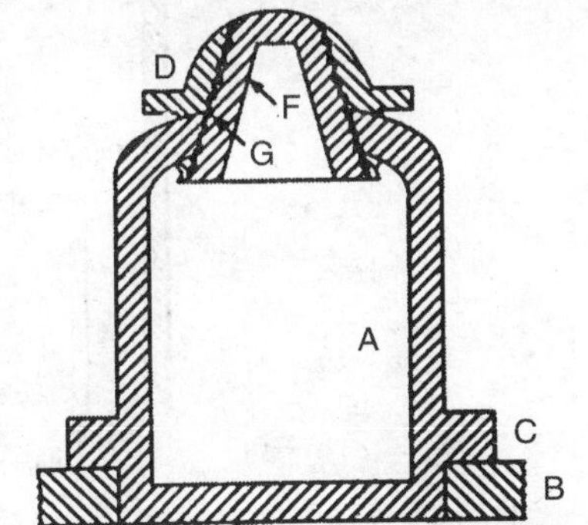

Fig. 7.17 Fusible plug. *A*–Outermost plut, *B*–fire box crown plate, *C–Hexagonal collar on* A, D–Plug screwed to *A, E Hexagonal collar on D, F*–Innermost plug, *G*–Fusible metal annulus.

falls below the lower limit, the plug will be uncovered. So the fusible metal G melts down due to furnace fire and the plug *F* drops out. Thus passing the steam into the fire-box and extinguishing the fire and indicating a danger.

7.9 BOILER ACCESSORIES

Boiler accessories are the equipments, which although not mounted on the boilers directly, but are essential for the efficient working of the boiler. Following accessories are used on the boilers:

(*a*) Accessories for increasing thermal efficiency:

(*i*) Economiser
(*ii*) Air Preheater
(*iii*) Super heater
(*iv*) Soot Blowers
(*v*) Feed pump
(*vi*) Injector
(*vii*) Steam traps
(*viii*) Reducing valves

1. Economiser

Economiser is a feed water heater extracting heat from the flue gases discharged from the boiler (see Fig. 7.18). As the name indicates, it saves in the cost of fuel required. These receive water from the feed pumps and deliver it at a higher temperature to the steam-generating section. An economiser consists of a group of tubes located in the main flue between the boiler and the chimney. The waste flue gases flow outside the economiser tubes and heat is transferred to the feed water flowing inside the tube. Economisers can be kept in a state of high efficiency with the help of soot scrappers moving up and down for cleaning the soot which is a bad conductor of heat.

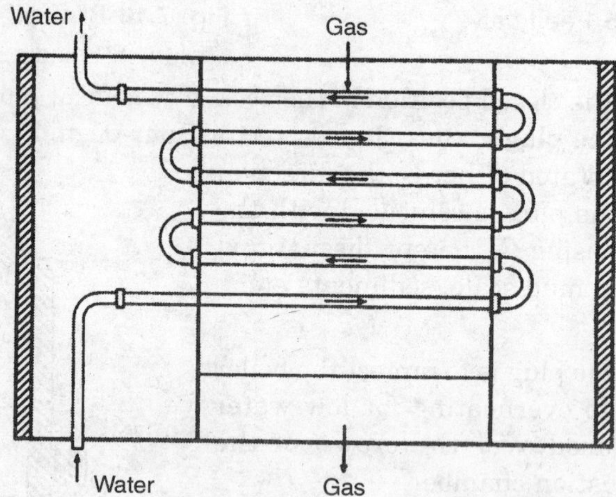

Fig. 7.18 Economizer.

Gain in efficiency by the economiser depends on the gas temperature out of the boiler and feed water temperature. Saving upto maximum of 20% in fuel cost can be achieved by using economiser. The maximum temperature to which the water can be heated in the economiser is 25°C below steam forming temperature in the boiler. Table 7.1 shows the percentage fuel saving by using economiser at different pressures:

Table 7.1 Percentage fuel savings by using economiser at different pressures.

Pressure	*Maximum economiser temp.*	*Percentage fuel saving with hot well temperatures*		
mPa	°*C*	*45°C*	*65°C*	*85°C*
0.8	150	17	14	10
1.3	169	20	17	14
1.8	183	23	19	16

The economisers may be classified in general as horizontal tube types, return, bend or continuous tube types, longitudinal or cross flow of gases, parallel or counter flow with regard to relative direction of gas and water flow.

Advantages of Economiser. Economisers have following indirect advantages in addition to those explained in preceeding paragraphs:

(*i*) It is easy to clean temporary hardness deposited on the tubes in the economiser as compared to that of boiler tubes. The water can be bypassed the economiser during cleaning without stopping boiler operation.

(*ii*) The economiser arrests the soot and flyash, which can easily be scrapped off from the tubes. Thus reduces the omission of soot and fly-ash.

2. Air Preheater

The heat carried with the flue gases coming out of economiser is further utilised for preheating the air before supplying to the combustion chamber (see Fig. 7.19). It has been observed that an increase of 20°C in the air temperature, improved the efficiency of boiler by about one per cent.

The high pressure boilers using solid fuels use preheaters. Preheated air, when used for combustion, accelerates ignition and burning of the fuel, and the drying of coal in pulverisers. Main advantages of the air preheaters are:

(*i*) Improved combustion.

(*ii*) Low grade fuels (with high ash content) can be used.

(*iii*) Saving in fuel combustion.

(*iv*) Increase in thermal efficiency.

(*v*) Increase in boiler capacity.

Air preheaters can be classified as:

(*a*) Tubular type. (*b*) Plate type. (*c*) Regenerative type.

In tubular type of air preheaters, while the flue gases pass through the tubes before going towards the chimney, the air is passed outside the tubes in a zig-zag way by providing baffles in its passage.

In plate type air preheaters, parallel plates are provided forming the chambers. Alternate chambers are used for flue gas and for air travelling in reverse direction.

In regenerative type preheaters, the chamber is dividing in two rotating compartments. The heating surfaces are made up of corrugated sheets. Hot flue gas moves upward on one side of such tubes, while air passes downward on the other side of the tube. With the rotation of the preheater at a slow speed, the heated surface comes in contact with combustion oil and cooled surface is heated by flue gases. This alternate heating and cooling of heating surface is known as regenerative heating.

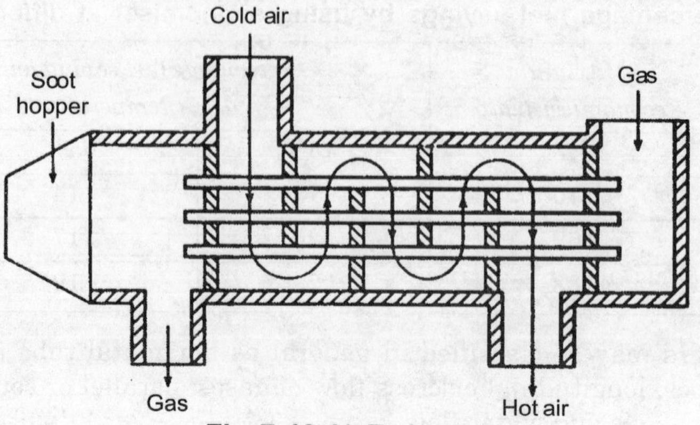

Fig. 7.19 Air Preheater

3. Superheaters

Addition of heat to steam, after evaporation, is accompanied by increase in temperature and enthalpy of the fluid. The superheaters are used to remove last traces of moisture from the saturated steam coming out of boiler and to increase its temperature sufficiently above saturation temperature. The superheating raises overall cycle efficiency as well as avoids too much condensation in the last stages of the turbine, when pressure reduces considerably, in order to avoid blade erosion.

Superheater consists of large number of parallel tubes in which steam flows, while hot gases come in contact flowing outside (see Fig. 7.20).

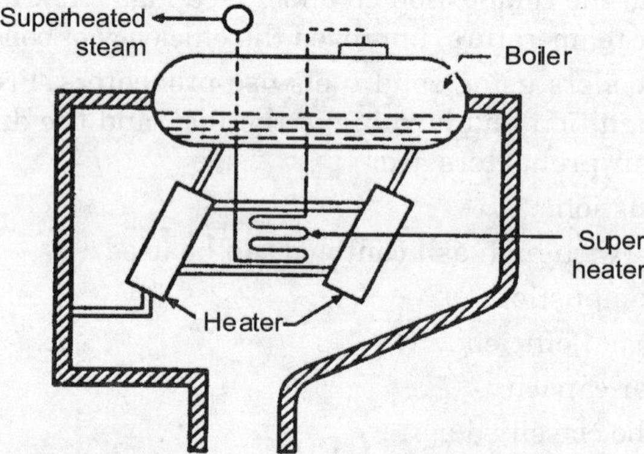

Fig. 7.20 Superheater for water tube boiler

4. Soot Blowers

Soot generated by the fuels in the boiler is deposited on the boiler tubes, economiser tubes and air preheater. Soot deposited on these heat exchangers drastically reduces heat transfer and significantly increases the amount of fuel required to produce heat equivalent to that generated by clean boiler surface. Therefore, the cleaning of these heat exchanger surfaces becomes essential (see Fig. 7.21).

The soot blowers therefore remove soot from heating surfaces by blowing with steam or compressed air. Soot blowers in addition to controlling soot build up and ash deposits that creates corrosive environment, improve heat transfer and controls draft loss. Soot blowers are of a variety of designs.

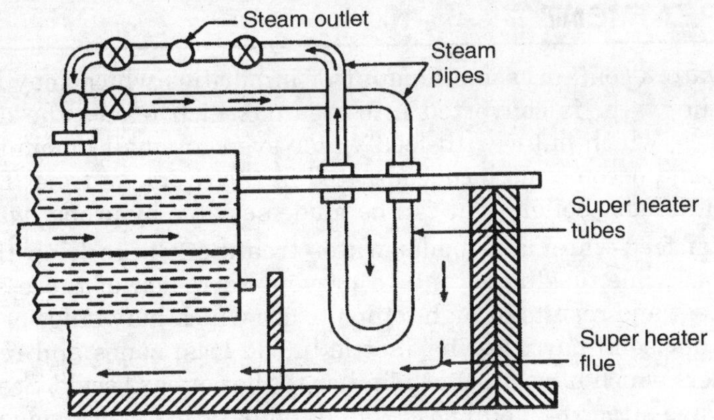

Fig. 7.20 (*b*) Superheater for Fire tube boiler.

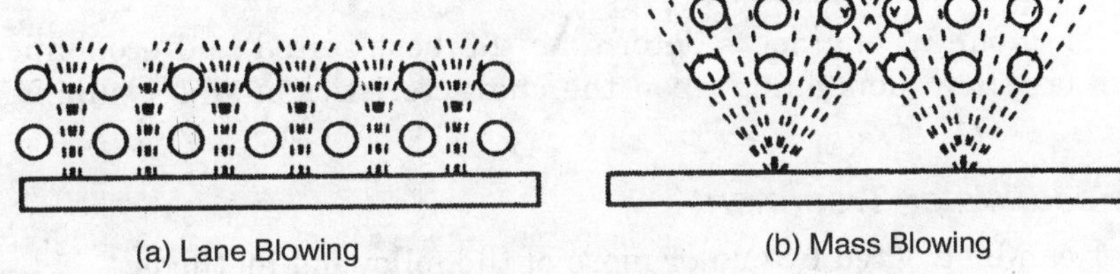

Fig. 7.21 Soot Blower.

5. Feed Pump

Water is fed to the boilers either by a feed pump or by an injector by raising its pressure, above the steam pressure in the boiler. Feed pumps may be either reciprocating type or centrifugal type. These may be driven either by a device driven by steam generated by boiler or by an independent device like an electric motor. Reciprocating pumps are driven by steam engine. Centrifugal pumps work on steam turbine or electric motors.

6. Injector

This device is also used for feeding water into the boiler drums against steam pressure. It raises the pressure of feed water from atmospheric to a pressure more than the pressure inside the boiler. Steam is used as a working medium in the injector. Injectors are mostly used for feeding the boiler as they occupy a very little space and also warms the feeding water. Injectors have high thermal efficiency, are easy to operate, and are compact. However they can not handle very hot water.

7. Steam Traps

The water formed by the condensation of the steam in the piping system, is entrapped in the steam traps which is later on disposed of. It is necessary to drain off the condensed water regularly, to prevent damage caused by this condensate in the supply-line.

8. Reducing Valve

Reducing valve is used to supply steam at a nearly uniform pressure to the steam engine or steam turbine. In fire tube boilers, it is easy to maintain uniform steam pressure, but it is difficult in case of water tube boilers. The water tube boilers are operated for higher steam pressure than required by the engine or a turbine. In such cases the steam is passed through the reducing valve to the engine, which maintains a constant reduced pressure on the engine side.

7.10 WATER TREATMENT

All natural waters contain, some amount of impurities which may be harmful in boiler operation. When the water is converted into steam, which leaves the drum in a very pure state. The impurities, which enter with feed water even in small amounts, are thus retained in the boiler water and become more concentrated in the course of time. Thus the water found in nature is not suitable for boilers, but can be made suitable by proper water-treatment.

The purpose of feed-water and boiler-water treatment is to keep the internal surfaces free from deposits of scale or sludge, and to prevent corrosion of those surfaces. Hard scale retard the flow of heat and results in high temperature which may bulge or rupture the affected parts. Sludge or other solid particles may settle in the local zones and restrict the flow. This form insulating layers which have an effect similar to that of hard scale. Scale or sludge deposits may increase corrosive attack of metal at elevated temperature. Corrosion weakens the general structure by removal of metal, but it more commonly appears in localised areas as cavitation and pets which may result in leakage.

Chemists or chemical engineers determine the best suited and economical way to treat for each individual plant, depending upon the characteristics of water supply and amount of make-up.

Methods of Feed-Water Treatment

Raw water can be treated by one or more of the following methods:

(*i*) Removal of suspended solids by settling or filtration.

(*ii*) Chemical treatment for removal of hardness.

(*iii*) Demineralisation for removal of dissolved solids.

This raw water after treatment is fed to the boiler with condensate. From this, the main concern is the removal of dissolved gases and preventing corrosion. The dissolved oxygen may cause pitting of steel surface in contact with water. Other gases dissolved in the water notably carbon dioxide, sulphur-dioxide, or hydrogen sulphide may result corrosion. The corrosion may be prevented by using a small percentage of alkali or ammonia or volatile amines.

7.11 DRAUGHT SYSTEM

The purpose of draught is to supply required quantity of air for combustion and remove the burnt products. To move the air through the fuel bed and to produce a flow of hot gases through the boiler, economiser, pre-heater, and chimney requires a difference of pressure equal to that necessary to accelerate the burnt gases to their final velocity and to overcome pressure losses.

Natural Draught (Chimney)

When the draught is produced with the help of chimney only, it is known as natural draught. The natural draught is obtained with the use of tall chimney which may be sufficient to overcome the losses in the system. This system of producing the draught is useful for small capacity boilers. A chimney is a vertical structure of masonry ; bricks, or steel or reinforced concrete built for enclosing a column of hot gases to produce the draught, and discharge the gases high enough to prevent air-pollution. The draught produced by the chimney is due to the temperature difference of hot gases in the chimney and cold air outside the chimney.

Artificial Draught

The draught produced by any means other than chimney is known as artificial draught. Since draught produced by chimney has no flexibility, poor efficiency and tall chimney. In order to have draught independent of atmospheric condition, and able to meet fluctuating requirement.

Artificial draught is generally produced by a fan and is known as fan (mechanical) draught. Draughts are of two types ; namely forced draught, and induced draught.

(*a*) Forced Draught. In this system, a blower is installed near the base of the boiler and air is forced to pass through the furnace, flues, economiser, air preheater and to the stock (see Fig. 7.22(a)). This type of draught is known as positive draught or forced draught, because the pressure of air throughout the system is above atmospheric pressure. In this case, function of chimney is only to discharge gases high in the atmosphere.

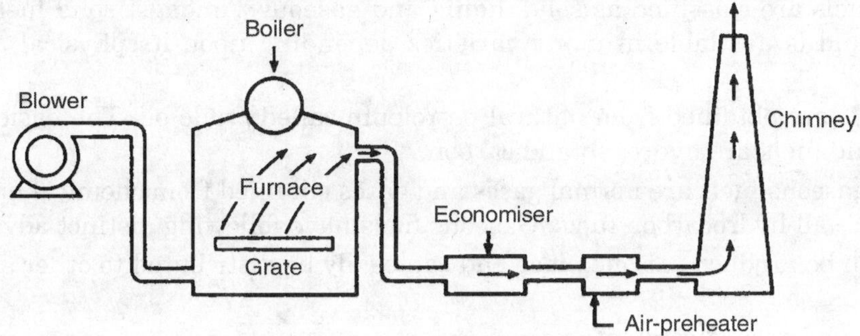

Fig. 7.22 (*a*) Forced Draught.

(*b*) Induced Draught. In this system, blower is located near the base of the chimney, and the air is sucked by reducing the pressure below atmosphere (see Fig. 7.22(b)). Induced draught fan sucks the burned gases from the furnace, resulting the drop of pressure inside the furnace, the induced atmospheric air to flow through the furnace. The action of the induced draught is similar to the action of the chimney. This type of draught is used generally when economiser and air preheater are introduced in the system. When chimney is also provided in this system total draught produced is the sum of the draught produced by the chimney and that by the fan.

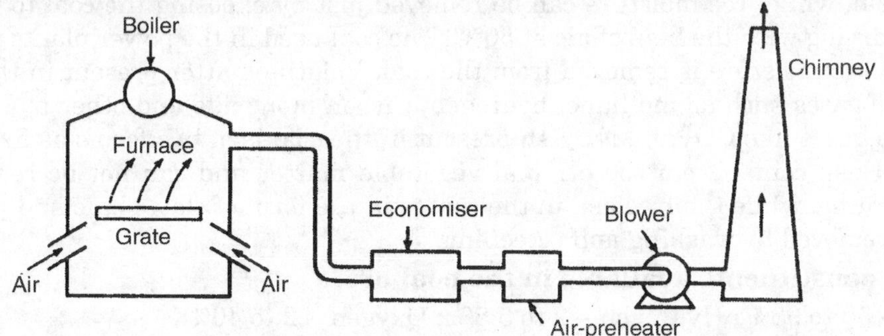

Fig. 7.22 (*b*) Induced Draught.

(*c*) Balanced Draught. It is always preferred to use a combination of forced draught and induced draught. In this system, the pressure of air below the grate is above atmosphere and it helps for proper and uniform combustion, whereas the pressure of air above the grate is below atmosphere and helps to remove the exhaust gases as quick as possible from the combustion zone.

Control of Air Supply

Quantity of air to be supplied changes with the change of load on the plant. Two methods are used to supply the variable quantity of air as per the demand.

(*a*) Damper control. By increasing the resistance of the system *i.e.* by partly closing dampers, the flow quantity of air can be reduced or *vice versa.*

(*b*) Speed control. The flow quantity of air can also be reduced *by* reducing the speed of the fan.

7.12 BOILER FUELS

A substance, which contains combustible elements which, when combined with oxygen, burns and generates heat, is called fuel. Combustible elements are generally carbon, hydrogen and sulphur. Fuels are classified as solid, liquid and gaseous. Amongst solid fuels, coal is the most common and is available in many varieties depending upon its physical and chemical properties.

Liquid fuels are obtained from natural petroleum called crude oil. The basic combustible elements of liquid fuels are hydrogen and carbon.

The most gaseous fuels are normal gases and gases obtained from chemical manufacturing process. They are all hydrocarbon fuels. Gaseous fuels have following distinct advantages:

(*a*) Gas can be produced at one place and can easily be distributed to other areas through pipes.

(*b*) Gas can be burned completely.

(*c*) Air pollution due to combustion of gaseous fuel is low.

(*d*) Disposal of smoke and ash is easier.

Coal is the oldest fuel for power generation. It is a fuel which brought about an industrial revolution and made a dramatic change in the life style of the people. It met most of our fuel requirement until oil was struck and even today it is a major source of energy.

Most coals have 3 to 30% moisture ; 3 to 50% volatile matter ; 2 to 30% ash ; 16 to 92% carbon. The coals have inherent as well as free moisture. Inherent moisture is held in the pores of the coal, while free moisture can be removed just by exposing the coal to the natural air flow or by drying with the help of air at 50°C. The coal used in the power plants is generally dried and the free moisture is removed from the coal. Volatile matter present in the coal may be combustible gases such as methane, hydrogen, carbon monoxide and other hydrocarbon or in combustible gases like CO_2 and N_2. Ash present in the coal is of two forms as fixed and free ash. The fixed ash comes from the original vegetable matter and can not be removed from coal before burning, while the free ash in the coal is in the form of clay, shale and pyrites, and which can be removed by washing and screening.

Various constituents available in the coal ore.

Carbon – 50 to 95% ; Hydrogen – 2 to 5.5% ; Oxygen – 2 to 40%;

Sulphur – 0.5 to 3% ; Nitrogen – 0.5 to 7% ; Ash – 2 to 30%.

Higher percentage of the carbon gives high heat value. Hydrogen exists in the coal in combined form with oxygen as inherent moisture. This inherent moisture is not desirable since it carries heat with flue gases without any contribution in combustion. The nitrogen has no heating value. The high percentage of sulphur is objectionable as it is responsible for clinkering, slagging, corrosion of air preheater, economiser and stacks, and air pollution. The presence of ash causes troubles during combustion, and it reduces the calorific value of coal, and reduces the rate of burning.

Cleaning of Coal

The coal is cleaned with the help of screens, coal drying equipments, washing equipment and dewatering systems.

The coal is passed through screens to separate the large coal above 8 cm in size from small coal. The coal below 8 cm in size is cleaned by dry cleaning plants (if moisture content is less than 3%) or by washing. The washed coal is dried by steam drying, oil dehydration and fuel gas drying. It is necessary to dry the coal as wet coal produces less heat and increases the transportation cost.

Storage of Coal at Plant Site

Coal is required to be stored as an insurance against complete shut down in case of failure of coal supply, and storage permits the date as to when it can be purchased to take advantage of market and other conditions.

The coal can be stored either in heaps or under water to avoid oxidation. The pulverised coal is preferred to be stored in bunkers.

Inplant Handling of Coal

The coal handling starts at the discharge end of the carrier system (rail head or harbour) and passes through various stages such as delivery to storage, reclamation and feeding to the combustion system. The inplant coal handling system should be designed in such a way that the inplant transportation should be minimum.

Coal Unloading. Unloading equipment depends upon the type of outplant handling mode, *viz.* rail, road or ship. Road transportation by tipping trucks is economical for small and medium capacity plants located within 30 to 50 kilometres of coal mines. Railroad movement of coal is adopted for large coal quantities and for longer distances where water transportation is not available. The railcars in which the coal is transported has gates in the bottom to drop the coal out into an unloading hopper or directly on to a conveyor hopper. When the coal is transported by sea, the unloading devices used are: coal towers, unloading bridges and self unloading boats. A travelling gantry cranes are used for unloading large coal carrying vessels.

Coal Preparation. Unsized coal brought to site is prepared before feeding to the combustion chamber. Coal preparation plant can either be located near the coal receiving point or at the point of actual use. The coal preparation plant includes ; (*i*) Crushers, in which the coal requires sizing is only crushed, and balance is bypass, (*ii*) Sizers, which removes unsized coal and feeds back to the crusher. (*iii*) Dryers, which removes the moisture from the sized coal, (*iv*) Magnetic Separators, which removes the iron scrap, before the coal is supplied to the storage hopper.

Transfer of Coal. The coal is carried from the unloading point to the storage site through different type of transfer equipments, namely ; (*i*) Belt conveyors, (*ii*) Screw conveyors, (*iii*) Bucket elevators, (*iv*) Plight conveyors.

For details readers are advised to refer the book "Construction Equipment and its Management" by S.C. Sharma.

Handling of Coal from Storage Point to the Firing Equipment. For simple stoker firing, only chutes are required to feed the coal, whereas for pulverised fuel firing system chutes, pulverised mills, feeders and belt weighing equipment are used for this handling.

Burning of Fuels

Burning of coal is generally done by either stoker firing, or by pulverised fuel firing. In stoker firing, the coal is generally fed over a travelling grate stokers or uniformly spread over a stationary spreader stoker. The pressurised air from forced draft fan is allowed to enter under the bottom of the grate. The gases leaving the upper surface of the fuel bed contain N_2, CO_2, CO, H_2 and H_2O. Additional secondary air is supplied at the top of the fuel bed to burn the remaining combustible gases *i.e.* volatile matter + CO + H_2.

The coal is fed from a hopper by gravity. The rate of fuel feeding depends on the requirement of steam generation, and which is controlled by the gate opening for fuel entry on the grate, and also by the speed of the grate. For efficient burning of coal it is necessary to burn the volatile matters completely, and to prevent unburnt carbon going with ash. For achieving this object adequate quantity of secondary air and its entry with high turbulence at proper location above the grate is essential.

Pulverised Fuel. Since stoke firing is not very suitable to take the fluctuating loads on large capacity boilers, the pulverised fuel firing systems are now-a-days used for large capacity boilers. This also enables the use of low grade fuels. In this system, coal is reduced to a fine powder and fed to the combustion chamber with the help of hot air current.

Ash Handling

Disposal of Ash produced from a large capacity power plant is very important as it is more than 10 to 20% of the coal used. Ash handling systems are of following types:

(i) **Mechanical Ash Handling System.** These are used for low capacity plants. The hot ash coming out of boiler furnace is cooled by water seal while falling on a belt conveyor, which is then carried continuously to the dumping site or overhead bunker. From the ash bunker, the ash is carried to the dumping site with the help of tipping trucks.

(ii) **Hydraulic Ash Handling System.** This system carries the ash with the flow of water with high velocity, through a channel and finally dumped into the sump. The water is separated from ash when it reaches to the sump, and is recycled.

(iii) **Pneumatic Ash Handling System.** This system is useful for handling abrasive ash as well as fine dusty materials such as flyash and soot. The ash carried by the air is separated into the separators working on cyclone principle, and clean air is discharged into the atmosphere.

Ash Disposal

Various methods are used for the disposal of ash from the collected hoppers to the discharge site which is located far away from the power plant. The method employed may be vacuum extraction system or water ejector system or steam ejector system or by mechanical conveyors like screw, and belt conveyor.

The ash and dust is transported either using wet system or dry system. In the wet system, ash is transported to ash ponds in the form of slurry. Ash settles in the pond, the part of the water evaporates and the remainder is either recycled or impounded. This system requires installation of pipelines and construction of embankments. In dry system, the ash is transported in a relative dry state (in the form of paste). Water is added only to compact the ash, compacted ash surfaces are covered with top soil and seeded.

Atmospheric Pollution

Fuel gas cleaning has become a problem of prime importance as the atmospheric pollution is highly restricted. The major emission from coal fired boilers (and furnaces) are fly ash, carbon ash (known as cinder), smoke, dust and irritating vapours like CO, SO_2, and nitrogen oxides. It has bad effects on the human and animal healths as well as bad effects on crops and destroying, spoiling and damaging actions on the public property. Cleaning of gas before discharge to the atmosphere is essential.

The devices used for cleaning the gas may be dust collectors (cyclonic collectors, fabric filters electronic precipitators). Scrubbers (for controlling SO_2), NO_2 formation is controlled by flue gas recirculation.

7.13 CHIMNEY (STACK)

Chimneys are made up of steel or bricks and concrete. Concrete chimneys are more popular. The average life of concrete chimneys is 50 years and that of steel chimneys is about 15 years depending upon the care taken to prevent corrosion. The net area of chimney depends on the following factors:

(*i*) Volume of gases to be discharged when the boilers operate at maximum rating.

(*ii*) Draught to be produced.

Chimney generally denotes brick concrete construction whereas stack means steel construction.

Fig. 7.23 (*a*) shows a brick chimney and Fig. 7.23 (*b*) shows reinforced chimney. Chimney is provided with a lightning conductor, aircraft warning lights, and various means of access and inspection.

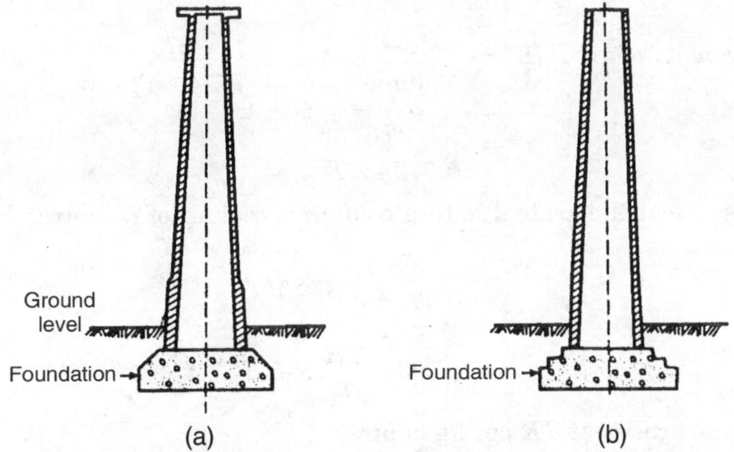

Fig. 7.23 (*a*) Brick and (*b*) Reinforced chimney.

The various advantages of steel chimneys over masonry chimneys are as follows:

(*i*) Lower overall cost

(*ii*) Slightly high efficiency,

(*iii*) Easier construction,

(*iv*) Requires lesser space,

(*v*) Lighter in weight.

R.C.C. chimneys are also becoming popular.

7.13.1 Calculation of Chimney Height

Air is required to burn coal or fuel in the furnace. We know

$$C + O_2 = CO_2$$

This shows that volume of chimney gases produced by the complete combustion of 1 kg of coal or oil is same as that of air required to support combustion if the temperature of flue gases and air is same.

Let H = Height of chimney above grate level in metres.

W = Weight of air required in kg per kg of fuel.

T = Average absolute temperature of chimney gases in °K

T_1 = Absolute temperature of air outside the chimney in °K

Weight of chimney gas produced

$$= (W + 1) \text{ kg of fuel burnt}$$

Volume of chimney gases at 273 K

$$= \text{volume of 1 kg air at } 273°\text{K}$$

$$= \frac{RT}{P} = \frac{287 \times 273}{1.01 \times 10^5} = 0.7734 \text{ m}^3$$

Therefore, volume of one kg of air T_1°K

$$= 0.7734 \times \frac{T_1}{273}$$

Thus volume of W kg of air at T_1°K

$$= \frac{0.07734 \times T_1 \times W}{273}$$

$$\text{Density of air } T_1°K = \frac{\text{Mass}}{\text{Volume}} = W \times \frac{273}{0.7734 \times T_1 \times W}$$

$$= \frac{273}{0.7734 \times T_1}$$

Therefore, pressure at the grate due to a column of cool air of H metres height

$$= \frac{273}{0.7734 \times T_1} \times H$$

$$= 1.293 \times \frac{273}{T_1} \times H \text{ P}_a \qquad \text{...}(i)$$

Volume of chimney gases at TK per kg of air

$$= 0.7734 \times \frac{T}{273}$$

Volume of $(W + 1)$ kg of chimney gases at T°K

$$= 0.7734 \times \frac{T}{273} \times (W+1)$$

Density of chimney gases at T°K

$$= \frac{\text{Mass}}{\text{Volume}} = (W+1) \times \frac{273}{0.7734 \times T \times W}$$

$$= 1.293\left(\frac{W+1}{W}\right) \times \frac{273}{T} = 353 \times \left[\frac{W+1}{W}\right] \times \frac{1}{T}$$

Therefore, pressure at the grate by a column of hot gases of height H metre

$$= \text{Density} \times \text{H Pa}$$

$$= 1.293\left(\frac{W+1}{W} \times \frac{273}{T}\right) \times H \text{ Pa} \qquad ...(ii)$$

Therefore, pressure P causing the draught

$$P = \left[\left(1.293 \times \frac{273 \times H}{T_1}\right) - 1.283 \times \left(\frac{W+1}{W} \times \frac{273}{T} \times H\right)\right]$$

$$= 353\, H\left[\frac{1}{T_1} - \left(\frac{W+1}{W}\right) \times \frac{1}{T}\right] \text{ Pa}$$

If this draught is h mm of water as measured by a U-tube water manometer.

$$h = 353\, H\left[\frac{1}{T_1} - \left(\frac{W+1}{W}\right) \times \frac{1}{T}\right] \text{ mm of water}$$

(We know 10 Pa = 1 mm of water column)

Let h' = height of column of hot gases.

$$h' \times 1.293 \times \left(\frac{W+1}{W}\right) \times \frac{273}{T} = P$$

$$= 353\, H\left[\frac{1}{T_1} + \left(\frac{W+1}{W}\right) \times \frac{1}{T}\right]$$

$$h' = \frac{353\, H\left[\frac{1}{T_1} - \left(\frac{W+1}{W}\right) \times \frac{1}{T}\right]}{1.293 \times 273 \times \left(\frac{W+1}{W}\right) \times \frac{1}{T}}$$

$$h' = H\left[\left(\frac{W}{W+1}\right) \times \frac{T}{T_1} - 1\right]$$

7.14 BOILER PERFORMANCE

7.14.1 Heat Flow

A power plant consists mainly of equipment for energy flow and transformation. The basic modes of heat transfer are as follows.

(a) Conduction (*b*) Convection (*c*) Radiation.

All the three forms of heat transfer are found in power plant engineering usually in combination with each other. The principal heat transfer calculations are associated with the following:

(*i*) Radiation from fuel bed and luminous gases to boiler tubes and water walls.
(*ii*) Radiation and conduction in heat insulators like refractories and pipe coverings.
(*iii*) Conduction of heat through boiler, economiser and air preheater surfaces.
(*iv*) Conduction of heat through condenser or heater tubes.
(*v*) Heat convection from combustion region to more remote boiler tubes, economiser tubes and air preheater surfaces.

Fig. 7.24 shows the flow of flue gas and feedwater and steam flow in the boiler plant. The heat of flue gases is utilised in heating the feedwater in economiser, in heating the air to be supplied to boiler furnace in air pre-heater and also in superheater.

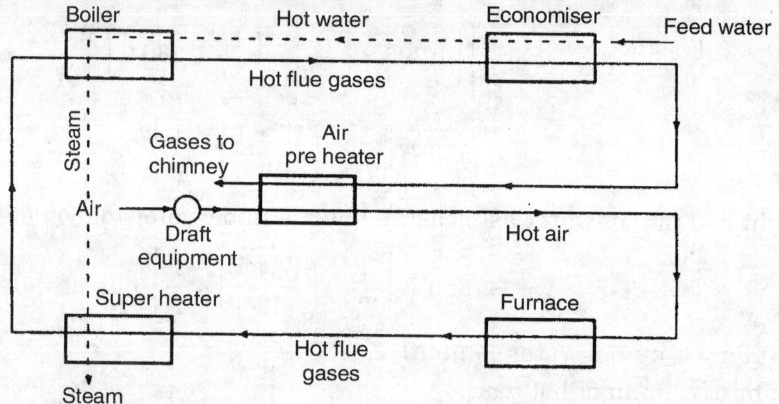

Fig. 7.24 Heat flow from flue gases.

7.14.2 Performance of a Boiler

The performance of a steam boiler may be expressed in terms of the following:

(*i*) Heat release per cubic metre of furnace volume.
(*ii*) Efficiency.
(*iii*) Heat transferred per square metre of the heating surface area per hour.
(*iv*) Rate of combustion in kcal per square metre of the grate area per hour for solid fuel.
(*v*) Amount (kg) of steam produced per hour.

Equivalent Evaporation

To compare evaporative capacity or performance of different boilers working under different conditions it is desirable to provide a common base so that water be supposed to be evaporated under standard conditions.

The standard conditions adopted are temperature of feed water 100°C and converted into dry and saturated steam at 100°C. As per these standard conditions 1 kg of water at 100°C requires 2257 kJ to get converted to steam at 100°C.

Therefore equivalent evaporation is defined as the amount of water evaporated from water at 100°C to dry and saturated steam at 100°C. However (*i*) the evaporation rate of boiler may also be expressed in terms of kg of steam per kg of fuel.

(*ii*) The capacity of a boiler may also be expressed in terms of total heat added per hour.

Factor of Evaporation (Fe)

It is defined as follows

$$Fe = \frac{H_1}{H_2}$$

where H_1 = Heat received by 1 kg of water under working conditions.

H_2 = Heat received by one kg of water evaporated from and at 100°C

Boiler Trial

Steam is generated in boilers under certain conditions of inlet water and exist steam while a certain rate of fuel is being burnt. In order to study the performance of boiler the experiments are conducted by operating the boiler for a certain length of time and recording the data. This procedure is known as boiler trial. A heat balance sheet is prepared which indicates the heat supplied, heat utilised through various sources and heat wasted.

The main objects of boiler trial are as follow:

(*i*) To determine the thermal efficiency of the boiler when working at a definite pressure.

(*ii*) To draw up heat balance sheet for the boiler to check the performance of the boiler.

Heat Losses in a Boiler Plant

The heat losses which may take place in a boiler plant are as follows:

(*i*) heat loss due to incomplete combustion of fuel

(*ii*) heat loss due to unburnt fuel

(*iii*) convection and radiation heat loss

(*iv*) heat lost in flue gases.

The various causes of heat loss in a boiler are as follows:

(*i*) Loss due to heat carried away in chimney gases.

(*ii*) Loss due to incomplete combustion of fuel. This can be due to insufficient air supply or fuel bed being in poor conditions.

(*iii*) Loss due to moisture formed by combustion of hydrogen in fuel.

(*iv*) Loss due to moisture in coal.

(*v*) Loss due to radiation and convection from boiler.

(*vi*) Heat loss due to unburnt fuel which is passed on to the ash

Thermal Efficiency of Boiler

Thermal efficiency of a boiler is defined as the ratio of heat actually utilised by water in generation of steam to the heat supplied during combustion of fuel.

$$\eta_{boiler} = \frac{W(H_2 - H_1)}{W_1 \times CV}$$

where Q = Weight of steam produced, kg

H_1 = Enthalpy of feed water, kJ/kg

H_2 = Total heat of steam, kJ/kg at the generation pressure

W_1 = Weight of fuel burnt, kg

$C.V.$ = Calorific value of fuel, kJ/kg.

Although the major portion of heat produced during combustion of fuel is utilised to heat the feed water for conversion into steam but some portion of heat produced is lost in flue gases leaving the chimney and some amount of heat is lost due to radiation from furnace to the outside. The boiler efficiency is nearly 75% without heat recovery equipment and it varies from 85 to 90% with heat recovery equipment.

Boiler performance can be improved by taking into account the following factors:

(*i*) Proper supervision should be carried out.

(*ii*) Too much excessive air should not be supplied.

(*iii*) Boiler, superheater, economiser and air heater surfaces should be cleaned at proper intervals.

(*iv*) Type of fuel used also controls boiler efficiency. Fig. 7.25 compares the performance in burning bituminous coal, oil or natural gas in the boiler.

Efficiency (η) indicates steam generating unit efficiency and C represents percent of rated capacity.

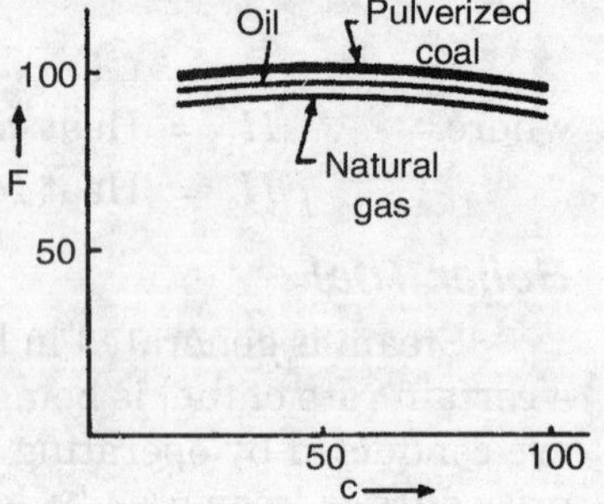

Fig. 7.25 Efficiency with different fuels.

(*v*) Proper supply of excess air.

(*vi*) Minimising combustible in refuse.

(*vii*) Minimising combustible in flue gas.

(*viii*) Maintaining required cleanliness of heat transfer surfaces.

Efficiency of a boiler depends on the following factors:

(*i*) Steam pressure

(*ii*) Calorific value

(*iii*) Effectiveness in burning of fuel. Indication of the effective burning of fuel is given by the percentage of the carbon dioxide (CO_2) in the flue gases. The percentage of CO_2 to give maximum efficiency and satisfactory operation will depend on:

(*a*) type of fuel such as oil, gas, coal. (*b*) method of firing.

(*iv*) Capacity of boilers

(*v*) Use of heat recovery apparatus such as air preheater, economiser etc.

7.14.3 Ways of Increasing Thermal Efficiency

The thermal efficiency of a steam power plant can be increased by the following ways:

(*i*) An increase in the initial pressure of steam raises the thermal efficiency.

(*ii*) The thermal efficiency can be increased by raising the initial temperature of the steam without changing the pressure.

(*iii*) Intermediate reheating of steam improves the thermal efficiency of the plant. As already mentioned an increase in the initial pressure of steam improves the thermal efficiency steam power plant but the wetness fraction of such steam also grows at the end of expansion. The wetness is so high that it causes wear of the blades of the last stages of a steam turbine. This shortcoming is

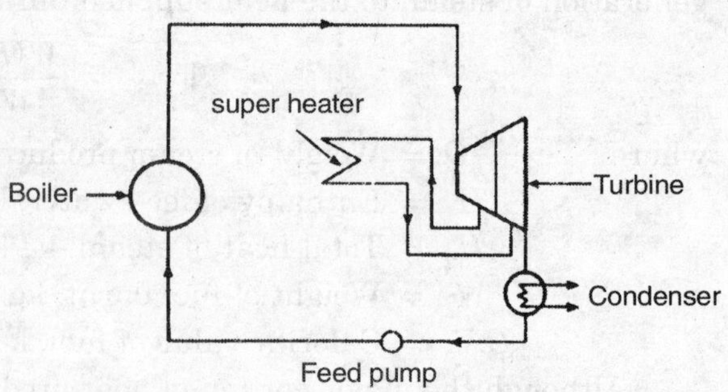

Fig. 7.26 Reheating of the steam.

eliminated by intermediate reheating of the steam. The steam from boiler after partial expansion in the first stage turbine is fed into super heater, where it is reheated at a constant pressure either to the temperature of live steam or to a temperature slightly below that one. The reheated steam then flows to the last stages of the turbine where the process of expansion is completed (Fig. 7.26).

(*iv*) Thermal efficiency of steam power plant can be improved by carrying out regenerative heating of the feed water. Such heating of the water is carried out by using the heat of steam partly tapped from the turbine. In this method (Fig. 7.27) the steam from boiler flows into steam turbine. After partial expansion some of the steam is tapped from the first stage of the turbine and directed to feed water heater and then to feed tank. The remaining steam enters the second stage of the turbine where it continues to expand. At the outlet from the second stage some of the steam is directed into water heater and then to feed tank. The other part of steam goes to the third stage of the turbine and expands there to the final pressure and enters the condenser. The condensate is delivered by pump to feed tank.

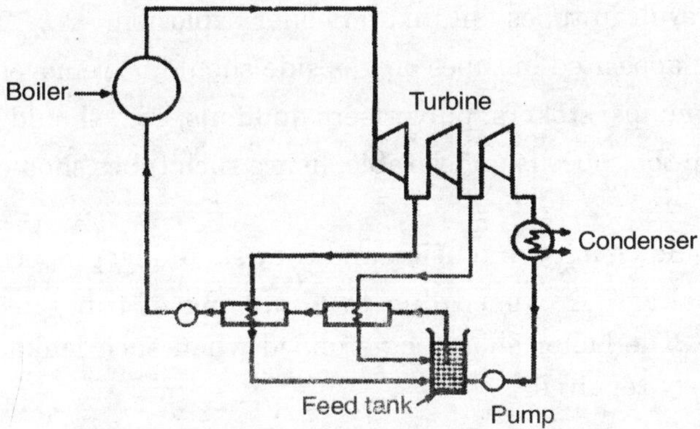

Fig. 7.27 Regenerative heating of feedwater.

7.15 BOILER INSPECTION AND MAINTENANCE

1. Boiler Inspection

The boilers are required to be tested regularly as per Indian Boilers Act. Before inspection, boiler should be made empty and all manhole doors, caps, mud drums, fire bars, fire arches, front end bridge plates, oil burners should be removed and boiler should be cleaned. All tubes, smoke boxes and external flues should be cleaned and checked. All the steam and water connections should be disconnected and then it should be made ready for conducting the following tests.

(*a*) **Hydraulic Test.** For this test, the mountings should be fitted tightly except the safety valves. The openings of the safety valves should be closed. All the joints should be made tight. Boiler is filled with water and inspector's gauges should be fitted to the boiler. Generally the boiler is tested at twice the working pressure of the boiler. During this test no part of boiler should deflect.

(*b*) **Steam Test.** This test is conducted to check whether the safety valves operate properly to relieve the steam when the steam pressure inside the boiler exceeds the rated pressure. This operation is tested keeping the feed valve and steam stop valve closed.

2. Boiler Maintenance

Regular maintenance and careful supervision of working of various components of boiler result in the efficient operation of the boiler. The various factors which should be carefully observed for proper maintenance of the boiler are as follow:

1. The combustion equipment should be so adjusted that the "temperature in the furnace does not exceed the designed value. The air supply to the furnace should be in correct proportion.
2. The water level in the boiler should not be allowed to fall beyond the minimum level.
3. The temperature should change slowly and uniformly in the various parts of the boiler. Rapid changes in temperature cause unequal expansion.
4. The water used for steam generation should be free from scale forming impurities because the scale if deposited on tube and boiler shell does not allow the heat transfer to take place effectively and sometimes causes over heating of tubes and boiler shell which may result in tubes failure and boiler explosion.
5. Soot and ash deposited in tubes on gas side should be removed regularly.
6. Bearings of pumps, stokers, pulverisers and fans, etc., should be lubricated regularly.
7. Oil or scale in the tubes is undesirable, hence such tubes should be located and problem is removed.
8. All stays should be in even tension and they should be properly adjusted.
9. The tube leakage is caused due to blistering of tubes as it is exposed to high temperature. The boiler should be stopped when such leakage is noted and it must be immediately repaired.

3. Boiler Tube Failure and Prevention

With the introduction of 200 and 500 MW capacity boiler, the outages of boilers due to tube leakage vary from 10 to 15% of the total outages. Therefore, it is necessary to combat this problem in the initial stages. The basic causes of the tube failures are corrosion, stress cracking, thermal fracture, stress rupture and creep distortion.

When the water and steam chemistry are not maintained within the limits recommended by the boiler manufacturer, corrosion or corrosion related damage may occur in feedwater heaters, economiser, and water wall. Over-heating damage may also occur in these tubes as well as in superheater and reheater tubes, if deposits are build up in them due to poor water treatment and improper boiler operation. Overheating of the boiler tubes is the main cause of their failure.

The problem of corrosion have already been discussed while dealing with feedwater treatment in earlier sections. The outer surface erosion of the water tubes is caused by an erosive action of ash particles in the gases. Erosion is enhanced by high flue gas velocities. Erosion creates spots, wall thinning and finally tube failure. These can be reduced upto some extent by controlling the excess air supplied to the combustion chamber, and by providing baffles, and tube shields can be welded in the maximum affected zones.

7.16 IMPORTANT FORMULAE FOR BOILERS

1. Boiler Efficiency:

The efficiency of a boiler is defined by,

$$\text{Boiler} \quad \eta = \frac{\text{energy gained by fluid}}{\text{energy input to boiler}}$$

$$= \frac{\text{steam quantity (steam head – feed water heat)}}{\text{quantity} \times \text{heating value of fuel}} \times 100$$

2. Calculation of Chimney Height:

$$H = h\left\{\frac{W}{W+1} \times \frac{T}{T_1} - 1\right\}$$

where
H = Height of hot gas column in metres.
h = Height of chimney in metres from the grate.
W = Weight of air used in kg per kg of fuel.
T = Mean absolute temp of chimney gases in °K.
T_1 = Absolute temp. of outside air in °K.

3. Power Required for Draught Fan:

$$\text{kW} = \frac{h.V}{3360 \times 60 \times \eta}$$

where,
h = Draught produced by fan in mm of water.
V = Volume flow of gas through fan in m^3/sec.
η = Efficiency of the fan.

SOLVED PROBLEMS

Example 7.1 *A chimney is 28 metres high and the temperature of hot gases inside the chimney is 320°C. The temperature of outside air is 27°C and furnace is supplied with 15 kg of air per kg of coal burnt. Calculate:*

(a) Draught in mm of water.

(b) Draught height in metres of hot gases.

Solution. We have

(*a*)
$$h = 353 \times H\left[\frac{1}{T_1} - \left(\frac{W+1}{W}\right) \times \frac{1}{T}\right]$$

$$= 353 \times 28\left[\frac{1}{27+273} - \frac{15+1}{15} \times \frac{1}{320+273}\right]$$

$$= 353 \times 28\left[\frac{1}{300} - \frac{16}{15} \times \frac{1}{593}\right]$$

$$= 353 \times 28\,[3.33 \times 10^{-3} - 1.8 \times 10^{-3}]$$
$$= 353 \times 28\,[1.53 \times 10^{-3}]$$
$$= 15.1 \text{ mm of water.} \quad \textbf{Ans.}$$

(b)
$$h' = H\left[\frac{W}{W+1} \times \frac{T}{T_1} - 1\right] = 28\left[\frac{15}{16} \times \frac{593}{300} - 1\right]$$
$$= \textbf{23.88 metres.} \quad \textbf{Ans.}$$

Example 7.2 *Determine the height of chimney to produce a static draught of 22 mm. of water if the mean flue gas temperature in a chimney is 290°C and ambient temperature in boiler house is 20°C. The gas constant (R) for air is 287 J/kg°k and for chimney flue gas is 257 J/kg °K. Assume barometer reading as 760 mm of mercury.*

Solution. Let

P = Absolute pressure of gas P_a (pascel)
V = Volume of gas (m²)
R = Gas constant
T = Absolute temperature of gas (°K)

Now
$$PV = RT \Rightarrow V = \frac{RT}{P}$$

Difference of pressure (ΔP) = Height of chimney (H) × ($\rho_{air} - \rho_{flue}$) kg/mm²

$$\Delta P = 22 \text{ mm of water} = 22 \text{ kg/m}^2$$

(As 9.8 Pa = 1 mm of water)

$$\rho_{air} = \frac{1}{V} = \frac{P}{RT}$$
$$= \frac{1.013 \times 10^5}{287 \times (273 + 20)} = 1.2 \text{ kg/m}^3$$

(P = Pressure = 760 mm of Hg = 1.013×10^5 Pa

$$\rho_{flue} = \frac{1}{V} = \frac{P}{RT} = \frac{1.013 \times 10^5}{257 \times (273 + 290)} = 0.69 \text{ kg/m}^3$$

$$\text{Height of chimney (H)} = \frac{\Delta P}{(\rho_{air} - \rho_{flue})} = \frac{22}{1.2 - 0.69}$$
$$= \textbf{43.1 metres} \quad \textbf{Ans.}$$

Example 7.3 *Determine the height of chimney to get net 'draught of 12 mm if the total draft losses are 4 mm. The temperature of air is 25°C and the temperature of chimney gases is 300°C. The mass of air used per kg of fuel used is 18 kg. One kg of air occupies a volume of 0.7734 m³ at N.T.P.*

Solution. Let

H = Height of chimney in metre
A = Gross sectional area of chimney in m²

$$\text{Density of air} = \frac{1}{0.7734} \times \frac{273}{273 + 25} = 1.24 \text{ kg/m}^3$$

Density of chimney gases at 0°C $= \dfrac{18+1}{18 \times 0.7734} = \dfrac{19}{18 \times 0.7734} = 1.365 \text{ kg/m}^3$

Density of chimney gases at 300°C

$$\frac{18}{18 \times 0.7734} \times \frac{273}{273+300} = 0.65 \text{ kg/m}^3$$

Mass of chimney gases $= A \times H \times 0.65$ kg

Mass of equal column of external air

$= A \times H \times 1.24$ kg

Difference in masses $= 1.24\,AH - 0.65\,AH = 0.59A \times H$ kg.

$$P = \text{Draft in Pa} = \frac{0.59\,AH \times 9.8}{A} = 0.59\,H \times 9.8 \text{ Pa}$$

$= 0.59\,H$ mm of water

(As 9.8 Pa = 1 mm of water)

$\therefore$ $0.59\,H = 16$ (as given draft = 12 + 4 mm)

$H =$ **27.1 metres. Ans.**

Example 7.4 *Determine the boiler efficiency if the boiler generate 23,000 kg of steam at* 1MPa *dry and saturated. The amount of coal used is 2100 kg and the calorific value of coal is 28500 kJ / kg. The feed water temperature is 105°C. Specific heat of water= 4.186 J / g°C.*

Solution. The heat of steam at 1 MPa = 2786 kJ/kg.

$$\eta_{\text{Boiler}} = \frac{23{,}000\,[2786 - (4.186 \times 105)]}{2100 \times 36{,}000} = 0.71 = \mathbf{71\%} \quad \textbf{Ans.}$$

Example 7.5 *Prepare the heat balance sheet for a boiler having the following data:*

Steam pressure = 1.3 MPa

Water evaporated = 300,000 kg

Coal fired = 30,000 kg

Superheat temperature = 300°C

Air temperature

(*a*) To air preheater = 45°C

(*b*) From air preheater = 90°C

Feed water temperature

(a) To economiser = 50°C

(b) From economiser = 120°C

Flue gas exit temperature = 115°C

Boiler house temperature = 25°C

Mean specific heat of the flue gases = 1 kJ/kg °K

Coal analysis (as fired)

C = 64%; H_2 = 4.5%; O_2 = 5.5%; H_2 = 1.5%; S = 0.7%;

Ash = 9.8 %

Total moisture = 14%

Calorific value of coal = 28500 kJ/kg

Duration of trial = 8 hours

Flue gas (dry) analysis

$$CO_2 = 12.5\% \text{ ; } CO = 0.4\%$$
$$O_2 = 6\% \text{ ; } N_2 = 81.1\%.$$

Solution. (*i*) To find weight of water evaporated,

Let, W = Actual weight of dry flue gas per kg of coal

$$W = \frac{11CO_2 + 8O_2 + 7(CO + N_2)}{3(CO_2 + CO)} \times \left(C + \frac{S}{1.83}\right)$$
$$= \frac{11 \times 12.5 + 8 \times 6 + 7(0.4 + 81.1)}{3(12.5 + 0.4)} \times \left(0.64 + \frac{0.007}{1.83}\right)$$
$$= 12.6 \text{ kg.}$$

W_1 = Total weight of flue gas per kg of coal = Actual weight of dry flue gas + Moisture evaporated from coal + Water produced from H = 12.6 + 0.14 + 9 (0.045) = 13.14 kg

Now ash content = 9.8%

W_2 = Weight of fuel burnt
$= 1 - 0.098 = 0.902$
W_3 = Weight of air supplied per kg of coal fired
$W_3 = W_1 - W_2 = 13.14 - 0.902 = 12.238$ kg
W_4 = Weight of water evaporated per kg of fuel fired
$$= \frac{300,000}{30,000} = 10 \text{ kg}$$

(*ii*) To Calculate Heat Absorbed:

H_1 = Heat absorbed by air preheater per kg of coal fired.
= Weight of air × Temperature rise × Specific heat
$= W_2 \times (90 - 45) \times 1$
$= 12.6 \times 45 = 567$ kJ

H_2 = Heat absorbed by boiler
= Weight of water evaporated per kg of coal × (Total heat in 1 kg of steam – Heat in feed water – Heat in air)
$= 10[666 - (120 - 0) - 132] \times 4.186 = 17330$ kJ

H_3 = Heat absorbed by economiser
= Weight of water evaporated × Temperature rise
$= 10[120 - 50] \times 4.186 = 2930$ kJ

H_4 = Heat absorbed by superheater
= Weight of water evaporated × [Super heat temperature – Saturated temperature] × Specific heat
$= 10[300 - 190] \times 2.09 = 10 \times 110 \times 2.09 = 2300$ kJ

(*iii*) To Calculate Heat Lost:

H_5 = Heat lost in dry flue gases = Weight of dry flue gases per kg of coal × Temperature rise × Specific heat

$$= \text{W} \times (115 - 25) \times 1$$
$$= 12.6 \times 90 = 1134 \text{ kJ}$$
$$H_6 = \text{Heat loss due to moisture in fuel}$$
$$= \text{w}(100 - t) + L + k(T - 100)$$

where w is the weight of moisture evaporated from coal

$$= 0.14[(100 - 25) \times 4.186 + 2256 + (115 - 100) \times 1.97]$$
$$= 0.14\,[314 + 2256 + 30] = 364 \text{ kJ}$$

(Assuming specific heat as 1.97)

H_7 = Heat lost due to water formed by combustion of hydrogen

$$= 9H\,[(100 - t) + L + k(T - 100)]$$

where H weight of hydrogen per kg of fuel, t is the atmospheric temperature, T is exist gas temperature and k is the specific heat of steam and L is latent heat of evaporation

$$= 9 \times 0.045\,[(100 - 25) \times 4.186 + 2256 + (115 - 100) \times 1.97]$$
$$= 0.405\,[314 + 2256 + 30] = 1053 \text{ kJ}$$

H_8 = Heat lost *due* to incomplete combustion of carbon

$$= \text{C} \times \text{C.V.} \frac{\text{CO}}{\text{CO} + \text{CO}_2}$$

where C is the weight of carbon per kg of fuel, and CO and CO_2 represent the per cent of volume of carbon monoxide and carbon dioxide in flue gases respectively and C.V. is the calorific value of coal

$$H_8 = 0.64 \times 28500 \times \frac{0.4}{12.6 + 0.4} = 561 \text{ kJ}$$
$$H_9 = \text{Heat unaccounted for loses}$$
$$= 28500 - [H_1 + H_2 + H_3 + H_4 + H_5 + H_6 + H_7 + H_8]$$
$$= 28500 - [567 + 17330 + 2930 + 2300 + 1134 + 364 + 1053 + 561]$$
$$= 28500 - 26239 = 2261 \text{ kJ}$$

Example 7.6 *A boiler with maximum continuous steam rating of 1.2 x 10^5 kg per hour is fired with bituminous coal having 10% moisture and 10% ash and a higher caloric value of 31400 kJ/kg. If it takes 4186 kJ to evaporate 1 kg of feed water entering the boiler and superheat it to final temperature and if overall efficiency of boiler is 80% determine the following:*

(*i*) Hourly coal supply.

(*ii*) Grate length, if building dimensions limit grate width to 6 metre. Given that the maximum heat produced is 12.56×10^6 kJ/m²/hr.

Solution. H = Heat required to convert water into steam

$$= 1.2 \times 10^5 \times 4186 = 5.023 \times 10^8$$

$$\text{Efficiency } (\eta) \text{ of boiler} = 0.8$$

$$\text{Heat to be produced by the fuel} = \frac{\text{H}}{0.8}$$

$$H = \frac{5.023}{0.8} \times 10^8 = 6.28 \times 10^8 \text{ kJ/hr}$$

$$\text{Coal required} = \frac{6.28 \times 10^8}{31400} \text{ kg}$$

$$= 2 \times 10^4 \text{ kg/hr}$$

$$\text{Grate area} = \frac{H}{12.56 \times 10^6} = \frac{6.28 \times 10^8}{12.56 \times 10^6} = 50 \text{ square metres}$$

$$\text{Grate length} = \frac{50}{6} = \textbf{8.33 metres.} \quad \textbf{Ans.}$$

Example 7.7 *The following data is supplied for a boiler plant:*

(*i*) Boiler

Mass of coal supplied = 220 kg/hr.
Calorific value of coal = 30140 kJ/kg
Mass of feed water = 2000 kg/hr
Enthalpy of steam produced = 2800 kJ/kg

(*ii*) Economiser

Inlet temperature of feed water = 18°C
Outlet temperature of feed water = 88°C
Atmospheric temperature = 19°C
Temperature of flue gas entering = 350°C
Mass of flue gases = 4150 kg/hr.

Determine the following:

(*a*) Efficiency of boiler (*b*) Efficiency of economiser
(*c*) Efficiency of whole boiler plant.

Solution. (*a*) W = Mass of coal supplied = 220 kg/hr

CV = Calorific value of coal = 30140 kJ/kg

H_1 = Heat obtained by combustion of coal

$= W \times CV = 220 \times 3140$

$= 6.63 \times 10^6$ kJ/hr

W_1 = Mass of feed water = 2000 kg/hr.

H_2 = Heat utilised in evaporation of steam

$= W_1 (h - h_1)$

where h = Enthalpy steam = 2.8×10^3 kJ/kg

h_1 = Enthalpy of feed water = 368 kJ/kg

$H_2 = 2000(2800 - 368)$

$= 4.86 \times 10^6$ kJ

$$\text{Boiler Efficiency} = \frac{H_2}{H_1} \times 100 = \frac{4.86 \times 10^6}{6.63 \times 10^6} \times 100 = 73.5\%. \quad \textbf{Ans.}$$

(*b*) T_1 = Inlet temperature of feed water = 18°C

T_2 = Outlet temperature of feed water = 88°C

H_3 = Heat utilised by economiser
$= W_1 (T_2 - T_1) = 2000 (88 - 18) \times 4.186 = 586 \times 10^3$ kJ
W_2 = Mass of flue gases = 4150 kg/hr.
T_3 = Temperature of flue gases = 350°C
T_4 = Atmospheric temperature = 19°C
C_p = Specific heat of flue gases = 1 kJ/kg/°K
H_4 = Heat supplied to economiser
$= W_2(T_3 - T_4) \times C_p = 4150(350 - 19) \times 1$
$= 1.37 \times 10^6$ kJ/hr

$$\therefore \quad \text{Economiser efficiency} = \frac{H_3}{H_4} \times 100 = \frac{586 \times 10^3}{1370 \times 10^3} \times 100 = \mathbf{42.5\%} \quad \textbf{Ans.}$$

(c) H = Total heat utilised

$H_2 + H_3 = 4.86 \times 10^6 + 586 \times 10^3$
$= 5.446 \times 10^6$ kJ/hr
η = Over–all efficiency

$$= \frac{H}{H_1} \times 100 = \frac{5.446 \times 10^6 \text{ kJ/hr}}{6.63 \times 10^6} \times 100 = \mathbf{82.3\%.} \quad \textbf{Ans.}$$

This shows that by installing economiser the efficiency of the boiler plant is increased.

QUESTIONS

7.1 Give flow sheet or layout of a steam power stations indicating major components.

7.2 Draw the layout of a big thermal power station showing all major components.

7.3 Draw a general layout of a thermal power plant and explain working of different circuits.

7.4 Define a steam boiler. State the uses of steam produced in the steam boiler.

7.5 What is a fire tube boiler? How does it function? Describe any one of the fire tube boilers with the help of a neat sketch,

7.6 Describe Babcock and Wilcox water tube boiler with the help of a neat sketch.

7.7 Discuss the advantages of water tube boilers over fire-tube boilers.

7.8 Draw and describe a Lancashire boiler.

7.9 Explain the terms (*a*) mountings of boilers, and (*b*) accessories of boilers.

7.10 Draw and describe the functions and working of: safety valves, pressure gauge, water level indicator, fusible plug, Steam stop valve.

7.11 Draw and describe the function and working of the following accessories:
Economiser, Air preheater, Superheater.

7.12 Define "Draught" in a boiler. How will you classify "draught".

7.13 Explain different types of draughts used in boilers.

7.14 Draw and describe "balanced draught"

7.15 Describe with a sketch the steam jet draught produced in a locomotive boiler.

7.16 Explain different types of fuels used in boilers.

7.17 Explain different systems of handling of ash produced by power plants. What are the important methods for disposal of ash?

7.18 How will you prevent atmospheric pollution through fuel gases?

7.19 How will you classify various types of boilers?

7.20 What is the difference between water-tube and fire-tube boilers? Describe the working principle of Cochron or Lancashire boilers.

7.21 What is a superheater? Describe different types of superheaters.

7.22 State the advantages of superheated steam. Explain the working of a superheater.

7.23 Write short notes on the following:

(*a*) Maintenance of a boiler.

(*b*) Efficiency of a boiler.

(*c*) Water tube failure and prevention.

7.24 State the advantages of high pressure boilers.

7.25 Write short notes on the following:

(*a*) Economiser (*b*) Hydraulic test of a boiler

7.26 What are the high pressure boilers and how they differ in construction and working principle from ordinary boilers? What are the advantages gained by forced water circulation?

7.27 Describe construction and working of any two of the following boilers giving suitable sketches: LaMont, Loeffler, Benson and Velox boilers.

7.28 Write short notes on the following:

(*a*) Advantages of high pressure boilers.

(*b*) Chimney for a steam boiler.

(*c*) Heat losses in a boiler.

7.29 Explain diffrent methods of increasing the thermal efficiency of a boiler.

THERMAL POWER PLANTS: 2—STEAM ENGINES, TURBINES AND CONDENSERS

Chapter 8

Learning Objectives

After studying this chapter, the students will learn:

- Steam engine and their construction and working.
- Steam turbines, its working principle and types.
- Compounding of steam turbines.
- Velocity compounding and velocity diagram of steam turbines.
- Steam cycles, namely carnot cycle, rankine cycle, reheat cycle and turbine efficiencies.
- Governing of steam turbine.
- Condensers; its functions, elements and efficiency.
- Types of steam condensers.
- Air extractors and cooling water system.

8.1 STEAM ENGINES

Steam Engine was invented by New common in the year 1706. Steam Engine is a heat engine in which working substance is steam, which is used to drive a tight fitting piston which moves a wheel. Thus steam engine converts the heat energy of steam into mechanical energy of piston.

In the initial days the utility of steam engine in industry and transport was so great that the invention of steam engine led to a historic revolution called industrial revolution. The steam engines were used to run trains, industrial machines, and water. With the introduction of steam turbines and I.C. engine, these have become obsolete, since steam engines are very heavy, take long time to start, and have very low efficiency.

The cylinder of the steam engine has two rectangular ports through which steam can enter and leave the cylinder. These ports lead to the steam chest, which is cast integral with the cylinder. This steam chest contains steam direct from the boiler. The cast iron piston having piston rings is pushed by the steam entering from one end of the cylinder to the other and *vice versa*. The piston rod which passes the cylinder end through a steam tight stuffing box, joins the piston to a cross-head, which slides between the guide bars in a straight line. This cross head is also joined by a connecting rod to the crank. The connecting rod converts the straight sliding motion of the piston and cross head to the rotary motion of the crank. Crank is attached to the crank shaft like an I.C. Engine for producing rotary motion.

The sliding valve provided in the steamchest controls the flow of steam into, and out of the cylinder. The valve is a rectangular box of *'D'* shape opening at its lower face and slides over a machined face of the side of the cylinder. On this face, three rectangular openings are provided out of which two are for the steam ports leading in the opposite ends of the cylinder, and one is for the exhaust port leading to the outside in the air or into a condenser. The movement of slide valve uncover each of the two steam ports in turn from where live steam enters from the steam chest to the cylinder and pushes the piston to move. Inside of the valve provides a passage for exhaust steam to travel from other steam ports to the larger exhaust port. This movement of piston converts the rotary motion of the crank shaft through piston shaft, cross-head and crank (see Fig. 8.1).

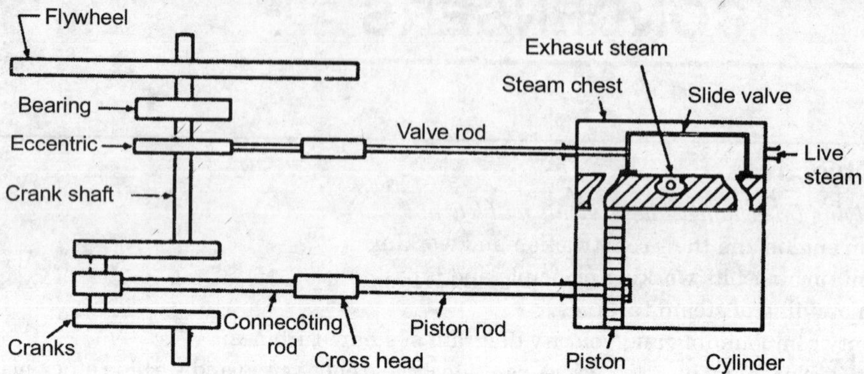

Fig. 8.1 Construction of Steam Engine.

8.2 CONSTRUCTION DETAILS OF STEAM ENGINES

(i) Cylinder. It is a cast iron cylinder inside of which is bored perfectly true. It consists of a steam chest connected to it by means of passages known as "steam ports". Over these ports a valve slides which is known as "D-slide valve". The passage which is called the "exhaust port".

Both the ends of the cylinder are closed by means of covers. The piston rod passes through one of the holes in the cylinder cover. The leakage of steam around the piston-rod is prevented by means of a stuffing box provided at the cover hole. When the cylinder wears out too much, then to replace the whole of the cylinder block is very costly. Therefore sometimes cylinders are fitted with "liner" made of cast iron or steel. An annular space is left between the linear and cylinder body which is fed with live steam from the boiler to prevent the condensation of steam in the cylinder. This space is called the "steam jacket".

(ii) Piston. It is a sliding plug which slides from end to end of the cylinder, under the pressure of the steam and through which the energy of the steam is converted into the motion of mechanism. The body of the piston is turned to a size which is an easy fit in the cylinder. To avoid the leakage of steam from the side of the piston, some spring rings are provided on the periphery of the piston. These rings are also known as compression rings.

Generally the pistons of steam engine are made of cast iron, which may be of box type or plate type. The piston rod is connected to the piston by means of a nut screwed on the end of piston rod.

(iii) Stuffing Box. A stuffing box is provided on the crank end of the cylinder. It ensures the free movement of the piston rod without allowing any leakage of the steam. It is also used on the steam chest where valve rod passes out of the valve chest.

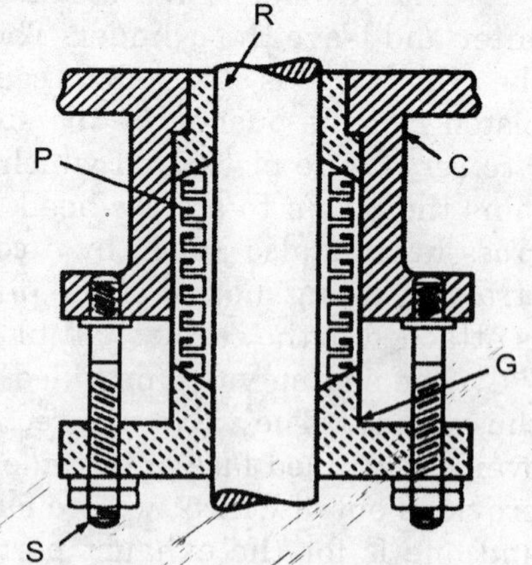

Fig. 8.2 Stuffing Box.

Referring Fig. 8.2, consists of a box C which is an extended portion of the cylinder. A gland G *is* fitted with the stuffing box by studs S. The piston rod R passes through the stuffing box and the gland. The space between the rod and the stuffing box contains the packings *P* which may be pressed by the gland thus to prevent leakage.

(iv) Connecting Rod. The connecting rod connects the cross-head with the crankpin on the crankshaft. Its function is to transmit the reciprocating motion to the piston to the crank of the crank shaft in which it is converted into circular motion. It has two ends, the end which connects the crosshead is known as small end and the crank end is known as big end. The connecting rod is generally made of forged steel of circular, rectangular or I-section rod.

(v) Crosshead. It is a H shaped block to which the ends of the connecting rod and the piston rod are joined. Usually the end of the piston rod is forged or attached rigidly with the cross-head and the small end of the connecting rod is connected with it by means of a cross-head pin. The sliding motion of the crosshead is controlled by means of guide bars provided parallel to the piston rod.

(vi) Crank Shaft. The function of a crank shaft is to convert the reciprocating motion of the piston to the rotary motion. The distance between the centre of the crank pin and the centre of the crank shaft is called as the radius of the crank. It consists of shaft, crank arm and crank pin. Generally the unit as a whole is a steel forging. To transmit reciprocating motion, the big end of the connecting rod is connected to the crank pin of the shaft.

Sometimes when the dimensions become sufficiently large, the different components are made in separate forgings which, after being machined, are assembled together to form a "built-up" crank shaft. The common types of the crank-shaft are as given under (see Fig. 8.3(a), (b) & (c).

(*a*) Overhung crank, (*b*) Centre crank, (*c*) Disc crank.

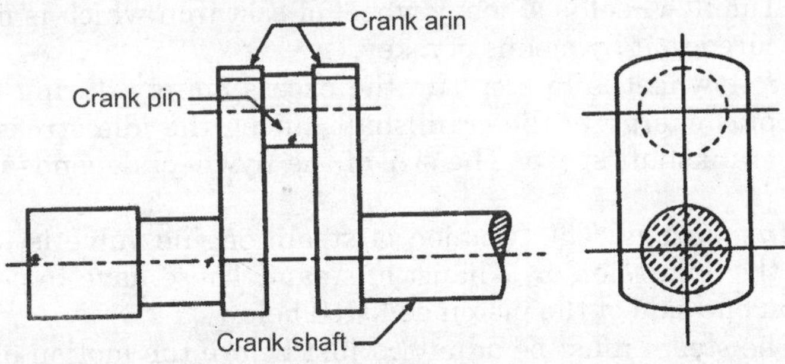

Fig. 8.3 (*a*) Overhung Crank.

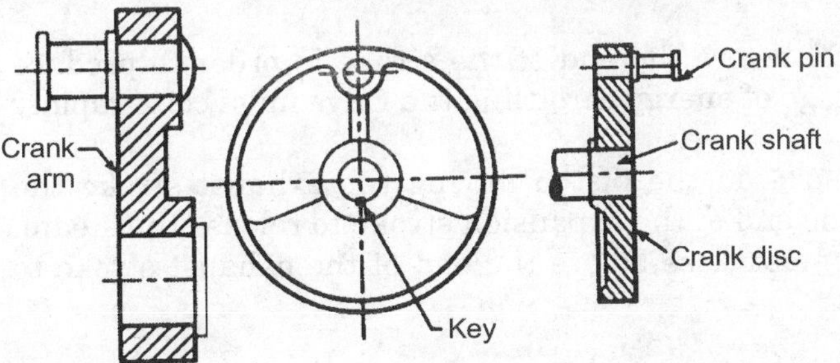

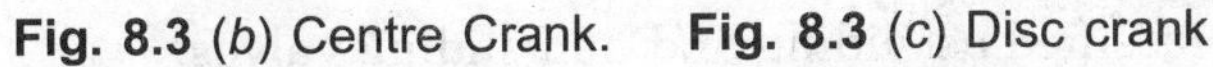

Fig. 8.3 (*b*) Centre Crank. **Fig. 8.3** (*c*) Disc crank

(vii) Eccentric. The function of the eccentric is just the reverse to that of the crank shaft. It converts the rotary motion of the crank shaft into the reciprocating motion of the slide valve.

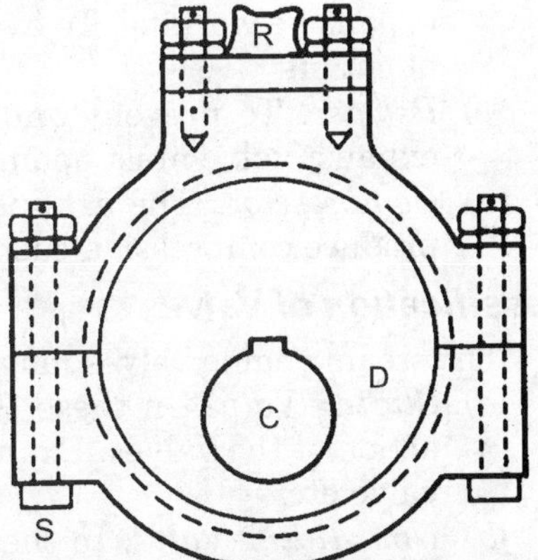

Fig. 8.4 Eccentric. D-Eccentric disc, C-Crackshaft, S-Strap, R-Eccentric rod.

Referring Fig. 8.4, it consists of a disc to which the crank shaft *C* is fitted out of centre. The disc *C* is free to rotate inside the two straps *S*. One strap is rigidly fixed with an end of eccentric rod *R*. When crankshaft *C* rotates then the disc *D* also rotate, with the result to provide reciprocating motion in rod *R*. The reciprocating motion of slide-valve fixed at the other end of the rod *R* just double that of the distance between the centre of the disc and the axis of the shaft.

(viii) Relief Valve and Cock. When the steam, inside the cylinder is condensed and the piston reciprocates continuously then sometime excessive pressure may be developed at the end of the piston which causes the bursting of the cylinder cover. To overcome this danger a safety device is provided at the cylinder cover which is called relief valve.

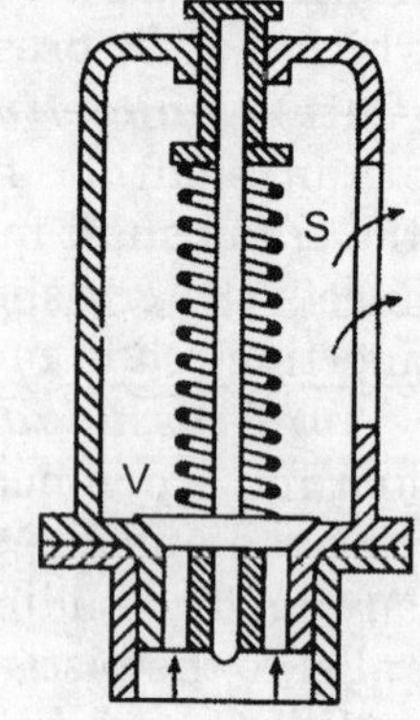

Fig. 8.5 Relief valve. S-Spring, V-Valve.

For working refer Fig. 8.5, when the cylinder pressure increases sufficiently during the return stroke of piston, the conical valve is lifted off its seat against the tension in the spring *S,* thereby permitting the condensed steam to escape out in the direction shown by the arrows.

Difference between Crank and Eccentric

(*i*) The crank converts the reciprocating motion of the piston transmitted through connecting rod into the rotary motion. The eccentric converts the rotary motion of the crankshaft into reciprocating motion of the slide valve.

(*ii*) The crank in a rod bent at different points. The eccentric is circular disc consisting of an axle out from the centre.

(*iii*) Kinematically both are identical but the crank has developed from the eccentric.

(ix) Fly-Wheel. The flywheel is a heavy wheel of cast iron which is mounted directly on the crankshaft and secured to it by means of a key.

The function of a flywheel is to store up the excess energy during the working stroke and to return the surplus energy to the crankshaft during the idle stroke. Thus minimizing the fluctuation of the crankshaft speed. The size of the flywheel depends upon the speed and type of the engine.

(x) Steam Engine Valves. The function of steam engine valve is to close or open the parts of cylinders for the admission or exhaust of steam. These have to perform four distinct operations on steam on one side of the piston as listed below.

(*a*) *Admission.* The steam must be admitted just before the motion of the stroke so that the beginning of the outward motion of the piston will be performed under the full pressure of the steam.

(*b*) *Cut-off.* The steam must be cut-off before the end of the stroke in order to produce expansive working. To avoid wastage of energy throttling the valve must close rapidly at cut-off.

(*c*) *Release.* To prevent undue pressure on the piston during the exhaust stroke the exhaust valve must open before the end of the expansion stroke to release the steam.

(*d*) *Compression.* The exhaust valve must close before the end of the exhaust stroke to produce compression cushioning.

Classification of Valves

The steam engine valves may broadly be classified as sliding and non-sliding valves.

(*a*) *Sliding Valve.* In these the control of opening in the cylinder is done by the sliding action of the valves. Example of sliding valve are D-slide valve, piston valve, Meyer's valve etc.

(*b*) *Non-sliding Valve.* In these the control of port opening is done by the rise and fall or by the twisting action of the valves. Examples of non-sliding valves are drop valves and the corliss valve etc.

'D' Slide Valves

This valve slide on the face of the cylinder. Its shape is that of English letter D and that is why it is known as D valve. It is made of iron, gun-metal or phospher bronze. It has ground and finished seats so that it may slide over the cylinder very smoothly. It is operated through the eccentric or valve rod from an eccentric mounted on the crankshaft. During every revolution of the crack, the valve completes one forward and backward stroke, thus opening and closing the rectangular ports in the cylinder head and performing desired operations.

8.3 CLASSIFICATION OF STEAM ENGINES

Steam engines may be classified into the following groups.

(*i*) According to the action upon the piston:

(*a*) Single action engine: Steam acts on the side of the piston.

(*b*) Double acting engine: Steam acts alternately on both the sides of the piston.

(*ii*) According to the nature of expansion of steam.

(*a*) No expansion: No expansion of steam takes place inside the cylinder.

(*b*) Expansive engine: There will be an expansion of steam inside the cylinder.

(*iii*) According to the range of steam:

(*a*) Simple steam engine: Total expansion of steam occurs in a single cylinder.

(*b*) Compound steam engine: Expansion takes place in more than one cylinder.

(*iv*) According to the method of exhaust of steam.

(*a*) Non-condensing steam: Steam is exhausted from the engine cylinder into the atmosphere.

(*b*) Condensing steam engine: Exhaust steam is discharged into a vessel known as condenser.

8.4 GOVERNING OF STEAM ENGINES

As the variation of speed during a cycle due to the variation of the pressure of the working fluids inside the cylinder is regulated by a fly-wheel, but it cannot control the variation of speed due to change in load. Therefore a steam engine is provided with a governor the function of which is to regulate the variation of speed due to the change of load.

The common types of Governors are as given under:

(*a*) Simple pendulum governor.

(*b*) Loaded governor (Porter governor).

(*c*) Parabolic and crossed-armed governor.

(*d*) Spring loaded governor.

Simple pendulum governor. Referring Fig. 8.6, it consists of a vertical spindle S to which two arms, A are hinged. Two weights B are fitted at the other ends of the arms A. There are two links K one end of which is hinged with the arms and the other end with the sleeve E.

When the power from the crankshaft is transmitted through a pair of bevel gears G to the main spindle, then due to centrifugal force the weights B fly away from the spindle axis, thereby moving the sleeve E upward direction, and when the speed decreases the sleeve moves in the downward direction. This movement of the sleeve through the lever L and L' controls the throttle valve T. Thus controlling the flow of steam passing through the pipe N and entering the engine cylinder. In this way the supply of steam is adjusted with the load and the engine maintains a normal speed.

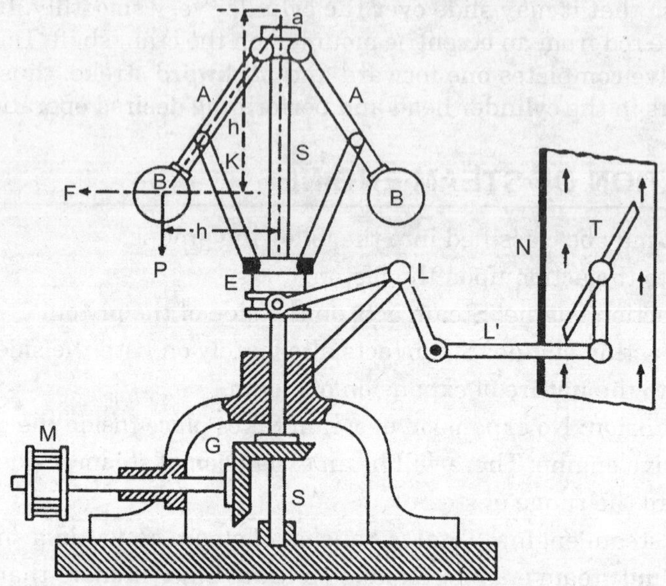

Fig. 8.6 Simple Pendulum Governor.
S-Spindle, A-Arms, K-Link, B-Fly-weight, E-Sleeve, L, L-Lever, G-Bevel gear, M-Pulley, N-Pipe, T-Throttle valve.

Methods of Governing

1. *Throttle Governing.* As already explained in the simple pendulum governor, the system consists of a throttle valve placed inside a pipe through which the steam may flow in the cylinder.

The function of the throttle valve is to control the flow of steam by varying the opening of the passage. Increase of engine speed causes the throttle valve to close the pipe and the decrease of the same causes the throttle valve to open the pipe.

2. *Cut-off Governing.* The system in which the action of governor alters the point of admission of steam to the engine cylinder by varying the point of cut-off is called "cut-off governing".

Difference between Fly-Wheel and Governor

(*a*) A flywheel controls only the variation in speed during one cycle of the engine, where the governor, exercises control on the mean speed of rotation of the engine over long period of several cycles.

(*b*) The flywheel cannot regulate the speed with variation of load on the engine. Where as the governor maintains fairly the same speed even with the increase or decrease of load.

(*c*) The governor control the mean speed of the engine by means of controlling the admission of the steam to the cylinder. Where as the cyclical fluctuation during a cycle is kept down by the flywheel.

(*d*) The action of the governor varies according to the load on the engine, where as the action of the flywheel remains continuous over the whole operation.

8.5 INDICATOR-DIAGRAM

The theoretical cycle on which the steam engine works is known as the Rankine cycle. It is an idle cycle with steam as a working fluid. Referring Fig. 8.7 the four operations of the cycle are as given below:

Operation 1—2. Let the unity amount of water is in the cylinder at a pressure p_1 and at the temperature of formation T_1. A hot body is applied to the end of the cylinder, thus warming the water to a temperature T_2, and at the same time let the pressure of the water rise *to* p_2. T_2 *is* the temperature of vaporation at pressure p_2.

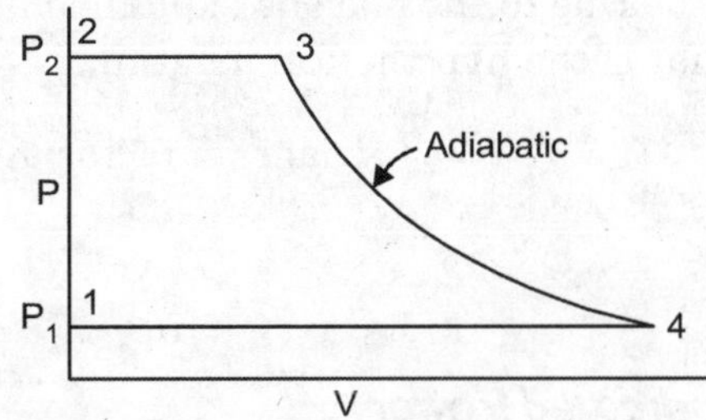

Fig. 8.7 Rankine cycle on *p-v* diagram.

Operation 2—3. The water is now evaporated at constant pressure p_2 to steam. The latent heat is absorbed from the hot body. Work is done on the piston.

Operation 3—4. The hot body is now removed and the steam is expanded adiabatically to a pressure p_1 and corresponding T_1. Work is done on the piston.

Operation 4—1. A cold body is now brought into contact with the cylinder end, and heat flows from the steam to the cold body, this condenses the steam into water, at constant pressure, its latent heat being rejected to the cold body. The cycle is thus completed.

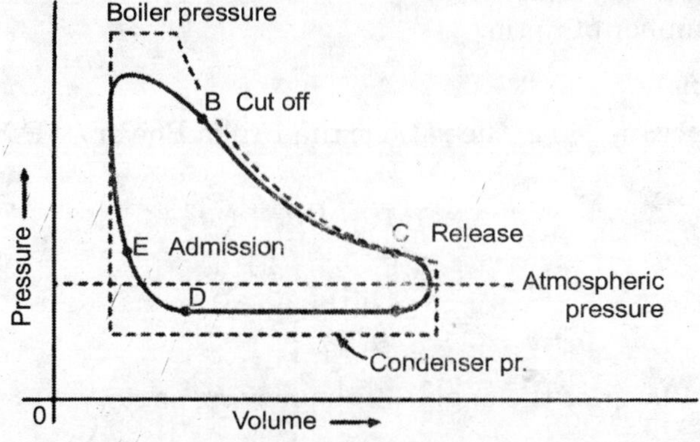

Fig. 8.8 Actual indicator diagram.

Actual Indicator Diagram

Indicator diagram obtained from an actual engine differs from the theoretical diagram due to following reasons (refer Fig. 8.8).

(*i*) There is always a clearance volume in an engine.

(*ii*) The pressure of the steam at entry to the cylinder is always less than the boiler pressure due to the pressure drop in the steam pipe and the throttling effect of the valves, and it also falls until cut-off occurs.

(*iii*) Cut-off does not take place as a definite point, since valve closure is not instantaneous.

(*iv*) Actual expansion curve is slightly lower than the theoretical line due to condensation of some of the steam on the cylinder walls.

(*v*) The steam starts releasing from the cylinder slightly before the end of the stroke, as some time is required for the steam to fall to the back pressure.

(*vi*) The back pressure is slightly higher than the condenser or exhaust pressure.

(*vii*) The exhaust valve is closed slightly before the end of the exhaust stroke, trapping some steam in the cylinder. This is necessary to provide a cushioning effect for the piston at the end of its stroke.

(*viii*) The live steam is supplied to the engine just before the end of the exhaust stroke.

Due to the reasons explained above, the area of actual indicator diagram is smaller than that of the hypothetical diagram.

$$\text{Diagram factor, } K_d = \frac{\text{Actual diagram area}}{\text{Hypothetical diagram area}}$$

$$= \frac{\text{Actual mean effective pressure}}{\text{Hypothetical diagram pressure}}$$

Indicated Power

Neglecting the area of piston rod.

Indicated power in kW

$$\text{I.P.} = \frac{2\, p_m\, LAN}{60{,}000}$$

where, p_m = Mean effective pressure in Pa

L = Length of stroke in m

A = Area of cylinder in m^2

N = Number of r.p.m.

Mechanical Efficiency

Mechanical efficiency ($\eta_{mech.}$) is the ratio of the Brake Power (B.P.) in kW to the Indicated Power (I.P.) in kW.

$$\eta_{mech} = \frac{\text{B.P.}}{\text{I.P.}} = \frac{\text{I.P.} - \text{F.P.}}{\text{I.P.}}$$

$$= \frac{\text{B.P.}}{\text{B.P.} + \text{F.P.}}$$

where, FP = Frictional Power of the engine of the moving parts.

Thermal Efficiency

Thermal efficiency ($\eta_{thermal}$) are either based on B.P. or I.P. of the engine, are known as Brake thermal efficiency or Indicated thermal efficiency respectively.

Indicated Themal Efficiency

$$= \frac{\text{Energy in I.P.}}{\text{Energy in steam supplied}}$$

$$= \frac{\text{I.P.} \times 60{,}000}{W \times (h_1 - h_w)}$$

where, W = Weight of the steam used per min.

h_1 = Enthalpy of the steam supplied.

h_w = Enthalpy returned in water.

Similarly,

$$\text{Brake Thermal Efficiency} = \frac{\text{B.P.} \times 60{,}000}{W \times (h_1 - h_w)}$$

8.6 COMPOUND STEAM ENGINES

Use of high pressure steam with the same back pressure requires a very long cylinder for large stroke. This also results in condensation of the steam entering the cylinder.

To overcome these difficulties a compound steam engine is evolved. The expansion of steam in compound engine is carried out in two difficult cylinders. High pressure steam is expanded to an intermediate pressure in first cylinder, and then this steam passes on to another low pressure cylinder where expansion takes place down to the back pressure.

Methods of Compounding Steam Engines

Steam engines can be compounded in two ways, namely, Tandem Compounding. Receiver type compounding. These are shown in Figs. 8.9 (a) and 8.9 (*b*).

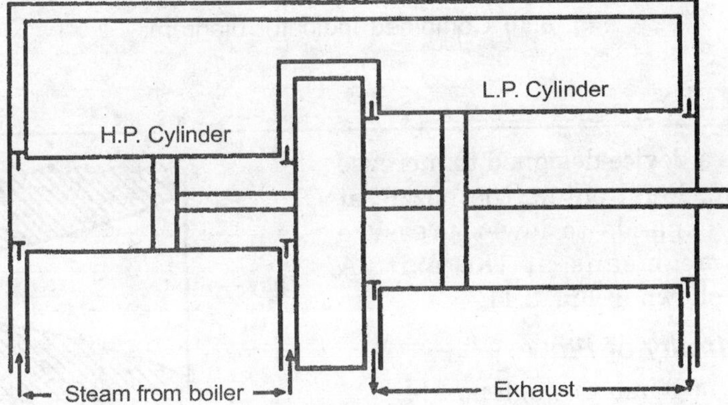

Fig. 8.9 (*a*) Tandem Compound steam engine.

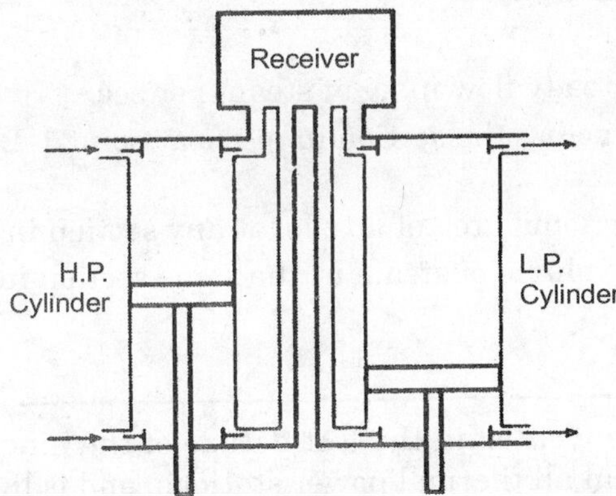

Fig. 8.9 (*b*) Receiver Type Compound steam engine.

Combined Indicator Diagram

In the testing of 2 cylinder compound steam engine, the indicator diagram for each of the cylinder is taken. These diagrams are then replotted between the same ordinates and to the same scale. This combined diagram represents the complete expansion of the steam as shown in Fig. 8.10.

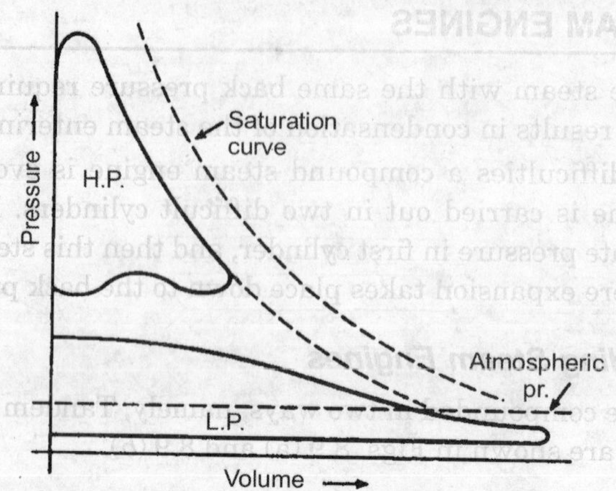

Fig. 8.10 Combined indicator diagram.

8.7 STEAM NOZZLES

Steam nozzle is a device designed to increase the velocity of steam. Fluid enters the nozzle at high pressure and expands to lower pressure maintained by some means at the exit. A converging nozzle is shown in Fig. 8.11.

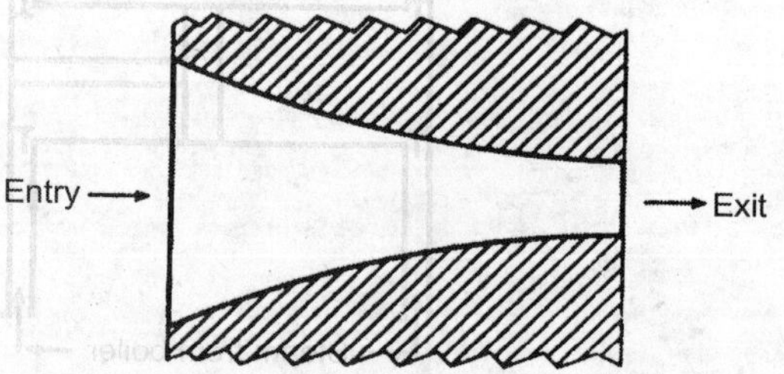

Fig. 8.11 Converging steam nozzle.

Equation for Continuity of Flow

For steady flow of steam through nozzle,

$$w = \frac{AV}{v_s} = \frac{A_1V_1}{v_{1s}} = \frac{A_2V_2}{v_{2s}}$$

where, w = rate of steady flow in kg of steam per sec.

V = velocity across the section in m/sec at point 1, 2 etc. along the axial length of the nozzle.

A = cross-sectional area of nozzle, at any section in m^2.

v_s = Specific volume of steam at the same section in m^3.

8.8 STEAM TURBINES

The steam turbine was invented in the last decade of the nineteenth century. These are mainly used as prime-mover in all thermal power stations, and is likely to continue to be used predominantly in the foreseeable future. Now-a-days, steam turbines upto 1000 MW capacity are being manufactured, and of capacity upto 1500 MW are planned for future.

Thus a steam turbine is a prime-mover which continuously converts the energy of high-pressure, high-temperature steam supplied by a boiler plant into shaft work with the low temperature steam exhausted to a condenser. The energy of the steam is converted into mechanical work in following two steps:

1. The high-pressure, high temperature steam first expands in nozzles and comes out of at a high velocity.

2. The high-velocity jets of steam coming out of the nozzles, impinge on the blades mounted on a wheel, get deflected by an angle and suffer a loss of momentum which is observed by the rotating wheel in producing torque.

In steam turbines, steam is allowed to expand in stationary nozzles or blades, where it acquires a high velocity. These nozzles or blades are mounted on the periphery of a fixed disc called diaphragm. These high velocity steam jets are then directed into a ring of blades fixed to the rim of wheel or rotor, which is free to revolve. The effective force of these jets, acting on these blades, rotate the wheel.

Main Components

Main parts of a steam turbine are as under:

1. A rotor on the circumference of which a series of blades or buckets are attached. The blades are so designed that they are able to withstand the action of steam and the centrifugal force caused by high speed. As the steam pressure drops the lengths and size of blades should be increased in order to accommodate the increase in volume.
2. Bearing to support the shaft.
3. Metallic casing which surrounds blades, nozzles, rotor, etc.
4. Governor to control the speed.
5. Lubricating oil system.

8.9 TURBINE PRINCIPLE

The steam turbine depends upon the dynamic action of the steam. The pressure of the steam is allowed to fall in a passage or nozzle in order to convert pressure energy into kinetic energy. High velocity steam when strikes the moving part of the turbine (known as blades), suffers a change in direction of motion and thus gives rise to change in momentum, and thereby a force, as shown in Fig. 8.12.

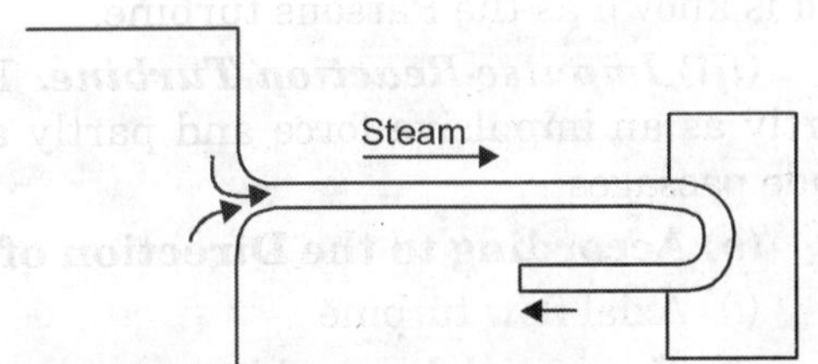

Fig. 8.12 Steam striking the turbine bucket.

The high velocity jets are converted to mechanical work in a rotating shaft by means of buckets.

The steam turbine obtains its motive power from the change of momentum of a steam jet flowing over a curved blade. In moving over the curved surface of the blade, the steam jet exerts a centrifugal pressure normal to the blade surface along the whole length of the blade as shown in Fig. 8.13. The resultant of these centrifugal pressures, plus the force due to change of velocity, is the motive force on the blade. It should be noted here that the blade obtains no motive force from the static pressure of the steam or from any impact of the steam jet, because the blade is designed so that the steam jet will glide on and off the blade without any shock.

In most modern steam turbines, steam is allowed to expand in stationary *nozzles* or *blades,* where it acquires a high velocity. These nozzles or blades are mounted on the periphery of a fixed disc called *diaphragm.* These high velocity steam jets are then directed on to a

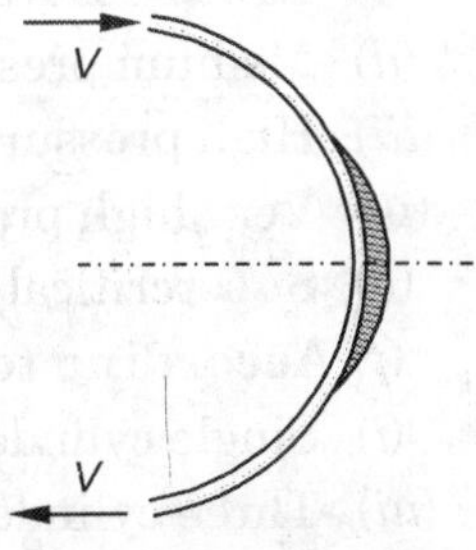

Fig. 8.13

ring of blades fixed to the rim of a *wheel* or rotor, which is free to revolve. The effective force of these jets, acting on these blades, rotate the wheel. Several of these wheels of moving blades are keyed to the same shaft, having a ring to fixed blades between each wheel of moving blades. These fixed blades face in the opposite direction to the moving blades, and are fixed to the turbine casing; their object is to receive the steam jet discharging from the moving blade ring and to divert it on to the next ring of moving blades by changing its direction.

A pair of blade rings consisting of a fixed ring of blades and a moving ring of blades is known as a *turbine pair or stage.*

8.10 CLASSIFICATION OF STEAM TURBINES

Steam turbines can be classified in several ways:

(*a*) According to the Action of Steam on the Blades:

(*i*) *Impulse Turbine.* The steam is expanded in nozzles only ; when passing over the fixed and moving blades its pressure remains constant. The steam jets pass over several rings of moving blades until its kinetic energy is expanded. The original steam turbine, the De Laval, was an impulse turbine having a single blade wheel. Other impulse turbines are the Curtis, Zoelly, and Rateau.

(*ii*) *Reaction Turbine.* In this turbine the steam is not expanded in nozzles, but expands as it flows over the fixed and moving blades. The passages between the blades are made convergent to control the rate of increase of the velocity of the steam, consequent on a pressure and heat drop through the blading. The reaction turbine was invented by Sir Charles Parsons, and is known as the Parsons turbine.

(*iii*) *Impulse-Reaction Turbine.* The impulse-reaction turbine obtains its motive force partly as an impulsive force and partly as a reactive force due to expansion of steam in the blade passages.

(*b*) According to the Direction of Steam Flow.

(*i*) Axial flow turbine
(*ii*) Radial flow turbine
(*iii*) Tangential flow turbine
(*iv*) Mixed flow turbine
(*v*) Helical flow turbine
(*vi*) Re-entry turbine.

(*c*) According to the Number of Stages:

(*i*) Single stage turbine
(*ii*) Multistage turbine.

(*d*) According to the Condition of Exhaust Steam:

(*i*) Condensing type turbine
(*ii*) Non-condensing type turbine.

(*e*) According to the Pressure of Steam at Inlet:

(*i*) Low pressure turbine. Pressure of 0.12 MPa to 0.2 MPa
(*ii*) Medium pressure turbines. Pressures upto 4 MPa.
(*iii*) High pressure turbines. Pressure 4 MPa and above.
(*iv*) Very high pressure turbines. Operation at pressures 17 MPa and above.
(*v*) Supercritical pressure turbines. Operating at pressures above 22.5 MPa.

(*f*) According to the Number of Cylinders:

(*i*) Single cylinder
(*ii*) Double cylinder
(*iii*) Three cylinder
(*iv*) Four cylinder.

(g) According to the General Arrangement of Flow

(i) Single flow
(ii) Double flow
(iii) Compound flow
1. Tandem 2. Cross compound 3. Vertical compound.

(h) According to the Method of Governing:

(i) Throttle governed (ii) Nozzle governed
(iii) By pass governed (iv) Combination of throttle and nozzle
(v) Combination of throttle and by-pass,

(i) According to the Heat Drop Process:

(i) High pressure non-condensing (ii) High pressure condensing
(iii) Back pressure turbine (iv) Topping or superposed turbine
(v) Mixed pressure turbine (vi) Regenerative or bleeder turbine
(vii) Passout or extraction turbine (viii) Reheat turbine
(ix) Low pressure or exhaust turbine (x) Turbine working on binary vapour cycle.

(j) According to the Usage:

(i) Constant speed stationary for power generation
(ii) Variable speed stationary for turbo-blowers.
(iii) Variable speed mobile turbines used for ships.

8.11 ADVANTAGES OF STEAM TURBINE

(i) *Thermal efficiency.* Thermal efficiency of steam turbine is higher than that of reciprocating steam engines (steam turbines 35 to 40% ; steam engines 15 to 20%).

(ii) *Work done per kg of steam.* In steam turbines, the steam can be expanded up to pressures as low as .0035 MPa whereas in reciprocating steam engines it is 0.2 to 0.3 MPa. Thus, more work is obtained per kg of steam in case of steam turbines for the same initial steam conditions.

(iii) *Speed.* Much higher speeds may be developed, and a far greater speed range is possible than in the case of reciprocating .steam engine. (*Engine speed about 150 r.p.m.*)

(iv) *Balancing vibrations foundation.* There are no reciprocating parts in steam turbines and therefore perfect balancing is theoretically possible. Due to balanced rotor, less vibration and consequently lighter foundations are made.

(v) *Capacity.* Steam turbines have been manufactured to develop 500 MW and even more.

(vi) *Flywheel.* Steam turbines generate power at the uniform rate (no cyclic fluctuation of energy) and require no flywheel.

(vii) *Internal lubrication.* Due to the absence of piston cylinder in case of steam turbines, no internal lubrication is required, which reduces the cost of lubrication and makes pure feed available for boiler.

(viii) *Steam Consumption.* Unlike reciprocating engine the steam consumption of the turbine does not increase with years of service.

(ix) *Loss due to condensation.* Because the parts of steam turbine are subjected to constant pressure and temperature at constant loads, there is no initial condensation of steam and consequently no loss.

(*x*) *Overloading.* Steam turbines can be overloaded considerably with only a slight reduction in overall efficiency.

(*xi*) *Mechanism.* Due to the absence of intermediate links between input and output like piston, piston rod, cross head, crankshaft etc., the mechanism is simple in case of steam turbines.

(*xii*) Turbines can be designed so that steam can be held at any pressure to be used for heating feed water (regenerative feed heating) or for any other industrial purpose.

In fact, when properly designed and constructed, the steam turbine is a very durable prime-mover. Steam engines can compete with steam turbines only in the small power range. Steam engines are used where frequent starting, stopping, reversing or change of speed is necessary.

8.12 COMPOUNDING OF STEAM TURBINES

If the entire pressure drop is carried out in single stage nozzle, then the velocity of steam entering the turbine shall be of the order of 1500 m/sec, and the blade velocity will be of the order of 30,000 rpm. Such a high velocity is dangerous for the blades due to excessive centrifugal stresses and such a high speed of the turbine rotor is also not useful for practical purposes. Therefore a reduction gear shall be required between the turbine and the generator. This high entering velocity also results in high exit velocity, which gives a considerable loss of kinetic energy (about 10 to 12%). Therefore, the velocity of the blades is limited to 400 m/sec. In order to solve these problems, number of stages are employed, and the combination of these stages is known as compounding.

1. Velocity Compounding

The drop in heat energy takes place only in the nozzle at the first stage and is converted into kinetic energy. This gain in kinetic energy of the steam is used by the rows of moving blades and finally exhausted from the last row of the blades. The function of fixed blades is merely to turn the steam into the direction required for entry into the next row of rotor blades without changing the pressure and velocity of the steam. (Refer Fig. 8.14)

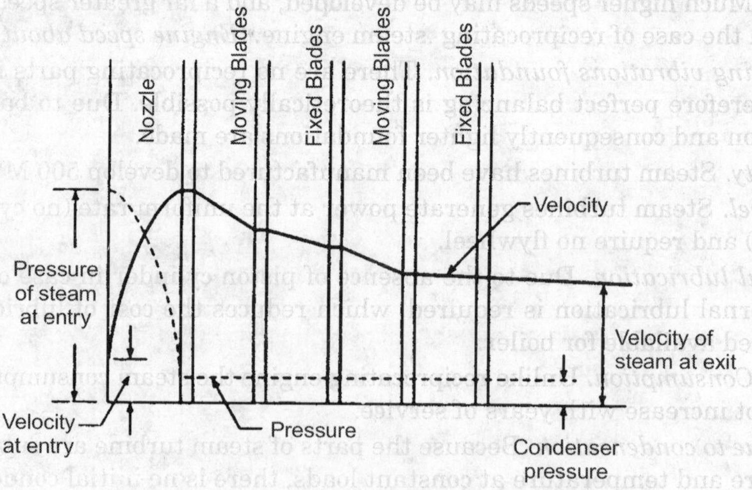

Fig. 8.14 Velocity Compounded turbine.

2. Pressure Compounding

In this compounding, the turbine is provided with one row of fixed blades (works as nozzles) at the entry of each row of moving blades. The pressure drop of the steam takes place in different rows of fixed blades, which work as nozzles. (Ref. Fig. 8.15)

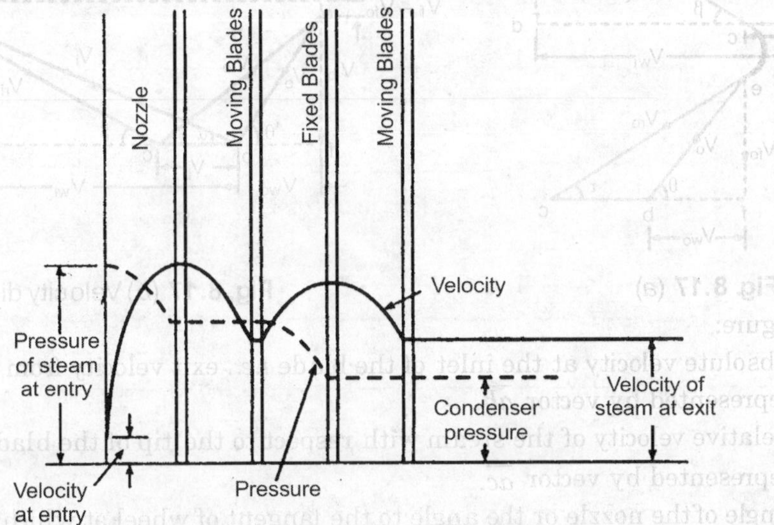

Fig. 8.15 Pressure Compound Steam Turbine.

3. Velocity and Pressure Compounding

This is a combination of pressure and velocity compounding. The drop of pressure of steam is compounded as in pressure compounding in each stage, whereas velocity obtained in each stage is also compounded as shown in Fig. 8.16.

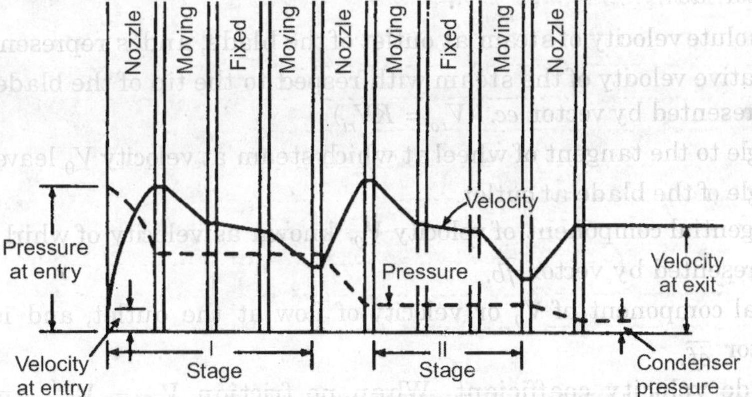

Fig. 8.16 Velocity and Pressure Compounding.

8.13 VELOCITY DIAGRAM

Velocity Diagrams at inlet and outlet are drawn at Fig. 8.17(*a*) to indicate the flow of steam through the blades of one row. Combined velocity diagram at inlet as well as outlet into single diagram is shown in Fig. 8.17(*b*). In this case both the velocity diagram are drawn on a common base representing the blade velocity V_b.

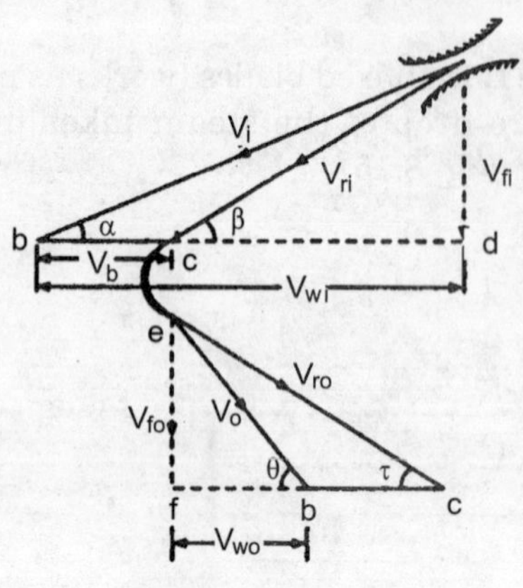

Fig. 8.17 (*a*)

Fig. 8.17 (*b*) Velocity diagram.

In the Figure:

V_i = Absolute velocity at the inlet of the blade *i.e.*. exit velocity from the nozzle and is represented by vector $\overrightarrow{ab}$.

V_{ri} = Relative velocity of the steam with respect to the tip of the blade at inlet, and is represented by vector $\overrightarrow{ac}$.

α = Angle of the nozzle or the angle to the tangent of wheel at which steam at velocity V_i enters.

β = Angle of the blade at inlet.

V_{wi} = Velocity of whirl at inlet *i.e.* tangential component of velocity V_i or the component responsible for rotating the turbine rotor and is represented by vector $\overrightarrow{db}$.

V_{fi} = Axial component of V_i or velocity of flow at inlet, and is represented by vector $\overrightarrow{ad}$..

V_o = Absolute velocity of steam at outlet of the blade, and is represented by vector $\overrightarrow{eb}$.

V_{ro} = Relative velocity of the steam with respect to the tip of the blade at outlet, and is represented by vector $\overrightarrow{ec}$. ($V_{ro} = KV_{ri}$)

θ = Angle to the tangent of wheel at which steam at velocity V_0 leaves the blade.

γ = Angle of the blade at outlet.

V_{wo} = tangential component of velocity V_0, known as velocity of whirl at outlet, and is represented by vector $\overrightarrow{fb}$.

V_{fo} = Axial component of V_0 or velocity of flow at the outlet, and is represented by vector $\overrightarrow{ef}$.

k = Blade velocity coefficient. When no friction $V_{ri} = V_{ro}$, and with friction $kV_{ri} = V_{ro}$.

From above, following formulae can be remembered:

(i) *Force*, $F = \dfrac{w}{g}(V_{wi} \pm V_{wo})$

where $w = \dfrac{\text{Mass } M \text{ in kg}}{\text{time } t \text{ in second}}$

(*ii*) Work done, $W = w(V_{wi} \pm V_{wo}).\ V_b$ J/s or W

and Power in kW = $w.\dfrac{(W_{wi} \pm V_{wo}).V_b}{1000}$

(*iii*) Blade efficiency, $\eta_b = \dfrac{2(V_{wi} \pm V_{wo}).V_b}{V_i^2}$

(*iv*) Axial Force = $w(V_{fi} - V_{fa})$

(*v*) Heat Energy due to friction = $H_{friction} = \dfrac{W}{2}.\dfrac{(V_{ri}^2 - V_{ro}^2)}{J}$ kJ/s.

where 'J' is a constant known as the Joule's constant or the mechanical equivalent of heat. If Q is expressed in kcal and W in kgf-m the value of J is 426.935 ≈ 427.

Also, $J = 4.1868\dfrac{kJ}{kcal} \approx 4.27\dfrac{kgf^{-m}}{kcal}$

Example 8.1 *A single row impulse turbine developes 135 kW at a blade speed of 175 m/sec using 2 kg of steam per sec. Steam leave the nozzle at 400 m/sec. Velocity coefficient of the blade is 0.9. Steam leaves the turbine blades axially. Determine the nozzle angle, blade angles at entry and exit, assuming no shock.*

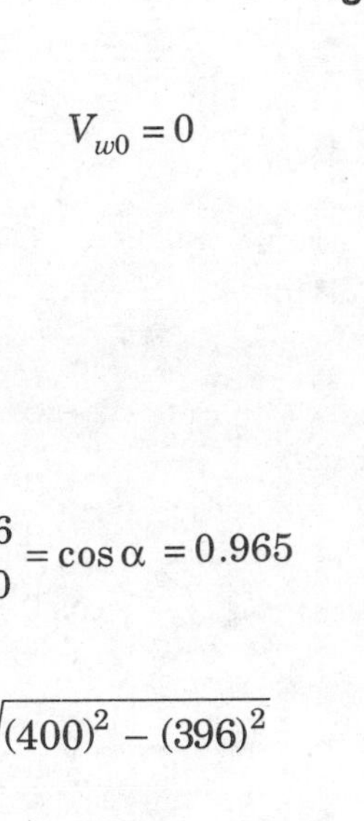

Fig. 8.18

Solution. U = 175 m/sec. V_1 = 400 m/sec.

W = 2 kg/sec. $K = 0.9$

As the steam leaves the turbine blades axially

$$\beta = 90°, \quad V_{f0} = V_0, \quad V_{w0} = 0$$

$$\text{Now power developed} = W\frac{U(V_{w1})}{100}$$

or
$$135 = \frac{2 \times 175 \times V_{w1}}{1000}$$

$$\therefore \quad V_{w1} = 386 \text{ m/sec.}$$

$$\frac{V_{w1}}{V_1} = \cos\alpha \quad \text{or} \quad \frac{386}{400} = \cos\alpha = 0.965$$

$$\therefore \quad \alpha = \textbf{18.9° Ans.}$$

$$V_{f1} = \sqrt{V_1^2 - V_{w1}^2} = \sqrt{(400)^2 - (396)^2}$$

$$\therefore \quad V_{f1} = 105 \text{ m/sec.}$$

$$\tan\theta = \frac{V_{f1}}{V_{w1} - u} = \frac{105}{386 - 175} = 0.49$$

$$\therefore \quad \theta = \textbf{26.5° Ans.}$$

$$V_{r1} = \sqrt{(V_{f1})^2 + (V_{w1} - u)^2} = \sqrt{(105)^2 + (386 - 175)^2}$$

or
$$V_{r1} = 235.7 \text{ m/sec.}$$

$$\therefore \quad V_{r0} = 0.9\, V_{r1}$$

$$V_{r0} = 0.9 \times 235.7 = 212 \text{ m/sec.}$$

$$\cos \phi = \frac{U}{V_{ro}} = \frac{175}{212} = 0.825$$

$$\therefore \quad \phi = \mathbf{34.4°} \quad \textbf{Ans.}$$

Example 8.2 *A simple impulse steam turbine has one ring of moving blades running at 150 m/sec. The absolute velocity of steam at the exit from the stage is 85 m/sec at an angle of 80° from the tangential direction. The blade velocity coefficient is 0.82 and the rate of steam flowing through the stage is 2.5 kg/sec. If the blades are equiangular, determine:*

(i) blade angles,

(ii) nozzle angle,

(iii) absolute velocity of steam issuing from the nozzle, and

(iv) the axial thrust.

Solution. Ref. Fig. 11.18.

$u = 150$ m/sec. $V_0 = 85$ m/sec. $\beta = 80°$ $V_{r0} = 0.82\, V_{r1}$ $w = 2.5$ kg/sec.

blades are equiangular, *i.e.*, $\theta = \phi$

$$V_{f0} = V_0 \,.\, \sin \beta = 85 \,.\, \sin 80°$$

$$V_{f0} = 83.71 \text{ m/sec.}$$

or $$V_{w0} = V_0 \cos \beta = 85 \,.\, \cos 80° = 14.756 \text{ m/sec}$$

$$V_{r0} = \sqrt{(V_{f0}^2)^2 + (u + V_{w0})^2}$$

$$= \sqrt{(83.71)^2 + (150 + 14.756)^2} = 184.8 \text{ m/sec.}$$

$$V_{r1} = \frac{V_{r0}}{0.82} = \frac{184.8}{0.82}$$

$$\therefore \quad V_{r1} = 225.36 \text{ m/sec.} \quad \sin \phi = \frac{V_{f0}}{V_{r0}} = \frac{83.71}{184.8}$$

$$\therefore \quad \phi = \mathbf{26° \ 56' = \theta.} \quad \textbf{Ans.}$$

$$V_{f1} = V_{r1} \sin \theta = 225.36 \,.\, \sin 26° \ 56'$$

$$\therefore \quad V_{f1} = 102 \text{ m/sec.}$$

$$\cos \theta = \frac{V_{w1} - u}{V_{r1}} \quad \text{or} \quad \cos 26° \ 56' = \frac{V_{w1} - 150}{225.36}$$

$$\therefore \quad V_{w1} = 351 \text{ m/sec} \quad \tan \alpha = \frac{V_{f1}}{V_{\omega 1}} = \frac{102}{351}$$

$$\therefore \quad \alpha = \mathbf{16° \ 12'.} \quad \textbf{Ans.}$$

$$V_1 = \sqrt{(V_{f1})^2 + (V_{w1})^2} = \sqrt{(102)^2 + (351)^2}$$

or $$V_1 = 365.52 \text{ m/sec.} \quad \textbf{Ans.}$$

$$\text{Axial thrust} = w(V_{f1} - V_{f0})$$

$$= 2.5 \times (102 - 83.71) = \mathbf{45.7 \ N.} \quad \textbf{Ans.}$$

Example 8.3 *Steam at a velocity of 450 m/sec enters an impulse wheel having a nozzle angle of 25°; the exit angle of the moving blade is 20°, and the relative velocity of the steam may be assumed to remain constant over the moving blade. If the blade speed is 180 m/sec, find the moving blade angle at inlet, the exit velocity and direction of the steam, the work done per kg of steam, and the diagram efficiency.*

If the turbine is supplied with 2 kg/sec of steam, find the power in kW of the wheel and the axial thrust.

Solution. Ref. Fig. 11.18.

$$V_1 = 450 \text{ m/sec}, \quad \alpha = 25°, \quad \phi = 20°, \quad V_{r1} = V_{r0},$$

$$u = 180 \text{ m/sec.} \quad V_{f1} = V_1 \sin\alpha = 450 \sin 25°$$

$$\therefore \quad V_{f1} = 190.17 \text{ m/sec.} \quad V_{w1} = 407.84 \text{ m/sec.}$$

$$\tan\theta = \frac{V_{f1}}{V_{w1} - u} = \frac{190.17}{407.48 - 180} = 0.8347$$

$$\therefore \quad \theta = \mathbf{39° 51'}. \quad \textbf{Ans.}$$

$$\frac{V_{f1}}{V_{r1}} = \sin\theta$$

$$\therefore \quad V_{r1} = \frac{190.17}{\sin 39° 51'} = 296.77 \text{ m/sec} = V_{r0}$$

$$\frac{V_{f0}}{V_{r0}} = \sin\phi$$

$$\therefore \quad V_{f0} = 296.77 \sin 20° = 101.5 \text{ m/sec.}$$

$$\frac{EB}{V_{r0}} = \cos\phi$$

$$\therefore \quad EB = 295.77 \cos 20° = 278.87 \text{ m/sec}$$

$$\therefore \quad EA = EB - AB = 278.87 - 180$$

$$\text{or} \quad V_{w0} = 98.87 \text{ m/sec}$$

$$V_0 = \sqrt{(V_{f0})^2 + (V_{w0})^2} = \sqrt{(101.5)^2 + (98.87)^2}$$

$$\therefore \quad V_0 = 141.7 \text{ m/sec.} \quad \textbf{Ans.}$$

$$\tan\beta = \frac{V_{f0}}{V_{w0}} = \frac{101.5}{98.87} = 1.0266$$

$$\therefore \quad \beta = \mathbf{45° 45'}. \quad \textbf{Ans.}$$

$$\text{W.D. per kg of steam} = u(V_{w1} + V_{w0})$$

$$= (407.84 + 98.87)\, 180 = 0.0912 \times 10^6 \text{ Nm/kg}$$

$$\text{Diagram } \eta = \frac{2u\,(V_{w1} + V_{w0})}{V_1^2}$$

$$= \frac{2 \times 180 \times (407.84 + 98.87)}{(450)^2} = \mathbf{0.9 \text{ or } 90\%}. \quad \textbf{Ans.}$$

$$\text{Power in kW} = \frac{\text{W.D. per kg} \times \text{kg/sec}}{1000} = \frac{0.0912 \times 10^6 \times 2}{1000}$$

$$= \textbf{182 KW} \quad \textbf{Ans.}$$

$$\text{Axial thrust} = w(V_{f1} - V_{f0})$$

$$= 2(190.17 - 101.5) = \textbf{178 N.} \quad \textbf{Ans.}$$

Example 8.4 *The nozzles of a De-Laval turbine deliver 1 kg of steam per sec at a speed of 730 m/sec to a set of blades moving at 180 m/sec. The nozzles are inclined at an angle of16° to the plane of wheel. Assuming a blade velocity coefficient of 0.72. Calculate:*

(*a*) *the blading η,*

(*b*) *the power developed by the blades, and*

(*c*) *the energy lost in the blades per sec.*

The blade angle at outlet may be taken as 25°.

Solution. Refer Fig. 11.18

$V_1 = 730$ m/sec $\quad u = 180$ m/sec $\quad \alpha = 16° \quad V_{r0} = k\,V_{r1}$

where $\quad k = 0.72 \quad \phi = 25°$

$$V_{w1} = V_1 \cos \alpha = 730 \cos 16°$$

or $$V_{w1} = 701.75 \text{ m/sec}$$

$$V_{f1} = V_1 \sin \alpha = 730 \sin 16°$$

or $$V_{f1} = 201.19 \text{ m/sec}$$

$$V_{r1} = \sqrt{(V_{f1})^2 + (V_{w_1} - u)^2} = \sqrt{(201.19)^2 + (701.75 - 180)^2}$$

or $$V_{r1} = 559.2 \text{ m/sec.}$$

$$V_{r0} = kV_{r1} = 0.72 \times 559.2$$

or $$V_{r0} = 402.6 \text{ m/sec.}$$

$$V_{f0} = V_{r0} \sin \phi = 402.6 \sin 25°$$

or $$V_{f0} = 170.14 \text{ m/sec}$$

$$EB = V_{r0} \cos \phi = 402.6 \cos 25°$$

or $$EB = 364.88 \text{ m/sec}$$

$$V_{w0} = EB - AB = 364.88 - 180 = 184.88 \text{ m/sec}$$

$$\text{Blading } \eta = \frac{2u(V_{w1} + V_{w0})}{V_1^2}$$

$$= \frac{2 \times 180(701.75 + 184.88)}{(730)^2}$$

or $$\eta = 0.5989 \text{ or } 59.89\% \quad \textbf{Ans.}$$

$$\text{Power developed} = w\frac{u(V_{w1} + V_{w0})}{1000}$$

$$= \frac{1 \times 180 \times (701.75 + 184.88)}{1000} = \frac{180 \times 886.6}{1000}$$

or $$\text{Power} = \textbf{160 KW} \quad \textbf{Ans.}$$

$$\text{Energy lost in blades/sec} = w\left[\frac{V_{r1}^2}{2} - \frac{V_{r0}^2}{2}\right]$$

or
$$\text{energy lost} = \frac{(559.2)^2 - (402.6)^2}{2} = \mathbf{75200\ Nm.} \quad \textbf{Ans.}$$

8.14 LOSSES IN STEAM TURBINES

In an ideal turbine the useful work done per kg of steam would be the same as that done on the Rankine cycle between the same temperature and pressure limits, *i.e.*, it would be equal to the *adiabatic heat drop*.

In practice many things reduce this work considerably, and their effects may be investigated by tracing the path of the steam from the stop valve to the condenser.

***(i)* Losses in Regulating Valves.** The first loss occurs at the governor valve where, for the sake of speed regulation, a throttling loss of 5 to 10% of the pressure at the main stop valve appears to be inevitable.

Although throttling is a constant enthalpy process, yet the drop available for work is reduced, (Refer Fig. 8.19).

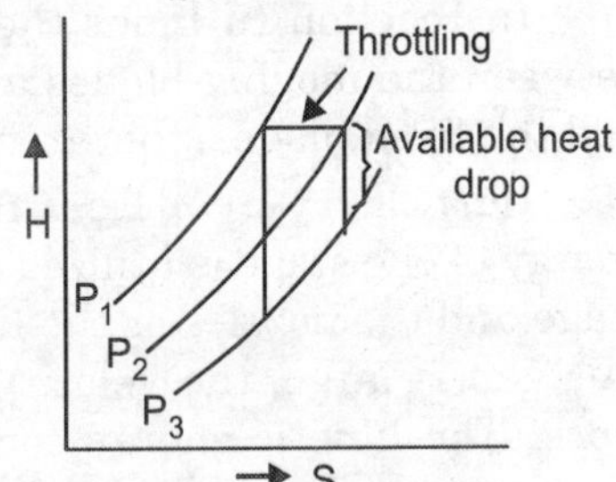

Fig. 8.19 Heat loss in throttling

***(ii)* Losses in the Nozzles.** Friction and eddies will cause various small pressure drops in the high pressure nozzle box, but the greatest loss occurs in the nozzle itself due to frictional resistance on the nozzle walls, viscous friction between steam particles, bending of the steam through considerable angle, growth of boundary layer, and the turbulence in the wake. This loss is accounted by the term nozzle efficiency.

$$\text{Thus, nozzle efficiency} = \frac{\text{actual enthalpy drop}}{\text{isentropic enthalpy drop}}$$

***(iii)* Losses in Moving Blades.** Following losses occur in the moving blades:

(*a*) *Trailing edge.* Steam jets, after leaving the nozzles and before entering the moving blades, mix with each other. This mixing results in the formation of eddies and loss due to turbulence. These losses are termed *wake losses.*

(*b*) *Impingement.* Steam jets attack the leading edges of the moving blades and cause a resistance in the flow and consequent loss of energy.

(*c*) *Leakage between the casing and the shroud.* Some of the steam leaks through the passage between the casing and the shroud without doing any work on the moving blades.

(*d*) *Friction.* Flow of the steam over the blades causes friction and loss of energy.

(*e*) *Turning of steam jet.* Due to curved surface of the blades, the path travelled by the steam is different which causes frictional loss. Presence of secondary flows also adds to this loss.

(*f*) *Shrouding.* Due to shrouding the height of flow of moving blades is larger than the nozzle height. This increases the thickness of boundary layer and losses.

The total losses in the moving blades are accounted by the term *velocity coefficient.*

***(iv)* Partial Admission, Fan Loss or Windage.** Small blade heights are undesirable from a thermal efficiency standpoint, but if these are increased, then the great density and limited quantity of high pressure steam prevent its admission over the whole periphery of the first stage, resulting *in partial admission.* This is unfortunate, because the blades which are not under the influence of the steam jets are idle churning the steam, with which the casing is filled, thereby causing a *Fan loss or Windage.* This loss may be reduced by fitting stationary shields around the moving blades which are not under the influence of steam.

***(v)* Disc Friction.** The high pressure turbine disc is itself running in a dense medium of steam that sets up a considerable fluid resistance, known as *Disc friction.*

(vi) Leakage Loss. The chamber containing the first stage is seated against leakage of steam by a gland at one side and diaphragm packing at the other side ; but since this seal is never perfect, some steam leaks from the chamber without doing useful work.

In reaction turbines there exists a radial gap between the guide blades and rotor and between the moving blades and stator. Since pressure drop occurs both in the moving blades and the fixed blades, there exists a leakage of steam through the radial gaps.

***(vii)* Carry-over Loss from One Stage to Another.** In multi-stage turbines the velocity energy of one stage is utilized in the following stage. The gap between the blades of the previous stage and the nozzles or the fixed blade row of the following stage should be kept minimum. If this gap is large, the velocity energy of the exit steam is completely lost without doing any work. The kinetic energy corresponding to the absolute discharge velocity V_0 of the steam, *i.e., carry over,* may be conserved by properly shaped and positioned diaphragm plates, with partial admission through angular positioning of the nozzles of successive stages in order to conserve the *carry over* is known as lead.

***(viii)* Loss due to Steam Wetness.** In the low pressure stages of the turbine near the exhaust, the steam is usually wet. Actually, no work is done by these water droplets, but some kinetic energy is lost in dragging them.

***(ix)* Terminal or Leaving Loss.** The exhaust steam enters the condenser with a considerable velocity, and the kinetic energy corresponding to this velocity represents a loss of energy.

***(x)* Radiation.** Loss of heat due to radiation, which, in a properly *lagged* casing amounts to about $\frac{1}{2}$ % of the adiabatic heat drop.

***(xi)* Friction in Bearings and Thrust Blocks.** Next to the glands we have bearings and the thrust block, which involve frictional losses.

Finally the power to drive the pumps as well as the governor gear must be deducted from the total power developed inside the turbine.

8.15 STEAM CYCLES

As discussed in the previous chapter, the energy relased by burning of the fuel is transferred to the water in the boiler to generate steam at a high pressure and temperature, which then expands in the turbine to a low pressure to produce shaft work. The steam leaving the turbine is condensed into water in the condenser where cooling water from a river or sea circulates carrying away the heat released during condensation. The condensate is then fed back to the boiler by the pump, and the cycle goes on repeating itself. The working substance thus follows along the same path of the cycle repetitively.

A cycle which uses heat in an ideal manner to produce work from the available energy, is called the *ideal cycle.* Actual heat cycles use heat in practical and imperfect processes of working fluids that develop less than the ideal available energy as work discard more energy to the receiver than the ideal cycle.

The efficiency of a cycle depends on the temperature of the heat flows into and out of the cycle. For a better understanding of the phenomenon of heat conversion into work, study of actual thermodynamic cycles is essential. The thermodynamic cycles used in the study of steam power plants are therefore being discussed here. The commonly used thermodynamic cycles are:

1. Carnot cycle
2. Rankine cycle
3. Reheat cycle
4. Regenerative cycle
5. Reheat – Regenerative cycle.

Although water does not satisfy all the requirements of ideal working fluid fully, but due to its low cost and general abundancy, it is used as the working medium in steam power plants.

8.15.1 Carnot Cycle

The Carnot cycle is an ideal but non-practical cycle giving maximum possible thermal efficiency for a cycle operating on selected maximum and minimum temperature ranges (see Fig. 8.20).

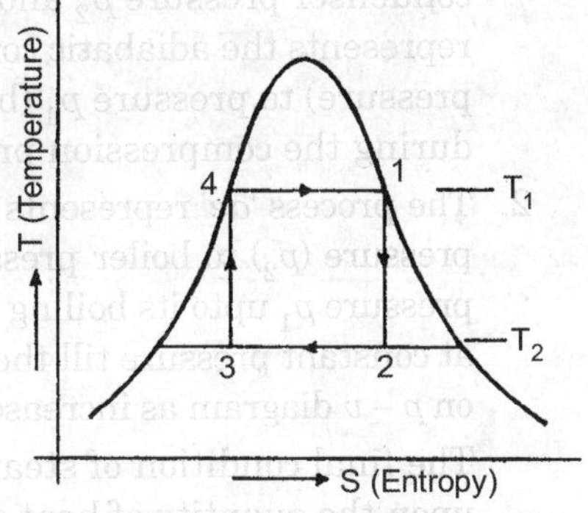

Fig. 8.20 Carnot cycle.

Thermal efficiency of Carnot cycle is given by

$$\eta_{Carnot} = \frac{\text{Work output}}{\text{Heat received}} = \eta_{max}$$

$$= \frac{T_1 - T_2}{T_1}$$

where, T_1 = Temperature of the heat source (higher temperature)

T_2 = Temperature of the receiver (lower temperature)

Being ideal cycle, the efficiency of the Carnot cycle is maximum and is treated as a standard of comparision for all other cycles.

8.15.2 Rankine Cycle

The basic cycle on which thermal power plant works is known as 'Rankine cycle'. The basic components of the power plant are shown in the line diagram given in Fig. 8.21 (*a*). These components are boiler, turbine, condenser and pump.

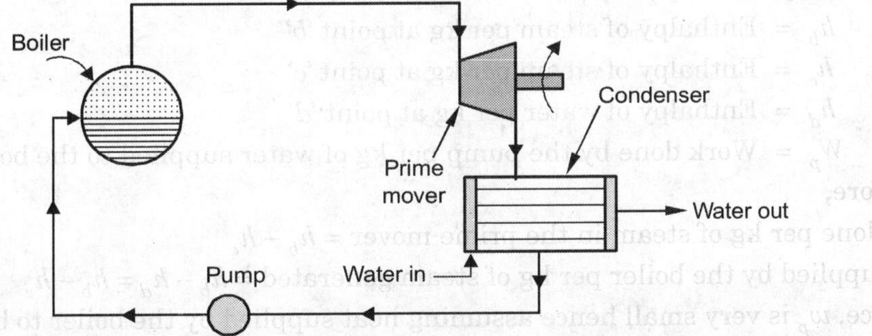

Fig. 8.21 (*a*) Components of steam power plant working on Rankine cycle.

The Rankine cycle used for steam power plant is shown in Fig. 8.21(*b*) and Fig 8.21(*c*) on $p-v$ and $T-S$ diagrams. Here S represents *Entropy,* which means heat (joules) per Kelvin.

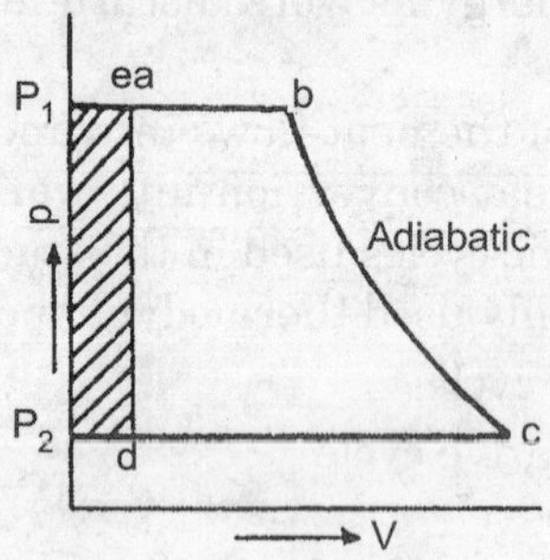

Fig. 8.21 (b)

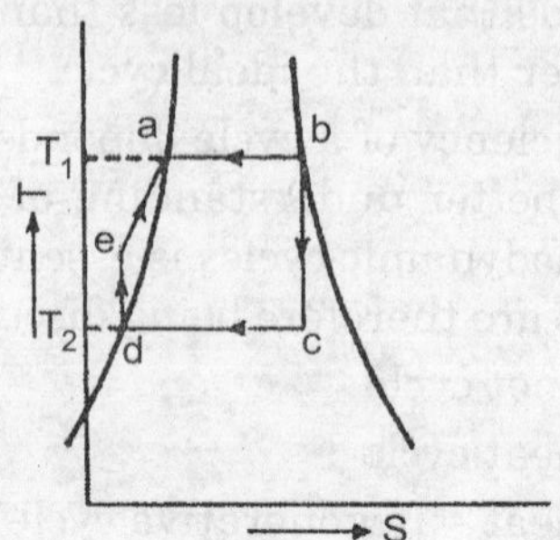

Fig. 8.21 (c)

Different processes of the Ranking cycle are described hereunder:

1. The point *'d'* represents the condition of condensate coming out of condenser at condenser pressure p_2 and corresponding saturation temperature T_2. The process *'de'* represents the adiabatic compression of water by the pump from pressure p_2 (condenser pressure) to pressure p_1 (boiler pressure). There is slight rise in temperature of water during the compression process *'de'*.
2. The process *'de'* represents adiabatic compression of water by the pump from condenser pressure (p_2) to boiler pressure (p_1). The process *'ea'* represents the heating of water at pressure p_1 upto its boiling point, and the process *'ab'* represents the generation of steam at constant pressure till the steam becomes saturated. The points *'e'* and *'a'* are the same on $p-v$ diagram as increase in volume of water during the heating process is negligible.

   The final condition of steam (b) may be wet, dry saturated or superheated depending upon the quantity of heat supplied by the boiler.
3. The process *'bc'* represents the expansion of steam isentropically in the turbine. During this expansion, external work is developed and the pressure of steam falls from p_1 to p_2 and its temperature will be T_2.
4. The process *'cd'* represents the condensation of steam coming out from the prime mover in the condenser at constant pressure p_2. During the condensation of steam, there is only change of phase from steam to water as the latent heat of steam is carried by circulating water in the condenser. This completes the cycle.
5. Again the 'process *'de'* represents the adiabatic compression of water by the pump from the pressure p_2 to p_1 and the cycle is repeated.

Let, h_b = Enthalpy of steam per kg at point *'b'*

h_c = Enthalpy of steam per kg at point *'c'*

h_d = Enthalpy of water per kg at point *'d'*

W_p = Work done by the pump per kg of water supplied to the boiler.

Therefore,

Work done per kg of steam in the prime mover = $h_b - h_c$

Heat supplied by the boiler per kg of steam generated = $h_b - h_d = h_b - h_e$

but since, w_p is very small hence assuming heat supplied by the boiler to be $h_b - h_d$.

Efficiency of the Rankine cycle is given by,

$$\eta_r = \frac{\text{Output}}{\text{Input}} = \frac{\text{Net work available}}{\text{Heat supplied}}$$

$$= \frac{(h_b - h_c) - W_p}{h_b - h_d}$$

Work done by the pump, W_p is neglected for all practical purposes as it is very small compared with other heat quantities.

$$\text{Therefore,}\ \eta_r = \frac{h_b - h_c}{h_b - h_d}$$

The Rankine cycle neglecting pump work is represented on T-S diagram as shown in Fig. 8.22.

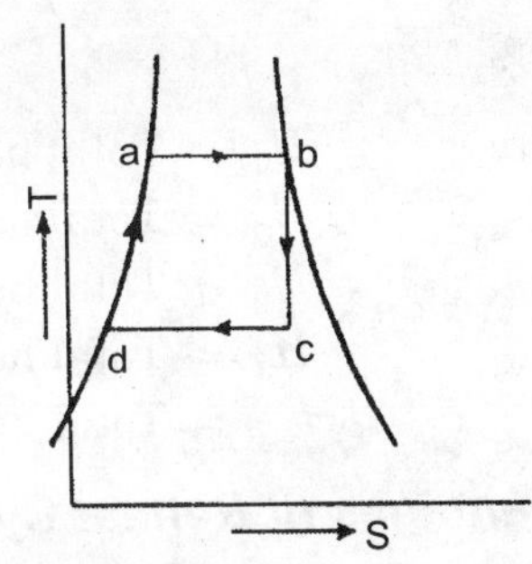

Fig. 8.22 Rankine cycle neglecting pump work.

8.15.3 REHEAT CYCLE

In a reheat cycle, steam after expansion upto a particular pressure, is reheated by passing it through reheater located in the combustion chamber of the boiler. Reheated steam is then passed through the second stage of the turbine and finally passed through the condenser.

The main purpose of the reheat is to reduce the moisture in the steam when passing through the last stages of the turbine and increase the output per kg of steam supplied to the turbine. This also reduces the blade erosion, and increases the nozzle and blade efficiency.

The reheat cycle is preferred for high capacity plants (above 100 MW) and when pressure of the steam is as high as 100 bar. It is not preferred for low capacity plants, as the cost of reheater in small plant, is not justified.

Fig. 8.23 shows flow diagram of reheat cycle, in which first turbine is high pressure (HP) turbine and second turbine is low pressure (LP) turbine. This cycle is shown on T-S (temperature-entropy) diagram in Fig 8.25.

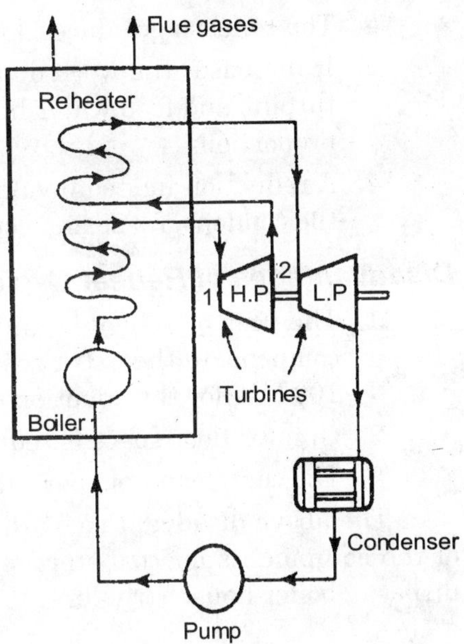

8.23 Reheat cycle.

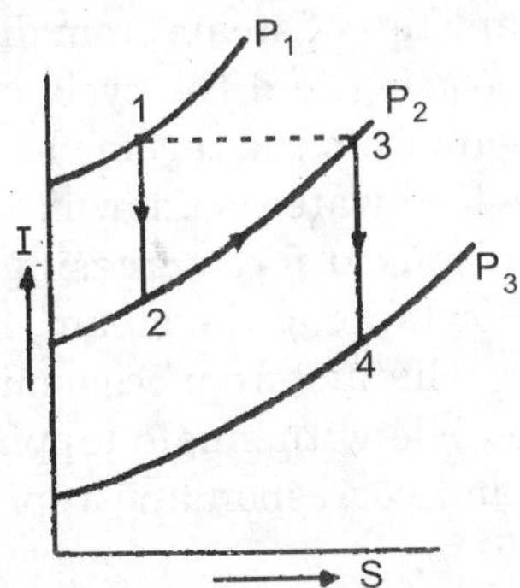

Fig. 8.24 Reheat cycle

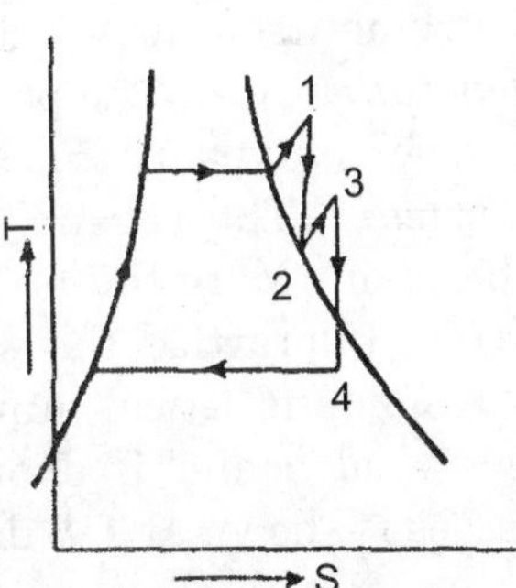

Fig. 8.25 T-S diagram for reheat cycle

Efficiency of the reheat cycle is given by;

$$\eta_{reheat} = \frac{(H_1 - H_2) + (H_3 - H_4)}{H_1 + (H_3 - H_2) - H_{w4}}$$

where, H_1 = Total heat of steam at 1.
H_2 = Total heat of steam at 2.
H_3 = Total heat of steam at 3.
H_4 = Total heat of steam at 4.
H_{w4} = Total heat of water at 4.

Advantages of Reheat Cycle

The advantages of reheat cycle over simple Rankine cycle are as under:

1. It avoids heavy blade erosion due to the impact of water particles carried with the steam and increase the life of turbine.
2. The reheating reduces 4 to 5% fuel consumption.
3. It increases the work done per kg of steam and this results in reduced size of boiler, turbine and feed water heating equipment. It also reduces the pumping power in that proportion.
4. A reduction in steam volume and heat to condenser is reduced by 7 to 8%. Therefore, the condenser size and cooling water requirement are also reduced accordingly.

Disadvantages of Reheat Cycle

1. The cost of reheater and long piping arrangements makes it more expensive. To compensate the extra cost with savings, the minimum capacity of the plant must be 100 MW for the adoption of reheat cycle.
2. Greater floor space is required to accommodate the longer turbine and reheat piping.
3. The complexity of operation and control increases with the adoption of reheat cycle.

The above disadvantages are compensated in new installations by the saving in the cost of the equipments for coal storage, coal handling, fuel burning, fans and dust collectors, ash disposal, boiler house structure.

8.15.4 Regenerative Cycle (Feed Water Heating)

When the feed water is heated by tapping small quantities of steam from different points in the steam turbine itself, it is known as *regenerative heating* and the cycle of operation is known as *Regenerative cycle.* The process of extracting steam from the turbine at certain points during its expansion and using this steam for heating the feed water is known as *Regeneration or Bleeding of steam.* The regenerative cycle is basically used for increasing the thermal efficiency of the plant. More the number of points used for tapping the steam, higher will be the cycle efficiency. In practice, the steam tapping points are limited from economic justification points of view to some practical number. A regenerative cycle with single tapping point with direct mixing type of heater is shown in Fig. 8.26 (*a*) and corresponding representation of different processes is shown at T-S diagram in Fig. 8.26 (*b*).

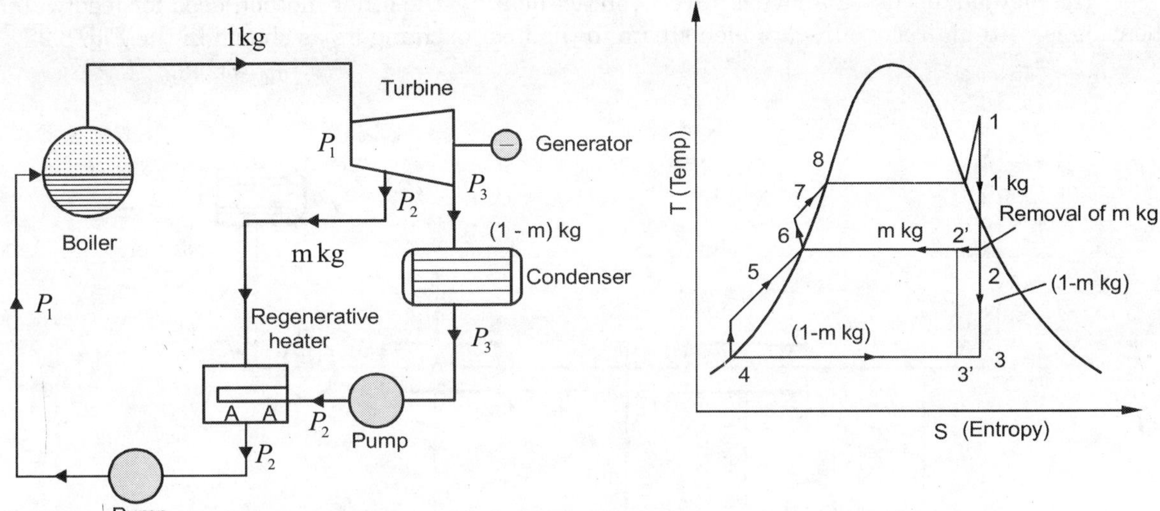

Fig. 8.26 (*a*) Regenerative cycle with single tapping point.

Fig. 8.26 (*b*) T-S diagram of Regenerative cycle.

If m is the steam extracted per kg of steam supplied to the turbine at pressure p_2, then the work done per kg of steam and cycle efficiency are given by,

$$W = (h_1 - h_2) + (1 - m)(h_2 - h_3) \quad ...(1)$$

$$\eta_r = \frac{W}{h_1 - h_{w_3}} \quad ...(2)$$

Neglecting the pump works.

The value of m is calculated considering the enthalpy balance at the regenerator. Heat gained by condensate of $(1 - m)$ kg

= Heat lost by m kg of steam.

$$\therefore \quad (1 - m)(h_{w2} - h_{w3}) = m(h_2 - h_{w2}) \quad ...(3)$$

In the equation (3), it is assumed that only latent heat of steam is used for heating the condensate from saturated temperature at pressure p_3 to saturated temperature at pressure p_2.

Different processes in this cycle, shown in Fig. 8.26(a) and (b) are as under:

1–2 = adiabatic expansion of 1 kg of steam.

2–3 = adiabatic expansion of (1–m) kg of steam.

3–4 = the condensation of steam at condenser pressure p_3.

4–5 = increase in pressure of condensate by first pump from pressure p_3 to p_2.

5–6 = the heating of (1–m) kg of condensate by using m kg of steam at condition 2 from saturation temperature corresponding to pressure p_3 to saturation temperature corresponding to pressure p_2.

6–7 = increase in pressure of condensate by second pump from pressure p_3 to p_2.

7–8–1 = the evaporation and superheating of steam in boiler.

The above discussion are for bleeding the steam from one point, for heating the feed water. The number of such points can be more, from economical considerations, these are generally upto 5 points

The method discussed above is 'direct contact heating' the other method used for feed water heating is with indirect contact of bled steam in the heat exchangers as shown in the Fig. 8.27.

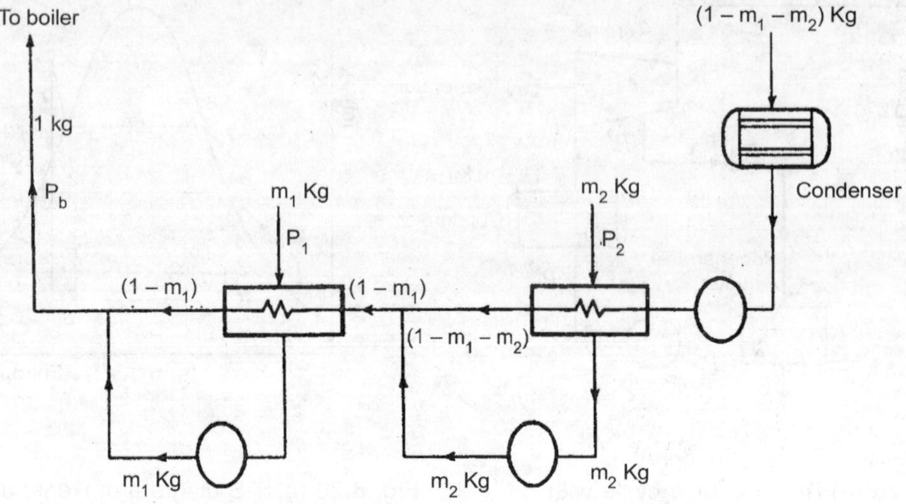

Fig. 8.27 Indirect contact heating in heat exchanger.

8.15.5 REHEAT-REGENERATIVE CYCLE

In all modern high capacity steam power plants, using high steam pressure, above 0.1 MPa bar reheat regenerative cycle, single stage reheating and regenerative heating at two to three points are generally used. This increases the total capacity of the power plant and also increases the overall efficiency of the plant. A simplified line diagram with single stage reheating and-with two extraction points for regeneration is shown in Fig. 8.28.

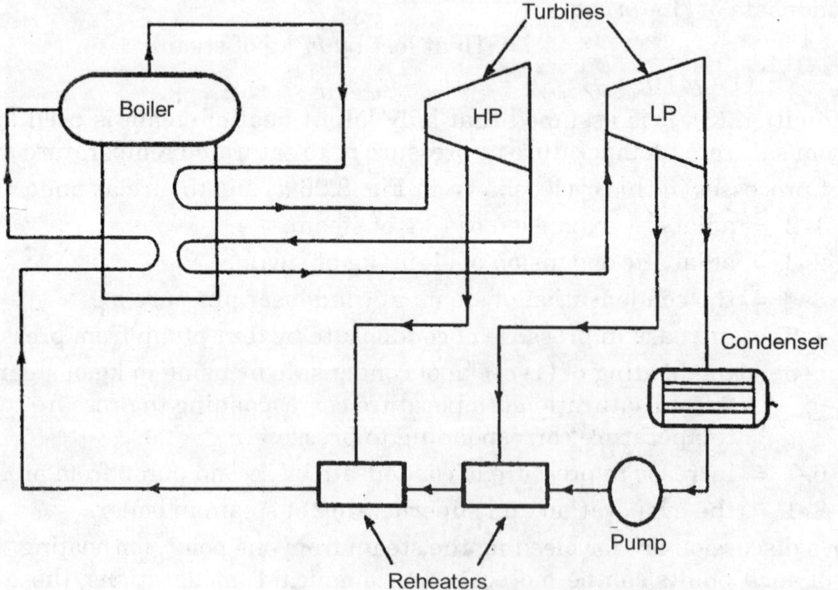

Fig. 8.28 Reheaters Reheat-regenerative cycle with single reheat and two tappings for regeneration.

8.16 EFFICIENCIES

***(i)* Blade or Diagram Efficiency.**

This has been given in para 8.13.

***(ii)* Stage Efficiency.** This covers all the losses on the nozzles, blades, diaphragms and discs that are associated with that stage.

$$\text{Stage } \eta = \frac{\text{Net work done on blades} - \text{Disc friction and windage}}{\text{Adiabatic heat drop per stage}}$$

***(iii)* Internal Efficiency.** This is equivalent to the stage efficiency when applied to the whole turbine, and is defined as

$$\eta_i = \frac{\text{Heat converted into useful work}}{\text{Total adiabatic heat drop}}$$

Assuming stage η constant for all stages, then,

$$\eta_i = \text{stage } \eta \times \text{reheat factor.}$$

***(iv)* Overall or Turbine Efficiency η_0.** This efficiency covers internal and external losses ; for example, bearings and steam friction, leakage, radiation, etc.

$$\eta_0 = \frac{\text{Work delivered at the turbine coupling in kcal}}{\text{Total adiabatic heat drop}}$$

***(v)* The Net Efficiency or the Efficiency Ratio.** It is the ratio

$$= \frac{\text{Brake thermal efficiency}}{\text{Thermal efficiency on the Rankine cycle}}$$

It is the overall or net efficiency that is meant when the efficiency of the turbine is spoken of without qualification.

8.17 GOVERNING

The purpose of governing is to regulate the supply of steam to the turbine to maintain as far as possible a constant mean speed of rotation under varying load conditions. This is necessary, because the turbines are designed to run at a particular speed for economy and better performance. Following are the methods of governing used in steam turbines.

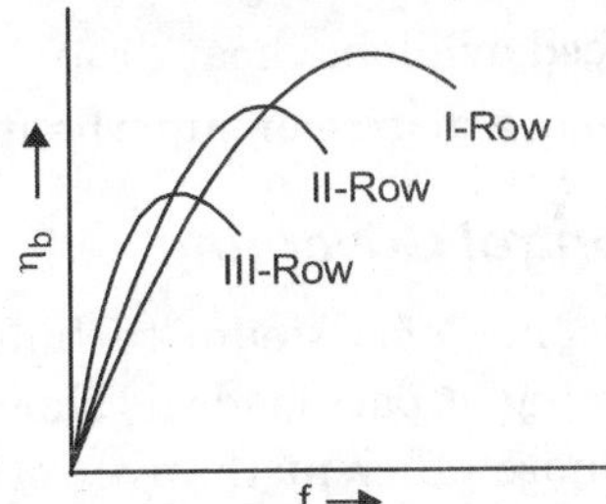

Fig. 8.29 Blade efficiency v/s speed ratio.

1. Throttle Governing

In this method a centrifugal governor (4), driven by worm gears (1, 2) at a speed less than that of the turbine, is usually employed to operate a double beat valve (8), which throttles the steam and reduces the quantity of steam flow (Refer Fig. 8.30).

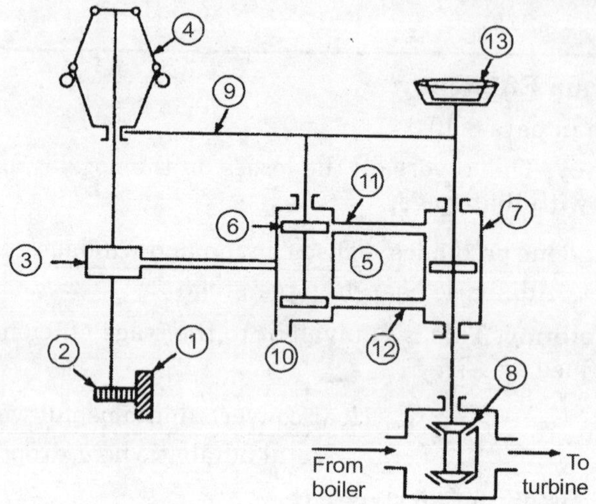

1. Worm connected to turbine shaft
2. Wheel connected to governor shaft
3. Oil pump
4. Centrifugal governor
5. Central chamber m
6. Regulating piston valve
7. Servo-motor
8. Double beat valve
9. Differential lever
10. Oil return pipe
11. Pipe to close valve
12. Pipe to open valve.

Fig. 8.30

In small turbines a mechanical linkage couples the governor to the valve, but in order to augment the effort of a small governor, to produce rapid response, and also to render the centrifugal governor applicable to large machines, an oil relay is used to replace the mechanical linkage.

As a protection against accident to the turbine in the event of abnormally low lubricating oil pressure, it is usual to connect the oil relay with the lubricating system, so that the relay acts as both a speed control and a safety device.

The effects of throttling are:

(*i*) reduced available heat drop,
(*ii*) reduced efficiency at part load,
(*iii*) increased degree of superheat,
(*iv*) delayed dew point.

2. Nozzle Control Governing

The efficiency of a steam turbine reduces considerably with throttle governing at part loads as shown in Fig 8.31. Nozzle control governing is more efficient than throttle governing and, therefore preferred for large steam turbines. Nozzle control governing is restricted to the first stage of the turbine, the nozzle area in the remaining stages remains constant. It is suitable for simple impulse turbine and large units which have an impulse stage followed by impulse-reaction stages. In pressure compounded impulse turbines, when some of the first stage nozzles are not supplied with steam, there will be some pressure drop at inlet to the second stage.

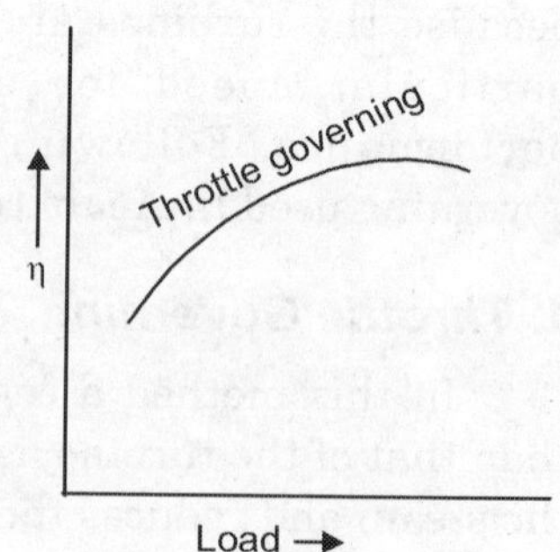

Fig. 8.31

In this method of governing, the nozzles are grouped together in 3 to 5 groups or even more, such as N_1, N_2, etc.. and supply of steam to each group is controlled by regulating valves such as V_1, V_2 etc., as shown in Fig. 8.32 (*a*) and (*b*). The valves are under the control of relay cylinder or pilot valve. There are several possible mechanical arrangements of regulating valves and nozzles. Two such arrangments are shown in Fig. 8.32 (*a*) and (*b*). In Fig. 8.32 (*a*) the groups of nozzles are arranged over the whole periphery of turbine and in Fig. 8.32 (*b*) nozzles are confined to the upper half of the turbine casing. At full load all the groups of nozzles are fed with steam through fully opened regulating valves. When the load on the turbine is reduced from full load, the supply of steam to one or more groups of nozzles is regulated depending upon the actual load on the turbine, thus keeping the speed of the turbine constant.

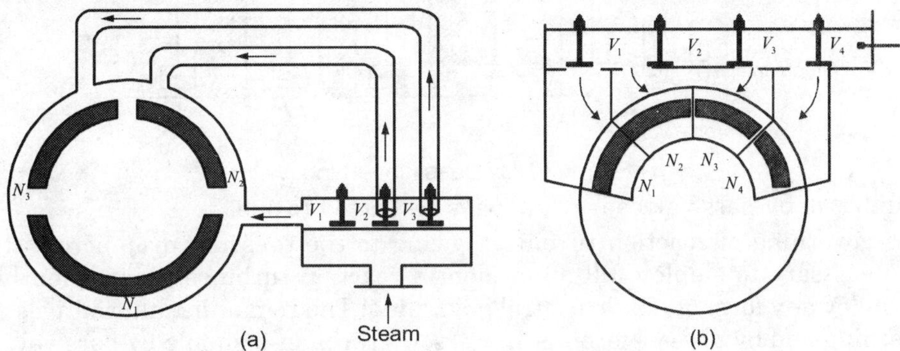

Fig. 8.32 Nozzle control governing.

3. By-pass Governing

In modern impulse turbines and specially those operating at very high pressures, the high pressure turbine comprises a number of stages of comparatively small mean diameter of wheel. All such turbines are usually designed for a definite load, termed the *Economic load,* at which the thermodynamic efficiency ratio of the turbine is a maximum. In the case of power station turbines, the economic load is made about *80% of the maximum continuous rating,* which is the maximum out put the generator is capable of giving continuously.

Due to the very small heat drop in the first stage it is not possible to employ nozzle control governing efficiently. Further more, it is desirable to have full admission in the H.P. stages at the economic load so as to eliminate the partial admission losses. Hence it is not possible to admit through additional nozzles in the first stage the 25 to 27% extra steam necessary to generate the full output. In fact, in most cases it would be quite impossible to do so because the steam pressure in the first stage wheel chamber would increase so much, due to the restriction at the second stage nozzles, that it would require to be greater than the initial pressure.

These difficulties of regulation are overcome by the employment of by-pass governing, the principle of which is illustrated diagramatically in Fig. 8.33. All the steam entering the turbine passes through a valve '*A*' (which. is under the control of the speed governor) and enters the nozzle box or steam chest '*B*'. In certain cases, *e.g.,* it would suffix for all loads up to the economical load, the governing being affected by throttling. For loads greater than the economic load a by-pass valve '*C*' is opened, allowing steam to pass from the nozzle of the 4th stage. The valve '*C*' is not opened until the lift of the valve '*A*' exceeds a certain amount, also as the load is diminishing the valve '*C*' closes first. The by-pass valve is under the control of the speed governor for all loads within its range. In later stages though there is increase in work output but the efficiency is low due to throttling effect.

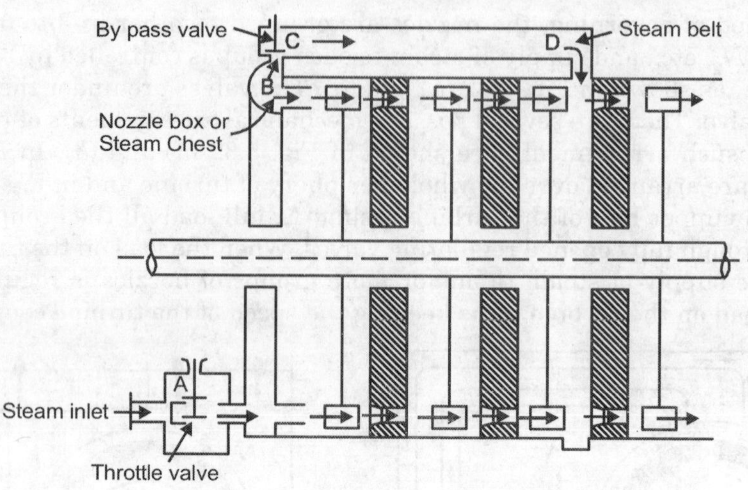

Fig. 8.33 By-gass governing.

The number of by-pass valves is not always restricted to one.

By-pass governing of reaction turbines: Owing to the pressure drop across the moving blades, it is necessary to employ full admission in reaction turbines. Consequently, it is not possible to employ any form of nozzle control governing. The remaining alternatives are simple throttle governing and by-pass governing. In Parson's turbines a simple by-pass valve is used.

Example 8.5 *(a) What methods are used in reducing the speed of turbine rotor.*

(b) The outlet angle of the blade of a Parson's turbine is 20° and the axial velocity of flow of steam is 0.5 times the mean blade velocity. If the diameter of the ring is 1.25 metre and the rotational speed is 3000 rpm. Determine:

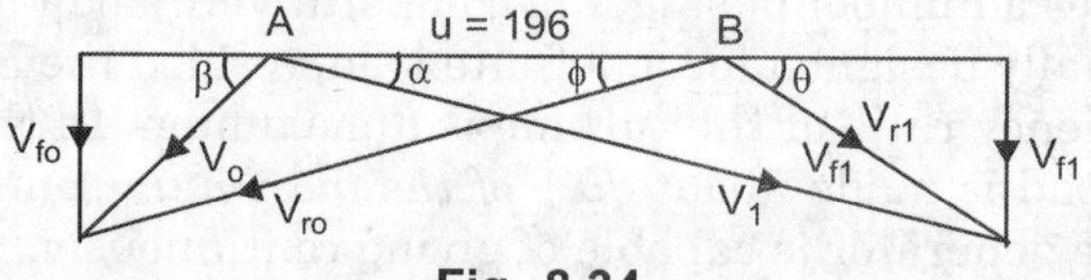

Fig. 8.34

(i) inlet angles of the blades.

(ii) the power developed, if dry saturated steam at 0.5 mPa abs. passes through the blade whose height may be assumed as 0.06 m cm. Neglect the effect of blade thickness.

Solution. For part (*a*) refer text.

(*b*) Refer Fig. 8.34

$$u = \frac{\pi DN}{60} \times \frac{\pi \times 1.25 \times 3000}{60} = 196 \text{ m/sec}$$

$$\alpha = \phi = 20° \text{ and } \beta = \theta \qquad \text{[Because blades are identical]}$$

$$V_{f1} = V_{f0} = 0.5\,u = 0.5 \times 196 = 98.0 \text{ m/sec}$$

Velocity diagram is drawn to scale and by measurement

$$\beta = \theta = 52° \quad V_{w1} = 77 \text{ m/sec} \quad V_{w0} = 268 \text{ m/sec}$$

$$\text{Mass flow rate in kg/sec} = \frac{\pi \,.\, D \,.\, h \,.\, V_f}{\text{specific volume}} = \frac{\pi \times 1.25 \times 0.06 \times 98}{0.3816}$$

[∵ sp. volume = 0.3816 from steam table]

$$= 60.5 \text{ kg/sec.}$$

$$\text{Power developed} = u(V_{w1} + V_{w0}) \times w$$

$$= \frac{196 \times (77 + 268) \times 60.5}{1000} = \textbf{4061 kW} \quad \textbf{Ans.}$$

Example 8.6 *(a) Distinguish between impulse and reaction turbines, (b) Explain the purpose of reheating in steam turbine, (c) In a stage of impulse turbine provided with single row wheel, the mean diameter of the blade is 1 m. It runs at 3000 rpm. The steam issues from the nozzle at a velocity of 350 m/sec and the nozzle angle is 20°. The rotor blades are equiangular. The blade friction factor is 0.86. Determine the power developed if the axial thrust on the end bearing of rotor is 12 kg.*

Solution. For parts (*a*) and (*b*) refer text.

(*c*)

$$D = 1 \text{ m} \quad N = 3000 \text{ rpm} \quad V_1 = 350 \text{ m/sec.}$$

$$\alpha = 20° \quad \theta = \phi$$

Axial thrust,

$$F_a = 12 \text{ kg}$$

$$u = \frac{\pi DN}{60} = \frac{\pi \times 1 \times 3000}{60}$$

or

$$u = 157 \text{ m/sec}$$

$$V_{f1} = V_1 \sin\alpha = 350 \sin 20° = 119.7 \text{ m/sec}$$

$$V_{w1} = V_1 \cos\alpha = 350 \cos 20° = 328.9 \text{ m/sec}$$

$$\tan\theta = \frac{V_{f1}}{V_{w1} - u} = \frac{119.7}{328.9 - 157}$$

$\therefore$

$$\theta = \phi = 35°$$

$$\frac{V_{f1}}{V_{r1}} = \sin\theta$$

$\therefore$

$$V_{r1} = \frac{119.7}{\sin 35°} = 208.68 \text{ m/sec.}$$

$$V_{r0} = 0.86\, V_{r1} = 0.86 \times 208.68 = 179.46 \text{ m/sec}$$

$$\frac{u + V_{w0}}{V_{r0}} = \cos\phi \quad \text{or} \quad \frac{157 + V_{w0}}{179.46} = \cos 35°$$

$\therefore$

$$V_{\omega 0} = -\,10 \text{ m/sec.}$$

(–) sign means that point *E* lies between *A* and *B*.

$$V_{f0} = V_{r0} \sin\phi = 179.46 \sin 35° = 102.94 \text{ m/sec}$$

$$\text{Axial thrust, } F_a = w(V_{f1} - V_{f0})$$

or

$$12 = w(119.7 - 102.94)$$

$\therefore$

$$w = 70 \text{ N/S}$$

$$\text{Power developed} = w \cdot u \frac{(V_{w1} - V_{w0})}{1000}$$

$$= \frac{70 \times 157(328.9 - 10)}{1000} = \textbf{350 kW} \quad \textbf{Ans.}$$

Example 8.7 *(a) Explain with the help of neat sketch, the difference between pressure compounding and velocity compounding. (b) The mean diameter of the blades of an impulse turbine with a single row wheel is 1.05 m and the speed is 3000 rpm. The nozzle angle is 18°, the ratio of blade speed to steam speed is 0.42, and the ratio of the relative velocity at outlet from the blades to that at inlet is 0.84. The outlet angle of the blade is to be made 3° less than the inlet angle. The steam flow is 8 kg per sec. Draw the velocity diagram for the blades and derive the following:*

(i) Resultant thrust on the blade ; *(ii) Tangential thrust on the blades:*
(iii) Axial thrust on the blades ; (iv) *Power developed in blades ;*
(v) Blading efficiency.

Solution. For part (*a*) refer text.

(*b*) Given

$$D = 1.05 \text{ m} \quad N = 3000 \text{ rpm} \quad \alpha = 18°$$

$$\frac{u}{V_1} = 0.42 \quad \frac{V_{r0}}{V_{r1}} = 0.84 \quad \phi = \theta - 3°$$

$$w = 8 \text{ kg/sec}$$

Refer Fig. 8.18.

$$u = \frac{\pi DN}{60} = \frac{\pi \times 1.05 \times 3000}{60} = 164.96 \text{ m/sec}$$

$$\therefore \quad V_1 = \frac{u}{0.42} = \frac{164.96}{0.42}$$

or $$V_1 = 392.76 \text{ m/sec}$$

$$V_{f1} = V_1 \sin \alpha = 392.76 \sin 18°$$

or $$V_{f1} = 121.36 \text{ m/sec}$$

$$V_{w1} = V_1 \cos \alpha = 392.76 \cos 18°$$

or $$V_{w1} = 373.55 \text{ m/sec}$$

$$\tan \theta = \frac{V_{f1}}{V_{w1} - u} = \frac{121.36}{373.55 - 164.96°}$$

$$\therefore \quad \theta = 30.1° \quad \text{and} \quad \phi = 27.1°$$

$$V_{r1} = \frac{V_{f1}}{\sin \theta} = \frac{121.36}{\sin 30.1°}$$

or $$V_{r1} = 242 \text{ m/sec}$$

$$V_{r0} = 0.84\, V_{r1} = 0.84 \times 242$$

or $$V_{r0} = 203.27 \text{ m/sec}$$

$$V_{f0} = V_{r0} \sin \phi = 203.27 \sin 27.1°$$

or $$V_{f0} = 92.6 \text{ m/sec}$$

$$\frac{u + V_{w0}}{V_{r0}} = \cos \phi \quad \text{or} \quad \frac{164.96 + V_{w0}}{203.27} = \cos 27.1°$$

$$\therefore \quad V_{w0} = 16 \text{ m/sec}$$

(*i*) tangential thrust, $F_t = w \times (V_{w1} + V_{w0})$

$$= 8 \times (373.55 + 16) = \mathbf{3116\ N} \quad \textbf{Ans.}$$

(*ii*) Axial thrust $F_a = w \times (V_{f1} - V_{f0}) = 8 \times (121.36 - 92.6)$

or $F_a = \mathbf{230\ N}$ **Ans.**

(*iii*) Resultant thrust $F = \sqrt{F_t^2 + F_a^2} = \sqrt{(3116)^2 + (230)^2}$

or $F = \mathbf{3125}$ **Ans.**

(*iv*) Power developed $= \dfrac{F_t \cdot u}{1000} = \dfrac{3116 \times 164.96}{1000}$

or Power $= \mathbf{515\ kW}$ **Ans.**

(*v*) Blading $\eta = \dfrac{2u(V_{w1} + V_{w0})}{V_1^2} = \dfrac{2 \times 164.96\,(373.55 + 16)}{(392.76)^2}$

or $\eta = \mathbf{0.83\ or\ 83\%}$. **Ans.**

Example 8.8 *(a) Distinguish between impulse and reaction turbines, (b) Steam with absolute velocity of 300 m/sec is supplied through a nozzle to a single stage impulse turbine. The nozzle angle is 25°. The diameter of blade rotor is 100 cm and it has a speed of 2000 rpm. Find suitable blade angles for zero axial thrust. If the blade velocity coefficient is 0.9 and the steam flow rate is 10 kg/sec. Calculate the H.P. developed.*

Solution. For part (*a*) refer the text

(*b*) Given: $V_1 = 300$ m/sec $\alpha = 25°$ $D = 1$ metre $N = 2000$ r.p.m.

For zero axial thrust $V_{f1} = V_{f0}$ $K = 0.9$ $w = 10$ kg/sec

Required H.P., blade angles

$$u = \frac{\pi \times 1 \times 2000}{60} = 104.73 \text{ m/sec}$$

Since, $V_{f1} = V_{f0} = V_1 \sin\alpha = 300 \sin 25°$

or $V_{f1} = 126.78$ m/sec

$V_{w1} = V_1 \cos\alpha = 300 \cos 25°$

$V_{w1} = 271.9$ m/sec.

$$\tan\theta = \frac{V_{f1}}{V_{w1} - u} = \frac{126.78}{271.9 - 104.73}$$

or $\theta = \mathbf{37.2°}$ **Ans.**

$$V_{r1} = \frac{V_{f1}}{\sin\theta} = \frac{126.73}{\sin 37.2°} = 209.7 \text{ m/sec}$$

$V_{r0} = 0.9\,V_{r1} = 0.9 \times 209.7$

or $V_{r0} = 188.72$

$$\sin\phi = \frac{V_{f0}}{V_{r0}} = \frac{V_{f1}}{V_{r0}} = \frac{126.78}{188.72}$$

$\because \quad V_{f1} = V_{f0}$, for no axial thrust

$\therefore \quad \phi = \mathbf{42°}$. **Ans.**

$$\frac{u + V_{w0}}{V_{r0}} = \cos\phi \quad \text{or} \quad \frac{104.73 + V_{w0}}{188.72} = \cos 42$$

$$\therefore \quad V_{w0} = 35.51 \text{ m/sec}$$

$$\text{Power developed } P = \frac{u(V_{w1} + V_{w0})\, w}{1000} = \frac{104.73(271.9 + 35.51)}{1000} \times 10$$

$\therefore \quad P = \mathbf{322\ kW}$ **Ans.**

Example 8.9 *In a reaction turbine the fixed blades and moving blades are of the same shape but reversed in direction. The angles of the receiving tips are 35° and of the discharge tips 20°. Find the H.P. developed per pair of blades for a steam consumption of 2.5 kg/sec. When the blade speed is 50 m/sec. If the heat drop per pair is 2.4 kcal/kg, find the efficiency of the pair.*

Solution. Receiving tips of moving blade and fixed blade have angles θ and β. Discharge tips of moving blade and fixed blade have angles θ and α

$$\therefore \quad \theta = \beta = 35° \quad \text{and} \quad \phi = \alpha = 20°$$

Since,
$$\tan\theta = \frac{V_{f1}}{V_{w1} - u}$$

But
$$V_{f1} = V_{w1} \tan \propto$$

$$\therefore \quad \tan\theta = \frac{V_{w1} \tan\alpha}{V_{w1} - u}$$

or
$$\tan 35° = \frac{V_{w1} \cdot \tan 20°}{V_{w1} - 50}$$

$$\therefore \quad V_{w1} = 102.97 \text{ m/sec}$$

$$V_{w0} = V_{w1} - u = 102.97 - 50$$

or
$$V_{w0} = 52.97 \text{ m/sec.}$$

$$\text{Power developed} = w \cdot \frac{u(V_{w_1} + V_{w_0})}{100}$$

$$P = \frac{2.5 \times 50(102.97 + 52.97)}{1000} = 0.125 \times (155.94)$$

or
$P = 19.5$ kW **Ans.**

$$\eta = \frac{W.D./kg}{J.H_d}$$

$$= \frac{2.5 \times 50(102.97 + 52.97)}{4186 \times 2.4 \times 2.5} = \mathbf{0.78 \text{ or } 78\%} \quad \textbf{Ans.}$$

8.18 CONDENSERS

The use of steam condensers in the steam prime-movers (generally steam turbines in power plants) is to improve the efficiency by decreasing the exhaust pressure of the steam below atmosphere. Another important advantage of the condenser is that the steam condensed may be recovered to provide a source of pure feed water to the boiler and reduces the water softening plant capacity requirements.

Since we know that the thermal efficiency of a steam turbine/engine is given by $(T_1 - V_2)/T_1$, where T_1 and T_2 are the supply and exhaust temperatures of the steam. This means that, efficiency can be improved either by increasing T_1 or by reducing T_2 or both. Increase in T_1 is difficult after a certain limit due to several reasons, but T_2 can be reduced by allowing the exhaust of the steam prime mover to take place below the atmospheric pressure (low exhaust pressure means low-exhaust temperature).

To allow the exhaust of the steam below atmospheric pressure, a vessel known as condenser is used, in which the pressure is maintained below the atmospheric pressure by condensing the steam with the circulation of cold water. Thus, a condenser is a closed vessel in which the steam is condensed by abstracting the heat from steam and the pressure is maintained below the atmospheric pressure.

Advantages of Condensers

(*i*) Steam condensed by the condenser is used as feed water for boiler. This reduces the cost of fuel as the condensate is supplied at higher temperature of the boiler. It also reduces the capacity of the feed water cleaning system.

(*ii*) Increases the efficiency of the prime mover, as the enthalpy drop $(T_1 - T_2)$ increases with the increase in vacuum in the condenser. This means reduction of steam consumption per unit of power generated.

(*iii*) Prevents the deposition of salt in the boiler by using the condensate instead of feed water from outer sources which contains salt. Deposition of salt in boiler reduces its efficiency and also reduces the life.

8.19 FUNCTIONS OF CONDENSERS

It is an equipment in which exhaust steam from the turbine is condensed. During condensing process some heat or exhaust steam is decipated to the cooling water known as "circulating water".

In other words, when steam is the working fluid, it may be returned to the boiler and used over again and again. This can be done in a condenser most conveniently and economically by first condensing it and then pumping the resulting water into the boiler.

Main functions of condenser are:

(*i*) The primary object of condensing exhaust steam in a condenser is to make it possible to remove it economically at a pressure less than that of the atmosphere after it has done its work in the turbine and thus enable the steam to expand to a greater extent and do more work.

This can be explained as follows:

Fig 8.35 indicates work done for a non-condensing steam plant while Fig. 8.36 indicates work done by a condensing plant of the same capacity as that of Fig. 8.35. Shaded area shows the extra work obtained.

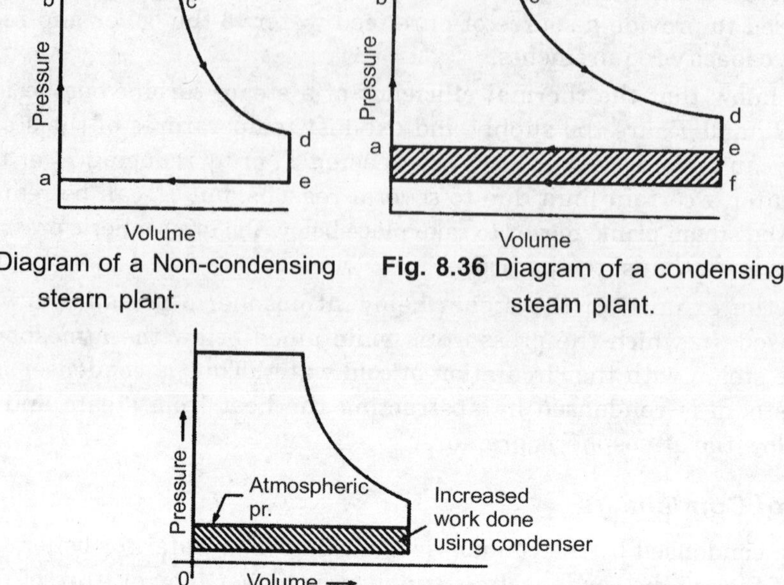

Fig. 8.35 Diagram of a Non-condensing stearn plant.

Fig. 8.36 Diagram of a condensing steam plant.

Fig. 8.37 Indicator diagram showing increased work done using condenser.

(*ii*) The secondary object of the condenser is to provide hot feed water for the boiler. The condensate is a pure hot water and therefore can be fed into the boiler directly without any treatment. Thus large savings are made and less fuel and time is required to raise the steam.

8.20 ELEMENTS OF A CONDENSING PLANT

It means in addition to condenser, all the essential equipment required for the proper functioning of a condensing plant. Such as:

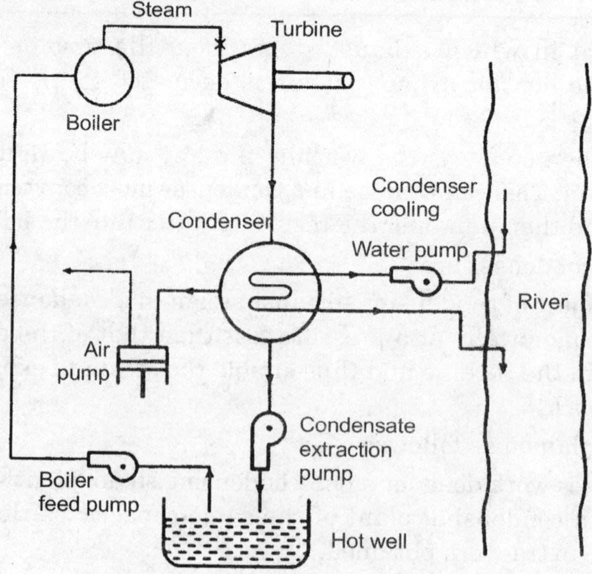

Fig. 8.38 Condensing System.

1. **A steam ejector or air pump,** which is needed to remove air and other non-condensable gases from the condenser. The air pump used to remove gases and air is known as **"Dry air Pump".**
2. **A hot well,** for collecting the condensate.
3. **An extraction pump,** to remove the condensate from the condenser hot well and then to feed it into the boilers or pre-heaters.
4. **A circulating water pump** to condense the exhaust steam of turbine in the condenser.
5. **Spray pond or cooling Tower** for cooling the circulating water of condenser, when there is only a limited quantity of cooling water available. If the circulating water arrangement is through a river having sufficient water then no other cooling arrangement is required. Thus cold circulating water is taken from the river.

8.21 TYPES OF STEAM CONDENSERS

Steam condensers are mainly classified as:

1. Mixing type or Jet Condensers.
2. Non-mixing type or Surface Condensers.

1. Mixing or Jet Type Condensers

In these condensers, the exhaust steam from prime-mover and cooling water come in direct contact with each other. The condensate coming out from these condensers can not be used as feed water to the boiler as it is not free from salt and pollutents. These condensers are generally preferred where the good water for feed to boiler is easily available in ample quantity. These condensers are rarely used in modern power plants.

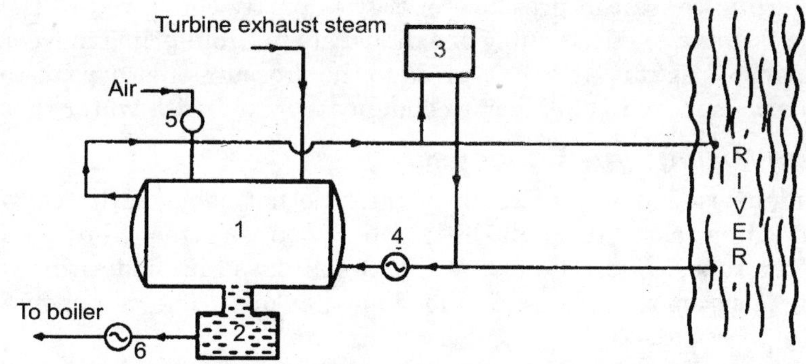

Fig. 8.39 Line Diagram of a condensing steam plant.

1 = Condenser

2 = Hot well

3 = Cooling Tower

4 = Circulating water pump

5 = Steam Air Ejector

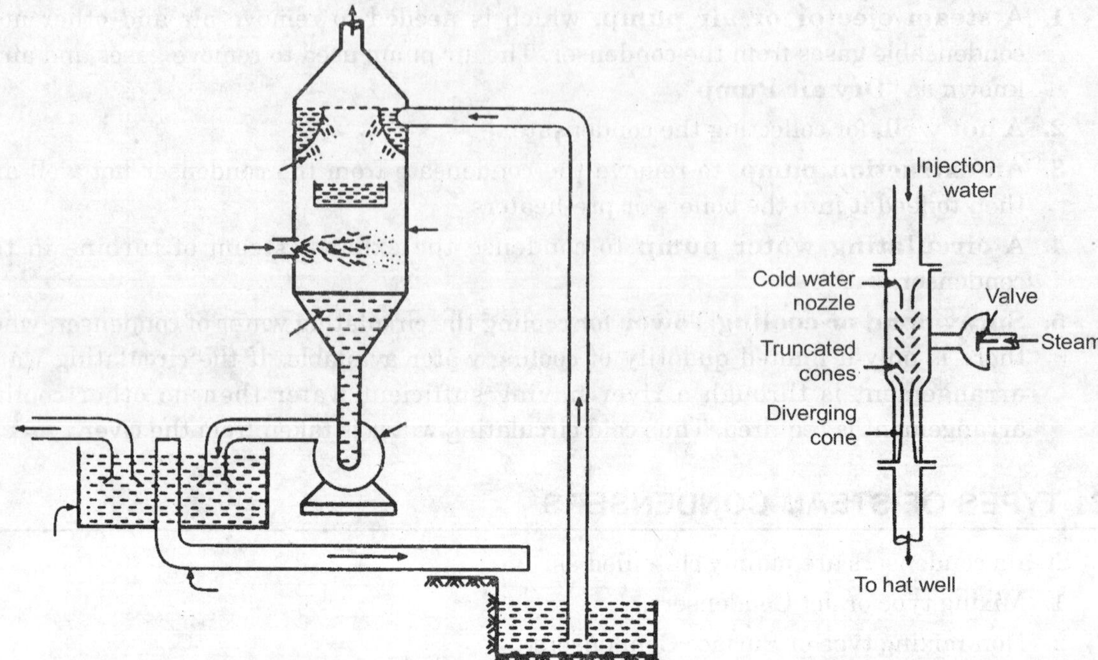

Fig. 8.40 Jet Condenser. **Fig. 8.41** Ejector Condenser.

The jet condensers may be either parallel flow (where steam and cooling water flow in the same direction) type, or counter flow (where steam and cooling water flow in opposite direction) type.

Ejector condensers also fall in the category of Jet condensers. The principle of operation of these condenser is that by discharging a smooth jet of cold water through a series of converging guide cones and finally leaves the condenser after passing through the diverging cone. In the converging cones the pressure energy is partly converted to kinetic energy and this reduction in pressure allows the entry of exhaust steam from prime mover. In the diverging cone the kinetic energy is again partly converted to the pressure energy so as to obtain pressure more than that of atmosphere to enable the condensate mixed with water to be discharged.

2. Surface or Non-Mixing Type Condenser

In these condensers, steam and cooling water do not come in direct contact with each other. The cooling water passes through the number of tubes attached to condenser shell, and steam surrounds the tubes. These types are universally used in modern power plants, as the condensate coming out from them is used as feed for the boiler. The arrangement of the surface condenser is shown in Fig. 8.42.

The steam gets condensed as it comes in contact with cold surface of the tubes in which water is flowing. The cooling water flows in one direction through the first set (lower half) of tubes and returns in the opposite direction through the second set (upper half) of tubes as shown in the figure. The cooling water coming out of the condenser is discharged into the river or pond, while the condensed steam is taken out by means of an extraction pump.

The surface condensers may be single or multiple pass, down flow or central flow type.

Evaporative Condenser. These condensers also fall in the category of surface condensers because, in this case also steam is not coming into direct contact of water when it is condensed, and the condensate can be used in boiler as feed water.

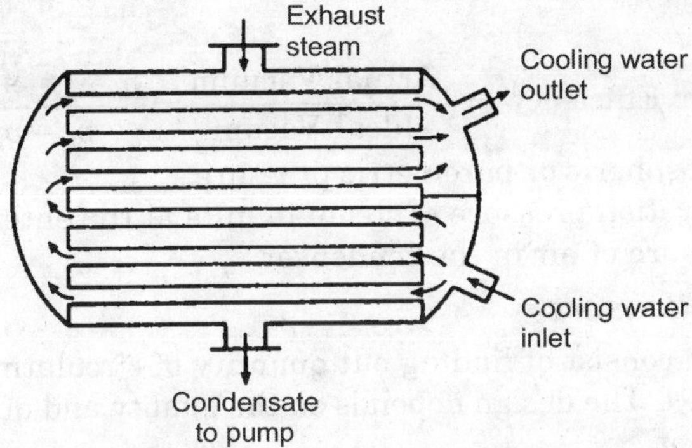

Fig. 8.42 Surface Condenser.

In this case, water is sprayed through the nozzles over the pipe carrying exhaust steam. The air is drawn over the surface of this pipe, which is bent like a coil, with the help of induced fan as shown in the figure. Air passing over the pipe carries the water from the surface of condenser coil in the form of vapour. The latent heat required for the evaporation of water vapour is taken from the water film formed on the pipes and drops the temperature of the water film, and this helps for heat transfer from the steam to the water. In this system cooling water requirement is only 10% of the requirement of surface-condensers. These condensers are preferred for small plants and where there is acute shortage of cooling water.

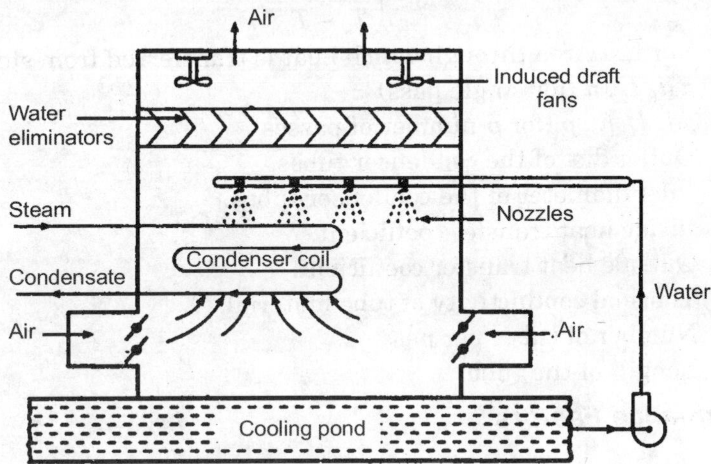

Fig. 8.43 Evaporative Condenser.

8.22 CONDENSER PERFORMANCE

Requirements of a good Surface Condenser

An ideal surface condenser should have following qualities:

1. The steam should be evenly distributed over the whole cooling surface.
2. There should not be under cooling of condensate. The quantity of water should be just sufficient to condensate the steam.
3. There should not be any air leakage in the condenser.

Vacuum Efficiency

$$\text{Vacuum Efficiency} = \frac{\text{Actual Vacuum}}{\text{Ideal Vacum}} = \frac{p_b - (p_a + p_s)}{p_b - p_s}$$

where, p_b = Atmospheric or barometric pressure.

p_s = Saturation pressure of steam in mPa at the entering steam temperature.

p_a = Pressure of air in the condenser

Design of Surface Condensers

Condenser designing consist of finding out quantity of circulating water and the surface area which comes in contact. The design depends on the quality and quantity of water available and condenser arrangement.

Surface Area Required in a Surface Condenser

Total heat lost by steam = Heat given by the steam to the water through tubes.

$$\therefore \quad Q = UA\theta.$$

where U = Overall heat transfer coefficient.

$$= \frac{d_o}{d_i} \cdot \frac{1}{h_i} + \left(\frac{d_o - d_i}{d_o + d_i}\right) \frac{d_o}{k} + \frac{1}{h_o}$$

$$\theta = \frac{(T_{wo} - T_{wi})}{\log_e \left(\dfrac{T_s - T_{wi}}{T_s - T_{wo}}\right)} = \text{logarathmic mean temp. difference}$$

and A = Surface Area through which heat is transferred from steam to water.

$= \pi\, d_0\, L\, .\, n$ (for single pass)

and A = $\pi d_0\, L, n\, .\, p$ (for p number of passes).

where, d_0 = Outlet dia. of the condenser tubes.

d_i = Inlet diameter of the condenser tubes.

h_i = Inside heat transfer coefficient.

h_0 = Outside heat transfer coefficient.

K = Thermal conductivity of tube material.

n = Number of tubes per pass.

L = Length of the tube.

Condenser Performance Factors

$$\text{Condenser Performance Factor, } K = \frac{M_s\, .\, H_s}{Q_m\, .\, A\, .\, V}$$

where, M_s = Mass of steam condensed in kg/hr.

H_s = Heat removed in kJ/kg of steam.

Q_m = Mean temp. difference in °C.

A = Total surface area in m^2

V = Velocity of water in m/sec.

8.23 AIR-LEAKAGE

The air enters in condenser from the following sources:

1. The dissolved air in exhaust steam can be liberated at low pressure.

2. Air can leak into the condenser from atmosphere through various parts or joints which are internally under a pressure less than that of atmosphere.
3. If the casing of the shell is porous or defective then air may also leak through them.
4. In jet condenser, air can enter with injection water.

Effects of Air Leakage or Causes of Inefficiency

(*a*) The presence of large amount of air increases back pressure on the prime mover, hence it will cause loss in heat drop or say plant thermal efficiency will reduce.

(*b*) Air leakage lowers the partial pressure of steam so there will be lower saturation temperature, thus latent heat of steam increases correspondingly. Therefore, exhaust steam will comparatively require more cooling water for condensation.

(*c*) Larger air leakage needs large capacity pump to remove air.

(*d*) As thermal conductivity of air is poor, the rate of heat transfer from vapour is less and hence for a given condenser duty, the surface area of the tubes has to be increased.

(*e*) The presence of air develops corrosion in condenser.

To Reduce Air Leakage

Air leakage can be minimised by the following means:

(*i*) The condensate is shut off from the atmosphere and is used over and over again, it will thus contain very small amount of air. Thus, there is an additional advantage of having to deal with less air in the condenser.

(*ii*) If condensate is not sufficient to the requirement of the boiler then additional water is to be fed into the boiler. This additional water is known as "Make up water" and contains certain amount of air. To remove this air, it is passed through steam evaporator while preheating. Thus, large quantity of air in the dissolved form is liberated in the evaporator.

(*iii*) Air leaking from the atmosphere can be reduced to a large extent by good design, accurate workmanship and careful attention during the life of the condensing plant from the joints.

Due to the high vacuum pressure in the condenser it is impossible to prevent air from leaking ; this can only be minimised. This air can be removed by the air extraction pump.

Air Leakage Allowance

The following relation is adopted to estimate the amount of air leakage allowance.

Air leakage = 5 kg per 10,000 kg of steam condensed for steam turbine plant

= 15 kg per 10,000 kg of steam condensed for steam engine plant.

Vacuum Measurement in Condenser

The difference between the atmospheric pressure and the absolute pressure in the condenser is called as a "vacuum in a condenser". To take measurement, it can be converted to correspond with the standard atmospheric pressure. **"The standard atmospheric pressure taken for vacuum measurement is the barometric pressure of 76 cm of mercury of 103 kg/cm^2".** The conversion of vacuum gauge reading to the standard is explained by Fig. 8.44.

Let "*A*" be the mercury barometer, "*B*" a mercury vacuum gauge, both dipping in an open bath of mercury "*C*". The vacuum gauge is connected to the condenser at "*D*" to measure vacuum. The column of mercury "*EF*" in "*A*" balances the atmospheric pressure acting on the surface of mercury in "*C*". The column of mercury "*EG*" in "*B*" plus the absolute pressure in the condenser connected to "B" at "D" acting on the top of the column "EG" also balances the pressure of the atmosphere. Therefore, the height "GF" gives the absolute pressure above the mercury in "B". The pressure in the condenser is independent of the atmospheric pressure, hence, if the reading of vacuum gauge corrected to standard is h, then

$$76 - h = EF - EG$$

or $$h \text{ (corrected vacuum)} = 76 - EF + EG \quad \ldots(1)$$

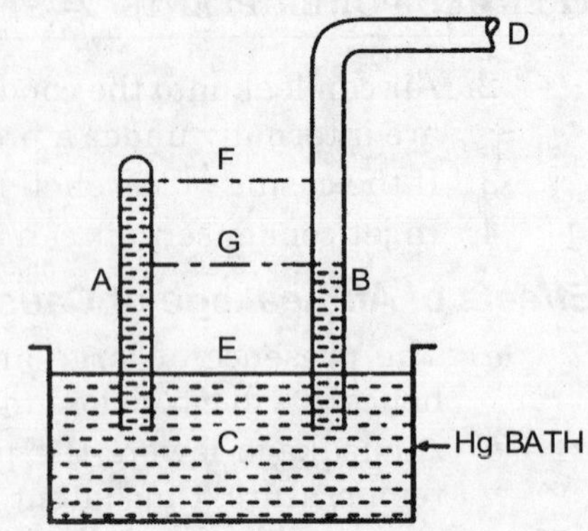

Fig. 8.44 Convertion of vacuum gauge reading to the standard.

Vacuum Efficiency

The purpose of an air pump on condensers is to remove air from the condenser so that the total pressure in the condenser may approach the partial pressure of the steam corresponding to the condensate temperature.

"Vacuum may be defined as atmospheric pressure minus absolute pressure and is generally given in millimeters of mercury".

Vacuum corrected to 760 mm barometer is = 760 – (Barometric height – Vacuum)

The high vacuum pressure in the condenser affects the steam consumption, the temperature of steam and its specific volume. A large variation of these factors occur for small variation of vacuum pressure,

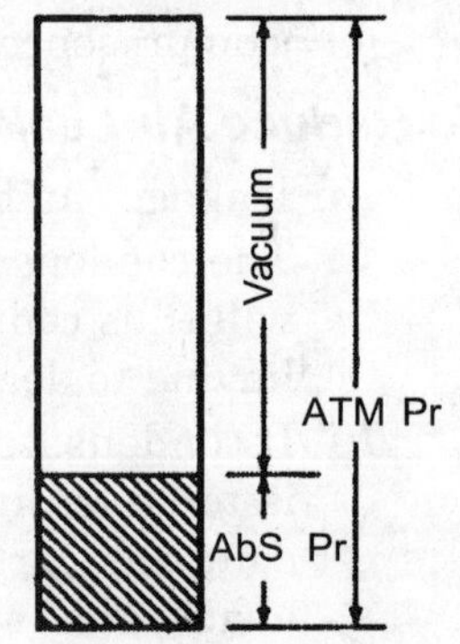

Fig. 8.45 Pressure distribution

$$\text{Vacuum efficiency} = \frac{\text{Actual vacuum}}{\text{Maximum obtainable vacuum}}$$

$$= \frac{\text{Actual vacuum at steam inlet to condenser}}{\text{Barometric pressure – Abs., pressure of steam}}$$

It is to note that vacuum efficiency depends upon the amount of air present in the condenser and it would be cent percent, if no air present in the condenser and such a condensing is called "perfect condensing plant".

The pressure equivalent of 1 mm of Hg

$$= \frac{10.325}{760} = 133 \text{ Pa}$$

Condenser Efficiency

Actually there is no standard method available to determine the condenser efficiency. The method used today was developed by M/s Parson & Co., a renouned manufacturer of steam turbines.

This is the ratio of the difference between the exit and the inlet temperatures of the cooling water to the difference between the temperature corresponding to the vacuum in the condenser and the entrance temperature of cooling water.

$\therefore$ Condenser efficiency = $\dfrac{\text{Rise in temp. of cooling water}}{\text{(temp. corresponding to vacuum in condenser} - \text{Inlet temp. of cooling water)}}$

Example 8.10 *The vacuum in a condenser is 720 mm when barometer is 770 mm. In another case the vacuum is 700 mm, when the barometer is 740 mm. Correct these vacuum readings to the standard barometer of 760 mm.*

Solution. (*i*) Here,

$$h = 760 - EF + EG$$

$$EF = \text{Barometer reading} = 770 \text{ cm}$$

$$EG = \text{Vacuum gauge reading} = 720 \text{ mm.}$$

Now

$$h = 760 - EF + EG$$

$\therefore$

$$h = 760 - 770 + 720 = 710 \text{ mm.} \quad \textbf{Ans.}$$

(*ii*)

$$EF = 740 \quad EG = 700$$

$\therefore$

$$h = 760 - 740 + 700 = \textbf{720 mm.} \quad \textbf{Ans.}$$

8.24 AMOUNT OF WATER FOR STEAM CONDENSATION

Suppose, steam entering the condenser = 1 kg

Latent Heat of entering steam = L kJ

Dryness fraction of entering steam = q

Temperature of entering steam = t °C

Sensible heat of entering steam = h kJ

Suppose, the condensate temp. in condenser = t_1 °C

Sensible heat of condensate = h_1 kJ

Then, heat given by the entering steam

$$= qL + h - h_1 \quad \text{...(2)}$$

Suppose, cooling water required to condensate 1 kg of steam

$$= W \text{ kg}$$

Entering temp. of cooling water to condenser = t_2 °C

Exit temp. of cooling water from condenser = t_3 °C

Sensible heat of cooling water at entrance = $h_2 = t_2$

Sensible heat of cooling water at exit = $h_3 = t_3$

Then, heat taken by cooling

$$= W(h_3 - h_2) \quad \text{...(3)}$$

Assuming, all the heat lost by the exhaust steam is given to the cooling water. Then equating Eq. (2) and (3)

$\therefore$

$$W(h_3 - h_2) = qL + h - h_l$$

or

$$W = \frac{qL + h - h_1}{(h_3 - h_2)} \quad \text{...(4)}$$

But

$$(h - h_1) = (t - t_1) \quad \text{and} \quad (h_3 - h_2) = (t_3 - t_2)$$

$\therefore$

$$W = \frac{qL + t - t_1}{(t_3 - t_2)} \quad \text{...(5)}$$

This equation (5) is employed in jet and surface condensers but in the case of jet condenser the final temperature t_3 of the injection water is equal to the temperature of the condensate. In surface condenser t_3 must be less than the temperature of the exhaust steam entering the condenser. The amount of cooling water required to condensate 1 kg of exhaust in surface condenser is greater than required in casė of jet condenser for the same condenser duty.

Example 8.11 *A surface condenser receives 0.833 kg/s of steam at 40°C and 0.8 dry. The rise of temp. of the cooling water to the condenser is 10°C. Estimate the quantity of cooling water in kg/min.*

Solution.

Steam consumed/min = 0.833 kg/s

Temp. of exhaust steam = 40°C

$q = 0.8$

Rise in temp. of cooling water = 10°C

Let, quantity of cooling water = W kg/s

∴ Heat lost by exhaust-steam = Heat carried by cooling water

or
$$x(h + qL) = W(t_3 - t_2) \quad \text{...(1)}$$

From steam tables of dry saturated steam

$$L = 2293 \text{ kJ/kg at } 40°\text{C}$$

and
$$h = t = 167.5 \text{ kJ/kg}$$

∴ By substitution in Eq. (1)

$$0.833\,(167.5 + 0.8 \times 2293) = W \times 10 \times J$$

or
$$W = \frac{0.833(167.5 + 1834.4)}{10 \times 4.186} = \textbf{40 kg/s} \quad \textbf{Ans.}$$

Exampley 8.12 *A surface condenser is designed to handle 2.78 kg of steam per sec. The steam enters at 0.008 mPa and 0.9 dry and the condensate leaves at the corresponding saturation temp. The pressure is constant throughout the condenser. Estimate the cooling water flow rate per hour, if the cooling water temp., rise is limited to 10°C.*

Solution. From steam tables at 0.008 mPa

$$h = 173.3 \text{ kJ/kg} \quad L = 2398 \text{ kJ/kg}$$

$$h_1 = 173.3 \text{ kJ/kg} \quad \text{...(Given)}$$

∴ Heat lost by exhaust steam = Heat carried by cooling water.

∴
$$(h + qL + h_1) = W\,(t_3 - t_2)$$

or
$$W = \frac{h + qL - h_1}{J(t_3 - t_2)} \text{ per kg of steam}$$

$$= \frac{173.3 + 0.9 \times 2398 - 173.3}{10 \times 4.186} = \frac{3398}{41.86}$$

$$= 51.6 \text{ per kg of steam} = 2.78 \times 51.6$$

$$= \textbf{143.3 kg of cooling water per second.} \quad \textbf{Ans.}$$

Example 8.13 *Exhaust steam of dryness fraction 0.9 enters a jet condenser at an absolute pressure of 0.012 mPa and is condensed to water at 35°C. The cooling water enters at 20°C. Compute the weight of injection water required per kg of steam.*

Solution. Heat lost by exhaust steam = Heat carried by cooling water

$$\therefore \quad (h + qL - h_1) = W(t_3 - t_2)\ J$$

From steam tables at 0.12 kg/cm^2

$$h = 205.1 \text{ kJ/kJ} \quad L = 2380 \text{ kJ/kg} \quad h_1 = 146.5 \text{ kJ/kg}$$

$$\therefore \quad W = \frac{205 + (0.9 \times 2380) - 146.5}{(35 - 20) \times 4.186}$$

$$= \textbf{35 kg of cooling water per kg of steam.} \quad \textbf{Ans.}$$

Example 8.14 *A barometric jet condenser consumes 1.25 kg 95% dry steam per second and records a vacuum of 700 mm of mercury when barometric pressure is 760 mm. The circulating water enters the condenser at 20°C and the mixed condensate and cooling water leave at 45.4°C. Calculate:*

(i) The minimum height of the tail pipe above the level of the hot well and

(ii) The quantity of cooling water required assuming no under-cooling.

Solution.

(*i*) Tail pipe length determination:

$$\text{Barometric height} = \frac{760 \times 13.6}{1000} = 10.3 \text{ m of water}$$

$$\text{Pressure in condenser} = (760 - 700) \times \frac{13.6}{1000} = 0.816 \text{ m of water}$$

$\therefore$ Required length of Tail pipe

$$= (10.3 - 0.816) = \textbf{9.484 m} \quad \textbf{Ans.}$$

(*ii*) Quantity of cooling water:

Since there is no under-cooling of the condensate, therefore, 45.4°C is also the temperature of the steam at condenser entrance.

From steam tables, corresponding to 45.4°C

$$h = 190 \text{ kJ/kg} \quad L = 2389$$

and $$x = 1.25 \text{ kg/s} \quad q = 0.95$$

Heat carried by cooling water = heat lost by exhaust steam

or $$W(t_3 - t_2)J = x(h + qL)$$

or $$W = \frac{1.25 \times (190 + 0.9 \times 2389)}{(45.4 - 20) \times 4.186}$$

or $$W = \frac{2925}{106.3} = \textbf{27.25 kg/sec} \quad \textbf{Ans.}$$

Example 8.15 *Steam enters the condenser at 35.8°C with barometer reading of 760 mm of Hg. A vacuum of 700 mm was recorded in the condenser. Determine the vacuum efficiency.*

Solution. Partial pressure of steam at 35.8°C

$$= 0.006 \text{ mPa}$$

$$= \frac{0.006}{0.000136} \text{ mm of Hg} = 44.1 \text{ mm of Hg}$$

$$= 760 - 44.1$$

or $$= 715.9 \text{ mm of Hg.}$$

$$\therefore \quad \text{Vacuum efficiency} = \frac{\text{Actual vacuum}}{\text{Maximum obtainable vacuum}}$$

$$= \frac{700}{715.9} = \mathbf{97.76\%}. \quad \textbf{Ans.}$$

Example 8.16 *The inlet and outlet temperatures of circulating water in a condenser are 286°C and 35.8°C respectively. The vacuum is 710 mm of mercury with barometer reading 760 mm. Estimate condenser efficiency.*

Solution. Condenser abs., pressure = (760 – 710) = 50 mm of Hg

= 0.00666 mPa

Temp. corresponding to 0.00666 mPa = 38° C of steam

$$\therefore \text{ Condenser efficiency} = \frac{35.8 - 28.6}{38 - 28.6} = \frac{7.2}{9.4} = 0.768 = \mathbf{76.8\%}. \quad \textbf{Ans.}$$

Example 8.17 *A surface condenser is designed to handle 4.44 kg of steam/second. The steam enters the condenser at 0.009 mPa abs and 0.88 dryness fraction and the condensate leaves the condenser at the corresponding saturation temperature.*

Determine the rise in the cooling water temp., if the cooling water flow rate is 249 kg/sec. Assume that the pressure is constant throughout the condenser.

Solution. Steam/second = 4.44 kg/second

Steam entrance pressure to condenser = 0.009 MPa

$$q = 0.88$$

$$W = \text{cooling water flow} = 2489 \text{ kg/s}$$

From steam tables at 0.009 mPa

$$h = 183 \text{ kJ/kg} \quad L = 2435 \text{ kJ/kg}$$

$$h_1 = 183 \text{ kJ/kg} \qquad \text{(Given)}$$

$$\therefore \text{ Weight of cooling water required to condense one kg of steam} = \frac{(h + qL) - h_1}{t_3 - t_2}$$

$\therefore$ For 4.44 kg of steam

$$249 = \frac{4.44(183 + 0.88 \times 2435 - 183)}{(t_3 - t_2) \times J} = \frac{4.44 \times 2142}{(t_3 - t_2) \times 4.186}$$

or

$$(t_3 - t_2) = \frac{9511}{249 \times 4.186} = \mathbf{9.12°C} \quad \textbf{Ans.}$$

8.25 AIR EXTRACTORS

This is also known as an "Air Pump" and its function is to always maintain high vacuum in the condenser by removing (*i*) air (*ii*) condensed steam (*iii*) and other non-condensable gases continuously all the time. The vacuum is created by the condensation of exhaust steam whereby a large volume of steam is converted into a relatively small volume of water. The **"Air Pump"** extract air that continuously enters the condenser. The air pump is operated by power and therefore the air entering the condenser should be least for minimum power consumption. The size of an air pump depends upon:

(*a*) The temperature of the air or the mixture of air and water which it has to extract.

(*b*) The volume of air to be removed.

It is important to see that air should be removed at a point where the air has the lowest temperature and hence the smallest volume. If the leakage of air is prevented the capacity of the pump required is reduced greatly.

Following two types of air pumps are used for this purpose.

1. **Dry Air Pump.** This removes air and other non-condensible gases only and may be of reciprocating, rotary or steam jet pump.
2. **Wet Air Pump.** This removes both air and condensate, hence called "Wet Air Pump"

This may be of reciprocating type or water pump.

8.26 COOLING WATER SYSTEMS

Heat rejected in a condenser is taken up by the water. As water comes out of the condenser, it gets heated and in case no natural source of water is available, the same water may be recirculated through the condenser (known as closed system), after cooling. The quantity of water required is considerably high. A 100 MW plant may require about 300,000 kg/hour of steam to be condensed in the condenser. Corresponding to this quantity of cooling water required will be something like 1000 tons/hr. Site conditions determine the method to be used for supplying the cooling water to the condensers, and may be one of the following systems:

1. River/sea or canal/lake water system.
2. Combined river and cooling water system.
3. Cooling tower or spray pond system.

As we have already discussed that a condenser is a device in which exhaust steam from steam turbines is condensed and the heat energy given up by the steam during condensation is taken up by the cooling water. The water coming out of the condenser is hot and is cooled in order that it may be recirculated through the condenser.

Availability of cooling water predominantly decides the plant site. The high cost of water makes it necessary to use cooling towers for the water coming out of condenser, in a closed system, before it is used again.

8.26.1 Types of Water Cooling Systems

1. Direct Use of River or Sea

Large power stations require enough quantity of cooling water per hour. Such plants are usually located near a river or sea. The water is constantly drawn from the river by the pump, filtered and circulated through the condenser. Hot coolant is discharged back into the river (Fig. 8.46)

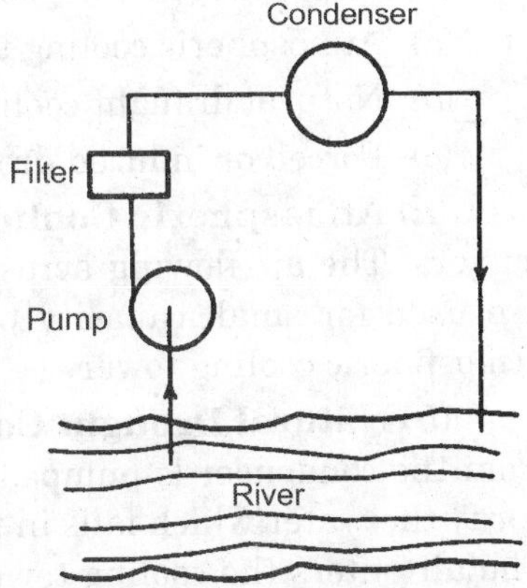

Fig. 8.46 Use of River water for Coolling

2. Cooling Ponds

In this system (Fig. 8.47) warm condensing water from the condenser is sprayed through nozzles over a pond of large area and cooling effect is mainly due to evaporation from the surface of water. In this system sufficient amount of water is lost by evaporation and windage.

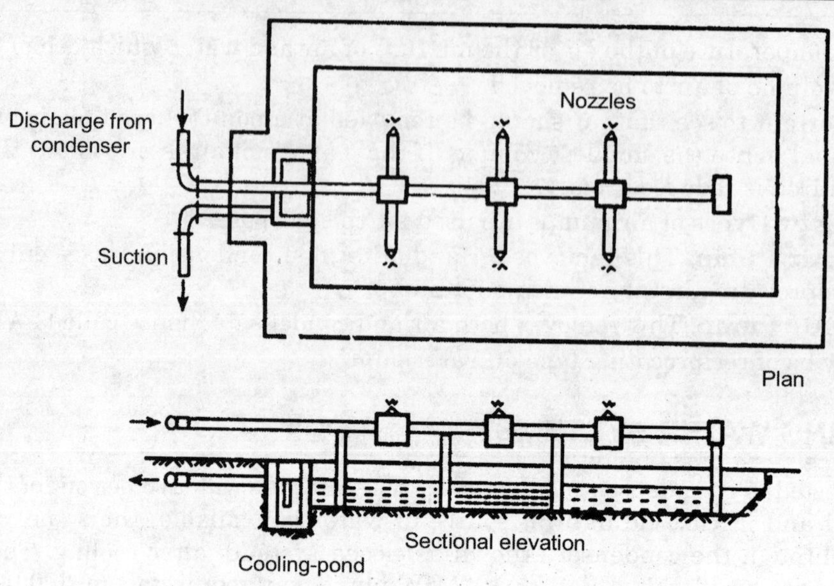

Fig. 8.47 Cooling with spray arrangment.

Some of the factors which influence the rate of heat dissipation from a cooling pond are as follows:

(*i*) Area and depth of pond
(*ii*) Temperature of water entering the pond
(*iii*) Atmospheric temperature
(*iv*) Wind velocity
(*v*) Relative humidity
(*vi*) Shape and size of water spray nozzles.

The ponds should be placed where the prevailing winds are not obstructed. The spray of water, as it comes in contact with atmospheric air, loses its heat and cold water is recirculated.

3. Cooling Towers

The different types of cooling towers are as follows:

(*a*) Atmospheric cooling tower.
(*b*) Natural draught cooling towers.
(*c*) Forced or induced drought cooling tower.

(*a*) Atmospheric Cooling Tower. In this cooling tower hot water is allowed to fall over louvers. The air flowing across in transverse direction cools the falling water. These towers are used for small capacity power plants such as diesel power plants. Fig. 8.48 shows an atmospheric cooling tower.

(*b*) Natural Drought Cooling Towers. In natural draught cooling tower, the hot water from the condenser is pumped to the troughs and nozzles situated near the bottom. Troughs spray the water which falls in the form of droplets into a pond situated at bottom of the tower. The air enters the cooling tower from air openings provided near the base, rises upward and take up heat of falling water.

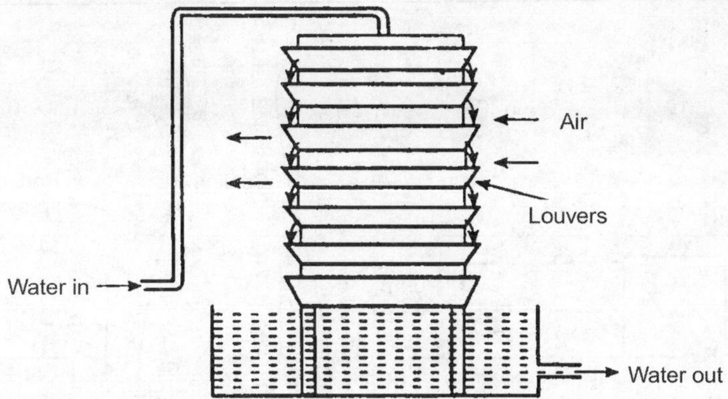

Fig. 8.48 Atmospheric cooling tower.

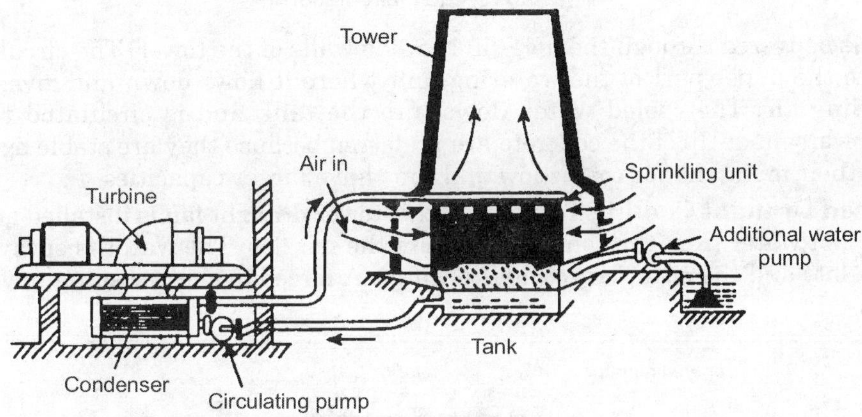

Fig. 8.49 Cooling tower with turbine.

Cooling towers may be made up of timber, concrete or steel. A concrete hyperbolic cooling tower is shown in Fig. 8.50. Fig. 8.51 shows the water circulation from the cooling tower to the condenser. Fig. 8.49 shows the cooling tower in which the position of turbine has also been shown. The system consists of turbine, condenser. Circulating pump, tank, additional water pump and sprinkling unit.

Hyperbolic towers offer greater resistance to wind pressure, eliminates shearing stresses and internal bracing, thus avoiding the formation of eddies. With the widening of the top of chimney permits condensation within the tower. It also avoids local fogging and warm air circulation. Chimney shape also creates its own draft, assuming efficient operation.

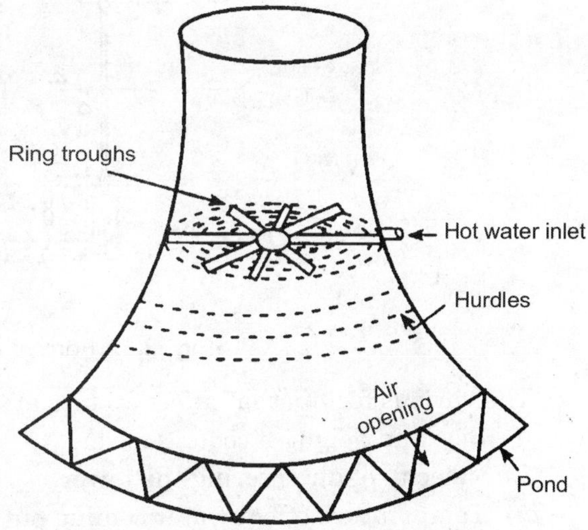

Fig. 8.50 Hyperbolic cooling tower.

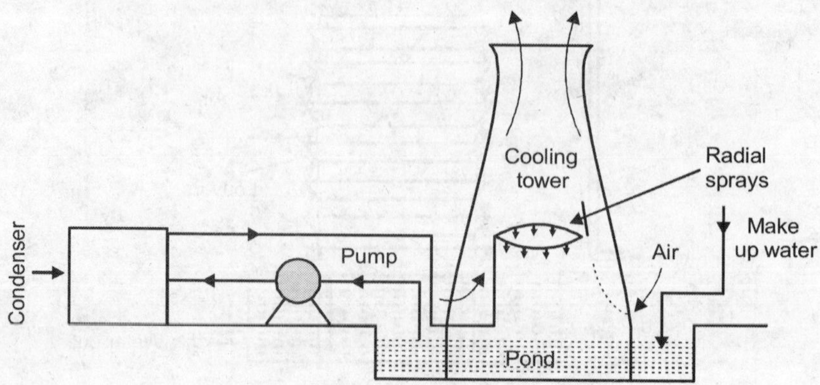

Fig. 8.51 Water circulation

The air is delivered through the holes in the side walls of the tower. The circulating water is delivered to the upper part of the watering unit where it flows down and gives its heat to the surrounding air. The cooled water flows into the tank and is circulated through the condenser. Towers made up of the concrete are preferred because they are stable against larger air pressure, their maintenance cost is low and they have larger capacities.

***(c)* Forced Draught Cooling Towers.** In this tower draught fan is installed at the bottom of tower. The hot water from the condenser enters the nozzles. The water is sprayed over the tower filling slats and the rising air cools the water. A forced draught cooling tower is shown in Fig. 8.52.

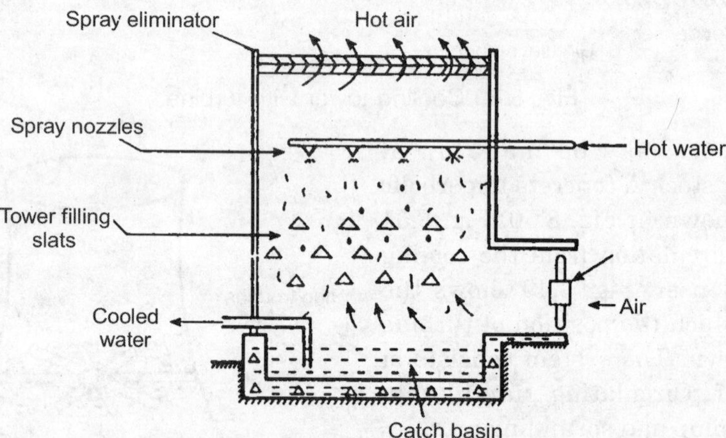

Fig. 8.52 Forced draft cooling tower.

The various factors that affect cooling of water in a cooling tower are as follows:

(i) Size and height of cooling tower.

(ii) Velocity of air entering the tower.

(iii) Temperature of hot water coming out of condenser.

(iv) Temperature of air.

(v) Humidity of air.

(vi) Accessibility of air to various parts of cooling tower.

Mechanical draft cooling towers may be forced draft cooling towers or induced draft towers. Fig. 8.53. shows an induced draft cooling tower. The hot water is allowed to pass through the air flowing upward in the counter direction. The draft fan installed at top of tower draw air through the large openings in the tower. The air moving the upward direction cools the water. Induced draft towers produce less noise.

In the cooling tower water cools by:

(*i*) evaporation. (*ii*) heat transfer to the air.

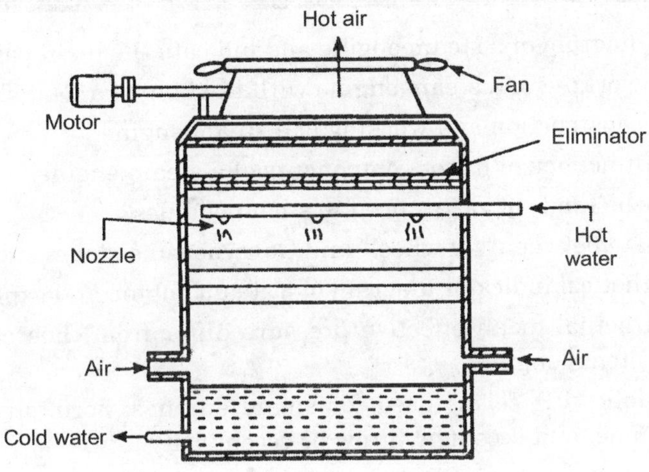

Fig. 8.53 Induced draught cooling tower.

Most of the cooling (aoout 75%) takes place by evaporation. Make up water should be continuously added to the tower collecting basin to replace the water lost by evaporation and spray carry over.

Maintenance of Cooling Towers. In order to achieve the desired cooling and to reduce the depreciation costs the regular maintenance of cooling towers is essential. The fans, motors housing etc. should be greased and gear boxes oiled. Any unusual noise or vibrations in them should be corrected immediately. At least once in a year motor's gear boxes should be checked for structural weakness. The circulating water should be tested for hardness and should be kept free from impurities to avoid scale formations and to avoid corrosive action of water. The water spraying nozzles should be inspected regularly for clogging.

4. Mixed System (Combined River and Cooling Tower)

This type of cooling system uses river water as well as cooling ponds or cooling towers, which overcome the difficulty of recirculation and meet the requirements of a fishery board on a fairly small river. The water from the condenser is discharged into the river at a point near the station, and after mixing with the river water an amount is drawn from the river and pumped to the cooling towers, where it is further cooled. Thus the part of the water from the condenser is discharged directly into the down stream of the river and part of the water is pumped to the cooling tower.

Some of the advantages of this system are:

1. The size of the cooling tower can be reduced.

2. A smaller quantity of cooling water is required.
3. The turbine plant efficiency is increased.

In this system, one pump set draws water from the river and feed to the towers and the other takes the water from the tower outlet and delivers it to the condenser. A third pumping set may be installed to draw water direct from the river, for delivery to the condenser.

QUESTIONS

8.1 Draw a neat diagram of a steam engine and indicate the main parts on it.

8.2 Describe the working of a steam engine with the help of a neat sketch.

8.3 Describe the construction and working of a steam engine.

8.4 Describe the functions of major components of a steam engine.

8.5 (*a*) What are the kinds of valves used in steam engines?

(*b*) Differentiate between "fly wheel" and Governor in a steam engine.

8.6 Draw a hypothetical indicator diagram of a steam engine and explain its use.

8.7 How does an actual mean effective pressure differ from theoretical mean effective pressure of a steam engine?

8.8 Explain the following terms used for steam engines: bore, stroke, swept volume, clearance volume, and eccentric throw or eccentricity.

8.9 (*a*) Give an expression for indicated power developed in a double acting and a single cylinder simple steam engine.

(*b*) Explain the reasons of actual indicated power being always lesser than hypothetical indicated power of a steam engine.

8.10 Write expressions for different efficiencies associated with a steam engine.

8.11 Explain governing of steam engine. Describe throttle governing and cut-off governing.

8.12 Explain the working of a compound steam engine.

8.13 Explain the principle of working of an impulse steam turbine with the help of a diagram.

8.14 Explain the reasons of compounding steam turbines. Describe the working of such a turbine.

8.15 Explain the principle of working of an impulse-reaction turbine.

8.16 State different losses in steam turbine. Explain different kinds of governing of steam turbine with merits and demerits.

8.17 (*a*) State the function of a condenser.

(*b*) Describe a jet condenser.

8.18 (*a*) Describe a surface condenser.

(*b*) State the advantages and disadvantages of jet condensers and surface condensers.

8.19 (*a*) Define vacuum and explain vacuum efficiency.

(*b*) Explain the effect of air leakage in a condenser. How the air leakage can be minimised? Explain any two methods.

8.20 Explain the working of following condensers:

(*a*) Ejector condenser

(*b*) Evaporative condenser.

8.21 Explain the term steam turbine performance in detail.

8.22 Write short notes on the following:

(*a*) Selection of a condenser.

(*b*) Sources of air leakage in a condenser.

8.23 What is a steam turbine? How are steam turbines classified?

8.24 Explain different methods of steam turbine governing.

8.25 What is a condenser? Name the different types of condensers.

8.26 Explain different types of cooling towers used in steam power plant. Discuss their specific advantages.

8.27 What are the advantages of steam turbines over steam engines?

8.28 Explain different methods of compounding of steam turbine stages.

8.29 List the advantages and limitations of velocity compounded steam turbines.

8.30 Discuss the merits and demerits of a hyperbolic cooling tower.

8.31 Write short notes on:

(*a*) Spray pond/cooling tower.

(*b*) Closed circuit cooling system.

8.32 State different losses in steam turbines.

Chapter 9

AIR COMPRESSORS

Learning Objectives

After studying this chapter, the students will learn:

- Compressors; types, theory and working.
- Cooling during compression.
- Reciprocating and rotary compressors, such as screw type, roots blower type, centrifugal, and slinding vane type.
- Components of air compressors.
- Mechanical and volumentric efficiencies.
- Applications of air compressors.

9.1 INTRODUCTION

Air compressors are used for producing high pressure air. Atmospheric air is drawn into a cylinder during the suction stroke and is compressed by the piston during the return stroke. The piston is driven by the electric power or by a diesel engine. It may be single acting or double acting depending upon whether air is admitted to one side of the piston only or to both sides of piston alternately. The air compressors are mainly of two types viz, reciprocating compressors and rotary compressors. The reciprocating compressors may be single acting or double acting.

Thus, air compressor is a machine used to increase the pressure of air by reducing its volume. A compressed air system consists of the air compressor with prime mover, suction line with air cleaner, inter-cooler, after-cooler, receiver and safety valve.

Air (or gases) are required at high pressure for many applications *e.g.* in chemical, industrial and construction field. It is used for operating small pneumatic hand tools in workshops, for fabrication works, and also for paint spraying, mining operations etc.

Pneumatic equipments commonly used in construction are: Road breakers. Jack hammers, Wagon drills, Drifters, Backfill tampers, Concrete vibrators. Compressed air is used for various other jobs like spray painting, pneumatic conveyors, pneumatic controls, riveting hammers, air wrenches, pneumatic grinders, sand blasting, running of hoists, grout pumps, drilling rock or other hard formations, operating air motors, hand tools, pile drivers, dewatering pumps etc.

Compressed air has been accepted as more useful power medium and replacing steam, because of following advantages.

(*i*) Compressed air can be transmitted from one point to another with ease.

(*ii*) Use of compressed air is safer.

(*iii*) Pneumatic tools are lighter than similar ones using steam.

(*iv*) The tools using compressed air do not overheat in use and are cheaper and easy to maintain.

Air Compressor Terminology

1. **Single Stage and Multi Stage Compressors.** *In single stage compressors,* the compression of air from the initial pressure (usually atmospheric pressure) to the final

pressure is carried out in one cylinder. *In multistage compressor,* air is compressed from initial stage to final stage in more than one cylinders, in a sequence, the delivery of one cylinder being the supply to the next.

2. **Single Action and Double Action Compressors.** In single acting reciprocating compressor, the suction, compression and delivery of the air take place on one side of the piston only. In double acting compressors, suction, compression and delivery takes place on both sides of the piston. Single acting compressor gives one delivery stroke per revolution of the crank, while double acting compressor gives two delivery strokes per revolution of the crank.
3. **Pressure Ratio.** It is the ratio of the discharge pressure to the inlet pressure, when both the pressures are expressed in absolute units.
4. **Displacement of the Compressor.** The volume swept by the piston in the cylinder (first cylinder of a multistage compressor) is known as displacement of the compressor.

9.2 CLASSIFICATION

Air compressors can be classified as follows:

(*i*) Positive displacement or non-positive displacement type.
(*ii*) Single stage or multistage.
(*iii*) Portable or stationary.

1. **In positive-displacement compressors,** air is pressurized with in a closed space. These are mainly reciprocating or rotary type. In the reciprocating air compressors air is pressurized by a piston in a confined space of a cylinder. Whereas in the rotary air compressors pressurising is done by positive action of rotating blades moving in a casing. The pressure developed by rotary compressors is limited to 75 Pa or less. These are used for larger volumes at medium pressures.
   Rotodynamic compressors are **non-positive displacement compressors** and centrifugal compressors are the main rotodynamic compressors. These centrifugal compressors may be radial flow, axial flow or mixed flow type. In these compressors the air is compressed by dynamic action of the rotating vanes or impellers which impart velocity and pressure to the air while flowing through the casing.
2. **Staging in the air compressor** is done for increasing the capacity (*i.e.* quantity of air) or the pressure rating. Volumetric capacity of the compressors can be increased by compounding the cylinders in parallel. The pressure rating is increased by compounding them in series, this is generally termed as multi-staging. Intercoolers are provided between the high pressure and low pressure units of the compressors when multi-stage compression is employed.
3. **Portable compressors** are mounted on trollies, having two wheel or four wheel, depending upon the size of the compressor, and can be hauled from one place to another place, whereas stationary compressors are installed on permanent foundations. These are used where large volumes of air is required for a considerably long period of time.

9.3 THEORY AND WORKING OF AIR COMPRESSORS

1. Single Stage Reciprocating Compressor

A reciprocating compressor consists of a cylinder, piston, crank shaft, connecting rod, inlet and outlet valves similar to that of an I.C. engine. Crank shaft is rotated by the prime mover.

This produces reciprocating motion of the piston in the cylinder. During the downward (i.e. suction) stroke, the pressure in the cylinder falls below the atmospheric pressure as a result of which the intake valve opens and air is drawn from the atmosphere into the cylinder. This continues till the end of stroke i.e. until when the piston reaches the bottom dead centre position. At the beginning of the upward or compression stroke the pressure in the cylinder will rise and the inlet valve closes. The cylinder pressure steadily increases until it is just greater than that in the delivery chamber. The delivery valve then opens, and discharge takes place from the cylinder to the receiver during the remaining upward motion of the piston. The receiver is a simple vessel which acts as a storage tank. Thus the suction, compression and delivery of air takes place within two strokes of the piston or one revolution of the crank. The suction takes place during one stroke and compression and delivery take place during the second stroke.

The thermodynamic cycle for a compressor is represented on the pressure-volume diagram as shown in Fig. 9.1.1-2 is a suction phase at constant supply pressure (p_s); 2-3 is a compression phase following the gas law, $pv^n = C$, 3–4 is a discharge phase at constant delivery pressure (p_d). The area of this cycle diagram represents the theoretical work required to drive the compressor.

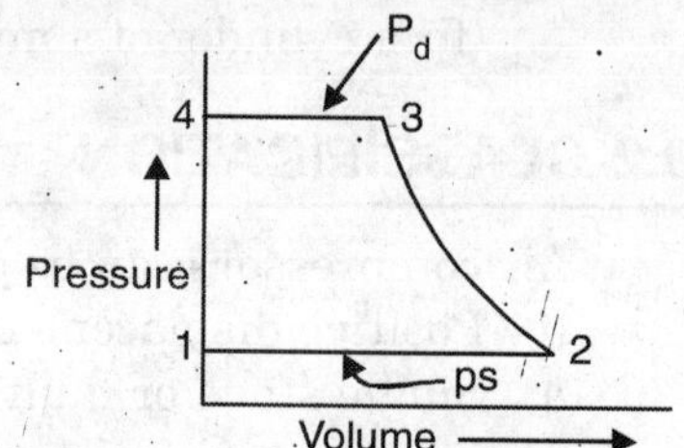

Fig. 9.1 Pressure-Volume diagram for a compressor.

When the process of compression is carried out at a constant temperature, it is known as esothermal and the air follows the law $pv = C$. And when the air is compressed rapidly or in a perfectly insulated cylinder, without any heat losses, the process is known as adiabatic and the air follows the law $pv^k = C$, where $k = 1.406$ for air.

In practice n is between 1 and k. The higher the value of n, the steeper the compression curve (as shown in Fig. 9.2) and the greater is the power requirement. Thus if isothermal compression is attained, there would be substantial power savings, as shown by the cross-hatched area of Fig. 9.2. (*a*).

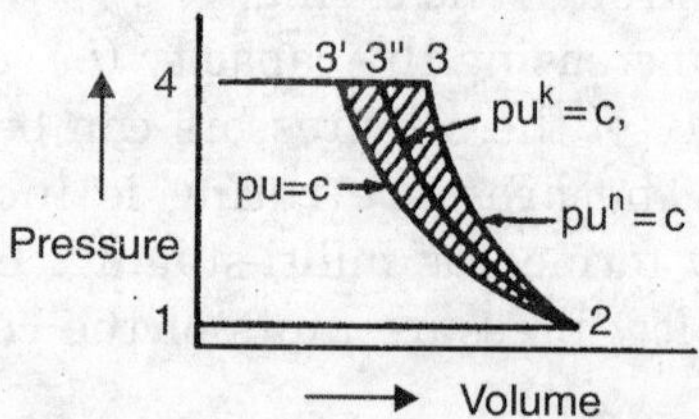

Fig. 9.2 (*a*) Pressure-volume diagram, indicates saving in power reqd, if isothermal compression is achieved.

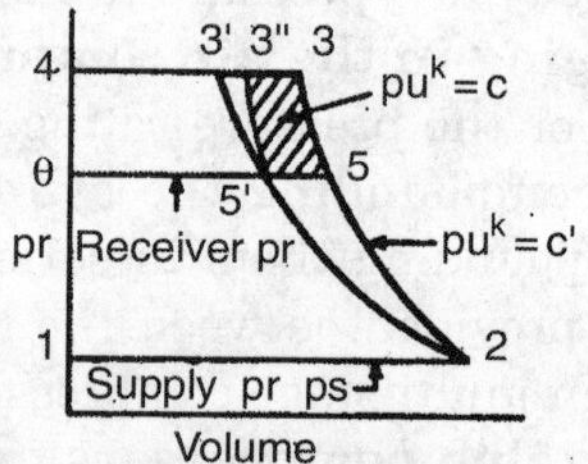

Fig. 9.2 (*b*) Pressure-volume diagram, 2-stage compressor with intercooling.

This multistage compression with intercooling result in, (*i*) a saving in power ; and (*ii*) improved volumetric efficiency.

Perfect cooling for two stage compression requires that the air be cooled back to the original temperature as shown by 5-5′ in Fig. 9.2 (*b*). Point 5′ is located on the isothermal compression line 2-3′. The power saving, over single stage adiabatic compression is shown by the cross-hatched area. The figure also indicates that the extent of saving of power is determined by the degree of intercooling and also by the pressure at which the intercooling affected.

2. Multi-stage Reciprocating Compressor

In this method, instead of compressing the air to high pressure in single cylinder, the air is compressed in several stages. The pressure of air is increased in each stage. In two stage air compressor, air is first taken into the low pressure (L.P.) cylinder. After compressing air to some desired pressure, it is passed through an inter-cooler. Finally the air is compressed in the high pressure (H.P.) cylinder and then discharged to the receiver. It is the cooling of the air between the cylinders that keep the compression very near to isothermal.

Advantages of Multi-stage Compressing

1. Reduction in power required to drive the compressor due to the compression being approximate to the isothermal.
2. A single stage air compressor must be strong enough to withstand the high delivery pressure, whereas in a multi-stage compressor, the low pressure cylinder is lighter.
3. Due to lower temperatures, lubrication is more effective.
4. Since pressure difference is on either side of the piston and valves is less, the loss due to leakage shall be less.
5. Better volumetric efficiency.
6. It is possible in this case to cool the air before being delivery to H.P. cylinder.
7. Better mechanical balance can be achieved with multi-stage compressor.
8. Higher volumetric efficiency for a given pressure ratio.

9.4 METHODS OF COOLING DURING COMPRESSION

Different methods of cooling the air during the compression are used for increasing the isothermal efficiency are discussed hereunder:

1. **External Fins.** Effective cooling for small capacity air compressors can be achieved by using the fins on the external surface of the cylinder.
2. **Water Jacketing.** The water is passed through the jacket provided around the cylinder and heat of air being compressed is carried away by the water.
3. **Inter Cooling.** Water jacketing is not very effective when the speed of compressor is high and pressure ratio required is also high. In such cases, the compression is carried out in two or more stages and the air coming out from the first stage is cooled in heat ex-changer known as inter - cooler before it is taken to the second stage.

Water jacketing as well as intercooling can be simultaneously used to reduce the work required. The jacket cooling reduces the work during compression in each cylinder while intercooling reduces the work to be supplied in the second (or next stage in multi - stage compression) stage.

Intercooling and Aftercooling

Cooling of air between the two stages of compression in a multi-stage compressor is known as intercooling (see Fig. 9.2(c). The objects of intercooling are,

(*i*) to reduce the temperature and thereby reducing the volume of air, which is increased during first stage of compression, which would otherwise need more power for compression.

(*ii*) to condense water vapour and thus removal of moisture from the air.

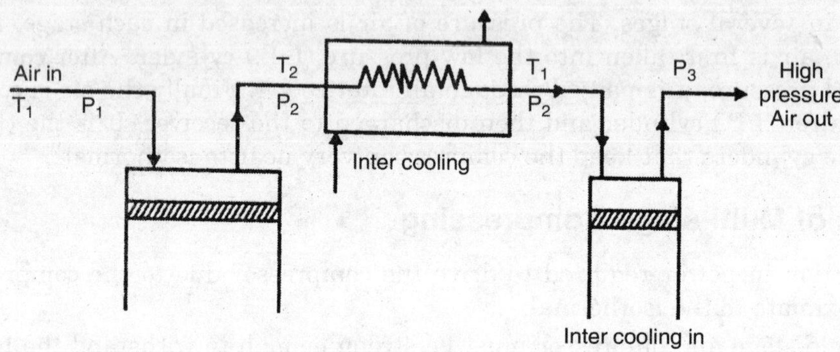

Fig. 9.2 (*c*) Arrangement of intercooling in a two-stage compressor.

Cooling of air after its final compression in the compressor is known as after-cooling. The aftercooling is done with the object,

(*i*) to remove moisture by condensing the vapour available in the air. This saves the pneumatic tools, from the harmful effects of wet air.

(*ii*) to cool the air before entering the pneumatic tools, otherwise, warm air is likely to cool and reduce volume during the distribution system causing a fall in pressure.

Construction

Intercoolers and aftercoolers are simple heat exchangers, where heat is removed from the air after it has been compressed. A simple line diagram of such a system is shown in the figure 9.3.

In this system coolant which is generally water, passes the tubes secured between two end plates, while air circulates over the tubes. For better contact between the tubes, air baffles are provided.

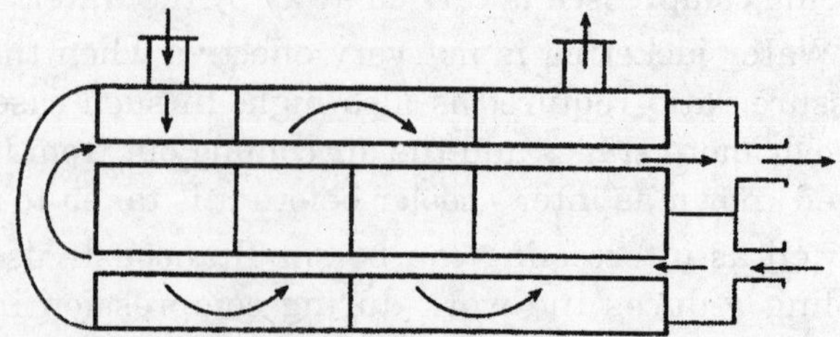

Fig. 9.3 Principle of construction of inter/after cooler.

9.5 TYPES OF AIR COMPRESSORS

As explained above air-compressors are of positive displacement type and non positive displacement type of compressor. Positive displacement compressors are mainly Reciprocating Air Compressors, Rotary Air Compressor and Screw Air Compressors. Whereas non-positive displacement type of compressors are centrifugal compressors and Axial-flow compressors.

The positive displacement compressors deliver air at high pressure, whereas non-positive displacement compressors give large volume of air at low pressure and are used in various industries as blowers. Since Construction Equipments require air at high pressure (generally at or above 7 kg/cm^2), positive displacement compressors are used for their operation. In this chapter we shall confine ourselves to positive displacement compressors only.

9.6 RECIPROCATING AIR COMPRESSORS

A reciprocating air compressor is a machine which increases pressure of air by compression process and delivers it at a higher pressure.

Earlier compressors followed steam engine practice of large bore and stroke, few cylinders, slow speed, double acting, horizontal construction. With the development of internal combustion engines, air compressors now-a-days are of small bore, short stroke, multi-cylinder, high speed, single acting, vertical construction.

The reciprocating air compressors resemble an Internal Combustion engine in construction. In this a piston moves in a cylinder through the motion of a crank shaft. Inlet and outlet valves are provided, as in the case of I.C. Engines. Internal lubrication is also done like I.C. Engines, with the result a small amount of oil is always present in the air delivered by the compressor. In multicylinder air compressors, cylinders are arranged in *V, Y* and opposed piston arrangements.

Prime-movers employed on air compressors are generally electric motor and Diesel engine. Electric motors are used where electric supply is available and compressor is stationary. Where as Diesel Engines are used to drive portable compressor and for stationary units when no electricity is available.

Reciprocating air compressors are available in single-or multi-stage. In single stage compressors, pressures between 0.65 to 1.05 MPa cm and capacities upto 0.14 m^3/s (300 cfm) are commonly available. While pressures upto 35 MPa cm have been attained in multi-stage compressors. For construction equipments, air compressors commonly used are 2-stage compressors giving pressure upto 1.5 MPa. Pneumatic equipments generally require a pressure of 0.7 MPa. Requirements of volume capacity for a compressor depends upon the number of pneumatic tools proposed to be operated with one compressor.

9.7 ROTARY AIR COMPRESSORS

Rotary machines which are used for supplying air or any other fluid are called fans or compressors. A fan is a machine which moves gaseous fluids with a pressure ratio of upto 1 : 15. A compressor is a machine which moves the gaseous fluids with a pressure ratio of more than 1 : 15. Rotary compressors have adiabatic compression. Their speed is very high and are small in size, as compared to reciprocating compressors, due to high speed and can be directly coupled to turbines, electric motors etc. Rotary compressors deliver more clean air and uniform delivery of air is obtained.

In recent years these have become very popular due to its light weight, uniform flow, variable output, carefree operation and long life.

1. Screw Type Rotary Air Compressors

Air Screw compressor is a positive displacement compressor and air is compressed by reducing the space progressively. This has two rotors, one male and other female. These rotors have a symmetric profile lobes encircling the rotors in the form of a thread (or an involute), the male has 4 lobes while female has 6. Both the rotors are housed in a closed housing and connected through external gears which are housed in a separate housing and filled with oil. The drive is given to male rotor while the female rotor just rotates on bearings with the lobes of the male rotor (see Fig. 9.4).

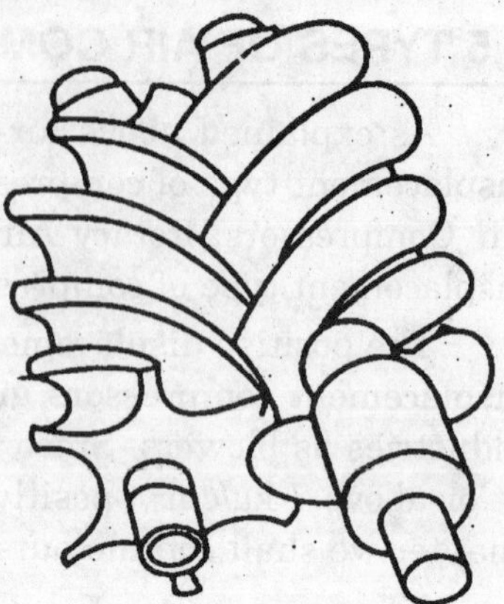

Fig. 9.4 Rotors of Screw Air Compressor.

There are small clearances between the two rotor lobes and between the rotor and the casing. Air enters the casing at one end, and is entrapped by the successive pairs of lobes as they rotate, and is compressed till the discharge port opens.

Air Screw Compressors have following advantages over reciprocating air compressors:

(*i*) Screw compressors are simple in construction, lighter in weight and very compact in dimensions.

(*ii*) Screw compressors has a few moving parts *i.e.,* gears and rotors. The life of screw compressor is therefore very long.

(*iii*) Since the screw compressors are of rotary design, therefore no unbalanced forces and hence no foundation is required. These can be directly put on a simple floor and can be started.

(*iv*) The screw compressors can run continuously round-the-clock for a long period without affecting the efficiency and life of the machine.

(*v*) These compressors give pulsation-free flow of air and therefore, no pulsation dampening devices are required.

(*vi*) Maintenance of screw compressor is very easy and requires a little time.

2. Roots Blower

These are positive displacement rotary compressors, and comprises two externally driven rotors. Such a compressor is shown in the figure. The shape of the two lobes is epicycloid or hypocycloid or involute to ensure a seal between high pressure region and the low pressure region at all angular positions (see Fig. 9.5).

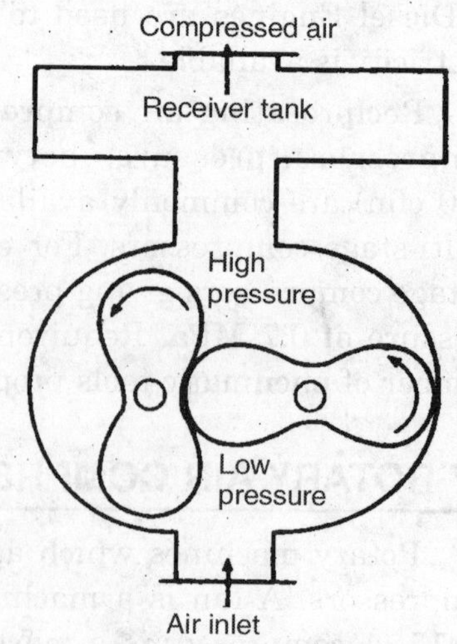

Fig. 9.5 Two lobe roots blower.

3. Centrifugal Compressors

The pressure in these compressors increases due to centrifugal force similar to that of centrifugal pumps. These compressors consist of a rotor, called impeller, provided with vanes and moving in a casing. The inlet of the air is at the hub and is provided with vanes to direct the air. The impeller gives a whirling motion to the air at high velocity and throws it out by centrifugal force. The high velocity is then converted to pressure rise in casing.

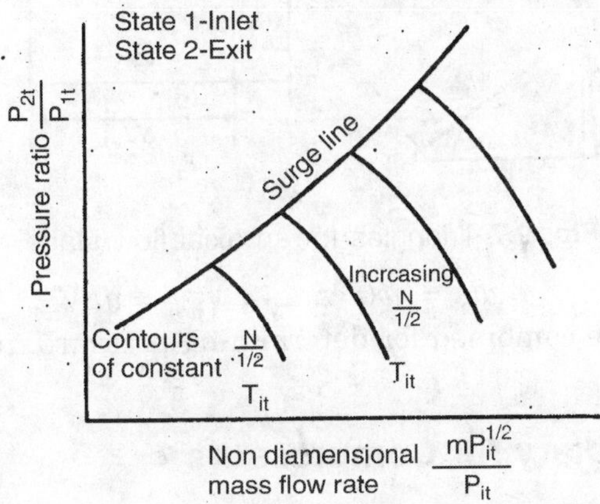

Fig. 9.6 General form of characteristics obtained from rotary compressors.

For higher pressure rise multistage centrifugal compressors are employed. Air compressed in one stage is fed into the next stage for further compression, and thus the pressure increases in each stage.

For pressure ratios less than 4, a centrifugal compressor may be used. High ratios may be achieved by using two or more in series but flow path is torturous. Hence, it is preferable to use an axial flow compressor, in which as many stages as required may follow each other.

Fig. 9.6 shows general form of characteristic obtained from rotary compressors. Points on the curves to the left of surge line are regions of unstable flow and compressor should not be operated in these regions.

Surge occurs when flow reverses temporarily in some part of the compressor and vibrations occur in the flow, which are audible and may cause damage to the compressor in extreme cases.

Work for compression is obtained by steady flow energy Eq. Compressors are ideally adiabatic and ideal specific work is given by

$$-w_x = C_p T_{1t}\left[r_{pt}^{(\gamma-1)/\gamma} - 1\right]$$

where T_{1t} = stagnation temperature at inlet (to account for inlet velocity), and

r_{pt} = stagnation pressure ratio.

Actual specific work is given by

$$-w_x = \frac{C_p T_u}{\eta_c}\left[r_{pt}^{(\gamma-1)/\gamma} - 1\right]$$

where η = process efficiency.

Fig. 9.7 shows triangles for an axial flow stage. Torque at the rotor = force × radius and work = torque × angular velocity.

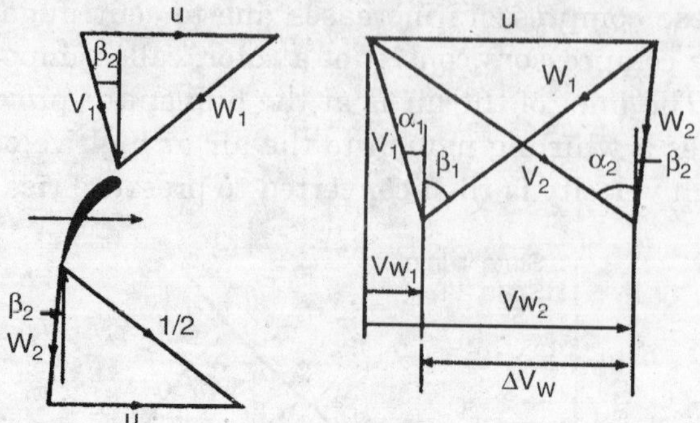

Fig. 9.7 Triangles for an axial flow stage.

Thus, $$w_x = w(r\,V_{w2} - r\,V_{w1}) = u\Delta V_w.$$

Above two equations be combined for determining pressure ratio, which will be given by a given blade shape.

4. Sliding Vane Type Rotary Air Compressors

The rotary sliding vane type air compressor, consists of cylinders, rotor and vanes. The rotor is mounted eccentrically to the cylinders. Non metallic sliding vanes inserted in slots made in the rotor, create sealed segments of varying capacity depending on eccentricity of the segment. Air is sucked into the stator bore, also known as cylinder, through inlet ports when the space between the rotor and stator is increasing. A slight further movement seals the filled segments from the intake port. Further rotation of the segment decreases its volume because of the eccentric location of the rotor. As the segment reaches the discharge port its volume is minimum, achieving the full compression. Slight rotation of the segment further, releases the compressed air through discharge ports and this cycle is repeated again and again (see Fig. 9.8).

This type of air compressor has following advantages over reciprocating air compressors:

(*i*) Due to high speed, its size and weight is less for equal capacities.

(*ii*) High speed also ensures smooth flow of air and vibration free running.

(*iii*) Since the starting torque is very low, this compressor enable the prime-mover to be directly coupled without any declutching device.

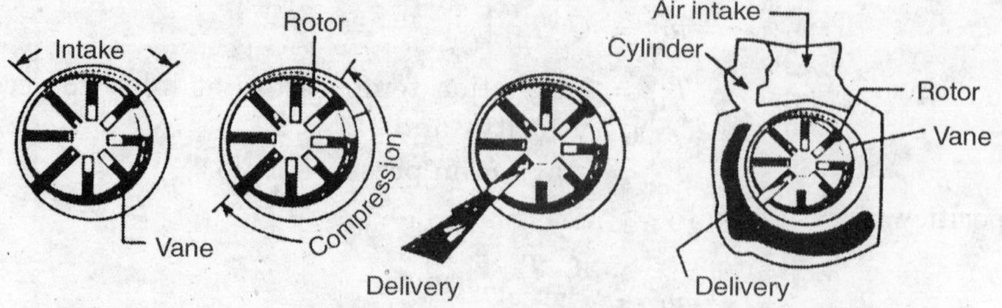

Fig. 9.8 Construction of sliding vane type rotary compressor.

(*iv*) The air delivered has much lower temperature than that of equivalent reciprocating air compressors.

(*v*) Replacement of blade is much easier and thus saves considerable down time and also the maintenance cost.

(*vi*) Since the rotary compressors have very few fast wearing-out parts, inventory of spare parts is also very less.

9.8 COMPONENTS OF AIR COMPRESSORS

9.8.1 Air Receivers

The compressed air must be delivered to a receiver:

(*i*) to absorb the pulsating effects of the compressor discharge.

(*ii*) to act as a strong vessel, keeping the air at a constant pressure, and

(*iii*) to remove moisture from the air by condensing it.

The receiver should be fitted with safety valves, pressure gauge, and drain valves.

9.8.2 Inter-cooler and After-coolers

These have already been discussed in detail in para 9.4

9.8.3 Prime-Movers for Compressors

Prime-movers for air compressors may be either an electric motor or a diesel engine. In electric driven compressors generally squirrel cage induction motors are used for high speed compressors, usually 90 kW and above, while direct connected synchronous motors are used for multi-cylinder compressors of 75 kW and below. The latter type has high efficiency at partial loads.

Diesel engines are common for portable compressors, and for stationary units where electricity is not available.

9.8.4 Oil Separators

An appreciable amount of oil is vaporised in the cylinders during compression. The oil vapours condens later on. Therefore, a suitable arrangement to trap all oil from the air is necessary to be installed in the system. Oil traps are used to separate oil from the air. An abrupt change of direction is affected as the air passes through the trap, and oil mist and moisture settles at the bottom of the trap. This device is very common being very economical.

In an improved system swirling motion of air causes centrifugal separation of large particles. The air is then directed on baffles where its further impingement produces more collection of oil and moisture.

Separators having filtering action are also employed. Filters having mats of felt, coconut fibre or wool are suitable for better results.

9.8.5 Compressed–air Distribution System

The air is required to be transmitted from the receiver of the compressor to the point of application at the pneumatic tools or pneumatic equipments. For a work likely to be continued for a long, a steel pipe line or for small work, a rubber hose is used for carrying the air. The air so transmitted should have a maximum pressure drop and leakage of air should be minimised and moisture in the air is eliminated for better performance and longer life of the tools. Provision for future extension should also be kept in view. Traps may be provided at frequent intervals along the pipe line to remove condensed moisture.

9.9 AIR ENGINES OR AIR MOTORS

These are the prime-movers which operate on air. Air at high pressure is supplied to the air engines or air motors and is expanded to perform the work, and then allowed to discharge at lower pressure. The indicator diagram for these engines or motors is similar to that shown in Fig. 9.9.

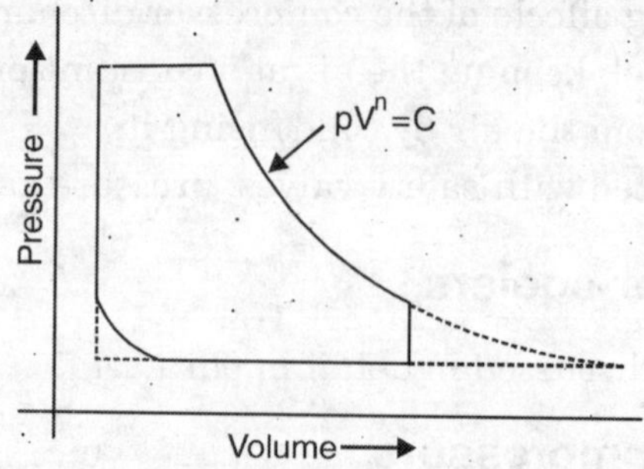

Fig. 9.9 Indicator diagram for air engine or air motor.

9.10 MECHANICAL AND VOLUMETRIC EFFICIENCIES

1. Mechanical Efficiency

$$\text{Mechanical efficiency. } \eta_{mech} = \frac{\text{Indicated power}}{\text{Shaft (brake) power}}$$

In case of compressor, Brake Power (B.P.) is the actual input to the compressor and Indicated Power (I.P.) is the energy which is actually used for compressing the air.

The I.P. of the compressor similar to engines is also given by,

$$\text{I.P.} = \frac{p_m LAN}{1000} \text{ kW.}$$

where, p_m = mean effective pressure (N/m^2)

L = Length of stroke (m)

$A = \frac{\pi}{4} D^2$ (D is diameter in metre)

N = Number of deliveries per sec.

Mean Effective Pressure

Mean effective pressure is determined from the indicator diagram plotted on the indicator card, by using the formula ;

Mean-effective pressure,

$$p_m = \frac{\text{Area of indicator card (mm}^2\text{)}}{\text{Mean length on the indicator card (mm)} \times \text{Vertical scale (MPa/mm)}}$$

Here, p_m is expressed in MPa.

If, A = Area of the cylinder in m^2.

L = Length of stroke of the piston in meter.

N = RPM of the compressor.

∴ The work done per cycle of the compressor.

$$= p_m LA \text{ (Nm)}.$$

∴ Work done per minute = $p_m LAN$.

∴ Indicated Power (I.P.) = $\frac{p_m LAN}{1000}$

2. Volumetric Efficiency

The volumetric efficiency of an air compressor is the ratio of the actual volume of free air inhaled during suction stroke to the swept volume of the piston. The volume of air inhaled must be reduced to the conditions of the free air or to a standard atmospheric conditions (S.T.P.) of (98.1 kPa)pressure at 15°C

$$\text{Volumetric efficiency} = \frac{\text{Volume of air inhaled at S.T.P.}}{\text{Swept volume of piston}}$$

The value of the volumetric efficiency varies between 65 and 90 percent according to the type of compressor. Reduction in volumetric efficiency of a compressor depends upon following factors:

(*i*) Leakage from piston and valves.
(*ii*) Very low speed.
(*iii*) Clearance volume being too large.
(*iv*) Obstruction at inlet valve.
(*v*) Inertia effect of air in the pipes.
(*vi*) Overheating of air due to contact with hot cylinder walls.

3. Compressors Performance

$$\text{Isothermal power} = 0.6 \times p_s\, v_s \log_e \frac{p_d}{p_s}$$

where p_s = supply pressure,

p_d = delivery pressure.

Isentropic of Adiabatic power

$$= 0.6 \times \frac{k}{k-1} \times p_s\, v_s \left[\left(\frac{p_d}{p_s} \right)^{(k-1)/k} - 1 \right]$$

Ratio of ideal useful capacity, v_s, to the piston displacement, D, is the "Conventional volumetric efficiency, e_v", and is given by the equation

$$e_v = \frac{v_s}{D} = 1 + C - C\,R_p^{1/n}$$

where C = clearance, expressed as a fraction of displacement,

D, R_p = ratio of pressures = $\frac{p_d}{p_s}$.

Multi-stage Compression

With inter-cooling between stages results in

(*i*) A saving in power, and

(*ii*) Improved volumetric efficiency of low-pressure cylinder, and therefore of the unit.

Perfect intercooling for two-stage compression requires that the gas be cooled back to the original temperature.

Work and power required for multi-stage isentropic standard with best receiver pressure and perfect intercooling are found by

$$W_{cycle} = 0.35 \times N\left(\frac{k}{k-1}\right) p_2 v_2 \left[\left(\frac{p_3}{p_2}\right)^{(k-1)/k} - 1\right]$$

and Isentropic horse power = $1.7 \times N\left(\frac{k}{k-1}\right) p_s\, v_s \left[R_p^{(k-1)/Nk} - 1\right]$

where p_s = supply pressure, MPa abs.

and v_s = capacity at supply pressure, p_s cm m/min.

Compression Efficiency

$$\text{Compression efficiency} = \frac{\text{Ideal power}}{\text{Actual power}}$$

Actual Capacity Indicated volumetric efficiency.

$$= \frac{\text{Indicated } v_s}{\text{Piston displacement}}.$$

$$\text{Actual volumetric efficiency} = \frac{\text{Actual metered capacity}}{\text{Piston displacement}}$$

$$\text{Slippage efficiency} = \frac{\text{Actual volumetric efficiency}}{\text{Indicated volumetric efficiency}}$$

Estimated actual volumetric efficiency

$$= \frac{\text{Indicated volumetric efficiency}}{1 + (0.25 \text{ to } 0.50)\left[\dfrac{\text{Abs. temp. air delivered}}{\text{Abs. temp. air supplied}}\right]}$$

9.11 APPLICATIONS OF AIR COMPRESSORS

1. Portable Compressors

Portable air compressors are generally used on construction jobs. These are self powered through prime-movers (normally a diesel engine) on a common frame. These are generally mounted on 2 wheel or 4 wheel trollies having rubber wheels and can be towed at 15 to 30 km per hr. Large size air compressors may be made self-propelled by installing it on a truck. The compressors may be single stage or multistage air cooled or water cooled.

While putting them in operation, they must be positioned at a level position. Air cleaner should be cleaned at an interval depending upon the atmospheric conditions *i.e.* whether the air is dusty or clean. In dusty atmosphere it should be cleaned frequently, may even be twice a day, and in clean atmosphere it may be cleaned after every 200 hours of working.

2. Capacity Factor

Number of pneumatic tools are connected to a compressor. Since all tools are not operated at any given time, hence extra tools can be connected, the total capacity of which may even exceed the total capacity of the compressor. For example, if 10 tools are connected, then not more than 5 or 6 will be consuming air at a given time Thus even if a compressor is capable of running 6 tools at a time, 10 tools can be connected to it. The ratio of average load to the maximum theoretical load that would exist if all tools are operating at the same time, is known as "Capacity Factor".

3. Air Operated Tools and Equipments

Air compressed by the air compressors is used to operate various pneumatic tools and equipments. These pneumatic tools are the means of converting air power into mechanical power. The pneumatic tools have various applications ranging from small work shop tools to the construction tools. Workshop tools include tools such as grinders, drills, hammers, screw drivers, rivetters, wrenches etc. Construction tools include tools for construction industry, quarrying operations, civil engineering works and comprising road breakers, rock drills, pumps etc. Air is also used for operating various other miscellaneous air operated equipment, like, winches, hoists, etc.

Pneumatic tools are generally connected by a 20 mm hose to a portable air compressor, which supplies the air power at 0.7 MPa pressure. The rotary tools, like drills are usually powered by rotary vane type air motors. Impact wrenches are used to tighten or remove nuts or studs through rotary and torsional impacts. Air percussion tools which are used for chipping, rivetting and ramming etc. have air operated pistons that delivers a series of blows to the tool held in a tool holder. The use of pneumatic tools for various construction activities are also common.

Advantages of Pneumatic Tools

Pneumatic tools have following main advantages over other types:

(*i*) Lighter in weight for same output.

(*ii*) Safer to work with these tools.

(*iii*) Easy to operate.

(*iv*) Can take considerable overload and do not overheat.

(*v*) Reliable in operation.

(*vi*) Initial as well as maintenance cost is low.

It is, however, necessary to maintain supply of adequate quantity of air at a specified pressure so as to get maximum efficiency (*i.e.* outlet) and keep the production cost low.

Air Pressure in Pipes or Hoses

The size of the pipe or hose should be selected after considering loss of pressure due to friction. Efficiency of the pneumatic tools or equipment depends upon the air pressure delivered to it. The pressure delivered to a pneumatic tool should not be less than that prescribed by the manufacturer. Therefore it is necessary to limit the pressure drop in transmission, by selecting a suitable size (dia and length) of the pipe or a hose. Otherwise the pressure of the equipment will drop, and the production will suffer.

9.12 TYPES OF AIR COMPRESSORS

Compressors used for different types of compressed air are of following types:

1. Positive displacement

(*i*) Reciprocating,

(*ii*) Reciprocating, wet.

(*iii*) Rotary, without liquid seals.

(*iv*) Rotary, with liquid seals.

2. Free compression, turbos

(*i*) Centrifugal fans

(*ii*) Centrifugal compressors

(*iii*) Axial flow fans

(*iv*) Axial-flow compressors.

3. Free compression, jet

(*i*) Hydraulic

(*ii*) Steam, air, or vapour.

Fields of Applications and Range of Performance

1. *Reciprocating Compressors.* Built in sizes as large as 2.5 to $5m^3$/s. piston displacement. Pressures range upto 1000 atm and vacua down to 12.5 mm of Hg. abs.
   Vacuum pumps are usually single stage, but positive-pressure compressors for large capacities and higher pressures require multistage with 0.7 MPa ; 0.4 to 2.5 MPa for two-stage ; 1.5 to 7.5 MPa for three stage ; and 20 to 30 MPa for four stage.
2. *Reciprocating Compressors running wet.* An old form particularly adopted to vacuum service where both air and water are to be removed. Piston displacement ranges upto 0.67 m^3 and vacua of the order of 550 to 700 mm of Hg. referred to 750 mm in barometer.
3. *Rotary compressors with liquid seals.* Available in sizes as 65 m^3/s. Suitable for moderate vacua not less than 375 mm of Hg abs, and for moderate pressures, 0.04 to 0.15 MPa. Some designs with sliding vanes can be carried to pressures of 0.7 MPa and to higher vacua, particularly if multistage.

4. *Rotary compressors with liquid seals.* Built in sizes upto 6.5 m^3/s and suitable for ratios of compression reflected in vacuum pump service to 700 mm in Hg and positive pressure service to 0.5 MPa gauge. If liquid is to be pumped simultaneously with the gas, this compressor is particularly favoured, as in wet vacuum service where 500 to 625 mm in Hg vacuum can be most satisfactorily met.
5. *Centrifugal fans.* Built in sizes of 0.04 to 330 m^3/s with pressure ranging upto 1.25 m or 1.5 m of water, positive or negative.
6. *Centrifugal Compressors.* Built in capacities of 0.3 to 65 m^3/s cm m/min and deliver pressures as high as 1 MPa, where 10 to 15 stages will be required. Lower pressure units deliver 0.005 to 0.15 MPa in a single stage. Light-weight high speed compressors for jet engines attain as much as 4: 1 pressures ratio in a single stage.
7. *Axial-flow fans.* Primarily for delivery of largest volumes under lowest pressures. If operated without a casing, static pressure is zero and all energy is kinetic. If a casing is included, heads upto 500 mm water can be developed with highest rotative speeds and with a single stage. Capacities are less than 65 m^3/s in commercial sizes.
8. *Axial-flow Compressors.* They extend the field of axial-flow fans by use of multi-staging so that pressures of 0.5 MPa can be developed with a 20 or 25 stage unit.

   Aircraft-type units use about one-half this number of stages. Capacities range from 2.5 to 125 m^3/s.
9. *Hydraulic-jet Compressors.* Primarily employed for vacuum pump service requiring 25 to 100 mm Hg abs pressure and capacities of 0.65 m^3/s at suction conditions.
10. *Vapour-jet Compressors.* With steam as usual actuating fluid are used for both exhauster and flower service. On the latter, delivery pressures run to 0.2 MPa kg/cm^2 and on the former, a suction pressure of 125 or 150 mm Hg abs can be maintained with two stage, and 6 mm of Hg with three stage.

Suction volumes of 0.3 to 1.33 m^3/s are common on vacuum pump service, but when developed for thermocompressor application, capacities range to 16.67 m^3/s.

QUESTIONS

9.1 Define 'reciprocating air compressor'. Mention its uses.

9.2 Describe a reciprocating air compressor with a neat sketch showing its main parts.

9.3 State the differences between reciprocating air compressor and rotary air compressor.

9.4 (*a*) What are uses of compressed air? (*b*) How do you classify air compressors?

9.5 Explain the principle of air compressor. Prove that the isothermal compression of air in an air compressor requires the minimum work input to drive it.

9.6 Explain "multi-stage compression" and mention its advantages.

9.7 Draw actual *p-v* diagrams for single acting single cylinder and two-cylinders reciprocating air compressors and state how they differ from ideal *p-v* diagrams.

9.8 Explain with suitable sketch the necessity of intercooling in between two stages.

9.9 Explain different types of rotary air compressors.

9.10 Explain the functions and working of intercoolers and aftercoolers.

9.11 Describe the working of an air motor with the help of *p-v* diagram and write its uses.

9.12 Write short notes on:

Air pressure in pipes or hoses, air operated tools and equipment.

9.13 List the advantages of pneumatic tools over other type of tools.

9.14 What are the rotary compressors? Classify them and state the differences between positive displacement type and non-positive displacement type.

9.15 (*a*) What do you understand by volumetric efficiency of a compressor?

(*b*) What are the effects of clearance volume and delivery pressure on the volumetric efficiency of a compressor?

9.16 Why cooling of air during compression is essential? What are the advantages of the same?

9.17 Explain the working of a centrifugal and axial flow compressors with a neat sketch.

9.18 List out the uses of compressed air in different fields.

9.19 In what ways a reciprocating air compressor differs from a rotary air compressor?

9.20 Define "Psychrometrics".

9.21 Define ; Atmospheric Air, dry air, vapour pressure, saturation pressure and dew point temperature.

9.22 Define "saturated air", and "specific humidity".

9.23 Explain the terms "humidification" and "dehumidification".

9.24 Write a short note on "Measurement of Flow".

9.25 Describe, flow of air through rectangular ducts.

9.26 Classify different types of air compressors.

9.27 What are the fields of applications for air compressors?

9.28 How will you define the compression efficiency?

INTERNAL COMBUSTION ENGINES (I.C. ENGINES)

Chapter **10**

Learning Objectives

After studying this chapter, the students will learn:

- Internal combustion engines, its classification, construction and working.
- Working cycles for I.C. engines, such as otto cycle, and diesel cycle.
- Number of strokes in I.C. engines, number and arrangement of cylinders.
- Various systems in I.C. engines, such as air supply system, exhaust system, lubrication system and supercharging.
- I.C. engine performance, heat balance sheet and governing.

10.1 INTRODUCTION

An engine is a device which transforms the chemical-energy of the fuel into thermal energy and utilises this to perform useful mechanical work. As the name suggests, Internal Combustion Engine is a heat engine in which the chemical energy of combustion is released inside the engine cylinder.

Any machine which derives heat energy from the combustion of fuel and converts part of this energy into mechanical work is known as **heat engine.** Heat engines are of two types namely, external combustion engines (such as steam engines), and internal combustion engines (popularly known as I.C. engines). I.C. engines are commonly used in automobiles, ships and locomotives, agricultural and earth moving machinery, many industrial applications and for power generation.

I.C. engines offer some special advantages over external combustion (steam engines) as mentioned below:

(*i*) The thermal efficiency of the I.C. engines is much higher (30 to 35%) than steam engines (15 to 25%).

(*ii*) The power developed by the I.C. engines per kg weight of engine is much higher, therefore, it is lighter and occupies less space.

(*iii*) These can be started quickly whereas a steam engine requires much more time as the steam has to be generated in the boiler, before starting the steam engine.

(*iv*) These are more compact and has monopoly in small and a portable power range.

10.2 CLASSIFICATION

According to the requirement, I.C. engines are designed in various varieties. General considerations for classification of I.C. engines are:

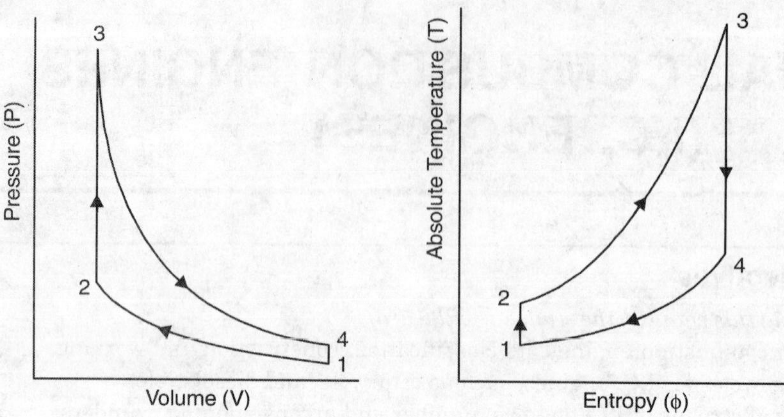

Fig. 10.1 Theoretical otto cycle.

1. Working cycle employed
2. Number of strokes
3. Fuel used
4. Number and arrangement of cylinders.
5. Valve arrangements
6. Method of cooling
7. Method of Ignition
8. Field of application.

10.3 WORKING CYCLE FOR I.C. ENGINES

10.3.1 Otto Cycle

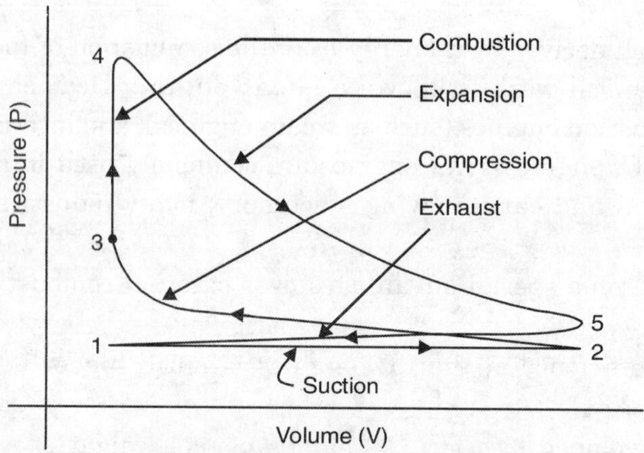

Fig. 10.2 Actual Otto Cycle.

This is the cycle on which gas, petrol and many other types of oil engines run. It is very common for spark-ignition engines. The theoretical otto cycle is shown in Fig. 10.1. The thermodynamic processes which take place during the cycle are given under.

(*a*) **Process 1-2, Adiabatic.** Compression or work done on the working medium by the piston.

(*b*) **Process 2-3, Constant Volume.** Heat supplied instantaneously.

(*c*) **Process 3-4, Adiabatic.** Expansion or work done by working medium on the piston.

(*d*) **Process 4-1, Constant Volume.** Heat rejected instantaneously. In this cycle heat is supplied at constant volume, therefore it is often called the "constant volume cycle".

The efficiency of this cycle is as follows:

$$\eta = 1 - \frac{1}{r^{\gamma - 1}}$$

where, r = compression ratio = $\frac{\text{volume at 1}}{\text{volume at 2}}$

γ = ratio of specific heat.

The **actual otto cycle** is slightly modified when used in I.C. engines. The diagram of actual otto cycle is shown in Fig. 10.2, which consists of four strokes as given under:

(*a*) **Suction stroke 1-2.** A charge is drawn into the cylinder by the outward movement of the piston.

(*b*) **Compression stroke 2-3.** The charge is compressed adiabatically by the return stroke of the piston, to the clearance volume of the cylinder.

(*c*) **Constant volume 3-4.** The compressed charge is ignited at point 3 by an electric spark thereby increasing pressure to 4.

(*d*) **Working stroke 4-5.** The burnt gas expands approximately adiabatically, thereby forcing the piston through the working stroke 4-5. This stroke provides power to the engine and also drives the piston through the other three strokes of the cycle. The excess energy during this process is stored in the fly-wheel.

(*e*) **Exhaust stroke 5-1.** During this process the burnt gases are pushed out through an exhaust port by the inward movement of the piston, thereby completing the cycle.

10.3.2 Diesel Cycle

Refer Fig. 10.3, the theoretical diesel cycle consists of two adiabatics, one constant pressure line, and one constant volume line. This cycle is very important and popular for the diesel engines. The basic difference between the otto and diesel cycles is in the heat supplied system. In the otto cycle heat is supplied at constant volume, while in the diesel cycle the heat is supplied at constant pressure.

The diesel cycle shown in Fig. 10.3 consists of the thermodynamic processes as given under:

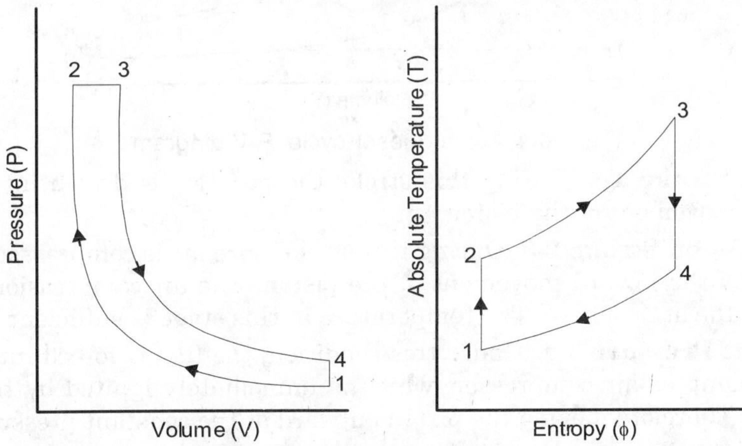

Fig. 10.3 Theoretical diesel cycle.

(*a*) **Adiabatic 1-2.** Compression or work done on the working medium by the piston.

(*b*) **Constant Pressure 2-3.** Heat supplied and part of work done.

(*c*) **Adiabatic 3-4.** Expansion or remaining work done by the working medium on the piston.

(*d*) **Constant Volume 4-1.** Heat rejected instantaneously.

In this cycle heat is supplied at constant pressure therefore it is often called "the constant pressure cycle". The air standard efficiency of this cycle is as given under:

$$\eta = 1 - \frac{1}{r^{\gamma-1}}\left[\frac{\rho^{\gamma-1}}{\gamma(\rho-1)}\right]$$

where,

$$r = \text{compression ratio} = \frac{\text{volume at 1}}{\text{volume at 2}}.$$

$$\gamma = \text{ratio of specific heats}$$

$$\rho = \text{volume ratio at cut off} = \frac{\text{volume at 3}}{\text{volume at 2}}.$$

The actual diesel cycle is slightly modified when used in I.C. engines. Because the engine uses heavy oil as a fuel which differs in ignition. The engine needs, an accessory, an oil pump or air compressor for forcing the oil fuel into the cylinder. Therefore the engine based on this cycle is called "compression-ignition engine". The diagram of actual diesel cycle is shown in Fig. 10.4, which consists four strokes as given under:

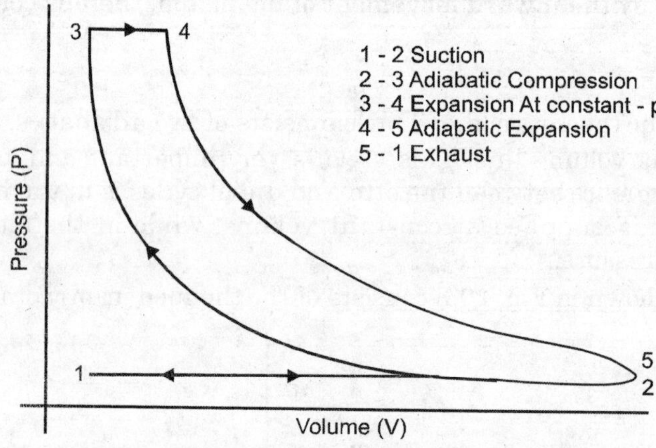

Fig. 10.4 Actual diesel cycle *P-V* diagram.

(*a*) **Suction Stroke 1-2.** During this stroke the pure air is drawn into cylinder by the outward movement of the piston.

(*b*) **Compression Stroke 2-3.** During this stroke pure air is compressed to the clearance volume by the inward movement of the piston. The air compression at this stage is 40 times the atmosphere. The temperature at this stage is sufficient to ignite the fuel.

(*c*) **Constant Pressure 3-4.** Under this condition, the fuel is forced into the cylinder by the oil pump or air compressor, which is immediately ignited by the hot air in the cylinder. Therefore moving the piston outward at the constant pressure upto the point of cut-off.

(*d*) **Working Stroke 4-5.** During this stroke the mixture of burnt fuel and air expands adiabatically to 5. Thereby moving the piston outward and supplying the power to engine crank-shaft. This power also drives the oil pump or air compressor as well to overcome the resistance of the other three strokes of the cycle.

(*e*) **Exhaust Stroke 5-1.** During this stroke the piston moves inward thereby forcing out the burnt gases of the cylinder, thus to complete the diesel cycle.

Example 10.1 *Two engines are to operate on Otto and Diesel cycles with the following data:*

Maximum temperature = 1227°C

Exhaust temperature = 427°C

Ambient condition 0.1 mPa and *27°C*

Find the compression ratio, maximum pressures and efficiencies.

Solution. (*i*) *Consider Otto Cycle*

Considering operation 4-1

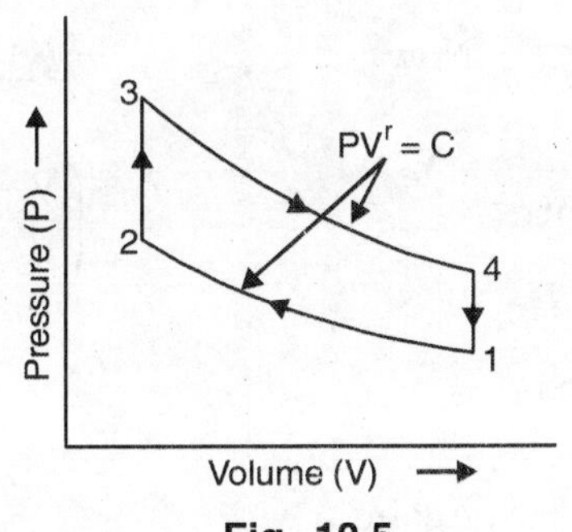

Fig. 10.5

$$\frac{P_4V_4}{T_4} = \frac{P_1V_1}{T_1}$$

$$\because \quad V_1 = V_4$$

$$\therefore \quad \frac{P_4}{T_4} = \frac{P_1}{T_1} \quad \text{or} \quad P_4 = P_1 \times \frac{T_4}{T_1}$$

Here

$$P_1 = 0.1 \text{ mPa}$$

$$T_1 = 273 + 27 = 300°\text{K}$$

$$T_4 = 273 + 427 = 700°\text{K}$$

$$\therefore \quad P_4 = 0.1 \times \frac{700}{300} = 0.234 \text{ mPa}$$

Considering operation 3-4

$$\frac{T_3}{T_4} = \frac{P_3V_3}{P_4V_4} \quad \text{and} \quad P_4V_4^\gamma = P_3V_3^\gamma$$

or $$\left(\frac{V_4}{V_3}\right)^\gamma = \frac{P_3}{V_4} \quad \text{or} \quad \left(\frac{V_4}{V_3}\right) = \left(\frac{P_3}{P_4}\right)^{\frac{1}{\gamma}}$$

or $$\frac{V_3}{V_4} = \left(\frac{P_3}{P_4}\right)^{-\frac{1}{\gamma}}$$

$$\therefore \quad \frac{T_3}{T_4} = \frac{P_3}{P_4}\left(\frac{P_3}{P_4}\right)^{-\frac{1}{\gamma}} \quad \text{or} \quad \frac{T_3}{T_4} = \left(\frac{P_3}{P_4}\right)^{\frac{\gamma-1}{\gamma}}$$

or $$\frac{1500}{700} = \left(\frac{P_3}{0.234}\right)^{\frac{1.4-1}{1.4}}$$

$$\therefore \quad P_3 = 3.37 \text{ mPa}$$

Now again $\frac{V_4}{V_3} = \left(\frac{P_3}{P_4}\right)^{\frac{1}{\gamma}} = \left(\frac{3.37}{0.234}\right)^{\frac{1}{1.4}} = 6.7$

∴ Compression ratio $= \frac{V_4}{V_3} = \frac{V_1}{V_2} = \mathbf{6.7.}$ **Ans.**

∴ Efficiency of the Otto cycle

$$= 1 - \frac{1}{(r)^{\gamma-1}} = 1 - \frac{1}{(6.7)^{1.4-1}} = 0.467 = \mathbf{46.7\%.}\ \textbf{Ans.}$$

(ii) Considering Diesel Cycle

Here $P_4 = 0.234 \quad P_3 = 3.37$

and $P_2 = P_3$

∴ Max. pressure = 3.37

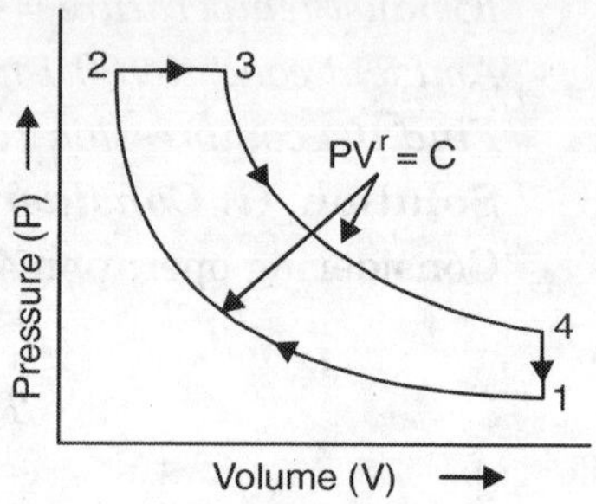

Now $P_1V_1^{\gamma} = P_2V_2^{\gamma}$

or $\left(\frac{V_1}{V_2}\right)^{\gamma} = \frac{P_2}{P_1}$ or $\frac{V_1}{V_2} = \left(\frac{3.37}{0.1}\right)^{\frac{1}{\gamma}}$

or $r = (33.7)^{\frac{1}{1.4}} = \mathbf{12.34.}$ **Ans.**

Again $\frac{T_2}{T_1} = \frac{P_2V_2}{P_1V_1} = \frac{3.37}{0.1} \times \frac{1}{12.34}$

or $T_2 = T_1 \times \frac{3.37}{1.234} = 300 \times \frac{3.37}{1.234} = \mathbf{819°K.}$

Fig. 10.6

Air standard efficiency $= \frac{\text{Heat supplied} - \text{Heat rejected}}{\text{Heat supplied}}$

∴ $\eta_{air} = \frac{C_p(T_3 - T_2) - C_v(T_4 - T_1)}{C_p(T_3 - T_2)} = 1 - \frac{1}{y} \times \frac{T_4 - T_1}{T_3 - T_2}$

$$= 1 - \frac{1}{\gamma} \times \frac{(700 - 300)}{(1500 - 819)} = 1 - \frac{1}{1.4} \times \frac{400}{681}$$

$$= 1 - 0.419 = 0.58 = \mathbf{58\%.}\ \textbf{Ans.}$$

Example 10.2 *The following observations were taken from a 4 stroke Diesel engine:*

Pressure at the end of suction	*= 0.1 mPa*
Temperature at the end of suction	*= 30°C*
Temperature at the end of constant pressure heat addition	*= 1500°C*
Compression ratio	*= 15*
Specific heat of air at constant pressure C_p	*= 1 kJ/kg° K*
Specific heat of air at constant volume C_v	*= 0.714 kJ/kg °K*

Find:

(i) % of stroke at which cut-off takes place

(ii) the temperature at the end of expansion stroke and

(iii) thermal efficiency.

Solution. $P_1 = 0.1$ mPa abs $\quad T_1 = 30 + 273 = 303°K$

$$T_3 = 1500 + 273 = 1773°K$$

$$\gamma = \frac{C_p}{C_v} = \frac{1}{0.714} = 1.4$$

Refer *P-V* diagram of Fig. 12.6 of Diesel cycle.

$$\frac{T_2}{T_1} = (r)^{\gamma-1} \quad \text{or} \quad T_2 = T_1(r)^{\gamma-1}$$

or
$$T_2 = 303(15)^{1.4-1}$$
$$= 303 \times 2.553 = 773°K$$

From the constant pressure line 2-3

$$\frac{V_2}{T_2} = \frac{V_3}{T_3} \quad \because P_2 = P_3$$

or
$$\frac{V_3}{V_3} = \frac{T_3}{T_2} = \frac{1773}{773} = 2.3$$

or
$$V_3 = 2.3\ V_2$$

$\therefore$ Stroke % at which cut off takes place

$$= \frac{\text{Vol. at cut off} - \text{Clearance Vol.}}{\text{Swept Vol.}} \times 100$$

$$= \frac{V_3 - V_2}{V_1 - V_2} \times 100 = \frac{2.3\ V_2 - V_2}{15\ V_2 - V_2} \times 100$$

$\because$
$$V_1 = 15\ V_2 = \frac{1.3\ V_2}{14\ V_2} \times 100 = \mathbf{9.3\%}$$

Considering adiabatic expansion 3-4

Expansion ratio,
$$\frac{V_4}{V_3} = \frac{V_4}{V_2} \times \frac{V_2}{V_3}$$

$$= \frac{V_1}{V_2} \times \frac{V_2}{V_3} = 15 \times \frac{1}{2.3} = 6.52$$

$$\frac{T_3}{T_4} = \left(\frac{V_4}{V_3}\right)^{\gamma-1} = (6.52)^{1.4-1}$$

$\therefore$
$$T_4 = \frac{T_3}{(6.52)^{0.4}} = \frac{1773}{2.12} = 836.3°K$$

Now heat supplied $= C_p(T_3 - T_2)$

$$= 1.0 \times (1773 - 773) = 1000 \text{ kJ/kg of air}$$

$$\text{Heat rejected} = C_v(T_4 - T_1)$$
$$= 0.714\,(836 - 303)$$
$$= 0.714 \times 533 = 380.6 \text{ kJ/kg of air}$$
$$\therefore \quad \text{Work done} = \text{Heat supplied — Heat rejected}$$
$$= 1000 - 380.6 = 619.4 \text{ kJ of air}$$
$$\text{Air standard efficiency} = \frac{\text{Work done}}{\text{Heat supplied}}$$
$$= \frac{619.4}{1000} = \mathbf{61.94\%}. \quad \textbf{Ans.}$$

10.4 NUMBER OF STROKES

Automotive engines are classified according to the number of strokes used in a cycle. They are operated on four-stroke or two-stroke principle as given under.

10.4.1 Four Stroke Engine

The four strokes of the complete cycle are as follows:

1. Suction or Induction Stroke. Refer Fig. 10.7 (*a*) where the inlet valve is open and exhaust valve is close. The piston connected with the connecting rod moves downward or outward thereby sucking the new charge of explosive mixture through inlet port from the carburettor. During this stroke the pressure inside the cylinder remains below atmospheric pressure which depends upon the throttle opening and engine speed.

2. Compression Stroke. Refer Fig. 10.7 (*b*), both the valves are in closed position. The piston moves upward or inward by the movement of the flywheel thereby compressing the charge into the combustion head of the cylinder. In petrol engines the compression ratio is usually between 6 and 9. The pressure at the end of the stroke is about 0.6 to 0.8 mPa. The compression ratio is the ratio of the total volume and the clearance volume of the cylinder.

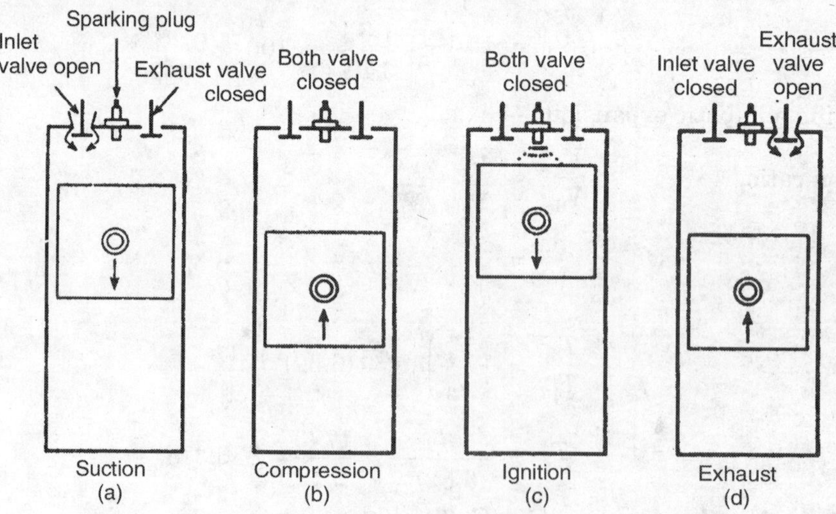

Fig. 10.7 Four-stroke engine.

3. Ignition or Explosion or Working Stroke. Refer Fig. 10.7 (*c*), both the valves are in closed position. Just before the end of the compression stroke, ignition takes place in the charge by means of an electric spark, thereby raising the temperature and pressure inside the cylinder. The explosion is completed while the piston is in rest or just to move downward, thereby expanding the hot gases and moving the piston downward or outward. The movement of the piston turns the crankshaft thereby propelling the vehicle against the external resistances. During this stroke the flywheel also stores the momentum lost during the idle strokes.

4. Exhaust Stroke. Refer Fig. 10.7 (*d*), the inlet valve is in closed and exhaust valve is in open position. The piston again moves upward or/inward by the momentum of the flywheel thereby discharging burnt gases through the exhaust valve. The exhaust pressure always remains slightly above the atmospheric pressure which depends upon the resistance created by the exhaust valve and silencer.

Therefore it is seen that crank-shaft faces four strokes in every two revolutions. Out of four strokes the ignition or working stroke is only the useful stroke and remaining three strokes are "idle" strokes. Therefore the engine based on the four stroke principle gives fuel economy.

10.4.2 Two-Stroke Engine

To reduce the proportion of idle strokes and to increase the useful strokes in a given time, the various forms of the "two-stroke" engine are designed. In two-stroke cycle the suction and exhaust strokes of the four-stroke cycle are eliminated. The cycle consists of only two strokes, compression and the working stroke. The entry of the charge under pressure and a high velocity, into the cylinder at the end of the working stroke scavenges the burnt gases thereby serving the purpose of suction and exhaust strokes of the four-stroke cycle.

The different stages of the two-stroke cycle are cleared in Fig. 10.8. Refer Fig. 10.8 (a) the exhaust port is in closed and inlet port is in open position. The piston moves upward by the momentum of the flywheel, thereby compressing the charge at the top of the piston and entering the new charge through the inlet port into the crank case due to creating the vacuum by the piston.

As soon as the piston reaches just at the top dead position the spark ignition takes place thereby increasing the temperature and the pressure inside the cylinder. Due to explosion, the hot gases expand and thus piston moves downward. Refer Fig. 10.8 (*b*), both the ports are in closed position and the piston is still in downward movement which compresses charge available in the crank case.

As soon as the piston moves slightly more downward, first exhaust port becomes open and then transfer port thereby scavenging the burnt gases and entering the new charge into the cylinder as shown by the directions of arrows in Fig. 10.8 (*c*). The process is completed by reaching the piston at bottom dead position. Then piston again moves upward and closes both transfer as well as exhaust port, while the inlet port is already in closed position as shown in Fig. 10.8 (*d*). Now the engine has completed one cycle and crank shaft has been turned through one rotation and the engine is ready just to start the next cycle. In this way every rotation of the crankshaft of this engine consists of one useful or power stroke thereby realizing double power than four-stroke engine of the same size.

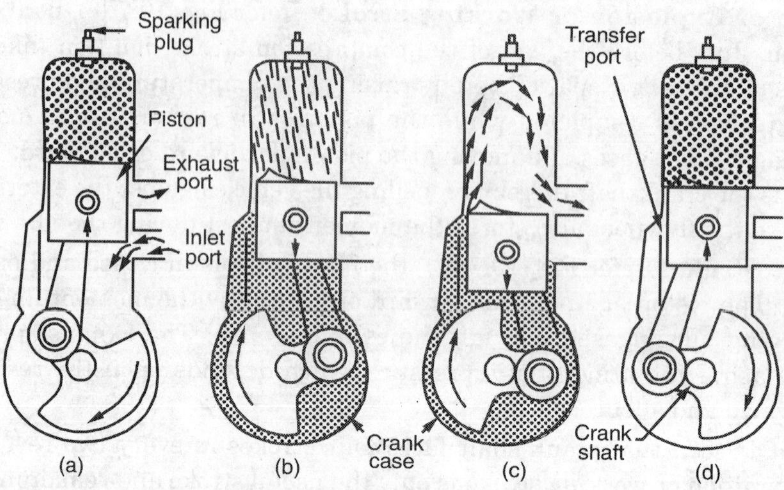

Fig. 10.8 Two-stroke engine.

Scavenging.

The burnt gases escape from the cylinder by virtue of the excess of their own pressure over that in the silencer or atmosphere. But the final clearing is done by the forcing of the air or the mixture of the charge into the cylinder which is known as "scavenging". There are three different methods of scavenging in the two-stroke cycle:

(*a*) **Mixture Compressing in Crank-case.** In this method the mixture of air and fuel is compressed in the crank-case by the piston during working-stroke. This method is adopted in petrol engines or engines working on constant volume cycle.

(*b*) **Pure Air Compressing in Crank-case.** In this method the pure air is compressed in the crank-case by the piston during its working stroke, which is utilized for scavenging purposes. This method is adopted in diesel engines or engines working on constant pressure and dual combustion cycle.

(*c*) **Pure Air Compressing by a Pump.** In this method pure air is compressed by a separate pump. The pump is driven by the crank-shaft of the engine. This method is adopted in both the petrol as well as diesel engines.

10.4.3 Four Stroke v/s Two Stroke Engines

The four-stroke engines are widely used as power plant for vehicles such as Ashok Leyland, Tata Mercedez Benz, Ambassador, Fiat, etc. The two-stroke engines are used for Motor cycles, scooters, model airplanes, motor boats, small pickups, minibikes etc. The general comparison between two is as given below:

Four-stroke engine	*Two-stroke engine*
1. It consists of one working stroke in every four strokes and crank-shaft completes two revolutions in a particular period of the cycle	1. It consists of one working stroke in every two strokes, and gives two working strokes in same period of cycles and crank shaft completes one revolution in each cycle. Thereby theoretically producing approximately double power than four-stroke engine of same size.

2. It is provided with inlet as well exhaust valves which need the special attention of wear, tear and lubrication.	2. Inlet as well as exhaust ports are provided instead of valves, which reduce the wear, tear and lubrication thereby also reducing the special attention.
3. It consists of three idle strokes in a complete cycle of two revolutions of the crankshaft.	3. It consists of two idle strokes in the same period of two cycles and two revolutions of the crank shaft.
4. It needs more space for producing equal amount of power.	4. The space required is less to produce the same amount of power.
5. It cannot be made reversible.	5. It can be reversible, if is of valveless.
6. Complete scavenging is not possible because the burnt gases remain in the clearance sgace.	6. The scavenging is more complete in low speed engines, as it does not leave the clearance space.
7. It needs heavier flywheel because there is more non-uniform torque on the crank-shaft.	7. It needs lighter flywheel because there is more uniform torque on the crank-shaft.
8. More power loss due to friction of idle, suction and exhaust strokes.	8. The power required to overcome the friction of the suction and exhaust strokes is saved.
9. Efficiency is more because none of the portion of the charge escapes with the exhaust gases.	9. Efficiency is lower because a portion of the fresh charge escapes with the exhaust gases while scavenging.
10. Due to working at half the speed, the wear, tear and the consumption of lubricating oil is less.	10. It needs more lubricating oil mixed with the fuel because the valve gear works at twice the speed.
11. Due to its low speed, fuel economy simplicity and heavier in construction is recommended for small and medium stationary power plants. It is also recommended for medium speed running vehicles.	11. Due to its lesser weight and high speed, is recommended as a power plant for small moving vehicles as well as for marine engines.

10.5 FUELS USED IN I.C. ENGINES

I.C. engines can be classified on the basis of fuel which they use. The general type of fuels used in I.C. engines are as given under:

1. The Petrol Engine

This engine uses a spirit fuel such as petrol, benzol or mixture containing alcohol. It works at high speed on the constant volume or otto cycle of the four-stroke or two-stroke type. The ignition is produced by means of an electric spark. The vaporization of the fuel is done by means of a carburettor or a pump driven by engine crank-shaft.

2. Heavy Oil Engine

This type of engine uses heavy oil as fuel. It may be of low or a high speed type. Its working cycle may be of a four-stroke or two-stroke type. It works on diesel or semi-diesel cycle. In this type the pure air is compressed in the cylinder then fuel is injected by air or solid method. An engine working on the constant pressure cycle is called "diesel engine."

Comparison between Petrol Engine and Diesel Engine

Petrol Engine	*Diesel Engine*
1. Lighter in weight.	1. Heavier in weight.
2. Requires regular maintenance of ignition system	2. Fuel injection does not require that frequent maintenance.
3. Requires change of spark plugs after every few thousand kms.	3. No spark plug is fitted.
4. Engine overhaul life is less.	4. Requires overhaul after long time.
5. Spares required for overhaul are less costly.	5. Spares required for overhaul are costly.
6. Initial cost of the engine is less.	6. Initial cost of the engine is more.
7. Due to low engine torque, the transmission is not made for heavy duty.	7. Due to high engine torque transmission is made for heavy duty.
8. Engine starting is easy even in cold weather.	8. Engine starting is difficult in cold weather and requires heater plugs.
9. Requires small battery as engine starting torque is less.	9. Requires high capacity battery as engine starting torque is more.
10. Due to less weight of engine and smooth working, vehicle chassis is not made very strong.	10. Due to heavy weight of engine and more vibration, chassis is made extra strong.
11. Produces less sound while running.	11. Produces more sound while running.
12. Radiators used are light in weight.	12. Radiators used in these engines are heavy duty.

10.6 NUMBER AND ARRANGEMENT OF CYLINDERS

On the basis of number of cylinders and their arrangements the automotive engines are classified as given under:

1. Single Cylinder Engine

As shown in simplified sketch Fig. 10.9 this engine consists of a single cylinder in which piston slides. The piston is connected to a crank-shaft by means of a connecting rod. The crank-shaft rests on main bearings and having a flywheel attached at one end to store the energy for idle strokes. Its maximum swept volume is about 0.0006 m^3 because with higher capacity it becomes very difficult to balance the engine. It is used generally for motor cycles and scooters. It may be two-or four-stroke cycle engine. It can be mounted in vertical, horizontal or inclined position on two-wheeler vehicles. It is generally air cooled type engine.

2. Two Cylinder Engines

For getting more power, better balance, uniform torque and compact size and shape, engines for motor cycles, scooters and light cars are designed of twin cylinder type. The volume capacity of the twin engines ranges from 0.0005 to 0.001 m^3.

3. Four Cylinder Engines

These engines are very common in the use of a small, light and medium size vehicles. These engines work on the four-stroke cycle principle. This type consists of two-working strokes per revolution thereby obtaining more uniform torque, as compared to a single cylinder engine. The crankpins are arranged at 180°, therefore a better balancing is obtained. The cylinder arrangement of four cylinder engines also can be straight or inclined type, Vee type or opposed type, such as in case of two cylinder engines.

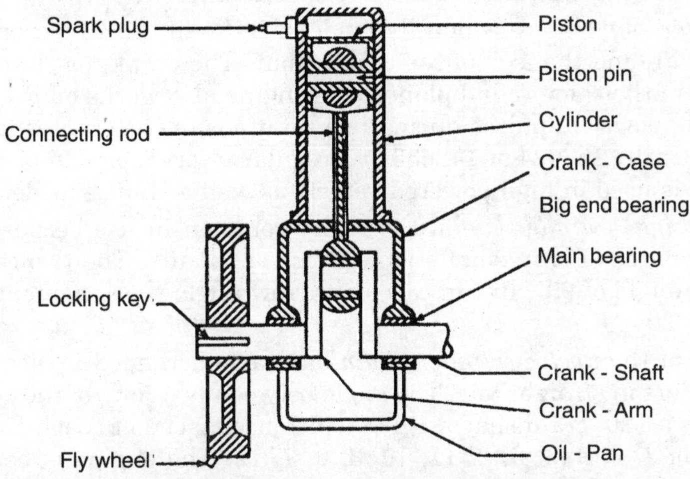

Fig. 10.9 Single cylinder engine.

Straight or In-line Type. In this type all the four cylinders are arranged in a straight line as shown in Fig. 10.10. The arrangement consists of two pairs of the pistons. When the pair of pistons, first and fourth moves upward then at the same time the other pair of pistons, second and third moves downward. Therefore in this way the reciprocating forces remain approximately in balance. The firing interval in the system is regular thereby obtaining one power impulse after every 180° rotation of the crankshaft. The crankpins are not exactly opposite to each other therefore reciprocating forces remain slightly unbalanced, thereby vibrating the engine at high speed. But the two-rocking couples neutralize each other. The firing order in this type of arrangement is generally 1, 3, 4, 2, or 1, 2, 4, 3, where the cylinder nearest to the radiator is numbered first.

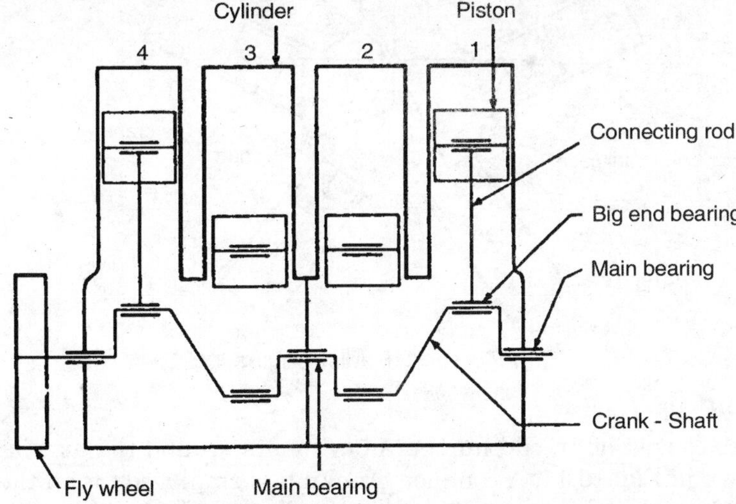

Fig. 10.10 Four cylinder in-line engine.

4. Six, Eight, Twelve and Sixteen Cylinder Engines

For obtaining higher power, better dynamic balance and smoother torque an engine is designed of six, eight twelve or sixteen cylinders type vehicle. The cylinders of these engines can also be arranged in straight line, vee or opposed manner.

The *six cylinder engines* are generally made of in-line type. The crankpins are designed at an angle of 120° around the axis of the crank-shaft. The crankpins 1 and 6, 2 and 5; and 3 and 4 are designed in the same radial plane. This engine also works on the basis of four-stroke cycle and having six power impulses during two rotations of the crank-shaft. The firing order of this engine is generally 153624 or 142635 at a regular interval of 120° of crank-shaft rotation. This type of engine is used in high powered vehicle as well as in the modern cars.

The Vee-type *eight cylinder engines* are very common in use because they are compact and having short lengths of crankshafts as well as cam-shafts. The two rows of the cylinders are arranged at an angle of 90°. But in case of narrow angle type engine, this arrangement is made at an angle of 20°.

The cylinders of the *twelve-cylinder engine* are also arranged in Vee manner. Each row consists of six cylinders in straight line. The angle between two rows of the cylinders is generally arranged 60°. It has a 120° crankshaft with two pistons attached to each crankpin. This engine has a firing order of 1, 4, 9, 8, 5, 2, 11, 10, 3, 6, 7, 12. The engine is better in balance and provides smoother torque. It is used in certain cars as well as in heavy duty trucks.

The cylinders in case of *sixteen-cylinder engine* are arranged in two rows at a Vee of 135°. The arrangement of individual row is in-line type. All the cylinders and crankcase are made in a single casting. This engine is perfectly balanced at all the speeds. This engine provides eight power impulses at an equal and regular interval for each revolution of the crank-shaft. This engine has a firing order of 1, 4, 9, 12, 3, 16, 11, 8, 15, 14, 7, 6, 13, 2, 4, 10. The brake power of this engine is 130 kW at 3,600 r.p.m.

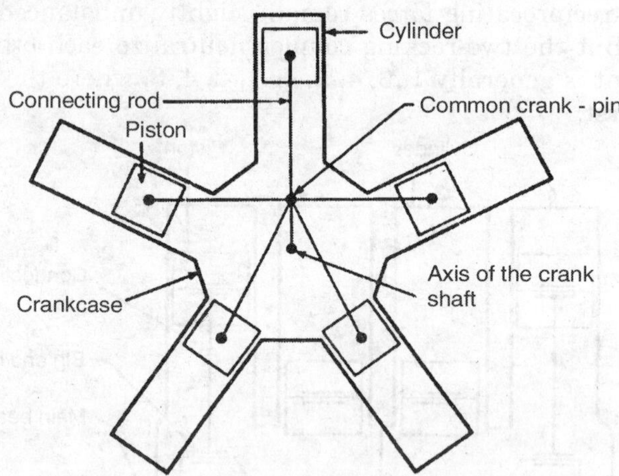

Fig. 10.11 Radial engine.

5. Radial Engines

These engines consist of an odd number of cylinders around the axis of a single common crankshaft as shown in Fig. 10.11. A common crankpin is employed for all the connecting rods. The advantage of the odd number of cylinders is that it provides uniform firing intervals. The three and five cylinder engines are generally used for automobile purposes, whereas seven and nine cylinder engines mostly used in air crafts. These engines are simple, compact and deliver higher Brake Power even with low weight. These engines are air cooled type therefore have large cooling surfaces and large frontal areas. Therefore their use in automobile are limited because they create difficulty in streamlining the vehicle.

10.7 OTHER METHODS OF CLASSIFICATION

1. Classification based on valve arrangements

Automobile engines are also classified according to arrangements of their intake and exhaust valves provided in the cylinder head or blocks. The different arrangements of the valves are I-head or L-head engines.

2. Method of Cooling

Refer Chapter on 'Automobile Engineering'.

3. Method of Ignition

On the basis of method of ignition, I.C. engines can be classified as:

(*i*) **Spark Ignition Engines.** These work on principle of otto constant volume cycle and generally use petrol as a fuel and hence are known as petrol engines.

(*ii*) **Compression ignition Engines.** These work on the principle of diesel cycle or the constant volume cycle. Since they run on higher pressure, these engines are heavy.

4. Field of Application

(*i*) Stationary Engine.
(*ii*) Marine Engine.
(*iii*) Automobile Engine.
(*iv*) Aero Engine.
(*v*) Locomotive Engine.

10.8 CONSTRUCTIONAL DETAILS

Description of main components of an I.C. Engine is given here-under as shown in fig. 10.12:

1. Cylinder Block

It is the basic frame work of the engine and consists of round cylinders in which the piston slides up and down. The cylinder head is attached at the top, and the crankcase at the bottom. The valve seats, ports, guides, and water jacket is also built in it. Cylinder blocks are casted from grey iron or cast iron alloys with nickle or chromium and aluminium.

2. Piston

A piston is a cylindrical plug which moves up and down in the engine cylinder. It is attached to the small end of the connecting rod by means of a piston pin. Its diameter is slightly smaller than that of cylinder bore. The space between the piston and the cylinder wall is called the 'piston clearance'. The material used for the piston is aluminium-alloy. Grooves are cut in the piston for providing piston rings.

Piston Rings. Piston rings are provided to (*i*) prevent leakage of the compressed and expanding gases above the piston into the crank case, (*ii*) control and provide the lubricating oil between the piston skirt and cylinder walls, (*iii*) scrap excessive lubricating oil from cylinder walls. Piston rings are usually made up of cast iron plated with chromium, cadmium or tin to reduce wear.

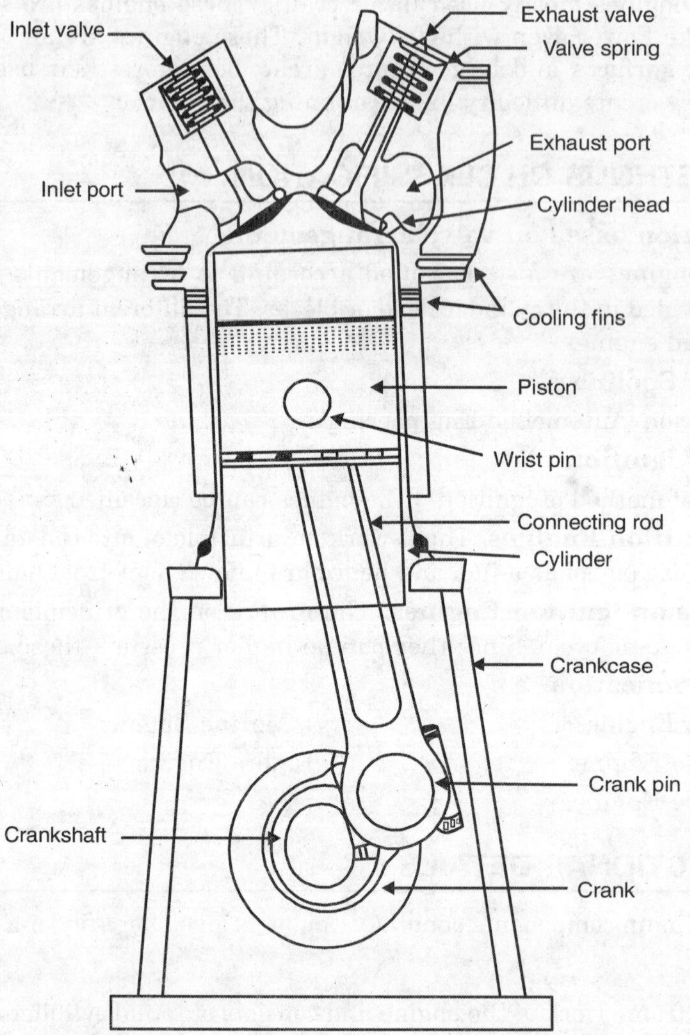

Fig. 10.12 Engine showing main components.

Piston Pins. These are also known as 'gudgeon-pins', or 'wrist pins'. These are case hardened steel pins used to connect the piston to the small end of the connecting rod. To reduce the weight it is made of tubular section. It passes through the piston bosses and the small end of the connecting rod.

3. Cylinder Liners

After long use when the cylinder itself worn out, then to replace it by new one is very costly. Therefore instead of replacing the complete cylinder block, a sleeve called 'liner' is fitted in the block. Material used for liners is generally iron alloy containing manganese, silicon, nickel and chromium. These are hardened by means of chromium plating or nitriding.

4. Cylinder Head

Cylinder head is a separate casting and is bolted to the top of the cylinder block, and is made of grey iron or aluminium alloy. To provide a gas and water tight seal between the cylinder block and the head, a gasket of copper-asbestos-copper is fitted in between. Cylinder head has cooling water jacket, provision for spark plug or diesel injector, and pockets for operating valves and combustion chambers.

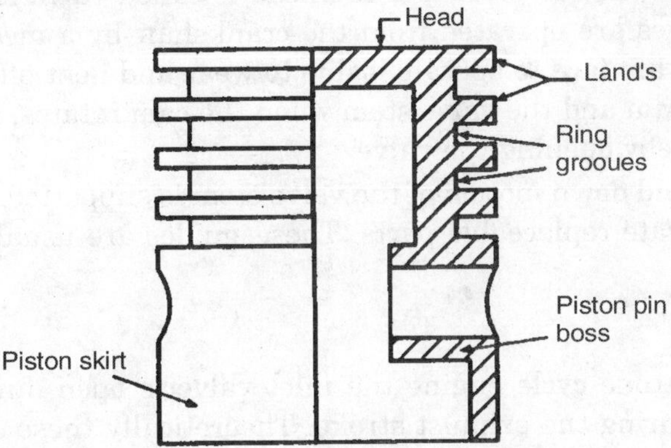

Fig. 10.13 Piston and its parts.

5. Crank Case

Crank case is a support and enclosure or the moving parts of the engine. It provides support for crankshaft and camshaft in suitable bearings. This is made of ferrous alloy or semi steel. This is fitted below the cylinder block and above the oil sump.

Oil Sump. It is a steel chamber bolted to the lower flange of the crank case and encloses the bottom of the casting. It also acts as a reservoir for the storage, cooling and ventilation of engine lubricating oil,

Engine Bearings. The bearings are provided between the relative moving parts of the engine. The bearings are provided to minimise the friction between two moving parts and to increase their life. The bearings may be sleeve or bush bearings, ball bearings, roller bearings, or split sleeve bearings.

Bearings provided at the crank pins and main journals of the crank shaft are called "Main Bearings". These are split sleeve type, Oil holes and annular grooves are provided in the main bearings for lubrication. A bush bearing is provided at the small end of the connecting rod. Big end of the connecting rod is connected to the crank shaft pin by a split type bearing and has a separate cap. The cap is secured by means of big end bolts.

6. Crank Shafts

Crankshaft is a rotating component of the engine which converts the reciprocating motion of the piston into rotary motion. Crank shafts has Crankpins, main journals, balance weights, webs and flywheel flange. The crank shaft may be single piece as well as built up type.

Connecting Rod. This provides connection between the piston and the crankshaft. It is made by drop forging from the steel or duralumin, in I section.

Cams and Camshaft. Cam is used to convert the rotary motion of the crankshaft in the straight line motion. The cams are provided to open and close the intake and exhaust valves. Camshaft consists of a number of cams integrated to it. For each engine cylinder, two cams are provided, one each for exhaust and suction valve. Cams and camshafts are made of alloy steel or alloy cast iron. Camshaft receives the drive from crankshaft, by means of gearing, and rotates at half the speed of crankshaft.

7. Valves and Valve Tappets

Valves are provided, in order to admit air fuel mixture into the cylinder of the engine during suction stroke, and allow burnt gases to be exhausted from the cylinder. Valve has 'head' and 'steam' as main parts. Lower portion of the head is called valve face and is ground at an angle of 45° or 30°. Valves are operated from the crankshaft by a mechanism. Valve stem is not directly kept on the cam face to operate it due to wear and heat effect. Therefore tappet is interposed between the cam and the valve stem when the cam rotates, it lifts the tappet which lifts the valve stem, thereby opening the valve.

Valve Guide. Up and down motion of the valve stem is supported and guided in the valve guides provided as separate replaceable parts. These, guides are usually made of cast iron or case-hardened steel.

Valve Timing

We know in four stroke cycle engine the inlet valve is open during the suction stroke and the exhaust valve during the exhaust stroke. Theoretically these actions of the valves as the stroke may be, should be performed either at the T.D.C. or at B.D.C positions of the piston. But in practice due to inertia of the airfuel mixture and quiet operation of the valves, more time is taken for the intake and exhaust processes to complete.

Refer Fig. 10.14. The inlet valve opens 5° before the piston starts downward to suction stroke, because the inlet valve takes time in the opening. As the piston moves downward air-fuel mixture enters the cylinder due to vacuum inside it. As the piston reaches to the B.D.C, still there is vacuum inside the cylinder and the charge tends to enter it. Now the piston moves upward and inlet valve remains open till the vacuum inside the cylinder comes to the atmospheric pressure. This timing is 44° after the B.D.C., position of the piston. In this way the inlet valve remains open for total 5° + 180° + 44° = 229° during the suction stroke. Then the inlet valve be closed.

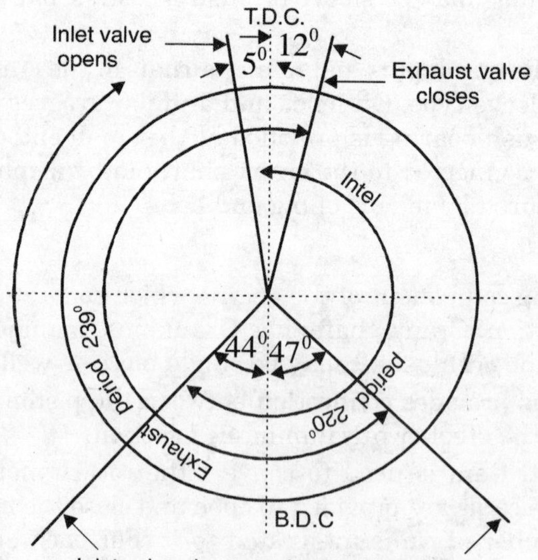

Fig. 10.14 Valve timing diagram.

Now-a-days generally both the valves are in closed position and the piston moves upward (however, value timing can vary from cyl. to clinder and it is an general example only). As soon as the piston completes compression stroke an electric spark takes place just few degrees before the T.D.C. In this way explosion takes place and piston moves downward, while both the valves are in close position.

The exhaust valve opens 47° before B.D.C. position of the piston thereby rushing on the burned gases with a slight pressure. Now the piston moves upward and draws out the burned gases fully. The exhaust valve remains open till the piston passes T.D.C. position by 12°. Then 12° after T.D.C. the exhaust valve be closed. In this way the exhaust valve remains open for total

$$47° + 180° + 12° = 239°$$

during the exhaust stroke. From the figure it is quite clear that there are timings when both valves are open simultaneously. This is called 'valve overlap'. In this diagram valve overlap is 17°.

8. Intake or Exhaust Manifold

These manifolds are hollow tube type castings of cast iron or aluminium. Intake manifold carries the air in diesel engine, and air-fuel mixture from the carburettor in petrol engine. The outlet ports of these intake manifolds are attached to the intake ports of the engine. The exhaust manifold carries the burned gases away from the engine cylinders. Inlet ports of these exhaust manifold are clamped to the outlet ports of the exhaust valves by means of studs and clamps. The gaskets are provided between the seats of the engine block and manifold ports to prevent the leakage of gases.

9. Silencers

If the exhaust gases from exhaust manifold are delivered directly to the atmosphere, a loud noise is produced, due to large pressure difference between the expanding exhaust gases at the atmosphere. Therefore to reduce the pressure, the exhaust gases are passed through a device called silencer or muffler. A good muffler permits the gases to expand to nearly atmospheric pressure before they are discharged into the air.

10. Flywheel

Since, the power supplied to the crank shaft is not uniform, it is liable to speed up when more power is delivered, and slows down when less power is received. In order to avoid fluctuation in speed flywheel is used. Flywheel is a heavy wheel mounted at the rear end of the crankshaft. Due to inertia it resists change in speed. Flywheel absorbs power when more power is supplied, and delivers a part of it to the crankshaft when it receives less power. Its rearface serves as the driving member of the engine clutch. Flywheel has a gear teeth ring around its outer rim meshing with the cranking motor drive pinion during the cranking of the engine for starting.

10.9 AIR SUPPLY SYSTEM

Air supply system is required to supply clean air to the engine. As atmospheric air contains dust and causes wear of piston rings and cylinder liner, an air cleaner is fitted to remove the dust from the air before supplying to the engine. The air cleaners may be either of dry type or wet type. The dry type air cleaner contains a pad of wool felt, steel wool or cloth. The dust in air passing through this cleaner is removed. Wet air cleaners, also known as oil bath air cleaners, are used in high capacity, high speed engines. In these, air is first passed through the oil bath and then through a steel wool pad.

10.10 EXHAUST SYSTEM

The function of the exhaust system is to discharge the exhaust gases to the atmosphere without creating much noise. It consists of exhaust manifold, exhaust pipe and silencer. The exhaust manifold collects the exhaust gases from all the engine cylinders to lead to a common exhaust pipe. Then all these exhaust gas pass through a silencer to the atmosphere reducing its velocity so its disagreeable noise is also reduced.

10.11 LUBRICATION SYSTEM

The main parts of an engine which need lubrication are as given under:

(*i*) Main crankshaft bearings
(*ii*) Big-end bearings.
(*iii*) Small end or gudgeon pin bearing
(*iv*) Piston rings and cylinder walls.
(*v*) Timing gears
(*vi*) Cam shaft and cam shaft bearings.
(*vii*) Valve mechanism
(*viii*) Valve guide, valve tappets and rocker arms.

The system which ensures the supply of lubricant to the different bearings and rubbing parts of the engine, to minimise the friction and provide free movement to them is called "lubrication system". The main factors considered in a good lubrication system are as given under:

(*i*) The lubrication system should be reliable.
(*ii*) The system should ensure the proper lubrication of all the working parts.
(*iii*) There should be minimum consumption of lubricant because the excessive oil provides friction and gummy action on the working parts of the engine.
(*iv*) The system should be provided with the proper oil filters.
(*v*) The system should be provided with pressure gauge or oil indicator so that engine man may know the circulation of lubricant through different working parts.
(*vi*) The system consists of lubricant of proper quality.
(*vii*) The system should be provided with proper oil sealing and drain plugs to replace the oil.

10.11.1 Types of Lubrication Systems

The common types of lubrications systems used in I.C. engines are as under:

1. Oil Mixed With Petrol. Under this system 3 to 6% amount of the lubricating oil is mixed with petrol itself. When charge reaches in the cylinder then oil available in it lubricates the main bearings, big end bearings, piston pin, piston rings and the cylinder walls. This system is very simple and does not require the oil pump, filter etc. This system is used in two stroke spark ignition engines which are generally used in scooters and motor cycles. The main disadvantage of this system is that during combustion the major part of the lubricant is burnt out thereby depositing carbon on different parts of the engine, which needs quick service of the engine.

2. Splash Lubrication System. In this system, splash troughs are provided under the connecting rods of the engine. Oil is filled in these troughs from the oil sump by means of oil pumps as shown in Fig. 10.15(*a*).

A scoop or dipper is made in the lowest part of each connecting rod. The dipper is hollowed out purposely to catch the oil. When the engine runs the dipper or scoop dips into the trough

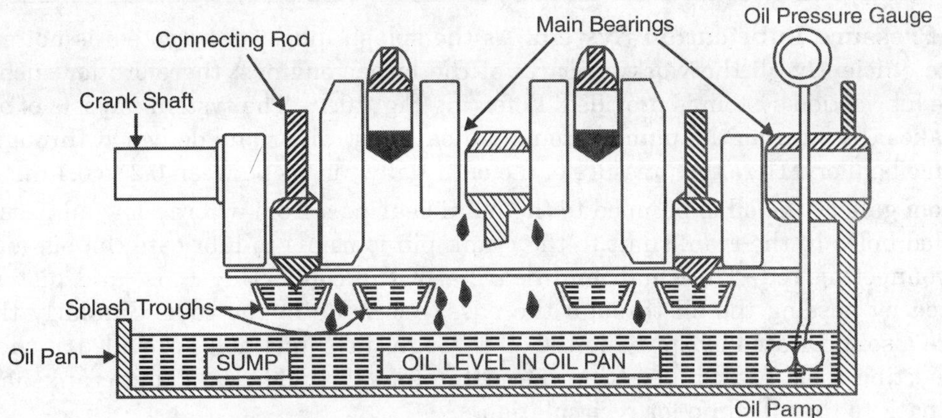

Fig. 10.15 (*a*) Splash lubrication system.

which is full with oil, then oil is filled into the hollow part of the dipper and is thrown by centrifugal force to all the engine parts such as cylinder walls, pistons, main bearings, gudgeon pin, big end bearings, valves, timing gears and cams as shown in Fig. 10.15(*b*), thereby lubricating all these working parts of the engine. In some designs pockets are provided above the various bearings to catch the splashing oil and deliver it by gravity to the bearings below them. The overflow oil from the splash troughs and other parts of the engine is returned to the sump by gravity force circulation.

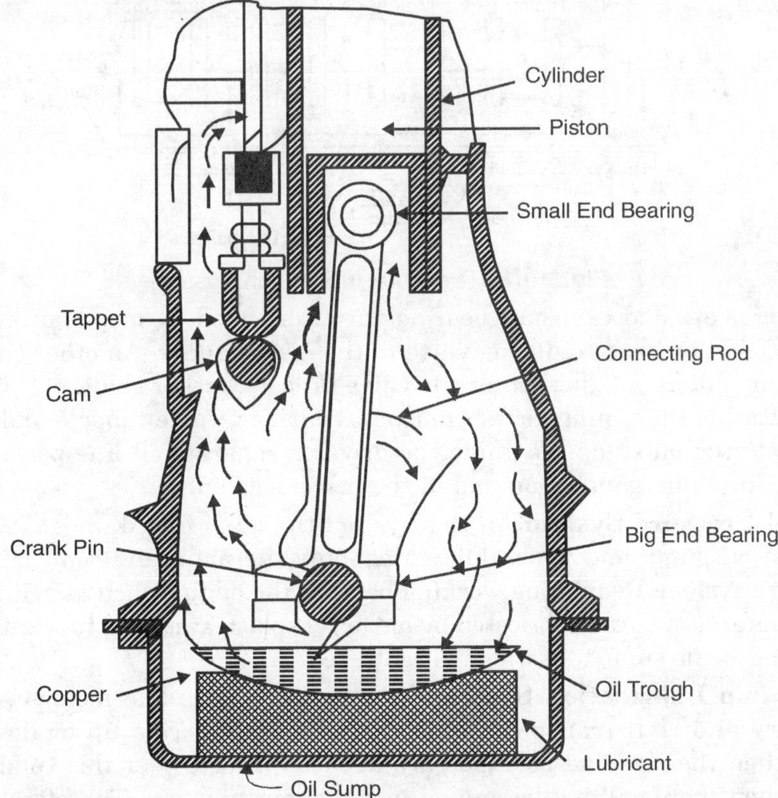

Fig. 10.15 (*b*) Splash lubrication system.

3. Pressure Lubrication System. As the splash lubrication system is not sufficient to lubricate efficiently all the working parts of the larger engines, therefore for such engines a pressure lubrication system is provided. Referring Fig. 10.16. The system consists of an oil pump which takes the oil from the pump through a floating strainer and deliver it through a filter to the main oil gallery. Here the pressure of the oil usually ranges between 0.2 to 0.4 mPa.

From gallery the oil is pumped to the main bearings, from where some oil passes through the drilled holes in the crankshaft to the crank pin journals to lubricate the big end bearings of the connecting rod. The small end bearing of the connecting rods, and piston pins are lubricated by passing the oil through the passages made in the rods. Similarly the cylinder walls are also lubricated by spraying the oil through the squirt holes provided at the big end of the connecting rods. After lubricating the piston pins, cylinder walls and crankshaft pins the oil falls back to the oil sump for recirculation.

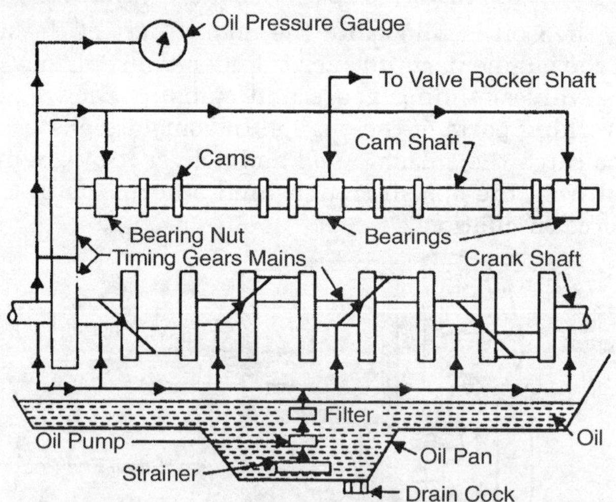

Fig. 10.16 Pressure lubrication system.

The timing gears and camshaft bearings are lubricated by means of a separate oil line connected through a pressure reducing valve to the main gallery. An other circuit which leads oil from the main gallery and lubricates the valve tappets, rocker shaft, etc. There is provision that after lubricating the timing gears, camshaft bearings, valve tappets and rocker shaft the oil comes back to the oil sump. To guide the driver a separate oil line passes from the main gallery to the oilpressure gauge mounted in the driver's cabin.

4. Partial Pressure System. In this system the oil is forced under 0.04 mPa pressure by means of an oil pump and which lubricates, only the main and connecting rod bearings, like the pressure system. Remaining working parts of the engine such as cylinder walls, piston walls, piston pins and cams are lubricated by means of splash system of lubrication. The overflow of the oil returns to the sump and which is reutilised.

5. Dry Sump Lubrication System. When the vehicles like motor cycles, sports cars, certain military and civil transport vehicles and aircrafts ride up or down the gradient continuously, then their engine positions are also changed. Under this condition a situation may develop when there will not be any oil in the sump to pump. Therefore to overcome this difficulty, a dry sump lubrication system is employed in certain vehicles.

Referring Fig. 10.17 the system consists of two oil pumps instead of one as in 'wet' system. The first which is called scavenging pump is installed in the lower-most portion of the inclined oil sump, from where it pumps the oil through a filter, to the main supply tank. The other pump which is called pressure pump is installed at the supply tank, which supplied the oil through the oil cooler to the bearings and working parts of the engine. The oil pressure is maintained at 0.4-0.5 mPa for the main and big end bearings. The oil pressure is reduced to about 0.05 mPa, like as in case of pressure system to lubricate the timing gears and camshaft bearings, rocker shaft etc. Care should be taken that the pumps should never be run dry, otherwise they would get defective.

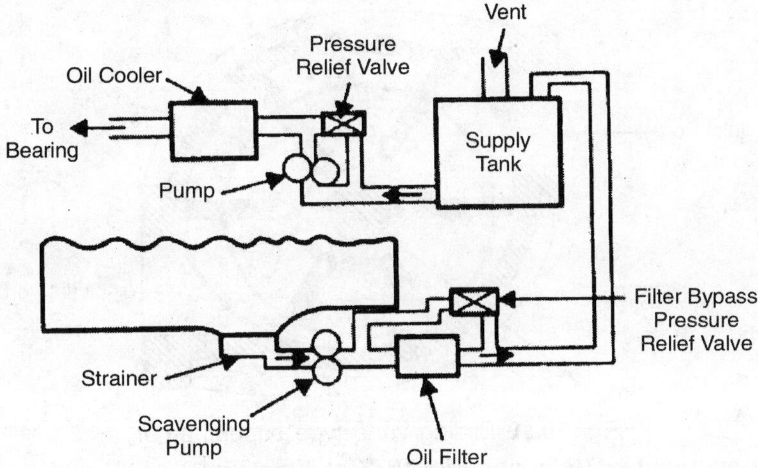

Fig. 10.17 Dry sump lubrication system.

10.12 SUPERCHARGING

When air-fuel mixture or charge from the carburettor is supplied to the engine at a pressure above the atmospheric then this process is called supercharging. The advantages of the supercharging are as under:

(*a*) With supercharging a densed charge enters the engine thus increases weight of the charge for the same swept volume. The power output of an engine is directly proportional to the weight of charge entered per minute to the engine, thereby increasing the engine output.

(*b*) At higher attitude the reduced air density reduces the rate of air weight flowing to the engine thus reduces the engine output. But the use of the superchager reduces this fall in the power output due to altitude.

(*c*) The use of the supercharger reduces the fuel consumption per Brake Power hour of the engine.

(*d*) In few engines the pressure of the exhaust gases is utilised to drive the supercharger thus saving the wastage of the engine power.

The supercharging device or supercharger is mounted between the carburettor and the inlet manifold of the engine. They may be gear, shaft or exhaust gases driven.

Following are most common **type of superchargers.**

1. Sliding Vane Type Supercharger. Referring Fig. 10.18 this type of supercharger consists of a rotor of drum shaped. The rotor contains a number of vanes mounted around it in such a manner that they can slide in or out against the spring force. The rotor is mounted eccentrically in a casing. The vanes always remain in contact with the inner surface of the casing. When the rotor is rotated then the space among the two vanes and casing, decreases continuously till it faces outlet port. In this way the air-fuel mixture from inlet port is taken between two vanes and delivered to the outlet port with a desired pressure. The outlet port is connected with inlet manifold hence begins the supercharging.

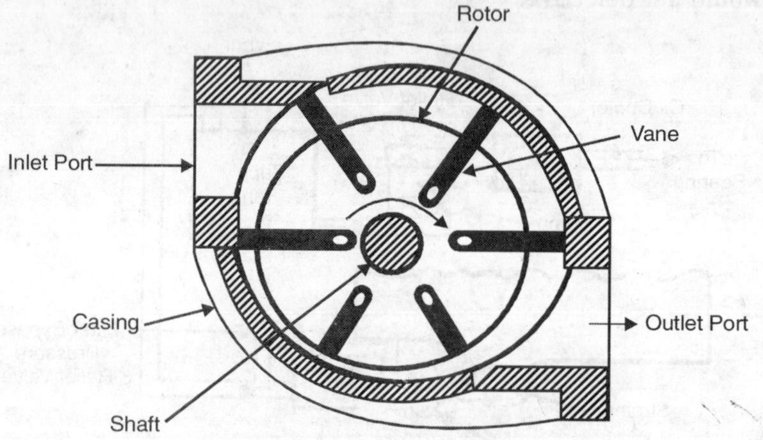

Fig. 10.18 Sliding vane type supercharger.

2. Root's Spercharger. Referring Fig. 10.19 this type of supercharger consists of two impeller mounted on their individual shafts. The shafts are driven by two equal sized gears so that the two impellers may revolve at the same speed. The shape of the impellers is epicycloid type. Both the impellers always remain in contact with each other as well as with inner sides of the casing. The working principle of this supercharger is same as that of a gear pump. The air-fuel mixture from carburettor enters at the inlet port and is delivered to the inlet manifold from the outlet port with a desired pressure. The construction of this supercharger is very simple, therefore, requires least maintenance and has longer life.

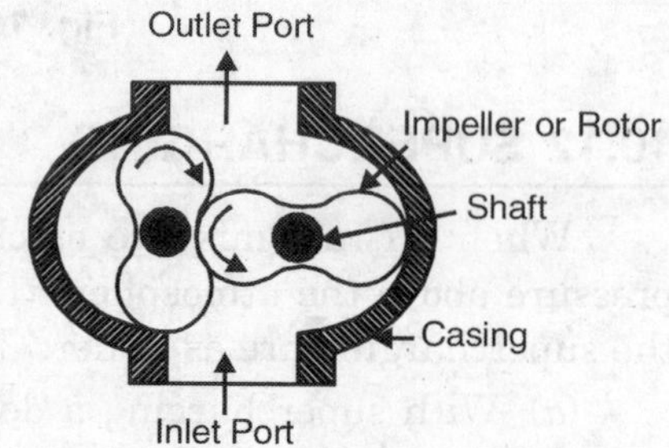

Fig. 10.19 Root's Supercharger.

3. Centrifugal Type Superchager. Referring Fig. 10.20 this type of supercharger consists of an impeller mounted on a power driven shaft. The impeller rotates in a casing which contains diffusers inside, around the periphery. The working principle of it is also same as that of centrifugal water pump. The air-fuel mixture enters at the Centre along the shaft of the impeller. When impeller is rotated at high speed then due to centrifugal action kinetic energy is developed in the entered charge. Then mixture

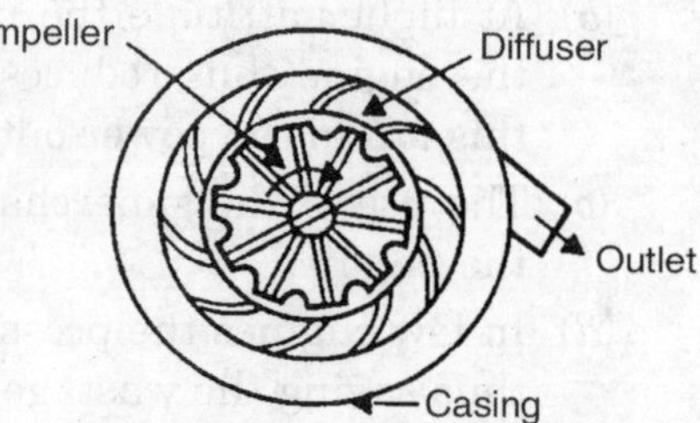

Fig. 10.20 Centrifugal type supercharger.

at supercharged pressure passes through the diffuser to the inlet manifold of the engine, which is connected at the outlet port of the supercharger. The disadvantage of this type is that it gives poor performance at lower speed.

10.13 ENGINE PERFORMANCE

10.13.1 Engine Power

The engine power may be stated in terms of indicated horse power (*i.h.p.*) or brake horse-power (b.h.p.). The indicated horse-power means the power developed inside the engine cylinder which is measured with the help of indicator diagram. The brake horse-power is the power available at the crank-shaft of the engine which is measured by means of a dynamometer. The difference of these two power measurements is known as friction horse-power (f.h.p.). The details of the power measurements are as discussed under:

1. *Indicated Power (I.P.):* The indicated power in kW of an engine is stated by means of the following formula:

$$I.P. = \frac{p_i \times l \times A \times n}{1000}$$

where, p_i = Indicated mean effective pressure in mPa (N/m^2)

l = Length of stroke in metre.

A = Area of cross section of cylinder in m^2.

n = No. of working strokes or impulse/second.

In two-stroke engines n is equal to N.

In four-stroke engines n is equal to $\frac{N}{2}$

where, N = Revolutions per second of the engine (*r.p.s.*)

The indicated mean effective pressure is calculated with the relation,

$$p_i = \frac{\text{Area of indicated diagram} \times \text{spring strength}}{\text{Length of diagram (in mm.)}}$$

2. *Brake Power (B.P.):* It is the real power taken from the engine crankshaft. Because it is measured with the help of brake test therefore named so. It is stated by means of following relations.

$$\text{B.P. in kW} = \frac{2\pi N \times (T_1 - T_2)R}{60{,}000} = \frac{2\pi NT}{60{,}000}$$

where, N = r.p.m. of the crank shaft

T_1 = Tension in the tight side of brake band in N.

T_2 = Tension in the slack side of the brake band in N.

R = Radius of brakewheel in metres.

T = Resisting torque.

If *brake mean effective pressure* is the mean effective pressure on the piston then,

$$B.P. = \frac{p_b \times l \times A \times n}{1000}$$

Therefore brake mean effective pressure (*b.m.e.p.*)

$$p_b = \frac{BP \times 60{,}000}{l \times A \times N} \text{ mPa.}$$

3. *Friction Power (FP):* The brake power is always less than the power developed inside the cylinder of the engine owing to frictional losses at the working surfaces.

Therefore, FP = IP – BP

10.13.2 Engine Efficiencies

1. *Thermal Efficiency:* It is termed on the basis of *IP* as well as on the B.P. as explained under:

(*a*) Indicated Thermal Efficiency,

$$\eta_i = \frac{IP \times 1000}{w_f \times C.V.} = \frac{IP. \times 1000}{w_f \times C.V.}$$

where, w_f = fuel supplied/second in kg.

C.V. = calorific value of fuel in kJ/kg.

∴ Indicated specific fuel consumption.

$$i.s.f.c. = \frac{1000 \times 60}{C.V. \times \eta_t} \text{ kg/IP hr}$$

(*b*) Brake Thermal Efficiency,

$$\eta_b = \frac{BP \times 1000}{w_f \times C.V.} = \frac{BP \times 1000}{w_f \times C.V.}$$

∴ Brake specific fuel consumption.

$$\therefore \quad b.s.f.c. = \frac{1000 \times 60}{C.V. \times \eta_b} \text{ kg/BP hr.}$$

2. *Mechanical Efficiency:* The ratio of the b.h.p. and the i.h.p. of the engine is known as the mechanical efficiency.

$$\eta_{mech} = \frac{B.P.}{I.P.} \text{ or } \frac{\eta_b}{\eta_t}$$

$$\text{or} \quad = \frac{b.m.e.p.}{i.m.e.p.}$$

3. *Volumetric Efficiency:* The ratio of the volume of charge admitted at N.T.P. to the volume swept by piston is called the volumetric efficiency of the engine.

$$\eta_v = \frac{\text{Volume of charge admitted at N.T.P}}{\text{Volume swept by piston}}$$

$$= \frac{\text{Volume of charge admitted at N.T.P.}}{\pi r^2 \times l}$$

where, r = Radius of the piston in metre

l = Length of stroke in metre

4. *Air Standard Efficiency:* Air in the alternatively cylinder absorbs and rejects heat during the cycle, and the engine is assumed to be working as a hot-air engine. This whole conception is treated theoretically only, therefore the efficiency obtained is known as the ideal or air standard efficiency of the cycle. The air standard efficiency for different cycles is given as under:

(*a*) For otto cycle, $$\eta = 1 - \frac{1}{(r)^{\gamma - 1}}$$

(*b*) For Diesel cycle, $$\eta = 1 - \frac{1}{(r)^{\gamma - 1}} \times \frac{1}{\gamma}\left(\frac{\rho^{\gamma - 1}}{\rho - 1}\right)$$

(*c*) For dual cycle, $$\eta = 1 - \frac{1}{(r)^{\gamma - 1}}\left[\frac{(\alpha\rho^{\gamma} - l)}{(\alpha - 1) + \alpha\gamma\,(\rho - 1)}\right]$$

where, r = compression ratio,
ρ = cut-off ratio
α = pressure or explosion ratio,
γ = adiabatic compression and expansion index for air

5. *Relative Efficiency:* The ratio of the thermal efficiency and the standard efficiency is known as the relative efficiency. As the thermal efficiency may either be indicated or brake thermal efficiency. Therefore relative efficiencies are.

(*a*) Indicated relative efficiency = $\dfrac{\text{Indicated thermal efficiency}}{\text{Air standard efficiency}}$

(*b*) Brake relative efficiency, $\dfrac{\text{Brake thermal efficiency}}{\text{Air standard efficiency}}$

10.13.3 Morse Test

It is a more accurate method to determine the mechanical efficiency of the multi-cylinder engine. The test is conducted with the help of a dynamometer. First the BP is measured with all cylinders operative. Then BP is measured with each cylinder rendered inoperative one by one. Under each case of measurement of BP the speed of the engine is kept constant by reducing the brake load on the engine. In this way the IP of the individual cylinder will be the difference of the first reading and the respective reading. Morse test is conducted for multi-cylinder engines only.

Let the engine consists of 4 cylinders,

B = B.P. of the engine with all cylinders operative
B_1 = B.P. of the engine with cylinder no. 1 inoperative.
B_2 = B.P. of the engine with cylinder no. 2 inoperative.
B_3 = B.P. of the engine with cylinder no. 3 inoperative.
B_4 = B.P. of the engine with cylinder no. 4 inoperative.

F_1, F_2, F_3 and F_4 be the FP of cylinders, 1, 2, 3 and 4 respectively.

Similarly, I_1, I_2, I_3 and I_4 be the IP of cylinders, 1, 2, 3 and 4 respectively.

$\therefore$ $$B = (I_1 + I_2 + I_3 + I_4) - (F_1 + F_2 + F_3 + F_4) \quad ...(i)$$

When cylinder no. 1 is rendered inoperative then its IP. will be zero and friction between cylinder and piston will be still there.

$\therefore$ $$B_1 = (I_2 + I_3 + I_4) - (F_1 + F_2 + F_3 + F_4) \quad ...(ii)$$

Now from the equation (*i*) and (*ii*) we get, $I_1 = (B - B_1)$

Similarly considering the remaining three measurements we get,

$$I_2 = (B - B_2), \quad I_3 = (B - B_3), \quad I_4 = (B - B_4)$$

∴ Total I.P. of the engine, $I = I_1 + I_2 + I_3 + I_4$

∴ Mechanical efficiency of the engine,

$$\eta_{mech} = \frac{B}{I}.$$

10.13.4 Heat Balance Sheet

To prepare the heat balance sheet an internal combustion engine is to run at a constant speed and load for a particular period. During this period an indicator diagram is drawn with the help of an indicator. For the same period consumption of fuel, its calorific value, internal and outlet temperature of fuel, gases and water and their amount is also noted. Engine i.h.p. and b.h.p. is also calculated. Now heat is calculated from the different divisions as explained under.

(*i*) Heat in fuel supplied $= wf \times C.V.$ kJ

where wf = weight of fuel supplied/second in kg.

$C.V.$ = Lower calorific value of fuel

(*ii*) Heat in IP per second $= IP \times 1000$ kJ.

(*iii*) Heat rejected to cooling water

$$= W_w (t_2 - t_1) \text{ kJ}$$

where, W_w = Weight of cooling water supplied/min.

t_1 = Inlet temperature

t_2 = Outlet temperature

(*iv*) Heat carried away by exhaust gases,

$$= W_e \times C_p (t_f - t_e) \text{ kJ.}$$

where, W_e = Weight of exhaust gases/min. in kg

C_p = Mean specific heat of the exhaust gases at constant pressure in kJ/kg °k

t_f = Temperature of the exhaust gases in degree centigrade

t_e = Engine-room temperature in degree centigrade.

∴ Heat Balance Sheet:

S.No.	*Particulars*	*Heat unit (kJ)*	*Percent*
(*i*)	Heat supplied in fuel per second		
(*ii*)	Heat in IP second		
(*iii*)	Heat carried away by cooling water		
(*iv*)	Heat carried away by exhaust gases		
(*v*)	Heat unaccounted for (by difference)		
	i.e., [S. No. (*i*) minus (*ii*) (*iii*) and (*iv*)		
	Total		

If the BP of the engine is also known then the heat lost in the friction can be calculated from the difference of the IP and BP.

10.14 GOVERNING OF I.C. ENGINES

In order to maintain constant speed of an I.C. Engine according to the load on the engine, governing of the supply of fuel (Diesel engine) or mixture of fuel and air (petrol engine or gas engine) is needed.

Governing of IC Engines can be achieved by following methods:

1. **Quantity Governing.** In this method, a throttle valve is kept in the supply pipe to regulate the quantity of the mixture of air and gas or petrol vapour according to the power needed.
2. **Quality Governing.** In this method, quality of mixture of air and gas or petrol vapour is varied by keeping the mass of air as constant while varying the mass of fuel supplied according to the power developed.
3. **Combined Governing.** In this method, governing can be done by combining the quantity and quality methods of governing.
4. **Hit and Mis Governing.** In this method, fuel supply is cut off completely during one or more number of cycles when the speed of the engine increases. This method is not suitable for heavy engines as it will require a heavy flywheel to minimise the variation of speed.

10.15 ABNORMAL COMBUSTION OF THE FUEL

Sometimes under certain operating conditions, the air fuel mixture has inherent characteristics thereby facing various forms of abnormal combustion. Important types of abnormal combustions are:

1. Pre-ignition

When any part of the combustion chamber such as spark plug, exhaust valve or carbon deposited, is overheated, this part ignites the fresh charge. This process is called pre-ignition, and tends to raise the temperature and pressure thereby producing "hot spot" in the chamber and increasing the rate of pre-ignition. It creates a pressure in the cylinder before the piston reaches the top dead centre. It reduces the total output of the engine and may also cause serious damage to the engine.

2. Detonation

Detonation, pinking or knocking is the violent sound pulsations within the engine cylinder. Under certain conditions of temperature, pressure and density when air-fuel mixture ignites without the help of a spark, then such an event is called as "auto-ignition". Cylinder walls combustion chamber and piston, face a violent blow caused due to the auto-ignition of a portion of the fuel. This loud pulsating noise is known as "pinking or knocking". Sometimes pressure wave under this condition breaks the piston of the engine. The main factors causing detonation are: chemical nature of fuel, shape of the chamber, relative position of the spark plugs, ratio of combustion of the fuels, and spark advance.

Certain chemicals are added in the fuel to reduce the knocking tendencies. Octane number rating, indicates the ability of a fuel to resist knock in an engine. A high octane fuel is highly resistant to knock.

10.15.1 Fuel Rating

Different fuel rating systems are utilised for different types of fuels used in spark ignition as well as in compression ignition engines. The important fuel rating systems are as given under:

1. Highest Useful Compression Ratio. For a particular engine operating conditions, the temperature towards the end of combustion is a function of compression ratio. As the temperature ratio is increased then the knock tendency of the fuel is also increased. When we continuously increase the compression ratio for a particular fuel, then a state comes where the detonation starts. This compression ratio is called as a highest useful compression ratio. This ratio is different for different engines.

2. Octane Number. Some fuels are more knock producing than others. The knocking tendencies are always undesirable characteristics in an engine. Therefore, certain chemicals are added in the fuel to reduce them. Thus, the octane number rating in this is an expression which indicates the ability of a fuel to resist knock in an engine, which is termed as octane number.

A high-octane fuel is highly resistant to knock. A low octane fuel knocks rather easily. A fuel which is very resistant to knocking is called "iso-octane" and is given an octane rating of 100. Similarly an other fuel which knocks very easily is called "heptane" and is given an octane rating of 0. Therefore, a mixture of 80 percent iso-octane and 20 per cent heptane would have an octane number rating 80.

3. Cetane Number. This number indicates the easiness by which the fuel ignites. With a high cetane number, the fuel ignites easily at low temperature. On the other hand, the fuel with low cetane number ignites relatively at higher temperature. The lower Cetane number tends to produce more knock.

As the fuel is sprayed or injected into the engine cylinder it does not ignite quickly and tends to accumulate there. But when the ignition takes place the accumulated fuel burns suddenly and produces a combustion knock. Therefore a fuel with suitable cetane number will ignite and begin to burn as soon as the injection spray starts, thereby producing an even combustion pressure rise and avoiding the knocking in the engine.

Example 10.3 *A petrol engine working on otto cycle consumes 8.2 litres of petrol per hour when developing 26 kW I.P. The specific gravity of petrol is 0.75 and its calorific value is 46000 kJ/kg. Determine the indicated thermal efficiency of the engine.*

Solution. Fuel consumed/min.,

$$w_f = \text{volume of fuel} \times \text{specific gravity of fuel}$$

$$= \frac{8.2}{60} \times 0.75 = 0.1025 \text{ kg/min}$$

$\therefore$ Indicated thermal efficiency,

$$\eta_t = \frac{\text{Indicated work done in heat unit/min.}}{\text{Heat supplied/min}}$$

$$= \frac{I.P. \times 60}{w_f \times C.V.}$$

$$= \frac{26 \times 60}{0.1025 \times 46000} \times 100\% = \mathbf{33.1\%.\ Ans.}$$

Example 10.4 *An eight-cylinder automobile engine of .086 mm bore and .083 m stroke with a compression ratio of 7 is tested at 4500 r.p.m. on a dynamometer which has a 535 mm. arm. During a 10 minute test a dynamometer scale beam reading of 420 N, 4.4 kg of gasoline for which the calorific value is 46000 kJ/kg are burned, and air at 22°C and 0.01 mPa is supplied to the carburettor at the rate of 5.5 kg per min, find (i) the b.p. delivered, (ii) the b.m.e.p. (iii) the b.s.f.c, (iv) the brake specific air consumption, (v) the brake thermal efficiency, (vi) the volumetric efficiency and (vii) the air-fuel ratio.*

Solution. (*i*)

$$\text{B.P.} = \frac{2\pi NT}{60000}$$

$$= \frac{2\pi \times 4500 \times 420 \times 0.535}{60000} = \textbf{105.8 kW Ans.}$$

(*ii*) Since,

$$\text{B.P.} = \frac{8\, p_b\, l\,.\,A\,.\,\frac{N}{2}}{60000} \quad \text{(8 is number of cylinders)}$$

∴ Brake mean efficiency pressure b.m.e.p.,

$$p_b = B.P. \times \frac{60,000}{4 \times l \times A \times N}$$

$$= \frac{105 \times 60,000}{4 \times 0.083 \times \frac{\pi \times (0.086)^2}{4} \times 4500} = \frac{6.35 \times 10^6}{8.674}$$

$$= 0.732 \times 10^6 \text{ Pa} = 0.732 \text{ MPa} \quad \textbf{Ans.}$$

(*iii*) Fuel consumed per minute,

$$w_f = \frac{4.4}{10} = 0.44 \text{ kg}$$

∴ Brake specific fuel consumption,

$$b.s.f.c = \frac{w_f \times 60}{B.P.}$$

$$= \frac{0.44 \times 60}{105.8} = \textbf{0.249 kg/kW-hr. Ans.}$$

(*iv*) Air consumption per minute,

$$w_a = 5.5 \text{ kg.}$$

∴ Brake specific air consumption

$$b.s.a.c. = \frac{w_a \times 60}{B.P.}$$

$$= \frac{5.5 \times 60}{100.8} = \textbf{3.12 kg/kW-hr. Ans.}$$

(*v*) Brake thermal efficiency,

$$\eta_b = \frac{B.P. \times 60}{w_f \times C.V.}$$

$$= \frac{105.8 \times 60}{0.44 \times 46000} \times 100\% = \textbf{31.3\%. Ans.}$$

(*vi*) Volume displaced per piston per cycle,

$$= \frac{\pi}{4}(0.086)^2 \times 0.083 = 4.82 \times 10^{-4} \text{ m}^3\text{/cycle.}$$

$\therefore$ Volume displaced at 4500 r.p.m. and for four-stroke, eight cylinder engine

$$= \frac{4.82 \times 10^{-4} \times 4500 \times 8}{2} \text{ m}^3\text{/min.} = 8.678 \text{ m}^3\text{/min.}$$

Volume of air used at intake conditions,

$$= \frac{w_a \times R \times T}{p} = \frac{5.5 \times 287 \times 295}{0.1 \times 10^6} = 4.65 \text{ m}^3\text{/min.}$$

$\therefore$ Volumetric efficiency, $\eta_p = \frac{4.65}{8.678} \times 100\% =$ **53.6%. Ans.**

(*vii*) The air fuel ratio, $\frac{A}{F} = \frac{5.5}{0.44} =$ **12.51. Ans.**

Example 4.5 *The following were noted for a 4-cylinder 4-stroke engine:*

Diameter	*= 100 mm.*
Stroke	*= 115 mm*
Speed	*= 1650 r.p.m.*
Fuel consumption	*= 0.2 kg / min*
Calorific value of fuel	*= 46000 kJ / kg*
Difference in tension on either side of the brake pulley	*= 400 N*
Brake circumference	*= 3.3 m*
Assume a mechanical efficiency	*= 80%*

Calculate:

(*i*) *Brake Thermal Efficiency*
(*ii*) *Indicated Thermal Efficiency*
(*iii*) *Mean Effective Pressure of the cylinder*
(*iv*) *Petrol consumption per b.p. - hr.*

Solution. $b.p. = \frac{2\pi NT}{60,000} = \frac{(2\pi R)NW}{60,000}$ $(\because T = WR)$

$$= \frac{3.3 \times 1650 \times 400}{60,000} = 36.3 \text{ kW} \quad (2\pi R = 3.3 \text{ m})$$

(*i*) Brake thermal efficiency, $\eta_b = \frac{b.p. \times 60}{w_f \times C.V.}$

$$= \frac{36.3 \times 60}{0.2 \times 46000} \times 100\% = \textbf{23.6\%. Ans.}$$

(*ii*) Indicated thermal efficiency,

$$\eta_i = \frac{\eta_b}{\eta_{mech}}$$

$$= \frac{0.236}{0.80} \times 100\% = \textbf{29.6\%. Ans.}$$

(*iii*) ∴ For 4 cylinders $i.p. = \dfrac{4p_i \times l \times A \times \dfrac{N}{2}}{60000}$ $\left(\because \eta_{mech} = \dfrac{b.h.p.}{i.h.p.}\right)$

or $\dfrac{b.p.}{\eta_{mech}} = \dfrac{2p_i \times l \times A \times N}{60000}$

∴ Indicated mean effective pressure,

$$p_i = \frac{60000 \times b.p.}{2 \times l \times A \times N \times \eta_{mech}}$$

$$= \frac{60000 \times 36.3}{2 \times 0.115 \times \frac{\pi}{4}(0.1)^2 \times 1650 \times 0.80}$$

$$= \frac{60000 \times 36.3}{0.115 \times \pi \times .01 \times 825 \times 0.80} = \frac{2.178 \times 10^6}{2.832}$$

$$= 0.914 \times 10^6 = \mathbf{0.914\ mPa} \quad \textbf{Ans.}$$

(*iv*) Brake specific fuel consumption

$$\text{b.s.f.c.} = \frac{w_f \times 60}{b.p.}$$

$$= \frac{0.2 \times 60}{36.3} = \mathbf{0.33\ kg/kW\ hr} \quad \textbf{Ans.}$$

Example 10.6 *A 4 cylinder engine running at 1200 r.p.m. gave 19 kW. The average torque when one cylinder was cut out was 100 Nm. Determine the indicated thermal efficiency if the calorific value of the fuel is 42,000 kJ/kg and the engine uses 0.25 kg of petrol/b.p. hour,*

Solution. (*i*) b.p. of 4 cylinders = 19 kW

(*ii*) b.p. of 3 cylinders, if one cylinder cut out

$$= \frac{2\pi \times 1200 \times 100}{60000} = 12.56 \text{ kW}$$

∴ i.p. of one cylinder = 19 – 12.56 = 6.44 kW

∴ i.p. of 4 cylinders = 4 × 6.44 = 25.76 kW

∴ Indicated thermal efficiency $= \dfrac{i.p. \times 60}{w_f \times C.V.}$

$$= \frac{25.76 \times 60}{0.25 \times 42{,}000} = 0.147 = \mathbf{14.7\%}. \quad \textbf{Ans.}$$

Example 10.7 *A petrol engine uses per kW hour 0.35 kg of fuel of calorific value 44000 kJ. The mechanical efficiency is 78% and compression ratio is 5.6 to 1. Calculate: (i) Brake thermal efficiency and (ii) Indicated thermal efficiency and (iii) the ideal air standard efficiency. Take γ = 1.4.*

Solution. (*i*) *Brake Thermal Efficiency*

$$\eta_b = \frac{b.p. \times 60}{w_f \times C.V.}$$

$$= \frac{1 \times 60}{\frac{0.35}{60} \times 44000} = \frac{3600}{0.35 \times 44000}$$

$$\eta_b = 0.234 = \mathbf{23.4\%.} \quad \textbf{Ans.}$$

(ii) Indicated Thermal Efficiency

Here $\eta_{mech} = 0.78$

$$\therefore \quad \eta_i = \frac{\eta_b}{\eta_{mech}} = \frac{.234}{0.78}$$

$$= 0.299 = \mathbf{29.9\%.} \quad \textbf{Ans.}$$

(iii) Air standard efficiency of petrol engine

$$\eta_{air} = 1 - \frac{1}{(r)^{\gamma - 1}}$$

$$= 1 - \frac{1}{(5.6)^{1.4-1}} = 1 - \frac{1}{(5.6)^{.4}}$$

$$= 1 - \frac{1}{2} = 0.50 = \mathbf{50\%.} \quad \textbf{Ans.}$$

Example10.8 *A 4 cylinder, 2 stroke cycle petrol engine develops 30 kW at 2500 r.p.m. The mean effective pressure on each piston is 0.8 mPa and the mechanical efficiency is 80%. Calculate the diameter and stroke of each cylinder if stroke to bore ratio is 1.5. Also calculate the brake specific fuel consumption of the engine, if brake thermal efficiency is 28%. The calorific value of the fuel is 44000 kJ/kg.*

Solution.

$$\eta_{mech} = \frac{b.p.}{i.p.}$$

or

$$i.p. = \frac{b.p.}{\eta_{mech}} = \frac{30}{0.80} = 37.5 \text{ kW}$$

$$\therefore \quad \text{i.p./cylinder} = \frac{37.5}{4} = 9.375$$

$$\frac{l}{d} = 1.5 \quad \text{where } l \text{ m = Stroke length, } d \text{ m = dia of piston}$$

$$\therefore \quad l = 1.5d$$

Now

$$i.p. = \frac{p_i.lAN}{60000} \quad \text{Here } n = N = 2500 \text{ being 2 stroke engine}$$

or

$$9.375 = \frac{0.8 \times 10^6 \times 1.5d \times \frac{\pi}{4}(d)^2 \times 2500}{60000}$$

or

$$d^3 = \frac{9.375 \times 4 \times 6000}{0.8 \times 10^6 \times 1.5 \times \pi \times 2500} \text{ m}^3$$

$$= \frac{37.5 \times 60000}{9420 \times 10^6} = 238.8 \times 10^{-6} \text{ m}^3$$

$\therefore$ d = **62 mm. Ans.**

$l = 62 \times 1.5 =$ **93 mm. Ans.**

Now, brake specific fuel consumption

$$\text{b.s.f.c.} = \frac{w_f \times 60}{C.V.\eta_b}$$

$$= \frac{0.35 \times 60}{44000 \times .28} = \mathbf{0.215 \text{ kg/kW hr.}} \quad \textbf{Ans.}$$

Example 10.9 *The following particulars refer to a 2-stroke Diesel engine:*

Bore	*= 100 mm.*
Stroke	*= 150 mm.*
Piston speed	*= 300 m/min.*
Torque developed	*= 58 Nm*
Mechanical efficiency	*= 81%.*
Indicated thermal efficiency	*= 40%.*
Calorific value of the fuel used	*= 44,800 kJ/kg.*

Determine :

(i) I.P. *(ii) Indicated mean effective pressure*

(iii) Fuel consumption per kW hour.

Solution. Piston speed = $2lN$ = 300 m/min

or $$2 \times \frac{150}{1000} \times N = 300.$$

$\therefore$ $$N = \frac{300 \times 1000}{2 \times 150} = 1000 \text{ r.p.m}$$

Now in 2 stroke engine N

= n. *i.e.*, No. of working strokes

= 1000 r.p.m.

$$b.p. = \frac{2\pi N.T.}{60000}$$

where T = Torque developed = 58 Nm

$\therefore$ $$b.p. = \frac{2 \times \pi \times 1000 \times 58}{60000} = 6.07 \text{ kW}$$

$$\eta_{mech} = \frac{b.p.}{i.p.} \quad \text{or} \quad 0.81 = \frac{6.07}{i.p.}$$

(*i*) $\therefore$ $i.p.$ = **7.5 kW. Ans.**

Now $i.p. = \dfrac{p_i \times l \times A \times n}{60,000}$ or $p_i = \dfrac{i.p. \times 60,000}{l \times A \times n}$

$$p_i = \frac{7.5 \times 6000}{0.15 \times \frac{\pi}{4} (.1)^2 \times 1000} = 382165 \text{ Pa}$$

$$= \textbf{0.382 MPa.} \quad \textbf{Ans.}$$

(*ii*) Fuel combustion per b.p. hr. = ?

Indicated thermal efficiency = 40%.

$$\therefore \quad \eta_t = \frac{i.p. \times 60}{W_f \times C.V.} \quad \text{or} \quad 0.40 = \frac{7.5 \times 60}{W_f \times 44,800}$$

$$\therefore \quad W_f = \frac{7.5 \times 60}{0.40 \times 44800}$$

$$= \frac{450}{17920} = 0.025 \text{ kg/minute}$$

$$= \frac{0.025 \times 60}{6.07} \text{ kg/kW hr.}$$

$$= \textbf{0.247 kg/kW} \quad \textbf{Ans.}$$

Example 10.10 *A petrol engine having a compression ratio of 7 has a brake thermal efficiency which is 40% of the ideal air standard efficiency. The calorific value of the fuel used is 42,000 kcal/kg. Calculate the fuel consumption in kg per hour, if the engine delivers 7.5 kW.*

Solution. Ideal air standard efficiency

$$\eta_{air} = 1 - \frac{1}{(r)^{\gamma - 1}} = 1 - \frac{1}{(7)^{1.4 - 1}} \quad (\because \ r = 7)$$

$$= 1 - \frac{1}{2.178} = \textbf{0.54}$$

Brake thermal efficiency.

$$\eta_b = 40\% \text{ of } \eta_{air} = 0.4 \times 0.54 = 0.216.$$

Now $\eta_b = \dfrac{\text{B.P.} \times 60}{W_f \times \text{C.V.}}$

where W_f = Fuel used in kg/hr.

C.V. = Calorific value

or $W_f = \dfrac{\text{B.P.} \times 60}{C.V. \times \eta_b} = \dfrac{7.5 \times 60}{42000 \times 0.216} = 0.0496$ kg/min

$$= \textbf{2.97 kg per hour.} \quad \textbf{Ans.}$$

Example 10.11 *The following results refer to a test on a petrol engine:*

I.P.	*= 31 kW*
B.P.	*= 26.5 kW*
Engine speed	*= 1800 r.p.m.*
Fuel per b.p. hour	*= 0.35 kg*
Calorific value of the fuel used	*= 44000 kJ/kg.*

Calculate:

(i) The indicated thermal efficiency *(ii) Brake thermal efficiency*

(iii) Mech. efficiency.

Solution. Fuel used by engine/hr.

$$= 0.35 \times b.p. = 0.35 \times 26.5$$

or $$W_f = 9.225 \text{ kg/hr} = 0.155 \text{ kg/min}$$

(i) $$\eta_t = \frac{\text{I.P.} \times 60}{W_f \times C.V.}$$

$$= \frac{31 \times 60}{0.155 \times 44000} = 0.27$$

∴ Indicated thermal efficiency = **27%. Ans.**

(ii) $$\eta_b = \frac{b.p. \times 60}{W_f \times C.V.}$$

$$= \frac{26.5 \times 60}{0.155 \times 44000} = 0.233$$

∴ Break thermal efficiency = **23.3%. Ans.**

(iii) $$\eta_{mech} = \frac{b.p.}{i.p.} = \frac{26.5}{31} = 0.86 = 86\%$$

∴ Mech. efficiency = **86%. Ans.**

Example 10.12 *From the data given below, draw a heat balance sheet for a 2 stroke diesel engine runs for 20 minutes at full load:*

r.p.m.	=	*350*
m.e.p.	=	*0.3 mPa*
Net brake load	=	*650 N*
Fuel combustion	=	*1.5 kg*
Cooling water	=	*160 kg*
Water inlet temperature	=	*35°C*
Water outlet temperature	=	*60°C*
Air used/kg of fuel	=	*30 kg*
Room temperature	=	*20°C*
Exhaust temperature	=	*300°C*
Cylinder bore	=	*200 mm.*
Cylinder stroke	=	*280 mm.*
Brake diameter	=	*1 metre*
C.V. of fuel	=	*44000 kJ/kg*
Steam formed/kg of fuel in the exhaust	=	*1.35 kg*

Specific heat of steam in exhaust = *2.1 kJ/kg °k*

Specific heat of dry exhaust gas = *1.00 kJ/kg °k*

Solution. Heat supplied = W_f × C.V. = 1.5 × 44000 = **66000 kJ.**

Heat carried away by cooling water

= 160 (60 – 25) × 4.186 = 23442 kJ

$$\text{i.p.} = \frac{P_i\, l\, A\, n}{60000}$$

$$= \frac{3 \times 10^5 \times 0.28 \times \frac{\pi}{4}\left(\frac{200}{1000}\right)^2 \times 350}{60000} = \frac{0.923 \times 10^6}{60000}$$

($\because$ $n = N = 350$; 2 stroke engine)

= 15.386 kW

$$\text{b.p.} = \frac{2\pi NT}{60000}$$

$$= \frac{2\pi \times 350 \times 650 \times 0.5}{60,000} = 11.9 \text{ kW}$$

Heat equivalent of i.p. in 20 minutes

= 15.386 × 20 × 60 = **18643 kJ**

Heat equivalent of b.p. in 20 minutes

= 11.9 × 20 × 60 = **14280 kJ**

$\therefore$ Heat lost in friction = 18643 – 14280 = 4363 kJ

Total air used = 1.5 × 30 = 45 kg.

$\therefore$ Total weight of exhaust gases = 45 + 1.5 = 46.6 kg.

Weight of steam = 1.35 × 1.5 = 2.025 kg

$\therefore$ Weight of dry exhaust gases = 46.5 – 2.025 = 44.470 kg.

Heat carried away by exhaust gases

= 44.470(300 – 20) × 1.00 = **12516 kJ**

Heat in steam = 2.025 [2230 + 2.1 (300 – 99.1)] = 2.025[2230 + 421.9] = 5370 kJ

Heat Balance Sheet

S.No.	*Particulars*	*Heat units (kJ)*	*Percent*
1.	Heat in fuel supplied.	66000	21.6%
2.	Heat in b.h.p. in 20 minutes.	14280	6.6%
3.	Heat in f.h.p. in 20 minutes	4363	35.5%
4.	Heat carried away by cooling water	23442	19.0%
5.	Heat carried away by exhaust gases	12516	19.0%
6.	Heat in steam	5370	8.1%
7.	Heat unaccounted for (By difference)	6029	9.2%

Example 10.13 *A four-cylinder four-stroke diesel engine has a bore of 110 mm, a stroke of 125 mm and develops 48 kW at 1650 r.p.m. During a test lasting 15 minutes at 1650 r.p.m., 2.8 kg of fuel of calorific value 46000 kJ/kg were used. The air fuel ratio for chemically correct combustion is 14.4 and the volumetric efficiency referred to air condition of specific volume 0.802 m³/kg is 82%. Determine the brake thermal efficiency and the percent of available air used for combustion.*

Solution. Brake thermal efficiency,

$$\eta_b = \frac{b.p. \times 60}{w_f \times \text{C.V.}}$$

$$= \frac{48 \times 60}{\frac{2.8}{15} \times 46000} \times 100\% = 33.5\%. \quad \textbf{Ans.}$$

$$\text{Stroke volume per cylinder} = \frac{\pi d^2}{4} \times l = \frac{\pi(0.11)^2 \times 0.125}{4} = 1.187 \times 10^{-3} \text{ m}^3$$

$$\text{Volume swept during test} = \text{Stroke volume/cylinder} \times \text{No. of cylinders} \times \frac{N}{2} \times 15$$

$$= \frac{1.187 \times 10^{-3} \times 4 \times 1650 \times 15}{2}$$

$$= 58.7 \text{ m}^3$$

Actual volume of air sucked in during test.

$$= 58.7 \times \eta_v = 58.7 \times 0.82 = 48.2 \text{ m}^3$$

Hence weight of air sucked in during test,

$$= \frac{48.2}{0.802} = 60 \text{ kg.}$$

Weight of air required for chemically correct combustion,

$$= 14.4 \times 2.8 = 40.32 \text{ kg.}$$

Hence percent of available air used for combustion

$$= \frac{40.32}{60} \times 100\% = \textbf{67\%. Ans.}$$

Example 10.14 *A four-cylinder petrol engine has an output of 45 kW of b.p. at 1800 r.p.m. A morse test was carried out and the brake torque readings were 169, 163, 160 and 167 N respectively. For normal running at this speed specific fuel consumption is 0.222 kg/kW hr. The L.C.V. of the fuel is 42,000 kJ/kg. Calculate the mechanical efficiency and the brake thermal efficiency of the engine.*

Solution. b.p. developed when cylinder No. 1 is rendered inoperative,

$$= \frac{2\pi NT}{60000} = \frac{2\pi \times 1800 \times 169}{60000} = 31.84 \text{ kW}$$

b.p. developed when cylinder No. 2 is rendered inoperative,

$$= \frac{2\pi \times 1800 \times 163}{60,000} = 30.71 \text{ kW}$$

b.p. developed when cylinder No. 3 is rendered inoperative,

$$= \frac{2\pi \times 1800 \times 160}{4500} = 30.14 \text{ KW}$$

b.p. developed when cylinder No. 4 is rendered inoperative,

$$= \frac{2\pi \times 1800 \times 167}{4500} = 31.46 \text{ kW}$$

$\therefore$ i.p. developed in cylinder No. 1 = 45 – 31.84 = 13.16 kW

i.p. developed in cylinder No. 2 = 45 – 30.71 = 14.29 kW

i.p. developed in cylinder No. 3 = 45 – 30.14 = 14.86 kW

i.h.p. developed in cylinder No. 4 = 45 – 31.46 = 13.54 kW

Hence, i.p. developed by the engine = 13.16 + 14.29 + 14.86 + 13.54 = 55.85 kW

$\therefore$ Mechanical efficiency, $\eta_{mech} = \frac{b.p.}{i.p.}$

$$= \frac{45}{55.85} \times 100\% = \mathbf{80.57\%.}\ \textbf{Ans.}$$

Brake thermal efficiency, $\eta_b = \frac{b.p. \times 60}{w_f \times C.V.}$

$$w_f = \frac{45 \times 60}{42000} \times 100 = \mathbf{28.9\%.}\quad \textbf{Ans.}$$

Example 10.15 *In a trial on 4 cylinder, 4-stroke, petrol engine of 102 mm bore, and 126 mm stroke, the net dynamometer load was 200 N at a radius of 0.495 m when the speed was 2400 r.p.m. At this speed and throttle opening the engine required 4.8 kW to motor it with ignition switched off.*

(i) Calculate the mechanical efficiency and the indicated mean effective pressure.

(ii) During a 3 minute run at this speed and power, the engine used 0.612 kg of petrol of calorific value 44000 kJ kcal/kg and 23 kg of cooling water with a temperature rise of 54°C.

Draw a heat balance sheet for the test in kJ/min. Specific heat of water = 4.186 J/kg °k

Solution. $\quad \text{b.p.} = \frac{2\pi NT}{60000}$

$$= \frac{2\pi \times 2400 \times 200 \times 0.495}{60000} = 24.8 \text{ kW}$$

i.p. = b.p. + f.p. = 24.8 + 4.8 = 29.6 kW

(i) Mechanical efficiency, $\eta_{mech} = \frac{b.h.p.}{i.h.p.} = \frac{24.8}{29.6} \times 100\% = \mathbf{83.81\%}$ **Ans.**

$$\because \quad i.p. = \frac{4 \times p_i \times l \times A \times \frac{N}{2}}{60000}$$

∴ Indicated mean effective pressure,

$$p_i = \frac{60,000 \times i.p.}{2 \times l \times A \times N}$$

$$= \frac{60,000 \times 29.6}{2 \times 0.126 \times \frac{\pi}{4}(0.102)^2 \times 2400} = 359552 \text{ Pa}$$

$$= \mathbf{0.359 \text{ mPa}} \quad \textbf{Ans.}$$

(*ii*) Heat supplied/min. $= \frac{0.612 \times 44000}{3} = \mathbf{8976 \text{ kJ}}$

Heat carried away by cooling water/min.

$$= \frac{23 \times 4.184 \times 54}{3} = \mathbf{1733 \text{ kcal.}}$$

Heat converted into work/min. $= b.p. = 60 \times 24.8 = \mathbf{1488 \text{ kJ/min.}}$

Heat Balance Sheet kJ/min.

Heat in	*kJ*	*Heat out*	*kJ*	*%*
Supplied by fuel	8976	As b.p. ...	1488	16.6
		To cooling water	1733	19.3
		To exhaust etc. (By difference)	5755	64.1
Total	8976		8976	100.00

Example 10.16 *A six cylinder petrol engine, 100 mm dia. by 110 mm stroke running at 1600 r.p.m. uses a mixture of air petrol by weight of 14 to 1. Assuming that the air drawn into the cylinder per stroke and measured at atmospheric pressure and 82°C is equal to $\frac{7}{8}$ths of the swept cylinder volume, and that the thermal efficiency of the engine is 24 percent, the power developed at ground level where the barometer reads 760 mm. of mercury. The calorific value of the petrol is 42,000 kJ/kg.*

What would be the power developed at an altitude of 1600 metres? A drop of 26 mm in barometric reading may be assumed for each 300 metre rise in altitude.

Solution. Swept volume $= \frac{\pi}{4}(0.1)^2 \times 0.11 = 8.635 \times 10^{-4} \text{ m}^3$

∴ Volume of air drawn in per cylinder/stroke

$$= \frac{7}{8} \times 8.635 \times 10^{-4} = 7.56 \times 10^{-4} \text{ m}^3$$

Weight of air induced per cylinder/stroke

$$= \frac{p.v.}{RT}$$

$$= \frac{0.1013 \times 10^4 \times 7.56 \times 10^{-4}}{287 \times (273 + 85)} \quad (\because 760 \text{ mm. of Hg} = 0.1013 \text{ mPa})$$

$$= 7.516 \times 10^{-4} \text{ kg.}$$

Weight of air supplied to the engine per min.,

$$w_a = 7.516 \times 10^{-4} \times \frac{N}{2} \times \text{ no. of cylinders.}$$

$$= 7.516 \times 10^{-4} \times 800 \times 6 = 3.608 \text{ kg.}$$

Weight of fuel consumed/min.

$$w_f = \frac{3.608}{14} = 0.2577 \text{ kg}$$

Heat supplied to the engine/min. $= w_f \times$ C.V.

$$= 0.2577 \times 42{,}000 = 10823 \text{ kJ}$$

∴ Power developed at ground level.

$$= \frac{\text{Heat supplied per min} \times \text{thermal } \eta}{60}$$

$$= \frac{10823 \times 0.24}{60} = \mathbf{43.3 \text{ kW}} \quad \textbf{Ans.}$$

∵ Drop in barometric reading at an altitude of 1600 metres,

$$= \frac{1600}{300} \times 26 = 138 \text{ mm.}$$

∴ Power developed at 1600 metres altitude.

$$= \frac{43.3\,(760 - 138)}{760} = \frac{43.3 \times 622}{760}$$

$$= \mathbf{35.44 \text{ kW}} \quad \textbf{Ans.}$$

Example 10.17 *It is assumed that an automobile, engine can operate at a thermal efficiency of 24 per cent when operating conditions are as follows:*

Volumetric efficiency 78 per cent; mechanical efficiency of 80%, heat value of petrol 46000 kcal / kg; theoretical air per kg of petrol 14; excess air of 26%; petrol vapour has density twice the density of air and the mixture at the end of suction stroke is at a pressure of 0.838 kg / cm² and a temperature of 58°C. Gas constant for air is 287 Nm / kg °k.

Find the cylinder dimensions of six cylinder engine at the above conditions when the engine develops its rated power of 65 kW at a speed of 2000 r.p.m. Assume the stroke is 20% greater than the diameter.

Solution.

The rated horse power means brake horse power of the engine.

$$\therefore \quad i.p. = \frac{b.p.}{\eta_{mech}} = \frac{65}{0.80} = 81.25 \text{ kW}$$

Indicated thermal efficiency, $\eta_i = \dfrac{i.p. \times 60}{wf \times C.V.}$

∴ Weight of fuel consumed, $w_f = \dfrac{i.p. \times 60}{\eta_{\text{volume}} \times C.V.} = \dfrac{81.25 \times 60}{0.24 \times 46000} = 0.424 \text{ kg}$

$$\text{Swept volume} = \frac{\pi}{4} \times 1.2\, d^3 \times 2000 \times 6$$

$$= 11310\, d^3\ m^3. \text{ (where } d \text{ is in m.)}$$

Weight of theoretically air required,

$$= 0.424 \times 14 = 5.936 \text{ kg.}$$

Actual weight of air supplied, $w_a = 1.26 \times 5.936 = 7.48$ kg.

Volume of air supplied per minute at intake conditions,

$$V_a = \frac{w_a \times R \times T}{p}$$

$$= \frac{7.48 \times 21.29 \times 331}{0.888 \times 10^4} = 8.648 \text{ m}^3$$

Density of air, $\rho a = \frac{7.48}{8.648} = 0.865 \text{ kg/m}^3$

Density of petrol vapour $\rho_f = 2 \times 0.865 = 1.73 \text{ kg/m}^3$

Volume of petrol vapour present in the mixture supplied/min

$$V_f = \frac{w_f}{p_f} = \frac{0.424}{1.73} = 0.2451$$

Volumetric efficiency,

$$\eta_v = \frac{\text{Per minute volume of charge taken at intake conditions}}{\text{Total swept volume/mm.}}$$

$$0.78 = \frac{(V_a + V_f) \times 10^6}{11310 d^3}$$

$$= \frac{(8.648 + 0.2451) \times 10^6}{11310\, d^3} = \frac{8.893 \times 10^6}{11310\, d^3}$$

$$\therefore \quad d = \left(\frac{8.893 \times 10^6}{0.78 \times 11310}\right)^{1/3} = 100.2 \text{ mm.}$$

$$l = 1.2 \times 10.02 = 120.2 \text{ mm}$$

$$\left.\begin{array}{l}\textbf{Bore = 100 mm}\\ \textbf{Stroke = 120.2 cm.}\end{array}\right\} \textbf{Ans.}$$

Example 10.18 *A two stroke cycle, 210 mm bore × 280 mm stroke, single cylinder oil engine gives the following results on test.*

Speed	=	*350 r.p.m.*
Net brake load	=	*620 N*
Diameter of brake drum	=	*1 m.*
Oil consumption	=	*4.25 kg/hr.*

I.m.e.p.	=	*0.275 kg / cm²*
Heating value of fuel used	=	*44680 kJ / kg*
Air fuel ratio by weight	=	*32*
Temperature of air in test room	=	*20°C*
Temperature of exhaust gases	=	*370°C*

Calculate

(i) The i.p.

(ii) The b.p.

(iii) Indicated thermal efficiency

(iv) Brake thermal efficiency and

(v) Percentage heat lost to the exhaust gases.

Assume a mean C_p *of 1.00 for the exhaust gases,*

Solution. (*i*)

$$i.p. = \frac{p_m.L.A.n.}{60000}$$

$$= \frac{0.275 \times 10^6 \times 0.28 \times \frac{\pi}{4}(0.21)^2 \times 350}{60000}$$

$$= \mathbf{15.55\ kW} \quad \textbf{Ans.}$$

(*ii*)

$$b.p. = \frac{2\pi N.w.R.}{60000}$$

$$= \frac{2\pi \times 350 \times 620 \times 0.5}{60,000} = \mathbf{11.36.} \quad \textbf{Ans.}$$

(*iii*) Heat equivalent of i.p. per minute = $p_m.L.A.n.$

$$= 0.275 \times 10^6 \times 0.28 \times \frac{\pi}{4}(0.21)^2 \times 350$$

$$= 930 \text{ kJ/min.}$$

$$\text{Heat supplied by fuel per minute} = \frac{4.25 \times 44680}{60} = 3165 \text{ kJ/mm}$$

$$\text{Indicated thermal efficiency} = \frac{930}{3165} \times 100 = \mathbf{29.3\%} \quad \textbf{Ans.}$$

(*iv*) Heat equivalent of b.p. per minute

$$= 11.36 \times 60 = 681.6 \text{ kJ/min}$$

$$\text{Brake thermal efficiency} = \frac{681.6}{3165} \times 100 = \mathbf{21.5\%} \quad \textbf{Ans.}$$

(*v*) Exhaust gases per kg of fuel burnt

$= 33$ kg $\quad \because$ 32 kg air + 1 kg fuel

Heat carried away by exhaust gases per kg of fuel burnt

$$= m \,.\, C_p\,(T_2 - T_1) = 33 \times 1.0\,(370 - 20) = 11550 \text{ kJ}$$

Heat produced by 1 kg fuel = 44680 kJ

$$\text{Percentage heat lost} = \frac{11550}{44680} \times 100 = \mathbf{25.8\%.} \quad \textbf{Ans.}$$

Example 10.19 *The compression curve on the indicator diagram from a gas engine follows the law $PV^{1.3}$ = constant. At two points on the curve 1/4 stroke and 3/4 stroke, the pressure are 0.14 mPa and 0.36 mPa respectively. Determine the compression ratio of the engine. Calculate the efficiency and the gas consumption per I.P. hour if the relative efficiency is 0.4 and the gas has the calorific value of 18840 kJ/m^3.*

Solution. Refer Fig. 10.21

Let AB be the compression curve and (1) and (2) the two points, '*r*' be the compression ratio and clearance volume unity. The (*r* – 1) will be the stroke volume.

$$p_1 = 0.14 \text{ mPa}, \; v_1 = 1 + 3/4\,(r-1)$$

$$p_2 = 0.36 \text{ mPa}$$

$$v_2 = 8 + 1/4\,(r-1)$$

$$p_1 v_1^{1.3} = p_2 v_2^{1.3}$$

or $$\frac{v_1}{v_2} = \left(\frac{p_2}{p_1}\right)^{1/1.3}$$

or $$\frac{1+3r}{3+r} = \left(\frac{0.36}{0.14}\right)^{0.7692}$$

$\therefore$ $$r = \mathbf{5.5745} \quad \textbf{Ans.}$$

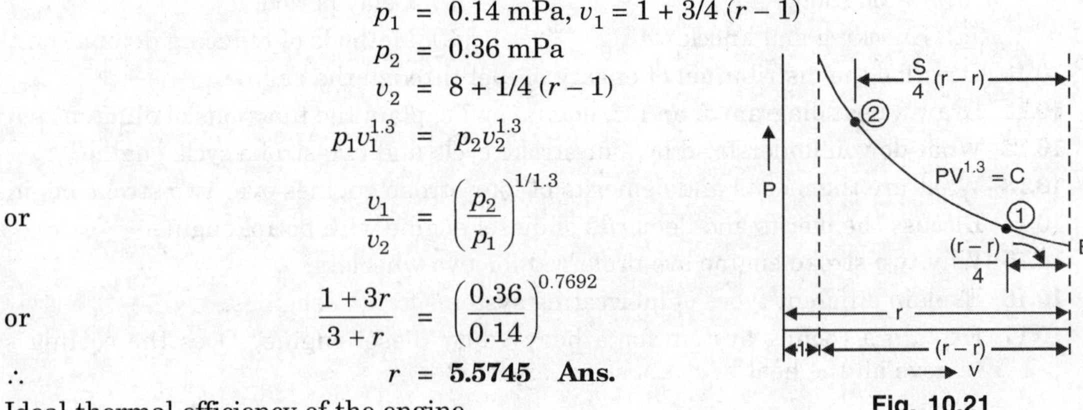

Fig. 10.21

Ideal thermal efficiency of the engine

$$= 1 - \frac{1}{r^{(r-1)}}$$

$$= 1 - \frac{1}{(5.5745)^{(1.41-1)}} = 1 - \frac{1}{2.03}$$

$$= 0.5054 \qquad \because r = 1.41 \text{ for air}$$

$$\text{Relative efficiency} = \frac{\text{Indicated thermal efficiency}}{\text{Ideal thermal efficiency}}$$

$\therefore$ $$\text{Indicated thermal efficiency} = 0.5054 \times 0.4 = 0.2022$$

$$\text{Now indicated thermal efficiency} = \frac{\text{heat equivalnt of I.P./hour}}{\text{Fuel/hour} \times \text{Calorific value}}$$

or $$0.2022 = \frac{1 \times 3600}{w_f \times 18840}$$

or $$w_f = \mathbf{0.945\ m^3/hour.} \quad \textbf{Ans.}$$

QUESTIONS

10.1 Differentiate between internal combustion engines and external combustion engines.

10.2 (*a*) How an internal combustion engine classified?

(*b*) Describe the main components of an I.C. Engine.

10.3 Explain 4-stroke and 2-stroke I.C. engines.

10.4 Draw a 4-stroke diesel engine diagram and explain its working.

10.5 Draw ideal and actual indicator diagrams for 4-stroke I.C. engine and explain their differences.

10.6 Explain the following in a Two stroke engine:
Scavanging, p-v diagram, advantages and disadvantages.

10.7 Draw and explain valve timing diagram for an I.C. engine.

10.8 Explain the "governing of an I.C. Engine" and describe different methods of governing.

10.9 Write short notes on ;

(*i*) Detonation, (*ii*) Delay period,

(*iii*) Knocking and knock rating (*iv*) Methods of reducing detonation.

10.10 Describe the distribution of energy of fuel through the engine.

10.11 Draw a neat diagram of an I.C. engine and explain the functions of different parts.

10.12 What do you understand by four-stroke cycle and two-stroke cycle engine?

10.13 What are the merits and demerits of four-stroke engines over two-stroke engines?

10.14 Discuss the merits and demerits of diesel engine with petrol engine.

10.15 Why two-stroke engine are preferred for two wheelers?

10.16 Explain different types of lubricating systems for I.C. engines.

10.17 Explain a cooling system for a heavy duty diesel engine? Does the cooling system remove all the heat of combustion, explain?

Chapter 11

FUEL SYSTEMS IN I.C. ENGINES

Learning Objectives

After studying this chapter, the students will learn:

- Fuel system in spark ignition system, fuel pumps carburettors, and air cleaners.
- Fuel system in compression ignition engines and its components including diesel filter, injection pump, fuel injector and injector nozzles.
- Multiple plunger fuel injection pump.

11.1 FUEL SYSTEM IN SPARK IGNITION ENGINES

In spark Ignition engines, fuel system is used to supply a proportionate combustible mixture of air and fuel to the engine. To suit the engine requirement under different operating conditions the mixture should be in varying proportions *i.e.* rich mixture when engine is cold and is being started, or for acceleration, or for high speed, or full load. A fuel system consists of fuel tank, fuel line, fuel filter, fuel pump, air cleaner, and carburettor etc.

Fuel Tank

It is a fuel container used for storing fuel. It is made of metal sheet and attached to the frame at the rear of the vehicle. The tanks are provided with baffle plates inside the tank, a cap at the top, fuel line near the bottom with a filtering element placed at the connection, and a vent pipe at the top. In modern vehicles a vapour recovery system is provided to condense the fuel vapours passing through the vent pipe and to reutilise. A drain plug is provided at the bottom for emptying the tank.

Fuel Line and Filters

Fuel line is a metallic tube through which fuel is sucked by the fuel pump from the fuel tank and delivered to the carburettor. The filter is fitted either inside the fuel tank or operates as a separate unit connected in the fuel line. The fuel filter used is generally a sediment bowl and a strainer screen. Bowl is made of glass or metal. The particles of dirt and water settle in the bottom of the bowl when fuel passes through it.

Fuel Gauge

This meters and cautions the driver of the vehicle or operator of the stationary engine about the quantity of fuel left in the tank.

11.2 AIR CLEANERS

If dust and dirt present in the air is not prevented from entering the engine through carburettor, it causes rapid wear and tear of the engine component, and blockage the passes of the carburettor. Therefore an air cleaner is provided at the carburettor. The air cleaners may be, oil bath type, dry type or oil saturated type.

***(i)* Dry Type Air Cleaner.** This type of air cleaner consists of a filtering element made of paper. This paper is arranged in the pleated form so that maximum filtering surface can be obtained. Inner surface of the element is connected to the carburettor or intake manifold and the outer surface is covered by a detachable cover. The edges of the element are sealed by means of tightening of the securing bolts. The air inlet pipe is connected with the crankcase breather to provide the extra air under pressure to the carburettor. This improves engine performance.

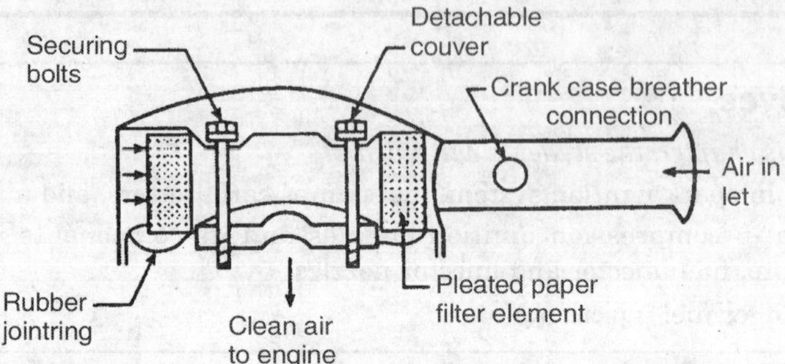

Fig. 11.1 Paper elemented dry air cleaner.

The filtering efficiency of this type of air cleaner is 99 to 99.7 per cent. The advantages of dry air cleaner is that it is lighter, compact and cheaper than other type of air cleaners (Refer Fig. 11.1).

***(ii)* Oil Bath Type Air Cleaners.** This type of air cleaner consists of a filter element made of mesh of metal wool. The filter element is suspended in a sheet metal shell. An oil sump is provided at the bottom of the shell. A passage way is provided through the filter and the oil sump to pass the filtered air to the carburettor. Filter element and passage way are covered by a top cover (see Fig. 11.2).

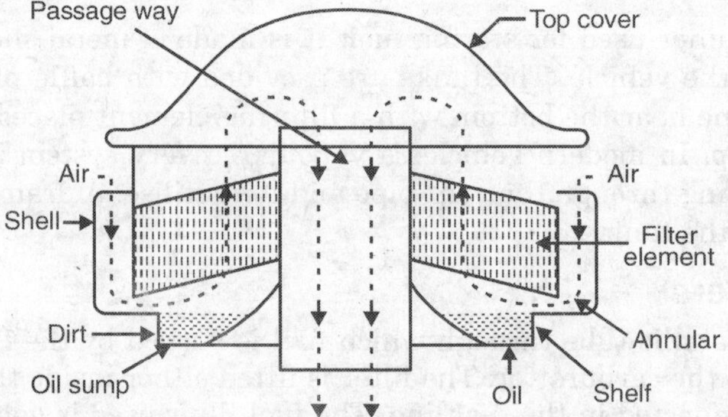

Fig. 11.2 A.C. oil bath air cleaner.

In operation the dusty air from atmosphere enters the shell as shown by the direction of the arrows. Then air passes through the angular passage and strikes the shelf, where it throws the dust particles into the oiled surface and changes direction of flow towards filter element. In this action oil is also carried with air, which is, returned back into the oil sump by the filter and only cleaned air passes through the filter element. Then this cleaned air passes through the passage way to the carburettor or inlet manifold of the engine. After certain period the unit is cleaned and the oil is replaced by the oil of particular viscosity.

***(iii)* Oil-saturated Type Air Cleaner.** This type of air cleaner is almost similar to the oil-bath type. It consists of an oil-saturated filter element to clean the air passing through it. The filter element is made of fine metal wool or polyurethene. The filter element is oiled to collect the dirt or dust particles. After certain period this element is cleaned by suitable solvent, then oiled and re-used.

11.3 FUEL PUMPS

It is a device which delivers fuel from the tank to the carburettor. Mostly it is mounted on the side of the cylinder block, at front of the engine. Fuel pumps are generally of two types say mechanical and electrical pump as described under.

***(i)* Mechanical Fuel Pump.** It consists of a rocker arm which is operated by means of an eccentric mounted on the camshaft. The inner end of the rocker arm is connected to the flexible diaphragm by means of a pull rod. The inlet of the pump is connected to the fuel tank whereas the outlet to the carburettor. Chamber of the pump consists of a suction or inlet valve and a pressure or outlet valve (Refer Fig. 11.3).

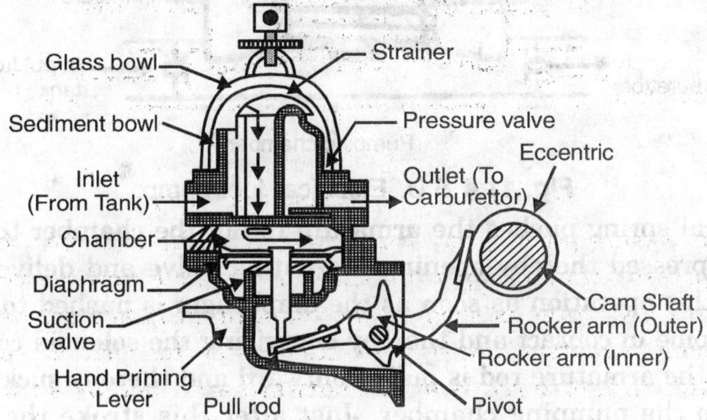

Fig. 11.3 A.C. Mechanical fuel pump.

As the camshaft rotates, the eccentric operates the rocker arm thus pushes the diaphragm up and down. The downward movement of the diaphragm creates a vacuum in the chamber thereby opening the suction valve and entering the fuel through the strainer into the chamber. The upward movement of the diaphragm closes the suction valve and opens the pressure valve thereby entering the fuel from the pump chamber to the carburettor float chamber.

When the float chamber of the carburettor is completely filled up then the needle valve of the float chamber closes so that no fuel can enter. Under this condition the fuel pump cannot deliver fuel to the carburettor. But in this case the rocker arm continues to rock and the diaphragm remains at its lower most position until the needle valve of the float chamber allows the fuel to enter the carburettor. Therefore, to avoid the building of excessive pressure and damaging of the pump, a flexible connection between the rocker arm and the pull rod is provided so that the diaphragm may not be operated all the time.

Dust and dirt particles which are seen through the glass bowl and are cleaned from the sediment bowl when filled excessive. A hand priming lever is provided to make the fuel available in the pump as well as in the carburettor in beginning of the engine start.

***(ii)* Electrical Pump.** This pump also consists of a diaphragm which is attached to the armature. The armature itself is attached to a bronze rod which passes through a solenoid winding. The outer end of the armature rod is hinged by means of a rocker which is provided a pair of breaker points made of tungsten. A compression type coil spring is provided to push the armature as well as diaphragm in the pumping chamber. The chamber consists of a suction valve and a delivery valve (see Fig. 11.4).

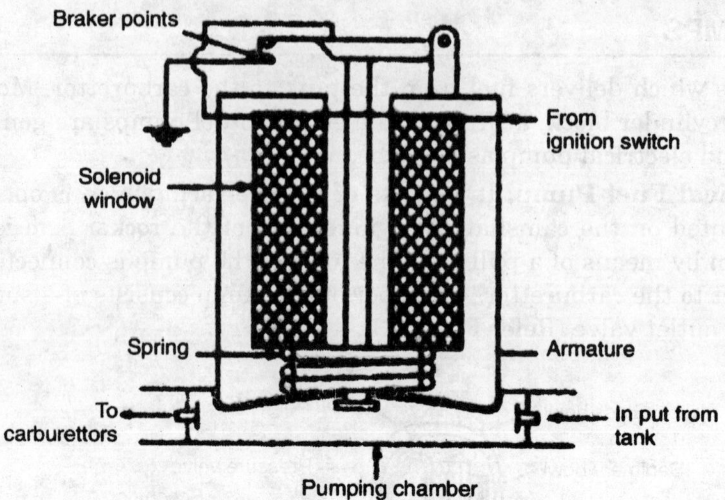

Fig. 11.4 S.U. Electrical fuel pump.

When the helical spring pushes the armature rod in the chamber then fuel available in the chamber is compressed thereby opening the out-let valve and delivering the fuel to the carburettor. Under this operation as soon as the diaphragm is pushed to its maximum limit, the breaker points come in contact and thereby energizing the solenoid coil. Therefore, due to this magnetic effect the armature rod is pulled outward and thereby sucking the fuel through the in-let valve into the pumping chamber. Just after this stroke the breaker points are disconnected automatically and the helical spring again comes in action for delivery stroke. Some times a condenser is provided at the contact points to reduce the sparking to its minimum possible.

11.4 CARBURETTORS

Carburettor is a device which vaporises and atomizes the fuel and mixes it with the air in varying proportions to meet the changing operating conditions of I.C. Engines. Here 'vaporization' means the change of the fuel from a liquid to a vapour, and 'atomization' is the braking-up of fuel by mechanical means, so that every small particle of the fuel is surrounded by air.

Simple Carburettor

A simple carburettor as shown in Fig. 11.5 is a round cylinder with a constricted section which ia called venturi. In this cylinder a fuel jet is provided at the venturi and a throttle valve is provided at the end which is bolted to the engine intake manifold. The other end of the tube remains open and acts as air entrance. The fuel jet supplies fuel from the float chamber in which fuel level is maintained at constant head by float and needle valve system.

When the throttle is turned to the vertical position then air from atmosphere flows through the venturi and produces a partial vacuum at the venturi. Therefore, due to this vacuum fuel jet delivers a spray of gasoline into the passing air stream. This air fuel mixture enters the engine cylinders via the intake manifold. The intake manifold remains hot thereby vaporizing any fuel particles left in the mixture. The level of the fuel in the float chamber remains the same as that in the jet. If the level tends to fall, the float drops and thereby opening the needle valve and admitting more fuel into the chamber. In beginning to provide the rich mixture to the engine a "tickler" is provided by depressing of which the float is also depressed thereby driping the fuel from the jet.

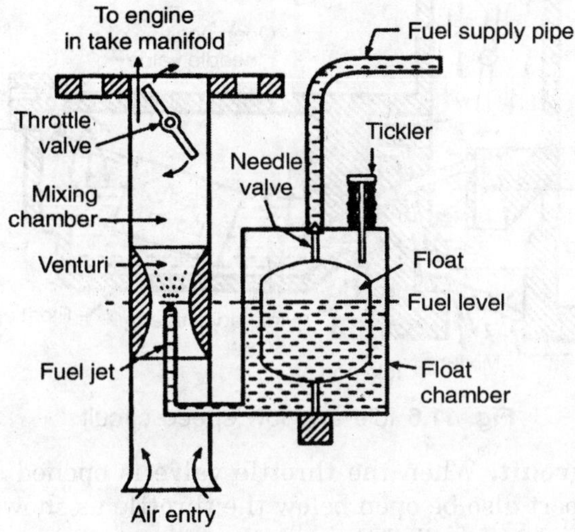

Fig. 11.5 Simple Carburettor.

Circuits of a Modern Carburettor

We know that under different operating conditions the engine demands a greatly varying air fuel mixture which is very difficult to provide by the simple tube carburettor. Therefore, modern carburettor is provided various circuits as described under, to meet the needs of the engine.

***(i)* Float Circuit.** This circuit generally consists of a float bowl or chamber, a float and a needle valve arrangement. The operation of float and needle valve system is to maintain a constant level of fuel in the float chamber. When the level of fuel in the float chamber falls due to consumption then needle-valve opens automatically and fuel enters the chamber. Similarly when the fuel enters the chamber faster than it is withdrawn then due to rise of fuel level the needle-valve closes automatically. The float tends to maintain the needle-valve partly closed thereby providing a balance supply of the fuel. The different mechanisms of float system are as given under.

***(ii)* Idle and Low Speed Circuit.** In this case, the throttle valve is fully closed or only slightly opened. Therefore no vacuum is developed in the venturi and no fuel is delivered by the main nozzle. Then under this position of the throttle valve the fuel is supplied by another circuit which is known as the "idle and low-speed" circuit, as shown by the direction of arrows in Fig. 11.6.

During this operation, high vacuum is created below the throttle valve and due to atmospheric pressure air and fuel mixture is pushed through the passages as shown in the figure. This rich mixture is controlled by idle adjustment screw and again mixes with the small amount of air past from the throttle valve edge. This final mixture is sufficiently rich to run the engine at idle and low speed. The idling speed of the engine can be adjusted by turning the adjustment screw in or out.

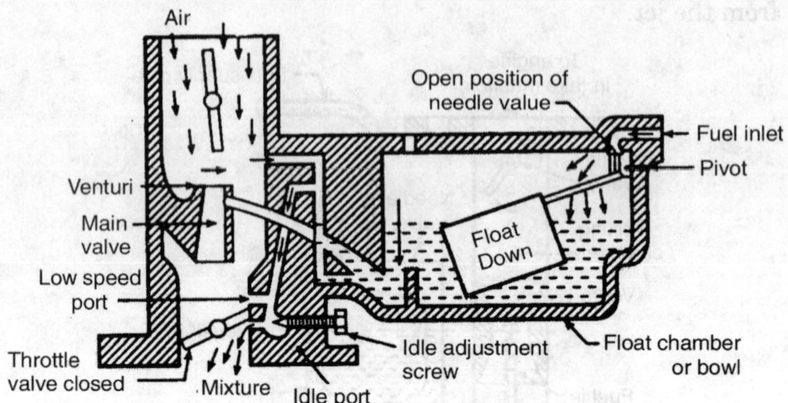

Fig. 11.6 Idle and low speed circuit.

***(iii)* Low Speed Circuit.** When the throttle valve is opened slightly then due to this movement the low-speed port also be open below the throttle as shown in Fig. 11.7. Therefore, an additional amount of fuel is supplied through the low-speed port to the engine. Under this position an additional air is also supplied from the sides of the throttle thereby obtaining the proper richness of the mixture. This additional air also improves the atomization of the fuel-received from the low-speed port thus running the engine smoothly at low speed.

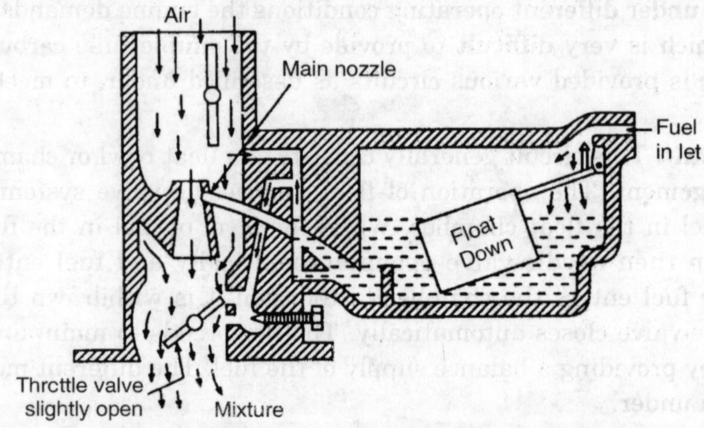

Fig. 11.7 Low speed circuit.

***(iv)* High Speed Part Load Circuit.** When the throttle valve is opened sufficiently or fully then there will be little difference in vacuum between the space above and below the throttle. Under this position most of the air passes through the venturi, which creates sufficient vacuum in the venturi to discharge the fuel from the main nozzle as shown in Fig. 11.8.

Under this condition a negligible amount of air fuel mixture passes through the low-speed port and major fuel is delivered from the main nozzle, therefore, the circuit from chamber to the main nozzle is called "high speed circuit". In this circuit a constant air fuel ratio is maintained by means of gradual operating of the throttle valve.

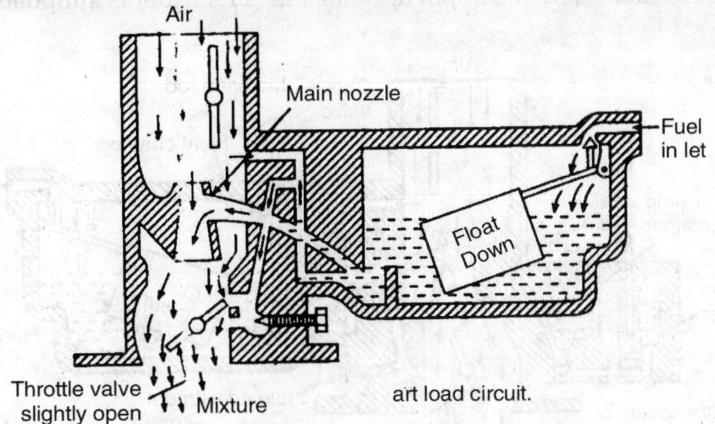

Fig. 11.8

(v) High Speed Full Power Circuit. In this case the throttle valve is full opened and the fuel is being fed through the main nozzle. But for high-speed full-power operation the mixture should be sufficiently rich, which is provided by incorporating an additional mechanical device in the carburettor.

This mechanical device is the metering-rod and metering-rod jet which are introduced in the high-speed circuit. The metering-rod consists of two or more steps of different diameters as shown in Fig. 11.9. The metering-rod is attached to the throttle valve by means of links. When the throttle valve is opened then the metering rod is also drawn from metering-rod jet as shown in Fig. 11.10, thereby flowing more fuel through the main nozzle. In this way more richer mixture is supplied by the carburettor. Similarly when the throttle is partly closed then part of the metering rod remains in its jet thereby restricting the flow of the fuel and controlling the speed of the engine at full load.

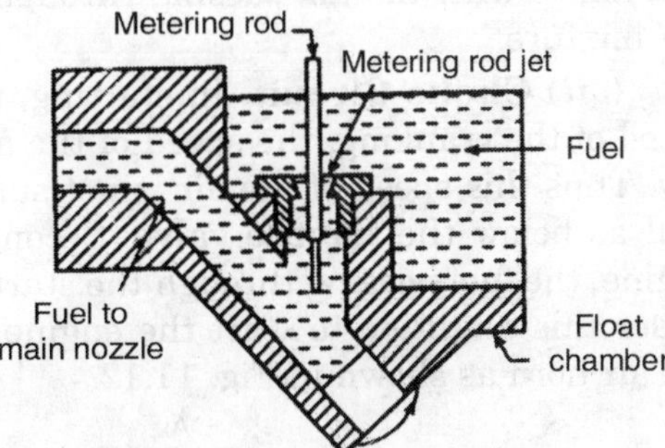

Fig. 11.9 Metering Rod and Metering Rod Jet.

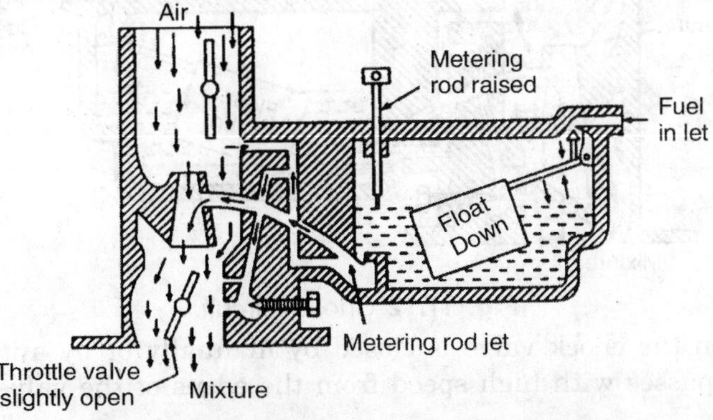

Fig. 11.10 High speed fuel power circuit.

(vi) Accelerator Pump Circuit. When the throttle valve is rapidly opened for acceleration then air enters into the carburettor suddenly. Therefore, under this condition to obtain a rich mixture, an additional fuel must be supplied. Refer Fig. 11.11 fuel is supplied by the accelerator pump circuit.

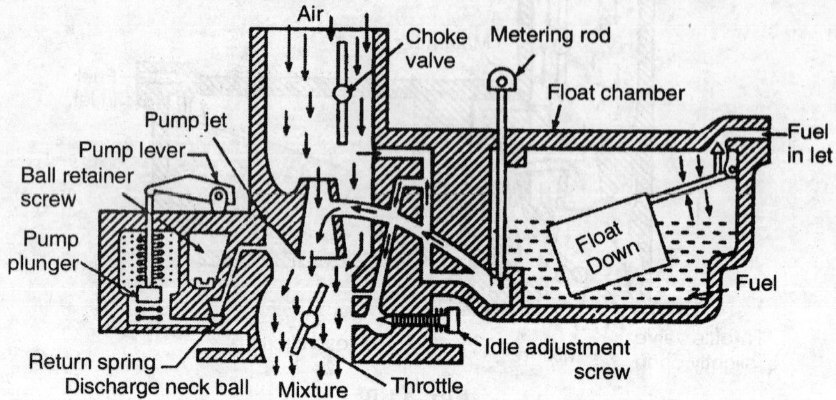

Fig. 11.11 Accelerator pump circuit.

The system consists of a pump plunger which can be forced downwards by means of a lever. The pump lever is linked with throttle valve. When the accelerator pedal is depressed to open the throttle valve then at the same time the pump lever also pushes the pump plunger downwards, thus forcing and delivering the additional fuel through this circuit. This additional fuel mixes with the air passing through the carburettor and ensures the desired richness of the mixture.

(vii) Choke Circuit. In starting, the engine needs a very rich mixture. But due to low speed of the cranking, the speed of the air passing through the carburettor horn also remains low. Thus this speed of the air is not sufficient to produce a desired vacuum at the venturi as well as below the throttle valve. Secondly due to cold condition of the carburettor and the engine, the fuel passing through the starting circuit is also not being fully vaporized. Therefore, under this condition to start the engine easily, carburettor is provided with a choke valve at the air horn as shown in Fig. 11.12.

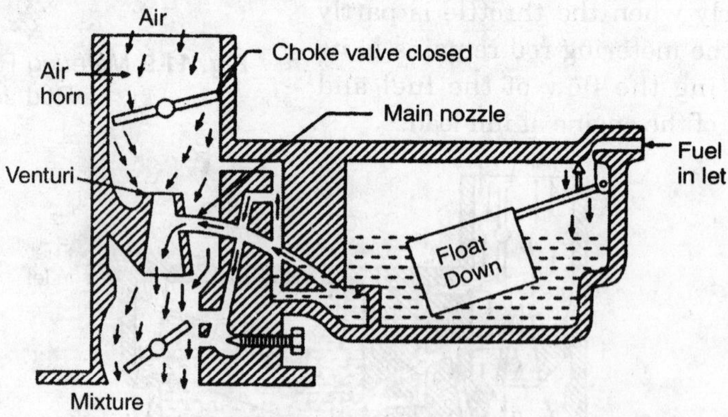

Fig. 11.12 Choke Circuit.

In starting when the chock valve is closed by manually or by automatic device then a small amount of air passes with high speed from the edges of the valve. This air develops a high vacuum in the air horn thereby discharging a heavy stream of fuel from the main nozzle as shown in Fig. 11.12. Therefore the fuel delivered by means of this circuit produces a direct air fuel mixture which is sufficient to start the engine in cold condition. After starting when the speed of the engine is increased then this choke valve must come to the off or open position automatically for obtaining a regular supply of air to the carburettor.

11.5 FUEL SUPPLY SYSTEM IN COMPRESSION IGNITION (C. I.) ENGINES

Fuel injection system is one of the important system of the C.I. engines. It supplies, meters, injects and atomize the fuel. Proper functioning of the fuel system greatly affects the engine performance.

The complete layout of diesel fuel system is clear from Fig. 11.13. The diesel fuel is stored in a main fuel tank which consists of a fuel-level gauge and in air-vented filter cap. In most of the vehicles this main fuel tank is installed lower than the engine level therefore a 'fuel-lift pump is necessary in the system. The fuel-lift pump is generally of mechanical type which functions by means of vacuum of a diaphragm.

The usual system is omitted the lift-pump and only it consists of a diaphragm-type 'feed pump' operated by the injection pump's camshaft. The feed-pump draws fuel from the 'main tank through a preliminary filter thereby obtaining the clean fuel. The feed-pump delivers the fuel to a 'main filter' which is provided with a relief valve. The relief valve acts as a safeguard against excessive pressure of the fuel and discharges the excessive fuel back to the main tank.

In recent installations the preliminary filter is replaced by a plug filter and the system consists of two fuel filters which are provided between the delivery side of the feed-pump and the inlet side of the 'injection-pump'. The primary filter takes out the greater part and the secondary filter takes out the finest particles of any dirt available in the fuel. Therefore, in this way the complete fuel system includes the main fuel tank, filters, fuel lift pump, fuel injection pump, injectors, low and high pressure pipe-lines etc. as shown in Fig. 11.13.

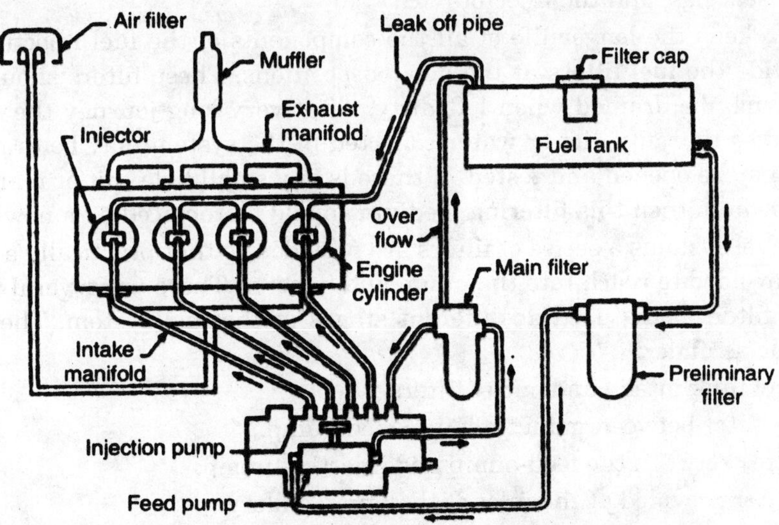

Fig. 11.13 The complete fuel system of a diesel engine.

11.6 COMPONENTS OF FUEL SYSTEM

Main components of a C.I. engine fuel system are: fuel tank, fuel filters, air cleaner, fuel injector pumps, nozzles, fuel gauges, and governors.

Fuel Tank

It is a metallic container mounted at a sufficient height above the fuel injection pump, for flowing the fuel by gravity. Where this is not possible, fuel lift pumps are used to create the gravity head.

Fuel Feed Pumps

These are used to supply fuel from the fuel tank to the injection pump. These are operated by a special cam on the injection pump's camshaft.

Fuel Filters

Fuel supplied to the fuel injection must be free from solid particles of dust and dirt, for good performance of the fuel system, otherwise it may adversely affect the fuel injection pump, nozzles, delivery valves and seatings etc. Therefore for long life of these components fuel filters are provided, and should be inspected from time to time and cleaned or replaced when necessary. There are several types and their combinations of filters are used at different stages in the fuel system.

11.6.1 Diesel Filter

As the fitting of the pump plungers and pump barrels is very precision, therefore, it is necessary that the fuel being forced by the pump must be free from the solid particles of dust and dirt. If greater plunger and barrel wear occurs in the pump, it means a breakage of the filter in the fuel system. The presence of dirt in the fuel is also harmful for the delivery valves, their guides and seatings, and the injectors.

Therefore to keep the longer life of all the components of the fuel injection system, it is essential to provide the fuel filters at the desired positions. These filters should be inspected timely and they may be drained when left dirty. After very long journey the sediment plugs should be opened so that any dirt or water collected in the system may be drawn off. On any doubt the filter may be opened and tested. If there is any small hole, cut or tear on the surface of the filtering medium then this filtering medium should be replaced by a new one.

In modern fuel systems a series of filters are provided so that practically a finest abrasive particle may be avoided to reach into the pump and injector. There are several types and their combinations of filters being used at different stages in the fuel system. The main types of fuel filters include as under:

(*i*) Main fuel tank intake and outlet filters.

(*ii*) Primary filter between main tank and feed-pump.

(*iii*) Main filter between the feed-pump and injection-pump.

(*iv*) Small filter provided at the inlet of the pump gallery.

(*v*) Small edged or disc-type filter provided at the inlet of the injector.

All the above type of filters are not provided in one type of fuel system but are used where felt necessary.

11.6.2. Microfilter for Diesel Fuel

Refer Fig. 11.14, this type of filter removes the finest particles that may have passed the primary filter. It consists of felt-element filter and mostly used on C.I. engines. The element is made of a series of felt-pads built upon a perforated support tube. As shown by the direction of arrow in Fig. 11.14 the fuel enters from the top left side and passes inside of the element and filtered fuel is taken out from the top right side connections.

The top-cover, body and filter element can be tightened by means of a wing nut provided at the top of the central stud. A felt ring is retained at the bottom of the perforated tube with the help of a spring retainer. A drain nut is provided at the bottom of the body to remove sludge when desired. The filter element can be cleaned with petrol, three or four times after that it should be replaced by a new one. When servicing the filter element always use a soft brush to remove the dirt. When assembling the unit check that all the exposed openings are sealed with clean plugs and corks.

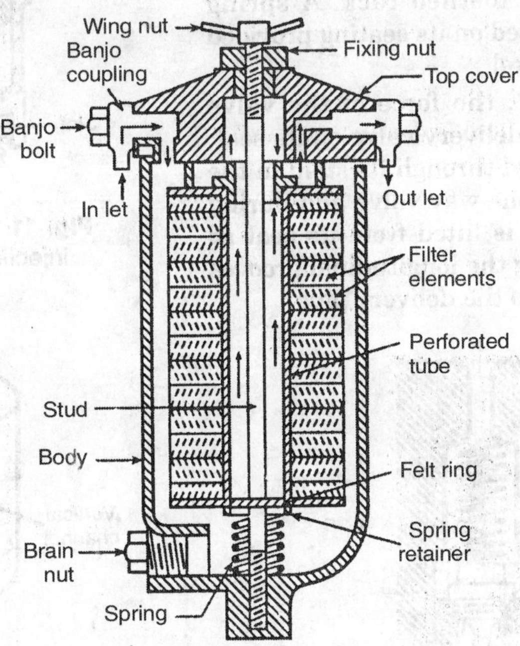

Fig. 11.14 Microfilter for diesel fuel.

11.7 FUEL INJECTION PUMP

The purpose of the injection-pump is to deliver the metered amounts of fuel at exactly the correct moments to the spring loaded fuel injector. The delivery pressure of this pump is of 150 to 175 atmospheres which is sufficient to open the spring loaded valve of the fuel injectors, thereby allowing the high pressure fuel to spray into the combustion chambers.

The mechanism and the most important parts of the single elements pump are shown in Fig. 11.15. It comprises a plunger and barrel ; delivery valve and its seating, which are always kept together and the individual components of which are not interchangeable. The barrel consists of two inlet parts through which the fuel enters from the gallery to the barrel. The plunger of the pump consists of a vertical channel extending from its upper edge to an annular groove, the top edge of which is cut in the form of a helix as shown in Fig. 11.16.

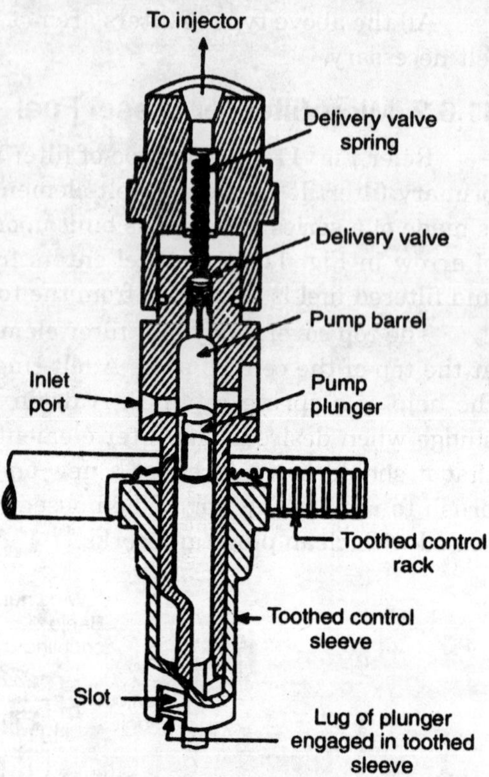

Fig. 11.15 Mechanism of an injection pump element.

In recent designs, a plunger consists of a central hole instead of the channel and a helical groove instead of helix and annular groove. The lug or lower part of the plunger is engaged in the slot of the toothed sleeve. The rotary movement of the toothed sleeve with the plunger is controlled by means of sliding movement of the toothed rack. A spring loaded delivery valve is fitted on its seating provided at the top of the pump barrel.

Referring Fig. 11.17, the force of the valve spring always presses the delivery valve on its seat. The delivery valve is guided through its stem in the valve holder. It acts as a one way valve and during the fuel delivery stroke it is lifted from its seat so that the fuel can flow along the longitudinal grooves and over the valve face into the delivery pipe.

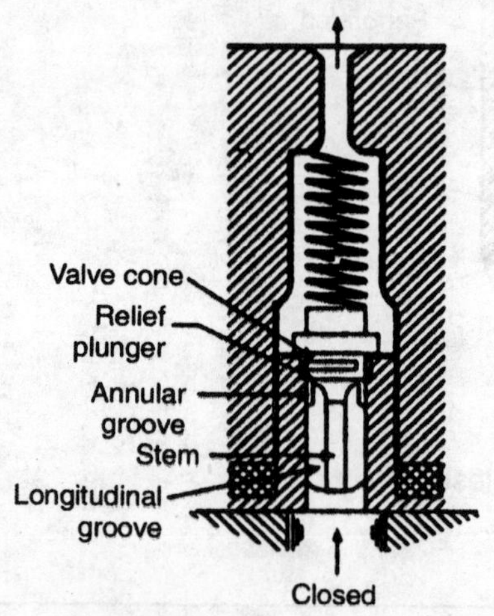

Fig. 11.16 Details of diesel pump plunger.

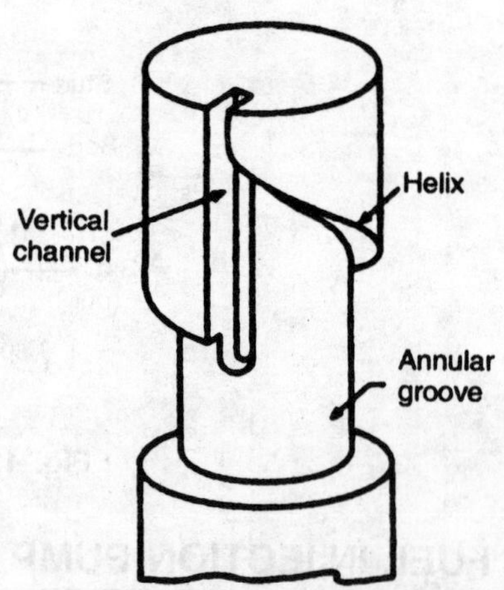

Fig. 11.17 Delivery valve.

For working principle of the pump, say the plunger is at its bottom dead centre starting position. Under this position the two fuel inlet parts are uncovered from the sides of the plunger thereby allowing the fuel to enter the barrel under feed-pump pressure. As the plunger moves upward the edge of it cuts off the fuel supply. Now the space above the plunger is full of fuel which, with the further upward movement of the plunger, is forced at a high pressure upwards, thereby lifting the delivery valve against the pressure of its spring and hence entering the fuel through the pipe to the injector. This delivery of the fuel from injection pump remains continue till the helix of the plunger opens the port in the barrel, thus communicating the fuel above the plunger to the gallery, via the vertical slot and thereby cutting off fuel delivery to the injector. When the fuel delivery to the injector is increased or decreased then with rack and pinion mechanism the movement of fuel cut-off by the plunger can be carried to any desired extent.

Similarly if the plunger is rotated in such a way that the vertical groove is opposite the right port, then there will be no pumping action because under this position the top of the plunger remains connected with the inlet port ; this corresponds to the zero delivery and this position is called 'stop' position of the pump.

11.7.1 Fuel Injector

The function of an injector is to convey the measured amount of fuel under the higher pressure from injection pump to combustion chamber in spray form. This unit consists of injection nozzle, its holder and pipe connections and is located in the head or side of the combustion chamber. All the joints of the unit are made leak-proof and a disc type filter is provided at the inlet port of the unit.

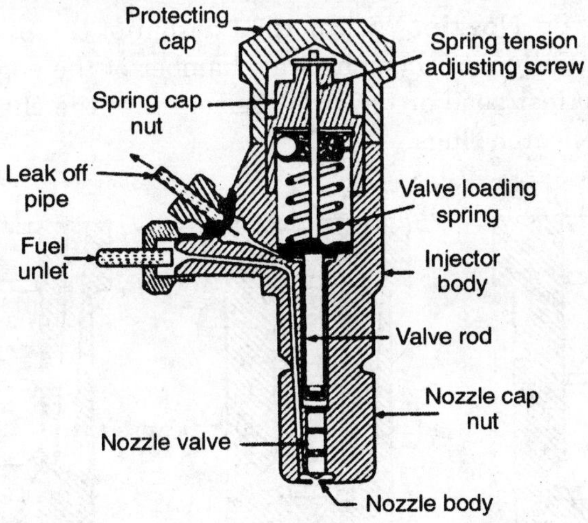

Fig. 11.18 Construction of fuel injector.

Referring Fig. 11.18, the unit consists of a spring loaded nozzle valve, which can slide by a small amount within its nozzle body and guide. The valve rod provided above the valve helps to hold the valve on its seating. The tension of the spring can be increased or decreased as per desired injection pressure by means of spring tension adjusting screw.

For working when the fuel from injection pump enters the lower-end of the nozzle valve through a vertical cavity shown on the left of the nozzle axis, then due to highest presence of the fuel, the valve lifts up from its seating thus the metered fuel escapes in the form of a spray, into the combustion chamber. When the injection pressure falls down then the spring load immediately closes the nozzle valve, now the nozzle is ready for the next injection. The small leakage of fuel upwards, past the valve and valve rod is separated by means of a 'leak-off fuel pipe.

11.7.2 Injector Nozzles

The injector nozzle is secured to the body of the fuel injector by means of a cap nut. The nozzles may be of various kinds according to their stem length, number of holes and their positions, shape of valves and their spraying angles. But the most common designs used in automotive engines are as given under:

***(i)* Pintle Nozzle.** The stem of the nozzle valve is extended to form a pin or pintle which protrudes through the mouth of the nozzle. The size and shape of the pintle can be varied according to the requirement. The spray cone in this type is generally kept at 60° angle.

***(ii)* Single Hole Nozzle.** In this type the mouth of the nozzle body consists of a single hole at its centre, which is closed by the nozzle valve. The diameter of the hole can be from 0.2 mm upwards. If this hole is made at an angle with the nozzle axis then the type is known as conical end type.

***(iii)* Multi-Hole Nozzle.** This type consists of a number of holes bored in the side of the valve seating. The number of holes may be up to seven. The number of holes, their size and angle depends upon the requirements of engine. The diameter of the holes can be from 0.2 mm upwards and the hole angle may be from 20° upwards.

***(iv)* Pintaux Injection Nozzle.** With normal design of the pintle or single hole nozzle the fuel is sprayed tangentially into the spherical chamber of the engine. In this way the fuel is not sprayed into the hottest zone or towards the centre of the chamber. Therefore for cold starting the engine needs heater plugs.

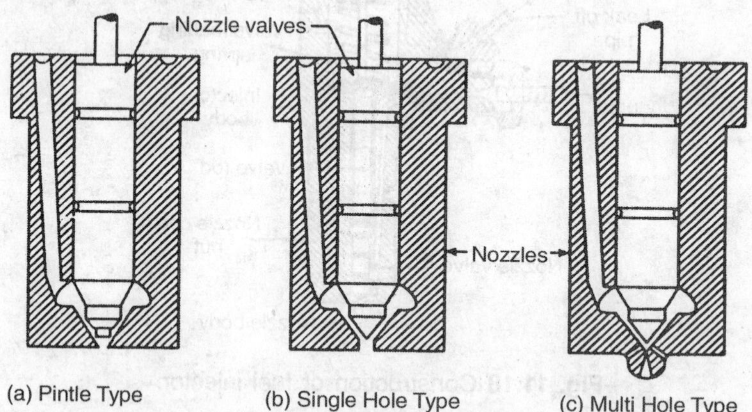

Fig. 11.19 Fuel injector nozzles.

Therefore to overcome the heating difficulty of the engine, the injection nozzles are designed of pintaux type as shown in Fig. 11.20. At starting, the nozzle valve is lifted slightly thus the pintle hole is not cleared and the fuel is only discharged through the auxiliary hole into the central hot zone, thereby obtaining better cold starting performance. When engine reaches to its normal speed then the needle valve is lifted from the pintle hole thereby allowing the whole of the fuel to pass through the pintle hole and entering the chamber tangentially.

In this way this type of nozzle provides both the advantage of cold starting as well as of normal running of the engine.

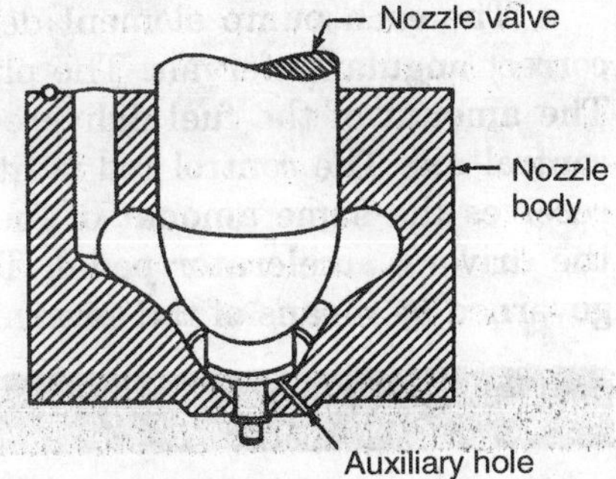

Fig. 11.20 Pintaux injection nozzle.

11.7.3 Injection Pump Governors

As the automotive engines run at speeds selected by the driver, but in order to prevent stalling when the vehicle is stopped and to ensure that the permissible top speed is not exceeded for which it is designed. Therefore all automotive diesel-engine injection pumps operate in conjunction with a governor which is required to control the injected quantity of fuel so that the engine neither stalls when idling nor exceeds the prescribed maximum speed. The governor automatically adjusts the control rod of the fuel injection pump and hence the amount of fuel injected also remains in accordance with variations in load (see Fig. 11.20).

In this way the functions of a governor are that the engine must idle smoothly, it must neither stall, race surge nor hunt, but must maintain a speed as nearly constant as possible.

In addition, it must never permit exceeding the maximum speed prescribed by the engine manufacturer.

11.8 MULTIPLE PLUNGER FUEL INJECTION PUMP (MPFI)

This pump is mostly used for commercial and other high-speed diesel engines. It consists of a separate pumping unit for each cylinder of the engine. All the units are grouped together in a common casting as shown in Fig. 11.21. The plungers are operated by a camshaft carried on roller bearings located in the casing. The movement of the cam is transmitted to the plunger through suitable roller tappet which is held in contact with its cam by a helical spring.

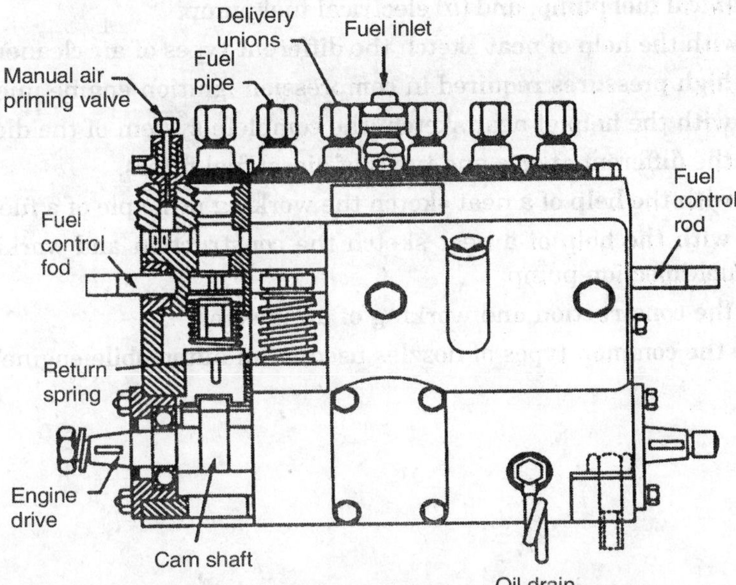

Fig. 11.21 Multiple plunger fuel-injection pump.

The each pump element delivers the fuel to the respective fuel injectors at exactly the correct angular intervals. The plungers are arranged in vertical position above the camshaft. The amount of the fuel delivered by each pump element is regulated by means of common control rod. The control rod is fitted with the pinions in such a manner so that each injector receives the same amount of fuel. The control rod is conrolled by suitable machanism from the driver's accelerator pedal. The limits of the maximum and minimum fuel delivery are governed by means of the governing device connected with the control rod.

QUESTIONS

11.1 (*a*) What are the different categories of I.C. engine fuels?
(*b*) What are the functions performed by the carburettor?

11.2 What is meant by lean mixture? a rich mixture? and how do they differ in the air-fuel ratio? Explain.

11.3 How does the throttle control the speed of the engine? Explain.

11.4 (*a*) Explain how a rich mixture is produced in a carburettor for starting purposes.
(*b*) What is the function of the idling system of a carburettor?

11.5 List the essential parts of the fuel-feed system of an automobile.

11.6 (*a*) Why do gasoline tanks have air vents?
(*b*) What prevents the fuel from returning to the tank on the return stroke of the diaphragm of a pump? Explain.

11.7 Why is a filter or fuel strainer attached to a fuel pump?

11.8 Why a clean air is required in the I.C. Engines? Explain different methods of cleaning the air, used in I.C. Engines.

11.9 Explain with the help of neat sketches the construction and working principle of, (*a*) mechanical fuel pump, and (*b*) electrical fuel pump.

11.10 Explain with the help of neat sketch the different types of air cleaners.

11.11 Why are high pressures required in compression ignition engine injectors?

11.12 Describe with the help of neat sketch, the complete system of the diesel fuel.

11.13 Describe the different stages and types of diesel fuel filters.

11.14 Describe with the help of a neat sketch the working principle of a fuel-injection pump.

11.15 Describe with the help of a neat sketch the construction and working of a multiple plunger fuel-injection pump.

11.16 Describe the construction and working of an injector.

11.17 What are the common types of nozzles used in an automobile engine?

AUTOMOBILE ENGINEERING

Chapter **12**

Learning Objectives

After studying this chapter, the students will learn:

- Various components of an automobile vehicle, such as frame, suspension system, steering, axles, wheels, tyres, engine, clutch, gear box, prolleler shaft and universal joints, differential, brake and braking systems.
- Cooling system and their types.
- Automobile electrical system including battery and its construction and working.

12.1 INTRODUCTION

First automobile tricycle propelled by an I.C. engine was developed in 1885 in Germany. Later on between 1890 and 1895 a car was developed almost simultaneously in America and France, but these came in production for sale after lot of improvements in around 1906.

An automobile is essentially a transportation equipment and is made up of a frame supporting the body, prime-mover and transmitting units. The frame is further supported by tyres and wheels through springs and axles. Power is generated by the engine and is delivered by the transmission and axles through the clutch or fluid coupling.

Automobile vehicles can be classified in 3 general categories, namely : Single unit vehicles (passenger vehicles or load carriers), Articulated vehicles, and the heavy tractor vehicles.

An automobile is a combination of large number of parts, and can be divided into 'chassis' and 'body' as the two main constituents. Chassis is a main machine portion and have frame, wheels, axles, springs, shock absorbers, engine, clutch, gear box, propeller shafts, differential, steering, brakes, battery, radiator, silencer etc. Body is the part where the passengers have their seats or the luggage and cargo to be transported is placed. The body is fixed over the frame and contains all the parts and accessories needed for the comfort of the human beings. In passengers vehicles the driver's seat is in the same compartment where passengers are sitting, whereas in trucks and other load carriers, the driver's compartment is separated from the main body carrying load and cargo.

12.2 CHASSIS

This is a main structure of a vehicle, and is generally used to denote the complete vehicle except the body for the heavy vehicles having a separate body. In light vehicles, the body is integral part of the chassis. Thus all the major units necessary to propel the vehicle, direct its motion, stop it, and to run smoothly over varying road surface are provided in the chassis (see Fig. 12.1). It is a main machine portion and includes :

1. Frame
2. Suspension
3. Steering mechanism.
4. Axles, Wheels, and Tyres
5. Engine, clutch and gear box.
6. Propeller shaft and Universal Joint.
7. Cooling system.
8. Differential.

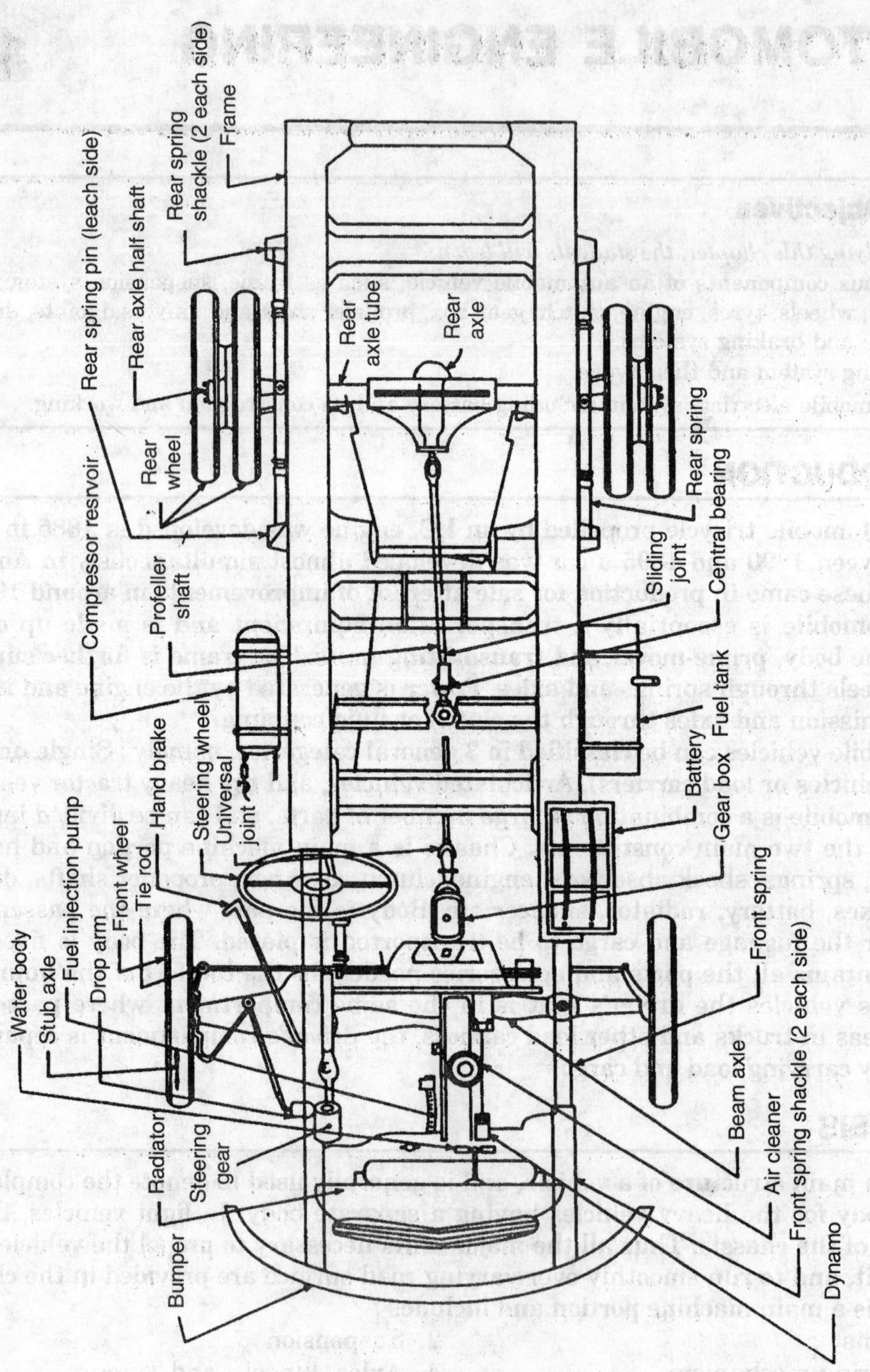

Fig. 12.1 Layout of a Chassis.

9. Brakes and braking system.
10. Fuel tanks and pipes.
11. Storage battery, starting and Generating System
12. Silencer

12.3 FRAME

It is the structural centre of the vehicle for carrying load, to furnish supports for transmission system, the body, and other units such as shock absorbers, springs, shackles. It also maintains the correct relationship between the other related parts in order to secure their normal function and freedom from strain and wear caused by operation in misaligned condition.

Frame has following functions to perform

(*i*) To support the power plant (engine), body, transmission system, springs, shock absorbers and some other miscellaneous units.
(*ii*) To maintain the correct spacing and relationship between the other units.
(*iii*) To withstand all the dynamic and static loads without undue twist and deflection.
(*iv*) To get normal functions of supported units and freedom from stress, strain and wear caused by operation in misaligned condition.

Frame should be able to take load of

(*i*) Impact loads due to sudden collision of the vehicle.
(*ii*) Inertia loads due to applying brakes of the vehicle.
(*iii*) Vertical loads due to crossing the bump or a hollow, by the vehicle.
(*iv*) Loads for long periods due to taking turn and applying the brakes at a time.

Construction

Frame consists of two mild-steel longitudinal members of channel or box section cross connected by channel, box or tubular section cross-members by bolting, riveting or welding system. The longitudinal-members are brought closer together at the front to give an adequate steering lock, and at the rear upswept to clear the axle movement. The depths of section of the longitudinal-members are frequently increased at the centre where the greatest bending loads occur.

On the sides of the frame some brackets and dumb irons are provided to support the body, springs, bumpers and bearings for spring shackles.

The engine is mounted at the one end of the frame, usually on the front end, and supported on the frame at the three places with the help of rubber blocks to avoid the effects of road-shocks on the engine and engine vibrations on the body.

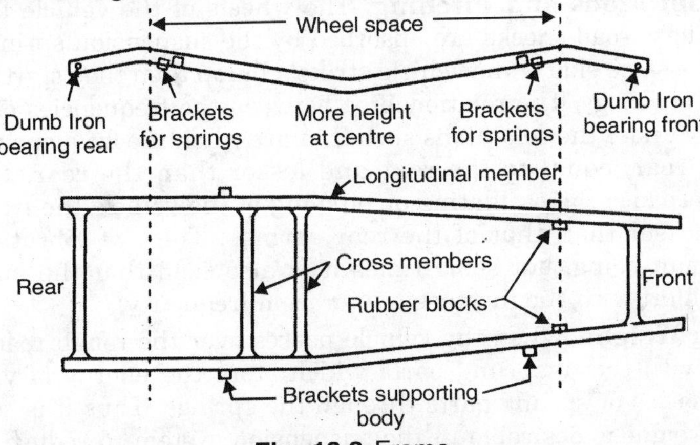

Fig. 12.2 Frame.

The different cross-sections used for the members of the frame are as given under ;

(i) Channel sections. (ii) Box sections. (iii) Tubular sections.

12.4 SUSPENSION SYSTEM

The frame as well body of the vehicle is attached to the rear axle and the front axle by springs. These springs damp the road shock transmitted to the body structure by the wheels when they travel over the road. In this way the springs are the protecting units supported directly by the frame of the vehicle. Therefore all the parts which perform the function of protection are collectively called a suspension system. These springs are generally of the laminated leaf type but may be of any other special type according to the need. These springs provide a best suspension system to the vehicle thereby protecting the passengers and load from jerks. The suspension system of vehicle is divided into (i) the front-end suspension and (ii) rear end suspension.

Objects of Suspension

(i) To provide the comfort and safeguard the occupants from road shocks.

(ii) To prevent the transmission of the road shocks resulting into stresses to the vehicle components.

(iii) To give the stability to the vehicle in case of rolling or pitching, while in motion.

(iv) To provide the particular height to body structure and to bear the torque and braking reactions.

Rate and Frequency

The flexibility of leaf spring depends upon the number of leaves containing it. When the pressure is applied to the ends of the spring then the leaves slip on each other to come back to the original radius of curvature. Therefore in this way the flexibility of a spring depends upon a property called "rate". The stiff spring has higher rate than soft spring. Thus the "rate" of a spring is the weight applied to deflect it by unit measurement. The other comparison of the springs is "frequency". The spring which returns quickly to its original position rapidly after being deflected is said to have higher frequency than which returns slowly. To avoid the friction, lubricant must be placed between the spring leaves.

Basic Requirements

(i) **Vertical Vibrations and Pitching.** The wheels of the vehicle fall or rise with the road undulations. These road shocks are absorbed by the suspension springs. When front and the rear wheels of the same side of the vehicle strike a bump then they start vibrating according to their time gap of strikings. The relationships between the frequencies of the front and rear suspensions can vary under the conditions such that are when the frequency of front suspension is higher than the rear, equal to the rear and lesser than the rear. It is seen from the experiments that to reduce the oscillating or pitching of the vehicle the frequency of the front springs should be lesser than that of the rear springs. The low vibration frequencies are comfortable for human being, but some times it is also seen that the maximum amplitude decreases the discomfort with the increase of vibration frequency.

***(ii)* Unsprung Weight.** When the vehicle passes over the rough road surfaces then the wheels rise or fall with the unsprung parts which store the energy of vibrations and then transmit it to the chassis or sprung parts through the springs. Thus it is seen that the lowest possible unsprung weight is desirable in the suspension system to reduce the vibrations and the inertia forces of wheels and axles.

***(iii)* Rolling.** When a vehicle turns the curve then the centrifugal force acts at the centre of gravity of the vehicle and also away from the centre, of the curve. The road resistance acts to the inside at the wheels. Therefore due to this force the vehicle tends to roll about the longitudinal axis. This type of turning of the vehicle is called rolling. The position of the axis of rolling depends upon the types of springs and the manner in which the vehicle is sprung.

***(iv)* Brake Dip.** When the brakes are applied to a vehicle then the front side or nose of the vehicle tends to be lowered or to dip. This brake dip depends upon the height of the centre of gravity of the vehicle from the ground as well as the wheel base and the quality and type of the suspension springs.

Leaf Spring

The details and the construction of the laminated leaf spring is shown in the Fig. 12.3. The longest or main or master leaf of the spring is coiled round at the ends to form eyes. These eyes embrace the shackle pins, and are also bushed with bronze bushes. Some times the second leaf also partly embraces the eye of the main leaf as a safe guard.

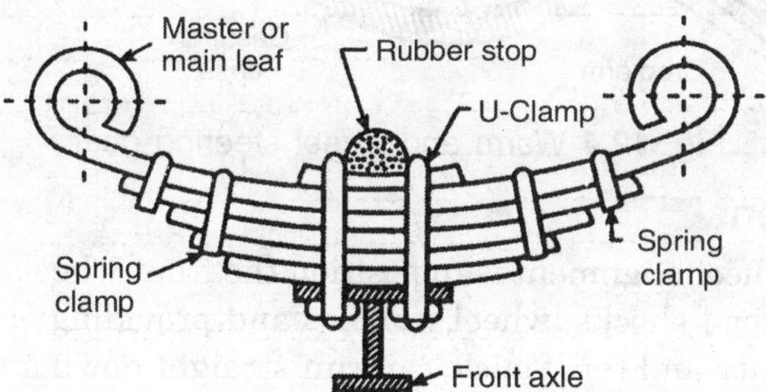

Fig. 12.3 Leaf spring.

12.5 STEERING

For effective control of the vehicle during entire range of speed on wide variety of road surfaces providing bumps and bounces, proper steering is necessary vehicle should be steered straight and maintained at that position and should be deviated only at the driver's will without much effort. The steering system provides directional changes to the moving automobile in such a way that the wheels undergo true rolling motion without slipping or scuffing. Moreover, the steering must be light, and stable with a certain degree of self-adjusting ability. Steering mechanism converts the rotory motion of the steering wheel into the angular turning of the front wheels, and also multiply the driver's effort with leverage or mechanical advantage for turning the wheels fairly easily.

Worm and Wheel Steering Gear Mechanism

This mechanism is shown in the simplified sketch Fig. 12.4. The system consists of a worm wheel which is carried in bearings in a cast iron case. The case is made in halves. The outer end of the worm wheel is fixed to a drop arm which is having a ball end to connect the side rod. The side rod is connected to the steering arm which is fixed to the stub axles. The worm which is keyed on to a steering shaft have a mesh with the worm wheel. The steering shaft is carried in bearing in the casing. The steering wheel is mounted at the upper end of the steering shaft. When driver rotates the steering wheel then drop arm moves either backward or forward direction by means of worm and worm wheel. This motion results in motion of the stub, axles.

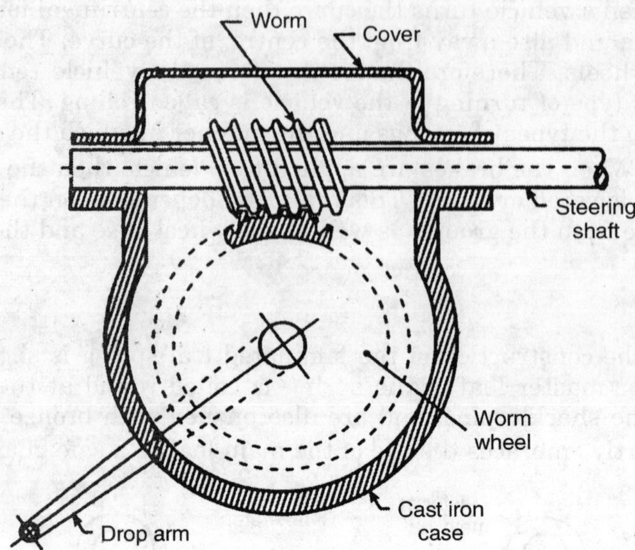

Fig. 12.4 Worm and wheel steering gear.

12.5.1 Wheel Alignment

The purpose of the wheel alignment is to position the front wheels and steering mechanism to reduce the tyre wear, road shocks, wheel wobbles and providing ease in steering control. It provides directional stability and the vehicle can run straight down a road under all conditions and at all loads. The other factors those effect the wheel alignment are as frame alignment, position of the rear axle, spring condition and condition of the shock absorbers. Therefore for smooth running of the vehicle, it is very necessary to consider and understand the factors which influence the stability and control of it. The balance of wheels, inflation of tyres and brake adjustment are also considerable in cases of wheel wobble, change of rolling radius and to pull the vehicle to one side respectively. For proper alignment, camber, toe-in, and kingpin inclination must properly be set.

Front End Geometry

The angular relationship between the front wheels, front wheel attaching parts and the vehicle frame are concerned with this term. It consists of the tilt of the front wheels from vertical, kingpin away from vertical and the toe in of the front wheels. The factors of the front end geometry are considered for the steering ease, steering stability, tyre wear and the riding qualities of the vehicle. The front end geometry depends upon the following factors :

1. Camber. The camber is the tilt of the front wheels from the vertical plane. When the tilt is outward at the top then the camber is positive and when the tilt is inward at the top the camber is called negative. The wheels are given a positive camber to start so that when the vehicle is loaded and rolling along on road, they automatically come to a vertical position. The positive camber is shown in the Fig. 12.5 which is measured in degrees from the vertical plane. The camber angle should not exceed 2°, while running any amount of camber, positive or negative tends to cause more rapid tyre wear because under this condition more of the load is shared on one side of the tyre. Therefore it is clear that the tyre life will be maximum when the camber angle in running condition is zero with average load.

Similarly unequal camber in the front wheels causes the vehicle to roll in the direction of the wheel having the greater camber. With excessive camber a tyre does not come is correct contact with the road which causes it to wear only on the side.

2. Kingpin Inclination. The kingpin inclination is the inward tilt of the kingpin from the vertical plane as shown in Fig. 12.5. The advantages of this inclination are, first it tends to keep the wheels straight ahead and helps the return of the wheels to the straight position after a turn has been completed, second it reduces tyre wear, and third, it reduces steering effort when the vehicle is stationary. The kingpin inclination, in combination with caster, is used to provide directional stability in the vehicle. Because this inclination causes the vehicle to move up and the front wheels are swing away from straight ahead, then the weight of the vehicle tends to return the wheels back to straight ahead after turn is completed. The amount of this inclination is generally kept 7 to 8 degrees.

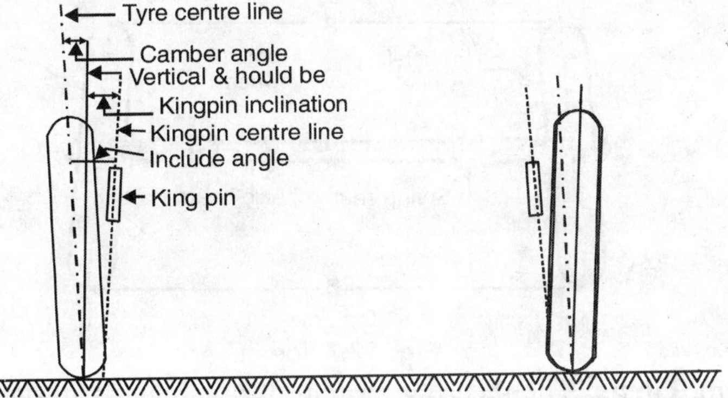

Fig. 12.5 Positive camber angle and kingpin inclination.

3. Caster. In addition to being tilted toward the centre of the front axle for kingpin inclination, the kingpin may be titled forward or backward from the vertical and in the plane of the wheel. The backward tilt from the vertical as shown in the Fig. 12.6 is called positive caster and for negative caster the kingpin is tilted forward in the same plane.

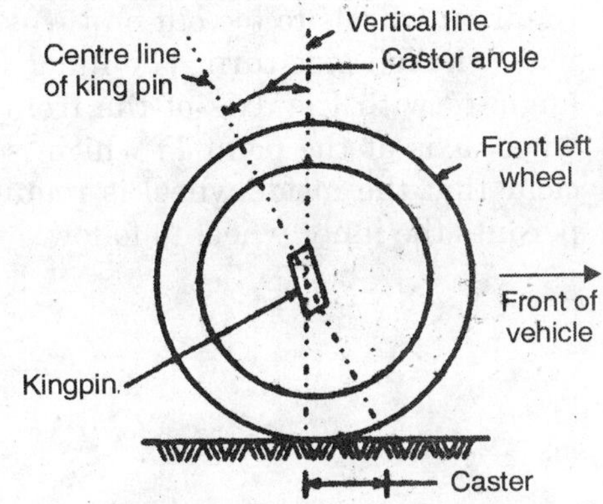

Fig. 12.6 Positive caster.

This tilt of kingpin from the vertical is called the caster angle in degrees. For good results the caster angle is kept about 3°. The positive caster on the front wheels provides directional stability to the vehicle. Another effect of the positive caster is that the vehicle tends to roll out or lean out on turns. But with negative caster the vehicle tends to bank or lean in on turns.

With positive caster when the right turn is made, the left side of the vehicle is lowered and the right side of the vehicle is lifted, then the vehicle rolls. Thus the positive caster helps the centrifugal force in rolling out the vehicle which is not desired. With negative caster, the left side of the vehicle would lift during a right turn while the right side of the vehicle would drop. Thus the negative caster tends to roll in, the vehicle.

The positive caster, causes a tendency in the wheels to toe in because the body of the vehicle is lifted as the front wheels toe out and it is lowered when the toe in. With negative caster, the wheels would try to toe out.

The positive caster provides directional stability and tries to keep the wheels straight ahead. But more amount of positive caster increases the effort required to steer. Therefore modern heavy-duty trucks, tend to have a negative caster which makes steering easier.

4. Toe-in. The toe-in is the turning in of the front wheels from the straight head position. Referring Fig. 12.7, the wheels are viewed from the top, the distance between the front wheels is less at A than at B. The amount of toe-in generally does not exceed 3 mm. The purpose of it is to move the wheels perfectly straight ahead under running condition of the vehicle, to stabilize steering, to prevent sideslipping and excessive wear of tyres. Due to rolling resistance the small deflections come in the wheel support system, when the vehicle is moving. But the toe-in of the wheels in standing position tends to roll them parallel when the vehicle is moving.

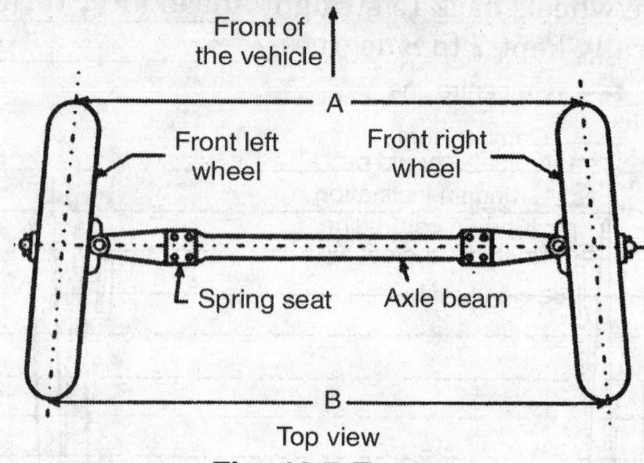

Fig. 12.7 Toe-in.

5. Toe-out on Turns. The steering system of a vehicle is designed to turn the inside wheel through a larger angle than the outside wheel when making a turn. This action causes the front wheels to toe-out on turns. Referring Fig. 12.8 where the front wheels are steered to make the left side turn. The inner wheel turns at an angle θ° and the outer wheel turn at an angle ϕ° with the axis of the front axle beam. The axes of the front wheels meet the rear wheel axis at the point *P*, which is called the instantaneous centre of the vehicle. Thus it is clear that the inside wheel is required to turn to a greater angle than the outer wheel. This permits the inner wheel to follow a shorter radius than the outer wheel.

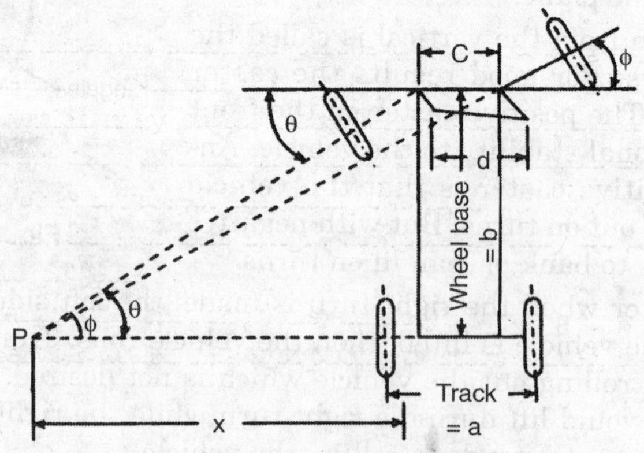

Fig. 12.8 Toe-out on turns.

Therefore in this way the steering angle of the inner wheel may have a maximum value of about 44°. This extreme turning position of the wheel is called lock position.

12.5.2 Wheel Balancing

The balancing of the front wheel is an important factor to make the steering easier. The unbalanced assembly of the front wheel produces tramp while running the vehicle. The vibrations of the rear wheels may affect the front suspension system. The unbalance of the wheel may cause of uneven tyre wear. For high speed vehicles the weight of the air pipe in the tube affects the wheel balance. Therefore tube tyres are marked a point where the tube outlet should be placed for proper balance.

The wheel balance can be classified in two ways, "static balance" and "dynamic balance". The wheel can be said statically unbalanced if its weight is not evenly distributed around the hub. Therefore when this wheel is rotated on a low-friction axis the heavy side comes to the bottom of it. If the weight around the hub is not evenly distributed in the same plane of rotation then a statically balanced wheel may be dynamically unbalanced as shown in the Fig. 12.9. Therefore dynamic unbalance of the wheel permits to wobble it, above the particular speed. Therefore it is necessary to provide the static and dynamic balance to each wheel.

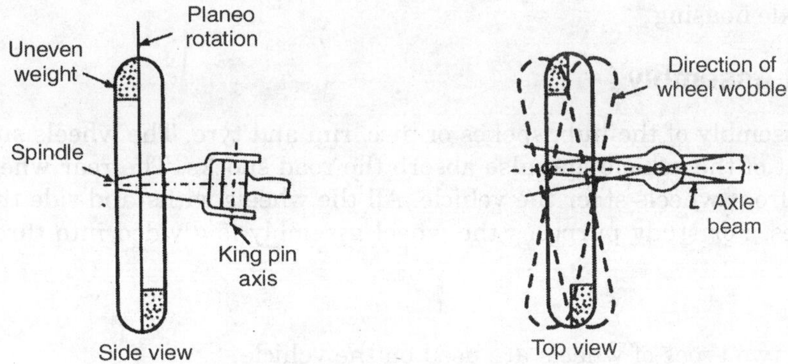

Fig. 12.9 Dynamic unbalance.

Power Steering

Power steering is used to reduce the steering wheel turning effort, turn sharp corners easily, negotiate winding roads, and manoeuvre the vehicles in confined space and at the same time to offer some resistance to make the driver feel and retain road feel or vehicle steering response. This solves the problem of loaded commercial vehicles in particular. For power steering a booster arrangement is provided for turning the steering wheel shaft. In most of the vehicles, hydraulic pressure operated power steering mechanism is adopted.

Power steering system basically consists of a oil pump, a valve, double acting power cylinder, an oil reservoir and steering gear box. The system becomes operative when the effort at the steering wheel exceeds a predetermined amount resulting the opening of valve, which directs fluid to the appropriate side of the cylinder for steering the wheels in the required direction.

12.6 AXLES, WHEELS AND TYRES

12.6.1 Axles

These carry the weight of the automobile.

Front Axle. These carry the weight of the front part of the automobile and then transmit it from the springs to the front wheels. In case of 4 × 4 vehicles, a live front axle as compared to dead axle has the additional function of transmitting the driving power taken from a transfer gear box to the front road wheels. The live front axles although resembling the rear axles have some difference at the axle half shaft ends where the wheels are mounted. In case of a dead front axle, axle beam of *I* or *H*-section made of alloy forged steel for rigidity and strength is used. These dead axles do not rotate with the engine in contrast to live axles.

Rear Axle. Rear axle or driving axles are used for transmitting power from the differential to the driving wheels. It not only acts as a bearing for the road wheels but provides attachments for springs. This supports the weight of the rear of the vehicle. Rear axles are of two types, namely, dead axle and live axle.

Dead axle do not rotate with the wheels but the wheels rotate on it. In case of front wheel drive car the rear axles are the dead axles which remain stationary with the wheels only rotating on them.

Live axles are attached to the wheel such that both of them rotate together. In this axle, the inner end of the axle shaft supports the differential *i.e.* the bearing is on the axle shaft supported by axle housing.

12.6.2 Wheel Assembly

It is an assembly of the hub, spokes or disc, rim and tyre. The wheels support the whole load and weight of the vehicle and also absorb the road shocks. The rear wheels transmit the power and the front wheels steer the vehicle. All the wheels withstand side thrusts and resist braking stresses. For study purposes the wheel assembly is divided into three parts–wheels, rims and tyres.

1. Wheels

Generally two types of wheels are used on the vehicle.

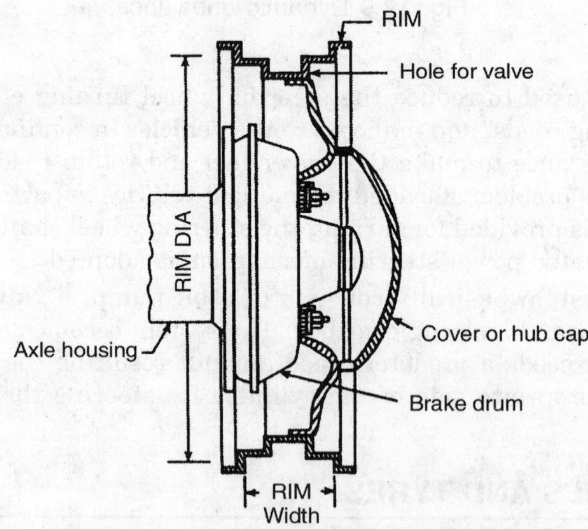

Fig. 12.10 Pressed-steel disc wheel.

***(i)* Wire-wheel.** It is very light wheel therefore generally used in sports cars. This wheel can be fitted and removed very easily. It has a separate hub which is attached to the rim by a large number of wire spokes. The spokes are arranged at different angles so that the combined tensions enable the forces due to weight, braking and the drive itself to be transmitted from hub to the rim. This construction also results an excellent cooling of the brake drums. Another advantage of this type is that wheel can be changed very quickly since there is only one nut to deal with. The hub is provided internal splines to mount on the axle shaft.

***(ii)* Disc wheel.** This type of wheel consists of a pressed steel disc instead of wire spokes. The disc may be rivetted or welded to the inner periphery of the rim. The disc also consists of narrow slots near the rim to encourage the cooling of the brake drum. This assembly is bolted to the brake drum as shown in the Fig. 12.10. The cover or hub cap is mounted on the disc by spring clips. This type of wheel is very simple, strong and cheap, therefore being used in most of the vehicles.

2. Rims

The rim of the wheel is so designed by which the mounting and dismounting of the tyre can be performed very easily. A "demountable rim" is one that can be removed easily from the wheel. The rims for motor vehicles are of four types (see Fig. 12.11) as given below.

***(i)* Drop-centre Rim.** The central portion of it is rolled to form a well. The well permits easy removal or mounting of the tyre. When tyre is removed the bead of one side is dropped into the well, while the opposite side of the other is pulled over the flange with tyre levers.

***(ii)* Flat-base Rim.** This type of rim is made in three pieces, main rim, side ring and locking ring. This rim permits the assembly of the modern tyres since their beads have little or no elasticity. When wheel is assembled then the tyre is pushed on the main rim and then the side ring is mounted on the bead of one side. The lock ring is fitted over the side-ring in the groove made for locking purpose on the main-rim. This type of rim is used on buses, trucks and other heavy duty vehicles.

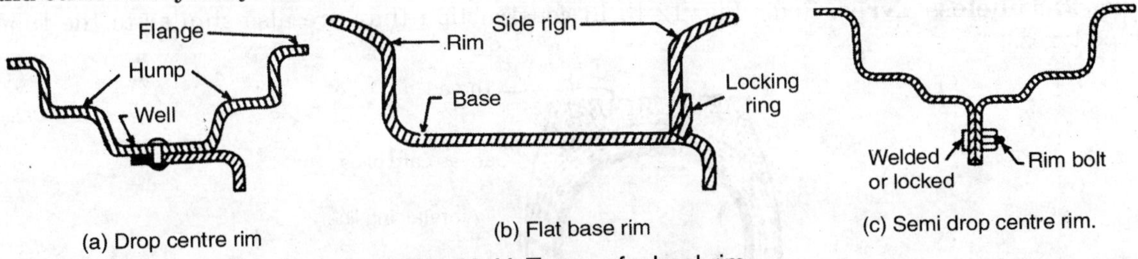

Fig. 12.11 Types of wheel rim.

***(iii)* Semi-drop Centre Rim.** This type of rim is made in two split type pieces. One piece contains four or six bolts welded or locked to it and other having holes to pass the bolts. The tyre can be removed from the wheel by pressing the beads into the well and then separating the two half rims. This type of rim is used on light duty vehicles.

***(iv)* Divided Type Rim.** This type of rim is also having flat base but is made in two pieces which can be bolted together. It is mainly used on military vehicles.

3. Tyres

The tyres are placed and designed around the rims of the vehicle to cushion it and its load and the passengers from the shocks and vibrations due to the road inequalities. The assemblies mounted above the axles can be protected by the springs and shock absorbers but the unsprung parts and units are protected only by the resiliency of the tyres. In addition to the comfort and protection the tyres provide fractional contact between the wheels and the road, thereby a good traction is secured. Tyres are of basic two types, solid and pneumatic. The solid tyres have very limited usage on automobile. The pneumatic tyres are mainly of two types, tubed tyres and tubeless tyres.

Types of Tyres

The pneumatic tyres are classified as given under.

1. Tube Tyres. Refer Fig. 12.12 where sectional view is shown of the tubed-tyre type wheel. This type of tyre consists of an inner endless tube of rubber, containing a valve through which air is filled. The air is retained under pressure in the tube so that it may act as a cushioning medium. Both the tube and the tyre are mounted on the wheel rim. For saving the life of the tube a rubber flap is used between the contact of the rim and the tube.

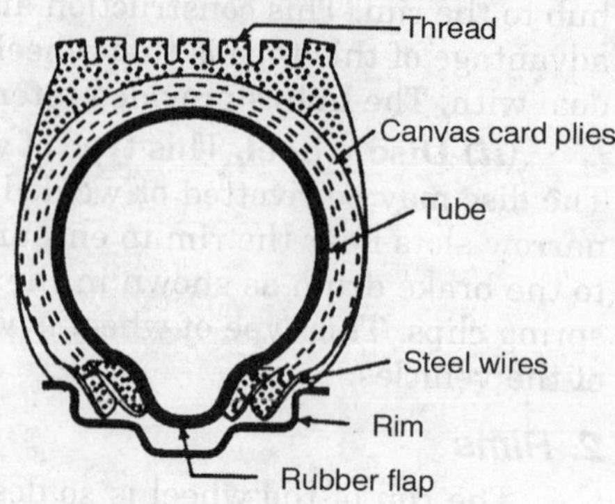

Fig. 12.12 Tubed tyre.

According to the pressure and volume of air in the tube, the tyres, are further classified into three categories.

(*a*) **High pressure tyres.** The maximum pressure in these tyres is up-to 5.6 kg/cm^2. They are used in commercial vehicles and are heavier in construction. They have more plies and constructed in square section with flat tread.

(*b*) **Balloon tyres.** The pressure range of these types of tyres is from 1.6 to 2.8 kg/cm^2. These types have larger sections than the highpressure tyres. They are generally used on cars.

(*c*) **Super-cushion tyres.** They are provided pressures ranging from 1 to 1.6 kg/cm^2. They are larger in section than balloon tyres. They are used on passenger cars and provide more soft ride.

2. Tubeless Tyres. Refer Fig. 12.13 in construction thay are also similar to the tubed type tyres.

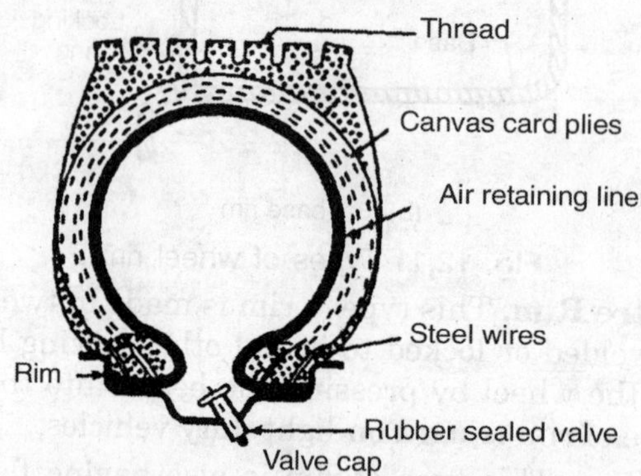

Fig. 12.13 Tubeless tyre.

12.6.3 Tyre Care and Maintenance

Following are the important factors which affect tyre life and performance of tyres :

(*a*) Inflation pressure
(*b*) Tyre Load
(*c*) Speed
(*d*) Heat on tyres
(*e*) Road conditions
(*f*) Vehicle maintenance—Mechanical irregularities
(*g*) Driving habits
(*h*) Maintenance for maximum tyre life.
(*i*) Tyre and Tube storage.

1. Proper Inflation. Inflation pressure of the tyres including spare tyre should be checked at regular intervals. Frequency of checking is decided on the basis of actual pressure losses experienced for that operation, but tyres must not be left unattended for more than a week. Tyre pressure should be checked and adjusted when it is cold, never bleed (reduce tyre pressure) during or immediately after the run when a tyre has warmed up.

If during a run, tyre pressure has been found to have increased by more than 20% it indicates over-loading, over-speed, under-inflation or combination of these factors, and needs to be immediately corrected so as to save the tyre from further damage.

Under Inflation. It can result rapid tread wear at the shoulders, over-flexing and excessive heat generation, causing premature tyre failure.

Over-Inflation. Over-inflation results over-stressing, cannot absorb road shocks or impacts satisfactorily thereby increasing its liability to impact damage. Over-inflated tyre is more prone to cut/snag/puncture and rapid tread wear at crown.

2. Proper Loading. It is vital for long tyre life to ensure that load is properly distributed, and prevent over loading of tyres on one side. Sometimes concentration of load on one side, over load tyres on that particular side. Over-load puts more strain on cords and weakens their inherent properties. Over-load creates severe flexing which results in heat generation, and abnormal heat deteriorates tyres cords. This weakened cord cannot withstand the stress imposed on it by overload. Hence cords start breaking at the side-wall or shoulder resulting in permanent tyre damage.

3. Effect on Speed. High speed is always dangerous and destructive to tyre. All tyres are designed for speeds upto certain range. On exceeding this range on a particular kind of surface, the result will be fast tread wear and excessive heat build up. Hence loss in tyre life.

4. Effect of Heat on Tyres. Heat generation in a tyre is directly proportional to the frequency of the flexing, friction of road surface and atmospheric temperature. Flexing is the result of tyre deflection and can be regulated by implementing recommended inflation pressure and load. Heat generation due to friction depend upon the speed. Therefore, by maintaining load, inflation, and speed, the tyre killing heat can be regulated.

Cord once weakened due to heat remains weak for ever, because they will not regain its original strength even if the tyre is cooled, and will result in a "Blow-out". Similarly due to high heat the tread rubber becomes soft and wear more rapidly resulting in low tread mileage. Also, this softened rubber reduces the resistance to cut.

To summarise, following factors promote over-heating :

(*i*) Improper or over loading
(*ii*) Over speeding
(*iii*) Under inflation
(*iv*) Quick stop and starts
(*v*) Insufficient spacing between dual tyres

5. Road Conditions. Tyre life is greatly affected by the type of road on which it runs. Fast wear is found in those areas where there are frequent speed changes on abrasive roads with severe curves, and up and down hills. However, a correct choice on tread pattern for any particular type of route and careful driving helps in obtaining good tyre performance.

6. Mechanical Irregularities. Regular inspection of tyre for signs of abnormal or spotty wear will reveal that these are due to certain Mechanical Irregularities. These irregularities cause more rapid wear on certain places than that at other places, the remedy is to correct the condition causing this wear. Most common vehicle problems affecting tyre performance are:

(*i*) Excessive toe-in or toe-out in the tyre causes feathery edges on the tread pattern.

(*ii*) Too much camber results in uneven wear on one side of the tyre.

(*iii*) Loose king pin results in fast and uneven tread wear.

(*iv*) Incorrectly and unequally adjusted brakes or oval brake drums cause tyre tread to wear rapidly specially in a single spot. Oval brake drum usually causes two spots of rapid wear to develop, one located opposite to the other, over circumferential length of the tyre tread.

7. Driving Habits. To get higher tyre life, driver should avoid speeding over potholes, stone etc. or running over avoidable obstacles and irregularities. Quick start and harsh braking would wear off a tyre tread faster due to rapid slippage or scraping of the tyre tread over the road surface. High speed operations result in increased tyre running temperatures, which would also cause the rate of tread wear to be faster. Low speed on rough roads and for more loads is desirable. Skidding of tyres is very harmful for life.

8. Maintenance for Maximum Tyre Life. Tyre life can be increased by adopting following maintenance practices :

(*i*) **Ample tyre-to-vehicle clearance.** Carefully check tyres of vehicles to make sure that they do not touch against any part of the vehicle, either on straight way driving or on turns, since this may result in pre-mature tyre failure.

(*ii*) **Wedged Stones.** Stones or other objects forming wedge between dual tyres or between tyres and vehicle should be removed promptly to prevent serious tyre damage.

(*iii*) **Proper matching of dual tyres.** A new tyre should never be mounted along side a considerably worn tyre because of difference in diameters. Tyres should not be mounted side by side in dual set if over all diameter vary by more than 6.5 mm for tyres upto and including 8.25 × section or 12.5 mm for tyres of 9.00 × section and larger. If these olerances are not observed, larger tyre will be over-loaded and is likely to fail pre-maturely while smaller tyre scuffs and wears out faster. Inflation pressure should never be changed just to increase or decrease tyre diameter when matching duals.

(*iv*) **Repair injured tyres promptly.** Prompt detection of cuts, snags or other damage enables proper repairs in time to prevent blow-out. Water, dirt or other material enter cuts or other injury which extend to or into the casing, resulting tread or ply separation possibly a blow-out.

(*v*) **Keep tyres free from oil, grease and gasoline.** Rubber quickly absorbs oil, grease or gasoline and then swells and becomes soft and spongy. This damage is fatal. Never clean tyres with gasoline or other petroleum products or allow tyres to stand in puddles of (or areas saturated with) petroleum products. If a petroleum product does get on tyres, promptly flush of or wipe with water.

9. Tyre and Tube Storage. For storage of tyres principle to be remembered is to prevent light, heat, air in motion, ozone, oils or grease, dust and dirt, and water from inside of tyres, as this accelerates deterioration in storage. Store them away from electrical devices such as motors, or switches, since they are an active source of ozone.

Used tyres should be stored after cleaning and carrying out necessary repairs to prevent absorption of moisture. If necessary to store tyres while mounted on machine, block up so that the weight does not rest on the tyres, and release the air from the tyre. Where this is not possible, check air pressure frequently and maintain proper inflation.

New tubes should be kept in original package, and store in dry cool and draft-free storage area.

10. Tyre Rotation. It has been noticed that some tyres wear more rapidly than others, also out side tyre in a dual assembly wears faster than the inside tyre or there is a difference in the wear of the tread pattern (uneven wear) due to different working positions. Therefore, in order to utilise full potential life of tyre, a planned periodic rotation of tyres is necessary so as to maintain equal rate of wear. Sometime uneven wear of tyres is noticed, in that case rotation is advisable even before the recommended interval. Since this uneven wear may be due to some mechanical irregularities, tyres would be changed from right to left and left to right immediately after rectifying the deficiency in the vehicle.

Protection while Mounting and Demounting Tyres

(*a*) Protection of Rims

(*i*) Always use right size of rim to avoid severe tyre damage and premature tyre failure because—

(*a*) Unnecessary stress on plies will prevent the tyre from proper deflection.

(*b*) If spacing becomes small in dual tyres because of incorrect rims, side walls will start rubbing and result in excessive heat build up and cord deterioration.

(*ii*) Rim and rim parts should be thoroughly cleaned, and damaged portions, such as bents, dents or rust are rectified before mounting a tyre on it.

(*b*) Protection of Tyres

(*i*) Examine thoroughly the tyre internally and externally for cuts, cracks, trapped stones or any other damage.

(*ii*) Inside of the tyre should be cleaned properly.

(*iii*) Inflate tyre to approximately 75 Psi pressure, (50 Psi for grader) then completely deflate to remove buckles and uneven stresses from the tube and flap before reinflating to correct operating pressure.

(*iv*) Use proper tools for mounting and demounting of tyres in order to protect them from any cut or damage of bead.

(*c*) Protection of Tubes

It is the air confined in the tube that carries the entire load and not the tyre. Hence a tube is constructed to prevent loss of confined air. Therefore maintenance of tube is more important. Following precautions should be taken in case of tubes :

(*i*) Always put new tube in a new type, because used tube will have stretched and grown in service. If a used tube is refitted in a new tyre it will crease and ultimately result in both tyre and tube failure.

(*ii*) Always use correct size of tube. If larger size tube is fitted in a smaller size tyre, the tube will crease.

(*iii*) Never fit an under size tube because it may result.

(*a*) Excessive stressing and thinning of tube causing substantial loss in air.

(*b*) Reduced resistance to punctures

(*c*) Excessive strain on valve base

(*d*) Reduction in life.

(*iv*) Mount the tubes correctly. Careless fitting may damage the valve stem, tyre bead, or rim may be damaged.

(*d*) Protection of Flaps

Flap is to protect the tube from damages that may be caused by rim, tyre bead or excessive brake drum heat. Since the flap can cause costly tube and tyre failure, hence care should be taken while mounting the flaps. Flaps should be properly centered between beads to prevent failure. Flaps which have once become twisted, creased, cracked in the rim, or folded over at edges should not be used again.

Daily Tyre Maintenance Programme

1. Check inflation pressure daily.
2. Check if under inflation is due to defect and loose valve etc., if so, rectify it and then top up pressure to the recommended range.
3. Check for proper seating and tightness of valve core and cap. If cap is missing, fit another tightly.
4. Remove damaged or distorted rim and rim parts for repair or replace.
5. Check for foreign material trapped between dual tyres. Inspect whether the spacing between twins is satisfactory.
6. If any damage on the tyre is noticed, remove the tyre for repairing, since a negligible score or cut in the initial stage may develop in a subsequent trips and promote tread or side-wall separation.
7. Check for uniform wear. If there is uneven wear, change the position of the tyre.
8. Check for irregular wear cuts, bulging, trapped stone etc. If there is irregular tread wear, change the tyre to suitable position. Clear off trapped stones, nails etc. and repair the spot.
9. If the tyre has been worn smooth, remove it from the vehicle for retreading.

Tubeless Tyres

These tyres do not require an inner tube, and are mounted on the rim while air is retained between the rim and casing (tyre). The tyres are provided with special lining on inside made of special air retaining material. The inner lining acts as seals at the edges of the tyre against wheel rim, thereby preventing air leakage.

Following precautions be taken in case of tubeless tyres :

1. Care should be taken while fitting or removing the tyre that the bead is not damaged.
2. Never lift rim by valve hole.
3. Never drop, tumble, or roll rim parts.
4. Since tubeless rim also functions as air seal, proper care must be taken not to distort the rim parts.
5. Do not lift tyre by beads with hooks or forks. Sharp hooks or forks may tear, cut, or snag the tubeless tyre beads and result in leakage at these points.
6. Foreign material or moisture should be removed from inside tyre while mounting on the rims.
7. Use babbit or lead hammers, not the sledge hammers while assembling rims.
8. "O" ring seals should be carefully stored in a cool, dry place where they are not damaged. During use, these should be periodically checked, and change if found damaged.
9. Unmounted tyres should be stored and transported in vertical position. Horizontal storing may compress the beads resulting initial inflation difficult.

12.6.4 Tyre Retreading

Tyre has a date to change for retreading. Driving of worn smooth tyre can be fully dangerous and destructive because a smooth tyre has no grip on road surface and skidding would inevitably result. However, when hard brakes are applied the vehicle would tend to slip. A worn smooth tyre has a very thin layer of rubber remaining, which provides less protection for casing and at this stage tyres cannot withstand cut and impact. Most of the tyre casings are rendered unfit for retreading by attempting to run further on smooth tyre. Hence tyres should be removed for retreading when they are worn smooth. Timely change for retreading prevents un-necessary purchase of new tyres.

12.7 ENGINE, CLUTCH, AND GEAR-BOX (TRANSMISSION)

12.7.1 Engine

The engine or the power unit is the source of power that makes the wheels go around thereby moving the vehicle on the earth. In automobile, generally the power unit is an internal combustion engine which burns the fuel inside. When the engine burns gasoline is known as gasoline engine. The I.C. engines also burn heavier oils but out of them the diesel engine has come into the widest use. When the fuels burn inside the I.C. engine, the chemical energy stored in the fuels is converted into heat energy. The heat energy is converted into mechanical energy by the expansion of gases against pistons. The movement of the pistons is carried by connecting rods to the engine crankshaft. The crankshaft thus made to rotate. The rotary motion is carried through the power train to the vehicle wheels thereby rotating the wheels and moving the vehicle.

I.C. Engines have been dealt separately in details,

12.7.2 Clutch

The clutch is located between the engine and the transmission. The flywheel of the engine is an integral part of the clutch mechanism. The clutch disconnects the engine from the power transmission system at the will of the driver by operating a foot pedal, resulting the engine to run without driving the automobile. The engine is required to be disconnected when starting or shifting gears or stopping or idling. The clutch also permits the gradual taking up of the load.

A clutch has two members positively driven by the engine (one is pressure plate bolted to flywheel, and other cone or disc or plate) attached to the main shaft of the transmission. When these members are separated by pushing down the clutch pedal, the engine will run without turning the transmission shaft, thus permitting the gears to be shifted, or the vehicle to be stopped without stalling the engine. The friction surfaces of the clutch are designed so that the driven member slips on the others when the pressure is first applied. As the pressure is increased (by releasing the clutch pedal slowly), the driven member is brought gradually to the speed of the driving member. When the speed of all the three members (flywheel, clutch or transmission shaft) becomes equal, slippage ceases entirely, and they make firm contact.

This drive between three members is accomplished by the friction between first two members and depends upon the materials in contact, and the pressure forcing them together which is generally by the spring pressure. On pushing down the clutch pedal, these springs are compressed, thus freeing the engine from the transmission line.

Types of Friction Clutches

Following types of friction clutches are common in automobile industry.

1. Cone clutch
2. Single plate clutch
3. Multiplate clutch
4. Centrifugal clutch
5. Hydraulic clutch.

1. Cone Clutch. As shown in the Fig 12.14 line diagram of cone clutch, both the driving and driven components are made of cone shaped. While fully engaged the male cone remains fully inside the female cone so that the friction surfaces may remain incomplete contact. This engagement of the cone remains all the time by means of pressure of springs.

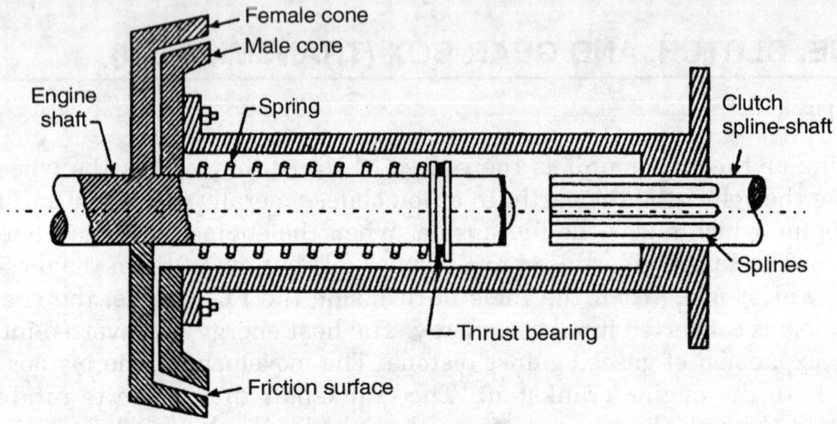

Fig. 12.14 Cone clutch.

When the clutch is in engaged position, the power is transmitted from engine to female cone to male cone to the gear box through the spline shaft. When the clutch system is disengaged the male cone is pulled out by means of the system and then both the contact surfaces are separated in this way.

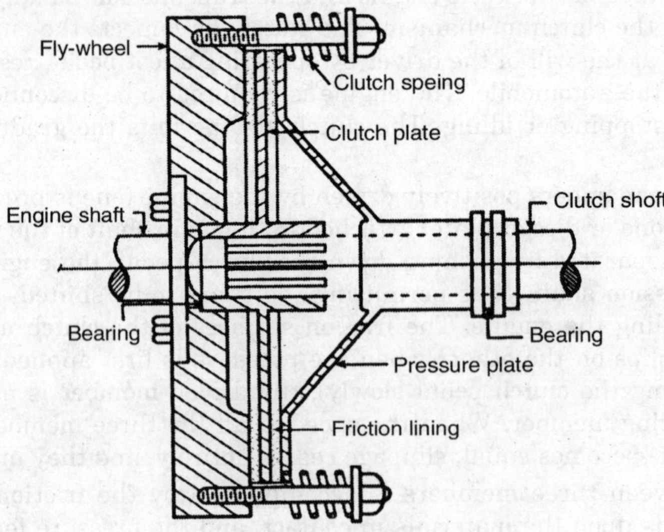

Fig. 12.15 Single plate clutch.

2. Single Plate Clutch. As shown in the Fig. 12.15 simplified sketch the friction plate is held between the flywheel and the pressure plate. The axial pressure is applied on the pressure plate with the six helical springs mounted circumferentially. The friction plate is mounted on the one end of the clutch shaft having splines. To increase the friction surface the friction lining are provided on both sides of the clutch plate. For disengaging the clutch a pedal is provided to pull the pressure plate against the springs to make loose the friction plate from both sides. While there is no effort on the clutch pedal, and clutch remains always in engaged position.

3. Multiplate Clutch. In Fig. 12.16. It is similar to the single plate clutch but only the difference is this that it has more number of frictional and metal plates. In this way due to increasing the number of frictional plates the frictional surface is also increased which increases the capacity of the clutch to transmit torque, the size remaining fixed. To transmit the same torque the diameters of plates are reduced in comparison to the single plate clutch.

The construction of the multiplate clutch is clear from the simplified diagram. Almost construction and system is same as that of single plate clutch except the number of frictional plates. It contains two sets of frictional plates one set of plates slides on splines of the pressure plate and the one slides in grooves on the flywheel. The both types of frictional plates are arranged in alternate manner one on the pressure plate and other on the fly wheel and so on. The construction of the fractional plates is made clear from the Fig. 12.17 (*a*) and (*b*).

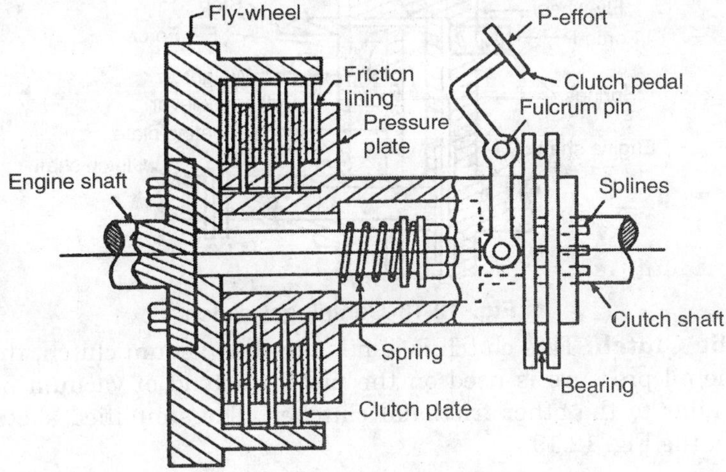

Fig. 12.16 Multiplate clutch.

4. Centrifugal Clutch. In the purely centrifugal type of clutches, the clutch pedal and the springs are eliminated altogether and only the centrifugal force of weight is used to apply the required pressure on the floating plate as well as on pressure plate for keeping the clutch in engaged position. In this system the operation of the clutch is automatic and depends upon the engine speed. There is no need to disengage the clutch and vehicle can be stopped in gear without stalling the engine. In this system the vehicle is controlled by pressing the accelerator and vehicle is put into the gear at the starting only. This makes the driving operation very easy and convenient.

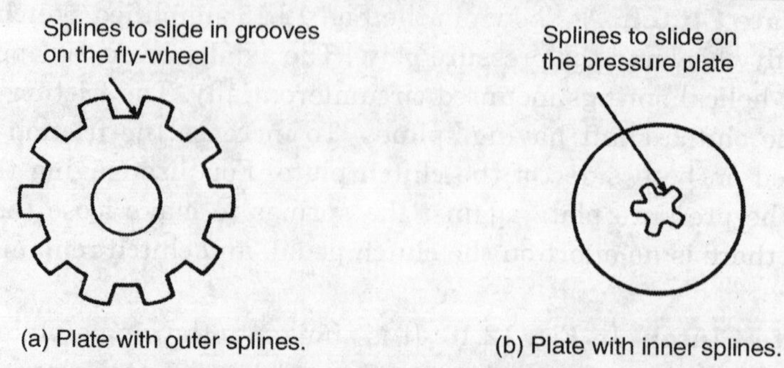

Fig. 12.17 Friction plates of a multiplate clutch.

The constructional details of the centrifugal clutch are as shown in the Fig. 12.18. When the speed increases, the weight flies due to centrifugal force, thereby operating the bell crank lever and the floating plate is pressed. Then force of floating plate is transmitted to the pressure plate through the helical spring S_1. The pressure plate containing friction lining also, gets pressed on the flywheel to complete the transmission, thereby the clutch is engaged. The spring S_2 keeps the clutch in disengaged position at low speed say about 500 r.p.m. and the stop above the weight, limits the amount of the centrifugal force. If even the speed is increased over this limit the pressure on the plates will be constant.

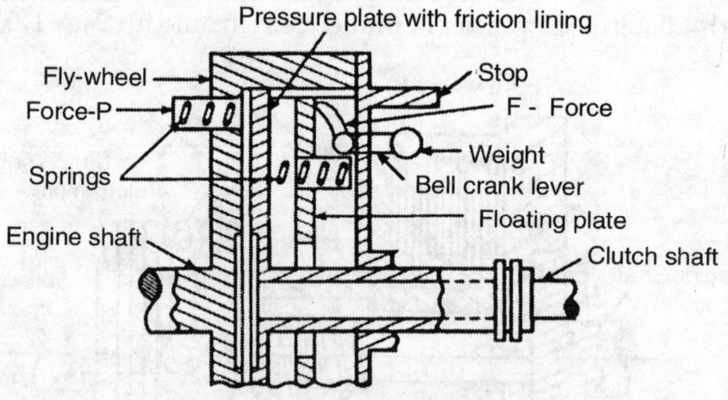

Fig. 12.18 Centrifugal clutch.

5. Hydraulic Clutch. This clutch is similar to the vacuum clutch, the difference is only this that in it the oil pressure is used on the piston instead of vacuum pull. The remaining clutch unit is similar to the other frictional clutches. The simplified sketch of the hydraulic clutch is shown in the Fig. 12.19.

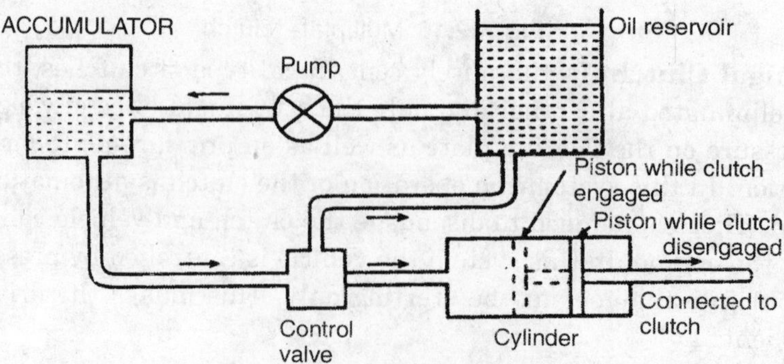

Fig. 12.19 Hydraulic clutch.

It contains an oil reservoir from which the oil is pumped into the accumulator tank. The pump is driven by the engine itself. Then the under pressure oil is sent to the cylinder through a control valve which is electrically operated by a switch connected with gear change lever. When driver operates the lever to change the gears the control valve opens automatically admitting oil under pressure to the cylinder. Then this pressure on the piston causes it to move and the movement disengages the clutch. When the driver leaves the lever in the geared position the switch control valve closes and in this way clutch is again engaged.

Clutch Components

(*i*) Clutch withdraw mechanism. (*ii*) Clutch plate.

(*iii*) Clutch lining.

***(i)* Clutch Withdraw Mechanism.** To put and hold the clutch in the disengaged position, the clutch withdraw mechanism is necessary. This mechanism including foot pedal and clutch housing, levers etc. are also seen some times from outside also.

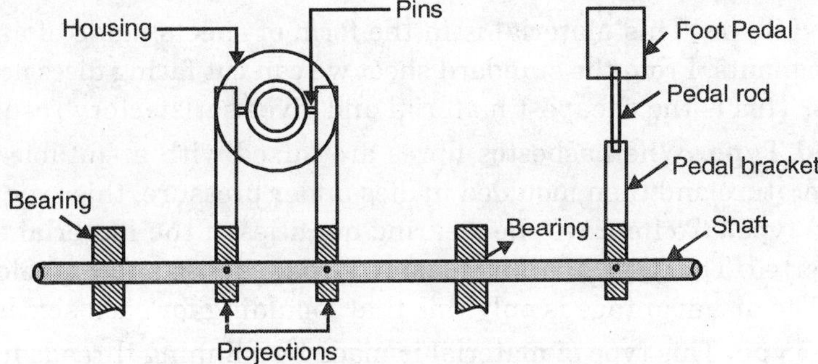

Fig. 12.20 Clutch withdraw mechanism.

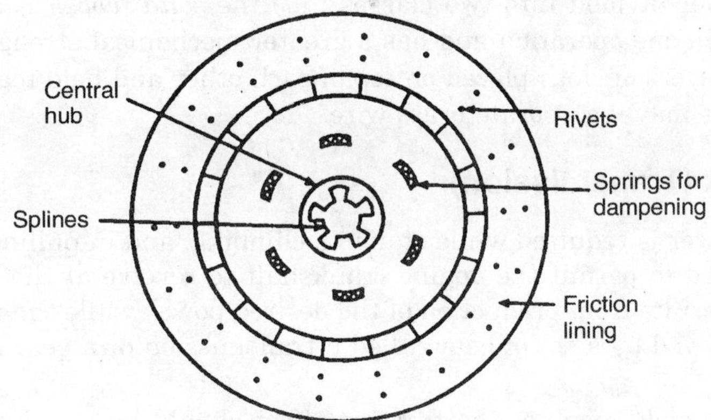

Fig. 12.21 Clutch plate.

A cross shaft type withdraw mechanism is as shown in the Fig. 12.20. When the foot pedal is pressed by the driver the shaft having the two projections and supported at two bearings is turned thereby moving the housing with the pressures of the pins. In this way this motion of the housing disengages the clutch. When the manual pressure from the foot pedal is removed the housing comes in its original position thereby the clutch again engaged.

***(ii)* Clutch Plate.** A clutch plate is made by the segments of spring-steel in series around the central hub and the friction facings are riveted to the segments. The helical springs are provided in between the relative movement to damp out the undesired vibrations and thus they serve as dampening devices. The friction lining is attached to waved cushioning springs, as the clutch is engaged, the waves compress slightly to provide cushioning effect. For the better friction, the friction linings are replaced by the new linings and when there is defect in the dampening then helical springs are also changed.

***(iii)* Clutch Lining.** It is the frictional material used in between the driving and driven faces for transmitting power through it. Thin linings of frictional material are riveted on both sides of the clutch plate which is known as "clutch linings". The basic requirements of a good clutch lining are as given under :

(*a*) High co-efficient of friction.
(*b*) High resistance to heat.
(*c*) Good wearing properties.
(*d*) Pressure of a good binder in it.
(*e*) Cheap and easy to manufacture.

Friction Materials

The various types of friction materials are as given under:

***(i)* Millboard Type.** This material is in the form of sheets made of asbestos and treated with certain impregnants. From the standard sheet we can cut facing discs according to different size requirements. This is the cheapest material and gives satisfactory results.

***(ii)* Moulded Type.** When asbestos fibres are mixed with a suitable binder and heated to a definite temperature and then moulded in dies under pressure, this type of friction material is called "moulded type". To improve the wearing qualities of the material sometimes metallic wires are also inserted. This type of lining is more strong, dense and capable to transmit heavy loads through it. The disadvantage is only this that moulding for each size lining is very costly.

***(iii)* Woven Type.** This type of material is made by spinning threads from asbestos fibres, weaving this thread into a cloth and then impregnating it with a bonding material. This type of material is further divided into two classes ; (*a*) the *solid woven variety* is woven to the required thickness in one operation and has a greater mechanical strength, (*b*) the *laminated variety* consists of layers of cloth placed on top of each other and held together by the bonding material. Both types may incorporate brass wire.

12.7.3 Gear–Box (Transmission)

Since more power is required while starting, climbing, and in pulling a load, means must therefore be provided to permit the engine crankshaft to revolve at the relatively high speed which necessarily results from production of the desired power while the wheels turn at slower speeds. This is achieved by a set of gear called a transmission or a gear box, which multiplies the turning force.

To keep the vehicle moving, the tractive effort should be equal to the sum of all the opposing forces. If tractive effort is more than the sum of all the resistances, the vehicle will accelerate, and if it is less then the vehicle will retard. These resistances are resistance due to wind, gradient, rolling etc.

The vehicle needs more acceleration at the time of starting to gain the speed quickly. The maximum tractive effort is available in I gear, therefore, in beginning the vehicle is always moved in I gear. When it obtains necessary speed then is shifted to II gear and so on. Similarly when the vehicle is moved across a steeper gradient then it is always shifted to the I gear and is accelerated at very slow speed.

Sliding Mesh Gear Box

This gear box is very simple in construction as shown in the simplified sketches of Fig. 12.22. The power arrives from the engine to the clutch gear through clutch shaft and this clutch gear is always in mesh with a largest gear on the counter-shaft. All the gears on the counter shaft are fixed in the positions of reducing diameters while considering from input to output side of the gear box and as such they are all the time rotating when the engine is running and power is transmitted through clutch. With the help of selector mechanism and lever the gears can be moved on the main shaft to mesh in various positions with the gears of counter shaft to obtain different speeds when desired.

The positions of the gear box provides four forward speeds and one reverse as explained under:

(*i*) Gear in Neutral. In this position of the gear the engine power is not transmitted to the rear axle. As shown in Fig. 12.22 (*a*) the clutch gear is transmitting the power to gear *A* on the counter shaft and the counter shaft further not transmitting the power to main shaft. Therefore, the output of the gear box is disconnected with the input of the gear box.

(*ii*) First or Low Gear. As shown in Fig. 12.22 (*b*) the gear *D* is slid along the main-shaft and meshed with gear I on countershaft. The drive comes from the engine shaft through the constant-mesh gears *G* and *A to* the countershaft and transmitted to main shaft through gears *I* and *D,* then the gear ratio is:

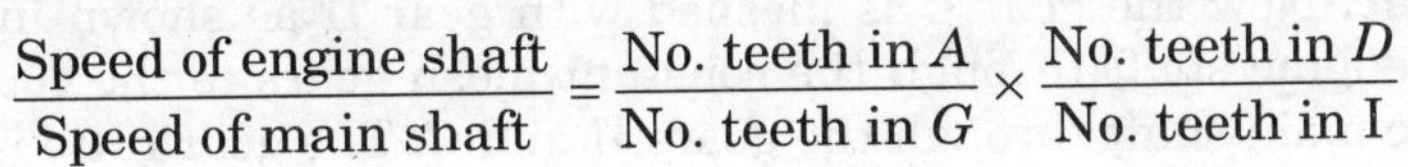

$$\frac{\text{Speed of engine shaft}}{\text{Speed of main shaft}} = \frac{\text{No. teeth in } A}{\text{No. teeth in } G} \times \frac{\text{No. teeth in } D}{\text{No. teeth in I}}$$

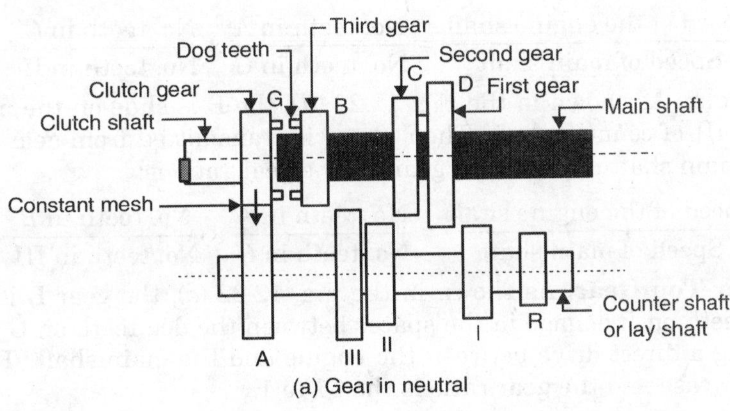

(a) Gear in neutral

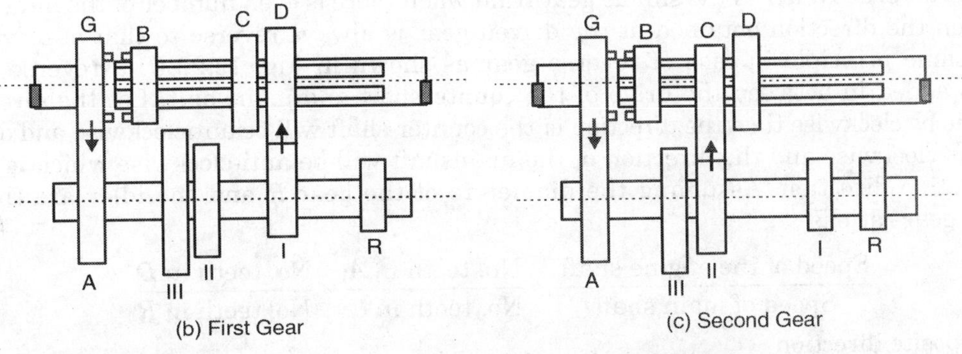

(b) First Gear

(c) Second Gear

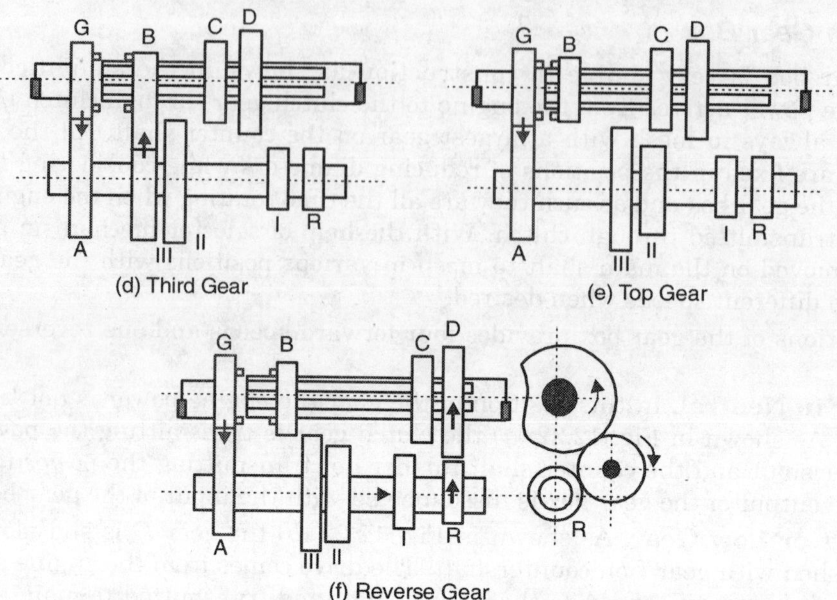

Fig. 12.22 Various gear positions of sliding mesh gear box.

***(iii)* Second Gear.** Now the gear *C* is meshed with gear II as shown in Fig. 12.22 (*c*). The drive is from the engine shaft through the constant mesh gears to the counter-shaft and then transmitted to the main shaft through the gears II and *C*. The gear ratio is:

$$\frac{\text{Speed of the engine shaft}}{\text{Speed of main shaft}} = \frac{\text{No. teeth in } A}{\text{No. teeth in } G} \times \frac{\text{No. teeth in } C}{\text{No. teeth in II}}$$

***(iv)* Third Gear.** As shown in the Fig. 12.22 (*d*), the B is slide on the main shaft and is meshed with gear III of counter shaft. Then power is transmitted from gear *G* to *A* and gear III to *B*, thus the main shaft rotates with gear *B*. The gear ratio is:

$$\frac{\text{Speed of the engine shaft}}{\text{Speed of main shaft}} = \frac{\text{No. teeth in } A}{\text{No. teeth in } G} \times \frac{\text{No. teeth in } B}{\text{No. teeth in III}}$$

***(v)* Fourth or Top Gear.** As shown in the Fig. 12.22 (*e*), the gear B is slid over to the left then the dog teeth on *B* engage in the spaces between the dog teeth on *G*, thus connecting *G* and *B* and giving a direct drive between the engine and the main-shaft. The counter-shaft revolves idle in this case and the gear ratio is then 1 to 1.

***(vi)* Reverse Gear.** In the simple gear train when there is even number of the intermediate gears then the direction obtained at the driven gear is always reverse to that of driver gear. On the same principle in case of reverse gear as shown in Fig. 12.22 (*f*) a reverse gear or "idler" is added in between the drive of the countershaft and mainshaft. Let the direction of the engine be clockwise then the direction of the counter-shaft will be anticlockwise and direction of idler as clockwise and the direction of the mainshaft will be anti-clockwise which is reverse to the engine direction. Assuming the diameters of the gear *R* and the idler are the same than the gear ratio is :

$$\frac{\text{Speed of the engine shaft}}{\text{Speed of main shaft}} = \frac{\text{No. teeth in } A}{\text{No. teeth in } G} \times \frac{\text{No. teeth in } D}{\text{No. teeth in } R}$$

in the opposite direction

Sliding Mechanism

The many type of mechanisms are used for sliding the selected gear to engage with the corresponding gear on the countershaft of the gear box. Generally they are divided into two categories, (*i*) sliding mechanism with ball-mounted lever and (*ii*) sliding mechanism of steering column gear change. The gear shifting system in these two types of the mechanism is similar but the difference is only in the external linkage system.

12.7.4 Formulae used in Transmission

1. Power for Propulsion

(*a*) $$P_v = \frac{R \times V}{1000}\left(\frac{1000}{60 \times 60}\right) = \frac{R \times V}{3600} \text{ kW}$$

(*b*) $R = (R_a + R_r)$ when vehicle moves along a level road.

(*c*) $R = (R_a + R_r + R_g)$ when vehicle moves up a gradient.

(*d*) $$P_{req} = \frac{P_v}{\eta_t} = \frac{R \times V}{3600 \times \eta_t} \text{ kW}$$

(*e*) $R_r = KW$ (K = 0.015 to 0.2 N for hard road; 0.2 to 0.3 for sand)

(*f*) $R_a = K_a A V^2$

(*g*) $R_g = W \sin \theta$

(*h*) % grade $= \tan \theta \times 100$

where, P_v = power required by the vehicle.
V = speed of the vehicle in km./hr.
R_a = air resistance in kN.
R_g = grade resistance in kN.
K = 0.006 for goodroods
K = 0.18 for sand, Average value 0.015
K = constant of rolling resistance.
A = projected frontal area in m².
$P_{req.}$ = engine b.p. required.
R = total resistance in kN.
R_r = rolling resistance in kN.
η_t = transmission efficiency.
W = total weight of the vehicle in kN.
K_a = 0.023 for steam lived cars
K_a = 0.031 for average cars
K_a = 0.045 for trucks

K_a = coefficient of air resistance in $\left(\frac{N \times hr^2}{m^2 \times km^2}\right)$

θ = road inclination.

2. Tractive Effort.

(*a*) $$T_E = \frac{60{,}000 \times P_E}{2\pi N} = \frac{9554\, P_E}{N}$$

(*b*) $$T_w = (ar. \times gr.) \times \eta_t \times T_E = G \times \eta_t \times T_E$$

(*c*) $$F = \frac{T_w}{r} = \frac{T_E \times \eta_t \times G}{r} \text{ kg.}$$

where, T_E = mean engine torque in kN. m.
P_E = engine brake power; N = r.p.m. of crank shaft.
T_w = torque at rear wheels.; *ar.* = back axle ratio.

gr = gear-box gear ratio.; η_t = overall transmission efficiency.
G = (gr × ar.) overall gear ratio.; r = radius of tyre in metre.

(*d*) Condition for acceleration, $F > R$.

where,
F = tractive effort.
R = total resistance,

3. Relation between Engine Revolutions and Vehicle Speed

$$\frac{2\pi \times r \times N}{G} = \frac{V \times 1000}{60} \quad \text{or} \quad \frac{N}{V} = \frac{8.33}{\pi}\left(\frac{G}{r}\right) = \frac{N}{V} = 2.65\left(\frac{G}{r}\right)$$

where,
V = vehicle speed in km/hr.
N = r.p.m. of the engine.
G = overall gear ratio.
r = radius of tyre in metre.

4. Performance of Vehicle in Different Conditions

(*a*) Acceleration.

$$\because \quad \text{Surplus Power} = m_e \times f \times \frac{V}{3600}$$

$$\therefore \quad f = (\text{surplus h.p.}) \times \frac{3600}{m_e}$$

$$= (P_e - P_{req.})\, \eta_t \times \frac{3600}{m_e} = \frac{(F - R)}{m_e}$$

(*b*) Gradiability

$$\because \quad \text{Surplus power} = \frac{W \times \text{Gradiability} \times V}{100 \times 3600}$$

$$\therefore \quad \text{Gradiability} = \frac{100}{W}(P_E \times \eta_n - P_V) \times \frac{3600}{V} = \frac{100}{W}(F - R)$$

(*c*) Drawbar pull

Maximum drawbar pull = $(F - R)$

where,
m_e = equivalent weight of the vehicle.
f = Maximum acceleration of the vehicle.
V = speed of the vehicle.
P_E = engine power in kW
$P_{req.}$ = engine brake power required.
F = tractive effort.
R = road resistance.
W = weight of the vehicle.
P_v = power required by the vehicle.

5. Equivalent Weight of the Vehicle

$$W_e = W + \frac{I_f \times \eta_t \times G^2 + I_s}{r^2}$$

where,
W_e = equivalent weight.
W = weight of the vehicle.

I_f = M.I. of faster rotating parts in N m^2.
I_s = M.I. of slower rotating parts in N m^2.
η_t = transmission efficiency.
G = overall gear ratio.
r = radius of the road tyre in metre.

6. Gear Ratio for Maximum Acceleration

$$G = \frac{b}{a} \pm \sqrt{\left(\frac{b}{a}\right)^2 + \frac{c}{d}}$$

where,

$$a = T_E \times \eta_t, \quad b = R \times r.$$

$$c = \frac{W}{g} \times r + \frac{I_s}{g \times r}, \quad d = \frac{I_f \times \eta_t}{g \times r}$$

7. Relations for Maximum Tractive Effort, Acceleration and Reactions for Different Drive

(*a*) Rear wheel drive :

$$F_R = \mu R_R, \quad W = R_F + R_R,$$

$$\frac{f}{g} = \frac{\mu(b - l)}{b - \mu h} \quad \text{and} \quad R_R = \frac{b - l}{b - \mu h} \times W.$$

$$R_F = \frac{l - \mu h}{b - \mu h} \times W.$$

(*b*) Front wheel drive :

$$F_F = \mu R_F, \quad W = R_F + R_R,$$

$$R_f = \frac{l}{b + \mu h} \times W, \quad \text{and} \quad R_R = \frac{b - l + \mu h}{b + \mu h} \times W,$$

where,
F_R = tractive effort on rear wheels.
F_F = tractive effort on front wheels.
R_R = total normal reaction at rear wheels.
R_F = total normal reaction at front wheels,
μ = coefficient of adhesion between the road surface and tyre surfaces.
W = weight of the vehicle.
f = maximum forward acceleration.
b = wheel base.
l = distance of C.G. from the rear axle.
h = height of the C.G. from road surface.

12.8 PROPELLER SHAFT AND UNIVERSAL JOINTS

1. Propeller Shaft

Propeller shaft transmits the power from the engine driven gear box to the driving axle. It is either fitted in a front or rear wheel drive or between the main gear box and an auxiliary gear box. It is made of a strong steel tube because it has to withstand torsional stresses of the transmitting torque and should also be light and well balanced to avoid vibrations at high speeds.

The gear box or transmission and the engine are attached to the vehicle chassis or frame and rise or fall as one piece with it. The rear axle housing is attached to the frame by springs. Due to uneven contour of the road, the springs compress or expand due to rear wheels recountering the irregular surface. The propeller shaft, therefore has to transmit power between two points at varying vertical and horizontal distances. Universal joints are incorporated with it for this purpose. These providing for variations in the angle of drive act as universal fittings. The slight variation in length is permitted either by one end of the shaft being free to slide on splines or by providing a slip joint.

2. Universal Joints

A universal joint is a form of mechanical connection between two shafts giving a positive drive whilst allowing angular movement of one or both of the shafts. The common types of universal joint is described here under :

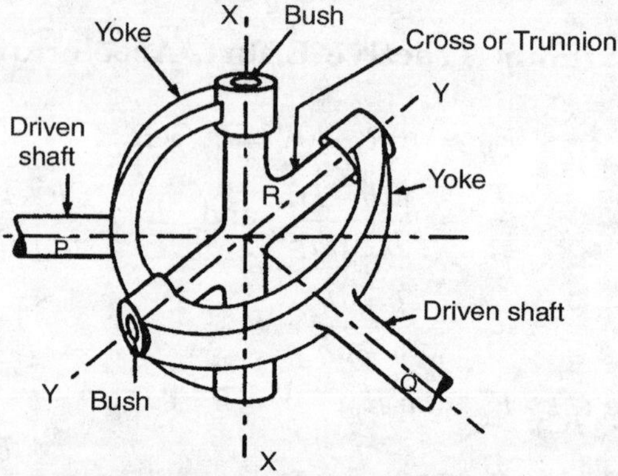

Fig. 12.23 Hooke's universal joint.

Hooke's Type Universal Joint. It is the most simple universal joint as shown in the Fig. 12.23. The axes of the driving shaft and the driven shaft P and Q respectively are intersecting. At the end of each of these shaft a yoke is made. The cross R which is called trunnion has four arms. The two opposite arms of the trunnion R are fitted in bushes or bearings in the yoke of shaft P, while the other two opposite arms of the trunnion are fitted in the yoke of shaft Q. Here the driving shaft P can have angular rotation about axis XX and the driven shaft Q about the axis YY. Therefore it is quite possible with this joint for the driving shaft P and driven shaft Q to have positive drive while allowing angular movement between the two.

To increase the efficiency of the joint in the improved form of the Hooke's joint, the needle roller bearings are used to support the trunnion in the yokes, instead of the uses of the bushes.

12.9 COOLING SYSTEM

Out of total heat generated due to burning of the fuel combustion chamber of an internal-combustion engine, only 25 per cent is converted into mechanical energy, 35 per cent losses through exhaust gases, 35 per cent passes through the cylinder walls and the remainder 5 per cent passes through the lubrication systems. The dissipation of heat from the cylinder walls is an important factor and which is removed by means of cooling system to the engine. If the cooling system is not provided in the engine then it would result in the pre-ignition of the charge, burned up of lubricant thereby scoring of cylinders, binding of pistons, bearings and warping of valves.

The too much cooling of the engine results decreased thermal efficiency, reduced vaporization of the fuel, increased viscosity of lubricant thereby decreasing overall efficiency of the engine. Therefore the duty of a cooling system is that neither it keeps the engine too hot nor too cold and allows it to run in a definite designed temperature range.

The cooling systems are broadly classified in two types, direct-air cooling and indirect or liquid cooling. But according to the need there are different cooling methods used in automobile engines which are described as under:

(*i*) Direct-air cooling.
(*ii*) Indirect or water cooling ; (*a*) Thermosiphon system, (*b*) Pump circulation system.
(*iii*) Liquid cooling. (*iv*) Pressure sealed cooling.
(*v*) Evaporative cooling or steam cooling.

1. Direct Air Cooling

This system is being used in light engines such as scooter, motor cycle, tractor and small aeroplane engines. The outer surface of the cylinders and cylinder heads consists of fins as shown in Fig. 12.24 which increases area of contact with the air thus radiates more heat to the atmosphere.

The rate of heat flow through the cylinder walls is expressed by law which may be stated as

$$Q = \frac{kA\Delta t\ M}{L}$$

where Q = rate of heat transfer.
k = average thermal conductivity.
Δt = temperature difference between the hot and cold surfaces.
A = surface heat transfer area.
M = mass flow rate of air.
L = length of the heat flow path.

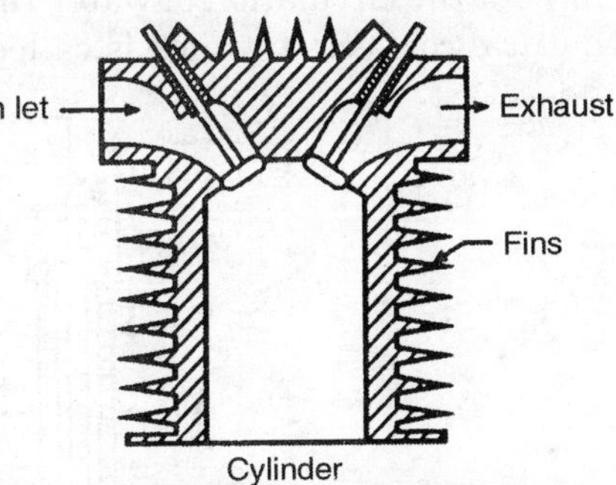

Fig. 12.24 Direct-air cooling.

Therefore for better air cooling, the surface area of the metal which is in direct contact with the air is increased by increasing the number of fins. The conductivity depends upon the metal used for cylinder and cylinder head, temperature difference depends upon the atmosphere and combustion temperature. The mass of air flowing over the fins can be increased by means of providing the fan or air blower with the engine fly-wheel. The flow of heat is also inversely proportional to the length of the heat flow path which depends upon the maximum pressure for which engine cylinder is designed.

Advantages

(*a*) As the air cooled engine does not consist of radiator cooling jackets, coolant etc. therefore it is lighter than others.
(*b*) Power lost in the cooling water circulating pump is saved in this system.
(*c*) The chances of freezing of water are omitted when used in cold climates or places.
(*d*) The air cooled engines are advantageous, where there is scarcity of cooling water.
(*e*) The cylinders and cylinder head prepared under this system are cheaper than any other types.

Disadvantages

(*a*) Air cooled engines are less efficient in cooling because the coefficient of heat transfer for air is less than that for water.

(*b*) As the cooling air can not provide the even cooling around the cylinder, therefore distortion in cylinders and cylinder head is possible.

(*c*) As the air cooled engines do not consist of water jackets therefore they are more noisy, because cooling water serves as sound insulator.

(*d*) It needs a fast running as well as bulky fan which produces more noise and absorbs more power.

2. Indirect or Water Cooling

In this system the cooling water flows through the jackets made around the cylinder, combustion chamber and valve ports. The heat from the cylinder is observed by cooling water which is circulated through the radiator. From radiator this heat is dissipated to the atmosphere by means of fins provided on the radiator. To maintain the proper working temperature the system is provided with a thermostat. The water cooling systems are of two types as given under:

(*a*) Thermosyphon System. Referring Fig. 12.25 this system does not consist of water pump but the circulation of water through the jackets and radiator is maintained only by means of natural convection which is called thermosyphon.

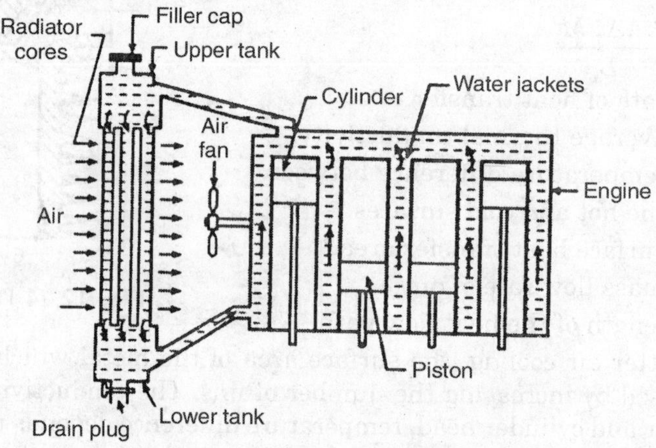

Fig. 12.25 Thermosyphon system.

This system is very simple and cheap but has some disadvantages by which it could not be much popular. Its cooling rate is slow. It always needs a particular or minimum level of water. It needs a big size radiator which requires more space. Therefore these are the main draw backs of this system.

(*b*) Pump Circulation System. This system is the modification of thermosyphon system. The system consists of a water pump provided between the lower tank of the radiator and the water inlet of the engine body. The pump is provided drive from the crankshaft by means of a belt. The water is circulated with the pump force thereby obtaining a rapid cooling.

As at the start the engine does not need cooling until the temperature exceeds the working range of the engine. Therefore in most modern engines the cooling system consists of a by-pass through which water circulates at the start. Referring Fig. 12.26. The system also consists of a thermostat which is opened automatically by means of high temperature whenever engine desires cooling. Thus obtained a rigid control over the cooling temperature under this system.

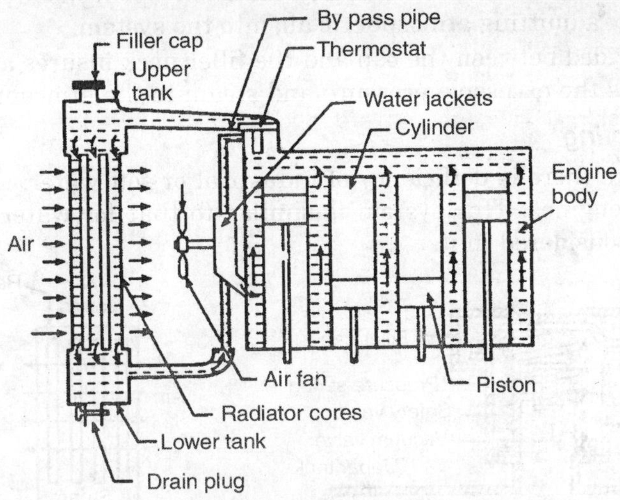

Fig. 12.26 Pump circulation system.

Advantages of Pump Circulation System.

1. The size of the water passage and jackets can be reduced thereby obtaining a compact structure of the engine.
2. Water can be circulated around the hottest spot of the engine such as sparking plug and valve seat bosses.
3. It needs less water and small size of the radiator.
4. As the water is circulated by means of pump force therefore, radiator can be placed at any where convenient to the designer.
5. A rapid cooling is obtained by this system.
6. To increase the cooling efficiency the system can act as pressurized system by minor alterations.

3. Liquid Cooling

The liquids such as glycerine and ethylene glycol are used as coolant instead of water because they have higher boiling points, this system is called liquid cooling. The advantages of this system are that a liquid of higher boiling point increases the heat carrying capacity of the system, reduces the weight of coolant as well as size of the radiator itself.

4. Pressure Sealed Cooling

The boiling point of the cooling water in the open atmosphere is always 100°C, which can be raised due to higher pressures by sealing the cooling system. Therefore the closed system reduces the weights of the coolant and the radiator and increases the thermal efficiency of the engine.

The pressure of the system can be raised by means of a special radiator cap which maintains pressure from 0.5 to 1 kg/cm^2. Referring Fig. 12.27 the cap consists of Pressure valve or blow off valve, vacuum valve over flow pipe and gasket for sealing purposes. When the pressure in the system exceeds the predetermined value then the spring loaded pressure valve be opened automatically. Similarly when the engine is stopped then the pressure falls below that of the outside atmosphere which creates a vacuum in the system. As the hoses and other thin parts of the cooling system can be collapsed due to creation of vacuum inside the system. Therefore to overcome this difficulty an other spring loaded vacuum valve opens automatically thereby admitting atmospheric air into the system.

The gasket provided between the cap and the filler neck insures an air-tight seal and the overflow pipe releases the excessive pressure and steam to the atmosphere.

5. Evaporative Cooling

The places where there is defficiency of industrial or soft water, this system of cooling is preferred in vehicle engines. This system is similar to that for water cooling but the latent heat of the steam is considered in it.

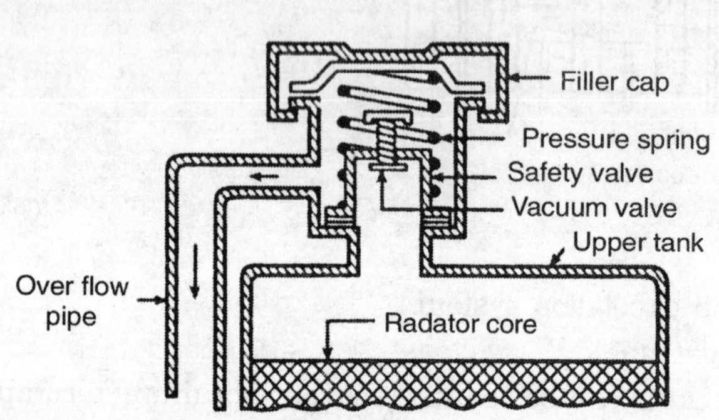

Fig. 12.27 Pressure sealed cooling system.

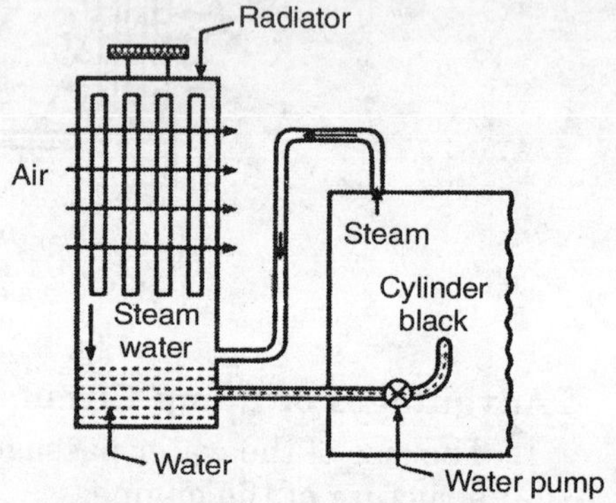

Fig. 12.28 Evaporative cooling system.

Referring Fig. 12.28, the system consists of a radiator which acts as a condenser. The water from the radiator is entered into the jackets of the engine block by means of a water pump. The water in the jackets is converted into the steam which flows out at the top of the engine blocks and enters at the bottom of the radiator. This steam is condensed in the radiator by means of air flow. The water formed is collected in the bottom of the radiator and recirculated by the same pump. The main advantages of this system are that it needs very little quantity of coolant at the start, the engine is also warmed up quickly.

Components of Water Cooling System

The main components of water cooling system are as given under.

(*i*) Radiator (*ii*) Water pump

(*iii*) Fan (*iv*) Thermostat

(*i*) Radiator. The purpose of a radiator is to cool the cooling water passing through it very efficiently. Referring Fig. 12.29, it consists of upper tank which is connected to the water outlet of the engine, and the lower tank which is connected to the water inlet of the engine. Water of both the tanks is connected by means of cores.

The radiator cores are of two types, the tubular type and the cellular type. In case of tubular type, the cooling water passes through the vertical or zigzag tubes and air passes around them. In case of cellular type, the air passes through the tubes and water passes in the spaces in between them. All the tubes, fins and tank walls of the radiators are constructed with the copper or brass because of their high heat conductivity and their joints are fixed together by means of solder.

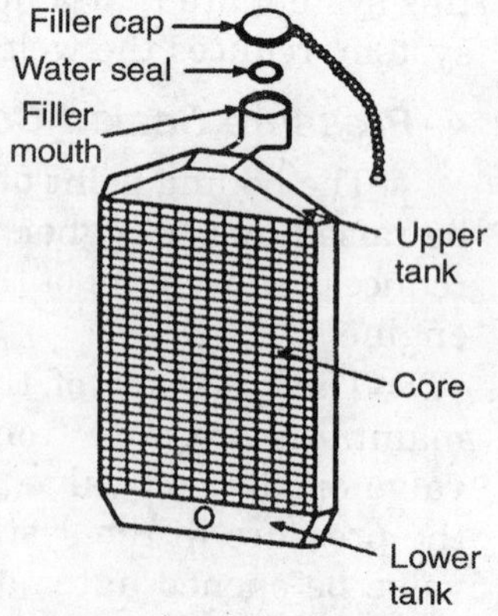

Fig. 12.29 Radiator.

2. Water Pump. The purpose of it is to increase the circulation of the coolant in the cooling system of the engine. It is mounted at the front of the engine in between the engine block and the radiator. It consists of a casing which contains inlet and outlet ports of water The impeller which may be of rotor or disc type is mounted at the one end of the pump shaft and encloses in the casing. A pulley is mounted on the outer end of the shaft by means of hub. The impeller shaft itself is supported in the pump body by a prelubricated ball-bearing assembly as shown in Fig. 12.30.

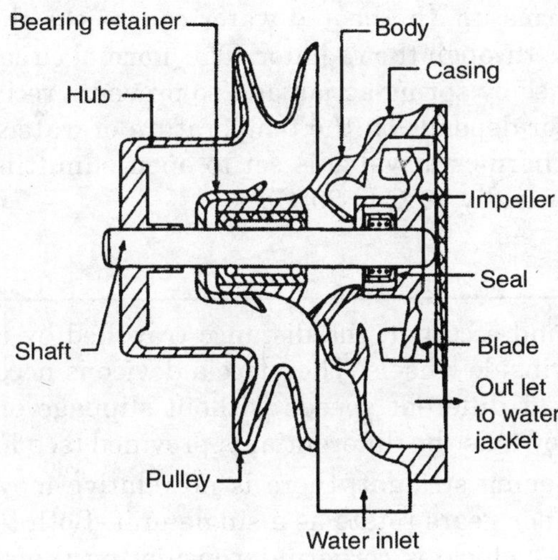

Fig. 12.30 Water pump.

When pulley gets drive from the crank shaft by means of a *V*-belt then the pump is also rotated within the casing. The water available at the impeller and its blades are forced out through the outlet to the engine water jackets by means of centrifugal force. Fresh water from the lower tank of the radiator always flows through the hose pipe to the pump. The leakage of water from the casing side towards bearing side is prevented by means of providing a water seal between two.

3. Cooling Fan. When the vehicle is moving under heavy load and at slow speed then the natural air-draft is insufficient to provide the desired engine cooling. Therefore the system consists of a cooling fan which provides a powerful draft of air through the radiator. The fan as shown in the Fig. 12.31 is made of sheet metal, moulded plastic, nylon or polypropylene material. It is mounted behind the radiator on the water-pump shaft and is driven by the same V-belt that drives the pump and the generator.

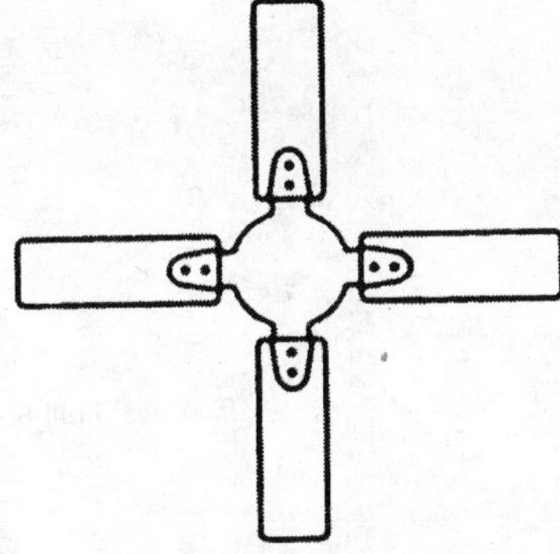

Fig. 12.31 Cooling Fan.

To improve the performance of the fan and to assure the passing of the whole cooling air through the radiator for which the system is equipped with a fan shroud. In modern vehicles to keep the cooling constant at different speeds and to save the unnecessary consumption of power in fan drive, the cooling systems are equipped with a thermostatically controlled power-booster fan that is driven through a liquid clutch which slips after reaching a set speed.

4. Thermostat. Thermostat prevents the water in the engine jackets from circulating through the radiator for cooling until its temperature has reached a level suitable for efficient operation, when the engine is first started and is cold, water from the engine water jackets is sent back to the pump for recirculation. This is done because overcooling results in deterioration of engine efficiency. Thermostat valve is closed and blocks off the flow of water to the radiator through the inlet hose. This results the pump to build up enough pressure to open the spring loaded bypass valve and allow the water to re-enter the pump through bypass, and allows the water to circulate between the pump and the water jackets.

When the engine warms up, the heated water causes the thermostat valve to open and allow the water to circulate through the radiator. This normal circulation, relieves the pressure on bypass valve which closes by spring action and so prevents recirculation. Quantity of water circulating through radiator depends on the temperature of water, which governs how far the thermostat valve opens. Thermostat valve is set to open admitting the water to the radiator at certain temperature (generally 70°C to 80°C).

12.10 DIFFERENTIAL

While travelling around a corner, the distance travelled by the outside wheel is greater than that travelled by the inside wheels. Therefore a device is necessary to permit the wheels, driving wheels to revolve at different speeds without slippage or other damage. To achieve this, a system of gears, known as the differential is provided (see Fig. 12.32).

When the vehicle is going straight, there is no relative movement among the different gears. The case and the inner gears rotate as a single unit. Both the half shafts revolve at the same speed. But when the vehicle is cornering, considering that the case is stationary, then both the sum gears will rotate but in opposite direction. To understand the principle, consider a vehicle with wheel speed N r.p.m. going straight. When it takes a turn towards left, at this

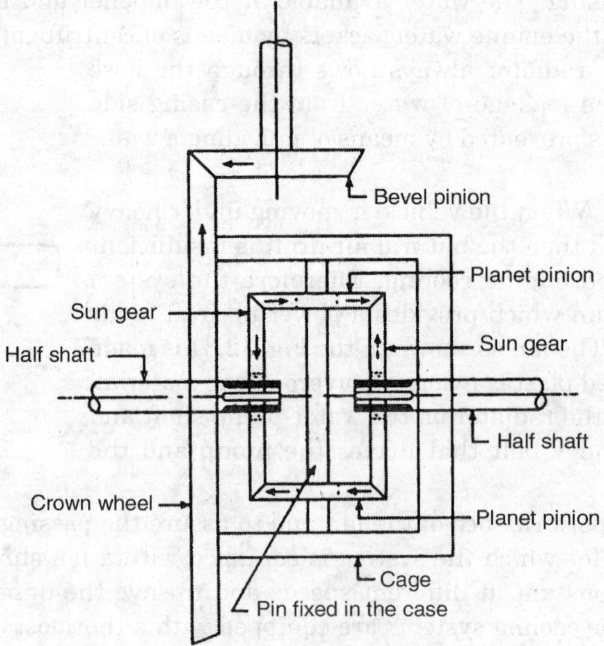

Fig. 12.32 Working of differential system.

time there will be resistance to the motion of the left wheel. Due to differential action, if the left wheel rotates back at n r.p.m., then the right wheel will rotate forward at n r.p.m. because the outer wheel has to cover more distance than the inner. This will give the resultant speed of the right wheel as $(N + n)$ and that of the left wheel as $(N - n)$ r.p.m.

12.11 BRAKES AND BRAKING SYSTEMS

1. Brakes

Brakes are the devices for slowing and stopping the vehicle. The braking operation is reverse of the acceleration to the vehicle. In the acceleration the heat energy of the fuel is converted into the kinetic energy of the vehicle whereas in the braking operation the kinetic energy of the vehicle is converted into heat, which is dissipated into the atmosphere.

When we drive the vehicle the torque of the engine produces a tractive effort at the peripheries of the driving wheels. When the brakes are applied the braking torque introduced at the brake drums produces a negative tractive effort at the peripheries of the braking wheels. In this way the acceleration and deceleration are limited by the friction available between the driving wheels and the ground.

The main purposes of fitting brakes on motor vehicle are as given under.

(*i*) In emergencies to bring the vehicle to rest in the shortest possible distance.

(*ii*) To control the vehicle when it is descending along the hills.

(*iii*) To keep the vehicle in desired position after bringing it in complete rest when there is no driver.

To fulfil the above needs two independent braking systems are provided in the vehicle, namely service brakes and emergency brakes.

2. Braking Systems

In general in a motor vehicle drum type brake system is preferred. If the shoes are provided to rub against the inside surface of the drum then the system is called “internal expanding system”. If the shoes are provided to rub against the outer surface of the drum then it is called “external contracting system”. In “transmission brake system” the brake drum is provided on the mainshaft of the gear box or on the level pinion or worm and in “wheel brake system” the drums are provided on wheels by one on each wheel. The wheel brakes are better from heat dissipation point of view as well require lighter design of parts through which the power is transmitted to the driving wheels.

In a vehicle when brakes are applied to all the wheels by operating a foot pedal then the system is called “foot brake”. When in emergency to park a vehicle, the rear wheels are braked through a separate connecting rod linkage by operating a hand lever then this system is called “hand brake”.

Methods of Actuating the Brakes

Different ways of operating the automobile brakes are as given below.

(*i*) **Manually Operated Brakes.** In this case the braking effort depends upon the physical strength of the driver. It is very common system, the brake pedal or lever may be connected to the actual brake either mechanically, by means of rods or wires or hydraulically, by means of a fluid in a pipe. This system is used on small or light vehicles.

(ii) **Servo Brakes.** When the weight or load of the vehicle is increased then it needs more effort for braking which can be beyond the physical strength of the ordinary driver to apply conveniently. Therefore for help an extra effort is added to the driver's effort in applying the brakes properly which is called "servo brake system". This system is used on a few large cars and to a considerable extent on lorries.

(iii) **Power Operated Brakes.** When driver does not apply any braking effort and operates the braking system by controlling a compressed air valve then this system is called "power operated brakes". This system is confined to heavy lorries especially those used with trailers.

Mechanical Brakes

When effort from the pedal is transmitted to the wheel brakes by means of cables, rods or shafts then this system is known as mechanical brakes. The forces between the brake pedal and brake shoes are developed or multiplied by means of different levers, linkages, cams, toggles or wedges to produce effective braking. The common terms and mechanisms of mechanical brakes are as explained under:

Hydraulic Brakes

When a fluid enclosed in a pipe is used to transmit the pedal effort to road wheel instead of the use of rod or cable, the system is called hydraulic brake. The outline and arrangement of the system is clear from the Fig. 12.33. The system consists of a master cylinder and piston which is connected by steel piping to hydraulic wheel cylinders. A hydraulic cylinder is provided at each wheel. A wheel cylinder consists of two pistons which are pivoted to brake shoes.

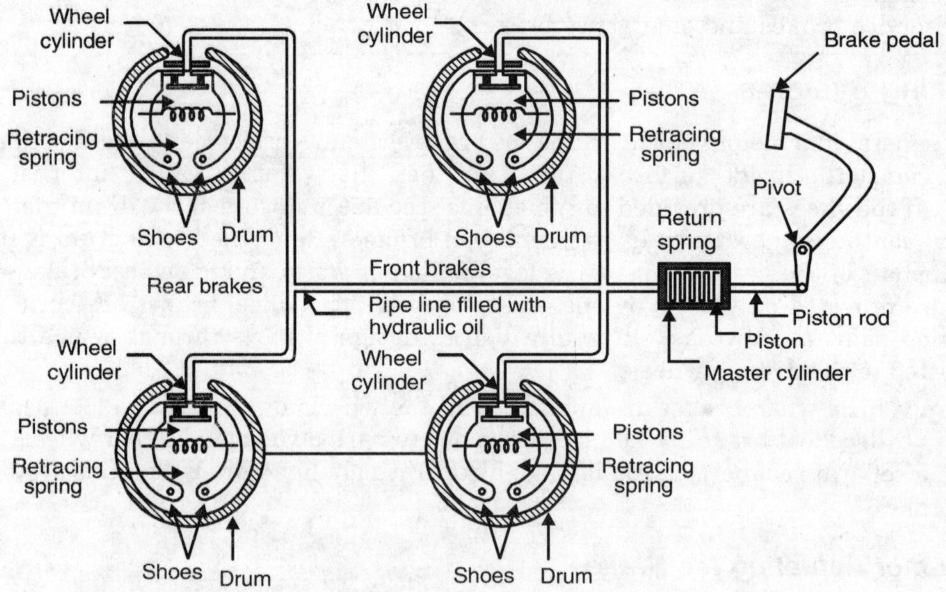

Fig. 12.33 Hydraulic brake system.

The air of the system is replaced by filling the brake oil under light pressure. The pedal lever is linked to the master-cylinder piston by means of a piston rod. When the pedal is depressed the master cylinder piston is forced into the cylinder thus increasing the pressure on the brake oil. Therefore due to this pressure the piston of the wheel cylinders are forced outwards and thus expanding the brake shoes against the drums.

When the brake pedal is released it reduces pressure on the hydraulic fluid. At the same time the retracing springs pull the brake shoes out of contact with the drums and force the wheel pistons back to their original positions. In this way the brake fluid is returned from the wheel cylinders to the master cylinder. Now the brake pedal is again ready for the next operation.

The main units and their constructions of this system are described as under:

***(i)* Master Cylinder.** It is a central unit provided near the brake pedal. It is either fixed on the frame of the vehicle or on the engine side of the fire wall as shown in Fig. 12.34. The piston of the unit is pivoted with the brake pedal by means of a link. This unit supplies fluid under pressure to the wheel cylinders and collects the excessive fluid of the lines when brakes are released.

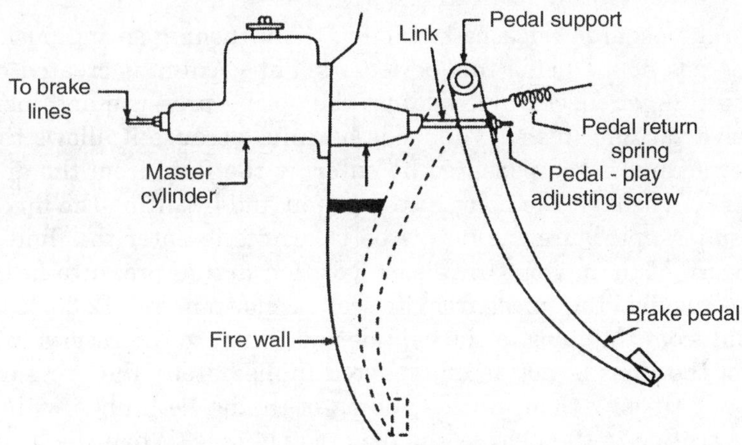

Fig. 12.34 Mounting of master cylinder.

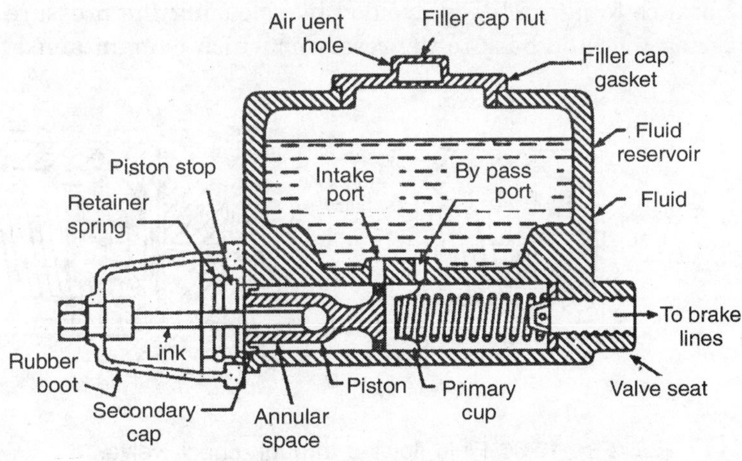

Fig. 12.35 Schematic view of master cylinder.

Regarding constructions and function of the master cylinder refer Fig. 12.35 where the schematic view of this unit is shown. The body of the cylinder is integral with the fluid reservoir which is filled with fluid in order to keep the cylinder submerged and to keep the air out of the unit. The fluid from reservoir is connected to the cylinder by means of intake port and by pass port. The top side of the reservoir is provided with a filler cap and air vent hole. The piston inside the cylinder is provided with primary cup and secondary cup for sealing purposes. The left side of the piston is connected to the brake pedal by means of a link and right side is provided with a return spring which retains the check valve on its seat. The left end of the cylinder is provided with a piston stop, retaining spring and rubber boot to protect the cylinder from the dirt and dust. The right end of the cylinder is connected to the brake lines.

When the brake pedal is depressed the link moves to the right against the force of the spring, till the piston covers the by-pass port. If the piston moves continuously then it builds up pressure in the spring side of the cylinder. This pressure allows to pass the fluid from cylinder to the brake line by depressing the lip of the check-valve as shown in Fig. 12.36 (*i*). This fluid under pressure enters the wheel cylinders and moves the pistons outwards thus pressing the brake shoes against the rotating drums.

When the brake pedal is released suddenly, the returning spring pushes the piston back to its left extreme position. During this period a slight vacuum is created in the cylinder due to delay in the opening of check valve against the spring pressure and inertia of the fluid in the lines. Excessive vacuum in the cylinder is harmful because it allows the air to enter into the system. This vacuum is compensated by entering the fluid from the annular space of the piston to the cylinder through the holes in the piston and by lifting the lip of the primary cup. Under the atmospheric pressure the intake port permits to enter the fluid from the reservoir to the annular space of the piston. In release position due to pressure in the brake lines the check valve is lifted against the spring from its seat as shown in Fig. 12.36. (*ii*). There-by entering the excessive fluid from the lines to the cylinder and then to the reservoir via bypass post. If this extra fluid of the lines is not accommodated immediately, the pressure in the lines will not be relieved fully. It may then, cause rubbing of the brake linings with drum. Due to this rubbing more generation of the heat at the drum is possible. When the heat is transmitted to the wheel cylinder then the fluid available in the cylinder by heating tends to expand the shoes, which causes the brakes to jam. This is avoided by releasing the pressure of the brake lines by means of delivering the fluid back to the reservoir which is maintained at the atmospheric pressure.

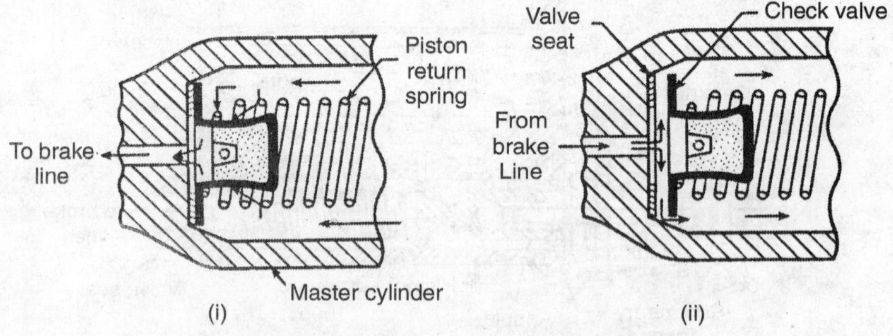

Fig. 12.36 Fluid flowing through check valve.

***(ii)* Wheel Cylinder.** It is an expander unit provided on each wheel by means of fixing on the back plate. Refer Fig. 12.37, where the sectional views of this unit are shown. The cylinder is kept open at both the ends. It consists of two pistons attached to the brake shoes by means of links. At the inner faces of the pistons two cup washers and a light spring between them are provided for sealing purposes. The ends of the cylinder are covered with the rubber boots which keep the cylinder free from the dust.

When from master-cylinder the fluid under pressure enters into the wheel cylinder it forces the pistons outwards, thereby pushing the brake shoes against the drum and applying the brakes.

In few vehicles to avoid the excessive expansion or vaporisation of the fluid, the wheel cylinder is mounted outside the drum, where it is better cooled. In some systems a girling-type expander is provided, which is operated by a hydraulic piston.

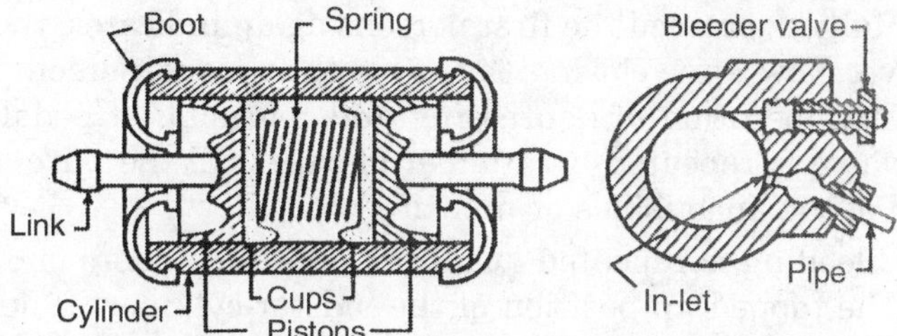

Fig. 12.37 Sectional views of wheel cylinder.

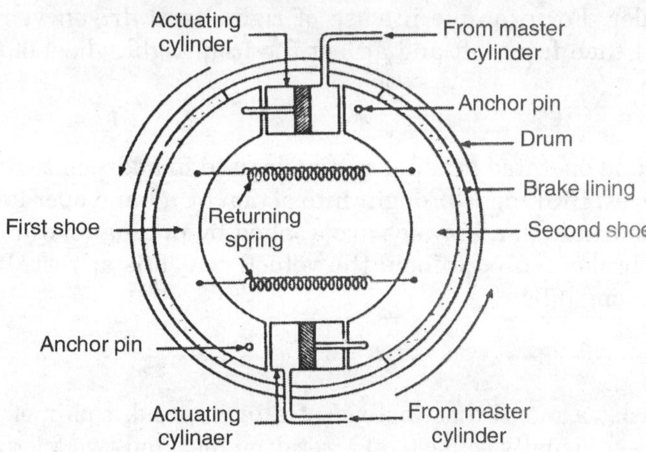

Fig. 12.38 Hydraulic, two leading shoe brake.

***(iii)* Hydraulic Two Leading Shoe Break.** The principle of this s-ystem is also similar to that of mechanically operated two leading shoe brake but it is provided with two single acting actuating cylinders instead of cams. One end of the each shoe is hinged on the fixed anchor pin as shown in simplified sketch Fig. 12.38.

Bleeding of Brakes

In hydraulic brakes the air enters into the system through the joints or when the fluid level has become low. The air is compressible therefore it eliminates the high braking pressure in the system and permits a soft or spongy pedal action. This results a poor braking action. Therefore it is necessary to remove any air available in the system. The procedure of removing air out of the braking system is called "bleeding".

For bleeding purpose a bleeder valve is provided at the highest point of the wheel cylinder. The bleeder valve consists a bleeding screw and a nipple which are provided at the back side of the anchor plate. The nipple is fitted with a rubber tube. The other end of the rubber tube is dipped into fluid contained in a clean glass jar. First be sure that all the bleeder valves of the cylinder are closed then top up the master-cylinder completely with the brake fluid.

The operation is done by two persons one sits at the driver's seat for operating the brake pedal while the other operates the respective bleeder screw at the wheels. The first person after pedaling, presses the brake pedal and indicates to the second. Then second person opens the bleeder valve thereby coming out some air from the line through the rubber pipe, in the form of bubbles produced in the glass jar.

When the pedal is fully depressed the first persons again indicates, then the second person closes the bleeding valve. Now after closing the valve the second person indicates to the first to release the pedal. This operation of depressing and releasing of pedal as well as opening and closing of bleeder valve is repeated by number of times till the pure fluid comes out from the rubber pipe instead of the air bubbles or mixture.

This procedure of bleeding is repeated at all the wheel cylinders one by one. Mean while it is essential to check the topped up position of the master-cylinder. If desired the fresh fluid may be filled in the reservoir or master-cylinder. For fully bleeding first bleed the furthestmost wheel cylinder, then second furthest, then third furthest and at last the nearest wheel cylinder from the master-cylinder. For example in case of right hand driven vehicle, first bleed the rear left then rear right then front left and at last the front right wheel of the vehicle.

The Hand Brakes

An independent hand operated braking system is used in addition to the hydraulic braking system in modern vehicles. Braking is brought into action by a hand operated lever with trigger ratchet and toothed quadrant. Hand brakes are applied by pulling forward the hooked handle of the lever. The hand brake is used to hold the vehicle which is at rest. It also helps the foot pedal when descending long hills.

Disc Brakes

These brakes consist a metal disc instead of a drum, and a pair of pads instead of the curved shoes. A disc brake usually consists of a rotating disc and two friction pads actuated by four hydraulic wheel pistons. The outer surface of the disc is fastened to the wheels. Pressure on the friction pads may be actuated by pull rod or by hydraulic system. In case of rear wheel brake the brake assembly is attached to the axle housing, while in the case of a front wheel brake, it is secured to the steering knuckle.

Brakes for Heavy Vehicles

Since heavy vehicles require heavier braking effort than can be furnished by the driver's foot pressure, powered brakes are used. These can be of following types:

***(i)* Air Brakes.** The brake valve controls brake operation by directing the flow of air from a reservoir against the diaphragms in the brake chambers during brake application, and from the brake chambers to the atmosphere during release. An air compressor, driven by the engine, furnishes compressed air to the reservoir whenever the air pressure in the reservoir falls a set value. These brakes are popular in very heavy to heavy vehicles.

***(ii)* Hydraulic Servo Brakes.** When extra oil under pressure supplied by a pump driven by the engine or some part of the transmission system is used to assist the braking effort the system is called *hydraulic servo brakes.* Refer Fig. 12.39, where the simplified mechanism of this system is shown. It consists of a pump which forces the fluid from a tank through the servo unit back to the tank as shown by the directions of the arrows, until the brake pedal is depressed. The inlet pipe is connected at *A* and the outlet at *B* to the servo unit. The servo unit consists a master cylinder with piston, which is provided with pipes for brakes. A reservoir is connected to the master cylinder for providing brake fluid. A plunger is provided in the right side portion of the unit, which is pivoted with the brake pedal rod. A space, *S* is provided between the master-cylinder-piston and plunger for building up the fluid pressure.

When brake pedal is at off position then the plunger rests at its right extreme position thereby flowing the fluid through space *S* from inlet *A* to the outlet *B* of the unit. When the brake pedal is depressed it results cut-off the fluid at outlet *B*. Therefore, the pressure build up in the space *S* pushes the master-cylinder-piston towards left thereby applying the brakes. When the pedal is released the pressure of fluid in space *S* is exerted on the left side of the plunger thereby pushing the plunger towards right and opening the passage at outlet *B*. The fluid available in the space, right side the plunger is forced gradually through the valve back to the space *S*. When the engine or vehicle is at rest, in the case the servo system is not operative and the complete unit functions just as an ordinary hydraulic brake system. In some systems an accumulator is provided to utilise the servo brakes at all the times.

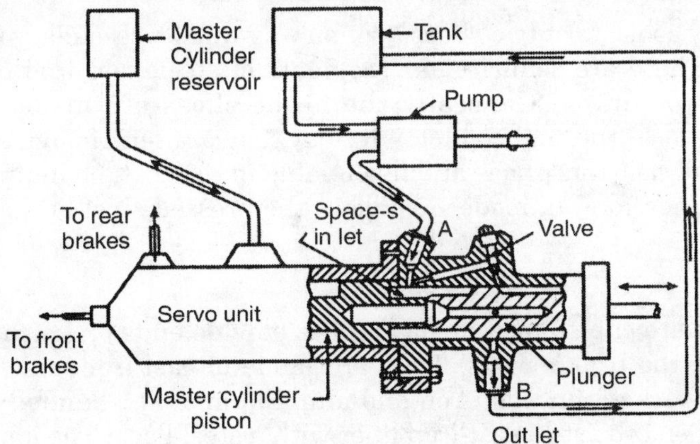

Fig. 12.39 Hydraulic servo brakes.

***(iii)* Vacuum Servo-brakes.** When the vacuum is obtained from the manifold of the engine or a separately driven exhauster used to assist the braking effort the system is called *vacuum servo-brakes.* Refer Fig. 12.40, where a simplified sketch of this system is shown. The system consists of vacuum reservoir connected through a non-return valve to the inlet manifold of the engine. Two connections from vacuum reservoir, one on each side of the piston of the servo cylinder is provided ; on left side the connection is through the control unit whereas right side is connected directly. The piston of the servo cylinder is further connected to the piston of the boost cylinder or the brake linkage. The control unit consists of a piston to which two valves are attached. The lower valve controls the connection between the reservoir and the right side of the servo cylinder piston. The upper valve controls the connection between the atmosphere and the left side of the servo cylinder piston. The other side of the piston of the control unit, is actuated by the pedal effort through a master cylinder.

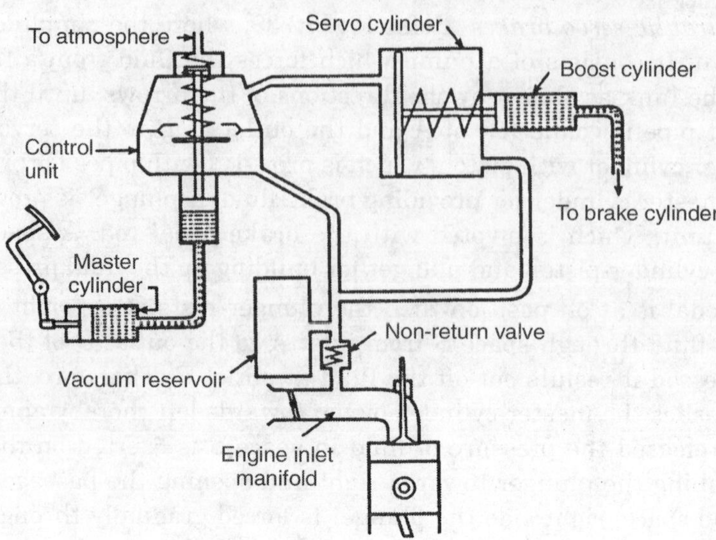

Fig. 12.40 Vacuum servo-brakes.

Brake Shoes

A wheel generally consists of two shoes, primary as well as secondary shoe. In some cases consists three shoes. These are semicircular segments of steel, cast iron or aluminium, with special treated asbestos lining. Generally the brake shoes are made of 'T' section. The aluminium shoes have good thermal conductivity but they are not enough strong to withstand the braking torque. The cast-iron shoes are also unable to resist the handling stresses induced due to braking force. Therefore, in modern vehicles the pressed steel brake shoes are used.

Brake Drums

The brake drums are cup shaped members, the outside ends of which are closed and the inside open to provide the brake shoes. They are made of cast-iron, steel, aluminium, cast-iron and steel, chrome-nickel iron, cast-iron and aluminium or steel and aluminium back. The most modern construction consists a cast-iron liner with a steel-back. The inner cast-iron surface provides a high coefficient of friction and dissipates the heat rapidly, while the steel back provides the strength to the drum. To dissipate the brake heat into the atmosphere, drums are provided with finnings or ribbens at their outer surfaces. The passenger-car brake drums are made between 200 mm and 375 mm in diameter. The inside diameters are made slightly greater than the diameter of shoelinings circle.

Brake Linings

As regard frictional materials is concerned these already have been discussed in the chapter of clutches, however the brake linings are required to provide friction in *dry* state. These are made in two main types as given under.

***(i)* Solid-Woven Type.** It is non-metallic lining made of asbestos spinned fibres. The coefficient of friction of asbestos is 0.4 upto about 250°C temperature. It can resist the temperature upto about 350°C without serious fade. This variety can be reinforced and strengthened by including brass or zinc wires and then impregnating it with a bonding material. The metallic wires help in dissipation of heat from the linings but provide a typical structure. Now these days the linings are made from entirely spinned asbestos threads, without using the metallic wires.

***(ii)* Moulded or Composition Type.** This type of lining is made by composing the natural asbestos fibres with the help of bonding materials. The mixed material is moulded in dies at elevated temperature thus converted into a solid mass. Wearing quality of the linings can be increased by including metallic wires. The temperature resistance of this variety varies from 400 to 450°C.

Brake Fluid

The oil which is used to transmit the braking force from the brake pedal to the brake drum in the hydraulic brake system is called *brake fluid.* The characteristic of this fluid are, that it should give a very flat viscosity curve over a very wide temperature range. It must not freeze or thicken unduly at the sub zero temperatures. It must have a wide margin against grassing at the elevated temperature. It must not cause chemical attack on the metallic parts as well as rubber seals. A special castor-oil based and compounded with suitable diluents fluid is usually used as brake fluid. In no case should a mineral oil be used for braking, since it causes immediate swelling of rubber seals and parts.

12.12 AUTOMOBILE ELECTRICAL SYSTEMS

Satisfactory operation of the electric equipment and system is most important in an automobile. Electrical equipment of an automobile as a whole consist of the six elements, namely ; generating circuit, starting circuit, battery circuit, ignition circuit, lighting circuit, and branch circuit.

Generating circuit or the charging circuit consists of a generator and its control equipment. Its purpose is to generate electricity for refilling or charging the battery when the motor vehicle is running. The starting system helps to turn the engine over until it fires for running under its own power. The battery system provides required amount of current for starting the engine and also operates most of the units. The ignition circuit, lighting circuit and the branch receives the current directly from the battery, but when the generator is operating the current is drawn by them directly from the generator. Branch circuit operates different accessories like horns, wind screen wipers, radio sets, heater, etc.

12.12.1 Automobile Battery

It supplies the current, when the engine is being cranked, and also when the generator is not charging. Thus it is the secondary source of electrical energy when vehicle is not operating the generator speed is insufficient to meet the full load requirements. The batteries are of 3 types : Lead acid battery, Nickel-alkaline battery, and Zinc-air battery.

Lead-Acid Battery

Modern vehicles mostly use these batteries, since these are capable of supplying larger amount of electrical current required by self-starter motor and is relatively cheap with longer life. The principle is that when two rods, one of pure lead and other of lead peroxide are placed in a solution of dilute sulphuric acid, then an electric voltage is generated between the two.

The lead peroxide, brownish in colour forms positive pole, while the pure lead is grey and form the negative pole. This type of simple cell generates 2 volts. Thus to form a 6 volt or 12 volt battery 3 or 6 cells are connected in series. To increase the surface area and current capacity of the battery, it is provided with plates instead of rod poles.

Battery Construction

The main components of the battery are (see Fig. 12.41):

***(i)* Container.** This is one piece moulded case made of hard rubber, plastic or bituminous composition. The material must withstand extremes of heat, cold, mechanical shocks and must be resistant to the absorption of acid. It is divided into compartments for different cells by partitions.

***(ii)* Plates.** A cell consists of two types of plates, the positive and the negative. Material for positive and negative plates are lead peroxide and spongy lead respectively.

***(iii)* Cell Element.** All the positive plates of the cell are welded to a post strap to form a positive group. Similarly negative plates are welded to another post strap to form a negative group. One positive and one negative group of plates are slid over each other with separators in between the plates. This assembly called an 'element' is inserted into battery compartment to form a cell.

***(iv)* Separators.** To avoid short circuiting of positive and negative plates, thin sheets of non-conducting porous material called separators are provided between them. These are made from chemically treated wood, micro-porous rubber, plastic, fibre glass sheets etc.

***(v)* Cell Covers.** Cell covers are provided to seal the battery on the top of the container. They are moulded from hard rubber and contain holes to protrude two terminal posts through them a vent and filler opening.

***(vi)* Cell Connector.** The cells of the battery are connected in series by cell connector.

***(vii)* Terminals.** The battery terminals are made slightly tapered, and are of standard dimensions (17.5 mm dia of positive, and 16 mm dia of negative terminals).

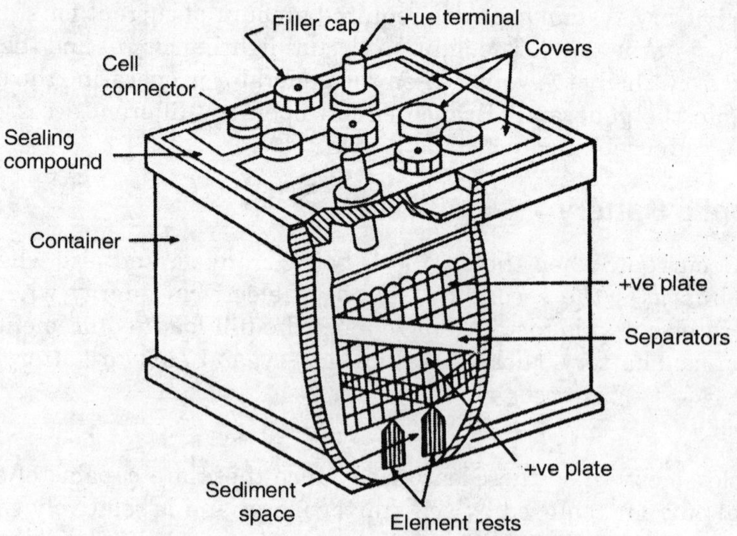

Fig. 12.41 Battery construction.

***(viii)* Electrolyte.** The solution of sulphuric acid is called the 'electrolyte' and covers the plates. It contains 31 percent sulphuric acid and 69% water by weight and has a specific gravity of 1.230 at 27°C. During discharge process, water is produced and thereby reduces the gravity of the acid, and ultimately fully discharged when gravity is 1.110.

Chemical Reactions

The battery cells converts chemical energy into electrical energy during discharge, and electrical energy into chemical energy during charge.

***(i)* Reactions During Discharging.** During discharge H_2SO_4 is broken up into H_2 and SO_4.

$$\underset{\text{Positive plate}}{PbO_2} + \underset{\text{Sulphuric Acid}}{2H_2SO_4} + \underset{\text{Negative plate}}{Pb} - \underset{\text{Lead Sulphate on positive plate}}{PbSO_4} + \underset{\text{Water}}{2H_2O} + \underset{\text{Lead sulphate on negative plate}}{PbSO_4}$$

The H_2 reacts with PbO_2 and forms PbO which further reacts with a part of H_2SO_4 and forms $PbSO_4$ and H_2O. Thus during discharging $PbSO_4$ arid H_2O are formed continuously and the available H_2SO_4 is diluted.

***(ii)* Reactions During Charging.** The discharged battery is recharged by passing the current through it in the opposite direction by means of an external source. The chemical reactions that takes place in the battery are just reverse of that of during discharging.

$$PbSO_4 + 2H_2O + PbSO_4 \rightarrow PbO_2 + 2H_2SO_4 + Pb.$$

Battery Rating

The voltage of battery depends upon the number of cells connected in series in the battery. But the amount of current being delivered by the battery depends on total volume and area of active plate material, and the strength and amount of electrolyte. The rated capacity of the battery is indicated by Ampere-hour and is denoted by AH.

Alkaline Battery

These have construction similar to that of the lead-acid battery. Nickle hydroxide is used as the active material on positive plates, while metallic iron or cadmium on the negative plates. Each cell in these batteries has 1.2 volts.

Zinc-Air Battery

It gets the energy from the zinc provided in the cell as well as from the atmospheric oxygen.

Battery Maintenance

In order to maintain the battery in proper working condition, the following measures should be taken:

(i) To make the loss of water, the level of electrolyte in the battery should be checked and brought to the required level by putting in distilled water. Correct level is, about 12.5 mm above the plates.

(ii) Top of the battery should be kept clean and dry. Corroded cable connectors should be scrapped, cleaned, and coated with anti-corrosive grease.

QUESTIONS

12.1 Draw a layout of a truck chassis indicating main components on it.

12.2 What are the functions of a frame? Describe the construction of a frame of an automobile.

12.3 What are the basic requirements of the suspension systems in a vehicle?

12.4 Write short notes on: Leaf springs, action of a telescopic type of shock absorber.

12.5 Discuss the terms: "Wheel-alignment", "wheel balancing".

12.6 (*a*) Describe the steering mechanism in an automobile.

(*b*) Describe the different types of steering gears.

12.7 Explain with the help of a neat sketch the construction of an automobile truck tyre. What care should be taken for the proper maintenance of the tyre?

12.8 Write short notes on:

tyre rotation, tyre chain, tubeless tyres, retreading or recaping of a tyre.

12.9 (*a*) List types of tyre wear, their causes and remedies.

(*b*) What are the factors influencing tyre life?

12.10 What is the function of a clutch? Explain the working of a single plate clutch with the help of a simple sketch.

12.11 (*a*) What, are the essential properties of a clutch lining material?

(*b*) Explain different type of clutches, in brief.

12.12 Describe with the help of a neat sketch the various gear positions in a sliding mesh gear box.

12.13 Explain the common faults encountered in gear boxes, their causes and suggest remedies.

12.14 What is the function of "universal joints" in power transmission? Explain with the help of a neat sketch the construction of a propeller shaft.

12.15 What is the necessity for cooling of an engine. Enumerate different methods of engine cooling.

12.16 Explain major components of water cooling system for an automotive engine.

12.17 Write short notes on:

(*a*) Utility of a thermostat in the cooling system.

(*b*) Mixing of anti-freeze with coolants.

(*c*) Causes of engine overheating.

12.18 Describe the function of a "differential gear".

12.19 What are the requirement of automobile brakes? Describe any one type of mechanical brake.

12.20 (*a*) Explain why mechanical brakes are used for parking and emergency brakes.

(*b*) Compare disc brakes with drum-type in regard to construction and operation.

12.21 (*a*) Describe with a neat sketch, the hydraulic brake system of a vehicle.

(*b*) What are the advantages of hydraulic brakes over mechanical brakes?

12.22 Write short notes on

(*a*) Bleeding of hydraulic brakes, (*b*) Master cylinder.

12.23 Explain with the help of a neat sketch the construction and working principle of a lead-acid battery.

12.24 What do you understand by discharging and charging of a battery? Explain the chemical reactions occurring during these two processes.

12.25 Explain the working principle and function of starter motor in an automotive.

12.26 Explain the principle of generator. How it differs from the starter.

12.27 Write short notes on:

(*a*) Bendix starting drive,

(*b*) Solenoid shift system

(*c*) Voltage regulator, current regulator and cut-out.

Chapter

13

GAS TURBINES

Learning Objectives

After studying this chapter, the students will learn:

- Gas turbines and their applications.
- Classification of gas turbine plants.
- Open cycle and closed cycle gas turbine plants and their efficiency.
- Components of gas turbine plants.
- Governing system for gas turbine plants.
- Fuels for gas turbine plants.
- Jet propulsion.

13.1 INTRODUCTION

The gas turbine is a rotary heat engine operating by means of a series of processes consisting of compression of air, increase of air temperature by the combustion of fuel, expansion of hot gases to the atmosphere, the whole being a continuous flow process. Thus, it is similar to the internal combustion, but is akin to the steam turbine in its aspect of the steady flow of the working medium. The compression and expansion processes are carried out in turbo-machines that is by means of rotating elements in which the energy transfer between fluid and rotor is effected by means of kinetic action.

Thus, in its simplest form a gas turbine consists of a compressor, a combustion chamber, and a turbine unit. Air acts as a working fluid and is compressed in the compressor and the energy is added to it in the combustion chamber. The high energy fluid is then expands in the turbine and thus mechanical energy is produced. The part of this energy is used up in driving the compressor, which is usually mounted on the same shaft as that of the turbine, and rest of the energy is utilised for various purposes. The net work is therefore the difference between the turbine work and work used-up in driving the compressor.

The gas turbines are the most economic type of power plants when they are used as intermittent or peak load plants in combination with the base load plants. These plants are cheaper in capital cost compared with steam plants of the same size. The fixed charges of these plants are also lower as compared to those for steam power stations.

These plants are highly reliable, flexible, quick starting and require less space. These are ideally suitable as peak load stations. Because of its low weight per unit power, the gas turbines are generally used to drive aviation systems. It is also being increasingly used in land vehicles like buses and trucks and also to drive locomotives and marine ships. However, the use of gas turbine in the power generation industry is more recent. Earlier, the turbine plants have been used in remote oil and gas fields, on offshore platforms and in gas-pipeline pumping stations. These were also used where fresh water is scarce, fuel supplies are cheap and space available was limited.

The rapid progress have been observed, in about last two to three decades, in the development of the gas turbine plants for electric power generation.

Applications of Gas Turbine Plants

Gas turbine plants have the following applications.

1. These may be used as peak load plants and stand-by plants to steam, diesel or hydro-power plants.
2. These are used in jet aircrafts and ships.
3. To work as combination plants.
4. To supply mechanical drive for auxiliaries.
5. These are also used in industrial applications ranging from petro-chemical, thermal process industries. Gas turbines are more economically used near oil fields because natural gas is available as by-product. They are also economically used in combination with steam plant because high temperature exhaust gases coming out from gas turbines can be used to generate steam which can be used in steam turbine.

Advantages and Disadvantages

Advantages and disadvantages of the gas turbine power plants (GTPP) are as follows:

(A) Advantages Over Steam Power Plants

The economics of power generation by gas turbines is proving to be more attractive, due to low capital cost, and high reliability and flexibility in operation. Quick starting and capability of using wide variety of fuels from natural gas to residual oil or powdered coal are other outstanding features of gas turbine power plants. Major progress has been made in three directions namely, increase in unit capacities of gas turbine units (50-100 MW), increase in their efficiency and drop in capital cost. Primary application of gas turbine plant is to supply peak load. However, gas turbine plants now a days are universally used as peak load, base load as well as stand by plants.

1. It is smaller in size and weight as compared to an equivalent steam power plant. For smaller capacities the size of the gas turbine power plant is appreciably greater than a high speed diesel engine plant but for larger capacities it is smaller in size than a comparable diesel engine plant. If size and weight are the main consideration such as in ships, aircraft engines and locomotives, gas turbines are more suitable.
2. The initial cost and operating cost of the plant is lower than an equivalent steam power plant. Civil engineering costs of these plants are very low.
3. The plant requires less water as compared to a condensing steam power plant.
4. The plant can be started quickly, and can be put on load in a very short time.
5. There is no stand by losses in the gas turbine power plant whereas in steam power plant these losses occur because boiler is kept in operation even when the turbine is not supplying any load.
6. The maintenance of the plant is easier and maintenance cost is low,
7. The lubrication of the plant is easy. In this plant lubrication is needed mainly in compressor, turbine main bearing and bearings of auxiliary equipment.
8. The plant does not require heavy foundations and building.
9. There is great simplification of the plant over a steam plant due to the absence of boilers with their feed water evaporator and condensing system.
10. It is very reliable.
11. A wide variety of fuels can be used.
12. It has fewer auxiliaries.
13. No coal handling and ash handling is required.

Disadvantages

1. Major part of the work developed in the turbine is used to derive the compressor. Therefore, net work output of the plant is low.
2. Since the temperature of the products of combustion becomes too high so service conditions become complicated even at moderate pressures.
3. Part load efficiency is low.
4. High air rate is required to limit the maximum inlet temperature, as a result of which the exhaust losses are high, unless the waste heat in it is utilised.
5. Air and gas filters have to be of very high quality so that no dust enters to erode and corrode turbine blades.

(B) Advantages Over Diesel Power Plants

A gas turbine has the following advantages over diesel power plants:

1. Its weight-power ratio is less.
2. The turbine is vibration-free as there are no imbalanced forces.
3. The operation cost is low as low grade fuel like kerosene or paraffin can be used.
4. The number of moving parts are less, therefore the mechanical efficiency is high.
5. Lubrication system is simple.

13.2 CLASSIFICATION OF GAS TURBINE PLANTS

Gas turbine plants may be classified according to the following criteria:

1. According to the Cycle:

(*a*) Open cycle plants. (*b*) Closed cycle plants.

2. According to the Type of Load:

(*a*) Peak load plants. (*b*) Standby plants,
(*c*) Base load plants.

3. According to the Application:

(*a*) Aircraft. (*b*) Locomotive,
(*c*) Marine. (*d*) Transport.
(*e*) Power station.

4. According to the Number of Shafts:

(a) Single shaft. (*b*) Multi-shaft.

5. According to the Fuel Used:

(*a*) Liquid fuel. (*b*) Solid fuel,
(*c*) Gaseous fuel.

13.3 OPEN CYCLE GAS TURBINE PLANTS

A simple open cycle plant is shown in Fig. 13.1. In this case fresh atmospheric air is drawn into by the compressor (C) continuously and heat is added by combustion of fuel in the working fluid (air) itself. The products of combustion are expanded through the turbine (T) and exhausted to atmosphere. For a given flow of working fluid there is a limit to the amount of fuel which can be burnt in the combustion chamber and the maximum amount of fuel which can be added to combustion chamber is governed by the working temperature of highly stressed turbine blades which should not be allowed to exceed to certain limit. This limiting value of temperature is dependent upon two factors namely creep strength of the material used in the construction of the turbine blades and the working life of the system.

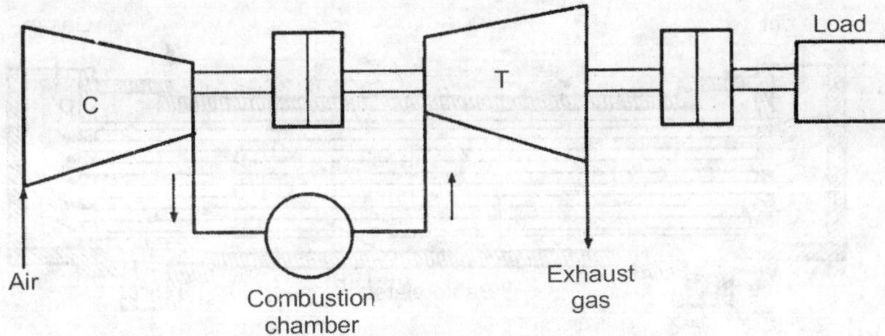

Fig. 13.1 Open cycle gas power plant.

Combustion of fuel takes place either at constant volume or at constant pressure. Theoretically the thermal efficiency of the constant volume cycle is more than the constant pressure cycle. But there are some practical difficulties in the case of constant volume cycle such as valves are necessary to isolate the combustion chambers from the compressor and the turbine. Thus combustion is intermittent and does not allow the smooth running of machine. Due to these defects this cycle has not been widely accepted. In case of constant pressure cycle, combustion is a continuous process and valves are not necessary. This system has proved to be quite satisfactory.

13.3.1 Refinement (increase in efficiency) of Simple Open Cycle

Refinement of open cycle is done by providing regenerator, intercooler and reheating combustion chamber (Fig. 13.2). This arrangement increases the efficiency by about 30%.

1. In simple open cycle system, the heat of the turbine exhaust gases goes as waste. To make use of this heat a *regenerator* (Fig. 13.3) is provided. In the regenerator the heat of hot gases coming from the turbine is utilised in preheating the air entering the combustion chamber. Air circulates all around the tubes of regenerator and the exhaust gases pass through the tubes. This arrangement saves some amount of fuel because less heat will be required in combustion chamber (C.C.) and consequently thermal efficiency is improved. However, there will be small reduction in power output because of pressure losses in this working fluid (air) while passing through the regenerator but it is negligible against the improvement in thermal efficiency.

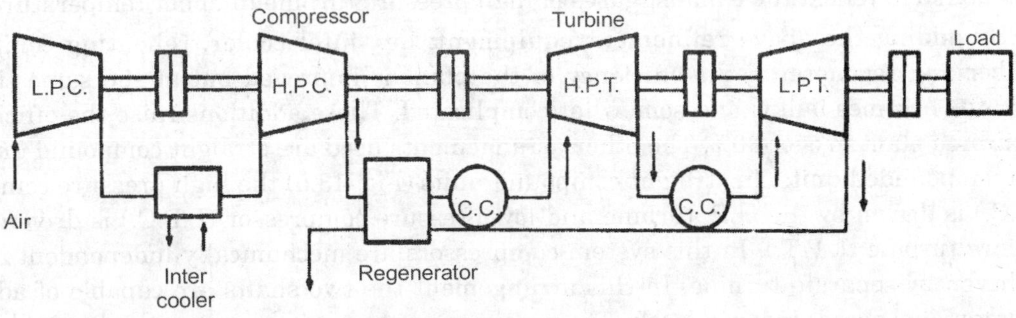

Fig. 13.2 Refined open cycle.

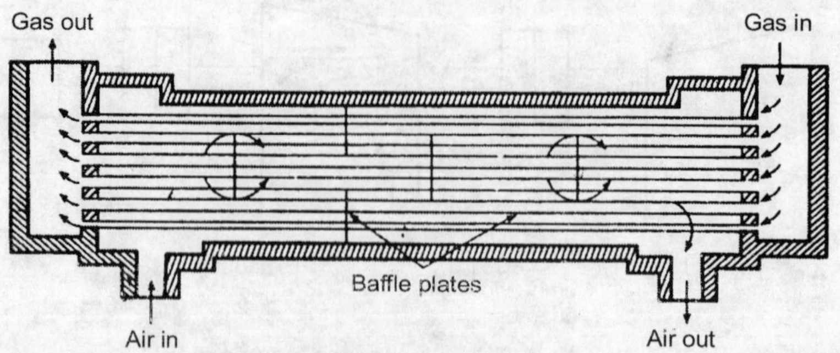

Fig. 13.3 Regenerator

2. For further improvement of the cycle, an ***intercooler*** is placed between low pressure compressor and high pressure compressor. In this case the compressor will have to do less work to compress the air (because of reduction in volume). Fig. 13.4. shows an intercooler. In the inter-cooler water is passed through the tubes and air circulated all round the tubes.

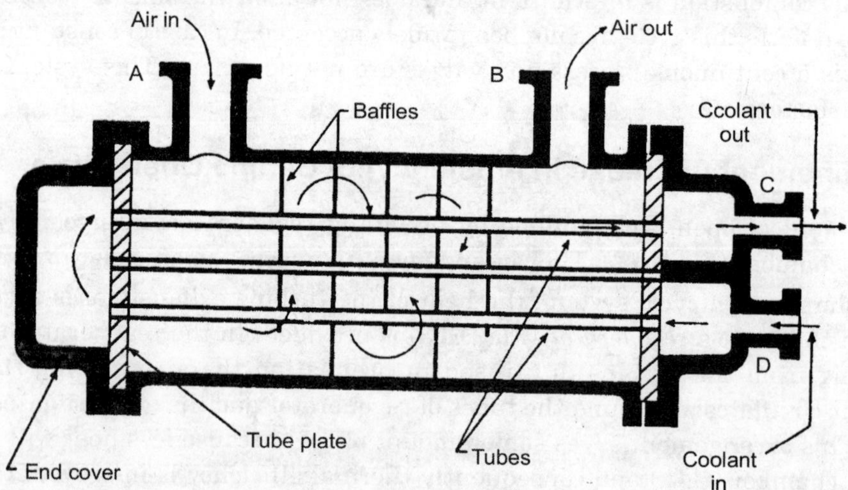

Fig. 13.4 Intercooler

3. The output can be further improved by providing a ***reheating combustion chamber*** between high pressure turbine and low pressure turbine. In reheating combustion chamber fuel is added to reheat the exhaust gases of high pressure turbine to upper temperature limit.

By adding the above refinement equipment, i.e., inter-cooler, reheating combustion chamber and regenerator, the efficiency of the plant is improved but at the same time the apparatus becomes bulky; and somewhat complicated. These additions raise the efficiency of gas turbine plant to over 30%. The other arrangements used are straight compound units and cross compounded units. In straight compound units (Fig. 13.5) the high pressure compressor (H.P.C.) is driven by the H.P. turbine and low pressure compressor (L.P.C.) is driven by low pressure turbine (L.P.T.). In this system compressors are mechanically independent as, they are driven by separate turbine. In this arrangement the two shafts are capable of adjusting their rotational speeds independently. The turbo-compressor unit may operate at variable speeds while turbo-generator unit can be maintained at constant speed.

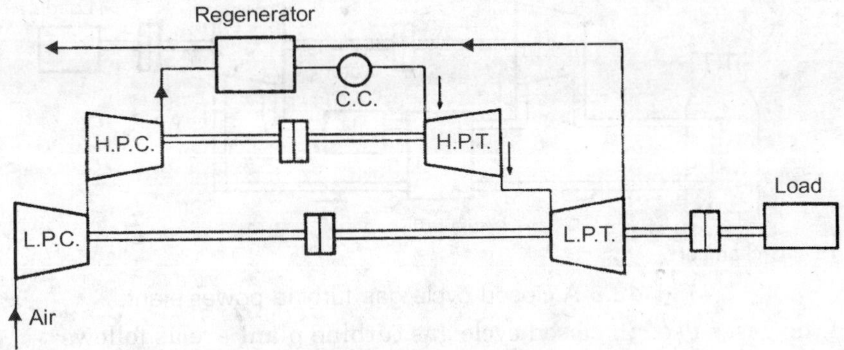

Fig. 13.5 A straight compound unit.

Advantages of Open Cycle

The advantages of open cycle are as follows:

1. **Simplicity.** The combustion chamber is lighter in weight and smaller in size with a high rate of heat release. Secondly the ignition system is simple only a spark is required for a short period to start the burning after which the combustion continues. The combustion chamber may be designed to burn almost any of the hydrocarbon fuels ranging from gasoline to heavy diesel oil including solid fuels.
2. **Vibrationless.** In this system the, moving or rotating parts being the rotor (consisting of turbine and compressor connected by a shaft) and the gear trains that drive the other auxiliaries. There being no unbalanced forces and therefore the engine is vibrationless.
3. **Cooling water.** Cooling water is not needed except in those turbines using inter-cooler.
4. **Low weight and size.** In this cycle the turbine has a lower specific weight and requires lesser space per horse power output.
   (Specific weight is the weight of engine per H.P. output). This property of producing more power output in a small space and low weight is quite useful in aviation engines.
5. **Warm up period.** The warm up period of the engine is negligible because after the engine has been brought up to the speed by starting motor and fuel ignited the engine then can accelerate from a cold start to a full load without warm up time. This property is quite advantageous in marine, aviation etc.

Disadvantages

1. Part load performance is low. This can be improved by using inter-cooler and reheater.
2. Reduction in component efficiencies lowers the thermal efficiency of the cycle.
3. The gas turbine in open cycle requires a large quantity of air.

13.4 CLOSED CYCLE GAS TURBINE PLANTS

In the closed cycle (Fig. 13.6) the same working fluid (air or some other stable gas) is constantly circulated. The fuel is burnt in the combustion chamber and the heat is transferred to the working medium through heat transfer surfaces. Thus the working medium does not mix with the products of combustion. The working medium is cooled in the water cooler before it enters the compressor. This minimises compressor work.

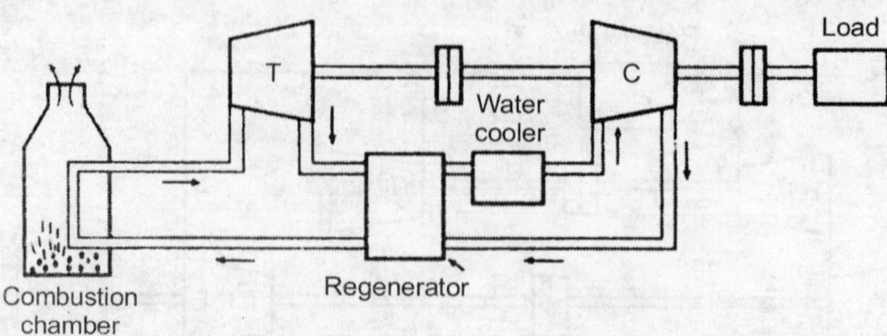

Fig. 13.6 A closed cycle gas turbine power plant.

Some of the gases used in closed cycle gas turbine plant are as follows:

(*i*) Helium (He)
(*ii*) Hydrogen (H_2)
(*iii*) Nitrogen (N_2)
(*iv*) Ammonia (NH_3)
(*v*) Ethylene (C_2H_4)
(*vi*) Krypton (Kr)
(*vii*) Oxygen (O_2)
(*viii*) Carbon-dioxide (CO_2)
(*ix*) Methane (CH_4)

Advantages of Closed Cycle Turbines

The various advantages of closed cycle gas turbines are as follows:

1. As the working fluid does not mix with the products of combustion, therefore, it is possible to use a gas of higher density and higher specific heat than air such as krypton, argon, xenon etc. This will reduce the size of various components. Helium has been, successfully used as working medium in closed cycle gas turbines. The specific heat of helium at constant pressure being nearly five times that of the air so heat drop and hence energy dealt per kg of mass flow is about five times in turbines using helium as working medium, as compared to gas turbine using air. Heat exchanger used in gas turbines using helium as working medium has a surface area of about 1/3 that of heat exchanger used in gas turbines using air as working medium. Therefore, the size of helium unit is comparatively smaller.
2. As the working medium does not mix with the product of combustion so there is no accumulation of carbon deposits on blades and nozzles of the turbine. The compressor remains free of dust as the working medium may be cleaned. Therefore, the periodic cleaning of components is not needed.
3. The system has improved part load efficiency as the output can be varied by withdrawing or admitting more working medium.
4. External heating can be done by inexpensive solid fuel such as coal.

Disadvantages

1. As the system is under an initial high pressure with a working medium other than air, therefore, the system should be gas-tight. This increases the cost of the system.
2. A large air heater is needed and this air heater is not as efficient as the combustion chamber used in open cycle.
3. As the system needs cooling water so the system cannot be used in aeronautical engines.

13.5 SOME OTHER POSSIBLE ARRANGEMENTS

Some other arrangements of open cycle system are as follows:

1. Simple Open Cycle - Twin Shaft. In this arrangement air enters the compressor (C) and after being compressed it flows to combustion chamber. The gases then expand in the turbines (*T*). The exhaust of burnt gases takes place in the atmosphere (See Fig. 13.7)

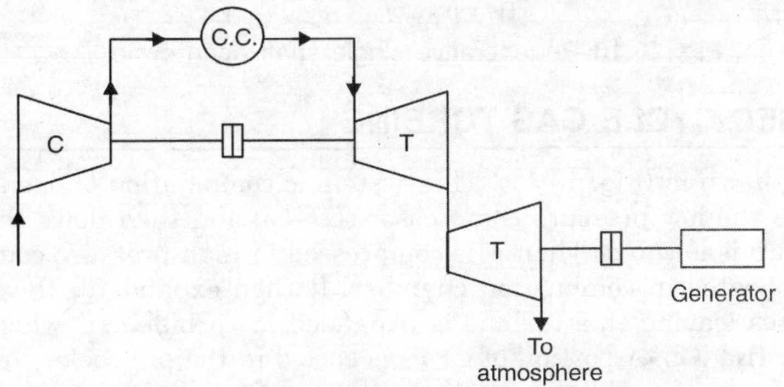

Fig. 13.7 Simple open cycle-twin shaft.

2. Inter Single Shaft Open Cycle. The air from low pressure compressor (L.P.C.) flows to inter-cooler and then to high pressure compressor (H.P.C.). The high pressure compressor delivers the air to combustion chamber (C.C.). The gases then expand in the turbine (Fig. 13.8).

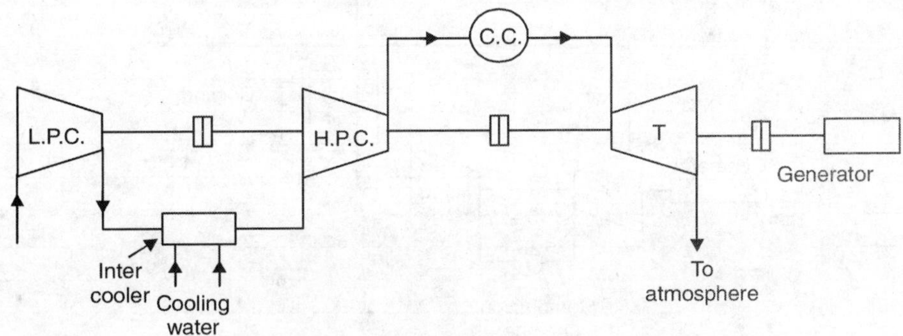

Fig. 13.8 Inter single shaft open cycle.

3. Reheat Single Shaft Open cycle. In this arrangement reheating is achieved by providing a combustion chamber between the two turbines (Fig. 13.9).

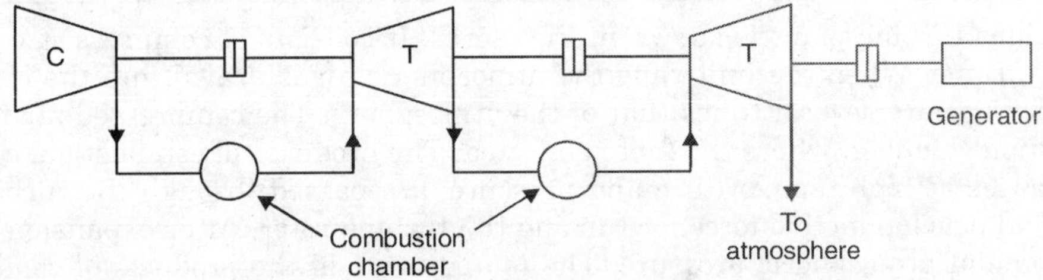

Fig. 13.9 Reheat single shaft open cycle.

4. Regenerative Single Shaft Open Cycle. In this system the heat of exhaust gases from turbine (T) is utilised in heating the compressed air (Fig. 13.10).

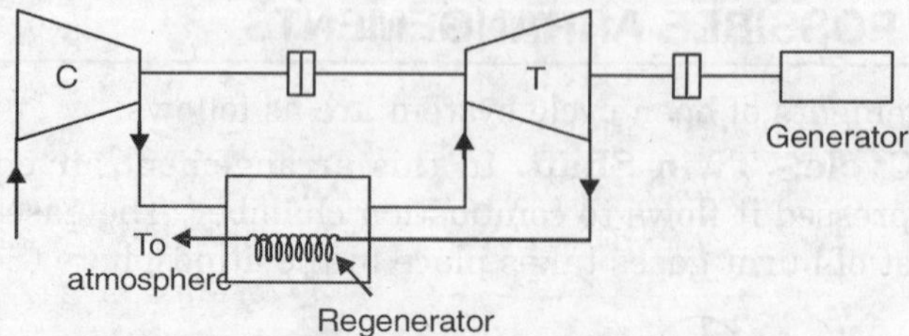

Fig. 13.10 Regenerative single shaft open cycle.

13.6 SEMI-CLOSED CYCLE GAS TURBINE

The system is shown in (Fig. 13.11). This system is combination of open cycle and closed cycle. The air enters the low pressure compressor (L.P.C.) and then flows through pre-cooler where its temperature is lowered. Then it is compressed in high pressure compressor (H.P.C.) and after that it is heated in combustion chamber. It then expands in the turbine (T_1) and some amount of gases leaving this turbine is expanded in turbine (T_2) which drives the low pressure compressor (L.P.C.). A portion of gases is cooled in the pre-cooler. In this system the heat exchanger has been omitted for simplicity.

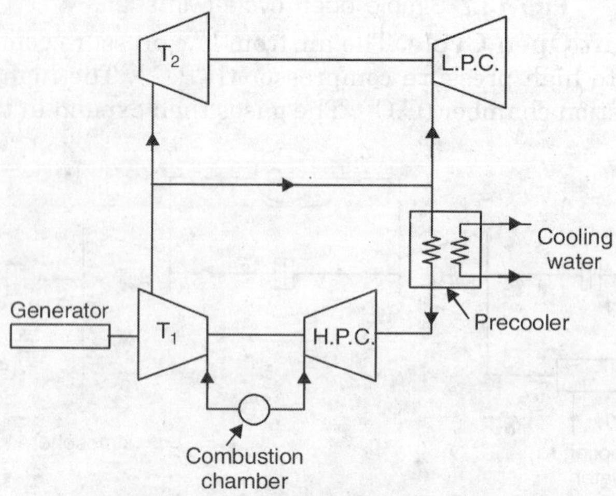

Fig. 13.11 Semi-closed cycle.

13.7 COMPONENTS (ELEMENTS) OF A GAS TURBINE PLANT

A simple gas turbine plant is shown in Fig. 25.12. It consists of compressor, combustion chamber and turbine. When the unit runs the atmospheric air is drawn into the compressor, raised to static pressure several times that of the atmosphere. The compressed air then flows to the combustion chamber, where the fuel is injected. The produces of combustion, comprising a mixture of gases at high temperature and pressure, are passed through the turbine where they expand and develop motive force for turning the turbine rotor. After expansion the gases leave the turbine at atmospheric pressure. The temperature of the products of combustion is nearly 1000°F to 1500°F. The temperature of the exhaust gases is in the range of 900°F to 1100°F. The compressor is mounted on the same shaft as that of turbine. Major portion of the work developed in the turbine is used to drive the compressor and the remainder is available as net power output.

1. Compressor

The various compressors used are reciprocating compressors, centrifugal compressors and axial flow compressors. The reciprocating compressors are not preferred due to the friction in sliding parts, more weight, less speed and inability to handle large volumes of air. For a gas turbine power plant of high output and efficiency generally pressure ratios of 10 : 1 or more is used. It is observed that when a single compressor with a pressure ratio not more than 4 : 1 is required the centrifugal compressor is the most suitable. It is quite rugged in construction, can operate more efficiently over a wide range of mass rate of flow of air than a comparable axial flow compressor. Centrifugal compressor is mainly used in super chargers and in jet aircraft plants, where lower pressure ratios and small volumes of air is needed.

For higher pressure ratios multi-stage centrifugal compressor does not prove to be as useful as an equivalent axial flow compressor. Therefore, when high pressure ratios are needed, axial compressor is advantageous and is always used for industrial gas turbine installations, Further it is desirable that more than one compressor should be used when the pressure ratio exceeds 6 : 1. Although the axial flow compressor is heavier than the centrifugal compressor but it has higher efficiency than the centrifugal compressor.

It is important that air entering the compressor should be free from dust. Therefore, air should be passed through a filter before it enters the compressor. Air filters are not needed in the closed cycle system.

(*a*) Centrifugal Compressor. It consists of stationary casing and rotating impeller. Impeller is provided with blades. When the impeller rotates the air enters axially and leaves radially. When the impeller rotates the pressure in the region R falls and, therefore, the air enters through the eye. The air then flows radially outward through the impeller blades. After that the air flows through the converged passages of diffuser blades and finally the air flows to compressor outlet (Fig. 13.12).

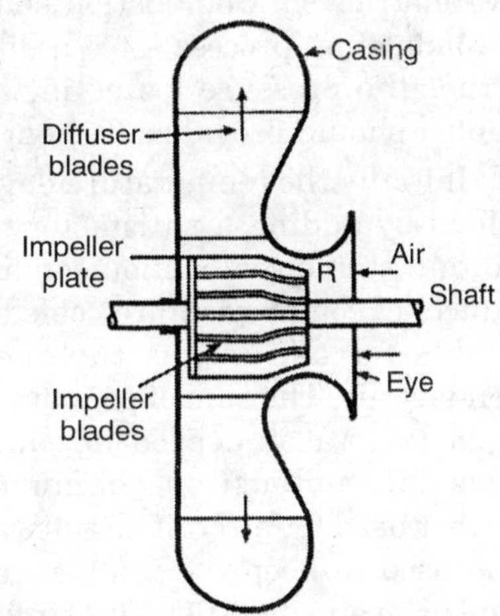

Fig. 13.12 Centrifugal compressor

(*b*) Axial Compressor. This compressor (Fig. 13.13) is quite commonly used in gas turbines. It consists of stator which enclose rotor (R). Both stator and rotor are fitted with rings of blades (RB -Rotor Blades, SB - Stator Blades). In this compressor the air flows in an axial direction from inlet to outlet. Air entering at one end as shown, flows through the alternatively arranged rings and gets compressed successively.

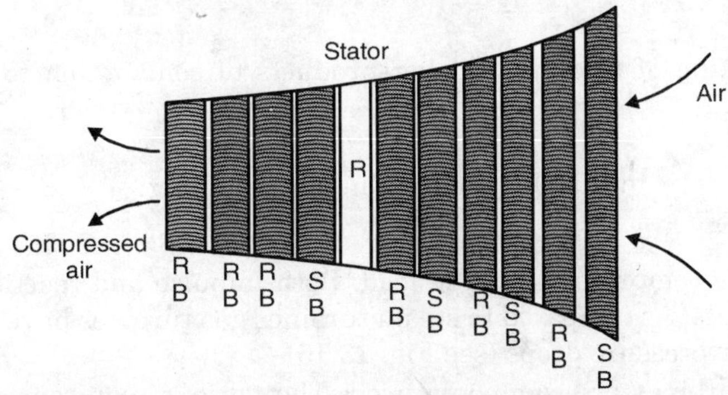

Fig. 13.13 Axial compressor.

2. Inter-coolers

These are used to cool the air between stages when multistage centrifugal compressors are employed. This reduces work input to the compressor. The cooling of compressed air is generally done with the help of cooling water. The water is circulated with the help of pump through the tubes. The compressed air is made to flow partly parallel and partly across the water flow with the help of baffles. The baffles are used to make the air to flow in contact for longer duration with heat transfer surface so as to increase the effectiveness of this heat exchanger.

3. Combustion Chamber

In the combustion chamber, combustion of fuel takes place. The combustion process taking place inside the combustion chamber is quite important because it is in this process that energy, which is later converted into work by the turbine, is supplied. Therefore, the combustion chamber should provide thorough mixing of fuel and air as well as combustion products and air so that complete combustion and uniform temperature distribution in the combustion gases may be achieved. Combustion should take place at high efficiency, because losses incurred in the combustion process have a direct effect on the thermal efficiency of the gas turbine cycle. Further the pressure losses in the combustion chamber should be low and the combustion chamber should provide sufficient volume and length for complete cobustion of the fuel.

Initially the temperature development in combustion chamber is too high. The difficulty is avoided by adding a satisfactory amount of air to maintain stable combustion condition and then the products of combustion are cooled to a temperature suitable for use in gas turbine by introducing secondary air. The sum of primary and secondary air supplied is total air needed for combustion. Fig. 13.14 shows the combustion chamber. In combustion chamber used for aircraft engines a large quantity of air is used to keep the temperature of combustion chamber to about 600°C. The air fuel ratio may be of the order of 60 : 1 in this case.

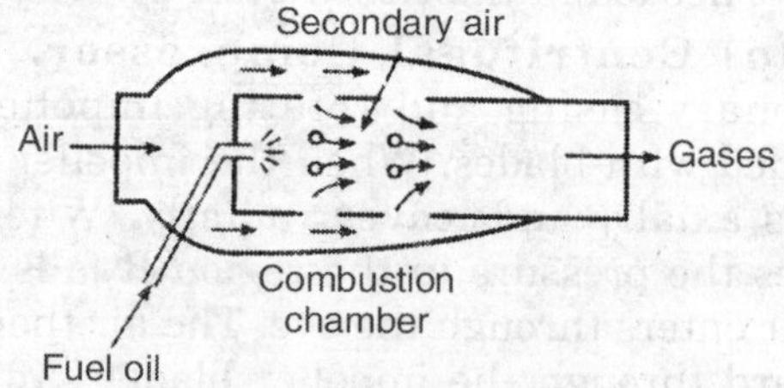

Fig. 13.14 Combustion chamber.

The requirements of a combustion chamber are as follows:

(*i*) Low pressure loss
(*ii*) High combustion efficiency
(*iii*) Good flame stability
(*iv*) Low weight
(*v*) Thorough mixing of cold air and hot products of combustion to generate uniform temperature
(*vi*) Reliability
(*vii*) Low carbon deposit in turbine, and combustion chamber.

4. Gas Turbine

Turbine drives the compressor and the load. Both impulse and reaction turbines can be used in gas turbine plants. As compared to steam turbines gas turbines have few stages because they operate on smaller pressure drops (see Fig. 13.15).

Axial flow type turbines are commonly used. The various requirements of turbines are as follows:

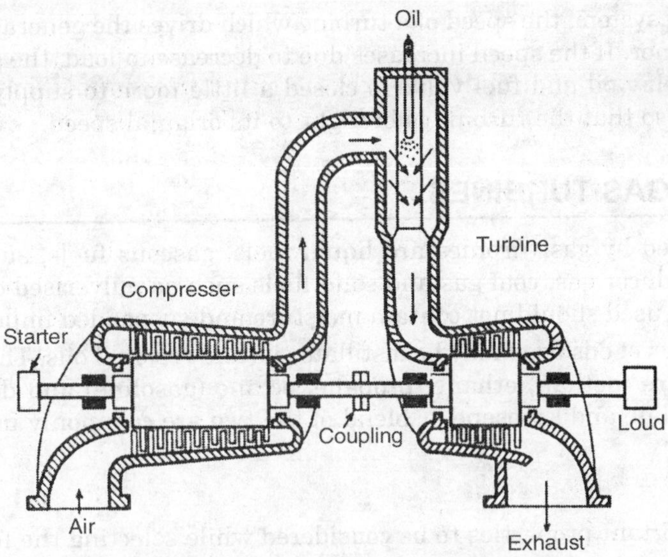

Fig. 13.15 A simple gas turbine plant.

(*i*) Light Weight (*ii*) High Efficiency

(*iii*) Reliability in operation (*iv*) Long working life.

5. Regenerators

The regenerators which are commonly used in gas turbine plants are of two types, namely, recuperate and regenerator. In a recuperative type of heat exchanger, the air and hot gases are made to flow in counter-flow direction as it gives high average temperature difference causing the heat flow effectively.

A regenerative type heat exchanger consists of a heat exchanging member which is exposed alternatively to the hot exhaust gases and cooler compressed air. It absorbs the heat from the hot gases and gives it up when exposed to the air. Though these regenerative type heat exchangers have high capital cost, it has principal advantages of lightness and smaller size for the same heat transferring capacity and low pressure drop.

6. Gas Turbine Auxiliaries

The gas turbines need additional equipment for serving the main components, these include: starting motor or engine, lubricating oil pump, fuel control system, oil coolers and filters, silencers, air and gas ducts and plant control panel.

13.8 GOVERNING SYSTEM FOR GAS TURBINE PLANTS

The efficiency of the gas turbine decreases with the decrease in load when plant is used as peak load plant. In all circumstances, it is necessary to maintain the speed of the turbine and alternator as constant. Governing system provide proper amount of fuel for starting the plant from cold, afterwards increase the fuel supply to accelerate the turbine to its normal speed. The provision must be made in the governing system to prevent overspeeding. A fuel control adjustment must be provided to take into account ambient air temperature and pressure changes. A control device is also included in the system to shut off the fuel for stopping the plant.

In this governing system, the speed of a turbine which drives the generator is kept constant by a centrifugal governor. If the speed increases due to decrease in load, the governor decreases the pressure of the relay oil and fuel valve is closed a little more to supply less fuel into the combustion chamber, so that the turbine is brought to its original speed.

13.9 FUELS FOR GAS TURBINES

Various fuels used by gas turbines are liquid fuels, gaseous fuels, such as natural gas, blast furnace gas, producer gas, coal gas and solid fuels such as pulverised coal. Care sould be taken that the oil fuel used should not contain moisture and suspended impurities.

The different types of oils used may be distillate oils and residual oils. The various paraffins used in gas turbine are methan, ethane, propane, octane (gasoline) and dodecane (kerosene oil). Out of these gasoline and kerosene or blend of the two are commonly used.

Qualities of Fuel

Some of the important properties to be considered while selecting the fuel for gas turbine are as follows:

1. Volatility. The properties has a major effect on starting and combustion efficiency of the engine particularly at low temperature and other adverse conditions. The volatility of the fuel should be such that it is conductive to a quick and successful restart blowout of flame. Highly volative fuels are also not desirable as they have the following disadvantages:

(*i*) They are more susceptible to fire (although they have less tendency to explode).

(*ii*) They are conductive to vapour lock and to excessive loss of fuel during flight because of evaporation of certain lighter hydrocarbons. Therefore, in case of aircraft gas turbines in which the quantity of fuel used is sufficiently high, the fuel wastage will also be more if the fuel is highly volatile.

2. Combustion Products. The products of combustion should not be in the form of solids because they tend to deposit on the combustion chambers, turbine blades and vanes and cause a loss in efficiency.

3. Energy Contents. Fuel should have greater heating value so that fuel consumption may be less.

4. Lubricating Properties. The fuel should provide a certain amount of lubrication of friction surface of fuel pumps.

5. Availability. The fuel selected should be available in large quantities so that it is cheaper.

Comparison of Kerosene Oil and Gasoline

Kerosene is quite commonly used in aircraft gas turbines. It is not as volatile as gasoline and, therefore, there is less possibility of vapour lock and fuel loss. But its combustion efficiency is low compared to gasoline. The lubrication properties of gasoline are poorer. About 5 to 20% of a barrel of crude may be refined kerosene whereas 40 to 50% of a barrel of crude oil be refined into gasoline which shows that gasoline can be available in large quantities.

Air Fuel Ratio

Air fuel ratio in the gas turbines is nearly 60 : 1.

13.10 AIR RATE

Air rate is the amount of air in kg needed per horse power hour.

Factors Affecting Air-Rate

1. **Effect of turbine inlet temperature and pressure.** With increase of turbine inlet temperature the air rate is decreased. As the pressure ratio is increased the air rate firstly decreases to a minimum value and then starts increasing.
2. **Effect of compressor inlet temperature.** With the increase of compressor inlet temperature the compressor work is increased and the net turbine output is decreased and therefore, air rate is increased.
3. **Effect of regeneration, inter-cooling and reheating.** If only regeneration process is added it increases the air rate to small extent. Addition of inter-cooling and reheating in open cycle reduces air rate especially at high pressure ratios.
4. **Effect of compressor and turbine efficiencies.** Air rate is decreased with increase in compressor and turbine efficiencies.

13.11 GAS TURBINE CYCLE EFFICIENCY

Gas turbines may operate either on a closed or on an open cycle. The majority of gas turbines currently in use operate on the open cycle in which the working fluid, after completing the cycle is exhausted to the atmosphere. The air fuel ratio used in these gas turbines is approximately 60 : 1.

The ideal cycle for gas turbine is Brayton Cycle or Joule Cycle. This cycle is of the closed type using a perfect gas with constant specific heats as a working fluid. This cycle is a constant pressure cycle and is shown in Fig. 13.16. On *P-V*diagram and in Fig. 13.17 on T-ϕ diagram. This cycle consists of the following processes:

The cold air at 3 is fed to the inlet of the compressor where it is compressed along 3-4 and then fed to the combustion chamber where it is heated at constant pressure along 4-1. The hot air enters the turbine at 1 and expands adiabatically along 1-2 and is then cooled at constant pressure along 2-3.

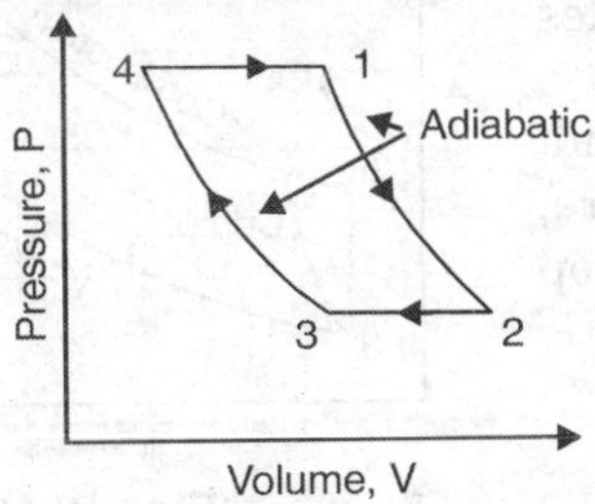

Fig. 13.16. Constant pressure cycle (P-V diagram)

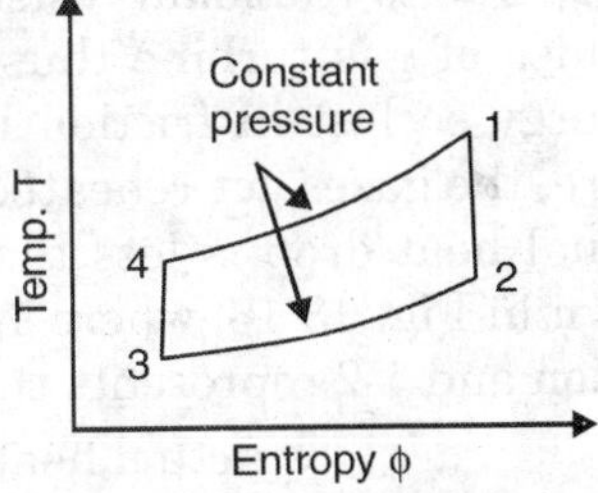

Fig. 13.17 T-ϕ diagram.

Heat supplied to the system = $K_p\,(T_1 - T_4)$

Heat rejected from the system = $K_p(T_2 - T_3)$

where K_p = specific heat at constant pressure,

Work done = Heat supplied – Heat rejected

$$= K_p\,(T_1 - T_4) - K_p(T_2 - T_3)$$

Thermal efficiency (η) of Brayton Cycle

$$\eta = \frac{\text{Work done}}{\text{Heat supplied}} = \frac{K_1[(T_1 - T_4) - (T_2 - T_3)]}{K_p(T_1 - T_4)}$$

$$\eta = 1 - \frac{T_2 - T_3}{T_1 - T_4} \quad \text{...}(i)$$

For expansion 1-2 $$\frac{T_1}{T_2} = \left(\frac{P_1}{P_2}\right)^{(\gamma-1)/\gamma}$$

$$T_1 = T_2\left(\frac{P_1}{P_2}\right)^{(\gamma-1)/\gamma}$$

For compression 3-4 $$\frac{T_4}{T_3} = \left(\frac{P_4}{P_3}\right)^{(\gamma-1)/\gamma} = \left(\frac{P_1}{P_2}\right)^{(\gamma-1)/\gamma}$$

$$T_4 = T_3\left(\frac{P_1}{P_2}\right)^{(\gamma-1)/\gamma}$$

Substituting the values of T_1 and T_4 in equation (1), we get

$$\eta = 1 - \frac{T_2 - T_3}{T_2\left(\frac{P_1}{P_2}\right)^{(\gamma-1)/\gamma} - T_3\left(\frac{P_1}{P_2}\right)^{(\gamma-1)/\gamma}}$$

$$\eta = 1 - \frac{T_2 - T_3}{\left(\frac{P_1}{p_2}\right)^{(\gamma-1)/\gamma}(T_2 - T_3)} = 1 - \frac{1}{\left(\frac{P_1}{P_2}\right)^{(\gamma-1)/\gamma}}$$

13.11.1 Effect of Blade Friction

In a gas turbine there is always some loss of useful heat drop due to frictional resistance offered by the nozzles and blades of gas turbine thus resulting drop in velocity. The energy so lost in friction is converted into heat and, therefore, the gases get reheated to some extent. Therefore, the actual heat drop is less than the isentropic heat drop as shown in Fig. 13.18, where 1-2′ represents the adiabatic expansion and 1-2 represents the actual expansion.

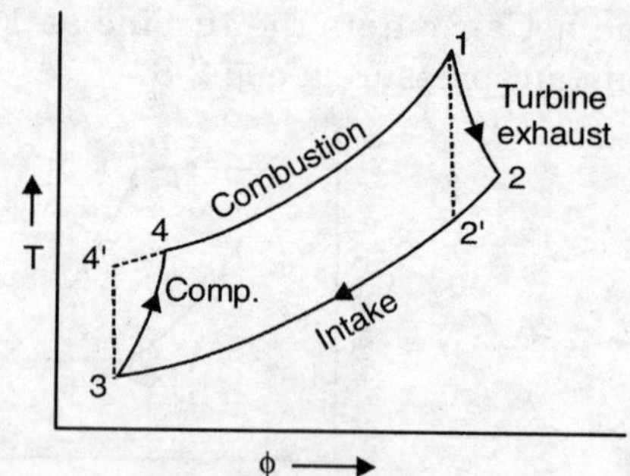

Fig. 13.18

$$\text{Actual heat drop} = K_p(T_1 - T_2)$$

$$\text{Adiabatic heat drop} = K_p(T_1 - T'_2)$$

$$\text{Adiabatic efficiency of turbine} = \frac{\text{Actual heat drop}}{\text{Isentropic heat drop}}$$

$$= \frac{K_p(T_1 - T_2)}{K_p(T_1 - T'_2)} = \frac{T_1 - T_2}{T_1 - T'_2}$$

For isentropic process 1-2′ $\dfrac{T_2}{T_1} = \left(\dfrac{P_2}{P_1}\right)^{\gamma-1/\gamma}$

In the compressor also reheating takes place which causes actual heat increase to be more than adiabatic heat increase. The process 3-4 represents the actual compression while 3-4′ represents adiabatic compression.

isentropic heat drop = $c_p(T'_4 - T_3)$

Actual heat drop = $c_p\,(T_4 - T_3)$

isentropic efficiency of compressor = $\dfrac{c_p(T'_4 - T_3)}{c_p(T_4 - T_3)} = \dfrac{T'_4 - T_3}{T'_4 - T_3}$

13.11.2 Improvement in Open Cycle

The open cycle for gas turbine is shown in Fig. 13.18. The fresh atmospheric air is taken in at the point in 3 and exhaust of the gases after expansion in turbine takes place at the point 2. An improvement in open cycle performance can be effected by the addition of a heat exchanger which raises the temperature of the compressed air entering the turbine by lowering exhaust gas temperature which is a waste otherwise. Less fuel is now required in the combustion chamber to attain a specified turbine inlet temperature. This is called a regenerative cycle (Fig. 13.19).

This regenerative cycle is shown on $T = \phi$ diagram in Fig. 13.20, where ϕ = entropy.

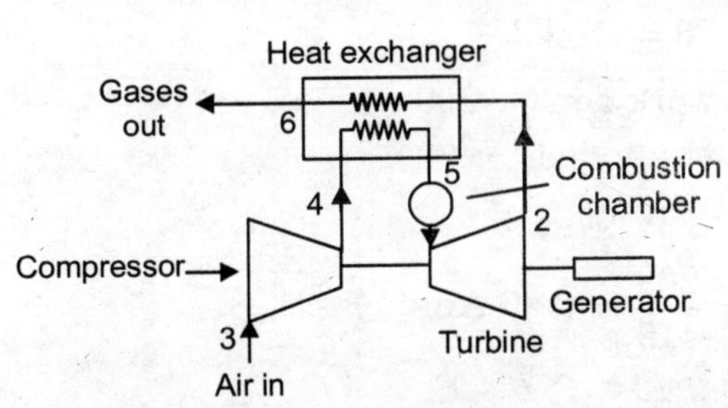

Fig. 13.19 Regenerative cycle.

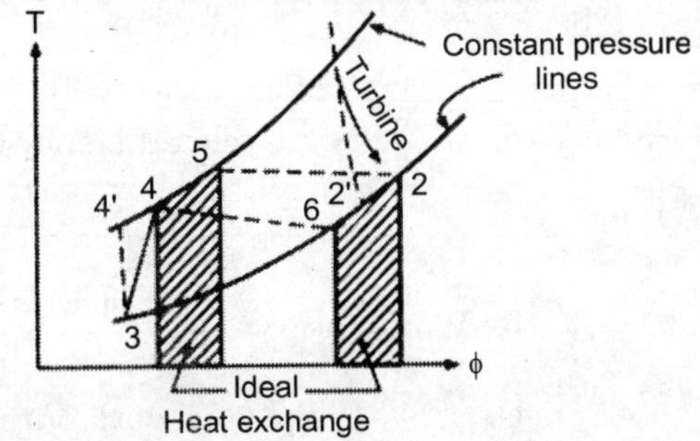

Fig. 13.20

Heat supplied = $c_p(T_1 - T_3) = c_p\,(T_1 - T_2)$

Heat rejected = $c_p(T_5 - T_3) = c_p\,(T_4 - T_3)$

(η) Thermal efficiency of theoretical regenerative cycle

$$\eta = \frac{c_p(T_2 - T_1) - c_p(T_4 - T_3)}{c_p(T_1 - T_5)}$$

For isentropic compression and isentropic expansion thermal efficiency is given by

$$\eta = \frac{c_p(T_1 - T_2) - c_p(T_4 - T_3)}{c_p(T_1 - T_5)}$$

Example 13.1. *In a gas turbine power plant working on Joule cycle, air is compressed from 0.1 MPa and 17°C through a pressure ratio of 6. It is then heated in the combustion chamber to 700°C and expanded back to a pressure of 0.1 MPa. Calculate the following:*

(*a*) *Cycle efficiency* (*b*) *Work ratio*

(c) *Specific work output of the plant.*

Solution. The cycle is shown in Fig. 13.21 on $T = \phi$ diagram

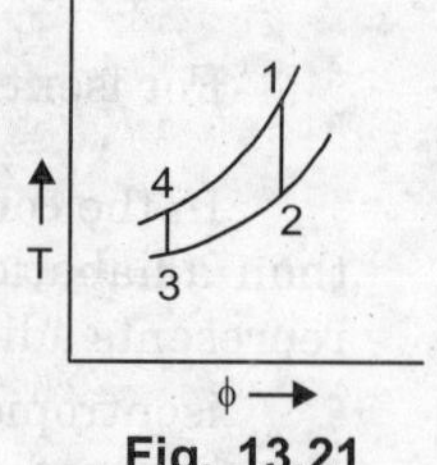

Fig. 13.21

$$T_3 = 273 + 17 = 290°K$$

$$T_1 = 273 + 700 = 973°K$$

$$\left(\frac{T_4}{T_3}\right) = \left(\frac{P_4}{P_3}\right)^{(\gamma-1)/\gamma}$$

$$\gamma = 1.4 \text{ for air}$$

$$T_4 = T_3\left[\frac{6}{1}\right]^{(\gamma-1)/\gamma} = 290\left[\frac{6}{1}\right]^{(\gamma-1)/\gamma} = 500°K$$

$$T_2 = T_1 = \left(\frac{P_2}{P_1}\right)^{(\gamma-1)/\gamma} = 973\left[\frac{1}{6}\right]^{(1.4-1)/1.4} = 554°K$$

Compressor work = $c_p(T_4 - T_3) = 1\,(500 - 290) = 210$ kJ

(Assuming $c_p = 1.00$ kJ/kg.K)

Work done by turbine = $K_p(T_1 - T_2) = 1 \times (973 - 554) = 419$ kJ

Net work done = 419 – 210 = 209 kJ

Heat supplied = $c_p(T_1 - T_4) = 1 \times (973 - 500)$

$$= 1 \times 473 = 473 \text{ kJ}$$

a. $$\text{Cycle efficiency } (\eta) = \frac{\text{Net work done}}{\text{Heat supplied}} = \frac{209}{473}$$

$$= 0.44 = 44\%$$

b. Work ratio = $\dfrac{\text{Net work done}}{\text{Work done by turbine}} = \dfrac{209}{419}$ = **0.49 Ans.**

c. Specific work output = 209 kJ/kg

= 209 × 1000 = **209,000 Nm/kg Ans.**

as (1 kJ = 1000 Nm)

13.12 JET PROPULSION

Gas turbines are quite commonly used for the propulsion of aircrafts. The aircraft units are classified as follows:

1. Turbo-jet unit.
2. Turbo-propeller unit.
3. Ram Jet

1. Turbo-jet Engine. Its principle is based on open cycle gas turbine. It consists of a diffuser, compressor, combustion chamber, turbine and exit nozzle. The air enters the diffuser at a velocity equal to that of aircraft. The pressure of air rises above atmospheric pressure in the diffuser. The air is then compressed in the compressor to a pressure of about 3.5 kg/cm^2. The compressor used is generally radial or axial type. The air is then supplied to the combustion chamber. The liquid fuel is injected into the combustion chamber where its combustion takes place. The hot gases expand in the turbine. The turbine extracts enough energy to drive the compressor and the necessary auxiliary equipment. The hot gases are then expanded through exit nozzle and they leave the nozzle at high velocity and causes a momentum due to which the engine exerts a forward thrust (Fig. 13.22).

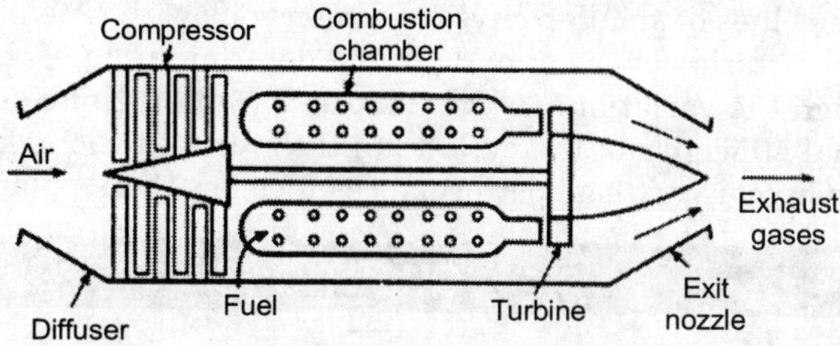

Fig. 13.22

Specific Thrust. The specific thrust of a turbo-jet is defined as the ratio of force to air flow-rate (sometimes it is defined as ratio of thrust force to fuel flow-rate).

2. Turbine-Propulsion. In the turbo propulsion system, the gases are partly expanded in the turbine and partly in the nozzle. The power developed by the turbine is used to run the compressor as well as propeller. The forward motion to the aircraft is given partly by propeller (which gives screw action to the craft) and partly by the jet produced by the nozzle. This combination gives higher efficiency even at low speed.

3. Ram Jet. It consists of only three main components, namely diffuser, combustion chamber and nozzle. The pressure of incoming high velocity air is increased by diffuser action. The fuel is injected into the high pressure air passing through the combustion chamber and is burned with the help of flame stabilizer. The high pressure and high temperature gases are further passed through the nozzle where pressure energy is converted into kinetic energy. The high velocity gases leaving the nozzle provide required forward thrust to the ram jet. The ram jet engines are used for supersonic speed aircrafts.

13.13 TERMS USED IN THE CHAPTER

(*i*) *Work ratio.* It is defined as the ratio of net-work output to the total work produced in the turbine.

(*ii*) *Thermal efficiency.* It is defined as the ratio of net-work output to the total fuel energy input. The higher the working temperature of the working medium the higher the thermal efficiency of the turbine.

(*iii*) *Air ratio.* It is defined as the amount of air (in pounds or kg) entering the compressor inlet per unit of network output of the turbine.

The size of gas turbine plant is dependent on rate of flow of air in relation to the useful horse power output. Lower the air rate smaller will be the size of plant. Inter cooling and re-heating reduces the air rate. An increase in the compressor and turbine efficiencies will decrease the air rate. An increase in compressor inlet temperature will decrease the net output of the turbine and hence will increase the air rate.

(*iv*) *Pressure ratio.* It is defined as the ratio of absolute pressure at the compressor outlet to the absolute atmospheric pressure at compressor inlet.

(*v*) *Compression efficiency or machine efficiency.* This term is related to the compressor, and is defined as the work required for ideal compression to the actually required by the compressor for a given pressure ratio.

Actual work required by the compressor is always more than the ideal work.

Air rate or Air ratio is a criterion of the size of the plant, *i.e.* the lower the air rate the smaller the plant. From the mechanical and metallurgical point of view the lowering of the air rate results in turbines of smaller physical dimensions with a nearly uniform temperature distribution. The work ratio acts as a guide in the determination of the size of gas turbine. If the work ratio is high the variations in the compressor and turbine efficiencies will have less effect on the thermal efficiency of the cycle than if the work ratio is low. A plant with a high work ratio will have higher part load performance efficiency than a plant with a low work ratio.

QUESTIONS

13.1 What are the essential components of a simple open cycle gas turbine plants? How Inter-cooling and regeneration help in improving thermal efficiency of the plant?

13.2 Describe a closed cycle gas turbine plant. What are the advantages of closed cycle?

13.3 What are the advantages and disadvantages of gas turbine power plant? Discuss the applications of such plant.

13.4 What are the various factors to be considered while selecting the site for gas turbine power plant?

13.5 Give the layout plan of gas turbine power plant.

13.6 What methods are used to improve the efficiency of gas turbine?

13.7 What are the various compressors used in gas turbine? State their relative advantages and disadvantages?

13.8 What are the quality requirements of fuel used in gas turbines?

13.9 Write short notes on the following:
(*a*) Pressure ratio (*b*) Semi-closed cycle (*c*) Work ratio

13.10 Compare a gas turbine with a steam turbine.

13.11 Define air rate. Does a low air rate increase or decrease the size of an engine?

13.12 List the five operating variables that strongly affect the thermal efficiency of an open cycle gas turbine.

13.13 (*a*) Explain what is meant by open or closed cycle gas turbine.
(*b*) What are the advantages and disadvantages of closed cycle over open cycle?

13.14 What are the different methods used to improve the thermal efficiency of the cycle? Explain them.

13.15 (*a*) Explain the effect of regeneration in a gas turbine plant.
(*b*) Discuss the effect of intercooling and reheating in a gas turbine plant.

13.16 Briefly describe the different fuels which can be burnt in gas turbine plants.

Chapter

14

REFRIGERATION AND AIR CONDITIONING

Learning Objectives

After studying this chapter, the students will learn:

- Refrigeration system and its performance.
- Vapour compression refrigeration system.
- Absorption refrigeration system.
- Refrigerants.
- Applications of refrigeration.
- Air conditioning and its classifications.
- Psychrometry.
- Different types of air conditioners and their applications.
- Air conditioning systems and selection of a suitable plant.

14.1 REFRIGERATION

Refrigeration is the process of cooling below the temperature of its surrounding and is defined as the science of providing and maintaining temperatures below that of surroundings. Refrigeration process can also be defined as the process of removing heat from a place where it is not wanted and transferring that heat to a place where it makes no difference (sink).

In order to maintain the low temperature in the refrigerated space, it is necessary to remove heat continuously equal to the amount of heat leaking into the system and reject the same to the surrounding atmosphere at higher temperature.

A refrigerator is a machine which produces cold and the process is known as refrigeration. The house hold refrigerator absorbs heat from the food products and releases this heat into the kitchen where it makes no difference. The amount of heat released is so small that it is not even noticed.

Reverse of a refrigeration is a ***heat pump.*** Heat pump is a device which performs the heating process. In the case of refrigeration, heat is pumped out from the system as against heat pump where heat from surrounding is pumped into the system continuously.

Mechanical Refrigeration

It may be defined as production of cold by machine. It requires mechanical power. It may be of the following three types.

(*a*) Air Refrigeration system.

(*b*) Vapour compression system.

(*c*) Vapour Absorption system.

Each mechanical system have 4 basic units.

1. An Evaporator—the cooling unit.
2. A pump or compressor.
3. A condenser—the heat disposer.
4. A liquid expanding device—Expansion valve or cylinder or capillary tube.

The refrigerant passes through these and gives cooling effect in the evaporator.

14.2 UNIT OF REFRIGERATION

As power unit of mechanical machines is given by H.P. For electrical equipment unit of power is kW or MW.

Domestic refrigerators are rated in kJ/sec or Btu/hr of cooling. For commercial and industrial refrigeration system in U.S.A. are rated in "tonnes of refrigeration (TR)". Historically one TR was defined as the energy removal rate that will freeze one short ton (2000 lb) of water at 0°C (32°F) in one day.

"One Tonne of Refrigeration", means machine is capable of removing heat from a body equal to the heat absorbed by one ton of ice when it melts from 0°C ice to 0°C water in 24 hours. The tonne is equal to 1000 kg. Then

$$\text{one } \textbf{tonne} \text{ of Refrigeration} = 1000 \times \frac{333.5}{24 \times 60} = \textbf{231.6 kJ/min = 3.85 kW}$$

Where 333.5 kJ/kg is the latent heat of ice.

Hence in MKS system, a machine having its capacity to produce cooling effect of 209 kJ/minute is called as a "one tonne-machine".

The value of one **tonne** refrigeration was approximately 3.505 kW.

It has now been conventionally redefined as exactly 3.517 kW or 3517 watts. It is commonly taken as 3.5 kJ/sec.

14.3 COEFFICIENT OF PERFORMANCE

In the case of an engine, the performance is judged by efficiency of the engine, that is

$$\text{Efficiency} = \frac{\text{Work done}}{\text{Heat supplied}}$$

and in case of refrigeration the process is first reversed that is we have to supply work to obtain lower temperature or say for the removal of heat. Thus performance of refrigeration is just reversed that is $\frac{N}{W}$

Where N = the rate of heat abstracted from the body in given time, technically called **"Net Refrigerating Effect"**

and W = Work done supplied

Thus
$$\text{C.O.P.} = \frac{N}{W} = \frac{\text{Net refrigerating effect}}{\text{Work done supplied}}$$

This is theoretical C.O.P.

and
$$\text{Relative C.O.P.} = \frac{\text{Actual C.O.P.}}{\text{Theoretical C.O.P.}}$$

14.4 REFRIGERATION SYSTEMS

Following are the commonly used refrigeration systems:

1. Air refrigeration system.
2. Vapour compression refrigeration system.
3. Vapour refrigeration absorption system.

In air refrigeration system air works as the working substance. The C.O.P. (Coefficiet of Performance) of the air refrigeration system is quite low. To get the C.O.P. of a refrigeration system close to ideal reversible cycle values, the vapour compression cycle is used. In "vapour absorption system" the solar energy, waste heat from process industry or the exhaust from power plants can be utilized for the purpose of heating the absorber.

14.4.1 Air Refrigeration System

In air refrigeration system, air is used as working fluid to remove the heat from the refrigerated space and discharge the same to the surrounding atmosphere. Air refrigeration system are of two types as discussed hereunder:

1. Reversed Carnot Cycle **2. Air Refrigeration Cycle**

1. Reversed Carnot Cycle

If the carnot forward engine cycle is reversed, it behaves like a refrigerating cycle, if driven by power from external source. The P.V. diagram will be as follows. During this cycle the working fluid is expanded adiabatically from point 1 to 2, thus causing fall of temperature from T_1 to T_2. After this expansion takes place from point 2 to 3 isothermally that is without any change in temperature during the process heat is removed from cold body (see Fig. 14.1).

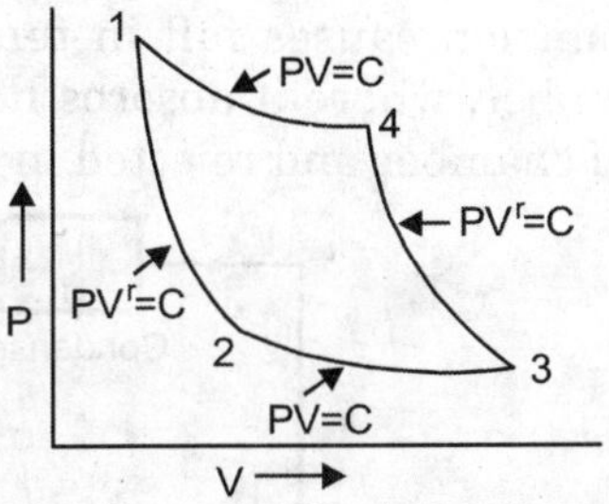

Fig. 14.1 Reversed Carnot cycle.

At point "3" the working fluid is compressed adiabatically, *i.e.,* external work is done raising temperature to T_1. The final operation is isothermal compression from point "4" to "1" during which heat is rejected to the hot body.

Thus, heat removed from cold body $= P_2 V_2 \log_e r$

$= RT_2 \log_e r$/kg of working fluid

Heat rejected to hot body $= P_1 V_1 \log_e r$

$= RT_1 \log_e \gamma$/kg of working fluid

Thus, work done supplied $= W = R(T_1 - T_2) \log_e r$

N = Net refrigerating effect = Heat removed from cold body

$$\text{Then C.O.P.} = \frac{N}{W} = \frac{RT_2 \log_e r}{R(T_1 - T_2)\log_e r} = \frac{T_2}{T_1 - T_2} \qquad r = \frac{P_1}{P_4} = \frac{P_2}{P_3} = \frac{V_3}{V_2} = \frac{V_4}{V_1}$$

Refrigerator working on a reversed Carnot Cycle is not a possible thing as in adiabatic process the machine has to run at higher speed whereas in isothermal process it has to run at much lower speed to avoid rise in temperature which is not possible. Further by lowering the higher temperature T_1 or raising temperature T_2, C.O.P. can be increased, as it is clear from above equation. But this is also not possible due to the following reasons:

(*a*) If T_1 lowers, denominator decreased but T_1 cannot be reduced below certain limit set by cooling medium temperature used in condenser. Suppose, if the condenser is air cooled then T_1 cannot be lowered than the ambient air temperature and further T_1 should be higher to have effective heat rejection so that there must be sufficient temperature difference.

(*b*) The temperature T_2 should be raised, this is again fixed according to the application *i.e.,* it must be lower than the temperature of substance or space to be cooled.

2. Bell Coleman Cycle or Air Refrigeration Cycle

Fig. 14.2 shows the Bell Coleman Cycle plant. This is one of the earliest type of refrigerating system.

It has 4 main components:

(*i*) Compressor (*ii*) Condenser
(*iii*) Expanding device (*iv*) Evaporator.

In its working, consider 1 kg of air is drawn into the cylinder during suction stroke 4-1 from the cold chamber and then compressed adiabatically to point "2". After this it is cooled into the condenser at constant pressure from 2-3 and temperature of air falls, but pressure remains high.

Now it enters in the expansion cylinder and expands adiabatically from 3-4. This adiabatic expansion causes fall in temperature. Now this low temperature air enters into the cold chamber, where it absorbs heat from the cold chamber. In this system heat is removed from cold chamber and rejected in condenser.

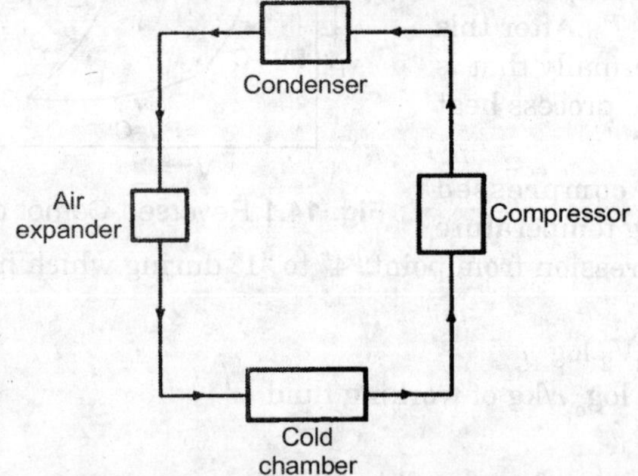

Fig. 14.2 Bell Coleman cycle or air refrigeration.

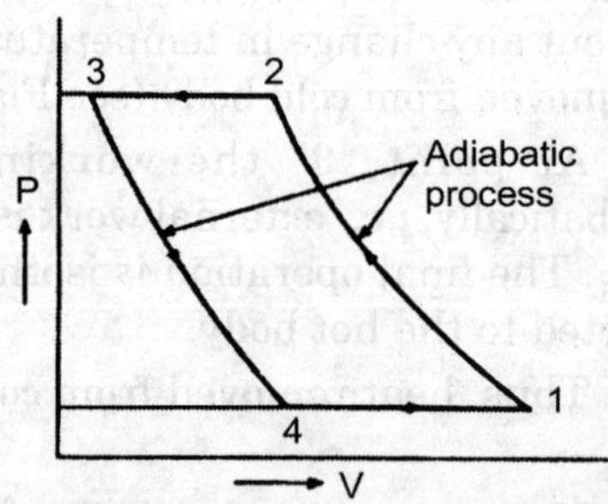

Fig. 14.3 Bell Coleman Cycle.

Let T_1, T_2, T_3, and T_4 be the absolute temperature at point 1, 2, 3 and 4 of Bell Coleman cycle. Consider 1 kg of air in operation.

Heat removed from cold chamber

$$= C_p(T_1 - T_4) \text{ per kg of air}$$

$$\text{Heat rejected to condenser} = C_p(T_2 - T_3) \text{ per kg of air}$$

$$\text{Work done} = \text{Heat rejected} - \text{Heat removed}$$

$$= C_p(T_2 - T_3) - C_p(T_1 - T_4)$$

$$\text{C.O.P.} = \frac{N}{W} = \frac{C_p(T_1 - T_4)}{C_p(T_2 - T_3) - C_p(T_1 - T_4)}$$

$$= \frac{(T_1 - T_4)}{(T_2 - T_3) - (T_1 - T_4)}$$

As process 1-2 is adiabatic, then

$$P_1V_1^{\gamma} = P_2V_2^{\gamma}$$

From this

$$\frac{T_2}{T_1} = \left(\frac{P_2}{P_1}\right)^{\frac{\gamma-1}{\gamma}}$$

Similarly during process 3-4

$$\frac{T_3}{T_4} = \left(\frac{P_3}{P_4}\right)^{\frac{\gamma-1}{\gamma}}$$

From Bell-Coleman Cycle, it is clear that

$$P_1 = P_4, \; P_2 = P_3$$

Thus $\quad \dfrac{T_2}{T_1} = \dfrac{T_3}{T_4} \quad$ or $\quad T_2 = \dfrac{T_1 T_3}{T_4}$

Putting value of T_2 in C.O.P. equation

Then $\quad$ C.O.P. $= \dfrac{T_1 - T_4}{\left(\dfrac{T_1 T_3}{T_4} - T_3\right) - (T_1 - T_4)}$

$$= \frac{T_1 - T_4}{\dfrac{T_3}{T_4}(T_1 - T_4) - (T_1 - T_4)} = \frac{T_4}{T_3 - T_4}$$

***(i)* Open Air System.** This system, which is called open air system, has some disadvantages, the air comes in contact with cold chamber or stuff to be preserved and collects moisture over such articles. This moisture freezes during expansion and there is likelihood of valve getting choked.

***(ii)* Closed Air System.** Most of the modern refrigeration work on this system. The air passes through the coil and absorbs heat from cold-chamber while coming in contact. Further in this system, the bulky size of compressor and expander can be avoided by supplying dense air. Therefore this system is also termed as "Dense Air System".

14.4.2 Vapour Compression Refrigerating System

Modern refrigerators work on this system. In this, the refrigerant is such that it can be readily evaporated and condensed. The refrigerant never leaves the plant but recirculates through the system. It removes heat from the cold body during evaporation and rejects latent heat during condensation.

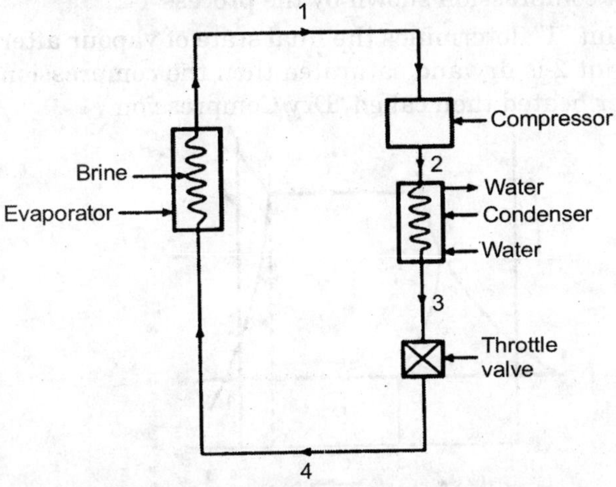

Fig. 14.4 Vapour compression system.

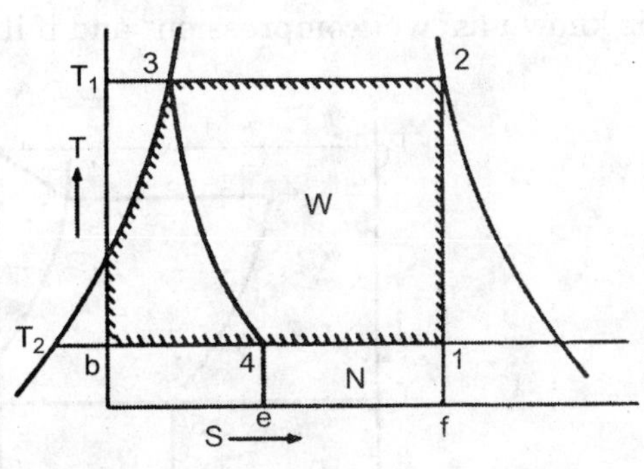

Fig. 14.5 Vapour compression cycle.

Fig. 14.4 shows the flow diagram of this system in which wet vapour is sucked from brine tank (evaporator) during suction stroke of compressor while in compression the vapour is compressed isentropically. Thus, completely evaporated sucked vapour is compressed and after this only dry vapour pass through the condenser. Its latent heat is rejected to the cooling medium and high pressure liquid vapour is obtained.

This high pressure liquid vapour is passed through the throttle valve, where through narrow opening, throttle expansion takes place, causing low pressure very wet vapour at a very low temperature nearly about – 10°C. This very wet vapour passes through pipes immersed in brine tank and evaporate after taking latent heat from brine. The refrigerant now leaves the evaporator somewhat fairly dry vapour, this completes the cycle.

Fig. 14.5 shows the cycle on the T-S diagram. At point "1" refrigerant is sucked into the cylinder and then compressed isentropically to point "2" raising temperature of vapour from T_1 to T_2. During process 2-3 the condensation takes place and throttling is done by the process 3-4. The removal of heat by very wet vapour is done during process 4-1.

Now from T-S diagram

Work done, W = Rankine area = area 1–2–3–4–1

$= H_2 - H_1$ per kg of refrigerant

Net refrigerating effect, N = area 1-4-e-f

$= H_1 - H_4$

During throttling process, there is no transfer of heat thus

$$H_4 = H_3$$

$$\therefore \quad N = H_1 - H_3$$

$$\therefore \quad \text{C.O.P.} = \frac{N}{W} = \frac{H_1 - H_4}{H_2 - H_1}$$

Dry and Wet Compression

If the vapour drawn into the compressor from evaporator is drier than the initial condition shown in Fig. 14.5 on the T-S diagram then while comprising between same pressure range, the vapour becomes super heated at the end of compression shown by the process 1-2.

The starting point of compression *i.e.,* point "1" determines the final state of vapour after compression. If the final state of vapour at point 2 is dry and saturated then the compression is known as "wet compression" and if it is super heated then called "Dry Compression".

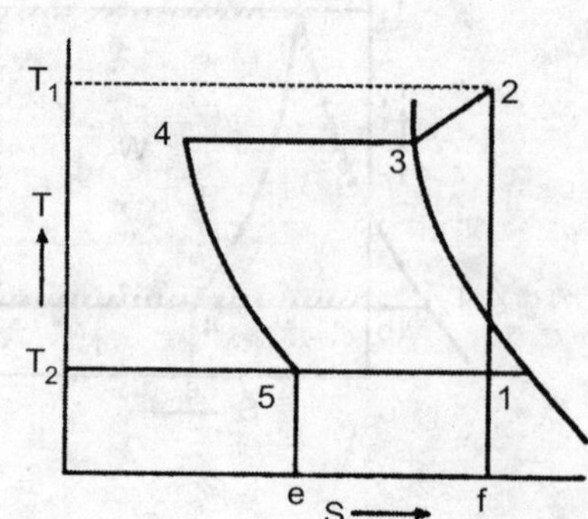

Fig. 14.6 Dry and Wet Compression.

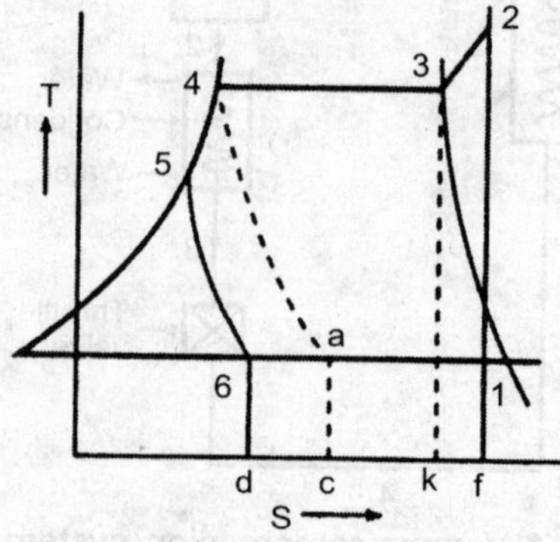

Fig. 14.7 Sub-Cooling Process.

The T-S diagram of Fig. 14.6 indicates that temperature after compression is T_{sup} and pressure P_1. When superheated vapour enters into the cylinder, cooling takes place in two stages at constant pressure P_1. That is first temperature falls from T_{sup} to T_1 and then latent heat is taken away. The process on the diagram is indicated by lines 2-3 and 3-4. Actual condensation starts at point "3" and completes at point "4". After this the high pressure liquid is throttled through the throttle valve.

$$\text{Now work done} = \text{Area } 1\text{–}2\text{–}3\text{–}5\text{–}1$$

$$= H_2 - H_1 \text{ per kg of refrigerant}$$

Net refrigerating effect, N = area 1-5-e-f.

$$= H_1 - H_5 = H_1 - H_4$$

$$\therefore \quad \text{C.O.P.} = \frac{H_1 - H_4}{H_2 - H_1}$$

From above it is observed that if after compression super heated vapour is more which means vapour removes more heat from evaporator for the same weight of refrigerant.

Further due to drier vapour, there is more specific volume hence greater amount of work done required to compress to higher pressure. It means with dry compression there is an increase in refrigerating effect as well as work done. But increase in refrigerating effect is less than increase in work done in respect to wet compression where no super heating of vapour is achieved. "Thus overall effect is that C.O.P. is reduced while superheating".

Under Cooling or Sub Cooling

The C.O.P. can be increased by under cooling or pre-cooling. It is the process where liquid refrigerant is allowed to cool below saturation temperature T_1 at higher pressure before throttling.

It is indicated on the T-S diagram, Fig. 6.7 by the process 4-5 along liquid line. Sub-cooling is brought about by circulating greater quantity of cooling agent, *i.e.*, water through the condenser or by using water colder than natural circulating water.

In under cooling the chances of vaporisation of liquid refrigerant are minimum during throttling process. Thus more liquid refrigerant is available in the evaporator. Further it is clear from T-S diagram that refrigerating effect is increased by an amount equal to area a.6–*d-C*. The T-S diagram in this case shows that the vapour is superheated as well as under-cooled before throttling. With both superheating and under cooling the refrigerating effect increases multifold.

Actual Vapour Compression Refrigeration Cycle

The actual cycle differs by the theoretical one as described below:

(*i*) Process 1-2-3 represents passage of refrigerant through evaporator at temperature "T_2" state "3" shows that vapour is of superheated condition. Superheating is due to large removal of heat from evaporator which is an advantage or due to heat picked up by vapour in suction pipe, which is a disadvantage.

2. *Process 3-4-5-6-7-8.*

(*a*) 3-4, 7-8—to operate suction and delivery valves which are due to pressure difference, the actual suction pressure and pressure should be lower and higher respectively than evaporating and condensing pressure. This is shown by 3-4 and 7-8 process lines.

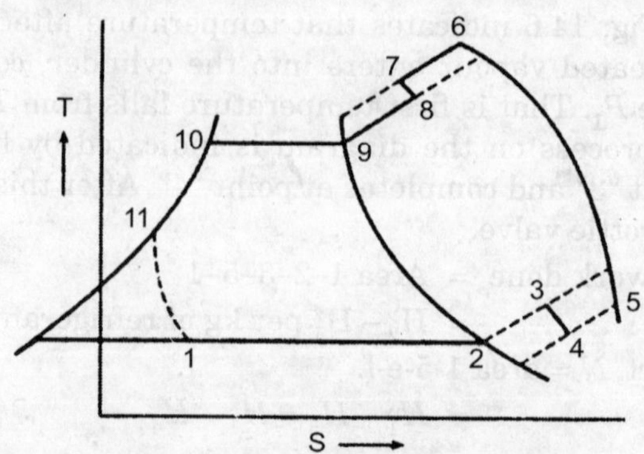

Fig. 14.8 Actual Vapour Compression Cycle.

(*b*) 4-5, 6-7—cold vapour enters into cylinder and is heated by hot cylinder walls shown by 4-5. Similarly cooling effect at discharge is represented by 6-7.

(*c*) 5-6—since the actual process of compression is neither adiabatic nor polytropic, therefore the process is represented by a curve 5-6.

3. *Process 8-9-10-11* shows path of refrigerant in condenser, 8-9 removal of superheat, 9-10 removal of latent heat and 10-11 sub-cooling.

4. *Process 11-1* is throttling of sub-cooled liquid so that state "1" can be obtained.

Advantage of Vapour Compression System

(*a*) Smaller size for a given refrigerating effect

(*b*) Higher C.O.P.

(*c*) Lower power consumption

(*d*) Less complicated in design and operation.

14.4.3 Absorption Refrigerating System

In vapour compression system to get condenser pressure there is a great change in volume during compression process. It is no doubt that only a small part of supplied energy is used to raise pressure and rest is utilized in negotiating the volume of gas. Thus, if this change in volume is avoided then even 95% of energy can be saved. This can be obtained by reducing the large volume of refrigerant, such as by using some absorbent which reduces the volume of refrigerating gas by liquid volume and while on demand refrigerant can be obtained by heating.

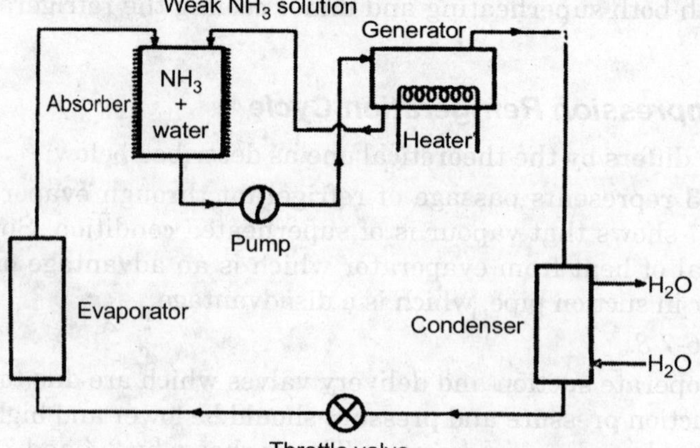

Fig. 14.9 Absorption Refrigerating System.

The basic difference between vapour compression and absorption cycle is the replacement of compressor unit by a set of equipment, which can solve the above said purpose. This consists of an absorber equipments such as condenser, evaporator, throttling valve, etc., remain the same.

Fig. 14.9 shows the diagram of this system. The most common used fluid in this is water as absorbent and ammonia as refrigerant. The vapour from evaporator is allowed to mix with water in absorber. The heat liberated due to chemical action can be used for raising the temperature of water which is further used for heating the aqua-ammonia (NH_3 + H_2O solution) in generator. Now this strong aqua-ammonia is sent to generator through a pump at condenser's pressure. In generator heat is given to aqu-aammonia where ammonia liberated being lower boiling temperature, thus vapour leaving the generator is only ammonia. This vapour is then sent to condenser before passing through throttle valve and evaporator. The weak solution left in generator is sent back to the absorber for repeating the cycle.

Practical Absorption Refrigerating Cycle

The replacement of the compressor by simple arrangement as shown in Fig. 14.9 is not very economical in practice. To have more economy some additional equipments are to be introduced. For example an analyser, a rectifier and two heat exchangers are as shown in Fig. 14.10.

(a) Analyser. Ammonia vapour leaving generator may contain some amount of moisture and for better performance the refrigerant must be free from moisture before passing through condenser and the expansion valve, otherwise moisture may freeze in the small valve passage and choke the flow. Thus function of an analyzer is to remove the moisture as far as possible.

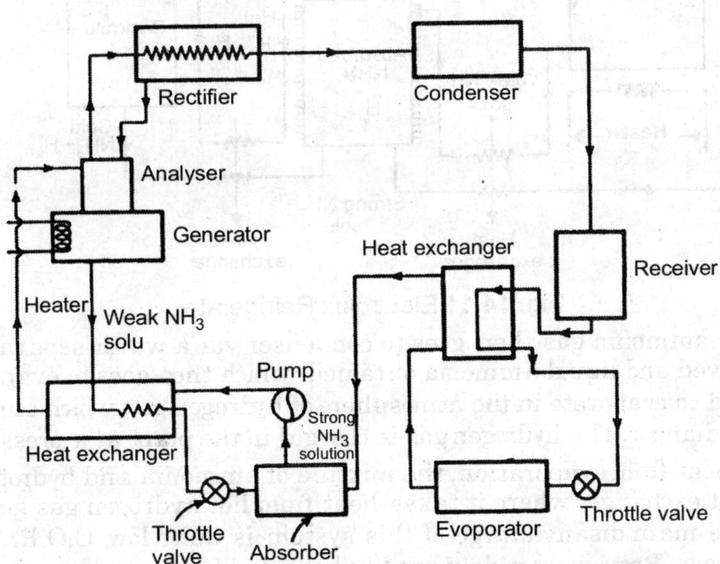

Fig. 14.10 Practical absorption refrigerating cycle.

(b) Rectifier. It is a closed type of cooler in fact a miniature condenser, where traces of water vapour left in ammonia vapour are removed by condensation. The cooling is done by circulating water similar to the condenser and the condensed liquid is returned to analyzer.

(c) Heat Exchanger. Two heat exchangers are fitted to transfer heat from high temperature liquid to low temperature liquid. One heat exchanger is fitted between liquid receiver and evaporator so that liquid is sub cooled and vapour heated up. The second heat exchanger is provided between generator and absorber so that strong aqua-ammonia solution is heated before entering into analyzer and weak aqua-ammonia solution is cooled before entering into absorber.

Electrolux Refrigerator (Ammonia Hydrogen Refrigerator)

This is an absorption type of Refrigerator. The main advantage is that it has no moving parts even compressor or pump. This can work on low grade energy *i.e.,* heat produced by kerosene or gas or electricity. In this aqua-ammonia solution hydrogen is used and the plant is charged at a pressure of 14 kg/cm^2. The principle behind this is that ammonia gas evaporates very readily in presence of hydrogen gas. The cycle works as under:

The strong aqua-ammonia solution enters into the boiler due to gravity from absorber via a heat exchanger where it takes heat from returning weak aqua-ammonia solution. The aqua-ammonia solution in boiler is heated up and thus ammonia gas is liberated from it and weak aqua-ammonia solution returns to absorber via heat exchanger (see Fig. 14.11).

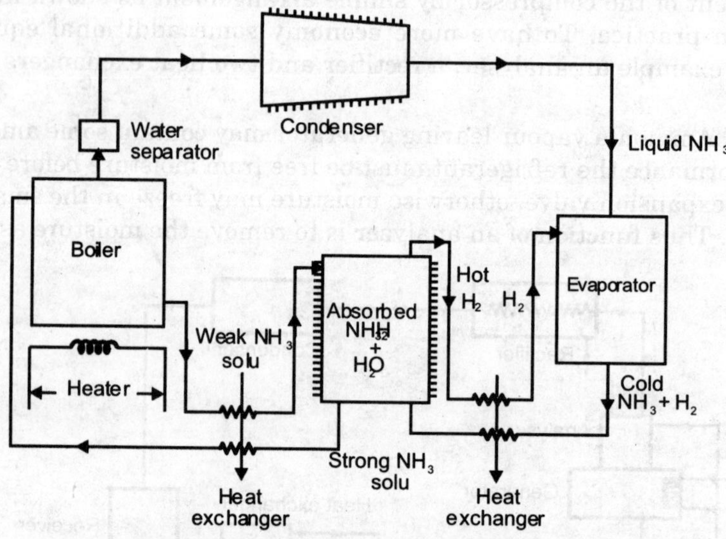

Fig. 14.11 Electrolux Refrigerator.

The liberated ammonia gas then goes to condenser via a water separator. In condenser latent heat is removed and liquid ammonia obtained, which then goes to evaporator. The liquid ammonia is allowed to evaporate in the atmosphere of hydrogen gas which comes from absorber via another heat exchanger. The hydrogen gas is charged in the plant at a pressure of 12 kg/cm^2.

After taking heat from evaporation, the mixture of ammonia and hydrogen is returned to absorber via a heat exchanger where it takes heat from hot hydrogen gas going to evaporator from absorber. The main disadvantage of this system is quite low C.O.P. The refrigerating effect obtained is poor. Previously widely used in household refrigeration.

14.5 REFRIGERANTS

Refrigerants absorb heat as they evaporate at low pressure and temperature in the evaporator and transfer this heat and heat equivalent of compressor work in the condenser.

Most of the refrigerants are vapours (gases) at atmospheric conditions but easily liquefied when pressure is increased. Every refrigerant has its own distinctive pressure and temperature characteristics. There are many factors which influence the choice of a refrigerant but most important are toxicity and relative ease with which it can be cycled from the vapour to the liquid and back again, and also on the suitability of its temperature range to the particular application.

These are fluids which act as cooling medium for absorbing heat from substances. These are of three types.

1. Primary Refrigerants
2. Secondary Refrigerants and
3. Cryogenic Refrigerants.

1. *Primary Refrigerants.* These are those working fluids which pass through the refrigerating cycle of evaporation, recovery, compression, condensation and expansion. For example, ammonia, freon, SO_2, methyl chloride and CO_2, etc.

2. *Secondary Refrigerants.* These are those working fluids which act as cooling medium but do not undergo the refrigerating cycle. A primary refrigerant undergoes the refrigerating cycle and this cools the secondary refrigerant in the evaporator. Air, brine and any other freezing solution are its examples.

3. *Cryogenic Refrigerants.* These are used for creating temperature below – 157°C, – O_2, H_2, fluorine and helium, etc., are its examples.

Properties of an Ideal Refrigerant

1. *High latent heat.* If it has high latent heat, it removes higher latent heat during evaporation, thus less quantity of refrigerant is required.
2. *Suitable evaporating pressure.* It should be according to the necessary temperature in the storage chamber and should be more than atmospheric pressure, if not, then air may enter in the refrigerator as well as it should not be too high, so that size becomes bulky.
3. *Low freezing point.* The evaporation temperature must remain between – 20°C to – 30°C.
4. *High critical temperature.* Critical temperature can be defined as the temperature above which the vapour can-not be condensed irrespective of any high pressure, so if high critical temperature is not higher than condensing temperature then liquifying of gas is not possible.
5. *High vapour density.* If it is not high, then more quantity of gas is required, causing longer stroke of compressor, which means bigger compressor size.
6. *Non-corrosive.* If corrosive then tubes used in the evaporator and condenser may be destroyed.
7. *Odourless.* Since odour is not good.
8. *Non-toxic.* Since toxic matters are harmful.
9. *Non combustible.* Refrigerant should be free from the danger of combustion.
10. *Low cost.* It should be easily available and at low cost.
11. *Leakage should be easily detectable.* Otherwise the refrigerant may leak without notice from the unit.

Important Refregerants

In each refrigerant all the properties are not available. Following are some of the *important refrigerants* which are commonly used.

1. Ammonia. Its boiling point at atmospheric pressure is –30°C and latent heat as 315 kJ/kg. It is toxic therefore rarely used in domestic refrigerator. It is mainly used in absorption system as it is highly soluble in water and easily expelled out when heated.

2. Freon - 12 (R-12). It is non-toxic and non-flammable and therefore more safe, hence it is most widely used refrigerant for many applications. Its boiling point at atmospheric pressure is –30°C.

3. Freon - 22 (R-22). It is superior to R-12 and is widely used for low temperature industrial and commercial systems as its boiling temperature at atmospheric pressure is –40°C. Its properties on safety aspects are similar to R-12. Its latent heat is 21 kJ/kg.

4. Carbon Dioxide (CO_2). Its boiling temperature at atmospheric pressure is –78.5°C. and latent heat is 274 kJ/kg.

14.6 APPLICATIONS OF REFRIGERATION

14.6.1 Household (Domestic) Refrigerator

The refrigerator is generally divided in three compartments as shown in Fig. 14.12 (*a*). The top compartment works as a freezer where the temperature of the refrigerent is minimum as the refrigerant first enters in this compartment. The other, middle and bottom, two compartments, known as cold chambers, are cooled by natural convection as the cooled air in the freezer descends and then goes up after absorbing the heat from the middle and bottom compartments.

Fig. 14.12 (*a*) Domestic Refrigerator.

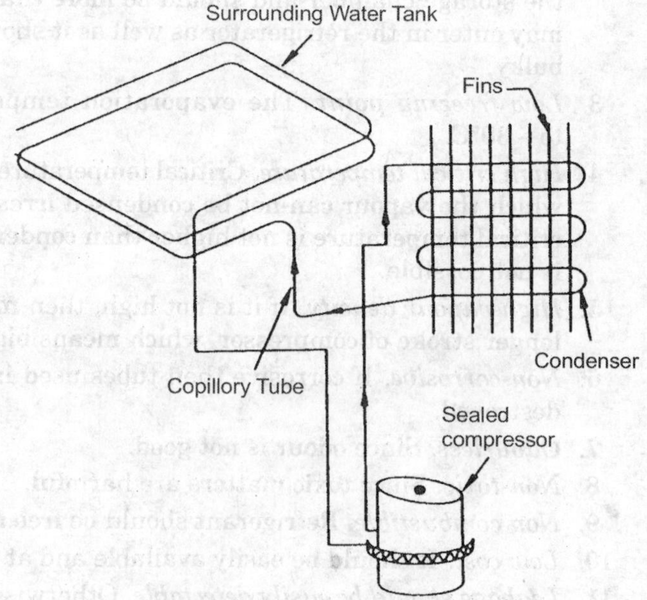

Fig. 14.12 (*b*) Refrigeration System.

The top compartment of the refrigerator is used for making ice, kulfi or ice-cream, while the middle and bottom compartments are used for preserving milk, curd, vegetables and fruits. The door of the refrigerator is also provided with space which is used for keeping the water bottles, cold drinks, sauce, and jam bottles. The top part of the door is used for putting the eggs. The temperature is controlled by on-off control which is operated automatically by a thermostat valve.

The refrigeration system is shown in Fig. 14.12 (b).

The domestic refrigerator works on the principle of vapour compression cycle and uses compressor, condenser, capillary and evaporator as basic components of the system. Since the compressor is directly coupled to an electric motor, and both are sealed in a box therefore this is known as sealed unit. This arrangement helps to reduce the noise of the motor, prevents the leakage of the refrigerant and motor is cooled by the refrigerant vapour as it is exposed to the same.

The condenser is generally made of a grid of copper tubes provided with fens, mounted behind the unit and cooled by natural air convection. Capillary is used as an expansion or throtling device for reducing the pressure and temperature of the refrigerant. The refrigerator is a box which forms the evaporator part of the system and is heavily insulated to reduce the heat flow from surrounding to the evaporator.

The R-12 is used as refrigerant in these refrigerators as this is most safe refrigerant.

14.6.2 Water Cooler

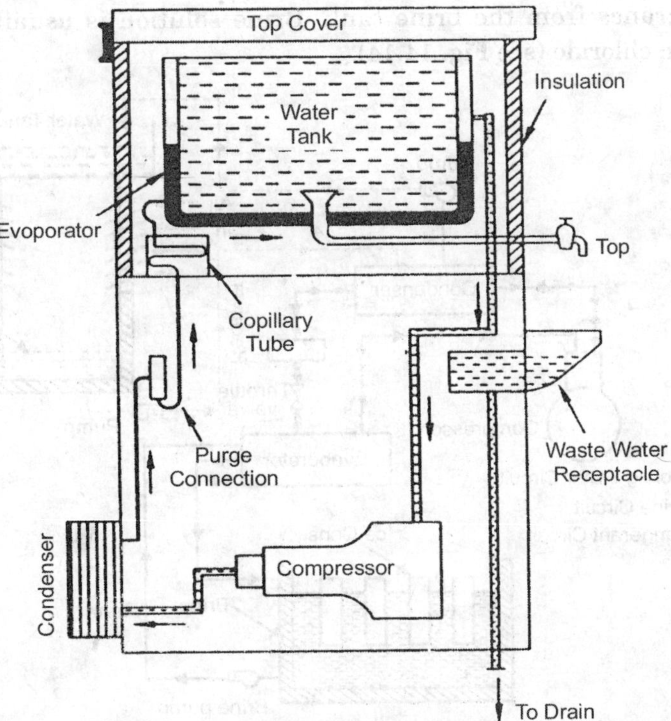

Fig. 14.13 Water Cooler.

It also uses vapour compression refrigeration system and has basic components of compressor, condenser, capillary and evaporator (see Fig. 14.13). Usually R-12 used as refrigerant is compressed in the compressor and is condensed passing through an air - cooled condenser where its heat is given to air. The condensed refrigerant is further passed through capillary where its pressure and temperature are reduced. Then the vapoure - liquid mixture enters into the hollow part of the water box where it absorbs its latent heat from the water and boils. As the latent heat for boiling the refrigerant is taken from water, the water is cooled. The water box is also provided with heavy insulation to prevent the heat flow from the surrounding. The cooled water is taken out from the water box through a water tap.

14.6.3 Ice-Plant

The refrigerant commonly used in ice-plant is ammonia (NH_3), as it has large latent heat compared to that of R-12 at atmospheric pressure. Further, it is very cheap. The only care that should be taken with ammonia is that the system should be perfectly air-tight. The vapour compression system is commonly used in ice-plants.

The throttled liquid ammonia at low-temperature and low pressure enters into an evaporator which is a coil dipped in brine. The brine is cooled by the low temperature ammonia and it gets evaporated. The cold brine from the evaporator is used for freezing water. The cold brine from the cold tank is circulated with the help of pump around the ice-forming cans which are also dipped in brine tank. The heat from the water in the ice cans is taken by the brine and the same is given to ammonia in the vaporator. Once the ice is formed, ice cans are lifted with the help of cranes from the brine tank. Brine solution is usually prepared through commercial calcium chloride (see Fig. 14.14).

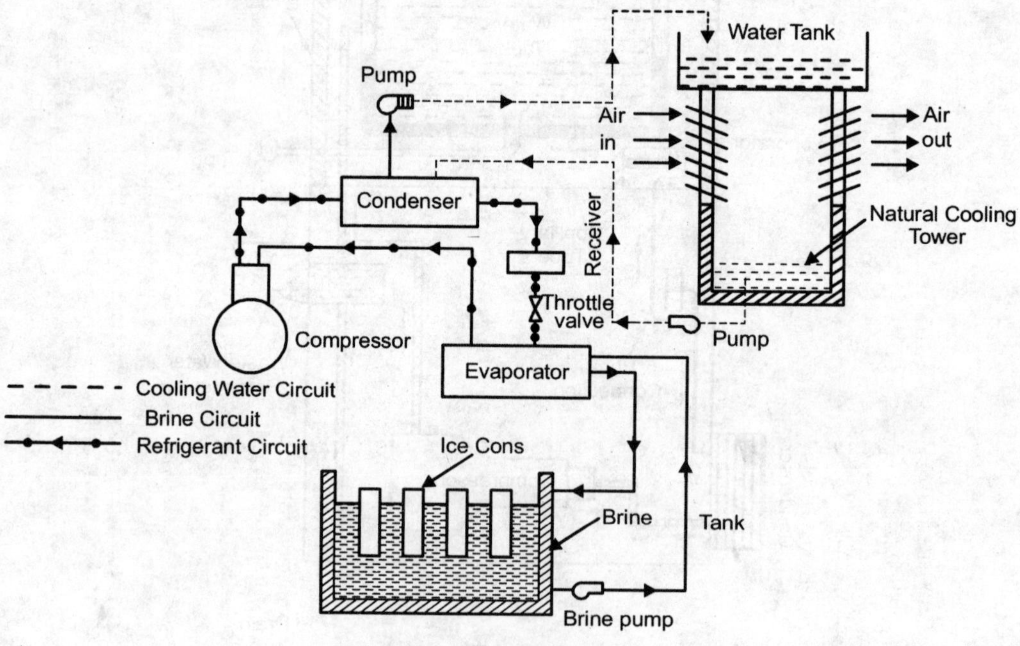

Fig. 14.14 Ice Plant.

Example 14.1 *The C.O.P. of a refrigerator operating on a Carnot cycle is 5.4 when it maintains the temperature in the evaporator – 5°C. Determine a condenser temperature and refrigerating effect of the refrigerator if the power required to run the refrigerator is 3 kW.*

Solution. The arrangement of the system is shown in the Fig. 14.15. In this particular system, evaporator is a source at temperature – 5°C and condenser is a sink at temperature T_1.

The C.O.P. of the refrigerator is given by

$$\text{C.O.P.} = \frac{Q_2}{W} = \frac{T_2}{T_1 - T_2}$$

where $T_2 = -5 + 273 = 268° \text{ K}$

$$\therefore \quad 5.4 = \frac{268}{T_1 - 268}$$

$$\therefore \quad T_1 = 268 + \frac{268}{5.4} = \mathbf{317.5\ K} \quad \textbf{Ans.}$$

$$Q = W \times 5.4 = 3 \text{ (kJ/s)} \times 5.4$$

$$= 16.2 \text{ kJ/sec} = \mathbf{16.2\ kW} \quad \textbf{Ans.}$$

Fig. 14.15

Example 14.2 *A reversible heat-pump is used for heating the building in winter season. The heat is absorbed from the earth at 0°C by a fluid circulating in buried pipe and delivered to the building to maintain the temperature at 23°C. How much heat is supplied to the building in kJ/hr if one kilowatt-hour of electrical energy is needed to operate the heat pump?*

Solution. The arrangement of the system is shown in Fig. 14.16. Here the source (building) and sink (earth) temperatures are given as 296 K and 273 K respectively.

$$T_1 = 296 \text{ K}, \quad T_2 = 273 \text{ K}$$

The C.O.P. of the system is given by

$$\text{C.O.P} = \frac{Q_1}{W} = \frac{T_1}{T_1 - T_2}$$

$$\therefore \quad \frac{Q_1}{W} = \frac{296}{296 - 273} = 12.9$$

$$W = 1 \text{ kW-hr} = 3600 \text{ kJ}$$

$$\therefore \quad Q_1 = 3600 \times 12.9$$

$$= \mathbf{46440\ kJ/hr} \quad \textbf{Ans.}$$

Fig. 14.16

Example 14.3 *Ice is formed at 0°C from water at 20°C. The temperature of the brine is – 10°C. Find the ice formed per kWh. Assume that the refrigeration Cycle used is perfect reversed Carnot cycle.*

Latent heat of ice = 334.5 kJ / kg.

Solution. Heat taken out per kg of water

$$= 4.18 \times (20 - 0) + 334.5 = 418 \text{ kJ/kg}$$

$$\text{One kWh} = 3600 \text{ kJ}$$

$$\text{C.O.P. of Carnot cycle} = \frac{T_1}{T_2 - T_1} \quad \text{where } T_1 < T_2$$

$$= \frac{-10 + 273}{20 - (-10)} = \frac{263}{30} = 8.77$$

Assume m_c is the ice formed per 1 kWh

$$\therefore \quad \text{C.O.P.} = 8.77 = \frac{Q}{W} = \frac{m_c \times 418}{3600}$$

$$\therefore \quad m_c = \textbf{75.57 kg of ice.} \quad \textbf{Ans.}$$

14.7 AIR CONDITIONING

Air conditioning is the process of altering the properties of air (primarily temperature and humidity) to more favourable condition. More generally air conditioning is referred to any form of technological cooling, heating, ventilation or disinfection that modifies the condition of air. In construction, a complete system of heating, ventilation and air-conditioning is referred to as "HVAC".

In true sense, air conditioning implies the simultaneous control of following four atmospheric conditions:

(*i*) Surrounding temperature,

(*ii*) Humidity (*i.e.* water content in the air)

(*iii*) Cleanliness or purity of air.

(*iv*) Air movement.

Seasonal extremities of heat and cost in our country are more than the human being can endure, it also affects the efficiency/performance of work. Air conditioning is used in residences, theatres, offices, hospitals, railway coaches, aeroplanes, vehicles, etc. The use of air conditioning for industrial purposes is also increasing with rapid industrial development.

Air Conditioning Units and Plants

In air conditioning system out door air would enter the system through some type of rain proof opening equipped with means to keep out rain and snow as the case may be. Air which contains dust particle, must be passed through some type of air filter.

After cleaning it may be necessary to heat the air for better humidification. A heater of proper capacity and size should be fitted. The heating should have proper automatic control. The out door air intake dust may be joined to the duct bringing air for recirculation. The two airs (fresh and recirculating) may join before the heating stage or afterwards.

The air is moved through the ducts by centrifugal fans or blowers. All fans cause certain amount of noise and therefore great care should be taken in selecting the fans so as to keep the noise with in permissible limits. The fans should be of sufficient capacity to overcome the resistance to the circulation of air. Ducts are prepared from sheet metal. The duct supplying conditioned air called "supply ducts". Conditioned air from ducts is supplied through out-lets known as "grills". Grills are available in large varieties of shape and design for the proper distribution of air.

14.8 CLASSIFICATIONS

14.8.1 Based on Location

Air conditioning systems may be classified as central system or unitary system. A central system is one in which the components of the system are all grouped together in one central room and the conditioned air is distributed from there to the space to be conditioned through extensive duct work.

A unitary system makes use of factory assembled units which are usually installed in or immediately adjacent to a zone or space to be conditioned.

The systems can be classified as:

(*a*) Cooling system or Summer air conditioning.

(*b*) Heating system or Winter air conditioning.

(*c*) Combination of both systems or year round air conditioning.

The advantages of the unitary system over the central system are:

(*i*) Savings in installation and field assembly labour.

(*ii*) Exact requirements of each separate room is met whereas a central system cannot meet the individual requirements of a separate room.

(*iii*) Zoning or distribution system or duct work is eliminated.

(*iv*) Only those rooms which need cooling will have their units running whereas the central plant will have to run all the time for the sake of only few rooms to be conditioned.

(*v*) Failure of any unit put the conditioner off in only one room. This disadvantage of central plant is many a time obviated by installing two plants instead of one of full capacity, so that if one is out of order, atleast few rooms can be served by the other.

(*vi*) Starting from one few rooms, the air conditioning may be progressively extended to other rooms without spending large sum to start with.

The advantages of the central plant system over the unitary system are:

(*i*) Lower initial cost

(*ii*) It can be housed quite away from rooms at a place which is not so useful otherwise.

(*iii*) Noise and vibration problems are not so actuate.

(*iv*) Easy maintenance.

14.8.2. Based on Applications

The air-conditioning is used in two types of applications; namely ; comfort applications, and process applications.

1. Comfort Applications

Comfort applications aim to provide a building indoor environment that remains relatively constant inspite of changes in external weather conditions or in internal heat loads. In comfort applications air surrounding human beings is maintained in such a condition that it is most suitable for their comfort and health.

Feeling of comfort in an air conditioned space depends on:

(*a*) Continuous supply of oxygen and removal of carbon dioxide.

(*b*) Removal of body heat dissipated by occupants and other appliances working in the area.

(*c*) Removal of body moisture dissipated by occupants. High humidity of air reduces comfort. The ventilation system must be capable of maintaining the relative humidity below 60%.

(*d*) Sufficient air movement and proper air distribution in conditioned space. A person may not feel comfort when air velocity is very high or very low. For human comfort air velocity should be around 8 m/sec.

(*e*) To maintain purity of air by removing odour and dust. The air must be free from odour, toxic gases such as smoke, and bacteria.

2. Process Applications

Process applications aim to provide a suitable environment for a process being carried out, regardless of internal heat and humidity loads and external weather conditions. It is the needs of the process that determine the conditions, not the human preference. The temperature and humidity conditions for industrial air-conditioning are always different from comfort air-conditioning.

Process applications include:

(*i*) Hospital operation theatres to reduce infection risks and humidity control, and to keep temperature in comfort range.

(*ii*) Pharmaceuticals where cleanliness of high level and control of temperature and humidity is required.

(*iii*) Textile manufacturing.

(*iv*) Power house control rooms.

(*v*) Chemical and biological laboratories.

(*vi*) Mining and other underground works.

(*vii*) Industrial environment

(*viii*) Food processing areas.

The air conditioning deals with supplying and maintaining a desirable internal atmospheric conditions irrespective of external conditions. The four essential factors required in a complete air conditioning system are:

(*a*) Temperature control

(*b*) Humidity control

(*c*) Air movement and circulation

(*d*) Air filtering, cleaning and purification.

A distinction should be made between "comfort cooling" and "Industrial cooling". Comfort cooling is applied for residences, stores, theatres and hospitals, etc., where the main object is to produce comfortable and healthy conditions for the occupants.

"Industrial cooling" is applied in industries where varying temperature and humidity conditions can affect the manufacturing processor product. Thus, here conditions which are essential for the best results are produced and maintained even, if they may not entirely agree with those required for human comfort. As used in tool room, textile mills, paper mills and photo processing plants, etc.

The 4 important factors in relation to human comfort are described as below:

(*a*) **Temperature Control.** This factor affects human comfort to a great extent. The temperature range for human comfort is 22.8°C to 25°C.

(*b*) **Humidity.** For human comfort the relative humidity should be in the range of 35% to 60% with air motion 5 metre/sec. to 8 metre/sec. Control of humidity is very essential for human comfort and it also increases the efficiency of the workers.

(*c*) **Air Purity.** Proper cleaning, filteration and purification of air is very necessary to keep it free from dust, dirt and such other impurities. We do not feel comfortable while breathing contaminated air.

(*d*) **Air movement and Circulation.** Air movement and its circulation is required to be controlled in order to ensure a uniform temperature and equidistribution of air throughout the space to be air conditioned.

14.9 PSYCHROMETRY

The study of air water mixture is known as psychrometry. Mixture of dry air and water vapour which may exist in superheated state or saturated state. Air is used as a medium for cooling the building in summer and heating in the winter, as discussed earlier, the complete process of air-conditioning is not only heating or cooling of air but also the control of water particles or humidity in air. Humidity or water vapour is an important property which affects the air-conditioning. The amount of water vapour present in the air affects the human comfort.

Some of the *terms used in the psychrometry* are:

(*i*) **Moist Air.** It is the mixture of dry air and water vapour.

(*ii*) **Humidity.** The mass of water vapour present per kg of dry air is expressed as humidity.

(*iii*) **Saturated Air.** When air contains all the vapour which it can hold, it is said to be saturated. Lower the temperature of the air, lower is the capacity of the air to hold the water vapour.

(*iv*) **Dry Air.** Air except water vapour is known as dry air.

(*v*) **Moisture.** The water vapour present in the air is known as moisture.

(*vi*) **Sensible Heat of Air.** The quality of heat which can be measured by measuring dry bulb temperature of air is known as sensible heat.

(*vii*) **Total Heat of Air.** It is the sum of sensible heat of dry air and sensible and latent heat of water vapour associated with dry air.

(*viii*) **Comfort.** The body of a man emits heat to the surrounding atmosphere. This heat should be carried out by the surrounding air otherwise it will be accumulated in the body and the man feels uncomfortable. The heat from the body is emitted in the form of sensible heat and latent heat. For carrying out the heat by the air from the body, the temperature of the air must be below the body temperature (36.4°C). The heat carried away from the body in the form of latent heat is due to the sweating from the body. The sweating maintains the softness of the body.

(*ix*) **Psychrometric Processes.** The summer air-conditioning generally requires dehumidifying and cooling the air. The winter air-conditioning generally requires heating and humidifying the air. The heating, cooling, dehumidifying by cooling coil and humidifying by water spray are the psychrometric processes.

(*a*) **Sensible Heating or Cooling.** The heating or cooling of air without the addition or subtraction of moisture is known as sensible heating or cooling. In both these cases the dry bulb temperature changes but specific humidity remains constant. During this the air remains in the unsaturated zone. For sensible heating air may be passed over the heating coil or steam, and for sensible cooling air is passing over evaporator coil.

(*b*) **Cooling of Air With Dehumidification.** The process of removing the water from air is known as dehumidification. This process results in a reduction of both dry bulb temperature and the specific humidity.

(*c*) **Humidification.** The process of adding the water to air is known as humidification.

(*x*) **Dry Bulb Temperature.** The temperature of air measured by ordinary thermometer is known as dry bulb temperature of air.

14.10 AIR CONDITIONER—TYPES

1. Window type room air conditioner
2. Packed air conditioner
3. Central air conditioning plant

1. Window Type Room Air Conditioner

This is unit or self contained type air conditioner having a complete refrigeration plant along with thermostat controls and filtering equipment. This is cooled by air and placed in a window or wall of the room to be air conditioned. No ducts are required.

The principle of working of this type is shown in Fig. 14.17. It has two compartments, *i.e.*,

(*a*) Outdoor compartment and (*b*) Indoor compartment.

The outdoor compartment consists of:

(*i*) Hermetically sealed motor compressor unit

(*ii*) Condenser and (*iii*) Motor driven fan

The Indoor compartment consists of:

(*i*) Filter (*ii*) Evaporator

(*iii*) Motor driven fan and (*iv*) Remote bulk refrigerant control

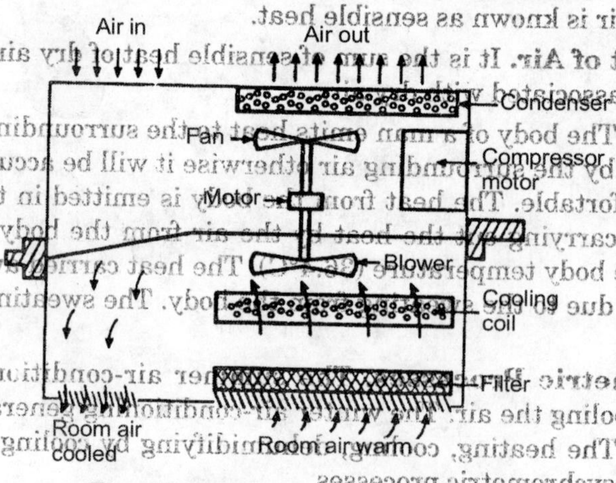

Fig. 14.17 Window Air Conditioner.

The outside air is forced over the condenser coils by a fan while in the indoor compartment there is second fan to draw air through the filter and forces it over the evaporator coils.

2. Packed Air Conditioner

This is also a unit or self contained type one and may be either air cooled or water cooled. It can be installed directly in the room to be conditioned or in a nearly room with a small lengths of supply ducts for the air. This is a factory assembled unit, in which the complete system is mounted in a cabinet.

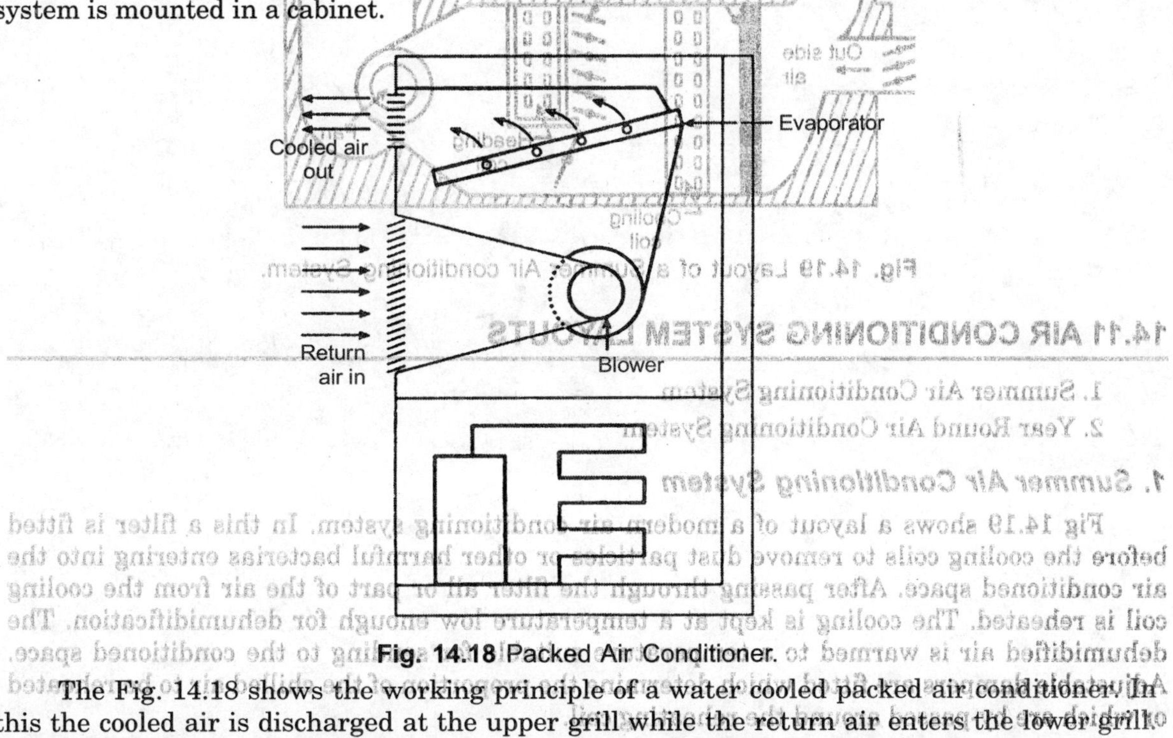

Fig. 14.18 Packed Air Conditioner.

The Fig. 14.18 shows the working principle of a water cooled packed air conditioner. In this the cooled air is discharged at the upper grill while the return air enters the lower grill. The unit can be used in restaurants, stores, banks and offices, etc. These units are available in capacities of more than 100 tonnes.

3. Central Air Conditioning Plant

The various factors affecting the selection of the equipment for designs and installation of the central air conditioning plant are:

(*i*) Capacity of the plant (*ii*) Filters
(*iii*) Refrigeration equipment (*iv*) Heating equipment
(*v*) Insulation (*vi*) Noise and vibrations.

This system is required to serve, a number of rooms required individual control of each room. Depending upon the type of "central air conditioning plant", the following components of the equipment may be needed:

(*a*) Filters (*b*) Heating coils
(*c*) Blower and driving motor
(*d*) Spray for cooling and humidifying or for washing air
(*e*) Coils for cooling and dehumidifying (*f*) Control devices and (*g*) Refrigeration plant.

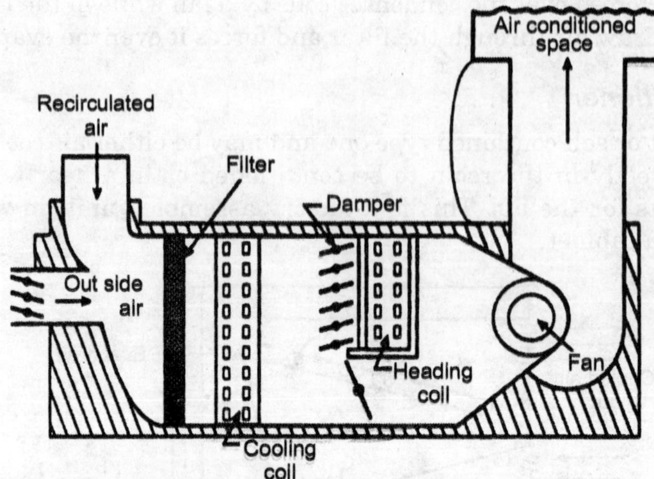

Fig. 14.19 Layout of a Summer Air conditioning System.

14.11 AIR CONDITIONING SYSTEM LAYOUTS

1. Summer Air Conditioning System
2. Year Round Air Conditioning System

1. Summer Air Conditioning System

Fig 14.19 shows a layout of a modern air conditioning system. In this a filter is fitted before the cooling coils to remove dust particles or other harmful bacterias entering into the air conditioned space. After passing through the filter all or part of the air from the cooling coil is reheated. The cooling is kept at a temperature low enough for dehumidification. The dehumidified air is warmed to a temperature suitable for sending to the conditioned space. Adjustable dampers are fitted which determine the proportion of the chilled air to be reheated or which are by-passed around the reheating coil.

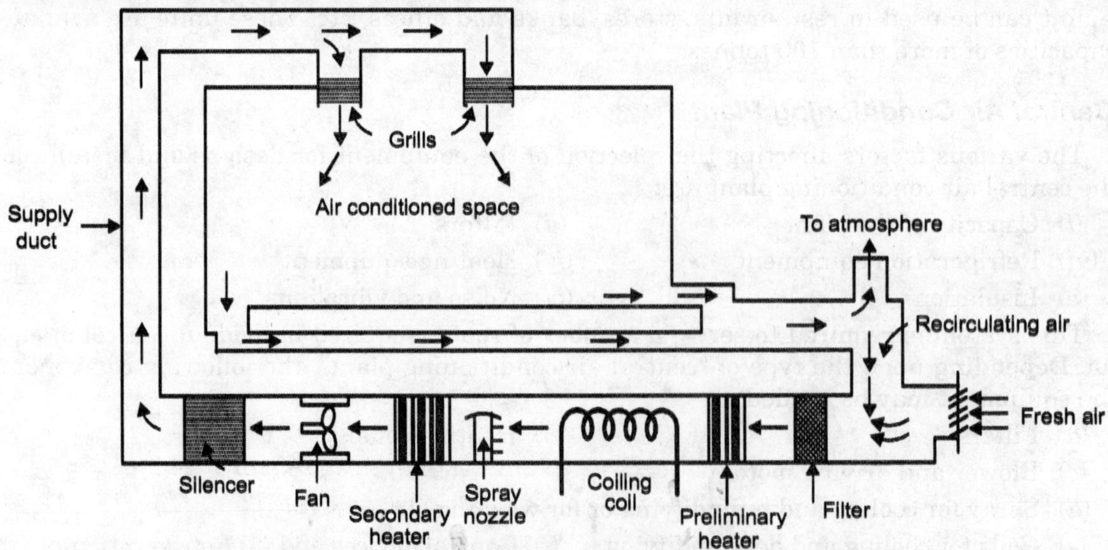

Fig. 14.20 Layout of year round air conditioning system.

2. Year Round Air Conditioning System

Fig. 14.20 shows the line diagram for the layout of a "Year round conditioning system". The mixture of recirculated air and fresh air first passes through a filter and cleaned air then goes ahead. For Summer air conditioning, the cooling coils cool the air to the desired temperature. Dehumidification is done by the cooling coils.

In Winter, the cooling coil is disengaged and only the heating coil functions to heat the air. To humidify the air in dry season, a spray type humidifier is also fitted.

14.12 SELECTION OF AIR CONDITIONING PLANT

Table 14.1 shows that which type of plant is the most suited for a given capacity range based on overall economy considerations:

Table 14.1 Type of plant suited for given capacity range.

	Capacity requirements	*Type of plant most often used based on economic considerations*	*Type of plant sometimes used*	*Type of plant seldom used*
1.	Upto 10 tonnes	Room cooler or packaged unit serving each room	Packaged air conditioning plant using duct work for distribution of conditioned air.	Built up central plant
2.	10 tonnes to 35 tonnes	Packged air conditioner and with duct work for distribution of conditioned air (Reciprocating compressor and direct expansion coils)	Central plant with direct expansion evaporators and reciprocating compressor.	Central system with centrifugal compressor and indirect expansion unit.
3.	35 tonnes to 100 tonnes	Central air conditioning plant with duct work for air distribution (Reciprocating compressor and direct expansion system).	Central plant with more than one reciprocating compressor or chilled water system.	-do-
4.	100 tonnes to 200 tonnes	Central system with more than one refrigeration plant, one or more compressors and most frequently with indirect or chilled water system.	One central plant with centrifugal compressor and chilled water system.	Multiple unitary system serving different zones and reciprocating compressor and direct expansion system or absorption plant or steam jet plant.
5.	200 tonnes to to 300 tonnes or higher capacity.	Central plant using centrifugal compressor and with indirect expansion chilled water system.	Central plant with absorption system or steam jet refrigeration plant.	Central plant using reciprocating compressor and direct expansion system.

14.13 AIR CONDITIONING APPLICATIONS

Following are some of the applications of air-conditioning which are most commonly used:

1. Industrial Applications

As discussed in process applications in previous sections, the main aim of their industrial applications is to control the temperature, humidity and air motion for maintaining the quality of the product. The function of air-conditioning is to control the temperature and humidity in chemical and biochemical reactions. Some of the common industries where air-conditioning is prominently used are:

(*i*) Photographic industry.
(*ii*) Food processing industry.
(*iii*) Textile industry.
(*iv*) Printing (paper and cloth printing) industry.
(*v*) Machine tool industry.
(*vi*) Hotel industry.
(*vii*) Pharmaceutical industry.
(*viii*) Oil refining. (ix) Ice factories.
(*x*) Electronic industries.

2. Commercial Applications

The comfort is the main consideration in commercial application of air-conditioning. Some of the common applications of commercial air-conditioning are:

(*i*) Theatres.
(*ii*) Hotels.
(*iii*) Hospitals.
(*iv*) Departmental stores.
(*v*) Banks
(*vi*) Offices.
(*vii*) Medical stores.
(*viii*) Places where electronic items are working (eg. computer centres, telephone exchanges, television centres, hospitals).

3. Transport Air-Conditioning

Now-a-days air-conditioning in moving vehicles is very common. The major problems encountered in design of air-conditioning system used in moving vehicles are the irregular variation of load due to change in speed, lightness and compactness of the equipment required. Transport air-conditioning include the following:

(*i*) Air-conditioning in automobile vehicles (cars, buses etc).
(*ii*) Air-conditioning of trains.
(*iii*) Air-conditioning in Aircrafts.
(*iv*) Air-conditioning in ships.

14.14 AIR CONDITIONING AND HUMAN COMFORT

Good working conditions produce a good effect on the workers' psychology in addition to greater efficiency. A worker working in an atmosphere of badly ventilated and hot conditions feels discomfort and fatigue. His efficiency will decrease and he will not be able to take interest in the work. Proper ventilation takes away the heat of human body, furnaces, boiler and other equipments thus reducing the effect of heat to some extent. Proper ventilation also removes dampness.

Atmospheric conditions have very important role in determining the efficiency and well being of the worker. Temperature, humidity, air quality and its circulation affect the health and comfort of the worker and hence the quality and quantity of production as well as general efficiency of the worker.

HVAC (Heating, Ventilation and Air-Conditioning) System

The main purpose of commercial HVAC systems is to provide the people working inside the buildings with "Conditioned" air so that they will have a comfortable and safe work environment. **Conditioned air** means that air is clean and odour free, and the temperature, humidity, and movement of the air are within certain comfort ranges.

Table 14.2 shows the ranges of ideal conditions for different environmental factors which are conducive to ideal working conditions.

Table 14.2 Range of ideal condition for different environment.

	Parameter	*Limit Range*
1.	Temperature	(*i*) Summer – 23–26°C (*ii*) Winter – 20–24°C
2.	Relative humidity	30 – 65%
3.	Air movement	0.22 m/sec
4.	Ventilation	0.5 to 2 m^3/minute/person depending upon type of space
5.	Particulate in cleaned HVAC system	1.0 μg/100 cm^3.
6.	Carbon dioxide	About 700 ppm over outdoor ambient
7.	Carbon monoxide	25 ppm.

QUESTIONS

14.1 Define Refrigeration. What is mechanical refrigeration?

14.2 What is the unit of refrigeration? How will you judge the performance of a refrigeration system?

14.3 Explain the reverse Carnot cycle.

14.4 Describe the Bell Coleman cycle for refrigeration.

14.5 Describe an ideal vapour compression refrigeration system.

14.6 Differentiate between Dry and Wet compression.

14.7 Differentiate between ideal and actual vapour compression refrigeration cycles.

14.8 What are the advantages of vapour compression system?

14.9 Explain "Absorption refrigeration system".

14.10 Explain Electrolux Refrigerators.

14.11 Name different types of refrigerants. Write important characteristics of refrigerants.

14.12 Write main applications of refrigeration and air conditioning.

14.13 What is the unit of refrigeration? Explain how it is defined.

14.14 How is the performance of a refrigeration system measured?

14.15 What is the difference between refrigerator and heat pumps?

14.16 What are the basic requirements of a good refrigerant? Discuss relative merits and demerits of R-12 and NH_3 as refrigerants.

14.17 Write short notes on: (*a*) water cooler ; (*b*) domestic refrigerator, and (*c*) ice-plant.

14.18 Classify air conditioning systems.

14.19 Write down advantages of unitary system over central air-conditioning system and *vice-versa*.

14.20 What are the essential factors required in a complete air conditioning system?

14.21 Differentiate between "comfort cooling" and "industrial cooling".

14.22 Explain different types of air conditioners.

14.23 How will you select an air conditioning plant most suited for given conditions?

Chapter

15

HYDRAULIC PUMPS

Learning Objectives

After studying this chapter, the students will learn:

- Reciprocating pumps and their rate of delivery, construction and working.
- Centrifugal pumps, its principle, construction, priming and workdone.
- Pump losses and their efficiencies.
- Miscellaneous types of pumps, their construction working and applications.

15.1 INTRODUCTION

Pump is a mechanical device to increase the pressure energy of a fluid. Broadly pumps can be classified as positive displacement or centrifugal pumps. Positive displacement pumps are reciprocating or rotary pumps. Rotary pumps include, gear type, screw type, lobe type pumps etc. Centrifugal pumps are either vertical or horizontal; multistage or single stage, submersible or dry type pumps.

Pumps are used for large variety of applications, like supply of drinking water, irrigation purposes, mine water drainage, handling mixtures of water and solids; for chemical liquids etc. Reciprocating pumps are used for developing very high pressures or heads; while centrifugal pumps are used for supplying large quantities at low and medium pressures with high discharges.

These pumps are of various metallurgical composition such as cast iron, cast steel, stainless steel, alloys of nickle, bronze, phosphorus, aluminium etc. Variety of linings are also used which include lead, rubber glass, enamel, PVC etc. The prime movers for the pumps are diesel engines or electric motors. The electric motors can be loosely coupled or close coupled to monoblock pumps.

15.2 RECIPROCATING PUMPS

Pump is a mechanical device to increase the pressure energy of a fluid. Generally pumps are used to lift the water from low level to high level. These are of two types, namely reciprocating pumps and centrifugal pumps. Reciprocating pumps are always used for developing very high pressures or heads. While centrifugal pumps are used for supplying very large quantities at low or medium pressures.

Construction and Working

Reciprocating pump is also known as a "positive displacement pump" because in it liquid is pushed out of the cylinder by the actual displacement of the piston or plunger.

A reciprocating pump is driven by power available from an external source and consists of a cylinder in which piston or plunger reciprocates. This movement of the piston or plunger creates alternately vacuum pressure and positive pressure in the cylinder which raises the water.

If the water acts only on one side of piston, the pump is known as "single acting pump". Fig. 15.1 shows this type of pump. These pumps suck the water into the cylinder during outward stroke. In this stroke atmospheric pressure acting on the surface of the water helps to push the water into the cylinder. During inward stroke the suction valve closes and the water is forced up in the delivery pipe. The construction and working of a single acting reciprocating pump is shown in Fig. 15.1.

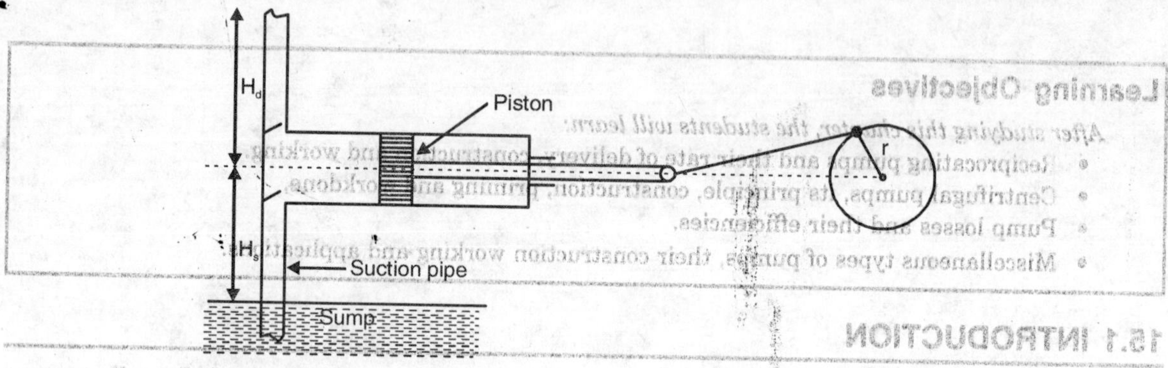

Fig. 15.1 Single acting reciprocating pump

The working of a reciprocating pump can easily be understood with the help of fig 15.1. The piston moves backward and forward in the cylinder due to the rotation of the crank. During outward (suction) stroke, piston moves to the right. This movement of the piston causes vacuum in the cylinder. The atmospheric pressure on the water surface forces the water into the cylinder through the suction pipe. This force of water opens the suction valve which is a non-return valve. On the inward stroke, pressure of water closes the suction valve and opens the delivery valve. In this stroke water is forced up the delivery pipe and so water is raised to the required height or pressure.

In "double action pumps" water acts on both the sides of the piston. Fluid is sucked on one side and delivered on the other side during each stroke. This is indicated in Fig. 15.2.

Types of Reciprocating Pumps

The classification may be:

(*a*) According to the action of water:
  (*i*) Single acting. (*ii*) Double acting.

(*b*) According to the number of cylinders:
  (*i*) Single cylinder, (*ii*) Double cylinder,
  (*iii*) Triple cylinder etc.

(*c*) According to the existence of air vessel:
  (*i*) With Air Vessel. (*ii*) Without Air Vessel.

Each type has been explained ahead.

15.3 WORK DONE BY PUMPS

1. Single Acting Pumps

These pumps have one suction and one delivery pipe, as shown in Fig. 15.1

Let A = Cross-sectional area of the piston in m^2

S = Piston stroke in m., N = Speed of crank in R.P.M.

H_s = Suction head in m., H_d = Delivery head in m.

w = 1000 g N = 9810 N/m^3

Then, average rate of flow, $Q = \dfrac{A.S.N.}{60}$ m^3/sec.

Neglecting losses, power required for the pump in kW,

$$= \frac{w.Q.H}{1000}$$

$$= \frac{w.(A.S.N.)(H_s + H_d)}{60{,}000} \quad \text{...(1)}$$

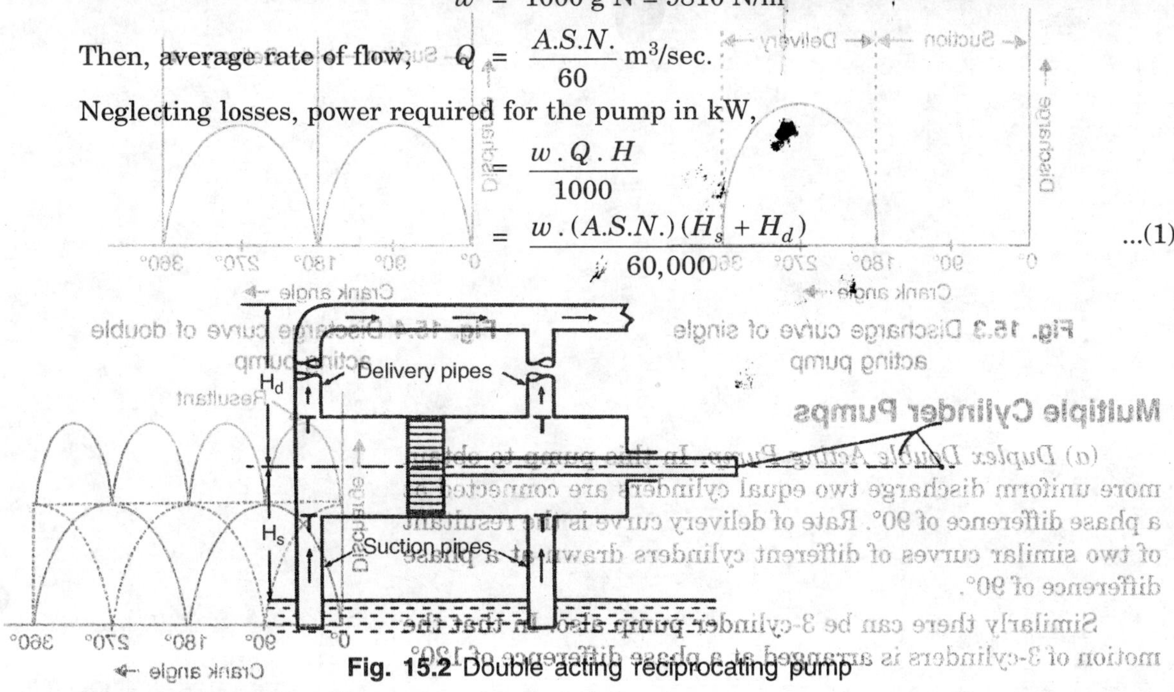

Fig. 15.2 Double acting reciprocating pump

2. Double Acting Pumps

These pumps have two suction pipes and two delivery pipes, one each on each side as shown in Fig. 15.2.

If a = cross sectional area of piston rod in m^2, Average rate of flow,

Q = Flow from cover side + Flow from crank side.

$$= \frac{A.S.N.}{60} + \frac{(A - a).S.N.}{60}$$

$$= \frac{(2A - a).S.N.}{60} \text{ m}^3\text{/sec} \quad \text{...(2)}$$

If area of piston rod is very small as compared to that of piston

$$Q = \frac{2A.S.N.}{60} \text{ m}^3\text{/sec} \quad \text{...(3)}$$

$$\therefore \text{ Power required by the pump } = \frac{w.Q.(H_s + H_d)}{1000} \text{ kW} \quad \text{...(4)}$$

15.4 RATE OF DELIVERY

(*a*) *Single Acting Pump.* During the first half revolution of the crank (*i.e.* crank angle 0° to 180°) only suction takes place and during the second half (*i.e.* crank angle 180° to 360°) only delivery takes place. As piston has its simple harmonic motion, therefore, discharge curve is a sine curve as shown in Fig. 15.3.

(*b*) *Double Acting Pump.* In such pumps each stroke is a suction stroke on one side while delivery stroke on another side. Shape of discharge versus crank angle curve is a two sine curves drawn at a phase difference of 180° as shown in Fig. 15.4.

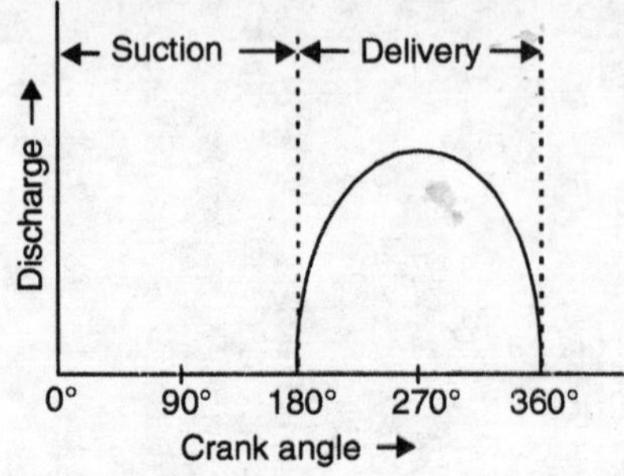

Fig. 15.3 Discharge curve of single acting pump

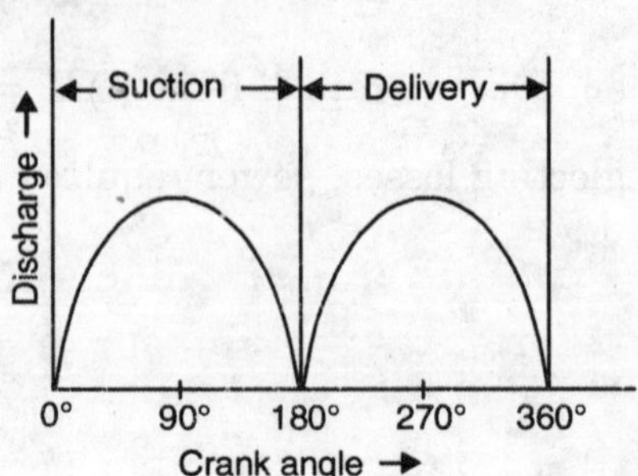

Fig. 15.4 Disctarge curve of double acting pump

Multiple Cylinder Pumps

(*a*) *Duplex Double Acting Pump.* In this pump to obtain more uniform discharge two equal cylinders are connected at a phase difference of 90°. Rate of delivery curve is the resultant of two similar curves of different cylinders drawn at a phase difference of 90°.

Similarly there can be 3-cylinder pump also. In that the motion of 3-cylinders is arranged at a phase difference of 120°.

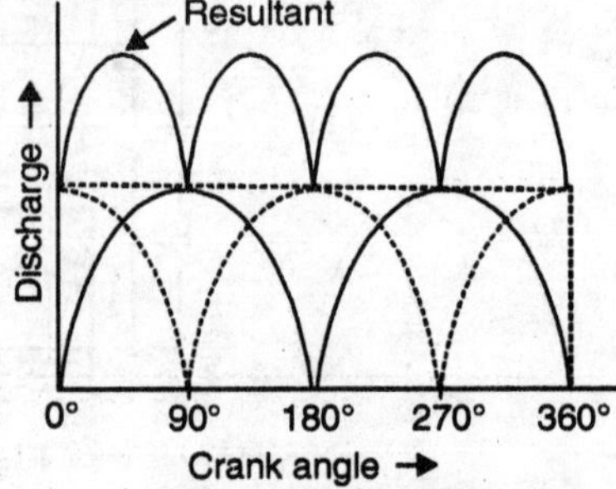

Fig. 15.5 Discharge curve of multi cylinder pump

Coefficient of Discharge

It is the ratio of the actual volume of water discharged to the volume swept by the piston or plunger. It is represented by C_d.

$$\therefore \qquad C_d = \frac{\text{Actual discharge/stroke}}{\text{Volume swept/stroke}}$$

Generally C_d is less than unity but it may be more than one, when delivery valve opens before the end of the suction stroke and thus a greater volume of water will be delivered than that swept by the piston or plungers.

Slip

It is the difference of volume swept by the piston and the actual discharge per stroke.

or Slip = Volume swept/stroke – Actual discharge per stroke.

Slip is generally positive but in case when coefficient of discharge is more than unity, slip will be negative.

$$\text{Percentage slip} = 100 \times \frac{(\text{Volume swept} - \text{Actual discharge})}{\text{Volume swept}}$$

15.5 AIR VESSEL

The air vessel is a cast iron chamber, which has an opening at the base, through which water can flow. One chamber is fitted on the suction pipe just near to the suction valve and one on the delivery pipe just near the delivery valve. Each chamber is connected through a small length of pipe. Such an air vessel is shown in Fig. 15.6.

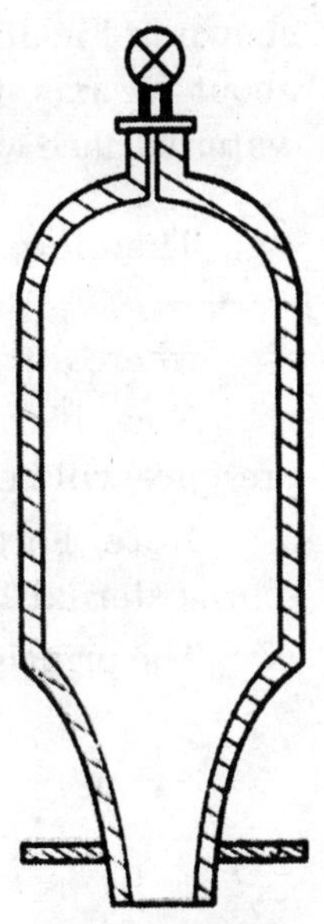
Fig. 15.6
Air vessel

Upper portion of air vessel contains air, which can compress or expand to absorb pressure fluctuations. When the level of water in the air vessel rises, the air in the air vessel is compressed. This compressed air forces the water out when the pressure in the pipe falls down.

Since the air slowly dissolves in the water at higher pressure, therefore, these air vessels are required to be charged or renewed periodically. In large pumps, it is done by a compressor while in small pumps an automatic valve known as shifting valve is provided. This valve draws the required amount of air during the suction stroke. Air vessel fitted on the suction pipe is sometimes also known as *Vacuum Vessel.* For efficient working, vacuum vessels should be 3 to 5 times the discharge/stroke and the air vessel on delivery side should be 6 to 10 times the discharge/stroke.

Working of Air Vessel

During the middle of the delivery stroke, when pump is forcing the water into the delivery pipe at a velocity greater than the average, excess water flows into the air vessel and compresses the trapped air in the upper portion of the chamber. At the end of this stroke when water flows into the delivery pipe at a rate less than the average, water flows out of the air vessel from the excess amount of water already stored to keep the discharge more uniform. This fluctuating water column causes the acceleration head to be reduced to that in between pump cylinder and the air vessel, which allows the pump to run at higher speeds. Thus in this way it saves the large amount of power lost in developing acceleration head.

On suction side, water first collects in the air vessel and then flows to the cylinder. On delivery side, water first goes to the air vessel and then flows with a uniform velocity. An air vessel provided in a reciprocating pump acts like a flywheel of an engine.

Functions

Following are the important functions of an air vessel:

1. Reduces the possibility of separation and cavitation.
2. Allows the pump to run at a higher speed.
3. Suction head can be increased by increasing the length of pipe below the air vessel.
4. Large amount of power is saved due to lesser acceleration head.
5. Uniform discharge.

15.6 CENTRIFUGAL PUMPS

15.6.1 Principle

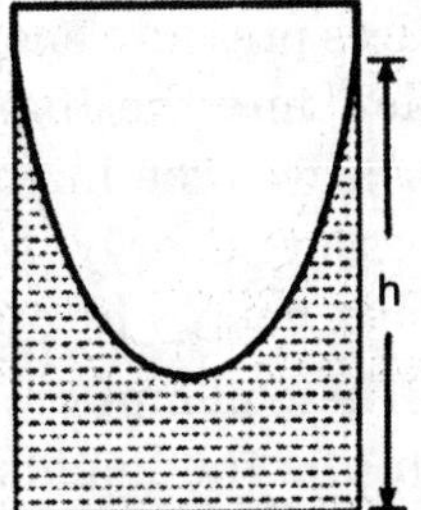

Fig. 15.7 Variation of pressure due to contritugal force

These are called "Centrifugal Pumps" because the centrifugal force or variation of pressure due to rotation is responsible for their operation or working. For example, as shown in Fig 15.7, if a container having water upto some height is rotated at a constant speed about its axis, the water will also tend to rotate at a speed as that of the container and the water surface will assume a parabolic shape.

Therefore, $$h = \frac{u^2}{2g}$$

where u = the linear velocity of container at radius r.

Now, if the water is confined in such a way that its surface cannot be changed then pressure will be produced by forced vortex action.

Note. Forced vortex action is developed when water is compelled to rotate by means of some external force.

The principle of Centrifugal Pump can also be demonstrated by Fig. 15.8.

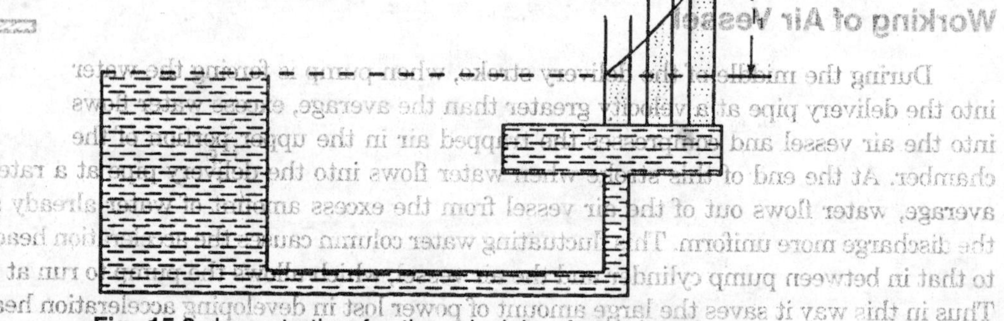

Fig. 15.8 demostration for the principle of contritugal pump

Suppose the water in a closed chamber is set in motion by means of a paddle wheel. The pressure in the outer part of the chamber will be greater than that in the centre by an amount $\frac{u^2}{2g}$. Suppose a piezometer tube is fitted in this chamber, then the water will tend to rise in it to a height, $h = \frac{u^2}{2g}$. Now, if the height of this piezometer tube is somewhat less than $\frac{u^2}{2g}$, water will flow out of the tube and we would have a Crude Centrifugal Pump. In Centrifugal Pumps mechanical energy is given up by the vanes of the Impeller to the water, which in turn produces pressure energy. The direction of flow through the Impeller is radially outward. During the flow into the Impeller, both the pressure and velocity of water are raised and when the water leaves the Impeller, it contains a large part of Kinetic or Velocity energy and small part as pressure energy. In an efficient pump this large part of kinetic energy is transformed into pressure energy by other means explained later.

These pumps are classified as rotor dynamic type, in which a dynamic pressure is generated which lifts the liquids from lower level to higher level. A Centrifugal Pump closely resembles a hydraulic turbine and its action is just the reverse of a radially inward flow reaction turbine. Thus in all the Centrifugal Pumps the water flows radially outward.

15.7 COMPONENT PARTS

The most important component parts of a Centrifugal Pump are:

1. Impeller

It is also known as rotor and on it series of curved vanes are fitted. Impeller is mounted on a shaft, which is driven by an electric motor. The function of Impeller is to force the water into a rotating motion and the shroud around the Impeller is to direct the water through the Impeller. The water coming out of Impeller is then led through pump casing under high pressure.

The Impellers may be classified as:

(*a*) Closed or shrouded Impeller. (*b*) Semi-open Impeller,

(*c*) Open Impeller.

(a) Closed or Shrouded Impeller. Fig. 15.9 shows this type of Impeller. It contains two shrouds in which curved vanes are inserted. This type is suitable for pumping liquids which are pure and comparatively free from foreign matter or debris. For example, ordinary water, oils, chemicals like acids etc. The material of the Impeller should be such that it can resist the reaction of chemical to be handled.

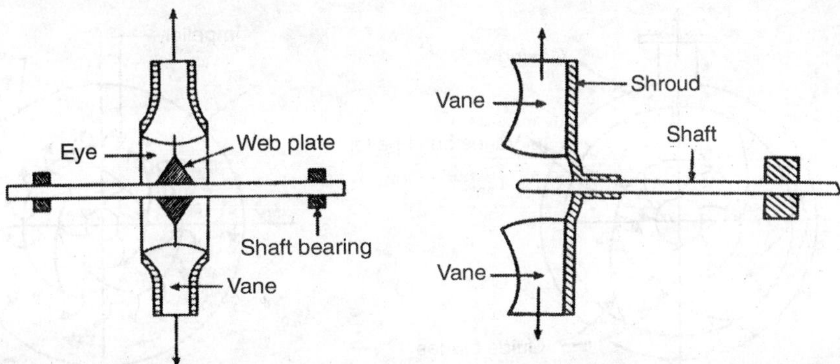

Fig. 15.9 closed or shrouded impeller **Fig. 15.10** semi-open impeller

(b) Semi-open Impeller. Fig. 15.10 shows this type. In this vanes are fixed on one shroud plate only. This type of Impeller is used for pumping liquids containing debris upto some extent only, such as sewage water, paper pulp etc. These Impellers have less number of vanes but have more length to avoid the clogging of the Impellers.

(c) Open Impeller. In this vanes are fixed from the centre with web plate as shown in Fig. 15.11. No crown and shroud plates are used. This is used for pumping liquids having coarse debris of large size such as mixture of water, sand, clay, and pebbles, in which the solid contents are in very large amount. As these impellers are required to perform very rough duty, these are generally made of forged steel. These Impellers have less life, as they perform very rough duty.

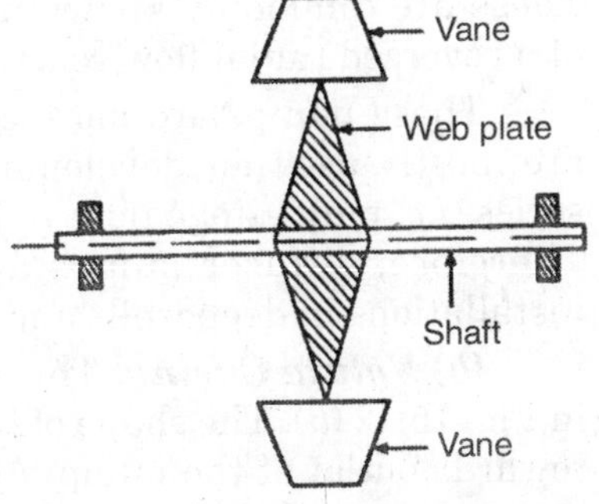

Fig. 15.11 open impeller

2. Casing

A casing is provided for housing the Impeller and supporting the bearing provided for housing the Impeller and supporting the bearing provided for the shaft carrying the Impeller. Two openings are provided in the casing for suction and delivery of water. As it handles the water at a higher pressure, therefore, a stuffing box is provided to prevent the leakage from the gap between the pump casing and shaft.

As the Impeller increases the kinetic energy of water, hence the casing shuld be such that least amount of energy should be lost due to eddy formation. Therefore, the efficiency of a pump also depends on the type of casing used. Following are the three common types of casing, which are generally used:

(*a*) Diffuser Ring

(*b*) Volute Casing

(*c*) Vortex or Whirlpool Casing

(a) Diffuser Ring. Fig. 15.12(*a*) shows diffuser ring type casing. In this, a diffuser ring having guide vanes surrounds the Impeller. These guide vanes are such that they provide gradually enlarged passage for the flow of water. These guide vanes are placed at such an angle that water enters without shock. It has been found that by using this type of casing, large amount of kinetic energy (upto 75%) is converted into pressure energy.

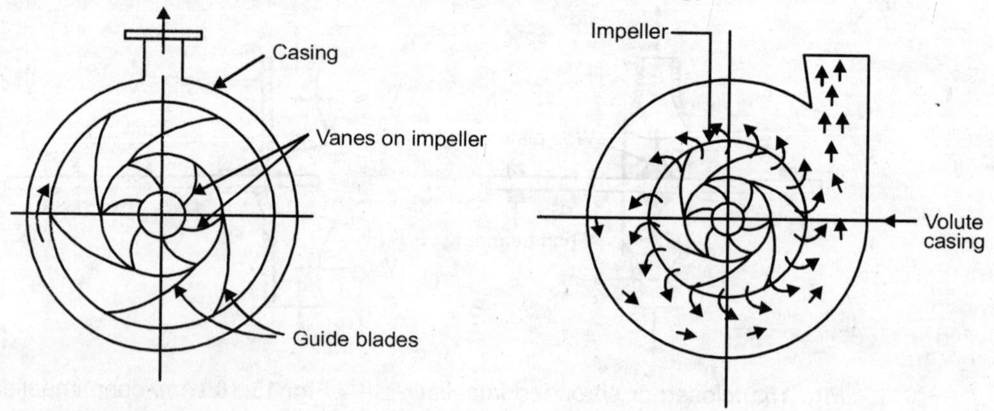

Fig. 15.12 (*a*) Diffuser casing. **Fig. 15.12** (*b*) Volute casing.

The pumps having diffuser casing are known as diffuser pumps or guide blade pumps. These are commonly known as "“Turbine Pumps” because in principle, they resemble to that of a reversed radial flow reaction turbine.

These pumps are more costly than that of the volute pumps, therefore, diffuser pumps are mostly used for developing high heads (60 m or more) by using number of Impellers in series (*i.e.* multi-stage pump). For single stage, volute or whirlpool casing pumps are preferred. Therefore, diffuser pumps are mostly used as vertical pumps and are very suitable for installations in deep wells and mines etc.

(b) Volute Casing. This is a spiral shaped casing which surrounds the Impeller as shown in Fig. 15.12 (*b*). The shape of the casing is such that its area of cross-section gradually increases towards outlet of the pump. As the flowing water progresses towards the delivery pipe, more and more water is being added from the outlet periphery of the Impeller. Increasing cross sectional area helps firstly to provide the space for the water coming out of the Impeller and secondly to reduce the velocity, thus converting kinetic energy into pressure energy.

These casings can convert only a small amount of kinetic energy into pressure energy and a large amount of kinetic energy is lost in eddies. Therefore, these pumps can produce comparatively low heads. These pumps are known as "Volute Pumps".

(c) Vortex or Whirlpool Casing. This is the improvement of volute pump. In such pumps an annular space known as vortex of whirlpool chamber is provided between the Impeller and volute casing. Thus the water enters the vortex chamber from the Impeller with a whirling motion and, therefore, water particles are having spiralling path. In the vortex chamber these water particles have constant energy because no work is done on the water in this chamber. Thus, when the water flows in this vortex chamber, the torque of the whirling motion also remains the same, *i.e.* free vortex is formed. Since for a free vortex velocity of whirl varies inversely as its radial distance from the centre; velocity of flow of liquid as it passes through the vortex chamber reduces. Thus, in this way reduction in velocity increases the pressure of water leaving the vortex. The water after leaving the vortex chamber passes through the volute chamber surrounding it. Thus it increases the efficiency of the volute chamber by converting the kinetic energy into pressure energy in the vortex chamber.

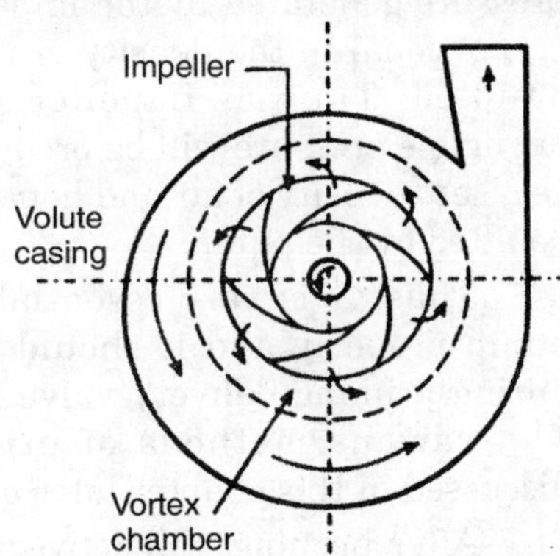

Fig. 15.12 (*c*) Vortex casing.

3. Suction Pipe

It is a pipe, whose upper end is connected with the pump on suction side and lower end is submerged in the source of water, *i.e.* sump. The lower end is fitted with a foot valve (one way valve) and a strainer. When water enters in the suction pipe, strainer keeps the debris (*e.g.* leaves wooden pieces and other rubbish materials) away from entrance to the pipe. Thus only clear water can enter the suction pipe through foot valve. The velocity of water in the suction pipe should be between 1.5 to 3.0 m/sec.

4. Delivery Pipe

Lower end of this pipe is connected to the outlet of the pump and takes the water upto the required head and then delivers it. The velocity of water in the delivery pipe should be between 1.5 to 3.5 m/sec. The velocity of water in the delivery is kept equal or slightly higher than velocity in suction pipe.

5. Delivery Valve

It is connected between the outlet of the pump and lower end of delivery pipe. It is used to regulate the quantity of water to be lifted by controlling the opening of the valve.

6. Prime Mover

It is used to rotate the Impeller of pump. Usually an electric motor is provided for this purpose. This motor is coupled with Impeller shaft.

Working

Fig. 15.13 explains in details the working of a centrifugal pump. Before starting a pump, priming is done. In priming, water is filled in the suction pipe, casing and into a portion of delivery pipe upto delivery valve. This is done for the removal of air. With the presence of air, it is not possible to create pressure because the pressure generated by the impeller is directly proportional to the density of the liquid to be handled. Thus, if Impeller is run in air, negligible pressure will be produced because of very less density of air and hence no liquid will be lifted by the pump.

Thus, it is very essential to prime the pump properly and it should be noted that during priming delivery valve is kept closed. The various methods of priming will be discussed in this chapter later on.

After priming, the delivery valve is kept still closed and prime mover (generally electric motor) is started to rotate the Impeller. The rotation of Impeller in the casing full of liquid produces centrifugal head in the liquid and thus pressure in whole of the liquid is increased.

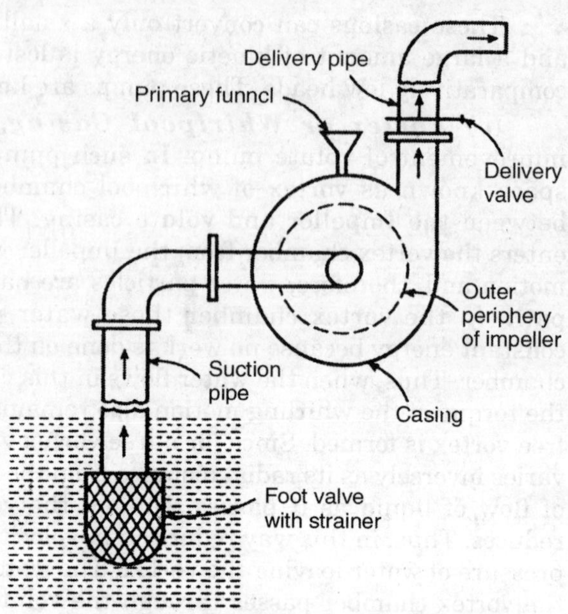

Fig. 15.13 Working of contritugal pump

After a short period, when pump attains a constant speed, delivery valve is gradually opened and thus liquid is allowed to flow in a radially outward direction through the Impeller vanes and attains a higher velocity and pressure at the outer periphery. With the help of special shape of casing (Volute, Vortex or Diffuser) this high velocity energy is converted into pressure energy and reduces high velocity. Now this water at higher pressure rises through the delivery pipe into a desired height. At the same time centrifugal action creates partial vacuum at the centre. Impeller known as EYE. The partial vacuum causes the water to rush through the suction pipe towards the Impeller eye, to take the place of water which has left the Impeller vanes, from the sump (where the pressure is atmospheric).

Thus, in this way water reaches and leaves the Impeller continuously and water is raised to a certain height at constant rate.

15.8 PRIMING

As explained earlier priming is the filling of water in the casing and suction pipe of a pump for the removal of air before starting it. If any air remains inside the casing, discontinuity of flow may be caused. Therefore, priming is very essential before starting a pump.

Reciprocating pumps do not require priming, if they are in good condition. In those priming is done only when the suction pipe is long. When the water sump is available above the pump level then priming is automatically done by simply opening the fitted valve in between the sump and pump suction pipe. If the pump is installed above the source of water then a foot valve is fitted at the bottom of the suction pipe submerged in water and priming is done by the following methods:

Pumps can be primed in several ways. These methods can be divided into the following three general categories:

(*a*) ***Manual Priming.*** Water is admitted to replace the air from the suction pipe and the pump casing. It may be done by hand or automatic means.

(*b*) ***Priming by Vacuum.*** In this air is removed from the suction pipe and pump by a vacuum producing device.

(*c*) ***Self-priming.*** In these methods pumps are fitted with self-priming devices to recirculate the water in the pump for the removal of air.

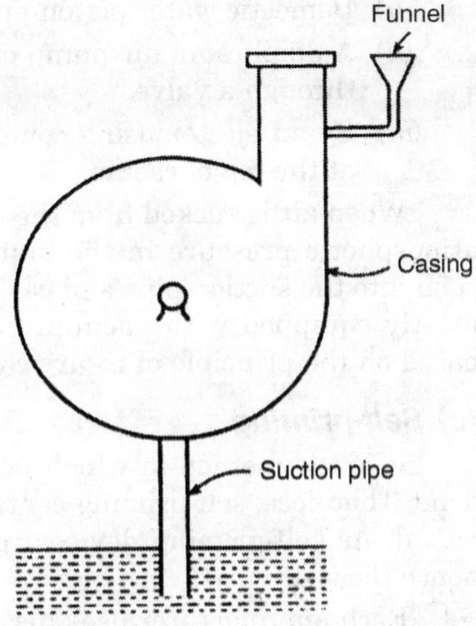

Fig. 15.14 Pouring water with tunnel

(a) Manual Priming

Some of the most common methods in this category are:

(*i*) Pouring water with funnel:

Fig. 15.14 shows this method. This is an ordinary method, in which water is poured through a funnel into the casing and thus replaces the air.

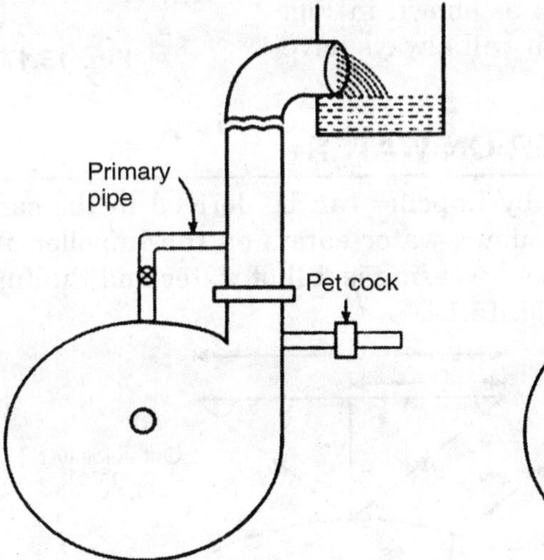

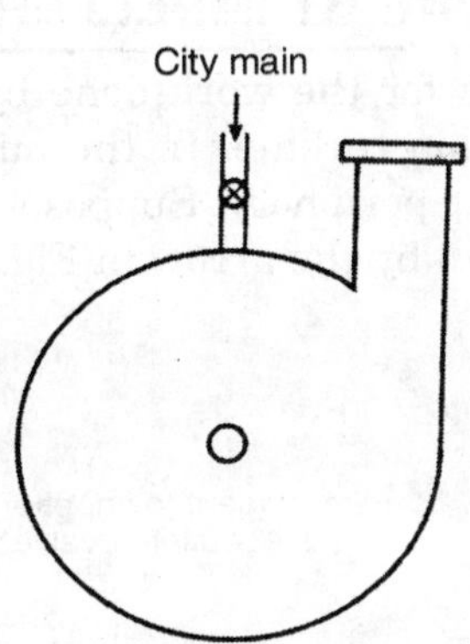

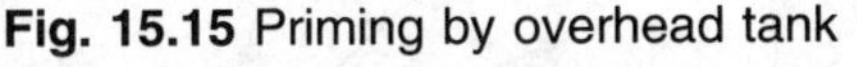

Fig. 15.15 Priming by overhead tank

Fig. 15.16 Priming by city water

(*ii*) When the pump is discharging into an overhead tank, then for priming purposes delivery pipe is connected with the casing so that when priming is to be done liquid from overhead tank can be used as shown in Fig. 15.15. If city water at some pressure is available then priming can be done by city water by connecting it with the pump as shown in Fig. 15.16.

Pet cock is provided at the top of the casing to allow air to escape from the casing while priming. After priming pet cock is closed.

(b) Priming by Vacuum

In these methods, the air from the suction pipe and pump is sucked by any of the devices, namely:

(*i*) Domestic water piston pump connected with the discharge line.

(*ii*) A diaphragm air pump operated by a handle connected with top of centrifugal pump through a valve.

(*iii*) By an ejector using compressed air or exhaust steam of an engine to create vacuum at the top of casing.

When air is sucked from the pump and suction pipe, atmospheric pressure in the sump forces the water to rush into the suction pipe and casing. Modern pumps are mostly equipped with vacuum pumps (electric driven) based on the principle of rotary compressors.

(c) Self-priming

It is the method in which priming is automatically done. Therefore, self-priming centrifugal pumps are very useful. As self-priming devices require large expenses, hence these are used with big size pumps only.

Each manufacturer uses his own design and patent for self-priming pumps. The device used in "Sini Pump" is very simple, in which the suction pipe and delivery pipe are connected at the top of the pump as shown in Fig. 15.17. When the pump comes to rest, it will always have some water for priming purposes.

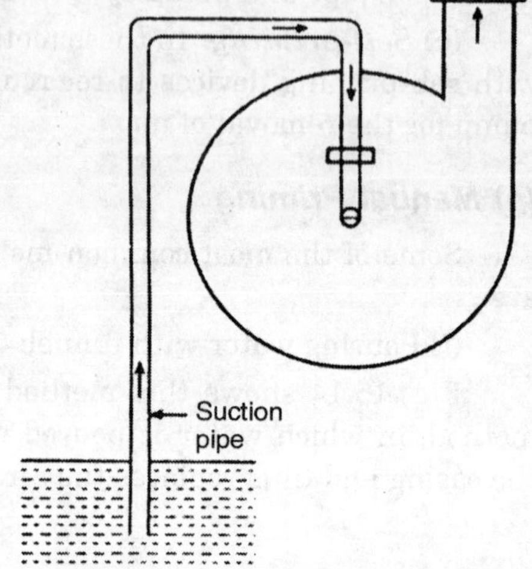

Fig. 15.17 Self priming

15.9 WORK DONE BY IMPELLER ON WATER

The equation for the work done by Impeller can be derived in the same way as for an inward flow reaction turbine. In the pumps water enters on the Impeller at inner periphery and leaves at outer periphery. Suppose the casing is full of water and the Impeller rotating in the direction shown by the arrow in Fig. 15.18.

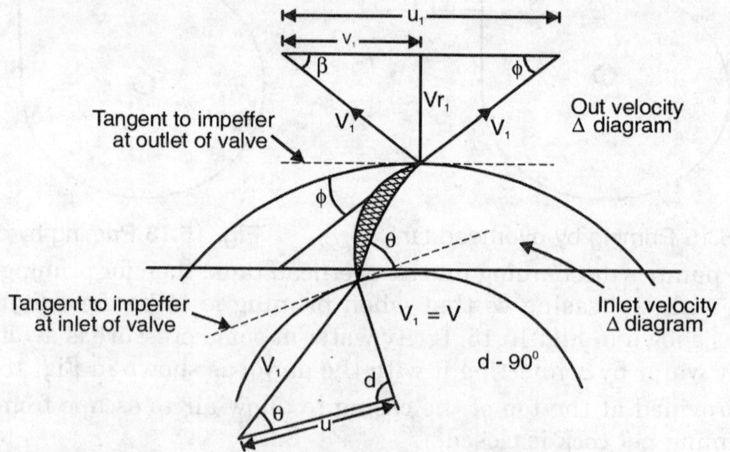

Fig. 15.18 Velocity diagram at inlet & outlet of pump

Velocity triangle diagrams at inlet and outlet are also shown in Fig 15.18, where

u = Peripheral velocity of Impeller at inlet.

u_1 = Peripheral velocity of Impeller at outlet.

V = Abs. velocity of water at entrance.
V_1 = Abs. velocity of water at exit.
V_r = Relative velocity of water at vane entrance.
V_{r1} = Relative velocity of water at vane exit.
V_f = Velocity of flow at entrance.
V_{f1} = Velocity of flow at exit.
V_w = Velocity of whirl at entrance.
V_{w1} = Velocity of whirl at exit.
$\alpha°$ = Angle which abs. velocity at inlet makes with the tangent at entrance.
$\theta°$ = Vane angle at entrance.
$\phi°$ = Vane angle at exit.
β = Angle with the abs. velocity at exit makes with the tangent of exit.
D = Diameter of Impeller at entrance.
D_1 = Diameter of Impeller at exit.
W = Quantity of water flowing through the Impeller in kg/sec.

For the best efficiency of pump, it is assumed that water enters the impeller radially *i.e.* $\alpha = 90°$ and $V_w = 0$. Like turbines, for shockless entry and exit to the vanes, water enters and leaves the vane tips in a direction parallel to their relative velocities at the two tips.

As it has already been stated that it is just reverse of a radially inward flow turbine, where the work done/sec by the water on the Impeller

$$= \frac{W}{g}(V_w\, u \pm V_{w1}\, u_1)$$

∴ In the case of centrifugal pump work done/sec by the Impeller on the water will be = $\frac{W}{g}(V_{wl}\, u_1 \pm V_w\, u)$. But for a radial flow at entrance as stated above $V_w = 0$.

$$\therefore \quad \text{Work done/sec} = \frac{W}{g}(V_{w1}\, u_1)$$

$$\text{Impeller power} = \frac{W \times (V_{w1}\, u_1)}{1000}$$

Fundamental Equation of Centrifugal Pump

As we have already discussed that work done per kg/sec.

$$= \frac{1}{g}(V_{wl}\, u_1 - V_w\, u)$$

In Fig. 15.19 from inlet velocity triangle

$$V_w = u + \sqrt{V_r^2 - V_f^2}$$

$$= u + \sqrt{V_r^2 - V^2 + V_w^2}$$

$$\text{or} \quad (V_w - u)^2 = V_r^2 - V^2 + V_w^2$$

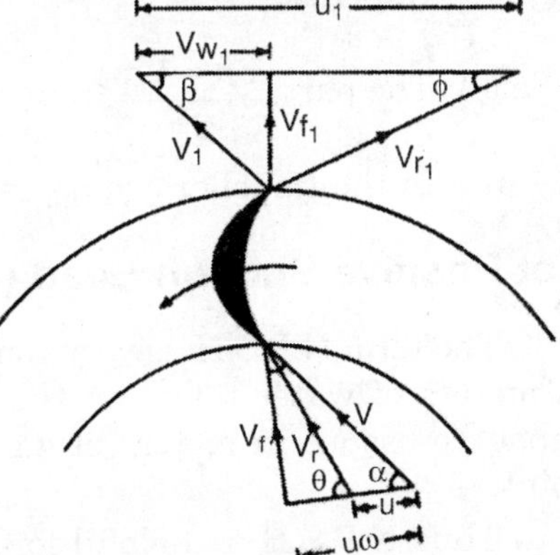

Fig. 15.19 Velocity triangle at inlet & outlet of centrifugal pump

$$\text{or} \quad (V_w^2 + u^2 - 2V_w u) = V_{r2} - V^2 + V_w^2$$

$$\text{or} \quad 2V_w u = V^2 + u^2 - V_r^2 \quad \ldots(1)$$

Similarly, from outlet Δ..

$$V_{w1} = u_1 - \sqrt{V_{rl}^2 - V_{f1}^2} = u_1 - \sqrt{V_{r1}^2 - V_1^2 + V_{wl}^2}$$

$$\text{or} \quad (V_{w1} - u_1)^2 = V_{r1}^2 - V_1^2 + V_{w1}^2$$

$$\text{or} \quad u_1^2 + V_{w1}^2 - 2u_1 V_{w1} = V_{r1}^2 - V_1^2 + V_{w1}^2$$

$$\text{or} \quad 2V_{w1u_1} = V_1^2 + u_1^2 - V_{r1}^2 \quad \ldots(2)$$

From equation (1) and (2)

$$\text{Work done per kg per sec} = \frac{1}{g}[V_{w1}u_1 - V_w u]$$

$$= \frac{1}{2g}\left\{\left(V_1^2 + u_1^2 - V_{r1}^2\right) - (V^2 + u^2 - V_r^2)\right\}$$

$$= \frac{1}{2g}\{(V_1^2 - V^2) + (u_1^2 - u^2) + (V_r^2 - V_{r1}^2)\}$$

$$\therefore \quad \text{Work done/kg/sec} = \frac{V_1^2 - V^2}{2g} + \frac{u_1^2 - u^2}{2g} + \frac{V_r^2 - V_{r1}^2}{2g}$$

Thus the above equation shows that the work done on the water by the pump consists of the following three parts:

(*a*) The part $\frac{V_1^2 - V^2}{2g}$ represents the change in kinetic energy of the liquid.

(*b*) The part $\frac{u_1^2 - u^2}{2g}$ represents the effect of centrifugal head or energy produced by Impeller.

(*c*) The part $\frac{V_r^2 - V_{r1}^2}{2g}$ represents the change in static pressure of the liquid, if the losses in the Impeller are neglected.

Net Positive Suction Head (N.P.S.H.)

The term N.P.S.H. is very common in the pump industry. It is defined as the gauge reading in meters of water taken on the suction side of the pump at the centre lines minus gauge vapour pressure in meters of water at the pumping temperature plus velocity head at that point.

Thus N.P.S.H. is helpful for designing a centrifugal pump and represents the minimum required difference between the suction head and vapour pressure head at the given capacity. Thus, if the N.P.S.H. is insufficient, cavitation may occur. Insufficient N.P.S.H. may occur due to very high suction lifts.

Manometric Head

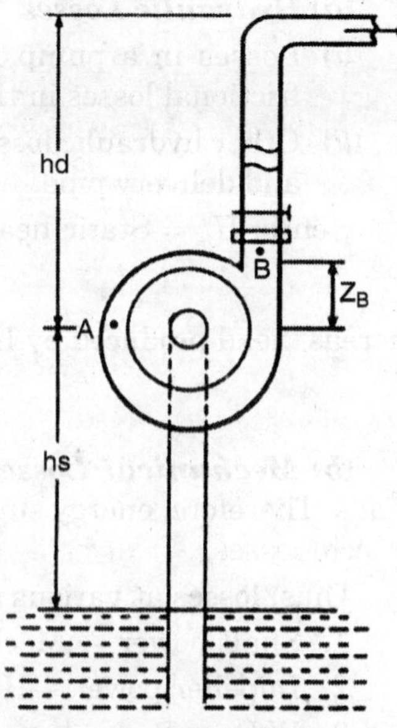

Fig. 15.20
Centrifugal pump

A. centrifugal pump with suction and delivery pipe is shown in Fig. 15.20, where

h_s = Suction lift
h_d = Delivery lift
V_s = Velocity of flow in the suction pipe.
V_d = Velocity of flow in the delivery pipe
h_{fs} = Head lost in friction in suction pipe
h_d = Head lost in friction in delivery pipe
A = Point at the outlet edge of Impeller
B = Point at the outlet edge of Impeller
H_A = Total head at point A
H_B = Total head at point B.

Then, (*i*) *Static Lift.* Static lift or total vertical lift is the height through which the water is lifted by the pump

$\therefore$ Static lift $h = (h_s + h_d)$.

(*ii*) *Suction Head.* Considering loss of head in suction pipe.

Suction head of a pump $= h_s + h_{fs} + \dfrac{V_s^2}{2g}$ neglecting head loss at entrance and bends etc.

(*iii*) *Delivery Head.* Delivery head of a pump $= h_d + h_{fd} + \dfrac{V_d^2}{2g}$; neglecting head lost at bends.

(*iv*) *Manometric Head* (H_m). It is the actual head against which pump has to work and is equal to static lift plus all the head loss occurring in the flow.

$$\therefore \quad H_m = h_s + h_{fs} + \frac{V_s^2}{2g} + h_d + h_{fd} + \frac{V_d^2}{2g}$$

$$\text{or} \quad H_m = (h_s + h_d) + (h_{fs} + h_{fd}) + \frac{V_d^2}{2g}, \text{ neglecting } \frac{V_s^2}{2g}$$

$$= h + hf + \frac{V_d^2}{2g} \quad \text{...(1)}$$

where h_f is the head lost due to friction in suction and delivery pipes.

As we know that manometric head developed by the pump is the difference of pressure heads between the outlet of the pump casing at B and inlet point A of the pump.

$$\therefore \quad H_m = H_B - H_A$$

$$= \left(\frac{P_B}{w} + \frac{V_B^2}{2g} + Z_B\right) - \left(\frac{P_A}{w} + \frac{V_B^2}{2g} + Z_A\right) \quad \text{...(2)}$$

15.10 PUMP LOSSES

Various losses occurring during the operation of a centrifugal pump can be grouped as follows:

(a) Hydraulic Losses. There are two types of hydraulic losses which may occur in a pump:

(*i*) Losses in a pump, these losses are eddy losses at entrance and exit of Impeller, frictional losses in the Impeller, frictional and eddy losses in the diffuser, if provided.

(*ii*) Other hydraulic losses, these are frictional losses and other minor losses in the suction and delivery pipe.

Hence, H_m = Static head + Friction and other minor head losses in suction and delivery pipe.

whereas, head produced by Impeller $= \dfrac{V_{w1}u_1}{g}$

$= H_m$ + Losses in the pump.

(b) Mechanical Losses. These are the losses due to friction of the main bearings and glands. Therefore, energy supplied by electric motor to Impeller = Energy produced by Impeller + Mech. losses.

Thus, losses at various stages can be understood by the following equations:

1. Shaft power — Mech. losses = Impeller power.
2. Impeller power —Hydraulic losses in the pump = W.P., *i.e.* power at casing exit.
3. W.P. — Power losses in suction and delivery pipes = Static power

15.11 PUMP EFFICIENCIES

Following important types of efficiencies are generally used for indicating the performance of centrifugal pump:

(*i*) Manometric Efficiency (η_{man}) (*ii*) Mechanical Efficiency (η_{mech})

(*iii*) Overall or Actual Efficiency (η_0).

(*i*)
$$\eta_{man} = \frac{H_m}{\dfrac{V_{w1}u_1}{g}} \quad ...(a)$$

$$= \frac{H_m}{H_m + \text{Hydraulic losses in the pump}} \quad ...(b)$$

$$= \frac{\dfrac{W \times H_m}{100}}{\dfrac{W \times V_{w1} \times u_1}{1000}} = \frac{\text{W.P.}}{\text{Power of Impeller}} \quad ...(c)$$

Note. Sometimes η_{man} is also known as hydraulic efficiency (η_{hyd}).

(*ii*)
$$\eta_{mech} = \frac{\text{Energy of Impeller}}{\text{Energy given to the shaft}} \quad ...(a)$$

$$= \frac{\dfrac{V_{w1} \times u_1}{g}}{\dfrac{V_{w1} \times u_1}{g} + \text{Mech. losses}} \quad ...(b)$$

$$= \frac{\text{Impeller Power}}{\text{Shaft Power of motor}} \quad ...(c)$$

(*iii*)
$$\eta_0 = \frac{\text{Manometric head}}{\text{Energy supplied to shaft/kg of water}}$$

$$= \frac{\dfrac{W \times H_m}{100}}{\text{Shaft power}} \quad ...(d)$$

If losses are not given, then instead of H_m, static head may be considered. The efficiency thus obtained is also known as actual or static efficiency.

Minimum Starting Speed of a Pump

When a pump is started, water will not flow until the pressure developed by the Impeller is large enough to overcome the manometric or gross head otherwise. Impeller will rotate in water but flow will not commence.

The water will start flowing only if the centrifugal head or pressure head raised by the Impeller $\frac{(u_1^2 - u^2)}{2g}$ be at least equal to external resistance, *i.e.* H_m.

Thus
$$\frac{u_1^2 - u^2}{2g} \geq H_m$$

Therefore, for limiting value,

$$\frac{u_1^2 - u^2}{2g} = H_m = \frac{V_{w1}\, u_1}{g} \times \eta_{max} \quad ...(i)$$

or
$$\frac{\left(\dfrac{\pi D_1 N}{60}\right)^2 - \left(\dfrac{\pi DN}{60}\right)^2}{2} = H_m \quad ...(ii)$$

From the above relation minimum speed (N) required to commence the flow can be found. $\frac{(u_1^2 - u^2)}{2g}$ is the churning pressure, *i.e.* until this head is obtained, the delivery valve should remain closed.

15.12 MISCELLANEOUS TYPES OF PUMPS

15.12.1 Axial Flow or Propeller Pump

This type of pump resembles to a reversed propeller turbine. This pump is fitted with an axial flow, Impeller in which water enters and leaves the Impeller axially. In such pumps no centrifugal head is created in the water but pressure is developed during the flow of liquid over the blades. Fig. 15.21 shows a line diagram of this type of pump. These pumps have vertical shafts.

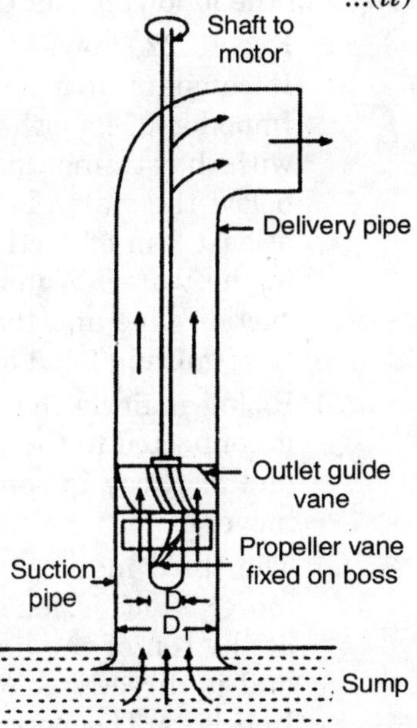

Fig. 15.21 Axial flow pump

As shown in the figure a small length of suction pipe at its upper end, ends in a casing in which a boss carrying about

4 vanes is provided. The boss is rotated by a motor connected at the top of boss shaft. Stucked liquid passes through a set of guide blades into the boss vanes, where it gets pressure energy. Then liquid passes through another set of guide blades into the delivery pipe.

The first set of inlet guide blades, guides the incoming water axially without any velocity of whirl into the Impeller. Guide blades at the outlet takes the whirl action again. (This whirl velocity is imparted by Impeller blades). Thus allowing the water to flow axially in delivery pipe.

These pumps are suitable for very large discharge against small heads (generally 9-12 m). As the speeds of these pumps are high hence danger of cavitation is more and, therefore, very low suction is kept.

These pumps are used in sewage purification, land drainage, flood control, condenser circulation in power houses and Irrigation etc.

15.12.2 Deepwell or Borehole Pumps

These pumps are used where water or any other liquid like oils etc. are required to be raised from a larger depth. These are generally used for pumping from depths more than 6 to 7 m. Special type of pumps can lift the oils even from a depth of 3000 m but generally ordinary deepwell pumps can lift upto 300 m. These pumps can have capacities upto 500 litre/sec and have pipe dia upto 0.6 m nearly.

These pumps can be plunger pumps, multi stage centrifugal pumps, jet pumps and air lift pumps etc. Out of these, plunger and multistage vertical type turbine pumps (centrifugal type) are most common. Plunger pumps are reciprocating pumps and are used for very high lifts.

Deepwell or turbine pumps are multistage centrifugal pumps and used for lifting the liquids upto 300 m and have a capacity varying from 2 to 500 litre/sec. The number of stage in the pump depends upon the head required. These pumps, having 20 stages in series are in use. These pumps consist of the following these mainparts.

(*a*) Fig. 15.22 shows the construction of this type of pump. It consists of a bowl *B*, which has suction case and Impellers. The lower end of the suction pipe is connected with the strainer at the foot valve and this end is placed below the water level. Generally all the Impellers and atleast 3 m of suction with strainer remains submerged in the water. Water enters the first. Impellers through suction pipe and then passes through all the Impellers, thus gaining head one after the other.

(*b*) Rising main is the second part, the lower end of which is connected to the outlet of the Impellers and the upper end of these is connected to the driving head (Prime mover).

(*c*) The third main part is the driving head or the prime mover which is generally an electric motor. This is placed at the top of the line shaft usually on the ground level and is directly coupled with the long vertical shaft of Impellers. The electric motor is properly protected against water.

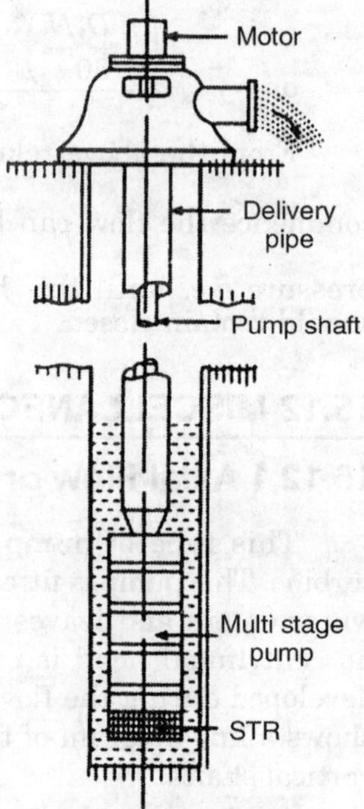

Fig. 15.22 Deepwel Pump.

The pump shaft and line shaft are coupled together and are enclosed in the rising main. These shafts are supported at suitable intervals with the rubber bearings. The bearings are lubricated by the delivery water passing through the rising main. Upper end of the line shaft is fitted with thrust bearing to carry the weight of all the Impellers at the shaft.

15.12.3 Submersible Pump

This is the modification on the deepwell pump explained above. As the name indicates in this electric motor and pump both are submerged in the water. By submerging electric motor, large economy can be made by avoiding long shaft, large number of a bearings and large sized rising main etc. The complications due to thrust (arising due to the weight of the shaft and Impellers etc.) are also avoided.

Since the motor is submerged in water, the only problem is to prevent the motor windings and other electrical connections to be spoiled by water coming in contact. For this purpose either of the following two solutions can be made.

(*i*) To use a normal type of motor but to prevent the water from reaching it by enclosing in a water-tight diving ball. This system of enclosing is known as "Dry motor" type system.

(*ii*) To allow the water to have more or less free contact with motor windings but to give them special protection by suitable type of insulation. This is known as "Wet motor" type system.

Construction of Submersible Pump

Submersible pumps are built together as single or multistage centrifugal pumps, directly coupled with squirrel cage electric motor, which is totally enclosed. The motor and pump shafts are supported in water lubricated plain bearings. A line diagram of a submersible pump is shown in Fig. 15.23.

These pumps are vertical centrifugal pumps with radial or mixed flow impellers. All the metallic bearings are water lubricated and protected against the sand. A non-return valve is fitted with flange at the top of the pump. The suction housing of the pump is situated between the pump and the motor, and provided with a perforated strainer.

Motor of the submersible pumps are wet squirrel cage type and are filled completely with water. The water cools the windings, insulated in a plastic impervious to water. To absorb the axial thrust, thrust bearings are provided. The pump shaft is connected to the motor shaft by a muff coupling. Very small submersible pumps also have common shaft for pump and motor.

These submersible pumps are installed below the lowest expected water level in the well. Gate valve *is* provided at the top of the pump, as a non-return-valve to discharge the water. These pumps can be usefully employed for installation in open wells, where frequent changes of water level make the working of a horizontal pump difficult. The total efficiency of these pumps is superior to conventional deep well pumps at much cheaper costs (including capital cost, running cost, maintenance cost, installation cost etc.).

Before a submersible pump is installed, the motor must be filled with clean non-acid water (not distilled water) free of sand for the purpose of priming. The pump must be started with the valve slightly open. As excessive amount of the sand in the water is harmful for the pump, hence discharge water must be checked for sand content. If there is an appreciable amount of sand found to be present, the pump must run with partly valve open only until the sand contents are reduced to acceptable quantity. The valve may then be opened gradually.

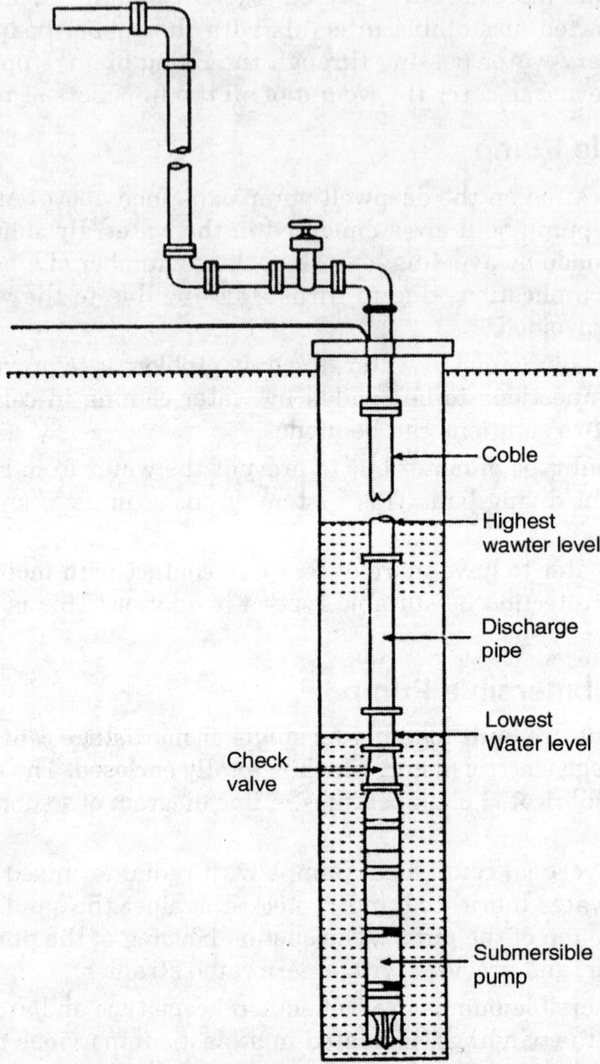

Fig. 15.23 Submersible pump.

Applications of the Submersible Pumps. These pumps can be used for:

1. Drinking water supply installations for cities and rural areas.
2. Service water supply installations for industry and commerce.
3. Lowering the ground water level on building sites.
4. Containing the ground water level in mines.
5. For irrigation purposes.

Advantages of Submersible Pumps

1. Economic in operation, because of high overall efficiency of these pumping sets.
2. Requires no maintenance, because of its glandless construction and its water lubrication.
3. Can be easily installed in crooked wells.

4. Due to submerged installations, priming of the pump is not required (after once it is installed), and since foot valve is not required, hence no suction trouble.
5. Low noise level, because of journal bearing.

15.12.4 Rotary Pumps

A rotary pump is a positive displacement type pump having circular motion. It is similar to centrifugal pump in outward appearance but differs in action. It continuously discharges liquid out of the pump chamber while centrifugal pump only imparts a velocity.

Classification. Mainly these are of two types:

1. *Constant Delivery Pump.* This type includes:

(*a*) Gear pump. (*b*) Screw pump.
(*c*) Vane pump. (*d*) Radial piston pump.

2. *Variable Delivery Pump.* This type includes:

(*a*) Rotary piston pump (sliding piston pump).
(*b*) Radial or seated valve piston pump.

"Constant" delivery pump is a pump in which there is continuous discharge of liquid at a uniform rate of flow whereas a "variable" delivery pump is a pump in which there is continuous discharge of liquid at varied rate of flow.

Some of the common and important type of pumps have been dealt as under:

15.12.5 Gear Pumps

Fig. 15.24(*a*) shows an ordinary external type of gear pump. It consists of a pair of meshed spur gears enclosed in a tight fitting housing, so that there is very little clearance between the gear teeths and the housing. This type of gear pump is the most commonly used pump in automobile engines. This is used to circulate lubricating oil, to various engine parts, under pressure.

Drive gear of the pump, (when used as oil pump) is securely attached to the crankshaft through suitable gearing, while other gear is free to revolve on its bearing. The pump is always submerged in oil in the crank case so that priming is not required. Both gears mesh together so that, if the driving gear is turned, the other gear is also turned, in opposite direction. The oil from the suction side of the pump is carried by the pockets formed by the spaces between the gear teeth to the delivery side. The close fitting of the meshing gear teeths in the centre prevents the passage of oil between the teeth from returning to the inlet side of the pumps.

It is clear that, the higher the speed of the gears, the more oil is pumped and the pressure in the system is greater. Generally the oil is delivered by the pump at a pressure of 0.3 to 0.5 MPa.

Fig. 15.24 (*a*) shows, how the fluid passes from the inlet to the discharge between the teeths of the gear and the casing. The meshing gear teeths at the centre seal the return of the fluid back into the suction. The gears are fitted in the casing with a slight clearance so that there should be minimum of friction. The discharge and the pressure from this type of pump is not steady. To get more uniform pressure, relief valve is provided in a by-pass passage between the suction and delivery side of the oil pump.

Fig. 15.24 (*b*) shows a gear type oil pump with pressure relief valve. Since pressure of the fluid in the pump becomes excessive due to high engine speeds, a pressure relief valve is provided as shown in the figure to prevent the pressure from rising above the desired values.

When the pressure of the pump rises above the value for which the spring of the relief valve is set, the spring is compressed, opening the valve and permitting some of the fluid to return to the suction side of the pump through the bypass, thus reducing the pressure.

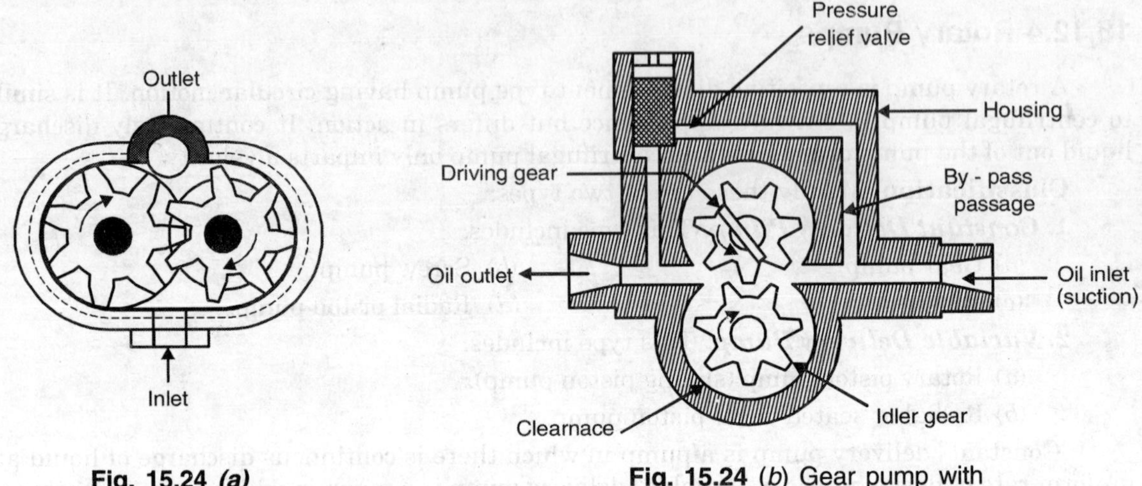

Fig. 15.24 *(a)*

Fig. 15.24 (*b*) Gear pump with pressure relief valve.

Fig, 15.24 (*c*) shows a rotary pump with lobed impeller. This lobed impeller has three teeth. This type of pump is used for relatively low pressures and has a high capacity because of large areas between the gears, called 'lobed impellers' and the casing. The working principle of these rotary pumps is same as that of gear pumps. These rotary pumps can also be made with two teeth.

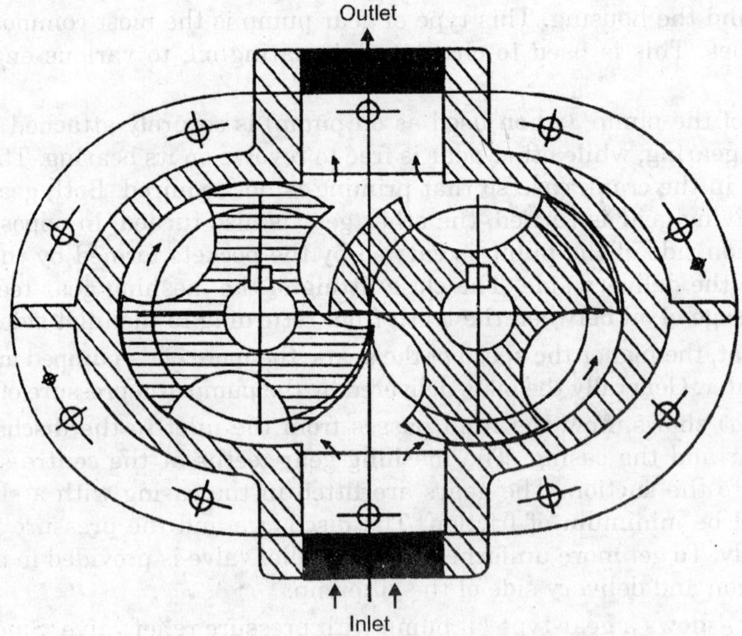

Fig. 15.24. (*c*) Rotary pump with lobed impeller.

15.12.6. Screw Pump

Another variation of the gear pump is the "screw pump" shown in Fig. 15.25. In this the gears are helical, screw threads meshed with each other at the centre and mounted in casing similar to the simple gear pump.

It consists of a rotor or screw to which the source of power say electric motor or steam turbine is directly connected. The screw may be single helical; *i.e.* one hand only or it may be double helical, *i.e.* right and left hand for axial balance. This fluid is carried forward to the discharge along the rotor in pockets formed between teeth and the casing.

In general, these pumps supply liquid, which is free from abrasive material and can be operated in either direction. This pump is suitable for operation at high speed by 3,000 or 4500 R.P.M. and can discharge liquids at 200 kg/cm^2 pressure.

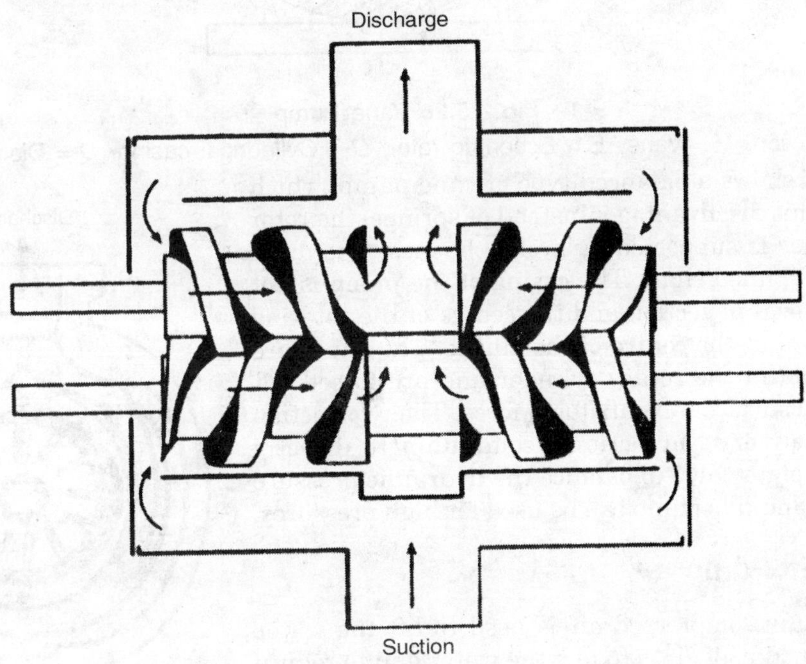

Fig. 15.25 Screw Pump.

15.12.7 Vane Pump

Fig. 15.26 shows a suction through a 'Rotary pump', sliding vane type. This has a rotor disc with the number of slots into which fit sliding blades or vanes.

The rotor rotates in a casing, which is bored eccentric to the driving shaft. The vanes are free to slide radially. The vanes form a seal between suction and delivery with the help of springs. When the rotor rotates, the fluid is sucked in from the inlet and carried around the discharge connection.

One disadvantage of this pump is that due to very great imbalance, it is not possible to develop high pressure.

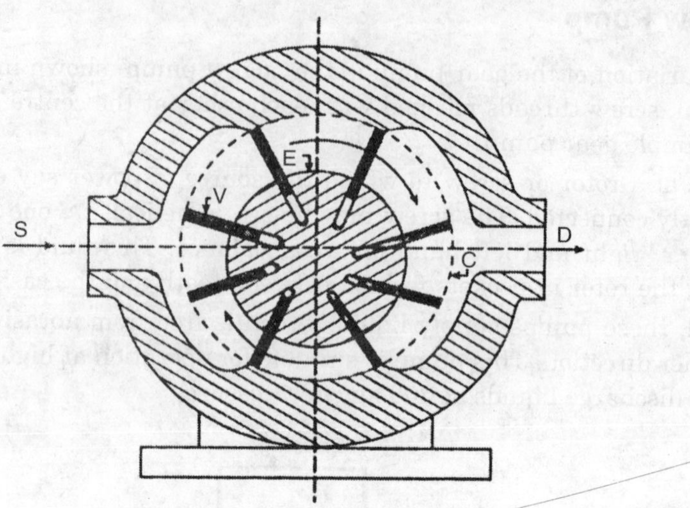

Fig. 15.26 Vane Pump

S = Suction, *V* = Vane, *E* = Eccentric rotor, *C* = Cylindrical casing, *D* = Discharge.

Fig. 15.27 shows a balanced type of vane pump, which does not have this disadvantage. Instead of springs, the rotor fitted with vanes is supported by bronze bushes. The drive is by means of splined shaft. The casing of the pump is not cylindrical but oval in section and the centre of the rotating element is placed at the centre of the ellipse. The vanes are mounted in slots in the rotary cylinder and are forced out against the oval surface by centrifugal force. The two suction and the two delivery connections, communicate directly opposite in the plate valve to balance the hydraulic pressure on the cylinder and this pump can be used for high pressures.

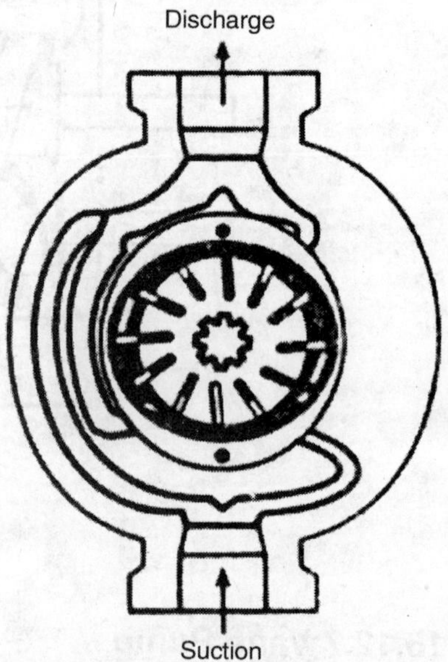

Fig. 15.27 Vene pump.

15.12.8 Air Lift Pump

In this pump compressed air is used to lift the water up. Here the function of air is to mix the water and to reduce the mixture density so that the weight of a given column of mixture is less than the weight of a much shorter column of'solid water'.

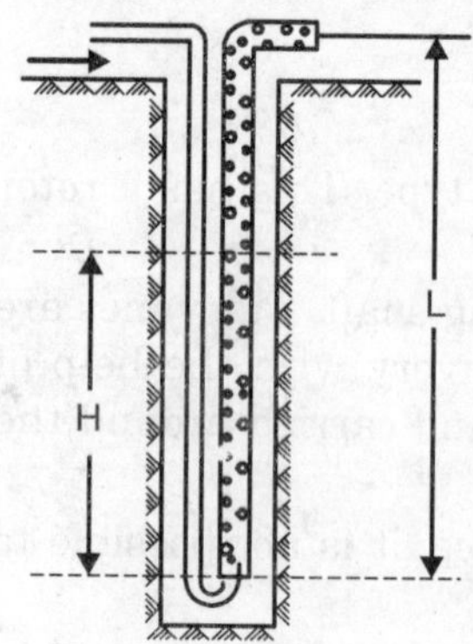

Fig. 15.28 Air Lift Pump

The air is passed through one or more nozzles at the bottom into the water at the foot of the rising main fixed in a wall. As soon as the column in the rising main has become impregnated with the bubbles, the pressure due to the column of length L is less than that due to the static pressure H, thus causing the flow of water and air issue from the top of the rising main so long as the supply of air is continued. Fig. 15.28 shows a line diagram of this type of pump.

To obtain better results the useful lift $(L - H)$ is less than the submergence H. The ratio $\left(\frac{L}{L-H}\right)$ vary from 4 when H is about 30 m to 1 when H is about 90 m.

Advantages

1. As it has no moving parts below water level, therefore, can not be damaged by solids suspension with the water.
2. It can raise more water through a bore-hole of given diameter than any other form of pump.

Limitation

Its efficiency is low say 20 to 40%.

15.12.9 Cooling Water Pumps

These water pumps are used in Automobile Engineering for circulation of water in the water cooled engines. These are generally a centrifugal or impeller pumps driven by a fan belt driven from the crankshaft pulley. Fig. 15.29 shows water pump widely used in automobile engines. In it water enters in the inlet pipe from the bottom of the radiator and is directed by a passage to the centre of the pump body. This water is caught by the rotary curved vanes of the impeller and is thrown outward with centrifugal force.

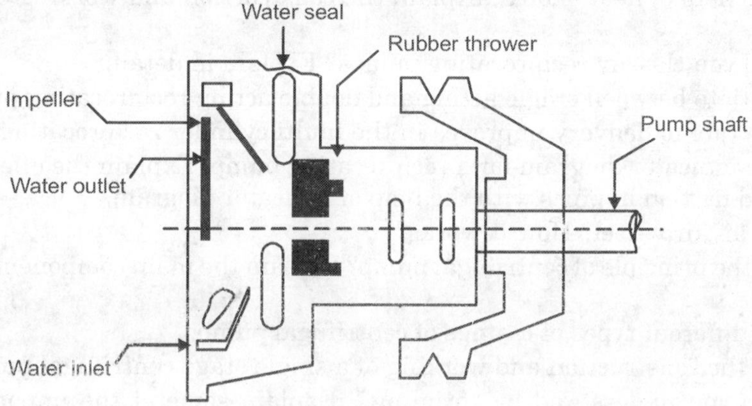

Fig. 15.29

Due to clearance between the impeller blades and the casing these water pumps do not build up dangerous pressures, if an obstruction occurs in the radiator or in the pipe line. Thus these pumps are non positive in action. When these pumps are held stationary for certain period of time, they do not block the passage of water through them, because of the clearance between the impeller blades and the casing and their wide spacing. This provides little resistance in case of impeller failure and allows water to flow due to natural thermosyphon action.

Since these water pumps remain submerged in water a seal assembly is provided around the shaft as shown in the figure to prevent the cooling water from leaking out around the shaft.

15.12.10 Jet Pump

Fig.15.30 shows a jet pump, used to lift the water in a greater quantity when a small quantity of water is available at more height. In this pump, water at a high pressure passes through a fine operate nozzle, thereby converting most of the pressure energy into kinetic energy. It results in a drop of pressure causing suction to take place.

These types of jet pumps are also operated by steam to feed the water in a boiler. With water at high pressure, a more perfect vacuum can be produced so that a suction lift upto 5 to 6 metre can be obtained. If large quantity of water is required to be lifted, then several jets are employed for lifting water.

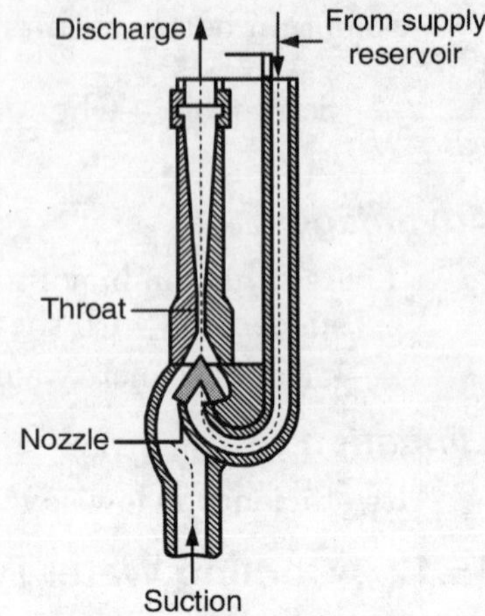

Fig. 15.30 Jet pump

If Q_1 = Rate of flow from supply reservoir.

Q_2 = Quantity of water per second sucked.

Then total delivery from pump, $Q = Q_1 + Q_2$.

If H_1 = Height of the supply reservoir.

H_2 = Suction head.

H_d = Delivery head.

$$\text{Then, efficiency of the jet pump} = \frac{\text{Work done}}{\text{Work consumed}}$$

$$\eta = \frac{Q_2(H_2 + H_d)}{(H_1 - H_d)}$$

QUESTIONS

15.1 With the help of neat sketch, explain the construction and working of a reciprocating pump.

15.2 How will you classify reciprocating pumps? Explain in detail.

15.3 Differentiate between single acting and double acting reciprocating pump.

15.4 How the rate of delivery improves in the multi cylinder reciprocating pumps?

15.5 Draw an indicator diagram for a reciprocating pump. Explain the effect of acceleration head and friction in pipes with the help of indicator diagram.

15.6 What is an air vessel? How it works?

15.7 What is the principle of centrifugal pump? Explain the main components of a centrifugal pumps.

15.8 Explain different types of casings of centrifugal pump.

15.9 Explain the construction and working of a single stage centrifugal pump.

15.10 What do you understand by "priming"? Explain some of the important methods of priming.

15.11 Explain the formulae used for calculating work done by impeller of centrifugal pump on water.

15.12 Explain following terms:
Manometric head, pump losses, pump efficiencies.

15.13 Explain the construction and working of a propeller or axial flow pump.

15.14 Why can the suction lift of a centrifugal pump not exceed a certain limit?

15.15 Describe "Deepwell pump" with the help of a sketch.

15.16 Explain the construction of a submersible pump with the help of a neat sketch.

15.17 Describe the function and working of "Air-lift pump".

15.18 Describe (*a*) Jet pump, and (*b*) Cooling water pump in detail.

15.19 Explain the construction and working of "gear pump".

15.20 What are the rotary pumps and how they are classified?

15.21 Explain the construction and working of (*a*) Screw pump, and (*b*) Vane pump.

Chapter 16

HYDRAULIC DEVICES

Learning Objectives

After studying this chapter, the students will learn:

- Hydro-static and hydro-kinetic devices.
- Hydraulic pumps and motors.
- Different hydraulic devices, such as hydraulic jacks, cranes, intensifier, lift, press, accumulator, hydraulic ram.

16.1 INTRODUCTION

Here we are discussing some of the commonly used hydraulic devices which are based on hydraulic systems following the Pascal's law.

Pascal's law states that, "pressure exerted on a confined liquid is transmitted undiminished in all directions and acts with equal force on all areas". This principle is utilised in different machineries for their different systems, like, hydraulic motors, torque converters, hydraulic lift cylinders, control valves, transmission system, engine clutch and steering clutch, brakes etc.

Hydraulic system is a circuit in which force and power are transmitted through a fluid, generally an oil. Hydraulic systems can be divided into following two groups:

1. Hydrostatic System.
2. Hydro-kinetic System.

1. Hydrostatic System. This is a system in which primary function of hydraulic fluid is the transmission of force and power by pressure. This system consists of (*i*) pumping unit to convert mechanical work into hydraulic energy, (*ii*) hydraulic motor to convert hydraulic energy into mechanic work, and (*iii*) a circuit connecting both these components. Here the pumping unit transmits fluid pressure and hydraulic motor, reciprocating or rotary type, receives force and power by means of fluid pressure. Hydraulic press is an example of hydrostatic system.

2. Hydro-kinetic System. This system also known as rotodynamic system, is based on change in velocity of working fluid. This differs with hydrostatic system as that is based on difference of pressure.

Hydro-kinetic system is used, to transmit power, and consists of (*a*) centrifugal pump or impeller mounted on driving shaft, (*b*) an oil turbine or runner mounted on the driven shaft, and (*c*) a closed hydraulic circuit consisting of both these elements. The power is transmitted from the driving shaft to the driven shaft through oil circulation between the impeller and the runner.

Since there is no mechanical connections between the driver and driven, impulsive shocks and vibrations are missing in the system. Hydraulic coupling and torque converters are the examples of hydrokinetic system.

Hydraulic Pumps and Motors

A pump converts mechanical energy into hydraulic energy. Input of the pumps is mechanical torque, while output is discharged with pressure. Reciprocating, centrifugal and other types of pumps discussed till now in the book are generally employed for lifting of large volumes of water. In the field of oil, rotary pumps are used which develop very high pressure. As discussed earlier rotary pumps are positive displacement pumps with a circular motion, delivers a given volume of fluid for revolution, and continuously scoops the liquid out of the pump chamber (see Fig. 16.1).

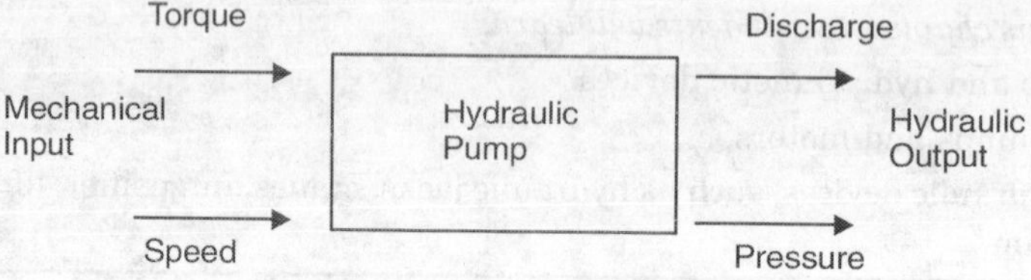

Fig. 16.1 Transformation through a Pump.

A Hydraulic Motor converts hydraulic energy into mechanical energy and is just reverse of hydraulic pump. Hydraulic motor is also a positive displacement motor, which accepts a given volume of fluid for every revolution (see Fig. 16.2).

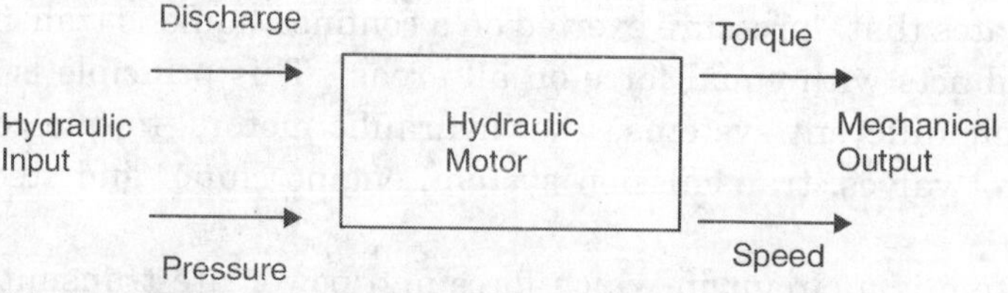

Fig. 16.2 Transformation through a Motor.

16.2 HYDRAULIC JACK

Hydraulic jack works on the same principle as that of hydraulic press. These are used to lift loads and are commonly used for lifting automobiles to facilitate cleaning and repair work. These are used to lift heavy loads through a short distance.

Fig. 16.3 shows a simple hydraulic jack. It is provided with plunger on one side. This plunger is reciprocated by a handle attached to it. Another side of the jack is provided with a ram, by means of which loads are lifted.

Up and down motion of the handle reciprocates the plunger. This movement of the plunger draws the water (or oil) from the reservoir and delivers to the cylinder fitted with the lifting ram through the delivery valve. Both the suction and delivery valves are the non return valves. Addition of high pressure water at the bottom of the ram causes it to move upward. Loads can be lifted if placed at the top of the ram.

Load is lowered with the help of a lowering screw. By unscrewing it, high pressure water from the bottom of the ram rushes to the reservoir, thus reducing the pressure of water below the ram and causes the ram and hence the load to come down.

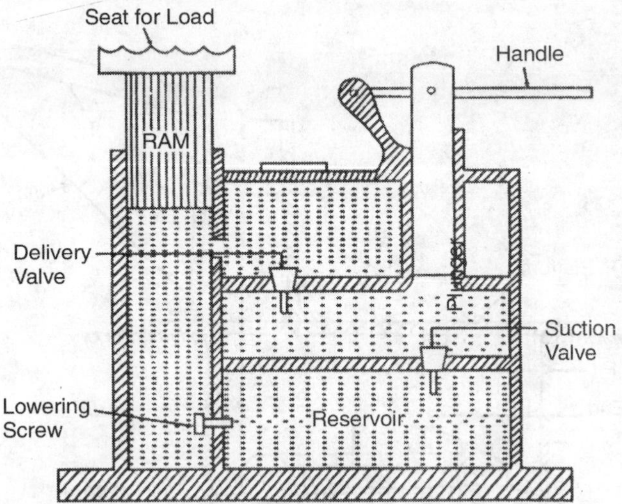

Fig. 16.3 Principles of a hydraulic jack.

$$\text{Pressure on ram} = \text{Pressure on Plunge}$$

$$\text{Theoretical force on the ram} = \text{Force on the plunger} \times \text{Ratio of areas}$$

$$\text{Ratio of Area} = \frac{\text{Area of ram}}{\text{Area of plunger}}$$

$$\text{Actual force on the ram} = \text{Theoretical force} \times \text{Efficiency}$$

$$\text{Velocity ratio of jack} = \frac{\text{Area of ram}}{\text{Area of plunger}} \times \text{Leverage}$$

16.3 HYDRAULIC CRANE

Hydraulic cranes are mostly used in docks, warehouses, sidings etc. for lifting and lowering heavy weights upto 250 tonnes. Its main advantage over electric crane is that these can be adopted to suit varying conditions. Hydraulic crane consist of two main parts (*i*) jigger and (*ii*) craning arrangement.

1. Hydraulic Jigger

It is an arrangement of pulleys for increasing the velocity ratio. It consists of a fixed cylinder in which a ram moves. A number of pulleys (2 to 6 fitted side by side) are fixed on the top of a ram and similar number is fixed at the end of the cylinder. A wire rope, with one end fixed to a point on the cylinder, is taken round all the pulleys on ram and cylinder alternately and finally to the other end is taken round a pulley fixed at the top of the mast (hollow vertical post) of the craning arrangement. It is then taken down after passing through a pulley fixed at the junction of jib and tie rods. At the free end a hook is provided for hanging loads. Jigger may be kept either horizontally or vertically.

2. Craning Arrangement

The crane consists of a hollow vertical post (known as mast) to which a jib and a tie is attached as shown in Fig. (16.4). The mast is supported on a bearing and can be revolved with a vertical axis, so that load from all directions can be handled. The jib can be raised or lowered together with the load.

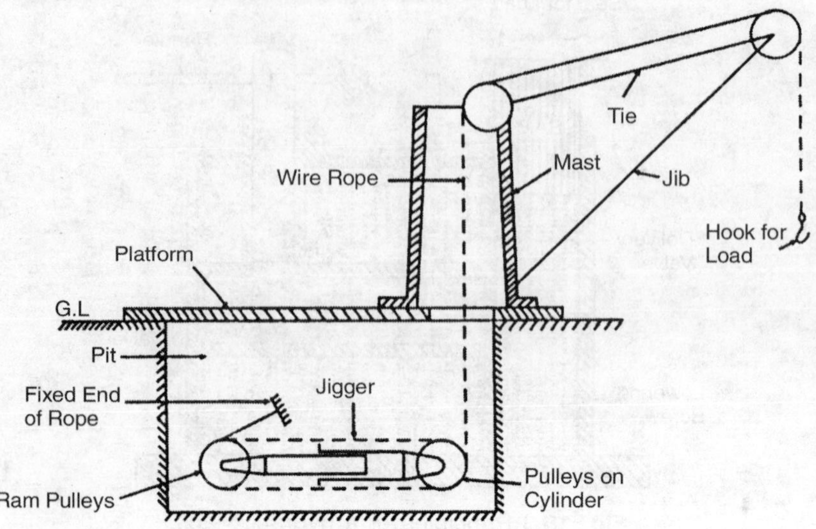

Fig. 16.4 Hydraulic Crane.

Working

When the ram is at the end of the stroke, water under pressure is admitted in the cylinder. This water forces the ram to go in outward direction. While ram is making outward stroke, the wire rope is pulled and the heavy weight hanged with the hook will be lifted upward. The mast can be revolved in a desired direction. Then for lowering the load, water from the cylinder is taken out and ram makes inward stroke. Thus the hook with load comes down on the ground.

Velocity Ratio

The velocity ratio of crane hook to the ram depends upon the number of sets of pulleys. If there be n pulleys on the ram and similar number of pulleys on the cylinder, then load will move through a distance which will be $2n$ times the travel of ram, *i.e.*, velocity ratio = $2n$.

16.4 HYDRAULIC INTENSIFIER

It is a device used for increasing the intensity of pressure by utilizing the energy of a large quantity of water available at low pressure. Its action is similar to that of step up transformer converting low voltage current to high voltage current. Mostly hydraulic process require very high pressure intensity which cannot be produced by pumps. Hence by the introduction of intensifier in between pump and press, high intensity pressure water can be supplied. It can develop pressures upto 16×10^7 N/m^2.

Principle

Its principle can be explained by Fig. 16.5 in which a load W *is* applied on a rod to two different areas, say A and a.

The intensity of stress in upper portion will be $\frac{W}{A}$ while in lower portion $\frac{W}{a}$.

As both the cross-sectional areas are subjected under same load W, therefore, the portion of smaller diameter will be subjected under greater stress, *i.e.* stress produced is inversely proportional to area, n. This, same principle is also applied in intensifier. Here low pressure fluid is acted in a cylinder of greater area while high pressure is developed in a smaller area cylinder.

Fig. 16.5

Construction and Working

In its construction, it consists of:

(*a*) Fixed cylinder with valve A.

(*b*) Sliding hollow ram with valve B.

(*c*) Fixed ram with valve C.

When the sliding hollow ram is at the bottom extreme position, at that time valve B remains open and is filled with low pressure water. Low pressure water coming in the fixed cylinder pushes the sliding hollow ram up and closes the valve B which in turn opens the valve C. Now gradually sliding hollow ram moves upwards and intensified water at high pressure passes out through a small dia pipe in the fixed ram.

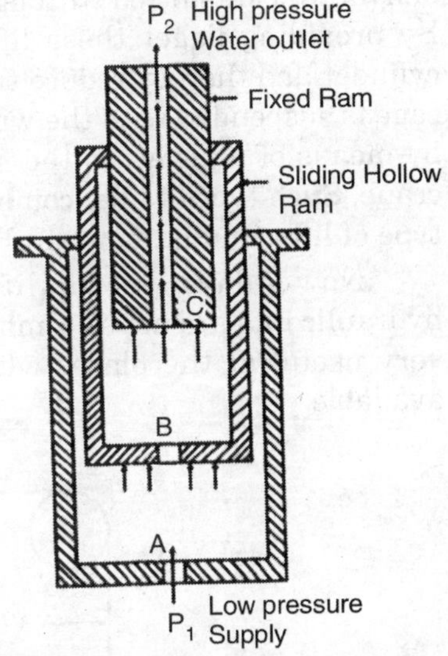

Fig. 16.6

As soon as sliding hollow ram reaches its topmost position, valves C and A are closed and B opens. The sliding hollow ram comes down by the opening of valve B due to its own weight. Valve A opens and low pressure water from fixed cylinder passes into the sliding hollow ram. In this way one cycle is completed and intensifier thus works itself.

If D_1 and D_2 be the dia at the base of sliding hollow ram and fixed ram respectively. Then

$$p_2 = \frac{p_1 \times D_1^2}{D_2^2} \text{ N/m}^2$$

where p_1 and p_2 are the initial and final pressures respectively.

16.5 HYDRAULIC LIFT

It consists of a cylinder and a ram. A platform or a cage is fitted at the top of the ram. Load or the persons are carried from one floor to another floor by means of these platforms or cages. Water under pressure enters the bottom of the cylinder and pushes the ram upward. Thus ram moves up along with a cage, in which load and persons may be there. In this type of lift (known as direct acting lift) stroke of the ram is equal to the lift of the cage. The cage moves between guides and can be stayed at any level so that one can get in or out of the cage to the floor. For lowering it high pressure water is gradually taken out. Such type of lift is shown in Fig. 16.7 in which dotted line indicate the initial position while firm lines indicate final positions.

Let, d = Ram dia in cm.

p = Pressure of water in N/m^2

W = Total load of ram, cage and its contents in Newtons

a = Cross-section area of ram = $\frac{\pi}{4} d^2$

Then, neglecting friction etc. $W = p \times a = p \times \frac{\pi}{4} d^2$.

Thus dia of ram is designed, considering the maximum weight likely to be lifted.

Direct acting lifts require deep wells for the ram and cylinder. To overcome this difficulty modified form of hydraulic lifts fitted with jigger are becoming common. By providing jigger these lifts require small ram and cylinder and thereby reduce the cost of construction. The cage is suspended with the wire rope and gets its motion by means of a jigger, in the same way as the hydraulic crane. Such type of lifts, commonly known as suspended type of lifts, and is shown in Fig. 16.8.

Now-a-days electric lifts are very common and hydraulic lifts are not so common. But hydraulic lifts are very useful at the places where electric supply is not available.

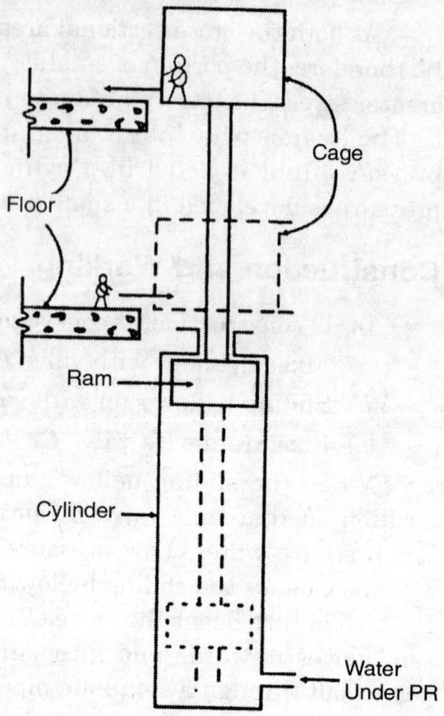

Fig. 16.7

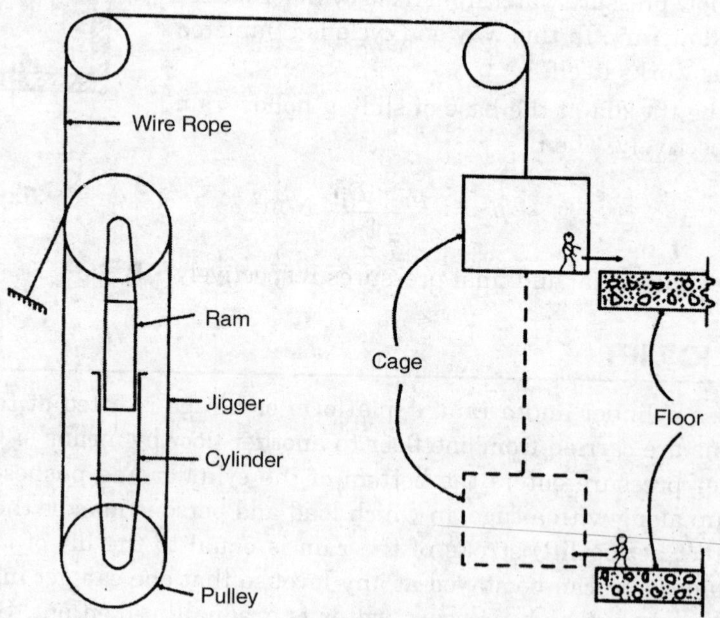

Fig. 16.8

16.6 HYDRAULIC PRESS

It is based on Pascal's law principle that pressure applied at any point on any confined liquid is transmitted equally to all other points. It was first built in 1785 by Ernest Bramah and, therefore, also called Bramah's press. With its help huge pressures can be applied. Therefore, it can be used for die sinking embossing, pressing cotton bales, forging hot bullets, shaping sheet metals, preparing moulds and castings of bakelite, vulcanising, packing, drawing, pushing rods and filter press etc.

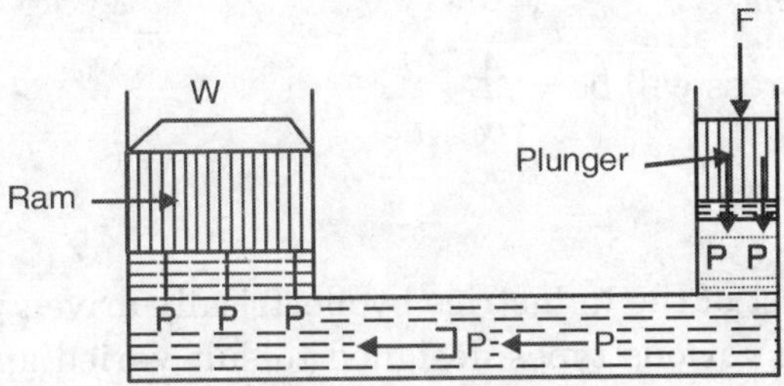

Fig. 16.9 *(a)*

In hydraulic press, mechanical advantage is obtained by applying Pascal's law of transmission of pressure.

In Fig. 16.9 *(a)*, $$p = \frac{W}{A} \quad \text{...(1)}$$

Fig. 16.9 *(b)*

where p = Pressure of fluid, W = Load to be lifted

and A = Cross-sectional area of ram.

Also $$p = \frac{F}{a} \quad \text{...(2)}$$

where F = Effort applied

and a = Cross-sectional area of plunger,

Equating equations (1) and (2), $p = \dfrac{W}{A} = \dfrac{F}{a}$

i.e., $$\frac{W}{A} = \frac{F}{a} \quad \text{or} \quad \frac{W}{F} = \frac{A}{a} \quad \text{...(3)}$$

But $$\frac{W}{F} = \frac{\text{Load}}{\text{Effort}} = \text{Mechanical Advantage (M.A.)}$$

$$\therefore \qquad M.A. = \frac{A}{a} \qquad ...(4)$$

Now by keeping the volume of fluid constant, the displacement of ram is proportionately smaller than the plunger.

In some cases effort is applied by a lever as shown in Fig. 16.9(*b*)

The M.A. of lever is $\frac{l}{m}$.

Hence total M.A. of press will be = $\frac{A}{a} \times \frac{l}{m}$ ...(5)

Construction

These are used in most of the industries by electrically driven pumps supplying oil under pressure. These may be of various types designed to suit varied applications. Mostly presses may be classified as:

(*a*) Direct type. (*b*) Inverted type.

***(a)* Direct Type.** In it pressure is applied to the ram from a pump either directly or through a small cylinder. By the application of pressure, the ram rises and moves up the movable plate, compressing any material which has been placed between the movable plate and a fixed plate.

***(b)* Inverted Type.** Fig. 16.10. shows inverted type press in which ram pushes the movable plate from top. Stationary plate is fixed at bottom. When pressure is released, the return weights come down pushing back the movable plate and ram to their initial position.

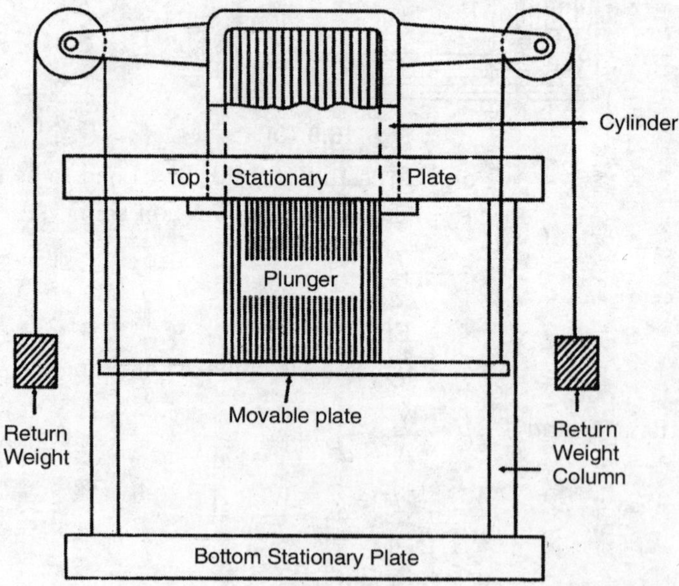

Fig. 16.10

16.7 HYDRAULIC ACCUMULATOR

It is a device used for storing or accumulating energy of water or any other fluid. Its actions are similar to storage battery used for storing electrical energy.

Some of the hydraulic machines such as lifts, cranes etc. require intermittent energy, *i.e.*, during lifting period they require large amount of hydraulic energy but during idle period very small amount of energy is required.

Hydraulic pumping machines supply energy continuously at a uniform rate. Therefore, an accumulator is placed between such machinery and pump device so that extra energy can be accumulated during idle periods and supplied at an increased rate when required during working periods.

By using this aid small capacity pumps can be used to supply greater amount of energy temporarily. This saves a lot of power consumption because of use of smaller size pumps.

Types

There are two types of accumulators,

(*a*) Direct type. (*b*) Inverted type.

(*a*) Direct Type. Fig. 16.11 explains the construction and working of direct type accumulator.

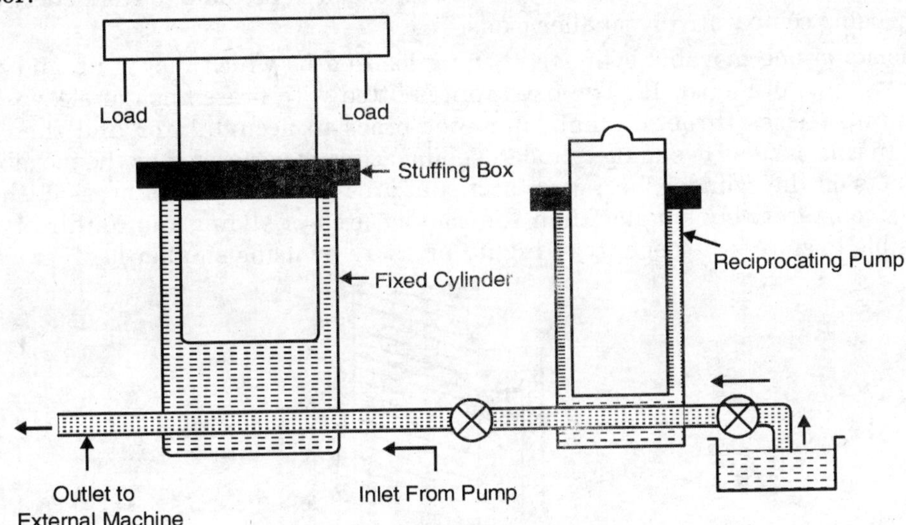

Fig. 16.11 Direct type accumulator.

In its construction it consists of:

(*i*) Plunger or sliding ram
(*ii*) Fixed cylinder
(*iii*) Weights
(*iv*) Stuffing box
(*v*) Stop (not shown in the figure) used to prevent the ram being thrown off from the cylinder due to very large pressure accumulation.

Working

Plunger moves up and down in the fixed cylinder. On the plunger heavy load is placed for developing pressure. A reciprocating or other pressure pump is used to supply water to external machine at constant high pressure. When less pressure energy is required by the driven machine, excessive energy is stored in accumulator by the upward movement of the plunger. When large energy is required by the driven machine in comparison to pump supply, the excessive required energy is supplied by the acumulator by descending ram. In this way pump supplies energy at constant rate.

The maximum energy that can be accumulated is the product of the total weight of plunger plus load on it with the maximum height of descent of the ram.

Let A = Area of ram base in sq.m

H = Maximum descent of ram in m

p = Intensity of input water by pump in N/m^2

W = Total weight of ram plus load.

Therefore, upward force acting on the ram = $W = p \times A$

Maximum work done by water in moving the ram through height H

$$= \text{Force} \times \text{Descent} = W \times H = p \times A \times H \text{ Nm}$$

$$\therefore \quad \text{Max. Energy Stored} = \text{Max. work done}$$

$$= p \times A \times H = p \times \text{Volume}$$

This is known as capacity of accumulator.

(b) Inverted Accumulator (Tweddell's Differential Accumulator).

This type is better than the direct type, as with this it is possible to store energy at higher pressure by using comparatively smaller loads.

It consists of one movable cylinder, (having load on it) which moves up and down on a fixed ram as shown in Fig. 16.12. The fixed ram is fitted with brass bush or sleeve. The water from the pump passes through small diameter pipes to accumulator and then to driven machinery. In this, instead of sliding cylinder, sliding action is performed by the movable cylinder and water acts on the cylinder for a cross-sectional area equal to that of brass bush or sleeve. This area is comparatively smaller than the contact area of sliding ram of direct type. This makes possible to store water energy at higher pressure by using smaller load.

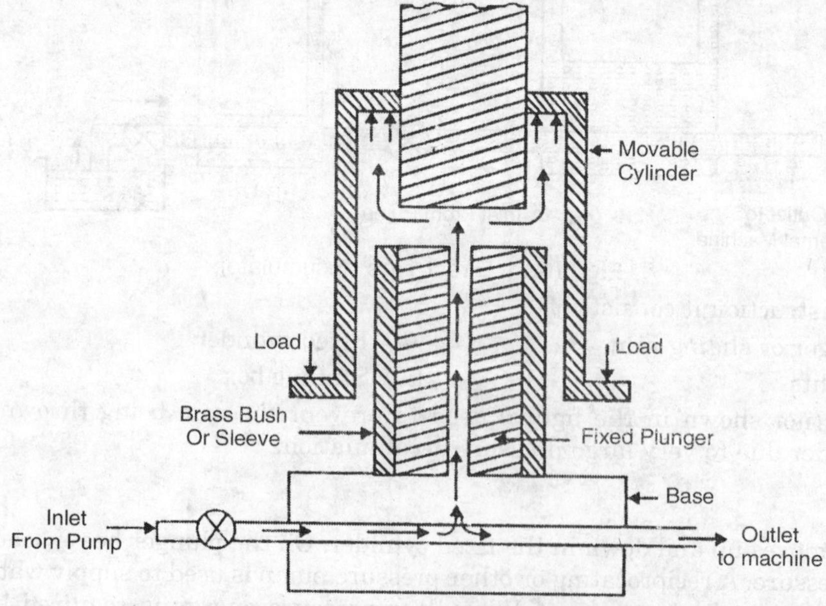

Fig. 16.12. Tweddell's differential accumulator

Let p = Pressure of incoming water in N/m^2

H = Max. descent in m

W = Wt. of loaded cylinder in N

Then, W = $p \times a$ N where a sq. m is the cross-sectional area of brass bush.

$\therefore$ Max. energy stored = W × H Nm.

Both of the accumulators may be dead load or variable load type. In dead load type, dead weights are placed to press the plunger but in variable load type, steam pressure is applied which can be varied at will. This type needs steam boiler and, therefore, applied where boiler is available such as ships.

16.8 HYDRAULIC RAM (HYDRAM)

It is a device like automatic pump used to lift water available in large quantities at low head to greater head in small quantities. In doing so, water hammer principle is utilized and no external power is required to run the ram.

Principle. The action of ram is due to the inertia force of the moving column of water in the supply pipe which produces water hammer effect due to sudden closer of impulse valve.

Application. This is applicable where some natural source of water such as river or spring at some height is available. For example hilly area, where no means of running the prime mover are there.

Construction. The following line sketch explains the construction of a Hydram (see Fig. 16.13). It consists of:

(*a*) Water source
(*b*) Supply pipe
(*c*) Gate valve
(*d*) Snifting valve
(*e*) Ram box or chamber
(*f*) Air chamber
(*g*) Delivery valve
(*h*) Waste valve
(*i*) Delivery pipe.

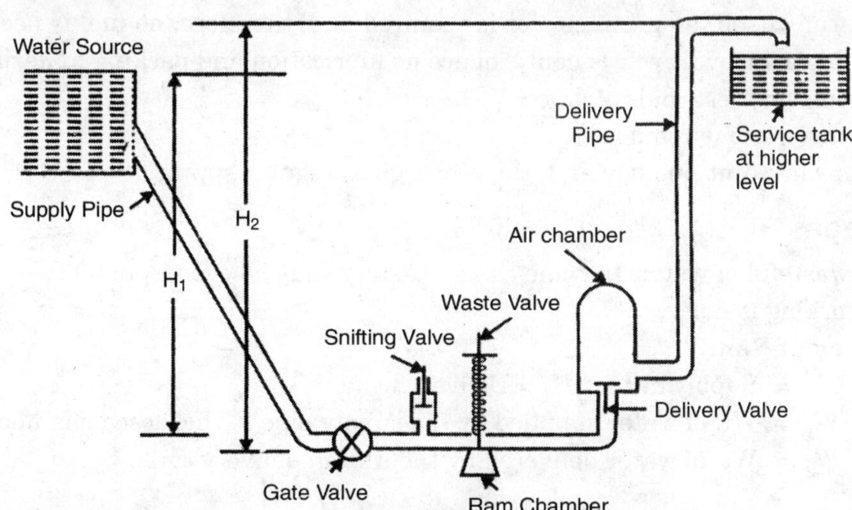

Fig. 16.13 Line diagram of Hydraulic Ram.

Working

At the time of starting, all the valves are closed. First of all, gate valve is opened and water from the supply source runs in the closed box striking the right side of box and rebounds. This rebound action of water causes partial vacuum, in the closed box and opens waste and air (snifting) valve together. Water passes into waste through the waste valve (Impulse valve), gradually increasing the velocity until it rapidly closes the waste valve. Rapid closer of the waste valve causes water hammer effect in the supply line by which tremendous pressure is created in the ram box to open the delivery valve in the air chamber. The water at high pressure passes through the air chamber and from it to the delivery tank through delivery pipe. Thus gradually pressure falls and delivery valve is again closed due to the back pressure of air and water in the air chamber.

In this way cycle of operations commences again and again running the ram automatically without any external driving source and supplying water to a greater height.

Precautions

(*a*) The most important part of ram is the impulse valve. This should not be made too large to reduce the rapidity with which it stops the flow nor too small to cause great frictional resistance to flow.

(*b*) In order to utilize the kinetic energy to a great extent, impulse valve must be suddenly operated.

(*c*) The snifting valve should open when the waste valve opens so that when the water flows through the delivery pipe it carries with it a small volume of air otherwise in a few strokes there will be vacuum in the chamber unless air is supplied through snifting valve.

(*d*) In order to develop maximum water hammer effect, supply pipe should be as long as possible. If the ram is placed near the source of supply, water hammer effect is less and small delivery head is developed.

Advantages

1. As it runs itself, no prime mover is required and, therefore, no fuel is needed.
2. As it has no moving components, hence no lubrication and packing material is needed.
3. It needs very less maintenance.
4. No attendance is required.
5. It can work continuously and, therefore, gives regular supply.

Disadvantages

1. It is wasteful of water. In some cases, wastage may be even upto 90%.
2. Its working is noisy.

Efficiency of Ram

Let H_1 = Supply head, H_2 = Delivery head

W_1 = Wt. of water supplied by the supply pipe to the closed chamber.

W_2 = Wt. of water delivered by the ram in delivery tank.

Then

$$\eta = \frac{\text{Output}}{\text{Input}} = \frac{W_1 H_1}{W_2 H_2}$$

SOLVED EXAMPLES

Example. 16.1 *A hydraulic jack has a ram of 100 mm dia and pump plunger of 20 mm dia. The leverage for working the pump is 8. When a force of 15 N is applied at the end of the lever, it lifts 2400 N on the ram. Determine (a) Velocity ratio; (b) Mechanical advantage: (c) Efficiency of jack.*

Solution. (*a*) As,

Area of plunger × Stroke of plunger

$$= \text{Area of ram} \times \text{Stroke of ram}$$

$$\therefore \quad \frac{\text{Stroke of plunger}}{\text{Stroke of ram}} = \frac{\text{Area of ram}}{\text{Area of plunger}}$$

Also,

$$\text{Velocity ratio} = \frac{\text{Distance moved by effort}}{\text{Distance moved by load (}i.e.\text{ ram)}}$$

$$= \frac{\text{Stroke of plunger} \times \text{Leverage}}{\text{Stroke of ram}}$$

$$= \frac{\text{Area of ram}}{\text{Area of plunger}} \times \text{Leverage}$$

$$\therefore \quad \text{Velocity ratio} = \left(\frac{100}{20}\right)^2 \times 8 = \mathbf{200.} \quad \textbf{Ans.}$$

(*b*)

$$\text{Mech. advantage} = \frac{\text{Load}}{\text{Effort}} = \frac{2400}{15} = \mathbf{160.} \quad \textbf{Ans.}$$

(*c*)

$$\text{Efficiency of jack} = \frac{\text{Mech. Advantage}}{\text{Velocity of Ratio}} = \frac{160}{200}$$

$$= \mathbf{0.80 \text{ or } 80\%.} \quad \textbf{Ans.}$$

Example. 16.2 *A hydraulic jack has a ram of 100 mm dia and a plunger of 25 mm dia. The mechanical advantage of the lever is 10 and the efficiency of the jack is 80%. Determine the weight which can be raised by it if a force of 20 N is applied at the end of the lever.*

Solution. As, velocity ratio of jack $= \dfrac{\text{Area of ram}}{\text{Area of plunger}} \times \text{Leverage}$

$$= \left(\frac{100}{25}\right)^2 \times 10 = 160$$

As, Mechanical advantage = velocity ratio × η = 160 × 0.80 = 128

Also, Mechanical advantage $= \dfrac{\text{Load}}{\text{Effort}}$

$$\therefore \quad 128 = \frac{\text{Load}}{20}$$

∴ Load which can be lifted = 128 × 20 **= 2560 N. Ans.**

Example. 16.3 *A hydraulic crane is supplied with water at a pressure, of 5 × 10⁶ N/m² and it consumes 0.025 m³ of water in order to lift a weight of 3 tonnes through a height of 3 m. Find the efficiency of the crane. Also determine the stroke of the ram, if the ram stroke and dia ratio is 6.*

Solution. Pressure of water $= 5 \times 10^6$ N/m²

Water consumption $=$ 25 litres $=$ 25 kg $= 25 \times g$ N

Load lifted $= 3 \times 1000 = 3000$ kg $= 3000 \times g$ N

Lift $=$ 3 m.

(*i*) Here, Output of the crane $= W \times H = 3000 \times 3 \times g$

Input of the crane $=$ Force on ram × Stroke of ram

$$= p \times A \times L$$

$$= 5 \times 10^6 \times \frac{\pi}{4} D^2 \times L \text{ Nm}$$

As, capacity of the crane cylinder $=$ 25 kg of water $= 0.025$ m³

∴ $0.025 = A \times L =$ Displacement of ram.

∴ Input to the crane $= 5 \times 10^6 \times 0.025 \times 10^4$ Nm

∴ $$\eta \text{ of the crane} = \frac{\text{Output}}{\text{Input}}$$

$$= \frac{3000 \times 3 \times 9.81}{5 \times 10^6 \times 0.025 \times 10^4}$$

$$= 72\%$$

(*ii*) Given, $$\frac{L}{D} = 6$$

∴ $$D = \frac{L}{6}$$

As $$0.025 = \frac{\pi}{4} D^2 L = \frac{\pi}{4}\left(\frac{L}{6}\right)^2 L = \frac{\pi}{144} L^3$$

∴ $$L^3 = \frac{0.025 \times 144}{\pi} = 1.15$$

∴ Stroke of ram $L =$ **1.05 m. Ans.**

Example. 16.4 *A hydraulic crane having 60% efficiency works at a pressure of 4 × 10⁶ N/m² to lift a load of one tonne with a speed of 20 m/min through a height of 10 m once in every two min. Determine (a) the volumetric capacity of the cylinder of the jigger of the crane (b) minimum power required for the pump (c) volumetric capacity of the accumulator.*

Solution. Efficiency of crane $= 0.60$

Pressure of water $= 4 \times 10^6$ N/m²

Speed of load $=$ 20 m/min.

Load $=$ 1000 kg $= 1000 \times 9.81$ N

Lift $=$ 10 m.

(*i*) Output $= W \times H = 1000 \times 10 \times 9.81$ Nm

Force of water on the ram $= p \times A$

$\therefore$ Input of crane $= 4 \times 10^6 \times A \times L$

$$\eta = \frac{\text{Output}}{\text{Input}}$$

$$\therefore \quad 0.60 = \frac{1000 \times 10 \times 9.81}{4 \times 10^6 \times AL}$$

$$\therefore \quad AL = \frac{10,000 \times 9.81}{0.6 \times 4 \times 10^6}$$

$$= 0.0416 \text{ m}^3 = 41.6 \text{ litres}$$

$\therefore$ Capacity of the jigger = **41.6 litres. Ans.**

(*ii*) As, Input of the crane $= 4 \times 10^6 \times A \times L = 4 \times 10^6 \times 0.0416$ Nm

= **166400 Nm. Ans.**

But this input is to be given to the crane cylinder after every 2 minutes. Therefore, the energy which pump should supply be at least = 166400 Nm.

$\therefore$ Minimum power required for the pump

$$= \frac{166400}{2 \times 60 \times 1000} \text{ kW} = \mathbf{1.39 \text{ kW}}. \quad \textbf{Ans.}$$

(*iii*) Work done by the pump during lifting of load by crane

$$= \frac{1.39}{2} = 0.695 \text{ kW} = 695 \text{ Nm/sec}$$

$\therefore$ Work done by the pump during lifting of load, by crane

= Work done/min × Time reqd. to lift by 10 m

$$695 \times 60 \times \frac{10}{20} = 20850 \text{ Nm}$$

The difference of work by which this value (20850) remain short is supplied by the accumulator.

Hence, minimum energy to be stored in accumulator

= 166400 – 20850 = 145550 Nm.

As we know that,

Energy stored in accumulator = Force on accumulator × Ram lift

$= (p \times A) \times H$

or $145550 = 4 \times 10^6 \times A \times H$

$$\therefore \quad A \times H = \frac{145550}{4 \times 10^6} = 0.0364$$

$\therefore$ Minimum capacity of accumulator

$= 0.0364 \text{ m}^3$ **Ans.**

Example. 16.5 *A hydraulic intensifier supplying water to a hydraulic press gets water from a supply main at a pressure of* 4×10^6 *N/m²* *and intensifies it to* 16×10^6 *N/m². The stroke is 1 m and capacity 0.015 m³. Find the dia of the two rams of intensifier.*

Solution. Given: $p_1 = 4 \times 10^6 \text{ N/m}^2, \quad p_2 = 16 \times 10^6 \text{ N/m}^2$

$$H = 1 \text{ m}$$

$$\text{Capacity} = 0.015 \text{ m}^3$$

a_1 and a_2 = Area of I and II ram in sq. mm respectively.

As,
$$p_1 a_1 = p_2 a_2$$

$$\frac{p_2}{p_1} = \frac{a_1}{a_2} = \frac{d_1^2}{d_2^2} \quad \text{or} \quad \frac{16 \times 10^6}{4 \times 10^6} = \frac{d_1^2}{d_2^2}$$

$$\therefore \quad \frac{d_1}{d_2} = 2$$

$$\text{Also volumetric capacity} = \text{Area} \times \text{Stroke of intensifier} = \frac{a^2}{10{,}000} \times H \text{ m}^3$$

or
$$0.015 = \frac{\pi}{4} d_2^2 \times \frac{H}{1000000} \quad \text{or} \quad d_2^2 = \frac{15000 \times 4}{\pi \times 1}$$

$$\therefore \quad d_2 = \mathbf{138 \text{ mm}} \quad \textbf{Ans.}$$

But,
$$\frac{d_1}{d_2} = 2$$

$$\therefore \quad d_1 = 2 \times 138 = \mathbf{276 \text{ mm}} \quad \textbf{Ans.}$$

Example. 16.6 *A hydraulic lift has a ram of 100 mm dia and is connected with a pipe of 40 mm dia in which pressure is 5 × 10³ kN/m². Find the maximum load that can be lifted steadily with a velocity of 1 m/sec. Also determine the maximum velocity with which this lift will descend steadily with an open exhaust.*

Solution. Total pressure on the ram = Pressure of water + Pressure due to velocity of water in the pipe

But, velocity of water in the pipe

$$= v = V \times \frac{A}{a} = 1 \times \left(\frac{100}{40}\right)^2 = 6.25 \text{ m/sec.}$$

$\therefore$ Velocity head of water in the pipe

$$= \frac{v^2}{2g} = \frac{(6.25)^2}{2 \times 9.8} = 2.0 \text{ m}$$

$$= 2 \times 10 \text{ kN/m}^2$$

$$\therefore \quad \text{Total pressure on the ram} = (5000 + 20) \text{ kN/m}^2 = \mathbf{5020 \text{ kN/m}^2}$$

$$\text{Hence load lifted by the ram} = p \times a = 5020 \times \frac{\pi}{4} \frac{(100)^2}{10^6} = 39.40 \text{ kN}$$

$$= \mathbf{3940 \text{ kg.}} \quad \textbf{Ans.}$$

Let,
$$V_d = \text{Velocity of descent in m/sec}$$

then velocity of water in 40 mm pipe

$$= \left(\frac{100}{40}\right)^2 V_d = 6.25\, V_d$$

∴ Velocity head of water in the 40 mm pipe

$$= \frac{(6.25\, V_d)^2}{2g}$$

But this velocity head is due to the pressure head of the ram.

∴ Pressure head of the ram in m of water

$$= \frac{5020}{10} = 502 \text{ m of water}$$

$$= \text{Velocity head in the 40 mm pipe}$$

$$\therefore \quad 502 = \frac{(6.25\, V_d)^2}{2 \times 9.8} \quad \text{or} \quad V_d^2 = \frac{502 \times 19.6}{6.25 \times 6.25}$$

$$\therefore \quad V_d = \textbf{15.8 m/sec.} \quad \textbf{Ans.}$$

Example. 16.7 *A hydraulic lift raises a load of 100 kN through a height of 10 m, once every two minutes. The speed of lifting being 1 m/sec. Lift works from an accumulator which is being continuously charged by a pump. The pressure of water is* 4×10^6 *N/m*^{2}*, find the power required to drive the pump and the minimum capacity of the accumulator, if efficiency of the lift is 75% and the efficiency of the pump is 80%. Take g = 10 and number of pulleys on ram 3.*

Solution. Theoretical work done by lifting the load

$$= 100 \times 10000 \times 10 \text{ Nm}$$

∴ Actual work required by the lift

$$= \frac{100 \times 1000 \times 10}{0.75} = 1330000 \text{ Nm}$$

This work is to be done in 2 min. by the pump

∴ Theoretical power for the pump

$$= \frac{1330000}{2 \times 60} = 11.1 \text{ kW.}$$

∴ Actual power required by the pump

$$= \frac{11.1}{0.80} = 13.88 = \textbf{13.88 kW.} \quad \textbf{Ans.}$$

Duration in which accumulator is charged

$$= 2 \text{ min} - \frac{10}{1 \times 60} = 2 - 0.166 = 1.834 \text{ min.}$$

But the energy stored in 1.834 min.

$$= \text{Energy required in lifting load through 10 m}$$

$$= \frac{1330000}{2} \times 1.834$$

If C *is* the minimum capacity of accumulator in m^3, then

$$\text{Energy stored} = \text{Volume} \times \text{Pressure of water}$$

$$= C \times 40 \times 100000 \text{ N}$$

$$\text{Also energy stored} = \frac{1330000}{2} \times 1.834$$

$$\therefore \quad C \times 40 \times 1{,}00{,}000 = \frac{13{,}30{,}000}{2} \times 1.834$$

$$\therefore \quad C = \mathbf{0.306 \ cu \ m. \quad Ans.}$$

$$\text{Velocity Ratio} = \frac{1}{2 \times \text{No. of pulleys on the ram}} = \frac{1}{2 \times 3}$$

$$= \mathbf{\frac{1}{6}. \quad Ans.}$$

Since, the Velocity Ratio = Mechanical Advantage, therefore, load lifted will be $\frac{1}{6}$ times the effort, if friction is neglected.

∴ Theoretical pull on the rope of the cage

= Force on the ram × Mechanical Advantage

$$= 500 \times \frac{\pi}{4} (8)^2 \times \frac{1}{6} = \mathbf{4180 \ N. \ Ans.}$$

$$\text{Stroke of the ram} = \frac{1}{6} \times \text{Distance travelled by the cage}$$

$$= \frac{1}{6} \times 15 = \mathbf{2.5 \ m. \quad Ans.}$$

Example. 16.8 *A hydraulic press has a plunger of 50 mm dia and ram of 400 mm dia. Find the weight that can be lifted by applying a force of 1000 N on the plunger.*

If the stroke of the plunger is 120 mm, how much load can be raised in 120 strokes? What is the work done in this period?

Solution. Here

Area of plunger, $a = \frac{\pi}{4} (50)^2 \ . \ \text{mm}^2$

Area of ram, $A = \frac{\pi}{4} (400)^2 \ \text{mm}^2$

Force applied, $F = 1000 \text{ N}$

(*i*) As, $\frac{W}{F} = \frac{A}{a}$ or $W = \frac{F \times A}{a}$

$$= \frac{1000 \times \frac{\pi}{4}(400)^2}{\frac{\pi}{4}(50)^2} = \frac{1000 \times 160000}{2500} = 64000 \text{ N}$$

$$= \mathbf{64 \ kN \quad Ans.}$$

(*ii*) Volume of water displaced by plunger in one stroke

= Area × Length of stroke

$$= \frac{\pi}{4} (50) \times 120 \ \text{mm}^3$$

Let S = Displacement of load in one stroke.

Then, $\frac{\pi}{4}(400)^2 \times S = \frac{\pi}{4}(50)^2 \times 12$

$\therefore$ $S = \frac{25 \times 120}{40 \times 40} = \frac{30}{16}$

Load moves in 120 strokes = $\frac{120 \times 30}{16}$ = 225 **mm. Ans.**

Work done = Load × Lift = $64 \times \frac{22.5}{100}$ kNm.

= **14.44 kNm. Ans.**

Example. 16.9 *In a hydraulic press dia of the plunger and the ram are 30 mm and 300 mm respectively. The leverage of the handle is 15:1. If a body weighing 5 kN be lifted on the ram by a force of 4 N applied on the lever, determine the efficiency of the press.*

Solution. Here,

Area of plunger, $a = \frac{\pi}{4}(30)^2 \text{ mm}^2$

Area of ram, $A = \frac{\pi}{4}(300)^2 \text{ mm}^2$

As, Actual Mechanical advantage

$$= \frac{15}{1} = \frac{\text{Load}}{\text{Effort}} = \frac{5000}{4} = 1250$$

But theoretical Mechanical advantage

$$= \frac{A}{a} \times \text{Leverage of handle} = \frac{\frac{\pi}{4}(300)^2}{\frac{\pi}{4}(30)^2} \times \frac{15}{1}$$

$$= 100 \times 15 = 1500$$

$\therefore$ η of the press $= \frac{\text{Actual Mech. advantages}}{\text{Theoretical Mech. advantage}} = \frac{1250}{1500}$

= **0.8334 or 83.34% Ans.**

Example. 16.10 *The ram and plunger of a hydraulic press are 150 mm and 25 mm in diameter respectively. With a plunger stroke of 300 mm the press is able to lift a load of 1 tonne through 1.5 m in a time of 25 minutes. What is (i) the load on the plunger, (ii) Power required to drive the plunger and (iii) No. of strokes done by the plunger.*

Solution. Given:

Dia of ram, D = 150 mm

Dia of plunger, d = 25 mm

Stroke of plunger, S = 300 mm

$A = \frac{\pi}{4}(150)^2 \quad a = \frac{\pi}{4}(25)^2$

Load, W = 1 tonne = 1000 kg

Load lift = 1.5 m in 25 min.

(*i*) Let F kg be the load on the plunger.

Then $$\frac{W}{F} = \frac{A}{a}$$

or $$F = \frac{W \times a}{A} = \frac{1000 \times \frac{\pi}{4}(25)^2}{\frac{\pi}{4}(150)^2} = \mathbf{27.78\ kg}$$

Hence required load on the plunger is **27.78 kg Ans.**

(*ii*) Let N be the total number of strokes of the plunger.

Then $N \times 300 \times \frac{\pi}{4}(25)^2 = \frac{\pi}{4}(150)^2 \times 1.5 \times 1000$

or $$N = \frac{(150)^2 \times 1500}{(25)^2 \times 300} = \mathbf{180.}$$

$\therefore$ Total number of strokes in 25 minutes.

$= \mathbf{180.\ Ans.}$

(*iii*) Power required by plunger $= F \times S \times$ Strokes/sec $\times g$ Nm/sec

$$= (9.81 \times 27.78) \times 0.30 \times \frac{180}{25 \times 60} \text{ Nm/sec}$$

$$= \frac{9.81 \times 27.78 \times 0.30 \times 180}{25 \times 60} = 9.78 \text{ W}$$

Hence power required to drive the plunger

$= \mathbf{9.78\ W\ Ans.}$

Example. 16.11 *The ram of a hydraulic accumulator is 200 mm in dia and 20 tonnes in weight. If the force required to move the cylinder along the loaded ram, against the resistance of friction is 2 tonnes, calculate the pressure of water when (a) Loaded ram is descending and (b) Loaded ram is ascending. Assume the velocity to be uniform.*

Solutions. Given:

Dia of ram, $D = 200$ mm

$\therefore$ Area $A = \pi(100)^2$

Weight $W = 20$ tonnes $= 20{,}000 \times g$

Frictional resistance $= 2$ tonnes $= 2000 \times g$

(*a*) *When ram is descending*

Gross load $= (20{,}000 - 2000) \times g = 18000\, g$

$\therefore$ $$\text{Pressure of water } p = \frac{\text{Gross load on ram}}{\text{Area A}} = \frac{18{,}000 \times g}{\pi \times (100)^2}$$

$$= \frac{18000 \times 9.8}{3.14 \times 10000} = \mathbf{5.61\ N/mm^2\quad Ans.}$$

(*b*) *When ram is ascending*

$$\text{Gross load} = (20{,}000 + 2000) \times g = 22000\,g$$

$$\therefore \quad \text{Pressure of water, } p = \frac{22{,}000 \times 9.81}{\pi \times (10)^2} = 687 \text{ N/m}^2 \quad \textbf{Ans.}$$

Example. 16.12 *Find the length of stroke required for an accumulator, if volume displaced by the plunger/sroke is 0.1 m³ litres and the dia of plunger is 250 mm.*

Solution. Given:

Dia. of plunger = 250 mm

Displacement of plunger/stroke = 0.1 m^3

As, displacement of accumulator is the volume displaced by the plunger/stroke.

∴ Displacement of accumulator = 0.1 m^3

Cross-sectional area of the plunger

$$= \frac{\pi}{4}(250)^2 = 49000 \text{ mm}^3 = 49 \times 10^{-6} \text{ m}^3$$

$$\therefore \quad \text{Length of stroke} = \frac{\text{Displacement of accumulator}}{\text{Area of plunger}}$$

$$= \frac{0.1 \times 10^6}{49}$$

$$= \textbf{2.04 m. Ans.}$$

Example. 16.13 *An accumulator having a ram of 200 mm dia has an effective stroke of 3 m and is loaded with a total weight of 40 tonnes. If the friction may be taken as 5% of the total load, find the total power delivered to the hydraulic machine. Ram falls steadily through its full stroke in 3 minutes, while during this period pump is delivering 500 litres/min.*

Solution. Given: Dia. D = 200 cm, Stroke = 3 m

Weight W = 40 tonnes = 40,000 kg = 40,000 × g-N

Friction = 5% of the total load = 0.05 × 40

= 2 tonnes = 2000 × g-N

Time of ram fall = 3 min

Pump discharge, Q_{pump} = 500 litres/min

Power supplied to the machine = ?

Machine gets work from two sources, firstly by a pump which is working continuously and secondly by the accumulator, the ram of which is falling steadily.

(*i*) Work supplied by the accumulator ram per minute

= Effective load × Distance moved/min.

$$= g\,(40{,}000 - 2000) \times \frac{3}{3}$$

$$= \mathbf{38000 \times g \text{ Nm}}$$

(*ii*) *Work supplied by pump*

$$\text{Pressure of water} \qquad p = \frac{\text{Effective load}}{\text{Area}} = \frac{38000 \times g}{\frac{\pi}{4}(200)^2} = 1.2 \text{ kg N/mm}^2$$

$\therefore$ Head due to this pressure, $H = \dfrac{p \times g}{w} = \dfrac{1.21 \times 10^6 \times g}{1000} = 1210$ m of water.

Hence, work supplied by the pump per min.

$$= gw.Q.H = g \times 500 \times 1210 \text{ Nm.} = 605{,}000\, g$$

$\therefore$ Total work supplied to the machine

$$= (38{,}000 + 605{,}000)\, g$$

$$= g \times 643{,}000 \text{ Nm/min}$$

$\therefore$ Power supplied to the machine $= \dfrac{643{,}000 \times g}{60}$ Nm/sec = **106 kW. Ans.**

Example. 16.14 *In an installation of 4 hydraulic cranes, the working cycle of each of which takes 2 minutes in hoisting and lowering. Each crane is supplied with water at the rate of 0.5 gN/mm² and is required to lift a load of 4 tonnes at a speed of 20 m/min through a height of 10 m. Jigger is giving a velocity ratio 4. Estimate the stroke and dia. of the ram, assuming its efficiency as 80%. Also calculate the minimum capacity of the pump and that of the accumulator, if all the 4 cranes perform working strokes at the same time.*

Solution. Given:

No. of cranes = 4, Time/cycle = 2 min.

Pressure of water p = 0.5 gN/mm², W = 4 tonnes = 4000 kg

Speed, v = 20 m/min, Height, h = 10 m

Velocity ratio = 4

η = 80%

As, Velocity ratio = Mechanical advantage

$$= \frac{\text{Load on each ram}}{\text{Load lifted}} \times \eta$$

$\therefore$
$$4 = \frac{\text{Load on each ram}}{4000} \times 0.8$$

or
$$\text{Load on each ram} = \frac{4000 \times 4}{0.8} = 20{,}000 \text{ kg}$$

$$\text{Area of ram} = \frac{\text{Load on ram}}{\text{Pressure, p}} = \frac{20{,}000}{0.50} = 40{,}000 \text{ mm}^2$$

or
$$\frac{\pi}{4} D^2 = 40{,}000$$

$\therefore$ Ram dia, D = **226 mm. Ans.**

$$\text{Stroke of ram} = \frac{\text{Height } h}{\text{Velocity ratio}} = \frac{10}{4} = \textbf{2.5 m. Ans.}$$

Volume swept by each ram = Ram area × Stroke

$$= 40000 \times (2.50 \times 1000) = 10^8 \text{ mm}^3$$

$\therefore$ Total volume supplied by pump to all the 4 cranes

$$= 10^8 \times 4 = \mathbf{4 \times 10^8 \text{ mm}^3}$$

∴ Minimum capacity of the pump

$$= \text{Amount of water supplied by the pump/min.}$$

$$= \frac{4 \times 10^8}{2} = 2 \times 10^8 \text{ mm}^3 = 0.2 \text{ m}^3$$

$$= \textbf{200 litres/min. Ans.}$$

$$\text{Operating time of crane} = \frac{\text{Height } h}{\text{speed } v} = \frac{10}{20} = \frac{1}{2} \text{ min.}$$

$$\therefore \quad \text{Idle time} = 2 - \frac{1}{2} = 1\frac{1}{2} \text{ min.}$$

As the pump feeds the accumulator for the whole time, therefore volume of accumulator

$$= 1\frac{1}{2} \times 200 = \textbf{300 litres = 0.3 m}^3 \textbf{ Ans.}$$

$$\text{Capacity of accumulator} = p \times \text{Volume} = 50 \times g \times 0.3 \text{ Nm}$$

$$= \frac{50g \times 0.3}{1000} \text{ Nm} = \textbf{147.15 kNm. Ans.}$$

QUESTIONS

16.1 Describe the working of the following with the help of sketches:
(*a*) Hydraulic Jack
(*b*) Hydraulic Crane
(*c*) Hydraulic Press
(*d*) Hydraulic Lift

16.2 Explain the construction and working of the following:
(*a*) Hydraulic Intensifier.
(*b*) Hydraulic Accumulator.

16.3 Describe a hydraulic ram with the help of a line diagram. Where this should be used? Define the efficiency of a hydraulic ram.

16.4 What is the principle of hydraulic press? Explain the contribution and working of a hydraulic press with the help of a line diagram.

16.5 Differentiate between hydraulic accumulator and hydraulic intensifier with the help of neat sketches.

16.6 Draw a neat diagram of hydraulic jack and explain its working.

16.7 Draw a neat diagram of hydraulic crane and explain its working.

16.8 Draw a neat diagram of hydraulic lift and explain its working.

Chapter

17

HYDRO-ELECTRIC POWER PLANTS

Learning Objectives

After studying this chapter, the students will learn:

- Hydro-electric power plants and their applications.
- Hydro versus steam power plants.
- Classifications of hydro-power plants.
- Selection of site and layout of hydro-power plant.
- Components of hydro-power plant.
- Case load and peak load plants.
- Pumped storage hydro-power plants.

17.1 INTRODUCTION

The word "Hydro" is a Greek word meaning water. The water is being used since long as the source of energy. The first designed water wheels were used in nineteenth century. Later on water wheels were further developed as water turbine. Afterwards, several types of turbines were build up, some of them are: Jonval turbine (in 1837), Girard turbine (in 1850), Francis turbine (in 1865), Kaplan turbine (in 1916).

When water strikes the blades fixed on the periphery of the turbine wheel, wheel starts rotating due to the force of the water. Mechanical energy thus produced is used for running the electric generator, which is coupled to the turbine shaft. The electric power thus produced is known as *'hydro-electric power'* or 'water power.'

Projects designed to produce electric power from water are known as "hydro-electric projects". Sometimes, these projects are made for controlling flood or irrigation purposes in addition to the electric generation, in which case these are known as "Multi-purpose projects". Examples of such projects are Bhakra Nangal, Tehri, Sardar Sarover etc., which is useful for power and irrigation purposes and Damodar Valley Corporation, which is mainly for flood control and also produces power.

For selecting the site for hydroelectric plant a large quantity of water at suitable head should be available. Regarding the quantity of water which could be available for the plant, a survey of past history of rainfall is done. From this it can be known that during past years what was the minimum and maximum rainfall. Maximum capacity of the plant should be decided on the basis of minimum quantity of water available and the storage reservoir should be constructed for the maximum quantity of water so that it can store the water even in the peak periods.

Initial cost of the hydroelectric plant is very high but operating costs are too low. Thus if the water is available in large quantity at a suitable head and near the load centre, hydroelectric plants are very advantageous. Now-a-days hydroelectric plants are being given much importance because of economical operation and no pollution.

The energy of water utilised for hydropower generation may be kinetic or potential. The kinetic energy of water is its energy in motion and is a function of mass and velocity, while the potential energy is a function of the difference in level of water between the two points, called the head. For the purpose of continuous availability of water, it is collected in natural lakes and reservoirs at high altitude may be utilised or water may be artificially stored by constructing dams across the flowing streams.

Electrical Energy produced can be written as,

$$P = g\rho QH\eta \text{ in watts}$$
$$= g\rho QH\eta \,.\, t \text{ in Wh}$$

where,

g = 9.81 m/sec² (The acceleration due to gravity)

ρ = water density (1000 kg/m³)

Q = flow or discharge, m³/sec

H = height of fall of water or head, m

η = efficiency

t = operating time in hours.

Therefore,

$$P = 9.81 \times 1000 \times Q \times H \times \eta \times t \text{ wH}$$
$$= 9.81\, QH\eta t \text{ kWh,}$$
$$= 9.81\, QH\eta \text{ kW}$$

Since, 1 metric Horse Power = 735 watts *i.e.*, 0.735 kW.

$$P = \frac{9.81}{0.735}\, QH\eta$$

17.2 APPLICATION OF HYDRO-POWER PLANTS

Hydro electric power plant can be used as independent power plant. But this will require large amount of water to be stored and at the times of low water flows the hydro power plant will not be able to meet the maximum load as otherwise the maximum capacity of the station has to be based on the maximum flow of water and this will not prove to be economical. Therefore, the present trend is to use hydro electrical power with a steam power station in an interconnected system. This will result in reduction in capital cost of hydro electric power plant as the size of reservoir may be reduced. In the interconnected system at times of low water flows the hydro-power plant can be used as peak load plant and base load should be supplied by steam power plant whereas during periods of high water flows the steam plants and hydro plants reverse their roles, the hydro power plant taking the base load and steam power plant supplying the peak load.

Following are the major applications of hydro plants (multi-purpose hydro power projects):

1. Generation of electric power
2. Storage of irrigation water.
3. To control the floods in the rivers.
4. Storage of drinking water supply.
5. Navigation.
6. Fish and wildlife development
7. Recreation etc.

When two or more of the above uses of water are combined together while designing a multipurpose project, increased benefits without a proportionate increase in the costs may be obtained, thus enhancing justification of the project.

17.3 ADVANTAGES AND DISADVANTAGES OF HYDEL POWER PLANTS

Advantages

The major factor which go in favour of hydro-electric power plant are as follows:

1. Water is the cheapest and reliable source of generation of electric power because it exists as a free gift of nature.
2. There are no ash disposal problems. Also the atmosphere is not polluted because no smoke is produced in this plant.
3. No fuel transportation problem.
4. It can take up the loads quickly and it is capable of meeting the variable loads without any loss in efficiency.
5. Its maintenance cost is low.
6. It requires less supervising staff.
7. Auxiliaries needed in the plant are less as compared to steam plant of equal size.
8. Running cost of the plant is low.
9. In addition to the power generation such plants are used for irrigation and flood control purposes also.
10. Hydro electric power plants have become economic competitors with steam power plants and will acquire more economic advantage with the rise in price of coal and oil.
11. The life of the plant is more and the effect of age is comparatively small on the overall efficiency of the plant.
12. These can be started and stopped very quickly.
13. Unforeseen outages are less frequent.
14. Plant has no standby losses.
15. As no fuel is required, no expenditure on material, material handling, storage and disposal of waste.

Disadvantages

The various disadvantages are as follows:

1. The power produced by the plant depends upon quantity of water which in turn is dependent upon the rainfall, so if the rainfall is in time and proper and the required amount of water can be collected, the plant will function satisfactorily otherwise not.
2. Hydro-electric plants are generally situated away from the load centres. They require long transmission lines to deliver power. Therefore, the cost of transmission lines and losses in them will be more.
3. Initial cost of the plant is high.
4. It takes fairly long time for the erection of such plants.

17.4 HYDRO ELECTRIC POWER PLANT *VERSUS* STEAM POWER PLANT

1. The initial cost (of land, dam, diversion works, water rights, rail, road and generating plant) of a hydro electric power plant is nearly twice (or more) as that of steam power

plant of equal size; but the labour, maintenance and repair costs of hydro-plant are much less than the steam power plant. In hydro-plant the cost of dam and water-ways constitute the major portion of plant costs. Depending upon the conditions the cost of a hydroplant vary from Rs. 5 to 7 Crores.

2. For preliminary estimates the fixed charges for the hydro-power station usually vary from 9 to 12%. The fixed chargeds for a steam power plant are higher and may be assumed from 12 to 14%. This is because of greater proportion of equipment and machinery costs and the depreciation charges are relatively high.
3. It is easy to locate steam power plant near the load centre and this eliminates the need for long transmission lines and thus transmission cost is less whereas power generated by a hydro plant is to be transmitted over longer distance and hence at greater cost.
4. The time to put the most steam power stations into operation is nearly 5 to 8 hours whereas a hydro electric power plant can be put into operation in very short time ranging from few seconds to 3 or 5 minutes. Thus where power is to be supplied both by hydroelectric power plant and steam power plant, the steam power plant should be used as base load plant and hydro-power plant should be used as peak load plant.
5. The efficiency of steam power plant drops as it gets older.
6. Steam power station can be operated as desired to suit the demand of the load system whereas power to be generated from hydro-plant is dependent on the quantity of water available in the storage.
7. The cost of transportation of fuel is quite high in case of steam power plant.
8. As compared to steam power plant the hydro-electric generation is the cleanest and most simple.

17.5 CLASSIFICATIONS OF HYDRO-POWER PLANTS

Hydro-power plants of large variety are in use. These are classified in following ways:

1. Classification Based on Purpose

Hydro power projects may be simple single purpose project for power production or multi purpose projects. Multi-purpose projects are for different functions, these may be water supply, irrigation, flood control, navigation etc. in addition to power production.

2. Classification Based on Hydraulic Features

These are based on basic hydraulic principles involved in the design of project. These can be classified as

(*i*) *Conventional hydro-plants.* These plants utilise available hydraulic energy of the flowing water of rivers.

(*ii*) *The pumped storage plants.* These utilise the same water by recycling using pumping.

(*iii*) *Tidal power plants.* These utilise tidal energy of sea waters.

3. Classification on the Basis of Operation

Hydro power plants can also be classified on the basis of actual operation in meeting the demand. In order to feed the power into the grid power station can be either (*i*) base load plant, or (*ii*) peak load plant.

In base load plants, a base load which is more or less constant is supplied. Its operating efficiency is usually high. The plants which meet the fluctuating peak loads are called peak load plants. Generally, hydro power is most suitable for peak load durations due to its quick starting and relatively easy in picking up the load.

4. Classification Based on Location and Topography

Hydropower plants can either be located in hilly areas or in plains.

5. Classification Based on Plant Capacity

(*i*) Micro hydel plants.......upto 100 kW.
(*ii*) Mini hydel plants.......From 100 upto 2000 kW.
(*iii*) Small hydel plants.......From 2000 upto 25000 kW.
(*iv*) Medium Capacity hydel plants.......15 MW to 100 MW.
(*v*) High Capacity hydel plants.......100 MW 1000 MW.
(*vi*) Super hydel plants.......above 1000 MW.

6. Classification Based on Constructional features

Hydel plants can also be classified based on the dominant constructional features of the plant. These may be:

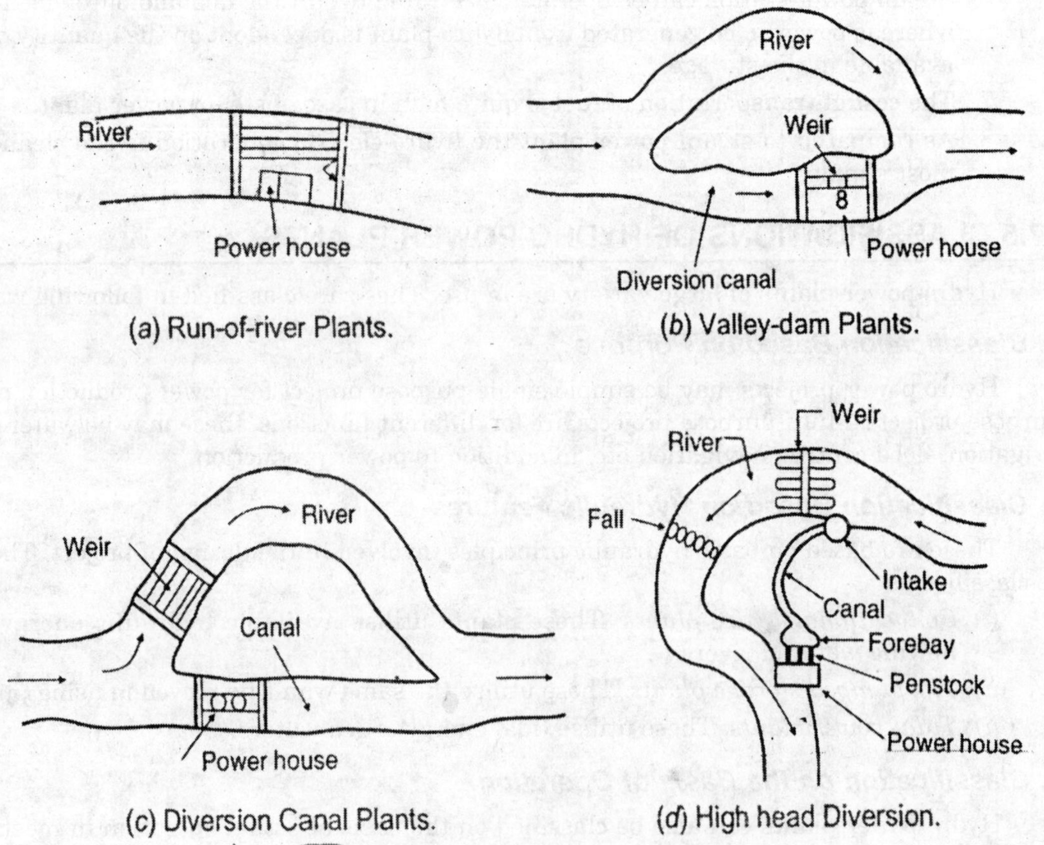

Fig. 17.1 Types of Hydro Power Plants.

(i) Run of River Plants. In these, the power house is located along the main course of the river. These are also of two types, namely without pondage, and with pondage. In run-of-river plants without pondage, the water is not stored and the plant makes use of running water. Generating capacity of these plants depends on the rate of flow of water. During rainy season, full quantity of water can not be utilised for power generation and a considerable quantity goes as waste. Due to this drawback this type is not common. Run-of-river plants with pondage are provided with a barrage to accommodate sufficient storage to take care of load fluctuations during the day [Fig. 17.1(a)].

(ii) Valley Dam (Storage) Plant. In these plants, a dam is constructed to create a storage reservoir, and power house is located at the toe of the dam. These are used for medium to high head plants with adequate storage. In this, water is stored during the period of rains and used when the flow in the river is less. Thus these plants are more reliable and are less affected by seasonal change of flow rates in the river [Fig. 17.1(b)].

(iii) Diversion Canal Plants. In these plants the water is diverted from the river through a canal to the power-house, located away from the diversion point. The water from the power-house is again brought back into the original river at a downstream point [Fig. 17.1(c)].

7. Based on Availability of Head of Water

According to the head of water available, the hydro-electric power plants can be classified as follows:

(i) Low Head Plant: Where operating head is less than 30 metres.

(ii) Medium Head Plant: When the operating head of water is between 30 to 80 metres.

(iii) High Head Plant: When the head of water exceeds 80 metres.

Pondage

The low head plants, that need to have a balancing reservoir to cater for short term fluctuations, are provided with a pond near the power house. In run-of-river plants, the main weir itself provides on its up stream side of the required pondage. While in diversion canal plants, the pond is required to be created at the end of the canal in the form of a forebay reservoir. Whereas, the storage reservoir provide storage of water for long term requirements, pondage take care of the hour-to-hour fluctuations over a day or occasionally.

(i) High Head Plants

In such plants water is usually stored in lakes on high mountains. This water is stored during the rainy season or during summer when ice melts-and this water is utilized throughout the year. Rate of flow should be kept in such a way so that water stored is sufficient for the whole year. Generally plants having head more than 200 m are known as high head plants. From these, lakes water is taken to smaller reservoirs (known as forebays) through tunnels. From forebays water is taken to turbines through penstocks.

(ii) Low Head plants

These plants generally have dam across a river. Such a plant layout is shown in Fig. 17.2. Let a river is having level at *A* as higher than the level at *B,* then a canal is constructed from A to B. Dam at *A* is built to divert the water in the canal. Sluice gates are provided to control the quantity of water in the canal. Near the end *B,* a power house is made from where water is discharged (from turbines) back into the river at *B*. Thus head available for the turbine is the difference between the levels at A and *B*. Generally the plants having head between 3 to 50 m are know as low head plants.

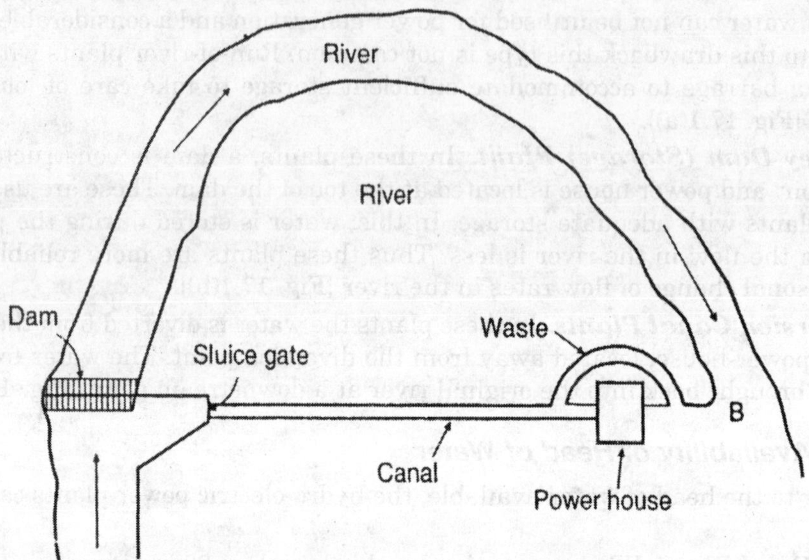

Fig. 17.2 Layout of a Low-Head Plant.

(iii) Medium Head Plants

If the head is more than 30 m then penstock is required for taking the water from reservoir to the turbine. The plants having head between 30 to 200 m are generally known as medium head plants.

The values of head in above cases overlaps *e.g.* for head between 30 to 50 m plants can be known as low head as well as medium head. Therefore, it is difficult to classify them only on the basis of heads and therefore, basis for specific speed (explained later) is also considered.

Here, head is the difference in elevation between two levels of water. In hydro-electric plants, it entirely depends on the topography. Head can be categorised as:

(*a*) Gross *Head.* It is the difference in elevations between the head race level at the intake and the tail race level at the discharge side. The gross head varies, as both the elevations of water do not remain the same at all the times.

(*b*) *Net or Effective Head.* It is the head obtained by subtracting from gross head, all the losses in carrying water from the head race to the entrance of the turbine. It is therefore, the pressure difference between the entrance to the turbine casing and the tail race water elevation.

17.6 SELECTION OF SITE

The ideal site will be one where the dam will have the largest catchment area to store water at high head and will be economical in construction.

The various factors to be considered while selecting the site for hydro-electric power plant are as follows;

1. ***Availability of Water.*** Geological, geographical and meteorological conditions of the site should be studied thoroughly. Daily, weekly and monthly flow water over a period of years should be recorded. Estimate should be made about the average quantity of water available throughout the year and also about maximum and minimum quantity of water available during the year. These factors are necessary to decide the capacity of the hydro-electric plant, setting up of peak load plant such as steam, diesel or gas turbine plant and to provide adequate spillways or gate-relief during the flood period.
2. ***Storage of Water.*** Water used in the hydro electric power plant is mainly dependent on the rain and since the rainfall is not regular throughout the year, therefore, it is essential to store water to afford a uniform output. Thus a storage reservoir is constructed at the site. During rainy season the excess water is stored in the reservoir and it is released to supplement low rates of flow during run-off periods to maintain the output. The reservoir should have large catchment area so that water in it should never fall below the minimum level.
3. ***Head of Water.*** Water in large quantities and at a sufficient head should be available. For a given power output an increase in effective head reduces the quantity of water required to be passed through the turbines.
4. ***Distance From Load Centre.*** It is desirable that the power plant should be set up near the load centres, so that costs of erection of the transmission lines and their maintenance are low. However, in hyclro-electric power plants it may not be possible because this plant will be located where sufficient quantity and head of water is available.
5. ***Accessibility to the Site.*** The site selected should have rail and road transportation facilities.
6. ***The Land Availability.*** The ideal site will be one where the dam will have the largest catchment area to store water at a high head and will be economical in construction. After the location has been chosen the exact position of the various structures are fixed by considering the following factors:
   (*a*) Details of foundation conditions
   (*b*) Requirements of head, flow demands of storage capacity
   (*c*) Arrangement and type of following:
      (*i*) Dam
      (*ii*) Intake system
      (*iii*) Conduits
      (*iv*) Surge tanks
      (*v*) Power house
   (*d*) Cost of dam and project
   (*e*) Transport facilities and accessibility of site.

17.7 SURVEYS NEEDED FOR HYDEL PLANTS

The following surveys are carried out for a hydro plant installation.

17.7.1 General Surveys

They constitute the following: .

1. **Topographical Surveys such** as
   (*i*) The dam site topographical survey plans including the area to accommodate dam, spillway, outlet works, diversion works etc.

(*ii*) The reservoir submergence plans

(*iii*) Contour plan of

(*a*) surge tanks (*b*) tunnels

(*c*) penstock (*d*) tail race channels.

(*iv*) River surveys

(*v*) For a barrage structure detailed survey maps to cover the area under the barrage.

2. **Geological and Foundation Investigations.** A study should be carried out for the following:

(*i*) Seismic condition of the region (*ii*) Water flow in reservoir

(*iii*) Sub-surface conditions (*iv*) Foundation investigations for dams.

3. **Meteorological and Hydrological Studies**

The following data should be collected:

(*i*) Precipitation, evaporation.

(*ii*) Maximum and minimum water flows.

(*iii*) Rainfall-run off relation data

(*iv*) Location of Rain gauge stations near the plant site.

4. **Construction Material Investigations**

Data should be collected for the following:

(*i*) Materials to be used for concrete and masonry dams, earth and rock fill dams.

(*ii*) Tests to be carried out for materials for their suitability.

5. **Transportation**

Data should be collected for road, water and rail routes.

6. **Environmental considerations**

Studies be conducted for impact assessment on social, forest and wild life, fish culture etc.

17.7.2. Special Surveys

Special surveys include

1. **Load Surveys.** These are carried out for assessing power station capacity, details of major loads to be served, peak load, load factor, future energy demand, inter connection with other power system.

2. **Layout Studies**

These include the following:

(*i*) Type of power plant such as storage type, run of river plant, pumped storage type etc.

(*ii*) Structural components, like

(*a*) Dams (*b*) Canals (*c*) Tunnels

(*d*) Penstock (*e*) Draft tube (*f*) Surge tanks

(*g*) Power house (*h*) Switch yard

17.8 GENERAL ARRANGEMENT (LAYOUT) OF HYDRO-POWER PLANT

As mentioned earlier, water flow during peak rainy season is stored in the reservoir and it is carried to the prime-mover (turbine) through a pipe known as penstock. The potential energy of the water is converted into mechanical energy as it passes through the water turbine. The mechanical energy of the water turbine is then used to generate electrical power. This electrical power is then transmitted through transformer and transmitting towers to the consumers.

The general arrangemet of a hydro-electric power plant is shown in Fig. 17.3 (*a*) and (*b*). These figures are showing main components of Hydro-power plants.

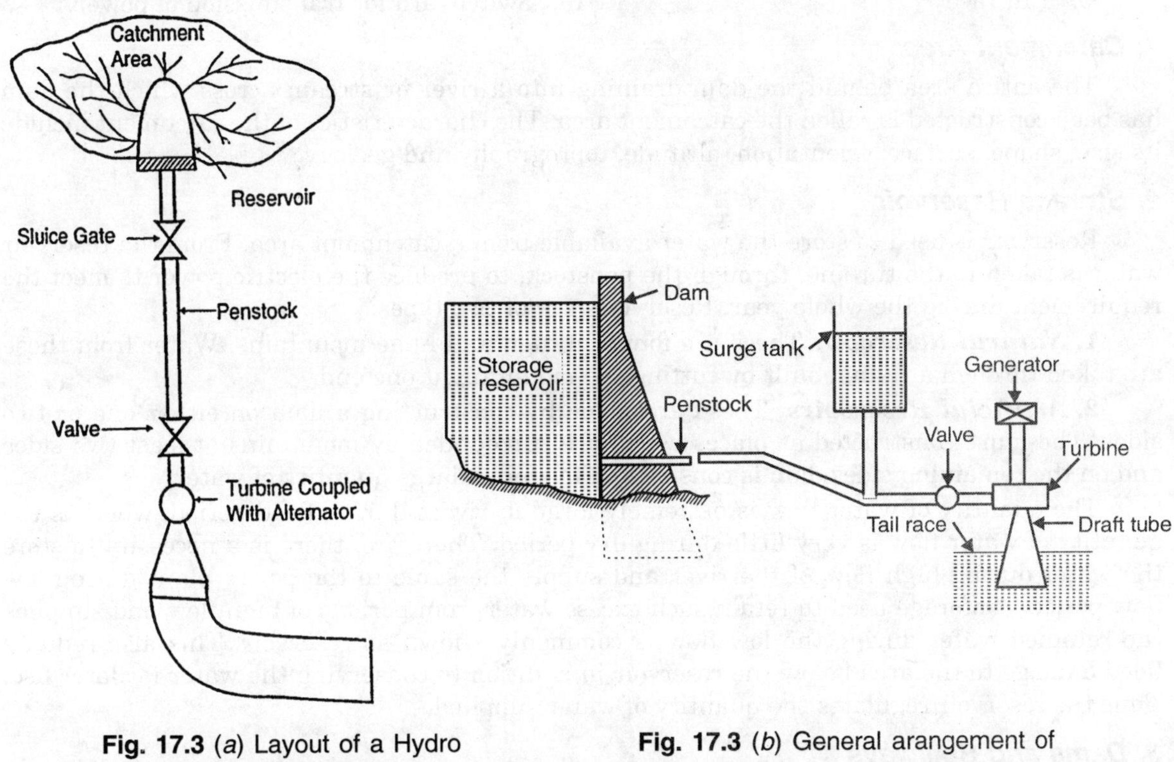

Fig. 17.3 (*a*) Layout of a Hydro electric plant (plan).

Fig. 17.3 (*b*) General arangement of Hydro-Electric plants (Elevation)

17.9 COMPONENTS OF HYDRO-ELECTRIC PLANTS

No two hydro-power projects are exactly alike and each will have its own unique problems of design and construction. The choice of a particular type of a plant at a given site depends upon various factors, such as topography of the area, available head, available flow, political influence in the area.

A hydro-electric development scheme ordinarily includes a diversion structure, head works and a conduit (penstock) to carry water to the turbines, and governing mechanism, generators, control and switching apparatus, housing for equipment, transformers, and transmission lines. In addition to these major components, trash racks at the entrance of the penstock, canal and penstock gates, a forebay, a surge tank etc. may also be required. A tailrace channel from the powerhouse back to the river must be provided, if the draft tubes cannot discharge water directly into the river.

The main components of hydropower plants are:

1. Catchment area
2. Storage Reservoir
3. Dams, and Spillway
4. Water ways : Canal, Forebay, Intake, Trash Rack, Gates
5. Penstock
6. Surge tank
7. Draft tubes
8. Powerhouse : Prime-movers, Generators etc.
9. Tail race
10. Switchyard for transmission of power.

1. Catchment Area

The entire area behind the dam draining into a river or stream across which the dam has been constructed is called the catchment area. The characteristics of the catchment include its size, shape, surface, orientation, altitude, toprography and geology.

2. Storage Reservoir

Reservoir is used to store the water available from a catchment area. From the reservoir water is taken to the turbine, through the penstock, to produce the electric power to meet the requirement during the whole year. Reservoirs are of two types:

1. *Natural Reservoir.* These are found in high and huge mountains. Water from these are taken through a tunnel built by cutting the mountain at one end.

2. *Artificial Reservoirs.* These are made by constructing a dam on either one or two sides. These are constructed at places which are surrounded by mountains at least two sides and on the remaining sides dam is constructed to store a large quantity of water.

The quantity of water flow is extremely large in rivers during flood period, whereas the quantity of water flow is very little during dry period. Therefore, there is a necessity to store the water during high flow of the river and supply the same to the power plant during low flow period. A storage used to retain such excess water from periods of high flow and supplies the retained water during the low flow is commonly known as reservoir. This also reduces flood damage to the area below the reservoir in addition to conserving the water for later use. Thus the reservoir regulates the quantity of water supplied.

3. Dams and Spillways

(A) Dams.

It is thick wall constructed at suitable places to store a large quantity of water and to create head. Dam may be built to make a artificial reservoir from a valley or to control the flowing water of a river, which may be for flood control or irrigation purposes. Dam may be either earthen dam, concrete dam or masonry dam. Concrete dams are more common.

A dam across the river is an important component, in most of the high head or medium head hydro-electric projects. Cost of this single component may be as much as 20 to 25 per cent of the total cost of the project.

The dam fulfils two fundamental functions, namely:

(*i*) It works as a reservoir which has a capacity to store water, and

(*ii*) It builds up head.

Classifications of Dams

(*a*) Based on Functions

(*i*) *Storage Dams.* To store water and use as and when required for various uses like power, irrigation, and water supply.

(ii) Diversion Dams. To raise water level and to divert river flow in another direction.

(iii) Detention Dams. To store flood waters.

(*b*) Based on Construction Materials. Earth, rock pieces and fragments, stone masonry, concrete or R.C.C. However concrete (plain or R.C.C.), earthen and rockfill dams are more popular.

Concrete dams, Earthfill dams and Rockfill dams are commonly employed depending upon the availability of material in the nearby areas. Earthfill dams, rockfill dams and their combination are becoming more common due to development in soil mechanics and heavy earth moving machinery. An earth and rockfill dam has high resistance to earthquakes because of its flexible character which permits considerable movements.

(*c*) Classification Based on Structural Design. These may be gravity dam, arch dam, or buttress dam.

(B) Spillways.

Some provision is made in the design of dam to allow the water to discharge in the downstream during the flood period. The provision so made is known as "spillway". Thus the part of the dam which discharges the flood flow to the downstream side is called as "spillway".

The spillway must have the capacity to discharge major flood without damage to the dam and other appurtenant structures and at the same time it must maintain some predetermined maximum level of water in the dam. In order to derive full benefit of the storing capacity of the reservoir, it is necessary that the spillway does not start discharging till the water reaches a predetermined level, called as Full Reservoir Level (FRL), Maximum discharging capacity of the spillway has to be such that the overflow over the spillway is at least equal to the instantaneous inflow. This ensures that the reservoir level will never cross. High flood level (HFL).

The spillway is an important part of the dam complex and is located either as a part of the main dam or separately at a suitable place near the dam. The different types of spillways used with different types of dams are:

1. Overflow spillways
2. Chute spillways
3. Side Channel spill ways
4. Shaft spillways
5. Siphon spillways

4. Waterways

A waterway is the passage through which water is carried to the turbine from the dam. Waterway may be canal, tunnel or penstock etc.

Various appliances used to control the flow of water are called *head works* or control works. Head works include booms, gates, valve and trash rack etc. Booms divert ice and floating logs from the intake. The gates are used not only at head works but also in penstocks at the turbine inlets. They also control discharge through sluices and conduits for flood–flow by-passes. *Trash rack* made up of steel bars is provided for preventing the debris from getting entry into the intakes from the dam. Manual or mechanical cleaning is used to remove the debris from trash rack.

(*a*) Canal.

The cross-section of the canal varies with the material through which the canal is constructed. In rock the section is approximately rectangular, while in the earth it would be trapezoidal. Economical section is one which requires minimum expenditure considering cost of construction per unit length, loss of head, maintenance etc. Canal should be located on an economically shortest route and lining may be done by stone paving or by concrete so as to (*a*) to reduce loss of head and (*b*) to prevent scour of banks and bottom.

Intakes. An intake structure is provided at the entrance of the canals or tunnels or pipes through which water is conveyed to the power house. Following are the functions of intakes:

(*i*) Control of the flow of water into the conveyance system. A gate or a valve is used for this purpose.

(*ii*) Smooth and turbulence free entry of water into the conveyance system. This is generally achieved by bell mouth entry.

(*iii*) Stopping entry of boulders, logs of wood etc. into the conveyance passage. Trash racks or steel grids are used for this purpose.

(*iv*) Not allowing heavy sediment load into the conveyance passage.

(*b*) Tunnel.

Tunnel is the passage by cutting the mountain to reduce the length of waterway between the reservoir and the power house. The tunnel may be circular or horse-shoe shaped constructed in the solid rock. In earth or unstable rock it may be lined with stone paving, reinforced concrete or steel depending upon the nature of rock, earth and topography.

5. Penstock

It is a pipe used to carry the water from reservoir to the turbine house. It should be laid in such a way that it follows the shortest route. While laying down the penstock it should be noted that, it should not rise above the hydraulic gradient at any point. Penstocks are made either of welded or rivetted steel plates or of reinforced cement concrete. Wooden or cast iron penstocks are also used for low heads and power houses.

When the distances between the forebay and the power house are short, a separate penstock for each turbine is preferred, while for long distances, a single penstock can be used to feed two or more turbines. The principles for the penstock design are the same as for pressure vessels and tanks, however consideration for water hammer pressures, caused due to governor control and turbine gate operation, is also taken into account during design.

Water Hammer

In long water conveyance systems, like penstock pipes, transient flow condition changes due to the governor operation of the turbine. In penstock pipes, governor operation of the turbine results in an altered value of discharge, this sends a high pressure wave which sweeps the penstock and may cause damage to the pipe. The design of the pipe or channel has to be safe against such transient phenomena. The most common extreme cases are:

(*i*) switching on the turbine directly to full load from a state of rest, and

(*ii*) sudden throwing off of the entire load by the turbine.

A gate or valve at the end of the penstock pipe controls the discharge to the turbine. Due to change in gate opening by the governor, the pipe flow has to be adjusted to the new magnitude of flow. This results in rapid pressure oscillations in the pipe, generally accompanied by a hammering like sound, known as 'water hammer'.

The pipe supplying the water must (*i*) withstand the high pressues caused by sudden closing of turbine (known as positive water hammer) due to sudden reduction of load on turbine; and (*ii*) there should not be any vacuum in the pipe line when the gate opens suddenly and turbine needs more water due to increased demand on generator, and therefore, water has to rush through the pipe (*i.e.* penstock).

Penstock Accessories

These are mainly safety devices for penstocks. Important penstock accessories are discussed here under.

(i) Air Vent. An air vent should always be installed just below the gates at the upper end of the penstock. It permits air to enter when the head gates are shut and water drawn out of the penstock.

(ii) Air Valves. These are provided at the sections of the penstocks where there are steep changes in the gradient or where the penstock line begins to pitch suddenly to turbine house. In cases of sudden load, the water is drawn out of the steep lower portion of the penstock, faster than the flow in flatter part at the upper portion, thus tending to cause a collapsing pressure on the pipe. (Refer Fig. 17.3).

(iii) Surge Tank. It is located on the penstock near the power house for the purpose of speed and pressure regulation. It protects the penstock against water hammer. These are very necessary when the length of penstock is more.

(iv) Forebay. These are enlarged body of water provided just infront of penstock. These are provided when head is low and length of penstock is short. Forebay provides a small balancing storage on the upstream of the powerhouse to store temporarily the water rejected by the turbine when the load is suddenly reduced and to meet the sudden demand of increased load, while the water in the penstock is being accelerated.

6. Surge Tanks

These are used when the water is carried through penstock. It is a reservoir fitted on the penstock near the turbine to receive the rejected water when the pipe line is suddenly closed by the governing arrangement. These help in reduction of water hammer in the pipe line.

Functions of Surge Tanks

The main functions of surge tanks are:

1. It reduces the distance between the free water level of the reservoir and the turbine and also reduces the intensity of water hammer. Thus, only relatively short length of conduit (*i.e* , penstock) below the surge tank need be designed to withstand the water hammer effects.
2. The surge tank acts as a relief valve when the load on the turbine is reduced and the pressure in the pipe suddenly increases by diverting the main conduit flow partly or wholly into this tank. The water level in the tank rises until it exceeds the level in the main reservoir. Thus retarding the main conduit flow and absorbing the surplus kinetic energy.
3. The tank acts as a temporary reservoir during increase of load on the turbine when it provides sufficient water to enable the turbine to pick up the increased load quickly and safely until the water level in the surge tank falls below its original level. Sufficient head is thus available to accelerate the flow in the penstock until it is sufficient to meet the new demand.

Types of Surge Tanks

Surge tanks may be classified based on:

(*a*) *The material of construction.* Concrete or steel etc.

(*b*) *Location with respect to ground.* Underground or overground.

(c) *Location in the hydraulic system.* Whether it is on the upstream side or downstream side of the power house.

(d) *Shape.* This is most common way of classification:

(i) *Simple Cylindrical Surge Tank.* These are cylindrical in shape and are provided between the penstocks and the pressure tunnels. These are not so common these days.

(ii) *Restricted Orifice Surge Tank.* This is a modification of simple cylindrical surge tank. In such surge tank with restricted orifice tanks, an orifice of considerably smaller diameter than the surge tank diameter is installed between the surge tank and the conduit system as shown in Fig. 17.4.

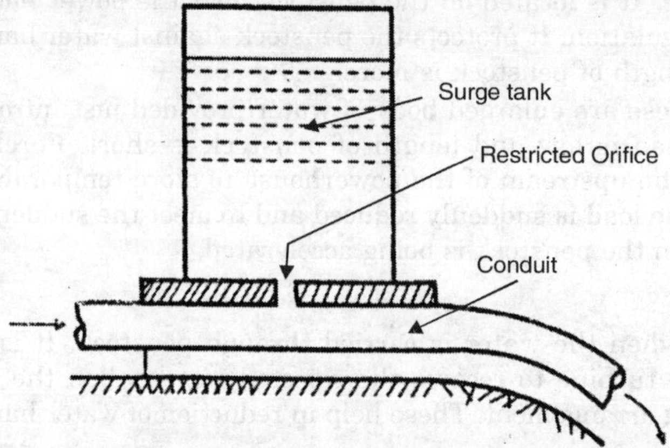

Fig. 17.4 Surge tank with restricted orifice.

When the load on the turbine drops, the rejected mass of water enters the surge tank through the restricted orifice and builds up quickly, a retarding head in the conduit.

(iii) *Surge Tank of Special Design.* Surge tanks of special designs are also in use. Some of these are shown in Fig. 17.5.

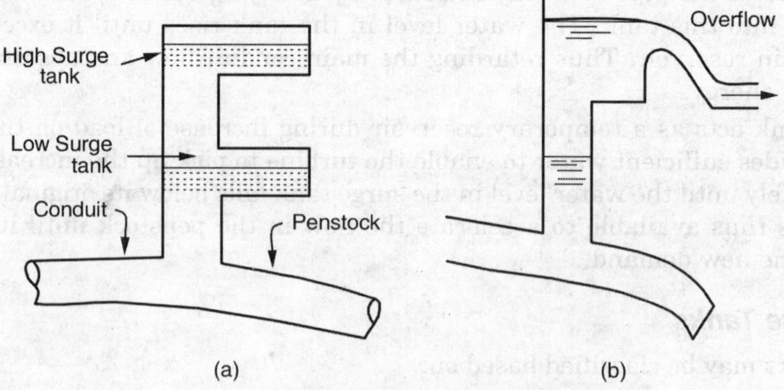

Fig. 17.5 Surge tanks of special design.

(*iv*) ***Differential Surge Tanks.*** These are much more efficient than simple surge tanks. This has an additional internal rise pipe provided with annular ports connected with the chamber as shown in Fig. 17.6. These annular ports open into the outer surge shaft, this interlinking the riser pipe and the surge shaft. When the load demand is higher, water falls in the internal riser quickly, thus, creating a large accelerating head on the conduit in a short time. The water falls slowly in the outer surge tanks being of larger area and water flows through the ports into the riser pipe and *vice-versa*.

In differential surge tanks, the head building function is achieved by the riser pipe, while the storage function is achieved through the outer shaft.

Surge tanks are dealt in detail in the chapter "Governing of turbines".

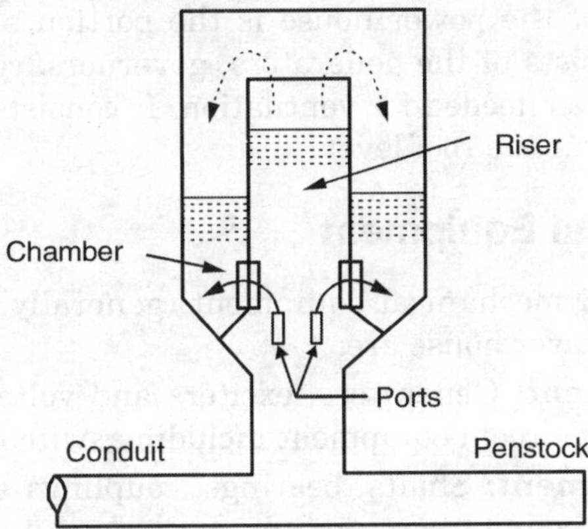

Fig. 17.6 Differential surge tanks.

7. Draft Tubes

Draft tubes have been discussed in detail in the chapter on "Hydraulic Turbines".

8. Power House

Power House consists of the main building of. hydro electric development where the conversion of energy of water to electrical energy takes place. Some important items of equipment provided in the power house are as follows:

(*i*) Turbines alongwith entrance (spiral casing) and exit draft tube.
(*ii*) Generators
(*iii*) Governors
(*iv*) Relief valve for penstock fittings
(*v*) Gate valves.
(*vi*) Transformers
(*vii*) Busbars
(*viii*) Oil circuit breakers
(*ix*) Storage batteries
(*x*) Outgoing connections
(*xi*) Cranes
(*xii*) Workshop and office.

Thus, power house is the main area of a hydro power plant and is designed to provide proper housing of electro-mechanical equipment.

The structural complex, where all the essential equipment needed are housed is known as "power house." The power houses are located either in a building above ground, called *surface power house* or in caverns excavated below the ground, called as *underground power house.*

The structures of surface power houses are broadly divided into three subdivisions, namely;

(*a*) Substructure (*b*) Intermediate structure

(*c*) Super-structure.

Sub-structure of the power house is that part of the structure which is situated below the turbine level, and includes Draft Tube, tail water channel, drainage galleries etc. It transmits the load of the structure above it to the foundation strata.

The **intermediate structure** extends from the top of the draft tube to the top of generator foundation. The turbine including its casing, the galleries for the auxiliary, machines and the governor servo-motor system are housed in this part.

The **super-structure** of the power house is the portion extending from the generator floor to the roof top and consists of the generators, governors, control room, the exciters and the auxiliary equipment like as needed for ventilation. It consists of walls and the roof with a main travelling gantry crane at the roof level.

Electrical and Mechanical Equipment

The main electrical and mechanical equipment (generally called as electro-mechanical equipment) provided in the power house are:

(*a*) **Electrical Equipment:** Generators, exciters and voltage regulators, transformers, switchgear and control room equipment including switch boards.

(*b*) **Mechanical Equipment:** Shafts, bearings, couplings etc. for the generator, the-oil circuits and pumps, compressors and air ducts, braking equipment, ventilation and air cooling system, cranes and other lifting equipment. Besides these, auxiliary equipment for the power house lighting and water supply and drainage *etc* are also provided.

9. Tail Race

It is a place where water from the turbine is disposed. In pumped storage plants (explained in next article) a storage in the tail race is also provided, to pump the water back into the head race (original reservoir).

10. Transmission of Electric Power

The transmission system delivers bulk power from the power stations to the load centres. As efficiency of transmission is quite high when the power is transmitted at high voltage, the power from power house is transmitted at a high voltage. Main components of overhead transmission lines are: (*i*) poles or towers, (*ii*) insulators, (*iii*) cross arms and clamps to support the insulators, (*iv*) conductors.

17.10 BASE LOAD AND PEAK LOAD PLANTS

As load required by the consumer does not remain constant with respect to time (hour, day or month) and fluctuates according to his requirements. The load curve indicates the load demand of consumer and shows how it varies with respect to time. This is useful in predicting the annual requirements of energy and capacity of the power plant required to take the peak load.

Different types of consumers *viz* industrial, commercial, domestic, agricultural, urban, municipal (street lights, drinking water pumping) have different type of activities and each have different load curve from the other. The load curve also varies from season to season. The demand load of a power plant is generally found out by adding all the loads mentioned above. Further, the changes in atmospheric temperature influence the electric load. Readers may refer the 'load studies' in the chapter on "Hydro Power Plant Economics".

Higher the peak load, the greater will be capacity of power generation if the total load of the system is taken by a single power plant. A plant designed to take base load as well as peak load will run at part load conditions during off peak hours and costs more. Because power plants give low efficiency of generation when they run at part load conditions. Therefore, the total load is generally taken by the base load and peak load plants. The base load plant is always allowed to run at full load condition to ensure maximum efficiency of generation and surity of return of the capital invested providing the cheapest energy to the public. The peak load plant may work at full load or part load as per the demand but only for few hours in a day (generally 3 to 5 hours.) during peak load demand and remain idle for the Remaining period of the day,

From the above it is clear that the cost of peak load generation will always be higher as the capital invested in peak load plants remain idle for 80% of the period. However, a part of the peak load demand can be nullified by diversifying the consumers demands, for example agriculture load and industrial load can be diverted to non-peak hours. The peak load demand can further be reduced by interconnecting the supply lines of different types of *power* plants (hydel, thermal, nuclear and gas turbines) and providing a common grid. Irrespective of all the measures, still peak load demands exist.

Thus base-load plants are those plants which are required to supply a constant power when connected to the grid and they run continuously without stop. Thermal and nuclear plants are run as base load plants. Run-off-river plants without pondage may also sometimes work as base bad plants.

Peak-load plants are those plants which supply the power during peak loads. Hydro power plants with pondage, gas-turbines and pumped storage plants used to run as peak-load plants. Peak-load plants have to fulfil the following main requirements:

(*i*) The response of the peak-load plant to load variations should be as quick as possible.

(*ii*) The capital cost of such plant should be as low as possible because they work only for about 20% of the time.

(*iii*) The voltage and frequency of generation must remain constant during the fluctuations of load.

17.11 PUMPED STORAGE HYDRO-POWER PLANT (PSP)

Pumped storage plant (Fig. 17.7) in combination with hydro electric power plant is used for supplying the sudden peak load of short duration. The water leaving the turbines of hydro electric power plant is stored in tail—race pond. This water is pumped back to the head race reservoir by means of reversible pump turbine sets and is used for power generation at the peak load time. Pump storage plants are generally inter connected with other plants such as steam power plants. The off peak capacity of steam power station can be used for pumping water in the head reservoir.

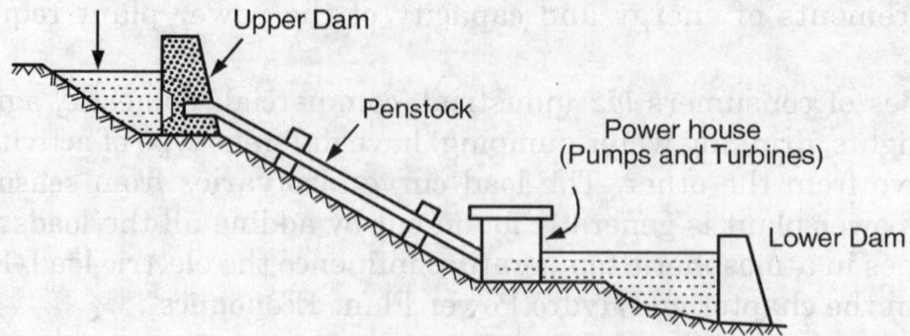

Fig. 17.7 Pump storage plant.

A 400 MW pumped stored plant is at design stage at Kadamparai in Tamil Nadu. This will be built with the existing upper Aliyer Reservoir at the lower pool, the higher pool being constructed at Kadamparai river. There will be four 100 MW reversible units in an underground power house.

Advantages of pumped storage plants are as follows:

(*i*) The capital cost of pumped storage plant is low as compared to other peaking units,

(*ii*) There is a great deal of flexibility in the operational schedules of the system.

(*iii*) A pumped storage plant can pick up load rapidly and is dependable.

(*iv*) They operate at higher load factors and improve the overall efficiency of the system.

They are used as peak load plants.

Fig. 17.8 shows the pumped storage power plant for peak load in conjunction with steam power plant as base load plant.

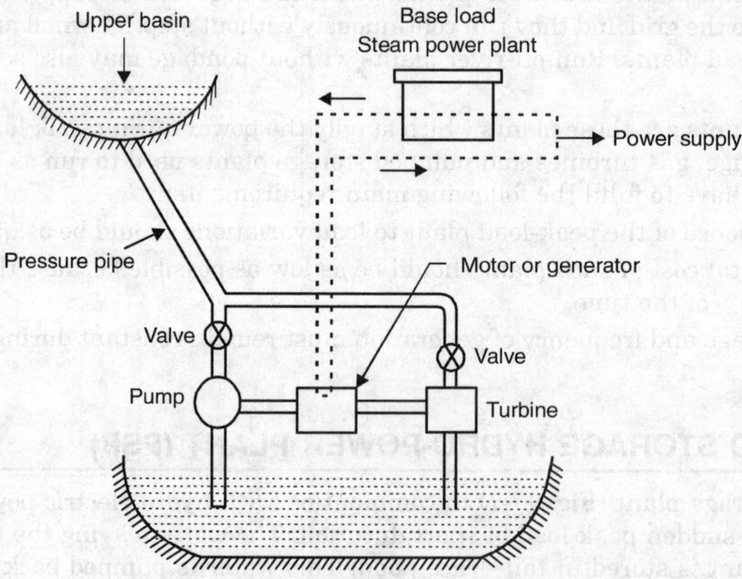

Fig. 17.8 Pumped storage power plant for peak load inconjunction with steam power plant as base load plant.

QUESTIONS

17.1 What are the different factors to be considered while selecting the site for hydro-electric power plant.

17.2 Discuss different surveys and investigations to be carried out while selecting the site for hydro-power plant.

17.3 Enlist the advantages and disadvantages of water power.

17.4 Explain different methods of classifying a hydropower plant.

17.5 Describe the working of a pumped storage plant, and describe its importance as a peak load plant.

17.6 Discuss various advantages of pump storage plant.

17.7 What are the multi-purpose hydel project?

17.8 What is the difference between run-off river plants and storage plants?

17.9 What is the difference between base load and peak load power plants?

17.10 Discuss various advantages of storage reservoir type power plants.

17.11 Give the names of major components of hydro-power plants. Explain any three of them.

17.12 Write short notes on:

(*a*) Catchment area (*b*) Storage Reservoir

(*c*) Penstock (*d*) Trashracks.

17.13 Explain different types of storage reservoirs.

17.14 How a surge tank helps in reducing the water hammer effect? Describe different types of surge tanks.

17.15 Describe a layout of a hydro-power plant.

17.16 Describe different types of dams used for hydro-power plants.

17.17 Explain the following in brief:

(*a*) Head works (*b*) Tail race

(*c*) Water ways (*d*) Draft tube

17.18 Write short notes on:

(*a*) Head race (*b*) Spillways

17.19 Enlist the various equipment provided in a power house.

Chapter 18

HYDRAULIC TURBINES

Learning Objectives

After studying this chapter, the students will learn:

- Classification of hydraulic turbines.
- Impulse turbines ie pelton wheel : construction and working.
- Reaction turbines : types, construction and working of francis turbine propeller turbine and kaplan turbine.
- Draft tubes, its functions and types.
- Carritation in turbines and its prevention.
- Governing of various turbines their construction and working.
- Turbine performance.
- Characteristic curves and selection of hydraulic turbines.

18.1 INTRODUCTION

Water turbines convert kinetic and potential energies of the water into mechanical power. Kinetic energy of the water depends upon the mass and the head of water. While the potential energy also depends on the quantity of water and the available head. This energy is available at natural sources, which can be **classified** as follows:

(*a*) Large quantity of water available at high heads.

(*b*) Small quantity of water available at high heads.

(*c*) Large quantity of water available at low heads.

(*d*) Small quantity of water available at low heads.

(a) Large quantity of water at high heads

In this category, hydraulic energy of natural sources is utilised by the Pelton turbine. As the large quantity of water is available, more machines with more nozzles per wheel can be provided to take the maximum energy.

(b) Small quantity of water at high heads

In such natural sources also, hydraulic energy can be converted into mechanical energy by Pelton wheel. As the quantity of water available is less, only a limited number of Pelton turbines are installed.

Such natural sources, where water is available at high heads, are available only at the top of mountains and hence this energy is available in J&K, Himachal Pradesh, North East Uttarakhand. etc. or the places of natural falls such as Jog Falls in Mysore.

(c) Large quantity of water at low heads

For low heads Francis turbines and Kaplan turbines are used. Where head available is less than 200 m, Pelton wheels are not much suitable. Because of fixed speed ratio at lesser heads, speed of the Pelton wheel will be small. If the Pelton wheel having less speed is connected with the alternator, then number of poles of alternator $\left(P = \dfrac{120f}{N}\right)$ will be more and this increases the size of the alternator and hence cost. Therefore, for low heads Pelton wheels are unsuitable.

For Francis and Kaplan turbines speed ratios are more and hence their speeds will be high even for low heads and hence can be coupled to the alternator economically and efficiently.

(d) Small quantity of water at low heads

In this case, energy available is very less. Therefore, such sources of energy are not utilised for hydro-electric powers.

18.2 CLASSIFICATION OF TURBINES

Water turbines can be classified according to:

1. The action of water on moving blades

On this basis water turbines are classified into following two categories:

(*a*) *Impulse turbine.* Whole of the pressure energy in this turbine is first converted into kinetic energy. The wheel revolves in open air at atmospheric pressure when water strikes on the series of buckets, *i.e.* in these there is no difference of pressure at the inlet and outlet of the runner. As these turbines convert the velocity energy into mechanical energy, these are known as Impulse or velocity turbines.

(*b*) *Reaction turbine.* These operates on the basis of difference of pressure at the inlet and outlet of the turbine, hence these are known as reaction or pressure turbines. In this case entering water has pressure as well as kinetic energy and when it moves over the blades both the kinetic energy and pressure energy provides turning moment of the wheel.

2. Name of the originator

Turbines can also be classified on the basis of names of their inventor or originator, *e.g.* Pelton wheel, Turgo, Girard, Jonval, Banki impulse turbines and similarly Francis, Kaplan and Thomson reaction turbines.

3. Head and quantity of water available

(*a*) High head (above 200 m) and small quantity of water requires impulse turbine such as Pelton wheel.

(*b*) Medium head (between 30 to 200 m) and medium quantity of water requires reaction turbines such as Francis turbines.

(*b*) Low head (below 30 m) and large quantity of water requires reaction turbines such as Kaplan or Propeller turbines.

4. Disposition of shaft

Turbines may have either horizontal or vertical shafts. Generally Pelton wheels have horizontal and Francis and Kaplan turbines have vertical shafts.

5. Direction of flow of water in the runner

(*a*) In radial flow turbines water flows in radial direction. The flow may be inward or outward.

(b) In tangential flow turbines, water strikes the runner in the direction of tangent on the depth of rotation of the runner. Pelton wheel is an example of tangential flow turbine.

(c) In axial flow turbines water enters and leaves the runner in a direction parallel to the axis of the turbine shaft.

(d) In mixed flow turbines water enters radially and leaves the runner axially.

6. Specific speed

Specific speed of a turbine is the speed of a geometrically similar turbines which would develop one kW under a head of 1 m. Specific speed $N_S = \dfrac{\sqrt{P}}{h^{1.25}}$ multiply by 0.2626 to convert to english units for power in HP and head in feet. On this basis some of the important turbines have following specific speeds:

(a) Pelton wheel has specific speeds varying from 10 to 25 for single jet and 25 to 50 for double jet.

(b) Francis turbine has specific speeds varying from 50 to 300.

(c) Kaplan turbine has specific speeds varying from 350–850.

18.3 IMPULSE TURBINE—PELTON WHEEL

Among different types of impulse turbines, Pelton wheel is the only turbine being used at present. It was discovered in 1880 by an American Engineer Lester A. Pelton. It operates under very high heads (upto 1800 m) and it requires comparatively lesser quantity of water.

From the head race in the mountains water is conveyed to the turbines installed in the power house through the penstocks. The lower end of the penstock is joined with a nozzle in the turbine casing. Water is delivered by the nozzle at a high velocity on the buckets. These buckets are mounted on the periphery of a circular wheel (also known as runner) which is generally mounted on a horizontal shaft. The quantity of water coming out of the nozzle or nozzles can be controlled by regulators (governing arrangement) in case of big installation and by hand wheels in the case of small installations (see Fig. 18.1).

The impact of water on the buckets causes the runner to rotate, thus develops mechanical energy. After doing work on the buckets water is discharged in the tail race. Being impulse turbine it must run at atmospheric pressure and, therefore, these are located above the tail race. The buckets are so shaped that water enters tangentially in the middle and discharges backward and flows again tangentially in both the direction to avoid thrust on the wheel (as shown in the line sketch). Actually the jet is deflected by 160°. To produce electric energy these are coupled with the electric generators.

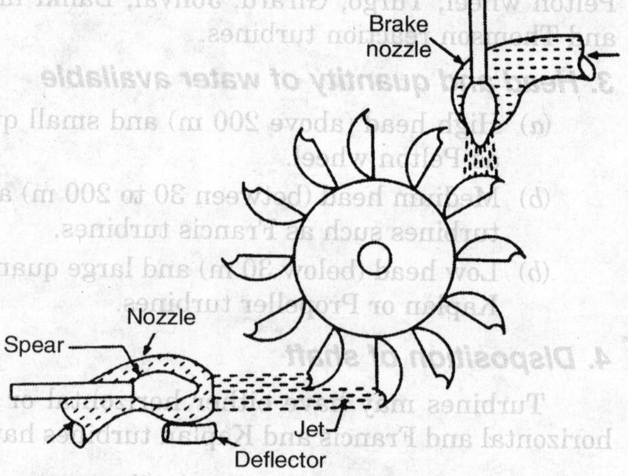

Fig. 18.1 Diagram to indicate Pelton wheel working.

18.3.1 Main Components of Pelton Wheel

1. *Nozzle.* As explained earlier the nozzle is provided at the end of penstock. The pressure energy of water in the reservoir while passing through the penstock is partly converted into kinetic energy and this kinetic energy is further increased by nozzle at atmospheric pressure in the casing. When the water strikes the buckets fitted on the periphery of runner mechanical energy is produced. For a small capacity turbine, single jet is sufficient and for the turbines producing large power, number of jets are to be provided. The nozzles are provided with a spear (*i.e. a* needle), which is connected with governing mechanism to regulate the quantity of water coming out from the nozzles, according to varying demand of power. A multi-nozzle arrangement is shown in Fig. 18.2.

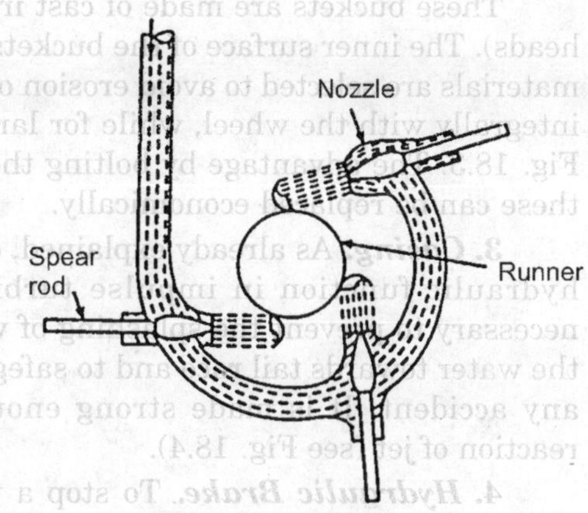

Fig. 18.2 Diagram to explain function of nozzles.

2. *Buckets.* The buckets of a Pelton wheel has a shape of a double hemispherical cup. Each bucket is divided by a common dividing straight edge, upon which the jet of water impinges as shown in Fig. 18.3. The incoming jet of water after impinging at the central common edge divides itself into two parts and after sliding over the inner surface of the buckets turns through 160° to 170° and leaves at the outer edge of the bucket. The two hemispherical cups give an advantage that the axial forces in both the cups neutralize each other and avoid the axial thrust.

As explained earlier, the shape of the bucket should be such that it turns the jet through 180°, so that full reaction of the out-going jet can be utilized. But practically it is not possible, because if the jet is deflected through 180°, it will strike the back of the bucket which is just following it, thus exerting retarding force. In actual practice the outlet tip angle of the bucket is kept between 10° to 20°, so that the outgoing jet of the previous bucket does not strike the following bucket.

To provide strength for the common straight edge, the inlet bucket angle is kept between 3° to 5° instead of 0°, but for all calculation purposes the inlet tip angle of the bucket is taken as 0°.

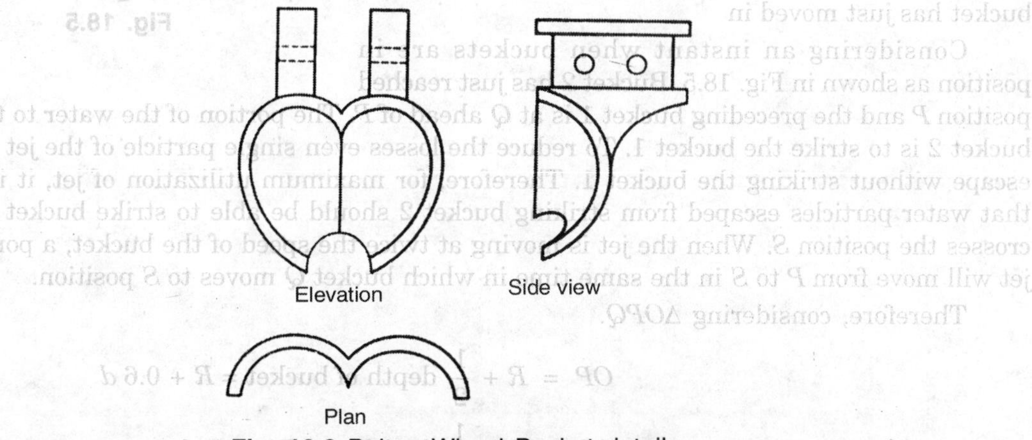

Fig. 18.3 Pelton Wheel Bucket details.

These buckets are made of cast iron (for low heads), cast steel or stainless steel (for high heads). The inner surface of the buckets is polished to minimise the frictional resistance. These materials are selected to avoid erosion of buckets. Buckets of small turbines are generally casted integrally with the wheel, while for large turbines these are bolted to the runner as shown in Fig. 18.3. The advantage by bolting the buckets is that, if the buckets wear or damaged then these can be replaced economically.

3. *Casing.* As already explained, casing has no hydraulic function in impulse turbines but are necessary to prevent the splashing of water, to lead the water towards tail race and to safeguard against any accident. It is made strong enough to resist reaction of jet (see Fig. 18.4).

4. *Hydraulic Brake.* To stop a turbine, even if the jet is closed, runner will go on revolving for a long time. Therefore, to stop the rotation quickly small brake nozzles are provided, which issue the water in the direction opposite to that rotation of the runner. When this brake nozzle is opened, it opposes the movement of the bucket by striking on the back of the bucket as shown in Fig. 18.4.

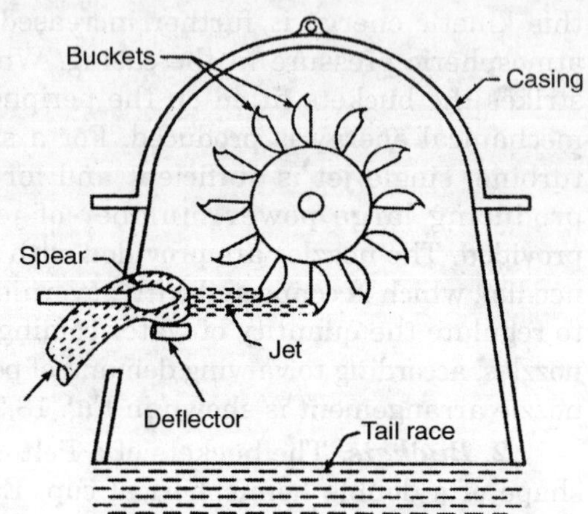

Fig. 18.4 Pelton wheel details.

18.3.2 Number of Buckets for Pelton Wheel

Let, Z = number of buckets required.

R = mean radius of bucket circle.

γ = angle between two adjacent buckets.

d = dia of jet.

Depth of bucket = 1.2 d (empirical relation).

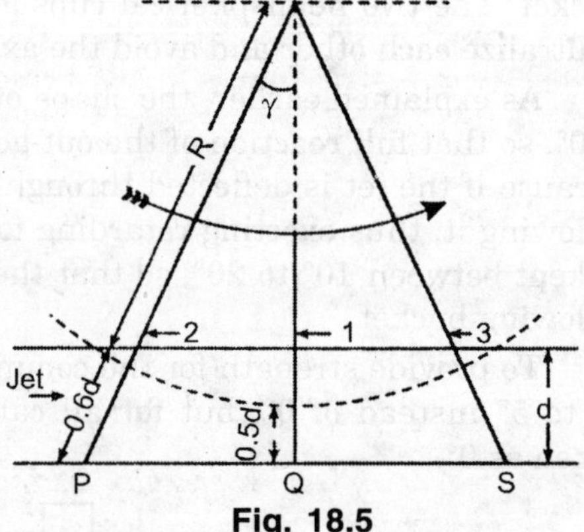

Fig. 18.5

Number of buckets can be obtained by arranging the buckets in such a way that the jet is always intercepted by the buckets. Therefore, one bucket will be just about to move out of the jet as the following bucket has just moved in

Considering an instant when buckets are in position as shown in Fig. 18.5. Bucket 2 has just reached position P and the preceding bucket 1 is at Q ahead of P. The portion of the water to the right of bucket 2 is to strike the bucket 1. To reduce the losses even single particle of the jet should not escape without striking the bucket 1. Therefore, for maximum utilization of jet, it is essential that water particles escaped from striking bucket 2 should be able to strike bucket 1 before it crosses the position S. When the jet is moving at twice the speed of the bucket, a portion of the jet will move from P to S in the same time in which bucket Q moves to S position.

Therefore, considering ΔOPQ.

$$OP = R + \frac{1}{2} \text{ depth of bucket} = R + 0.6\,d$$

$$OQ = R + \frac{1}{2} \text{ dia of jet} = R + 0.5\,d$$

$$\therefore \qquad \cos \gamma = \frac{R + 0.5\,d}{R + 0.6\,d} \qquad \text{...(1)}$$

From equation (1) angle γ can be obtained and then number of buckets,

$$Z = \frac{360}{\gamma} \qquad \text{...(2)}$$

In practice, Eq. (2) does not hold good because the number of buckets obtained by this equation is such a large which cannot be inserted on the wheel periphery. Therefore, actual number of buckets will be less than calculated by Eq. (2) (although it reduces efficiency).

Empirical Relation for Number of Buckets

Approximate number of buckets is generally found by using the following empirical relations:

(*a*)
$$Z = 0.5\frac{D}{d} + 15 \qquad \text{...(1)}$$

$$\left(\text{where } \frac{D}{d} \text{ is the jet ratio}\right)$$

(*b*)
$$Z = 5.4\sqrt{\frac{D}{d}} \qquad \text{...(2)}$$

18.3.3 Working Proportions

1. Peripheral velocity of wheel,

$$u = 0.44\sqrt{2gH} \text{ to } 0.46\sqrt{2gH}.$$

The ratio $\dfrac{u}{\sqrt{2gH}}$ is known as speed ratio

$$= 0.44 \text{ to } 0.46.$$

2. Outlet tip angle of the bucket,

$$\phi = 10° \text{ to } 20°$$

3. The ratio D/d is known as jet ratio and its value lies between 11 to 18.
4. Width of bucket = $3d$ to $5d$.
5. Depth of bucket = $1.2\,d$.
6. Number of buckets,

$$Z = 0.5\left(\frac{D}{d}\right) + 15.$$

18.4 REACTION TURBINES

In reaction turbines water enters the wheel under pressure and flows over the curved blades. When the water flows over the curved blades, the pressure head is transformed into velocity head. Water leaving the blade has a large relative velocity but small absolute velocity. Thus practically whole of the initial energy of water is given to the runner.

Let H = Total head of the entering water.

V_1 = Abs. velocity of the leaving water at exit.

Then, energy given to the wheel = $H - \frac{V_1^2}{2g}$.

In these turbines water leaves the runner at atmospheric pressure. The difference of pressure between the entrance and exit points of the runner is known as reaction pressure. This reaction pressure is responsible for the motion of the runner and hence, these turbines are known as 'Reaction' or 'Pressure' turbines.

In these turbines water comes down from the reservoir to the turbine casing through penstocks . Thus total head H consists partly of pressure head and partly velocity head. Thus water enters the runner (moving vanes) through the stationary vanes, which are fixed around the periphery of runner as shown in Fig. 18.6. As the water flows under pressure, the turbine must run full and hence may be entirely submerged below the tail race or it may discharge through a closed draft tube submerged in the tail race.

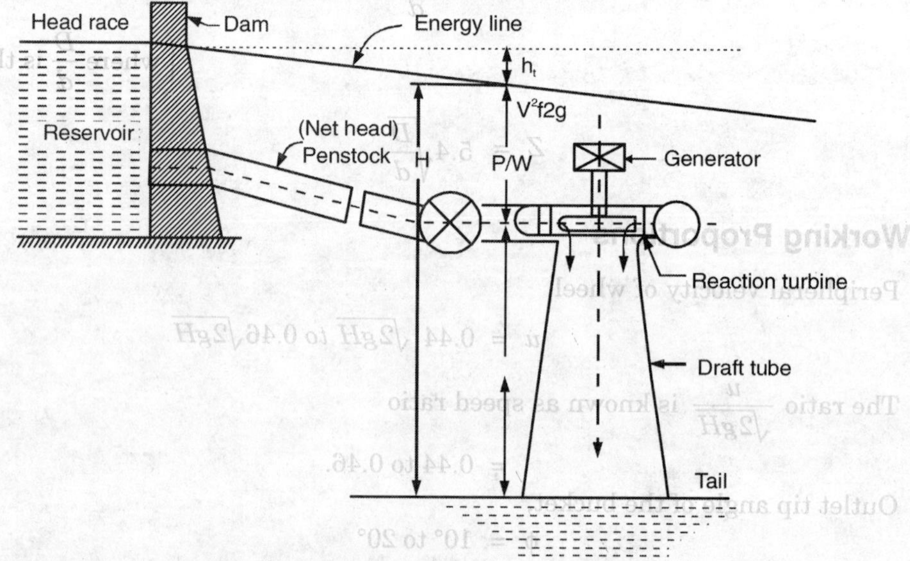

Fig. 18.6 Reaction Turbine Installation.

Classification

Reaction turbines can be classified as under:

1. Radial Flow Turbines

   (*a*) Inward flow Turbine. (*b*) Outward flow Turbine.

2. Axial Flow Turbine.

1. Radial Flow Turbines

(*a*) *Inward Flow Turbine.* In this type of reaction turbine, water enters the runner at the outer periphery and flows radially towards the centre and leaves the runner axially, *i.e.* in a direction parallel to the axis as shown in the Fig. 18.7.

The guide blades surround the runner externally and both are surrounded by an outer scroll casing. The advantage of this type of reaction turbine is that the centrifugal head produced on water by the runner acts against the radial flow of water. Thus any increase in speed of the runner due to the reduction in load, will try to reduce the flow of water from the runner, which results in reduction of power developed and thus makes governing easy.

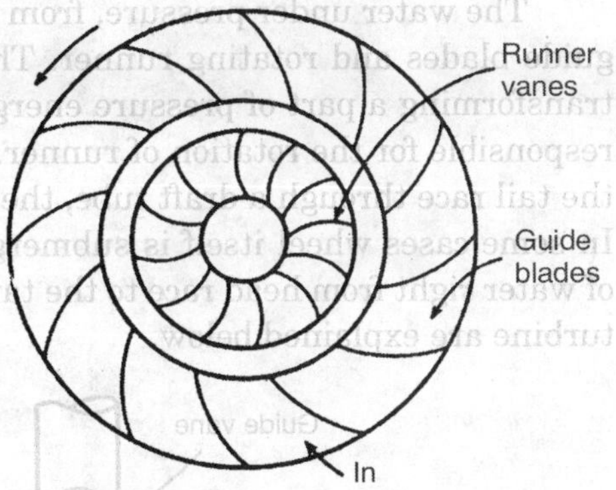

Fig. 18.7 Radial flow turbine

The maximum efficiency is produced when the discharge through the runner at exit is radial, as in this case velocity of whirl at outlet zero and the absolute velocity (V_1) at exit is as small as possible. Francis turbine is an example of modern type of such turbines, which will be explained in this chapter.

(b) Outward Flow Turbine. This is just reverse of the inward flow turbine. In this the water enters the turbine from centre through fixed guide blades into the runner blades as shown in Fig. 18.8.

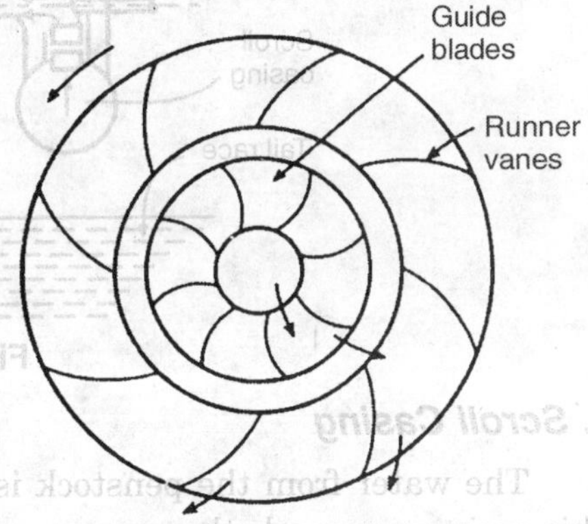

Fig. 18.8 Outward flow turbine

The purpose of guide blades is to direct the water to the moving vanes at an angle α. The water thus passes through the runner and discharges at the outer circumference. The disadvantage of this type of turbine is that the runner produces a centrifugal head on the water, which increases the relative velocity and consequently increases the quantity of water flowing through the runner. Because of this reason, even if there is a slight increase in speed, the centrifugal head, further increases its speed, which causes the runner to race, thus it complicates governing problem. Therefore, this type of turbine is unsuitable for modern installations and is obsolete.

2. Axial Flow Turbine

As shown in Fig. 18.9 in this type of turbine, water flows over the runner blades axially and leaves parallel to the axis of the shaft. These turbines are vertical type and also known as parallel flow turbines. Kaplan turbine is a modern type of axial flow turbine, which will be explained in this chapter.

18.5 FRANCIS TURBINE

Fig. 18.10 shows the outline plan and elevation of a modern Francis Turbine, which is a mixed flow inward reaction turbine. In these turbines water under pressure enters the runner through the guide blades radially in inward direction and leaves axially *i.e.* parallel to the axis of shaft. This turbine, as already explained, is a medium head (30 to 200 m) turbine and requires medium quantity of water.

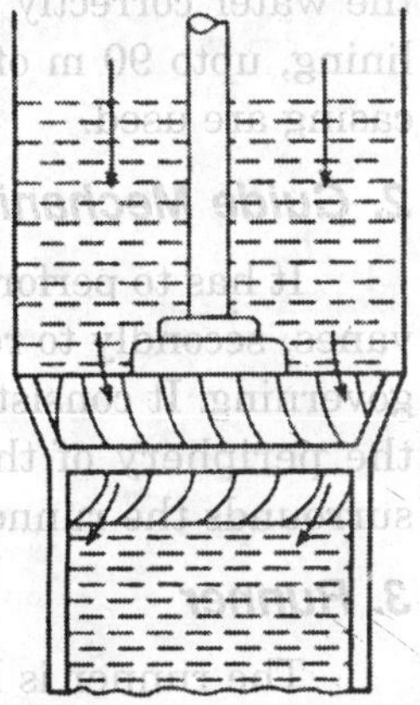
Fig. 18.9 Axial flow turbine

The water under pressure, from the penstock enters the casing which surrounds the fixed guide blades and rotating runner. These turbines generally use scroll casing, which helps in transforming a part of pressure energy into kinetic energy. Thus this difference of pressure is responsible for the rotation of runner. After imparting hydraulic energy water is discharged to the tail race through a draft tube, the lower end of which is submerged, as shown in the figure. In some cases wheel itself is submerged in the tail race. Being a reaction turbine it runs full of water right from head race to the tail race and is totally enclosed. Main components of Francis turbine are explained below.

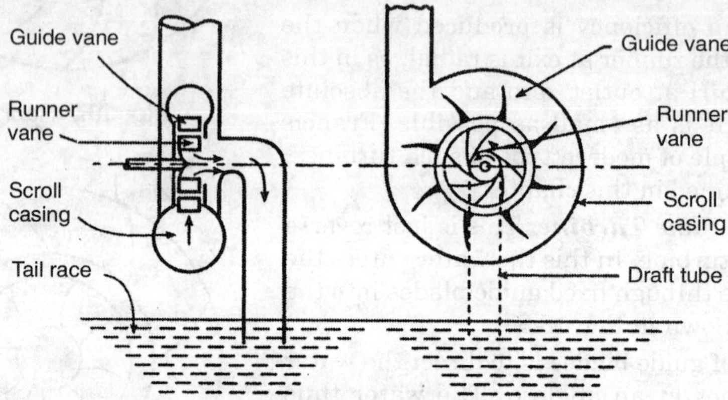

Fig. 18.10 Frana's turbine

1. Scroll Casing

The water from the penstock is supplied to the turbine casing which is spiral in shape. This casing surrounds the runner and guide blades. There should be no eddy formation during the flow from penstock to runner. Water is distributed around the guide blades by this special casing having cross-sectional area reducing gradually around the periphery of the guide blades, as the water passes from the entrance to the end.

In large installations, the inside circumference is fitted with several stay vanes to direct the water correctly over the guide vanes. Upto 30 m of head concrete casing with steel plate lining, upto 90 m of head welded rolled steel plate casing and above 90 m of head cast steel casing are used.

2. Guide Mechanism

It has to perform two functions, firstly to direct the water at correct angle α to the runner vanes, secondly to regulate the quantity of water according to load variations, as explained in governing. It consists of two rings in which several number of guide blades are pivoted around the periphery of these rings. The guide mechanism is held stationary in the casing and surrounds the runner. These guide vanes are generally made of cast steel.

3. Runner

The runner is keyed to the shaft which is coupled to the electric generator. The runners are made of C.I. for small turbines and of cast steel for large turbines. If the water contains chemical impurities then there is a danger of corrosion and in such cases it is made of some non-ferrous metals like bronze. It can also be made up of stainless steel.

In its construction ; it consists of a series of curved vanes uniformly fitted around the circumference in the annular space between the two rings. In Francis turbine the guide mechanism and the runners are concentric in such a way that the water first passes through the guide blades from the outer periphery and enters the runner radially in inward direction and leaves the runner axially from the centre.

4. Draft Tube

It is a tube or passage having gradually increasing cross-sectional area. Its upper end is connected with the exit of runner and lower end is sufficiently submerged in tail race. It helps in increasing the work done by the turbine by reducing pressure at the exit.

Working Principle of Francis Turbine

(*i*) Speed ratio = $\dfrac{u}{\sqrt{2gH}}$. Its value generally varies from 0.6 to 0.9.

(*ii*) Flow ratio = $\dfrac{V_f}{\sqrt{2gH}}$. Its value generally varies from 0.15 to 0.3.

(*iii*) Breadth ratio = $\dfrac{B}{D}$. Its value generally varies from 0.15 to 0.4.

18.6 AXIAL FLOW TURBINES

Following are the two common types of axial flow reaction turbines:

(*i*) Propeller Turbine (*ii*) Kaplan Turbine

18.6.1 Propeller Turbine

Since the power developed is proportional to the product of head and discharge. Hence, if head is less than discharge should be more. For low heads Francis turbine is inadequate to allow more water to pass over the blades because their passages are small. Thus the necessity of allowing more water over the blades can be met by providing purely axial flow runner. This axial flow reaction turbine known as propeller turbine has 4—8 vanes fixed over the boss as shown in Fig. 18.11. The shape of the runner is similar to the propeller of a ship and hence these turbines are known as "Propeller Turbines".

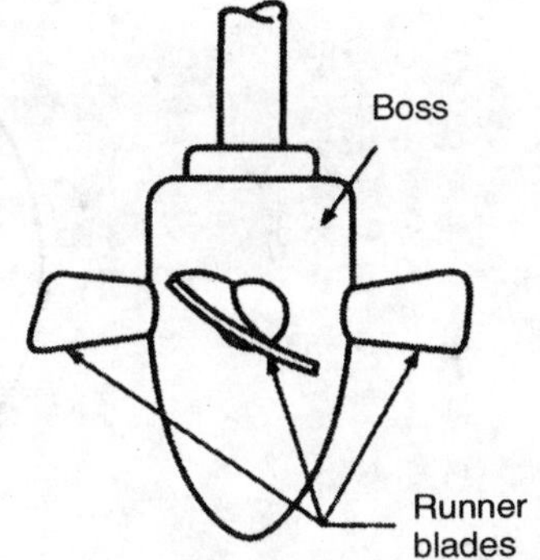

Fig. 18.11 Runner of a Propeller Turbine.

The runner is provided in the scroll casing and water goes to the runner blades through the guide blades. Water enters and leaves the runner axially and hence this turbine is also known as axial flow turbine.

9.6.2 Kaplan Turbine

Mr. V. Kaplan of Germany modified the propeller turbine and designed the turbine runner. The turbine using this type of runner is known as Kaplan turbine. The runner also has two vanes attached to a hub or boss and are mounted in such a way that their angle (θ and ϕ) can be adjusted while the turbine is in motion. These blades are usually adjusted automatically by means of a servo motor governing. With the change of angle of the runner blade at varying load conditions, quantity of water passing through the blades also changes to suit the varying load conditions. This type of turbine thus has a speciality that it gives good efficiency even at different load conditions and water flows over the blades without shock. This turbine is shown in Fig. 18.12.

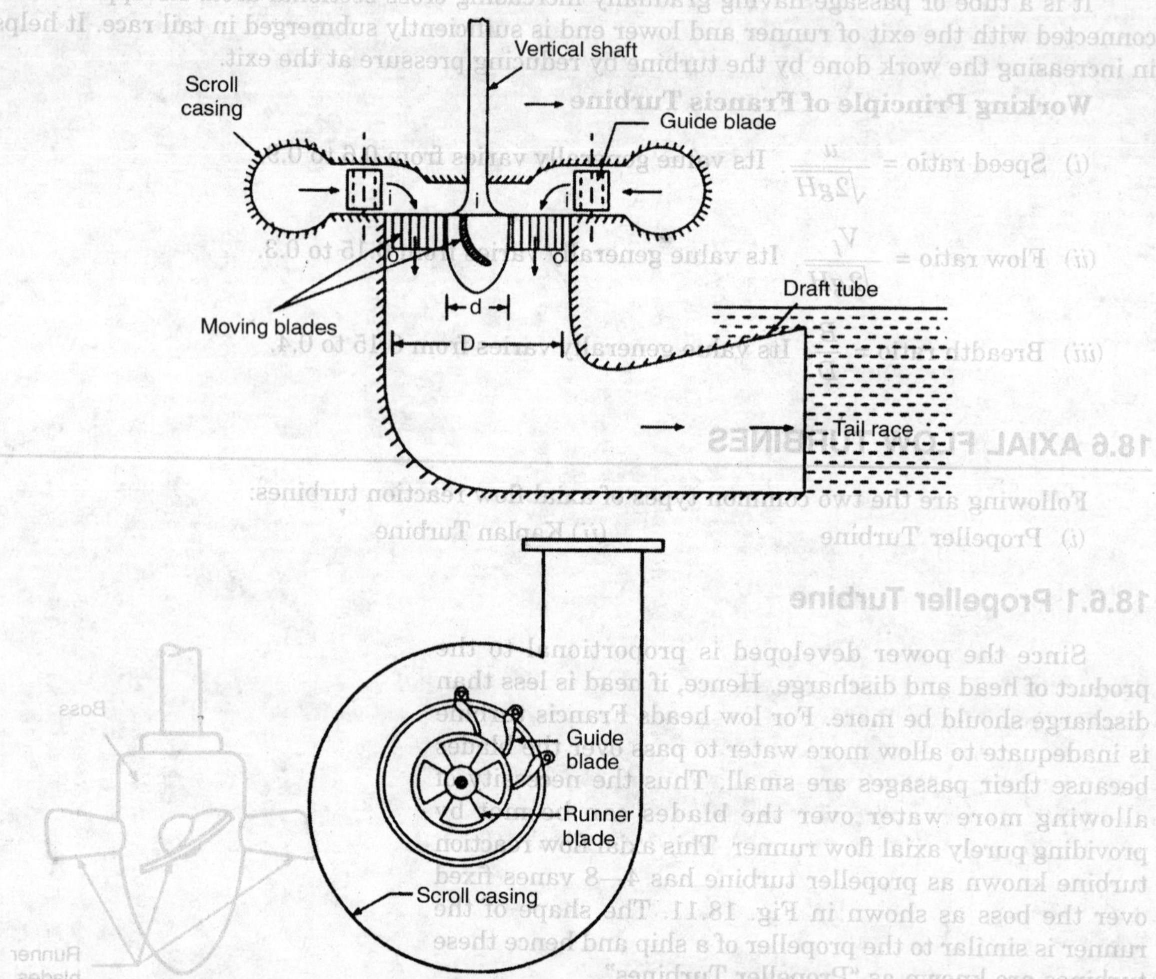

Fig. 18.12 Views of Kaplan Turbine.

In this turbine, similar to that of propeller turbine water passes through the scroll casing *via* guide blades and enters and leaves the runner blades axially. As shown in the figure water discharges into the tail race through draft tube.

This type of turbine is suitable for heads ranging from 3 to 30 m and have speeds from 60-220 R.P.M. These can give an overall efficiency upto 92%.

Main components of Kaplan turbine such as spiral casing, guide mechanism and draft tube are similar to that of Francis turbine.

Discharge Through Kaplan Turbine

The discharge through the runner,

$$Q = \frac{\pi}{4}(D^2 - d^2) \times V_f$$

where D = Dia of boss.

Working Proportions

$$\text{Speed Ratio, } \frac{u}{\sqrt{2gH}} = 2.09$$

$$\text{Flow Ratio, } \frac{V_f}{\sqrt{2gH}} = 0.69$$

$$\frac{d}{D} = 0.35 \text{ to } 0.60.$$

18.7 DRAFT TUBES

As already explained draft tube is a tube or a pipe which connects the outlet of reaction turbine with the tail race. Water after doing work on the runner passes to the tail race through this tube. This tube is either made up of rivetted steel plate pipe or cement concrete. Its cross-sectional area gradually increases towards the outlet (see Fig. 18.13).

Draft tube should be submerged by about 1 m in the tail race and made strong enough because the water leaving the runner vanes always has a tangential component which may damage the walls at entrance. Thus to prevent this danger, sufficient metal at entrance of the tube is allowed or otherwise it is made of concrete.

Functions

Draft tubes are required to perform the following two main functions:

(i) If water is discharged freely from the runner then turbine will work under a head equal to the height of head race water level above the runner outlet. But by providing an air-tight draft tube, which connects runner with the tail race it is made possible to establish a negative head at the outlet by amount equal to the height of the runner outlet above the tail race.

Thus, by providing a draft tube, it is possible to install the turbine above the tail race without loss of head, this makes the inspection and maintenance to turbine easy.

(ii) When water leaves the runner, it possesses high velocity and thus Kinetic energy of the turbine is lost, if it discharges freely. By using a draft tube of increasing cross-sectional area, the velocity is largely reduced at the exit of draft tube. Thus a kinetic head is gained, but at the outlet turbine only atmospheric pressure is required. Therefore, a further negative head at runner exit can be established by an amount equal the gain in velocity head.

Thus with the help of above two functions, negative head at the runner exit is increased, which result in the increase of net working head on the turbine and also output increases. In this way with the help of draft tube nearly 10% increase in efficiency is obtained.

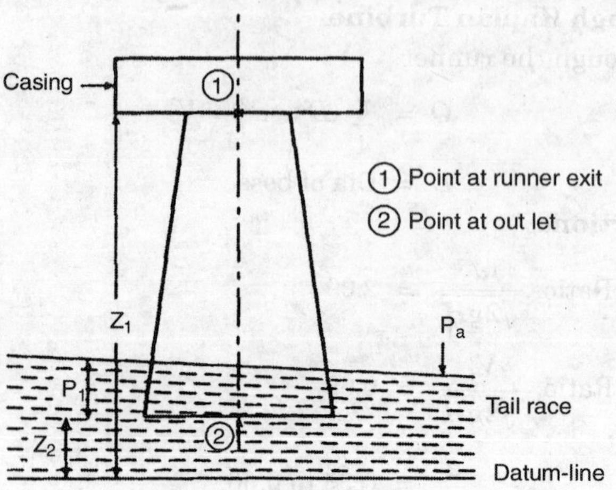

Fig. 18.13 Draft Tube

Types of Draft Tubes

The following are four types of draft tubes used with reaction turbines:

(*i*) Straight divergent tube (*ii*) Moody spreading tube

(*iii*) Simply elbow tube

(*iv*) Elbow type with circular cross-section at inlet and rectangular at outlet.

1. Straight Divergent Tube. The shape of this type of tube is similar to a frustum of a cone. The maximum cone angle of such tube is 8° ($\alpha = 4°$). If the cone angle is more than 8° water remains away from the inner walls of the tube and cavitation occurs, causing loss of head. This tube should be submerged by nearly 1 m in the tail race. This has an efficiency up to 90% and is most suitable for Francis turbines (see Fig. 18.14).

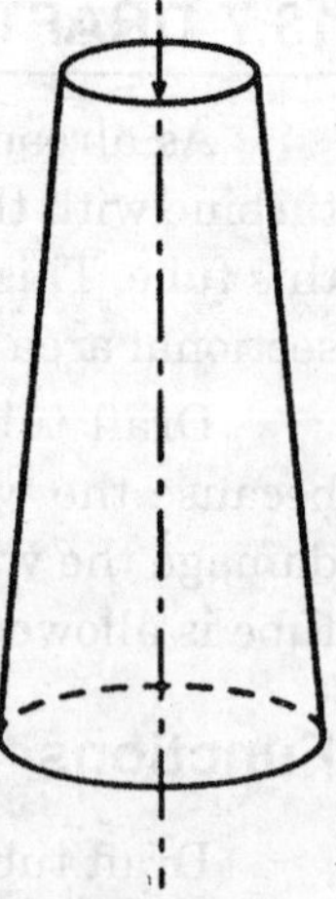

Fig. 18.14 Straight divergent tube

2. Moody Spreading Tube. As shown in Fig. 18.15 this type of the tube is a short vertical tube and with two long horizontal tubes. It is helpful in reducing whirling action of the water coming out from the runner and thus reduces eddy losses and also increases the life of the tube. Efficiency of such draft tubes is about 85%.

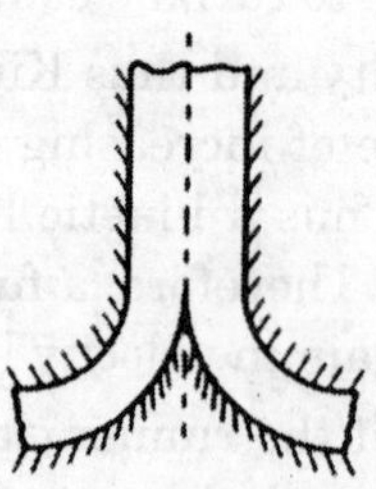

Fig. 18.15 Moody spreading tube

Fig. 18.16 Simple elbow tube

3. Simple Elbow Tube. As shown in Fig. 18.16 it has circular cross-section at the top as well as at the bottom. This type has an advantage over the above two types that this requires lesser excavation for the installation. Its efficiency is only about 60% and, therefore, rarely used.

4. Elbow type with circular cross-section at inlet and rectangular at outlet. This type of tube is shown in Fig. 18.17 and is useful where larger excavation is necessary. This has a larger cross section at outlet, which recovers more kinetic head at outlet. This is generally used with Kaplan turbine and has an efficiency of about 70%.

Fig. 18.17 Elbow type with circular cross-section at inlet & rectangular at outlet

18.8 CAVITATION IN TURBINES

As it has already been discussed that when the pressure in any conduit falls below the vapour pressure, cavitation may occur. Similarly, when the water flows through a reaction turbine and if pressure falls below the vapour pressure in any part of the turbine vapour traps in the liquid, thus forming the bubbles and when these bubbles are carried by the stream of liquid from lower pressure zones (where bubbles are generated) to the higher pressure zones vapour condenses and again is converted into water. This causes the bubbles to collapse. Collapsing bubbles produce very high pressure (upto 1000 atmospheres) which causes damage in the blades of the runner and draft tube etc. It causes small pits and cavities to be formed on the inside surface as shown in Fig. 18.18. This action on the metallic surface is known as pitting action and the phenomenon is known as cavitation because of the formation of cavities on the metallic surface.

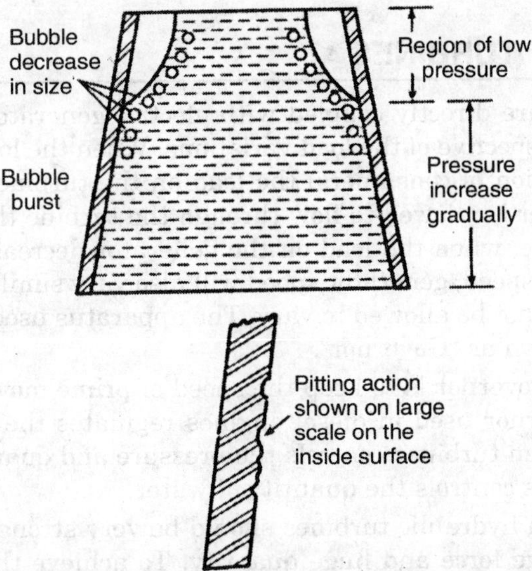

Fig. 18.18 Pits & cavities due to cavitation

Disadvantages Due to Cavitation

Cavitation causes the following disadvantages:

(*a*) It causes very great noise due to the bursting of the bubbles.

(*b*) When the bubbles burst, adjoining liquid rushes with a large velocity to occupy the space previously occupied by bubbles. It causes lot of shock and vibrations.

(*c*) Inside surface of the draft tube and runner blades etc. get pitted near the bubble collapsing range.

(*d*) The above factors affect badly on the overall efficiency of the turbine or pump.

How to Avoid Cavitation

Following are few methods to avoid cavitation in turbines.

(*a*) *Installation of turbine belows tail race level:*

From the equation of Thoma's eavitation factor,

$$\sigma = \frac{H_a - H_v - H_s}{H}$$

it is clear that if the value H_s is less, σ increases. This means, by reducing H_s, turbine may work in safe zone. In case of high specific speed turbines, like Kaplan turbine, runner exit velocity is very high, which lowers the value of pressure and thus brings the cavitation zone nearer. In order to keep pressure within cavitation limits, the value of suction head H_s is made negative. Negative H_s means that runner of such turbine is installed below the tail race level.

(*b*) *Use of Material*

Cast iron has less resistance to cavitation as compared to cast steel. Further cast steel parts can be easily repaired by welding the cavitation affected portions. However stainless steel is the best material to resist cavitation, but because of expensiveness turbine runners are made of cast steel and then coated with stainless steel. The turbine blades are then highly polished, since polished surfaces also avoid cavitation.

18.9 GOVERNING OF TURBINES

Hydraulic turbines are directly coupled with electric generators. The generators should run at constant speed irrespective of the load variations. When the load on the generator varies, (due to varying consumption of consumers) the load on the turbines also vary accordingly. If the same quantity of water is allowed to flow through the turbine the speed of the runner will either increase or decrease, when the load on the generator decreases or increases. Thus due to the variation in runner speed, generator speed will also vary similarly. But it is not desirable because frequency should not be allowed to vary. The apparatus used for controlling or keeping the speed constant is known as "Governor".

The function of the governor is to keep the speed of prime mover constant irrespective of load variations. The governor used in diesel engines regulates the quantity of fuel, governor in steam engines and steam turbines controls the pressure and quantity of steam whereas the governor in water turbines controls the quantity of water.

The governor used in hydraulic turbines should be very strong as it has to deal with the water coming at very large force and huge quantity. To achieve the above requirements, all type of turbines use oil pressure governor, which is desired in detail below:

Main Components of Oil Pressure Governor

1. Servomotor, also known as relay cylinder.
2. Distribution valve, also known as relay or control valve.
3. Actuator or Pendulum.
4. Oil Sump.
5. Oil Pump.
6. A system of oil supply pipes.

1. Servomotor

It consists of a cylinder in which piston reciprocates under the action of oil pressure. It is connected at both the ends with the distributor valve through the pipe lines, as shown in Fig. 18.19. The motion of the piston is transmitted to spear rod in Pelton turbines and the regulating rings in the reaction turbines.

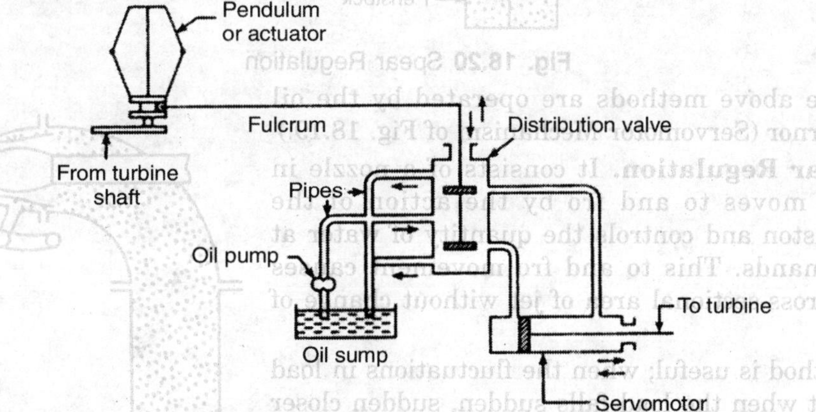

Fig. 18.19 Oil supply system in turbine governor.

2. Distribution Valve

It is similar to that of a piston valve used in steam engines. This valve slides whenever load changes due to the motion of the actuator and thereby allows the oil to go to either side of the servomotor as shown in Fig. 18.19. It receives oil from an oil pump. Its construction and connection of the oil supply system are shown in Fig. 18.19.

3. Actuator

It is a flyball mechanism connected to the turbines main shaft through belt or gears. It works on the principle of centrifugal force.

4. Oil Sump

It is a oil reservoir from where oil is sent to the distributor through oil pump. This oil comes back from the distributor after displacing the position of piston in the servomotor.

5. Oil Pump

It is generally a gear pump which receives power from the turbine shaft. It supplies oil at a pressure from 10 – 20 kg/cm^2.

18.9.1 Governing of Pelton Wheel

In Pelton wheel turbine, the quantity of water supplied by the nozzle and striking over the buckets can be regulated by any of the following methods:

(*a*) Spear Regulation. (*b*) Deflector Regulation.

(*c*) Combined spear and Deflector Regulation—known as Double Regulation.

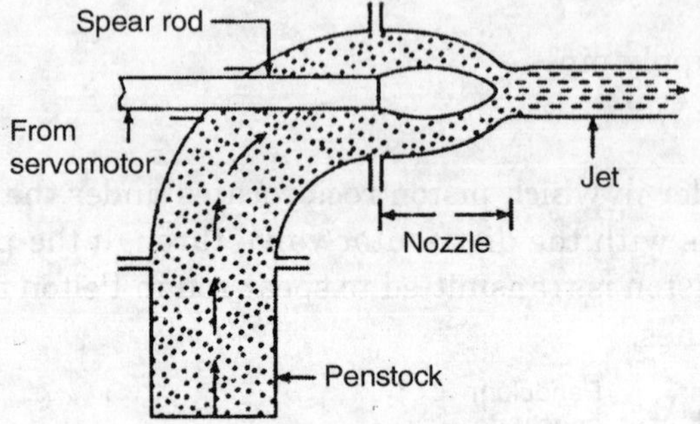

Fig. 18.20 Spear Regulation

All these above methods are operated by the oil pressure governor (Servomotor Mechanism of Fig. 18.19.)

(*a*) Spear Regulation. It consists of a nozzle in which spear moves to and fro by the action of the servomotor piston and controls the quantity of water at changing demands. This to and fro movement causes variation in cross-sectional area of jet without change of velocity.

This method is useful; when the fluctuations in load are small. But when the load falls sudden, sudden closer of the nozzle causes water hammer in the penstock. Therefore, simple spear regulation system is not used in modern turbines where fluctuations in the load are sudden.

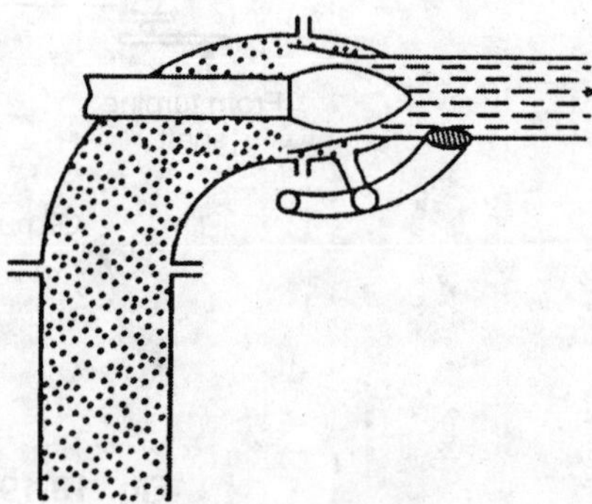

Fig. 18.21 Deflector Regulation.

(*b*) Deflector Regulation. It is a plate, pivoted just outside the nozzle and is connected to the oil pressure governor through levers. When the load drops, it deflects a part of the jet and thus controls the quantity of water striking the buckets. In this system large amount of water goes into waste but there are no chances of water hammer in penstock.

This system is also not mostly used due to large wastage of water and poor speed regulation.

(*c*) Double Regulation. Fig. 18.22 shows this method and is used by all modern Pelton wheels because it has the advantage of both spear as well as deflector regulation system. If load on a turbine drops suddenly, the speed tends to rise suddenly then deflector which comes into action immediately and obstructs a part of water reaching to the buckets. In the mean time, spear gradually comes forward to its new position (due to the servomotor) and thus the risk of water hammer is avoided. When the spear has moved to its new position to allow the required quantity of water, the deflector moves backward and allows the full jet to strike the buckets through spear nozzle.

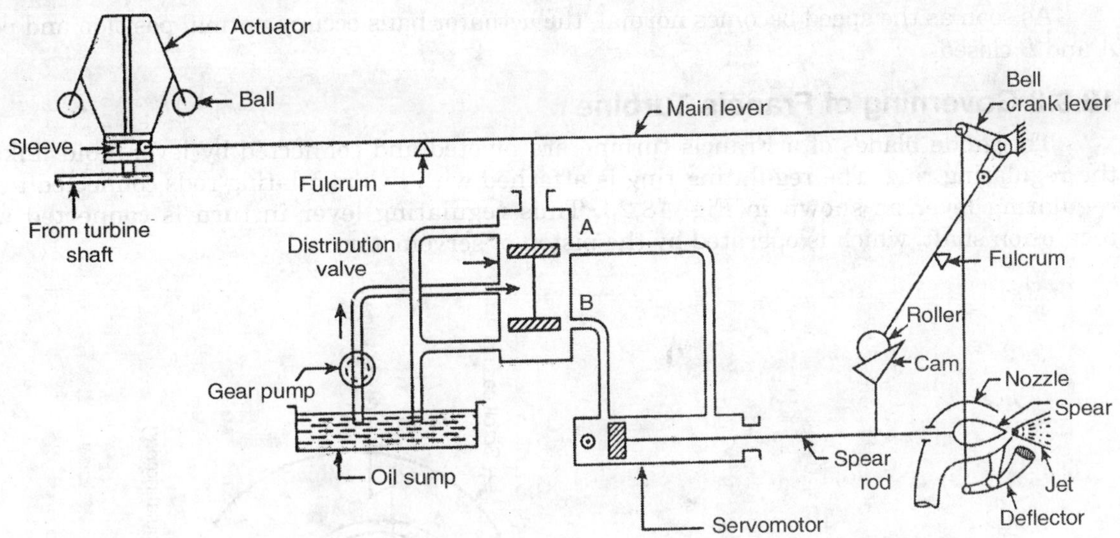

Fig. 18.22 Double Regulation,

As this type of governing controls the speed of turbine and pressure (*i.e.* water hammer) in the penstock by the combination of spear and jet deflector, this system is known as "Double Regulation".

Working

When the turbine is running at normal load, piston in the distribution valve and the actuator occupies their normal positions as shown in Fig. 18.23. In this position as the ports *A* and *B* are closed the oil supplied by the gear pump remains in the middle portion of the distributor valve and thus gears of the pump go on rotating without developing further pressure.

As the load on the turbine decreases, speed increases correspondingly. As the actuator is connected to the turbine shaft, its speed also increases and balls fly off. Therefore, sleeve moves upward the lever rod inclines and the bell crank lever moves downward. When the bell crank lever moves downward, the jet deflector will operate and divert whole or part of the jet away from the buckets. With the downward movement of bell crank lever, the roller on the cam rises. In the mean time distribution valve also works as follows:

When the lever rod inclines then sleeve control valve rod is pushed down in the distribution valve. Downward movement opens the port B and keeps port *A* still closed. As soon as port *B* opens, the oil from the middle port of distribution valve goes to the servomotor on the left side of the piston. This oil of high pressure forces the piston to move towards right side, thus forcing the spear to reduce the area of jet and hence the rate of flow. The spear occupies its final position, jet deflector occupies its original position and does not obstruct jet as the speed again becomes normal.

When the load on the turbine increases, the speed reduces. This causes the fly balls of the actuator to come down and thus sleeve also moves downward, causing the lever rod to move upward on the other side of the fulcrum. This causes the control valve rod to move upward and to open port *A* but to close port *B*. Thus the oil under pressure will go into the servomotor *via* port *A* to the right side of the piston. The piston moves towards left, causing the spear to move away from the nozzle and increases the flow, which in turn increases the amount of water striking the buckets and thus brings the speed to come to normal. In this case jet deflector moves up but deflector plate is not able to obstruct the jet at normal speed.

As soon as the speed becomes normal, the actuator balls occupy normal position and ports *A* and *B* closed.

18.9.2 Governing of Francis Turbine

The guide blades of a Francis turbine are pivoted and connected by levers and links to the regulating ring. The regulating ring is attached with two regulating rods connected to the regulating lever as shown in Fig. 18.23. Thus regulating lever in turn is connected with regulation shaft, which is operated by the piston of servomotor.

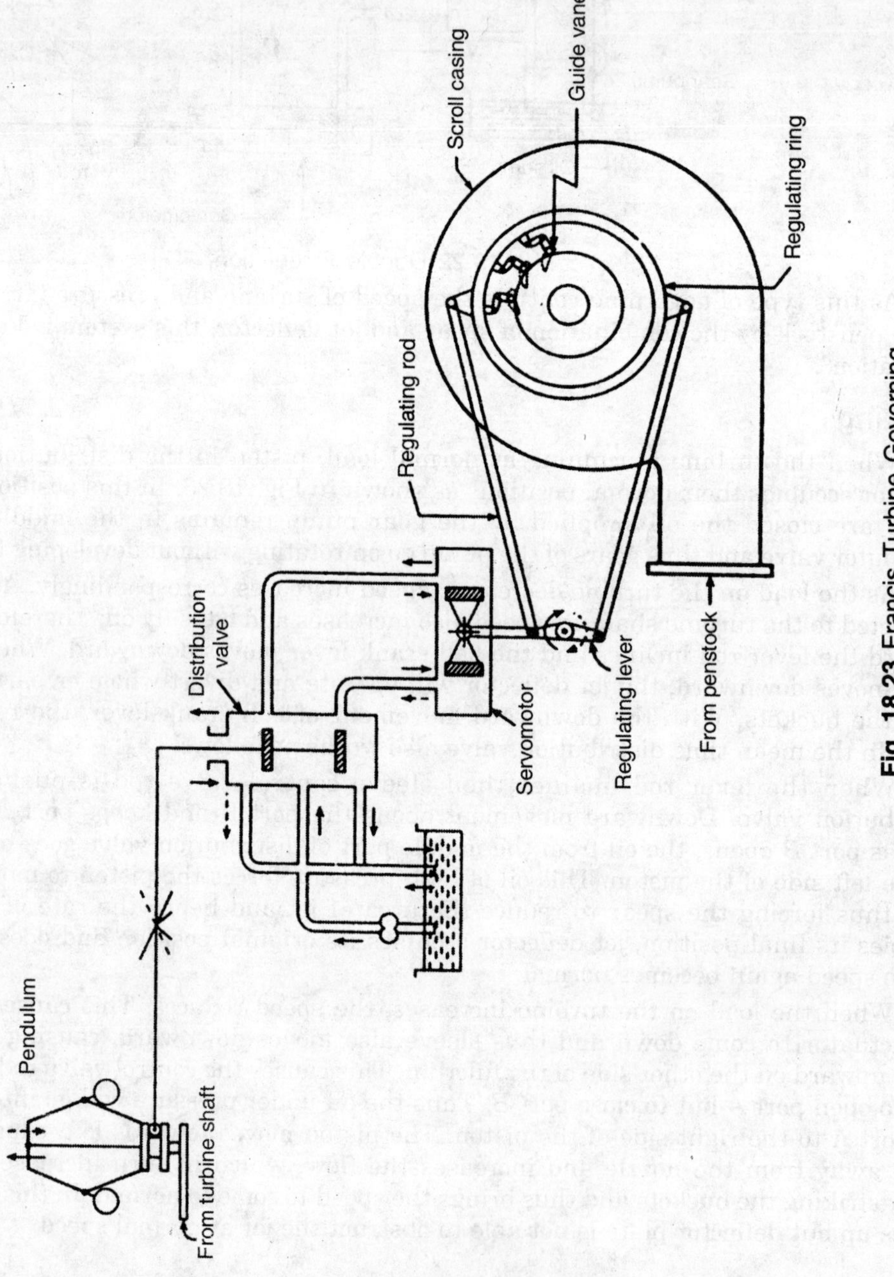

Fig. 18.23 Francis Turbine Governing.

When load on the turbine decreases, speed tends to increase, which moves fly balls of actuator away and thus raises the sleeve. Main lever on the other side of the fulcrum pushes down the control valve rod and opens the port *B*. With the opening of the port *B*, oil under pressure enters the servomotor from left and pushes the piston to move towards right. The previous oil at the right side of the servomotor is thus pushed back into the oil sump through port *A* and upper part of the distributing valve. When the piston of the servomotor moves towards right, regulating ring is rotated to decrease the passage between the guide vanes by changing guide vane angles. Thus quantity of water reaching the runner blades is reduced. Therefore, gradually speed comes to normal and then actuator, main lever and distribution valves attain their normal position.

Similarly when load on the turbine increases, exactly reversed action takes place. Regulating ring is moved in reversed direction to increase the passage between the guide blades thus allowing more water to meet the increased load demand.

Sudden reduction in the passage between the guide blades may cause water hammer which can be prevented by providing a relief valve near the turbine which diverts the water directly to the tail race. Thus it functions similar to that of jet deflector as in Pelton wheel.Thus double-regulation (speed and pressure) is also well performed in Francis turbines.

18.9.3 Governing of Kaplan Turbine

As explained earlier Kaplan turbine has guide vanes as well as runner vanes adjustments (see Fig. 18.24). In Francis turbine, only guide vanes are adjusted as per load requirements but in Kaplan turbine governor is required to operate guide vanes and runner vanes simultaneously.

The governor of this turbine is similar to that of Francis turbine as explained above except that it is also provided with a distribution valve for the runner vanes servomotor. Both the servomotor distribution valves (runner vanes and guide vanes) are interconnected. This interconnection facilitates to operate the runner vanes and guide vanes simultaneously in such a way that water passes through them without shock at all load conditions. This system is known as "Double Regulation" and gives higher efficiency.

As Kaplan turbine is low head one, hence the problem of water hammer is of less importance and, therefore, only speed regulation is necessary and no pressure relief valve is required.

As explained above, Kaplan turbine governing employs two servomotors. One for operating guide vanes which is similar to that described in Francis turbine. The second servomotor, operating runner vanes works as under:

This mechanism operates inside a hollow coupling fitted between the turbine and generator shaft as shown in Fig. 18.25. For more clarity it is shown in section.

Servomotor consists of a cylinder with a piston working under oil pressure (supplied by distribution valve) on either side. This piston is not shown and connected with an operating rod which moves up and down and passes through the turbine shaft, which is made hollow for this purpose. This operating rod actuates the cross-head which carries number of arms each connected with runner blades through small cranks. Runner blades are connected to the crank with the help of pivots casted integrally with the blades. This mechanism is enclosed in the runner hub leaving blades outside.

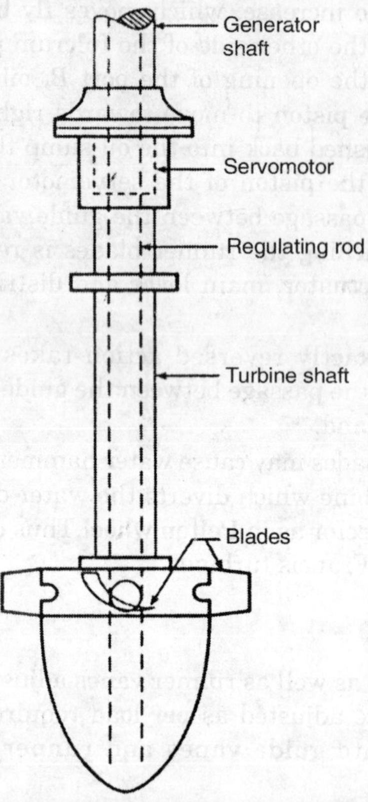

Fig. 18.24 Governing of Kaplan Turbine

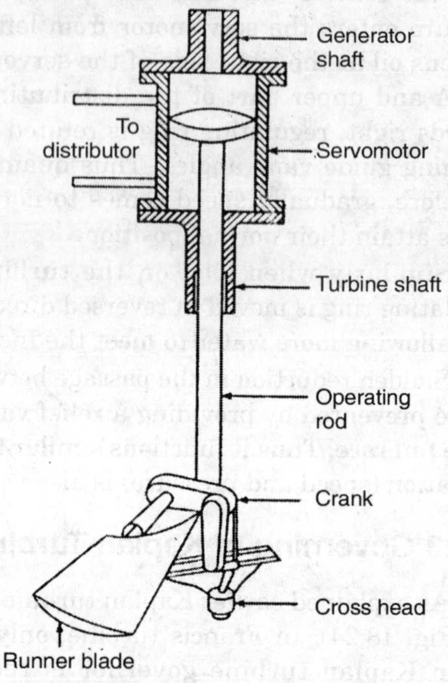

Fig. 18.25 Servomotor Mechanism operating Runner Vanes of Kaplan Turbine. Shown in section and on reduced scale.

Qualities of Governor

For a satisfactory speed regulation, a governor must have following qualities or performance characteristics:

***(i)* Sensitiveness.** Sensitiveness means the change in the position of the governor due to smallest change in speed at any turbine gate opening. It is the ratio of the difference between maximum and minimum speed to mean speed of action *i.e.*,

$$\text{Sensitiveness} = \frac{\omega_2 - \omega_1}{\dfrac{\omega_1 + \omega_2}{2}} \text{ or } \frac{2(\omega_2 - \omega_1)}{(\omega_2 + \omega_1)}$$

***(ii)* Rapidity of Action.** The action of the governor must be rapid. The governor must act quickly to open the gate if the load increases and *vice-versa*.

***(iii)* Stability.** Stable governor helps the gates to come to their normal position with the minimum possible oscillations. In other words stable governor requires least time between a change of load and setting down of the turbine to its proper speed with the new load.

***(iv)* Hunting or Racing.** If there is periodic speed fluctuations with steady load, the governor is said to hunt or race.

***(v)* Capacity.** The force required to move the turbine wicket gates, determines the capacity of governor which controls the flow of oil under pressure to the turbine servomotor.

***(vi)* Speed Regulation.** Speed fluctuations caused by sudden rise or fall of load, are regulated by the governor and normal speed is returned immediately.

***(vii)* Regulating Time.** Regulating time is the time required by the governor in seconds in order to move the turbine gates from the open position to close position or *vice-versa.*

18.10 TURBINE PERFORMANCE

Turbine Model. After preparing a design of a turbine, their performances are determined in the laboratory. For this purpose a similar model is made geometrically, over which experiments are performed and if this model turbine is found to be suitable under different heads, load and gate openings then actual turbines of larger sizes are manufactured. Geometrically similar turbines are the small turbines which are similar to the actual turbine but of smaller size, whose linear dimensions such as diameter, width etc. are proportional to the prototype and all the angles are equal.

Model turbines are also actual turbines but of smaller output, and are helpful to determine performance conveniently and economically and then from this model, the prototype (full size turbine) is manufactured.

The hydraulic tests are performed on this model in order to perdict what will happen in the prototype (or full-sized system) under similar operating conditions.

Performance of Water Turbine. Performance of a turbine means working of the turbine under different loading conditions. Although turbines are required to run at constant speed but this is difficult to achieve because of fluctuations in load. To accommodate varying loads the discharge is regulated.

The important parameters for any particular turbine are:

(*i*) Discharge (*ii*) Head

(*iii*) Efficiency (*iv*) Speed

(*v*) Power

It is desirable to study the behaviour of these quantities with respect to one another.

Such studies can be carried out under, unit quantities like unit power, unit speed and unit discharge.

18.10.1 Turbine Characteristics

The turbine characteristis like unit power, unit speed and unit discharge help in studying the performance of turbines.

(a) Unit Power (P_u). It is defined as the power produced by the turbine when running under unit head (1 metre).

We know
$$P = \frac{\omega H \times Q}{1000}$$

where P = Power produced by a turbine in kW

Q = Discharge in m^3

H = Net head in metres

ω = Unit weight of water.

Now $Q = A \times V = k_1 . D^2 . V$

where D = Diameter of the runner V = Velocity

$$V \propto = \sqrt{2gH} = k_2\sqrt{H}$$

$$\therefore \quad Q = k_2 . k_1 . D^2 \sqrt{H} = k . D^2 \sqrt{H}$$

$$P = \frac{\omega H}{1000} \times Q$$

$$= \frac{\omega k}{1000} H\sqrt{H} = k_3 H^{3/2}$$

For unit power (P_u)

as $H = 1 \quad \therefore \ P_u = K_3, P = K_3 H^{3/2}$

$$P_u = \frac{P}{H^{3/2}}$$

(*b*) **Unit Speed (N_u).**. It is the speed developed by a turbine when operating under a unit head (1 metre).

$$V = \frac{\pi DN}{60}$$

= speed of turbine in R.P.M.

Now $V = \sqrt{2gH}, \quad V = k_1\sqrt{H}$

$$\therefore \quad \frac{\pi DN}{60} = k_1\sqrt{H}$$

$$\therefore \quad N = K_2\sqrt{H}$$

For unit speed (N_u) $H = 1$

$$\therefore \quad N_u = \frac{N}{\sqrt{H}}$$

(*c*) Unit Discharge (Q_u). It is defined as the discharge of a turbine working under a unit head (1 metre).

$$Q = A.V. = k_1 D^2 V = k_1 D^2 \sqrt{2gH} = K_2\sqrt{H}$$

For unit discharge (ϕ_u) $\quad H = 1$

$$\therefore \quad Q_u = K_2$$

$$\therefore \quad \frac{Q_u}{Q} = \frac{1}{\sqrt{H}}$$

or

$$Q_u = \frac{Q}{\sqrt{H}}$$

(*d*) Specific Speed (Ns). It is the speed of a geometrically similar turbine running under a unit head and producing unit power.

Specific speed is quite useful as follows:

(*i*) It helps in selecting type of turbine to be used for a particular power station.

(ii) It permits to visualise the performance of turbine.

(iii) Specific speed being known the normal running speed can be determined.

$$P = \frac{\omega QH}{1000} \propto QH$$

As already proved $Q \propto D^2 \sqrt{H}$

$\therefore$ $P \propto d^2 H^{3/2}$ ...(1)

Also $V = \frac{\pi DN}{60}$ and $V \propto \sqrt{2gH}$

Therefore, $D \propto \frac{\sqrt{H}}{N}$

Substituting in equation (i),

$$P \propto \frac{H}{N^2} H^{3/2}$$

$$\propto \frac{H^{5/2}}{N^2}$$

$$N \propto \frac{\sqrt{H^{5/2}}}{\sqrt{P}} \propto \frac{H^{5/4}}{\sqrt{P}}$$

$$N = N_s \frac{H^{5/4}}{\sqrt{P}}$$

where N_s = specific speed

$$N_s = \frac{N\sqrt{P}}{H^{5/4}}$$

Table 18.1 shows the specific speeds for different turbines.

Table-18.1: Specific speed of different turbines

Turbine	*Classification*	*Specific speed*
Pelton turbine	Impulse	8.5 to 50
Francis turbine	Reaction	50 to 350
Kaplan turbine	Reaction	350 to 850

18.10.2. Turbine Efficiencies

Different types of efficiencies used in connection with hydraulic turbine are as follows:

(i) Volumetric efficiency

(ii) Hydraulic efficiency

(iii) Mechanical efficiency

(iv) Overall efficiency.

1. Volumetric Efficiency. Some of the water flowing in the turbine may leak through the joints or some water may not be effective in imparting its energy to the turbine. The leakage of water reduces the efficiency of the turbine.

$$\eta_v = \frac{Q - Q_L}{Q}$$

where η_v = Volumetric efficiency

Q = Discharge doing useful work

Q_L = Leakage from the turbine.

2. Hydraulic Efficiency. Hydraulic efficiency takes into account the loss of head that takes place in the turbine due to incomplete conversion of head over the blades into useful work. It is defined as follows:

$$\eta_H = \frac{H - h}{H}$$

where η_H = Hydraulic efficiency

H = Net head utilised

h = Head not effectively utilised.

3. Mechanical Efficiency. Mechanical efficiency takes into account the power loss due to friction.

$$\eta_M = \frac{P_1 - P_2}{P}$$

where P = Power produced by the turbine

P_1 = Shaft power; P_2 = Power loss.

4. Overall Efficiency (η). Overall efficiency of the hydraulic turbine is about 90% and is given by

$$\eta = \eta_v \times \eta_H \times \eta_M$$

18.11 CHARACTERISTIC CURVES

Turbines are generally designed to work under given conditions of head, discharge, speed, power and efficiency. But in actual practice, when turbine runs at some other different conditions, then the efficiency of the turbine will be reduced. For the study of behaviour of these turbines under varying conditions, some tests are generally conducted either on the actual turbine or on the model. The data obtained from the tests are graphically plotted and the curves thus obtained, are known as characteristic curves.

These characteristic curves are of the following three types:

1. Constant head or main characteristic curves.
2. Constant speed or operating characteristic curves.
3. Constant efficiency curves.

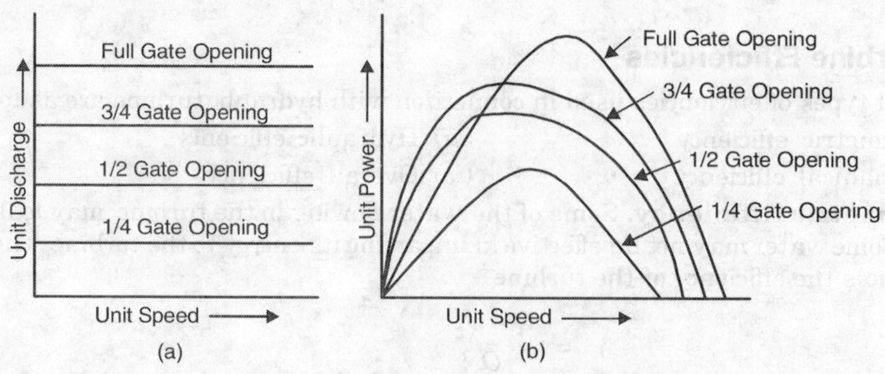

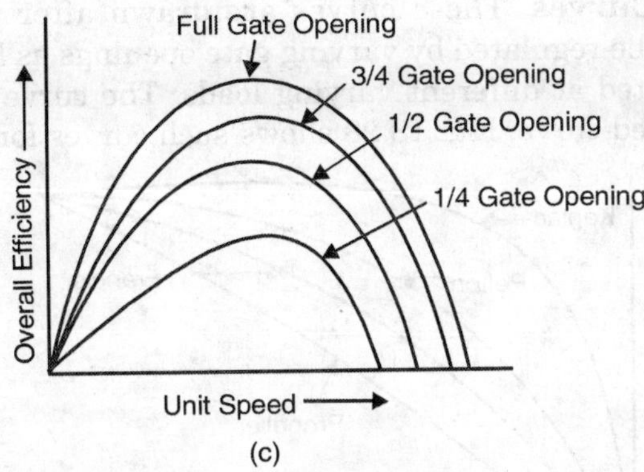

Fig. 18.26 Constant head characteristics of Pelton wheel.

1. Constant Head Curves. These curves are drawn after performing the tests on either actual or model turbines keeping the head and gate openings constant for all conditions and corresponding power, discharge and overall efficiency are measured at varying unit speeds. The curves are drawn keeping horizontal axis for varying speeds and vertical ordinates as power, discharge and overall efficiency. Such curves at various gate openings are plotted as shown in Fig 18.26 and Fig 18.27.

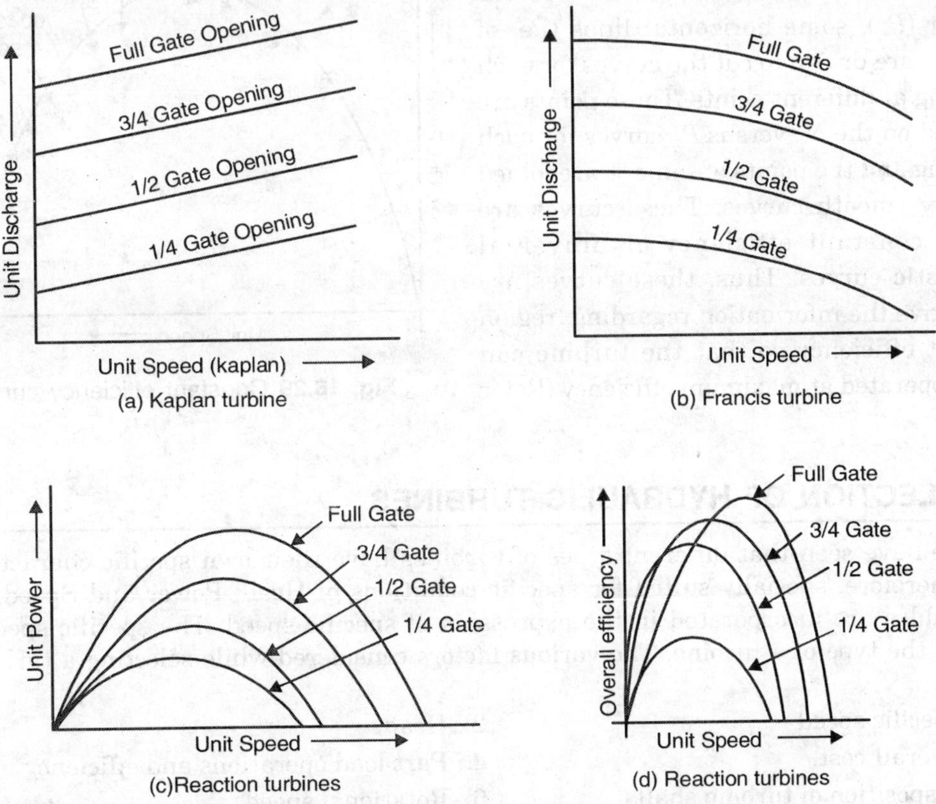

Fig. 18.27 Constant head characteristics curves of Reaction turbines.

2. Constant Speed Curves. These curves are drawn after performing the tests at constant speeds, which may be regulated by varying gate openings as load varies. The readings of overall efficiency are plotted at different varying loads. The curve thus obtained is known as operating or constant speed curve. Fig. 18.28 shows such curves for different turbines.

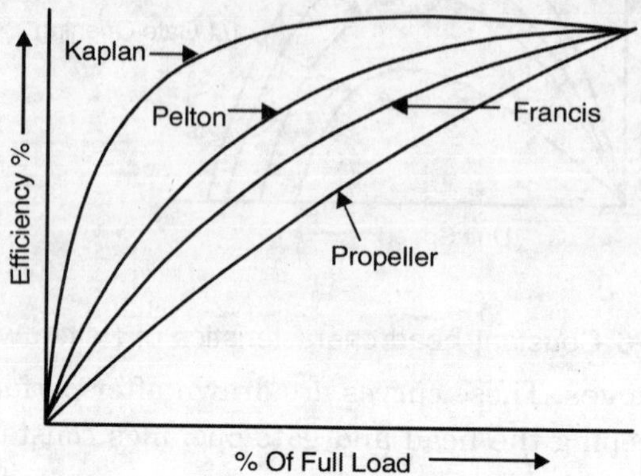

Fig. 18.28 Constant speed curves.

3. Constant Efficiency (Universal Characteristic) Curve. On constant head characteristic curves of unit speed (N_u) versus unit power (P_u), some horizontal lines (i.e. of constant η_0) are drawn to cut the curves for each gate opening at different points. These points are then plotted on the N_u versus P_u curves for each gate opening and the points of same % are joined together by smooth curves. These curves are known as constant efficiency or universal characteristic curves. Thus, these curves are helpful to give the information regarding region of constant efficiency so that the turbine can always be operated at maximum efficiency (Refer Fig. 18.29).

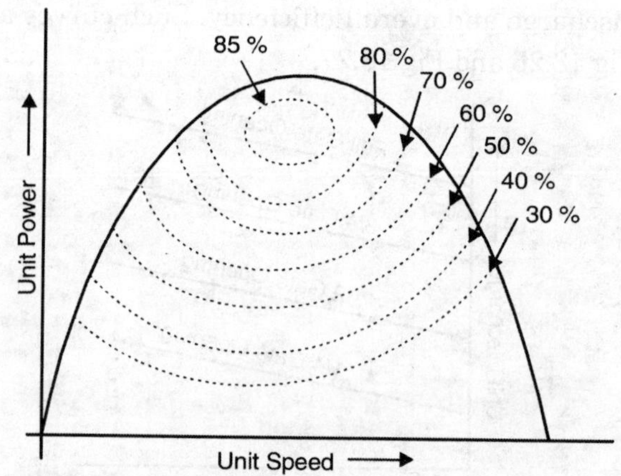

Fig. 18.29 Constant efficiency curves.

18.12 SELECTION OF HYDRAULIC TURBINES

As we have seen that different types of turbines have their own specific characteristics and are, therefore, specially suited for specific conditions of Head, Power and Speed. These three variables are incorporated in the expression of specific speed. The specific speed thus determines the type of a turbine. The various factors considered while selecting a turbine are as under:

1. Specific speed
2. Head
3. Overall cost.
4. Part load operations and efficiency
5. Disposition of turbine shaft
6. Rotational speed
7. Availability of water
8. Type of water.

1. Specific Speed. As already discussed each type of turbine has different range of specific speeds, which is most suitable for selecting a turbine.

Table 18.2 shows that which type of turbine is suitable for which range of specific speed.

Table 18.2: Range of spicific speed for different turbines.

S.No.	*Type of Turbine*	*Specific speed*
1.	Pelton wheel:	
	(*a*) Single jet	10 to 25
	(*b*) Multi jet	25 to 50
2.	Francis turbine:	
	(*a*) Slow speed runner	50 to 90
	(*b*) Normal speed runner	90 to 185
	(*c*) Fast runner	185 to 350
3.	Kaplan or propeller turbine	350 to 850

It is common practice that as far as possible a turbine with highest permissible specific speed should be selected because it will be cheap, small in size and have high rotational speeds, will require a generator (and hence power installations) of smaller size.

2. Head. Type of turbine also depends on the head of water available and the selection of turbine based on the head is shown in the following table:

Head above 200m - Pelton wheel

Head between 60 m and 200 m - Francis turbine

Head below 60 m - Kaplan or Propeller turbine.

The above table shows that impulse turbines are useful for high heads and reaction turbines are useful for low and medium heads.

3. Overall Cost. While selecting a turbine the cost is considered for two factors

(*i*) Initial cost, which includes purchase price and installation cost,

(*ii*) Running cost, it includes maintenance, repairs, depreciation, overheads and operation costs etc.

4. Part Load Operation and Efficiency. The load for which the turbine is designed, is known as full load. At this load the η_0 is not maximum and hence the turbine are not run at full loads. When the load on the turbine is such that it gives maximum η_0 such load is known as "normal load". This load occurs at nearly 80% of full load and generally the turbine runs at this load. When the load is less than the normal load, then the turbine is said to be working at "part load". At this load η_0 is less than the maximum. When load is more than the normal load, turbine is said to be working at "over load".

Performance curve shows that at part load η_0 is less than the maximum for all the turbines but the performance of Kaplan and Pelton turbines are better than those of Francis and Propeller turbines. Thus variability of load influences on the choice of type of turbine and, therefore, Pelton wheels may be used even for heads between 150 to 300 m, if the turbine is required to run at part loads. Similarly Kaplan turbine is preferred for heads between 30 to 60 m, if the turbine is required to run at part loads.

5. Disposition of Turbine Shaft. Turbine shaft can be either vertical or horizontal. A vertical shaft turbine requires deeper foundation and high building and is generally adopted for reaction turbines, whereas horizontal shaft turbine requires more area and is suitable for impulse turbines. Thus disposition of shaft is affected by the type of turbine.

6. Rotational Speed. Rotational speed of the runner of a turbine should be sufficiently large so that smaller size of turbine can meet the requirements. A large speed turbine will require less number of poles in the generator and therefore generator will be of smaller size. These smaller size turbines and generator will need less space, less installation cost and less excavation cost.

There is a definite limit for the increase of rotational speed because:

(*a*) Larger speeds give more specific speed of turbine, which may cause cavitation.

(*b*) Larger speed gives more centrifugal force in the revolving parts, thus causing very large stresses.

Thus the rotational speed is decided by considering the above factors and it should be such that the turbine can directly be coupled with the generator.

7. Availability of Water. If the water available is in less quantity and head is high then Impulse turbine is suitable. If more water at high head is available then more number of impulse turbines can be installed. If medium quantity of water is available at medium heads then Francis turbine may be used but if large quantity of water is available at low head then Kaplan turbine should be used. If less quantity of water at small head is available then installation of any type of turbine is not suitable.

Thus the availability of water affects to a large extent that which type of turbine and number of such turbines should be installed.

8. Type of water. If the water to be used has large amount of dirt and sand then Francis turbine cannot be used because its runner cannot withstand the erosion caused by such water. Thus for the upper limits of head (150-300 m) generally Pelton wheel and for the lower limits of head (30-60 m) generally Kaplan turbine is preferred.

SOLVED EXAMPLES

Example 18.1 *A turbine develops 13,500 kW working under a head of 520 m when running at 400 R.P.M. Calculate the specific speed of the turbine. Specify the type of turbine to be used.*

Solution.

H = Head = 520 m.

N = R.P.M. = 400 R.P.M.

P = Power = 13,500 k.W

$$\text{Specific speed} = N_s = \frac{N \cdot \sqrt{P}}{N^{5/4}}$$

$$N_s = \frac{400\sqrt{13,500}}{520^{5/4}} = \frac{400 \times 116.2}{2483} = 18.72 \text{ R.P.M.}$$

For this specific speed **Pelton turbine** is used. **Ans.**

Example 18.2 *Calculate the specific speed of a turbine and suggest the type of turbine required for a river having a discharge of 250 litre/sec with an available head of 50 metre. Assume efficiency of turbine as 80% and speed 450 R.P.M.*

Solution. Q = 250 litre/sec., H = 50 metre

η = efficiency = 80%, N = 450 R.P.M.

P = Power developed

$$= \frac{\eta \omega Q H}{75} = \frac{0.8 \times 9.8 \times 250 \times 50}{1000} = 98 \text{ kW}$$

(Taking ω = 9.8 N/litre)

Now, N_s = specific speed

$$= \frac{N\sqrt{P}}{H^{5/4}} = \frac{450\sqrt{98}}{50^{5/4}} = \frac{4454}{133} = 33.5 \text{ RPM}$$

For this specific speed **Pelton turbine** is suitable. **Ans.**

Example 18.3 *It is desired to build a hydro-electric power station across a river having a discharge of 30,000 litre / second at a head of 10 m. Assuming turbine efficiency 80% and speed ratio (K_u) as 0.83, determine the following:*

(*a*) Is it possible to use two turbines with a speed not less than 120 R.P.M. and specific speed not more than 350 R.P.M.?

(*b*) Specify the type of runner that can be used. Also calculate the diameter of runner.

Solution. (*a*)

$$P = \frac{\eta \omega Q H}{75}$$

where P = Total power available, ω = Specific weight of water

Q = Discharge, H = Head (metre), Assuming $g = 9.8 \simeq 10$ m/sec

$$P = \frac{0.8 \times 10 \times 30000 \times 10}{1000} = 2400 \text{ kW} = 2.4 \text{ mW}$$

Using two turbines each of capacity 1.2 mW H.P. the specific speed (N_s) of the turbine is calculated as follows:

$$N_s = \frac{N\sqrt{P}}{H^{5/4}} = \frac{120\sqrt{1200}}{10^{5/4}} = \frac{120 \times 34.64}{17.78}$$

$$= 234 \text{ R.P.M.}$$

For this speed **Francis turbine** is suitable. **Ans.**

(*b*) U = Tangential velocity of the runner.

$$= K_u \sqrt{2gH} = \frac{\pi D N}{60}$$

where K_u = speed ratio amd D = diameter of runner

$$D = \frac{60 \times K_u \sqrt{2gH}}{\pi \times N} = \frac{60 \times 0.83 \sqrt{2 \times 9.81 \times 10}}{\pi \times 120}$$

= **1.85 metres. Ans.**

Example 18.4. *A reaction turbine is supplied with 100 cu m of water per second and works under a maximum head of 120 m at 350 R.P.M. Assuming overall efficiency of the plant 80% and specific weight of water 1000 kg / m³ ; calculate the horse power developed and power in kW.*

Solution. H = 120 m, Q = 100 m³/sec, ω = 1000 kg/m³, η = 0.8

$\therefore$ Power developed $= \dfrac{\omega QH.\eta}{1000} = \dfrac{1000 \times 10 \times 100 \times 120 \times 0.8}{1000}$

$= 96{,}000$ kW **Ans.**

Example 18.5. *(a) Calculate the total energy in kWh which can be generated from a hydro power station having following data:*

Reservoir area = 2.5 sq. km

Capacity = 5×10^6 m^3

Net head of water at the turbines = 80 m

Turbine efficiency = 80%

Generator efficiency = 90%

(b) Also find by how many metre the level of reservoir will fall if a load of 20,000 kW is supplied for 5 hours?

Solution. (*a*) Work done/sec. $= 1000 \times 5 \times 10^6 \times 80 \times 10 = 4 \times 10^{12}$ Nm/s

Power Developed $= \dfrac{4 \times 10^{12} \times 0.8 \times 0.9}{1000} = 0.288 \times 10^6$

Energy produce kW sec $= 0.288 \times 10^6 \times 1 = 0.288 \times 10^6$

Energy produced (kWh) $= \dfrac{0.288 \times 10^6}{3600} = \mathbf{8 \times 10^5}$ **kWh. Ans.**

(*b*) Let the fall in level of reservoir = *h* metre

Time = 5 hours.

Area of reservoir = $(2.5 \times 1000 \times 1000)$ sq. m.

Discharge/sec. $= \dfrac{2.5 \times 1000 \times 1000 \times 10}{5 \times 3600} = \dfrac{10^5 \times h}{72}$

Work done/sec. $= \dfrac{1000 \times 10^5 \times h \times 80}{72}$

Power in kW produced $= \dfrac{1000 \times 10^4 \times h \times 80 \times 0.80 \times 0.9}{72 \times 1000}$

$= 0.8 \times 10^5 \times h$

But kW produced = 20,000 (given)

$\therefore$ $20{,}000 = 0.8 \times 10^5 \times h$

h = **0.25 metre. Ans.**

Example 18.6 *Calculate the power that can be developed from a hydro-electric power station having the following data:*

Catchment area = 100 sq.km., Average rainfall = 1.2 m

Run-off = 80%, Available head = 300 m

Overall efficiency of the power station = 75%.

Solution. Run-off available $Q = \dfrac{100 \times 10^6 \times 1.2 \times 0.8}{365 \times 24 \times 60 \times 60} = 3.4$ cu m/sec.

$\omega = 1000 \text{ kg} = 10{,}000 \text{ N} = 10^4 \text{ N}$

(as 1 cu m of water weight = 1000 kg) = 1000 g N and $g \simeq 9.81$)

H = Available head = 300 m, η = Efficiency = 0.75

$$\text{Power in kW} = \frac{10^4 \times 3.4 \times 300 \times 0.75}{1000} = 7{,}650$$

= **7.65 MW Ans.**

Example 18.7 *Determine the power that can be developed from a hydroelectric power plant with following data:*

Catchment area of reservoir = 3×10^8 sq. metre, Mean head = 40 m
Annual average rainfall = 1.1 m, Efficiency of turbine = 80%
Efficiency of generator = 85%, Load factor = 45%

Assume that 25% of the rainfall is lost due to evaporation and 4% head lost in penstock.

Solution. Quantity of water available per annum $Q = 3 \times 10^8 \times 1.10 \times (1 - 0.25)$

$= 2.475 \times 10^8$ cu m.

Quantity of water available per second

$$Q = \frac{2.475 \times 10^8}{365 \times 24 \times 60 \times 60} = 7.85 \text{ cu m/sec.}$$

Head H = 40 m

Overall efficiency (η) = Turbine efficiency × Generator efficiency × Penstock efficiency

= 0.8 × 0.85 × 0.96 = 0.65

$$\text{Power developed} = \frac{\omega Q H \eta}{1000 \times \text{load factor}}$$

$$= \frac{1000 \times 40 \times 7.85 \times 10 \times 0.65}{1000 \times 0.45} = \mathbf{4535\ kW\ \ Ans.}$$

Example 18.8 *A steel penstock carries 12 m^3/sec of water under a head of 240 m. The penstock diameter is 2.8 m and the allowable stress in penstock material is 14 kPa. Determine the following: (a) Water hammer pressure assuming velocity of pressure wave 1425 m/sec.*

(b) Thickness of penstock. *(c) Describe water hammer.*

Solution. (*a*)

Q = Discharge = 12 m^3/sec.

d = Diameter of penstock = 2.8 m

H = Head = 240 m

V_1 = Velocity of pressure wave = 1425 m/sec.

V_2 = Velocity of water, $Q = V_2 \times \frac{\pi}{4} d^2$

$$12 = V_2 \times \frac{\pi}{4} \times (2.8)^2, \quad V_2 = 1.95 \text{ m/sec.}$$

p_1 = Water hammer pressure

$$= \frac{V_1 \cdot V_2}{g} = \frac{1425 \times 1.95}{9.81} = \mathbf{283\ m.\ \ Ans.}$$

(*b*) f = Safe stress in penstock material

= 140 kPa

h = Total pressure head = H + P_1

= 240 + 283 = 523 m

p = Total pressure = *w.h.*

= 1000 × 523 kg/m² = 5.23 kPa

t = Thickness of penstock

$$= \frac{pd}{2f} = \frac{5.23 \times 2.8}{2 \times 140} = 0.0523 \text{ m}$$

= **52.3 mm Ans.**

(*c*) Water Hammer. When the load on the generator is suddenly reduced the values admitting water to the turbines are to be closed suddenly. This sudden stoppage brings the water near the valve to standstill but the along column of water in the pipe line is still moving. The momentum of this water causes sudeen increase in pressure in pipe section. This sudden rise of pressure above normal is called water hammer. Surge tanks help in releasing and dissipating the excessive water hammer pressure built up in a penstock.

Example 18.9 *The following data is supplied for a hydropower plant.*

Catchment area = *2200 sq. km*

Annual average rainfall = *1.5 m, Available head = 130 m*

Turbine efficiency = *0.86, Generator efficiency = 0.91*

Percolation and evaporation losses = 18%.

Determine

(*a*) Power development in kW taking load factor as Unity.

(*b*) Suggest type of turbine to be used if the speed of runner is to be maintained below 260 R.P.M.

Solution. A = Area = 2200 sq. km. = 2200 × 10^6 m²

h = Annual rainfall = 1.5 m.

y = Percolation and evaporation losses = 0.18

H = head available = 130 m

Q_1 = Quantity of water available for power generation per year

$= A \times h \times (1 - y)$

= 2200 × 10^6 × 1.5 × (1 – 0.18) = 2.7 × 10^8 m³

Q = Quantity of water available per second

$$= \frac{2.7 \times 10^8}{365 \times 24 \times 60 \times 60} = 8.56 \text{ m}^3\text{/sec}$$

η_t = Turbine efficiency = 0.86

η_g = Generator efficiency = 0.91

P = Power developed

$$\frac{\omega QH}{1000} \times \eta_t \times n_g = \frac{1000 \times 8.56 \times 1.3}{1000} \times 0.86 \times 0.91$$

= 8709 kW = **8.709 MW Ans.**

$$N = 260 \text{ R.P.M.}$$

$$N_s = \frac{N\sqrt{P}}{H^{5/4}} = \frac{260\sqrt{8709}}{130^{5/4}} = \frac{260 \times 93.32}{438.9} = \mathbf{55.23}$$

Single Pelton turbine with 4 jets can be used. Further since head available is large and discharge is low, **Pelton turbine** will work satisfactorily.

Example 18.10 *The available discharge for a hydro-electric power plant is 320 m^3/sec. under a head of 28 metre. The turbine efficiency is 90% and the generator is directly coupled to the turbine. The frequency of generation is 50 cycles/sec, and number of poles used are 26. Find the number of machines required when:*

(a) A Francis turbine with a specific speed of 280 is used.

(b) A Kaplan turbine with a specific speed of 500 is used.

Solution. N = Speed of generator

$$= \frac{120f}{p}$$

f = frequency = 50 cycles/sec.

p = Number of poles = 26

$$N = \frac{120 \times 50}{26} = 230 \text{ R.P.M.}$$

Since the generator is directly coupled to the turbine the speed of turbine used must be equal to the synchronous speed of the generator.

$$P = \text{Power} = \frac{\omega Q H}{1000} \times \eta$$

where ω = 1000 kg/m³ = 10^4 N/m³

Q = Discharge = 320 m³/sec, H = Head = 28 m

η = Efficiency of turbine = 0.9

$$P = \frac{10^4 \times 320 \times 28 \times 0.9}{1000} = 80.64 \times 10^3 \text{ kW}$$

(*a*) Let P_1 = Power capacity of each Francis turbine

$$N_s = \frac{N\sqrt{P}}{H^{5/4}}$$

$$280 = \frac{230\sqrt{P_1}}{28^{1.25}}$$

$$\sqrt{P_1} = \frac{280 \times 64.41}{230} = 78.4, \quad P_1 = 6168 \text{ kW}$$

Number of Francis turbines required

$$\frac{P}{P_1} = \frac{80.64 \times 10^3}{6148} = \mathbf{13.} \quad \textbf{Ans.}$$

(*b*) Let P_2 = Power capacity of each Kaplan turbine

$$N_s = \frac{N\sqrt{P_2}}{N^{5/4}}, \quad 500 = \frac{230\sqrt{P_2}}{28^{1.25}}$$

$$\sqrt{P_2} = \frac{500 \times 64.41}{230} = 140, \quad P_2 = 196 \times 10^2 \text{ kW}$$

Number of Kaplan turbines required

$$\frac{P}{P_2} \frac{80.64 \times 10^3}{196 \times 10^2} = \mathbf{4} \quad \textbf{Ans.}$$

Example 18.11 *A Pelton turbine is to satisfy the following requirements:*

Head = 340 m Power = 9.5 × 10³ kW

Speed = 740 RPM, Jet diameter = $\frac{1}{6}$ of wheel diameter

Overall efficiency = 87%

Determine

(a) Wheel diameter *(b) Jet diameter*

(c) Number of jets required

Given: Speed ratio = 0.48, $V_v = 0.986$

Solution.

H = Head = 340 m, P = Power = 9.5 × 10³ kW

C_v = Coeff. of velocity = 0.986

V_1 = Velocity of jet = $C_v\sqrt{2gH}$

$$= 0.986\sqrt{2 \times 9.81 \times 340} = 80.53 \text{ m/s}$$

V_2 = Velocity of wheel = Speed ratio × $\sqrt{2gH}$

$$= 0.48 \times \sqrt{2 \times 9.81 \times 340} = 39.2 \text{ m/s}$$

N = RPM = 740, D = Wheel diameter

$$V_2 = \frac{\pi D N}{60}$$

$$39.2 = \frac{\pi \times D \times 740}{60}$$

D = **1.01 m Ans.**

$$d = \text{jet diameter} = \frac{1}{6} \times D = \frac{1}{6} \times 1.01$$

Jet diameter = **0.168 m. Ans.**

n = Number of jets required

q = Discharge throughout jet

$$= \frac{\pi}{4} \times d^2 \times V_1 = \frac{\pi}{4} \times (0.168)^2 \times 80.53$$

$= 1.78$ m³/sec

η = Overall efficiency = 0.87

$$\eta = \frac{9500}{W.Q.H.}$$

where w = specific weight of water = 9.81 kN/m^3

Q = Total discharge

$$0.87 = \frac{9500}{9.81 \times Q \times 340}, \quad Q = 3.273 \text{ m}^3/\text{sec}$$

$$3.273 = n \times q = n \times 1.78$$

$$n = 1.838 = 2 \text{ (say)}.$$ **Ans.**

QUESTIONS

18.1 Classify different types of hydraulic turbines.

18.2 Describe major components of Pelton wheel.

18.3 Under what conditions will you select Pelton wheels for a hydro-electric project.

18.4 Draw and explain the inlet and outlet velocity diagrams of Pelton wheel.

18.5 What is the difference between Impulse turbines and Reaction turbines?

18.6 How will you classify the reaction turbines.

18.7 In what respects outward flow reaction turbine differs from inward flow reaction turbines? Which one is better and why?

18.8 Why Francis turbine is not used as a high head turbine with special reference to scroll casing and guide mechanism?

18.9 Why a reaction turbine should either be submerged in tail race or fitted in the draft tube?

18.10 Distinguish between Francis turbine, Propeller turbine, and Kaplan turbine.

18.11 What are the functions of a draft tube? How draft tube prevents the loss of head of a reaction turbine?

18.12 Describe various types of draft tubes with the help of sketches.

18.13 (*a*) Why does a Pelton turbine not have any draft tube?

(*b*) What is cavitation? Why it occurs? What are effects. Describe two methods to avoid cavitation in water turbines.

18.14 What is governing of a water turbine? What are the main components of oil pressure Governor? Describe them.

18.15 Explain governing arrangement suitable for Pelton wheel.

18.16 What qualities are expected in a governor?

18.17 Explain the servomotor mechanism for governing of Francis turbine.

18.18 Describe the mechanism for governing of Kaplan turbine.

Chapter 19

SOURCES OF ENERGY

Learning Objectives

After studying this chapter, the students will learn:

- Different sources of energy.
- Fuels and their types.
- Nuclear energy, hydro power, biomass energy.
- Non-conventional energy sources, such as solar, wind, ocean, tidal and geothermal.

19.1 INTRODUCTION

As we know that, energy is required for doing any type of work in life. It is required for cooking, heating, cooling, lighting, etc., in our homes. It is required to run our machines and other mechanical equipment in industries. It is also required for transportation. In fact almost all our development activities are directly or indirectly dependent upon the energy conservation.

The standard of living of a given society can be directly related to per capita consumption of energy. Energy is an important input in all sectors of any country's economy.

The energy is the primary and most universal measure of all kinds of work by human beings and nature. Every thing what happens in the world is the expression of flow of energy in one of its forms.

An energy source may be defined as the one that can provide us with adequate amount of energy in a usable form over a long period of time.

19.2 CLASSIFICATION OF SOURCES OF ENERGY

19.2.1 Primary and Secondary Sources

1. Primary energy sources are those which provide a net supply of energy. Coal, oil, uranium etc., are examples of this type. The energy required to obtain these fuels is much less than what they can produce by combustion or nuclear reaction. Their energy yield ratio is very high. The energy yield ratio is the ratio of energy fed back by the material to the energy received from the environment. Though these (primary fuels) contribute considerably to the energy supply, their supply is limited and hence it becomes very essential to use them sparingly.

Secondary energy sources do not produce net energy, though, these are necessary for the economy. Intensive agriculture is an example where, in terms of energy, the yield is less than the input. These sources require higher investment in terms of energy.

Coal, natural gas, oil and nuclear energy using breeder reactor are net energy yields and are primary sources of energy. Example of secondary sources are: solar energy, wind, water energy, etc. It may be necessary in future to develop the secondary sources like, solar, wind, tide wave, geothermal and ocean thermal sources.

19.2.2 Primary and Secondary Fuels

Primary fuels are those fuels which are directly available like wood, coal, petroleum or natural gas. Secondary fuels are those fuels which are manufactured/converted from primary fuels or obtained as by-products.

Examples of primary and secondary fuels are given in the table 19.1.

Table 19.1: Primary and secondary Fuels

Type of fuel	*Primary fuel*	*Secondary fuels*	
		Manufactured	*By-product*
Solid	Wood coal	Coke	Wood refuse
		Charcoal	biogas
		Briquettes	Coke breeze
		Pulverised coal	
Liquid	Petroleum	Diesel	Tar
		Petrol	Pitch
		Kerosene	Benzol
		Alcohol	Paper pulp mill waste
		Fuel oil	
		Naphtha	
Gaseous	Natural gas	Producer gas	Blast furnace gas
		Water gas	Coke oven gas
		Coal gas	Oil refinery gas
		Oil gas	Sewage gas.
		Gobar gas	
		Butane	
		Propane	
		Acetylene	
		Hydrogen	
		LPG	

19.2.3 Commercial and Non-commercial Sources

The *commercial sources* include fossil fuel (coal, oil and natural gas), hydro-electric power and nuclear power, while the *non-commercial sources* include wood, animal waste and agricultural wastes. In developed countries, most of the energy requirements are met from commercial sources, while in developing countries the use of commercial and non-commercial sources is almost equal. World's energy supply comes mainly from fossil fuels. The dependence of developing countries on non-commercial sources is likely to continue till it is replaced by other alternative sources of energy.

19.2.4 Major and Minor Sources of Energy

Major sources of energy are:

(*i*) Fossil fuel i.e., solid fuels (mainly coal including anthracite, bituminous, brown coals, lignite and peats), liquid and gaseous fuels including petroleum and its derivatives, and natural gas.

(*ii*) Water power i.e., energy stored in water

(*iii*) Energy of nuclear fission.

Minor sources of energy include: sun, wind, tides in the sea, geothermal, ocean thermal electric conversion, MHD, fuel cells, thermo-ironic converters, thermo-electric generators etc.

19.2.5 Renewable and Non-renewable Sources of Energy

The renewable energy sources are those which are perpetually available in nature (such as heat of the sun, power of the winds, power of tidal sea waves, thermal energy of the oceans, heat present in the earth's crust etc); or are generated continuously in nature (such as the potential energy of the stored water; and energy of the biomass such as wood, animal dung, agricultural waste, kitchen waste and leafy materials, and organic municipal solid wastes).

The *non-renewable energy sources* are those energy sources which have accumulated in nature over a long span of time (millions of years) and cannot be replenished even in hundreds of years. Therefore, these are the sources which are finite in quantity and cannot be reproduced. They are sure to be exhausted after some years. These energy sources include: fossil fuels such as coal, petroleum and natural gas and nuclear fuels like uranium-235.

In the age of development, mainly through industrialisation, and due to urbanisation these sources of energy are mainly used in the form of electrical energy. The electricity is the only form of energy which is easy to produce, easy to transport, easy to use and easy to control. Energy in the form of electricity is therefore used for transmission and distribution. As electricity is the main form of energy being used, its consumption per capita is considered as the index of living standard of people of a place or country. Electricity in bulk quantities is produced in power plants by using various sources of energy.

Thermal power plants generate about 80% of the total electricity produced in the world. Fossil fuels, viz. coal, fuel oil and natural gas are the energy source, and steam is the working fluid. Hydraulic power plants, besides generating power, also cater for irrigation, flood control, fisheries, afforestation, navigation, etc.

19.2.6 Classification According to Use of Fuel

Depending upon the usage, fuels are classified as domestic fuel, illuminating fuel, industrial fuel, rocket fuel, etc.

Domestic fuels supply heat for cooking and space heating. Illuminating fuels, e.g. Kerosene provide light in darkness. Industrial fuels supply heat for process heating, steam and electricity generation, etc. Rocket fuels e.g., hydrazine are used for producing enormous thrust for the propulsion of rocket.

19.2.7 Chemical Fuels and Nuclear Fuels

As per this classification, the fuels which generate heat by combustion are called 'chemical fuels' and the fuels which generate heat by nuclear fission or fusion are known 'nuclear fuels'.

Nuclear fuels are of two types: (*i*) fossil fuel, which release heat, by its fission by neutron bombardment e.g., uranium-233, uranium-235, plutonium-239. (*ii*) There are some nuclear fuels e.g. thorium-232 and uranium-238 which cannot be fissioned directly, but they can be 'converted into fissile material e.g., uranium-233 and plutonium-239 respectively. Such nuclear fuels are called fertile fuels.

19.3 FUELS

As discussed in the chapter on "Fuels and combustion of Fuels", fuel is defined as any material that combines chemically with oxygen and liberates heat.

The fuels are broadly classified as:

(*a*) *Solid fuels,* like wood and coal

(*b*) *Liquid fuels,* like petroleum products

(*c*) *Gaseous fuels,* like natural gas.

All these classes of fuels have been discussed in detail in the chapter on "Fuels and Combustion of Fuels."

19.4 NUCLEAR ENERGY

As we have discussed earlier, cheap and abundant power is essential to modern world. The rapid increase in industry and living standard of the people creates the pressure on conventional sources of power (coal, oil and gas), but these sources are rapidly depleting and will soon be unable to meet the rapidly increasing demand of power.

The discovery that energy can be liberated by the nuclear fission of materials like, uranium (U), plutonium (Pu), has opened up new sources of power of great importance.

It has been found that complete fission of one kilogram of U can produce as much energy as can be produced by burning about 4000 tonnes of coal or 2,000 tonnes of oil. This shows that nuclear energy can be successfully employed for producing low cost energy in abundance as required by the expanding and industrialising population. Therefore, nuclear energy should be used to as much extent as possible to solve power problem.

In a nuclear power station instead of a furnace, there is a nuclear reactor, in which heat is generated by splitting atoms of radioactive material under suitable conditions. The heat produced to fission of U or Pu is used to heat water to generate steam which is used for running turbo-generators. A nuclear power plant differs from a conventional power plant only in the steam generating part. The main parts of a nuclear power station are the nuclear reactor and a heat exchanger, together with the steam turbine, condensor and a generator similar to that of thermal power station. Heat is provided by fissioning or splitting of uranium atoms, in a reactor. A cooling medium takes up this heat and delivers it to the heat exchanger, where steam for the turbine is raised. The reactor and heat exchanger are equivalent to the furnace and boiler in a conventional steam plant.

The reasonably assured **uranium** reserves in our country are placed at about 34,000 tonnes of uranium oxide of which about 15,000 tonnes are considered to be economically exploitable. The established uranium resources are capable of supporting natural uranium reactors of about 8,000 MW of installed capacity. This shows that the share of nuclear power generation in the total power programme will remain modest. However, when the fast breeder reactor technology is fully developed and U-233 cycle used then there will be a large increase in nuclear power as there exist an estimated deposits of about 3,63,000 tonnes of thorium.

India has the largest, thorium reserves in the world. It would help development of nuclear power. The other countries having significant thorium reserves are USA and Brazil. Although conversion of thorium into uranium 233 a fissionable form of uranium on commercial scale has not been achieved but it is hoped that it would be possible to do so and thus thorium would be successfully used for power generation. This will solve India's power problems to a great extent. The nuclear reactors using thorium-uranium cycle are economical and have longer life. The utilisation of thorium in fast breeder reactors is quite successful.

Homi Jehangir Bhabha, the father of Indian nuclear energy programme has given a "three phased Indian Nuclear Fission Power Programme". In the first phase, the available uranium resource of about 70,000 tonnes is to be used in thermal reactors of pressurised heavy water type with a total generating capacity of 10,000 MW. In the second phase the plutonium obtained as a by-product from the first phase will be used to fuel fast breeders to sustain a generating capacity of 3,50,000 MW and at the same time irradiate naturally occurring thorium, U^{232} to convert to fissionable U^{233} variety in the third phase, breeder reactors using U^{233} as fuel will sustain a total generating capacity of 7,00,000. But to use this resource (thorium), it is necessary to go through the first two phases as described above. The two PHWR (Pressurised Heavy Water Reactors) of 235 MW capacity and the FBR (Fast Breeder Reactor) at Kalapakkam demonstrate the capacity of Indian nuclear engineers to build and operate reactors safety.

The major centre for research and development work in atomic energy in our country is the Bhabha Atomic Research Centre (BARC) at Trombay. The centre is the largest single scientific establishment in India. Besides BARC, three other national institutions associated with important aspects of atomic energy programme are as follows:

(*i*) Tata Institute of Fundamental Research, Bombay

(*ii*) Tata memorial centre, Bombay

(*iii*) Saha Institute of Nuclear Physics, Kolkata.

Advantage of Nuclear Energy

Nuclear energy is the most useful power available to mankind today. In large parts of the world it is becoming a predominant source of electrical power and a versatile tool for use in many areas of human endeavour. In India too atomic energy is being used to generate electricity and to bring about improvement in industry, agriculture, medicine and other fields through its varied applications.

The main advantages of nuclear power station compared to the conventional thermal power stations are:

1. It reduces demand for coal, oil and gas, the costs of which are tending to rise as their stocks are depleting. Greater nuclear power production leads to conservation of coal, oil etc, which can be used in other areas of utility.
2. The amount of nuclear fuel used in the plant is small. Therefore, no problems of fuel transportation, storage, etc. As stated earlier, one kg of uranium can produce, as much energy as can be produced by burning about 4,000 tonnes of high grade coal or about 2000 tonnes of oil.
3. Nuclear power plant requires less space compared to any other plant of the equivalent size.
4. Besides producing large amounts of power, the nuclear power plant can produce valuable fissile material, which is extracted when the fuel has to be renewed.

5. Nuclear plants create no air pollution, and are unaffected by adverse weather conditions.
6. Despite higher initial capital cost of a nuclear power plant, the lower fuel cost permits economical power production.
7. It minimises the ecological effects of power generation.
8. Siting of nuclear power station, generally in remote areas, would open a new field of general development in the area.
9. It does not have problems associated with mines safety.

Limitations

1. High initial capital required.
2. Difficulty in getting trained technologically competent man-power.
3. Danger of radioactivity, detrimental working conditions to health of workers, problem of disposal of radio active waste etc.
4. These can be used as base load plants only. They are not suitable for variable load operations as the reactors can not be easily controlled to respond quickly to load changes.

The biggest drawback of a nuclear power plant is that the uranium rods need to be changed periodically. This produces a lot of high level nuclear waste, which poses serious problems of disposal. When not disposed of properly, the radioactive nuclear waste produce a very adverse impact on the environment and human health. When human body is subjected to the radioactive radiations, large scale health hazards like cancers, shortening of life spans, deformations, and genetic changes (long term effects which take several years and even generations to cure) may occur. The safe disposal of nuclear wastes is an extremely difficult and costly exercise. Besides the continuous production of nuclear wastes from nuclear power plants, which need safe disposal, enormous problems do arise in decommissioning of old plants.

In addition to the above, another problem is for the disposal of hot waste water, which damage aquatic ecosystems. Possibility of a nuclear accident is most threatening and dangerous. As, a single nuclear accident can kill thousands of people, make many others seriously ill with incurable diseases, and destroy a wide area for decades by its radioactivity. Land, water and vegetables are also destroyed for long periods of time.

19.5 HYDRO POWER

The word "Hydro" is a Greek word meaning water. The water is being used since long as the source of energy. The first designed water wheels were used in nineteenth century. Later on water wheels were further developed as water turbine. Afterwards, several types of turbines were build up, some of them are: Jonval turbine (in 1837), Girard turbine (in 1850), Francis turbine (in 1865), Kaplan turbine (in 1916).

When water strikes the blades fixed on the periphery of the turbine wheel, wheel starts rotating due to the force of the water. Thus, hydraulic turbines are the devices which convert hydraulic energy into mechanical energy. Mechanical energy thus produced is used for running the electric generator, which are coupled to the turbine shaft. The electric power thus produced is known as hydro-electric power or water power. As water power is very cheap, it is now-a-days gaining the importance over thermal power.

Projects designed to produce electric power from water are known as "hydroelectric projects". Sometimes, these projects are made for controlling flood, drinking water supply, and irrigation purposes in addition to the electric generation, in which case these are known as "multi-purpose projects". Examples of such projects are Bhakra-Nangal project, Sardar Sarovar project, Hirakud, Nagarjun Sagar, Koyna, Rihand, Tehri project etc.

For selecting the site for hydroelectric plant a large quantity of water at suitable head should be available. Regarding the quantity of water which could be available for the plant, a survey of past history 'of rainfall is done. From this it can be known that during past years, what was the minimum and maximum rainfall. Maximum capacity of the plant should be decided on the basis of minimum quantity of water available, and the storage reservoir should be constructed for the maximum quantity of water, so that it can store the water even in the peak periods.

Initial cost of the hydroelectric plant is very high, but operating costs are too low. Thus, if the water is available in large quantity at a suitable head, hydroelectric plants are very advantageous.

The energy of water utilised for hydropower generation may be kinetic or potential. The kinetic energy of water is its energy in motion and is a function of mass and velocity, while the potential energy is a function of the difference in level of water between the two points, called the head. For the purpose of continuous availability of water, it is collected in natural lakes and reservoirs at high altitude may be utilised or water may be artificially stored by constructing dams across the flowing streams.

Advantages of Hydro Power

Advantages of hydroelectric power generation over thermal power generation are:

(*i*) Besides power generation, hydropower plants are quite useful for irrigation, flood control, and for supply of drinking water, afforesatation, navigation and aqua-culture.

(*ii*) Hydropower plants can be used both as base load plant and peak load plant.

(*iii*) Running cost of hydro plant is low as compared to steam power plant or nuclear power plant of same capacity.

(*iv*) The hydro plant system reliability is greater that that of other power plants.

(*v*) The hydraulic turbines can be put off and on in matter of minutes. Nuclear power plants and steam power plants lack this facility.

(*vi*) Modern hydropower plant equipment has a greater life expectancy which is about 50 years or more whereas nuclear power plant has an effective life about 30 years.

(*vii*) Steam power plants have problem of ash disposal where as hydropower plants have no comparable problem.

(*viii*) Modern hydropower plants give very high efficiency over a considerable range of load.

(*ix*) Maintenance requirements are very less as compared to fuel burning plant. Thus, in addition to lesser maintenance expenditure, there is lesser requirement of reserve capacity, leading to economy of investment cost.

(*x*) Ability of varying, their outputs is much higher and faster in hydro power plants.

(*xi*) Unforeseen outages are less frequent.

(xii) The problem of emission of polluting gases and particulates to the atmosphere also does not exist. Therefore, it does not cause any acid rain and emit obnoxious oxides of nitrogen.

(xiii) Being simple in design and operation, the hydropower plants do not require highly skilled workmen. Manpower requirement is also low.

(xiv) The water stored in the reservoir also helps in developing tourism and recreation spots.

(xv) The efficiency of the hydropower plant does not change with age, whereas there is considerable reduction in efficiency of thermal as well as nuclear power plant with age.

(xvi) The hydropower plant and its associated civil structures have a long life.

(xvii) It operates on water, which is a renewable energy source.

(xviii) The cost of generation of energy varies little with the time.

(xix) The machine used in hydel plants are more robust and generally run at low speeds (300 to 400 rpm) where the machines used in thermal plants run at a speed of 3,000 to 4,000 rpm. Due to low speed and low temperature, there are no complications of special alloys required for construction.

Due to above mentioned reasons hydel plant provides a stable and reliable operation of the system by taking fluctuations of all kinds, whether arising from system load or from forced outage of units. The hydel plants are best suited for peaking and is more economical in taking up variations of load generation that may occur through various causes and not merely covering the top of the load curve.

Disadvantages of Hydropower

The hydroelectric power plant have the following disadvantages:

(i) Initial cost of the plant including the cost of dam is high.

(ii) Hydro-station has special requirement of site which usually is an isolated area with difficult access.

(iii) Power generation by the hydel plant is dependent on the quantity of water available, which in turn depends on rain and vary from season to season.

(iv) The transmission of power from hydel stations to the load centre generally requires long transmission lines. Thus, the cost of transmission lines and losses in them are more.

(v) Hydel plants take long time for its construction as compared to thermal plants.

(vi) Large hydro-power plants disturb the ecology of the area, by way of deforestation, destroying vegetation and uprooting people. Strong public opinion against erection of reservoir type hydropower plants is a deterrent factor. The emphasis is now more on small and run-of-river hydel stations.

Positive Environmental Impacts of Dams

Although there are some negative impacts of dams, but they also help in improving the general environment of the area. Following, are the some important positive environmental impacts of dams:

(a) Overall increase in wood production due to flourinshing growth of trees and crops in the irrigated command area of the project.

(b) Excellent habitat for fisheries and water liking birds.

(c) Net improvement in public health due to availability of ample domestic water supplies leading to overall sanitation, cleanliness and better living conditions.

(d) Improvement in micro-climate caused in the adjoining areas due to evaporation from the open water surface of the reserviour and from the irrigated command area of the project. This also increases overall improved oxygen production.

(e) Development of sanctuaries and wild life becomes feasible, when adequate sweet water become available round the year by the implementation of the dam project.

19.6 BIOMASS ENERGY

Biomass is the organic matter produced by the plant or animals, and include wood; cattle dung; agriculture wastes such as crop residues; organic municipal solid wastes like paper, food wastes, and other non-fossil-fuel derived material like textiles, natural rubber, etc. Biomass is used on a large scale to obtain energy, either by directly burning the same or by converting it into more convenient form of: (a) *solid fuels* like manure, dung cakes, etc. or (b) into *liquid fuels* like ethanol, biodiesel, etc., or (c) into gas *fuels* like biogas etc.

Biomass has provided a major source of energy to human beings, since the early days of human civilization. While fossil fuels take millions of years to form, the biomass is an alternate fuel, the source of which may be considered renewable, since plant life renews and adds to itself every year. It is the solar energy stored by way of photosynthesis. The products of photosynthesis (or biomass), also called biofuels, are bulky and contain large amounts of water. Therefore, the stored energy in them must be utilized by direct burning in regions close to the source. The biomass can also be converted into variety of solid, liquid or gaseous fuels.

The fibrous waste of sugar industry is the world's largest potential source of biomass energy. Ethanol produced from sugarcane molasses is a good automobile fuel, and is now used in a third of the vehicles in Brazil.

Biogas is produced from animal dung, kitchen wastes, plants like hyacinth etc., and organic wastes from households and industries like fish processing, dairies etc. Since, in India, the biogas is primarily being produced from cattle dung (gobar), the biogas is called the *"gobar gas"*. Biogas primarily contains, methane (60 to 70%) and carbon dioxide (30 to 35%), along with small quantities of nitrogen (1 to 5%), and oxygen (0-0.1%). The use of biogas not only gives a cleaner fuel for generation of energy, but also provides an efficient method of disposal of society's wastes. Biogas plants are increasingly becoming popular in India, particularly in the rural sector and is used as a fuel for cooking, heating and lighting. The reduction in kitchen smoke by using the gobar gas has given a great relief to the ladies cooking food in houses. It also helps in reducing indoor air pollution to a great extent. Once the biogas is produced from the waste biomass, some residue is left, which can be used as an agricultural fertilizer (manure).

The biogas generation is a chemical process of decomposition of organic matter. This gas is highly combustible, and therefore can be used by burning for various applications.

As a result of energy shortages in the years to come, interest in the alternative fuel sources has increased considerably. Biomass is one of the such sources being considered. As discussed above, biomass as a source of energy has following advantages over dwindling fossil resources:

(i) It is renewable,

(ii) It is environmentally clean, and

(iii) It is easily adaptable.

Biomass is produced in nature through photosynthesis achieved by solar energy conversion.

Solar energy → Photosynthesis → Biomass → Energy generation.

Biomass resources are as follows:

(*a*) Concentrated wastes like; industrial wastes, municipal wastes and manure at large lots

(*b*) Dispersed waste residue like; crop residue, disposed manure

(*c*) Harvested biomass, standing biomass.

Energy from biomass is obtained in following three ways:

(*i*) Biomass in its traditional solid mass (wood and agricultural residue). This type of biomass is burnt directly to obtain the energy.

(*ii*) The biomass in non-traditional form (converted into liquid fuels). In this case the biomass is converted into ethanol and methanol to be used as liquid fuels in engines.

(*iii*) The biomass is fermented anaerobically to obtain a gaseous fuel called bio - gas.

19.7 NON-CONVENTIONAL ENERGY SOURCES

As already discussed conventional sources of energy *viz,* fossil fuels (coal, oil and natural gas), nuclear fuels and hydel power has very limited scope. The fossil fuels being the main fuels for thermal power, are likely to get exhausted in the current century. Therefore other systems based on non-conventional and renewable sources are being tried by many countries. These sources are solar, wind, sea, geothermal; biomass, etc. However, efforts are being made to find a suitable generic term to describe the whole range of technologies designed to tap the earth's natural energy flows. These technologies have variously been called 'alternative', 'appropriate', 'natural', 'new' sources of energy. However the term 'non-conventional' sources have gained the most wide spread acceptance.

Renewable energy sources include both 'direct' solar radiation intercepted by collector, and 'indirect' solar energy such as wind, hydropower, ocean energy (which includes tidal, seawaves, ocean thermal electric conversion, i.e., OTEC), biomass resources. Biomass being the traditional source have been discussed in separate section. Similarly hydropower have also been discussed separately. Geothermal fields tapped with present drilling technologies have a finite life, but being a new field, have been discussed in this Chapter.

Major obstacle to the implementation of non-conventional energy systems is the technical and economic uncertainties in comparison with conventional supply options and energy efficiency measures. In addition, there is a shortage of qualified personel to design, manufacture, market, operate and maintain such systems.

19.8 SOLAR ENERGY

Utilisation of solar energy is of great importance to India, since it lies in a tropical climate region of the world where sunlight is abundant for a major part of the year. Solar energy has the greatest potential of all the sources of renewable energy and if only a small amount of this form of energy could be used. It will be one of the most important supplies of energy specially when other sources in the country have depleted. Sun is regarded as an inexhaustible source of useful energy.

The 'solar power' (solar power is referred to the electricity generated by solar energy) would eliminate most of the serious environmental problems associated with fossil fuel and nuclear power.

There are certain major limitations to the extensive application of solar energy such as:

(*i*) The intermittent and variable manner in which it arrives at the earth's surface and

(*ii*) The large area required to collect the energy.

The solar energy is effective only during the day time, and if the power supply is to be made during night, then some reservoir of energy such as storage battery or heat accumulator should be used. Solar energy cannot be used during cloudy weather and rainy season. Its intensity during winter is very less. At present harnessing, storage and use of solar energy is much more expensive than using fossil fuel. It is believed that solar power can become economically feasible with the following aims achieved:

(*a*) Availability of better heat collectors

(*b*) Availability of improved materials and manufacturing techniques

(*c*) Better techniques for storage and cheap distribution of solar power.

India has the potentialities of solar and wind energy, but it needs the advanced technology to tap this potential.

Applications of Solar Energy

Traditionally, we have been using solar energy for drying clothes and food grains, for preservation of eatables, and for obtaining salt from seawater. However, in this modern age, we have already invented several techniques for harnessing the solar energy. Some of the important solar energy applications are:

1. Solar water heating
2. Heating and cooling of residential buildings
3. Solar drying of agricultural and animal products
4. Solar distillation on a small community scale
5. Salt manufacturing by evaporation of sea water or inland brines
6. Solar cooking
7. Solar pumping of water
8. Food refrigerator
9. Solar furnaces
10. Solar power generation
11. Solar photovoltaic (PV) cells
12. Bio-conversion, wind energy, ocean thermal energy conversion, tidal energy, sea waves etc., which are indirect sources of solar energy.

Electricity can be produced from the solar energy by photovoltaic solar cells, which convert the solar energy directly to electricity. This photovoltaic technology is also used for the energisation of pump sets for irrigation, drinking water supply, and rural electrification covering street lights, community TV sets, medical refrigerators, and other small power loads.

19.9 WIND ENERGY

Winds arise primarily from temperature difference of the earth's surface resulting from unequal exposure to (or absorption of) solar radiation. Air heated by contact with a warmer surface tends to rise, and cooler air from a less heated surface flows in, to take this place; the resulting air flow constituted wind. Wind energy is thus a form of solar energy. Like direct solar energy, wind energy is also highly variable.

It is the conversion of kinetic energy of the wind into mechanical energy that can be utilized to perform useful work or to generate electricity. Wind energy is a renewable source of energy. India with its long coastline, blessed with steady winds almost throughout the year has tremendous potential of wind power.

Wind energy which is an indirect source of solar energy can be utilised to run wind mills which in turn drives a generator to produce electricity. Wind mills may be either horizontal axis wind mills or vertical axis wind mills.

Some **characteristics** of wind energy are as under:

1. It is a free and inexhaustible renewable source of energy
2. Wind power systems are non-polluting, and has no adverse influence on the environment
3. It avoids fuel provision and transport
4. These are economical to the rural areas which are at far remote places from existing grids.

However, following **problems** are associated with wind power:

1. Wind energy available is unpredictable and fluctuating in nature
2. Wind energy systems are noisy in operation
3. Large areas are needed to install wind farms of electrical power generators
4. Wind mills usually kill the birds.

Wind being an intermittent source, cannot be used as the sole source of electricity, and requires some other back.

Wind power is a function of wind speed and therefore, the average wind speed of an area is an important factor for economically feasible power. As wind speed increases with height, therefore, power available 30 m above ground is 60% greater than at 10 m heights. Therefore wind mills need to be located at enough heights. As already discussed in the previous chapter, in India, high wind speeds are obtainable in coastal areas of Saurashtra, Maharashtra, Tamil Nadu, Kerala, Karnataka, Orissa, Andhra Pradesh, Western Rajasthan and some parts of central India. In these areas, there could be possibility of using medium and large sized wind mills for generation of electricity and feeding the same into the grid.

As a rough estimate, India has a potential of 20,000 MW of wind power. Traditionally, wind mills were used extensively in the middle ages to mill grain and lift water for land drainage and watering cattle. Wind energy converters are still used for these purposes today in some parts of the world, but the main focus of attention now lies with their use to generate electricity.

19.10 OCEAN ENERGY

Solar radiation is absorbed by the seas and oceans, which causes, like the wind, ocean currents and moderate temperature gradients from the water surface downward, especially in tropical waters. The oceans and seas constitute about 70 per cent of the earth's surface area, hence, they represent a large storage reservoir of the solar input. The conversion of solar energy stored as heat in the ocean into electrical energy by making use of temperature difference between the warm surface water and the colder deep water. The facilities proposed for achieving this conversion are commonly referred to as **OTEC** (Ocean Thermal Energy Conversion) plants.

Periodic rise and fall of the water level of the sea which are carried by the action of the sun and moon on the water of the earth is known as **tide.** The tidal range varies from few centimeters to about 8 to 10 m in some parts of the world. The potential energy of the tides can be trapped to generate power.

Ocean and sea waves are caused indirectly by solar energy like wind and OTEC. Wave energy drives from wind energy, which drives in turn from solar energy. Devices that convert energy from waves can therefore produce much higher power densities than solar devices. But the wave energy development is not nearly as far along as wind and tidal energy.

The hydrological cycle results in rain fall, which causes river flows, that can be trapped behind dams to even out the variations in river flows and thus become the source of **hydro-electric energy.**

Thus the **Ocean sources of energy** are:

(*i*) Ocean Thermal Energy Conversion (OTEC)

(*ii*) Tidal energy

(*iii*) Wave energy and

(*iv*) Hydrological cycle, which emanates from the sun - ocean system through surface water evaporation by solar heating.

19.11 TIDAL ENERGY

The water level in a sea rises and falls during tides, caused mainly by gravitational attraction of the moon and the sun on the water of solid earth and the oceans. About 70 per cent of the tide producing force is due to the moon and 30 per cent due to the sun. The moon is therefore is the major factor in the tide formation.

Surface water is pulled away from the earth on the side facing the moon, and at the same time the solid earth is pulled away from the water on the opposite side. Thus, high tides occur in these two areas (directly under the moon and diametrically opposite the earth's surface), with low tides on intermediate points. As the earth rotates, the position of a given area relative to the moon change, and so also do the tides. Thus, there are periodic succession of high and low tides. Therefore, the time between high tides and low tide at any given location is a little over 6 hours.

Thus, a full moon as well as no moon produce a high tide. In a period of 24 hours 50 minutes, there are therefore, two high tides and two low tides. The difference between high and low water levels is called the **range of the tide.** The periodic rise and fall in levels is in rhythms with daily cycle of rising and setting of sun and moon.

The potential energy developed by the water head created during high tide can be used to run the turbines, to generate electricity. Electricity produced by these tides is called *tidal power.* If the differential head between high tide and low tide could be utilized in operating a low head reversible hydraulic turbine, the tidal energy could be, covered into electical energy by means of an attached generator. During high tide, the sea water flows towards the land reservoir through the opeinings in the dam or the barrage, and turns the turbines, generating electricity. During the low tide, as the seawater level falls, the water from the upstream reservoir flows towards the downstream in seaward direction, again turning the turbine blades, to generate electricity.

The main problem in harnessing the tidal power is that, it is not possible to harness the tidal power along most of the coasts of the world. If the tidal range is less than about 8 metres or if narrow and enclosed bays are not available, then tidal power development becomes very uneconomical. Therefore, very few sites are available where tidal energy can be economically harnessed.

Following *characteristics* are to be noted regarding the .tidal phenomenon:

1. The tidal are periodical phenomenon and two tides in any cycle are alike
2. In any month, the tides on the full moon and no moon days are particularly higher than the rest
3. The mean tidal range varies from place to place
4. It is free from pollution
5. It is inexhaustible, and independent of the rains
6. Tidal power plants require less area of valuable land and are located on sea shore
7. The tidal range is highly variable and therefore, the turbines have to work on a wide range of head variation. This affects the efficiency of the plant
8. High chances of machinery being get corroded due to sea water.

19.12 GEOTHERMAL ENERGY

Geothermal energy is the energy which lies embeded within the earth as heat (i.e. thermal energy) in the earth's crust. The earth's crust constitutes a potentially useful and almost inexhaustible source of energy. This heat is apparent from the increase in temperature of the earth with increasing depth below the surface. The eruption of hot lava from volcanoes at the surface at about 1200°C temperature, clearly shows that the deep interior parts of the earth are very very hot, and may be in liquid or semi-liquid state.

From the extremely hot interior of the earth, a continuous upward current of heat flows towards the upper crust of the earth. Due to such currents, the upper 3 km portion of the earth's crust, also contains a huge amount of heat energy, which if removed is liable to be replaced again by the new heat coming from the heat currents from the interior portion of the earth. This source of heat energy is therefore, considered as a renewable energy source.

In the natural geothermal areas, water circulating systems are continuously transporting the heat energy to the surface, and this is observed as natural hot springs, called natural geysers. These natural eruptions of hot waters with steam with temperatures reaching upto about 350°C, may occur through the cracks existing in the crystal rocks.

The utilisation of geothermal energy can be made by drilling holes or wells into these natural circulation systems, and to top the geothermal fluids through these wells. The naturally erupting geyser can also be utilised for this purpose. These hot fluids (water + steam) can either be used for generation of electricity, or for direct heating purposes. For large scale use bore holes are normally sink with depth upto 1000 m, releasing steam and water at temperatures upto 200 or 300°C and pressures upto 30 mPa.

Geothermal energy is advantageous both from the point of view of conserving fossil fuel and of pollution control. The hot water coming out of the ground with steam (which has about 1/3 of the available thermal energy) is discarded because it contains about 30% dissolved salts and minerals, which can cause serious rust damage to the turbine.

QUESTIONS

19.1 Name and explain various sources of energy in brief.
19.2 What are the various types of fossil fuels? Explain.
19.3 What is meant by composition of fuel? Give percentage composition of some fuels.
19.4 Explain ultimate analysis to find various constituents of a solid fuel.
19.5 What are the advantages of liquid fuels and gaseous fuels?
19.6 What are non-conventional sources of energy? Explain.
19.7 Write short notes on the following:
(*a*) Proximate analysis. (*b*) Solar energy.
(*c*) Wind energy. (*d*) Geothermal energy.
19.8 What is wind energy? State the characteristics of wind energy.
19.9 What is tidal power? What is tidal range? State the advantages and disadvantages of tidal powers.
19.10 What are primary and secondary energy sources? Give two examples each.
19.11 Discuss "alternate energy sources".
19.12 What are the conventional and non-conventional energy sources? Describe briefly.
19.13 Discuss the adventures and limitations of solar power.
19.14 What is meant by renewable energy sources? What are the advantages and limitations of renewable energy sources?
19.15 Name the main applications of:
(*a*) Solar energy. (*b*) Wind energy
19.16 (*a*) What is the difference between biomass and biogas?
(*b*) What is wind mill?
19.17 (*a*) Discuss geothermal source of energy.
(*b*) What are the difficulties in tidal power developments?
19.18 (*a*) Explain how the tides are formed.
(*b*) What are the limitations of the tidal power stations?
19.19 What are the main characteristics of a good hydel plant site?
19.20 Discuss the advantages and disadvantages of hydro-electric power generation over thermal power generation.
19.21 What considerations have to be kept in view in siting nuclear plants?
19.22 Discuss the advantages and disadvantages over nuclear power plants as compared with conventional thermal power plant.

Chapter 20

ENGINEERING MATERIALS, STRESS AND STRAIN

Learning Objectives

After studying this chapter, the students will learn:

- Types engineering materials.
- Properties of materials, namely physical, mechanical, thermal, electrical and magnetic properties.
- Ferrous metals and alloys, such as pig iron, cast iron, wrought iron, steel, alloy steel and special alloy steels.
- Heat treatment of steel.
- Non ferrous metals and their alloys.
- Corrosion and its prevention.
- Miscellaneous engineering materials.
- Stress and strains in materials including strain energy, thermal stresses and strains, and stress-strain diagram.

20.1 INTRODUCTION

In engineering, successful design depends upon the selection of materials having the most desirable properties for a given function, whilst at the same time taking into consideration the cost of the materials and the problems connected with the bringing of these materials to the required shape. Therefore, an engineer must know about the properties of available materials before deciding on its use.

Engineering materials are mainly classified as :

1. Metals
   (a) Ferrous metals. (6) Non-ferrous metals.
2. Non Metals.

Metals are those materials which possess lustre (on polishing), strength and conductivity eg. iron, copper, aluminium, nickle, lead, tin, zinc, magnesium, gold, silver, etc.

Non-metals are those materials which mostly have low value of lustre, strength and conductivity eg. timber, rubber, plastic, carbon, sulphur, phosphorous, glass, etc.

In engineering practice the durability of the materials is an important factor. Therefore, study of properties of different materials and their standards is very essential. Metals used in engineering trades and construction are broadly classified into two groups:

(i) Ferrous metals and their alloys.

(ii) Nonferrous metals and their alloys.

The important ferrous metals in use are: cast iron, wrought iron, low-carbon steel or mild steel and high-carbon steel or hard steel. The more useful nonferrous metals are; aluminium, copper, lead, tin and zinc.

An **alloy** is an intimate mixture with metallic properties and is composed of two or more elements, of which one at least is a metal. The metal having the largest proportion is termed as the "bare metal" whereas the other elements are termed as "alloying elements". An alloying element may be a non-metallic material.

Sometimes it so happens that none of the available pure metals exactly satisfy all the requirements expected of a material for a particular purpose. In all such cases, the material, called alloy, is obtained by mixing or melting together two or more metals.

Some examples of commonly used alloys are cast iron, wrought iron, steel, brass, bronze, gun metal, white metal, duralumin etc.

Pure metals are generally too soft and weak for many commercial applications. By alloying a pure metal in proper proportions with other metals or non-metals desirable mechanical or other properties can be developed in the materials. For example, steel, being an alloy of iron and carbon, is stronger than pure iron and can also be heat-treated to produce desirable mechanical properties. Therefore, alloys are more widely used as engineering materials than pure metals.

Alloys are preferable to pure metals in the following respects :

1. Alloys are stronger than pure metals.
2. Alloys have lower melting points, therefore, these can be easily shaped by casting.
3. Alloys are harder and hence are more durable as they do not wear out so soon.
4. Alloys generally resist the action of air and acids more than pure metals.
5. Alloys offer more resistance to electricity than pure metals.

20.2 PROPERTIES OF MATERIALS

1. Physical Properties

(i) Specific Gravity. The specific gravity of a material is the weight of a given volume of the material as compared with the weight of an equal volume of pure water. It is usually measured at a temperature of 60°F (15.5°C).

(ii) Specific Heat. The heat absorbing capacity of the substance differs from material to material. Therefore, specific heat is the heat required to raise the temperature of unit weight of a material by one degree.

(iii) Fusibility and Fluidity. This property means that when heat is applied to a metal then a stage comes where it melts and tends to flow. It is a major factor considered while casting metal.

(iv) Weldability. It is the ability of uniting two pieces of metal by means of heat or pressure or both heat and pressure.

(v) Elasticity. It is that property of metal by virtue of which it behaves like a spring and regains its original dimensions on removal of the load.

A material is said to be *perfectly elastic* if the whole of the stress produced by the load disappears completely on the removal of load. The modulus of elasticity of Young's modulus (E) is proportionally constant between stress and strain for elastic materials.

(vi) Plasticity. Plasticity is the property that enables the formation of permanent deformation in a material. It is reverse of elasticity, a plastic material will retain exactly the shape it takes under load, even after the load is removed.

Upto a certain limit all the materials behave elastic but beyond that limit they are unable to gain their original dimensions and retain to their moulded shape. This property of the material is called "plasticity".

(vii) Porosity. When the materials are in plastic or molten state then they contain a certain amount of dissolved gases or steam, which are evaporated when they set, thereby forming gas holes and pores in the materials. This property is called porosity.

2. Mechanical Properties

(i) Strength. The strength of a material is a measure of its ability to resist the application of load without rupture. The application of load to a part of a structure causes internal balancing forces to be set up on that part. The intensity of such internal force is called "stress", which may be divided into three kinds: Compression, tension and shear.

Thus, the strength of metal is its ability to withstand various forces to which it is subjected during a test or in service. It is usually defined as tensile strength, compressive strength, shear strength etc. Strength of materials is a general expression for the measure of capacity of resistance possessed by solid to any cause tending to produce in them a permanent and disabling change of form or fracture.

The "ultimate strength" means the breaking load per unit of original section of the material. Therefore, the property which enables a substance to resist fracture under the action of tensile stress, is the ultimate strength in tension.

Under load, the length of material also increases, therefore measure of increase in length per unit length is called "strain".

For tensile and compressive stresses the ratio,

$$\frac{\text{Stress}}{\text{Strain}} = \text{Modulus of Elasticity } (E) \text{ of material.}$$

Similarly for shear stress the ratio,

$$\frac{\text{Stress}}{\text{Strain}} = \text{Modulus of Rigidity } (N) \text{ of material}$$

The values of E and N are naturally not the same for a given material because, the nature of stress and strain is different.

(ii) Hardness. It is the property by virtue of which a material resists penetration or scratching by other materials. For example, glass is very hard material but diamond is the hardest of all known materials. Therefore glass is cut by a diamond-pointed tool.

(iii) Hardenability. It is the property by virtue of which a metal is capable of being hardened by heat treatment. Steel can be hardened by heating and quenching in a cool medium like water, dry air, oil or molten lead.

(iv) Toughness. Some materials withstand great deformation together with high stress without fracture, this ability is called toughness. In other words it is the amount of energy a material can absorb before actual fracture takes place.

(v) Brittleness. It is the property of a material to fracture on receiving a blow or shock. Glass and cast iron are good examples of brittle materials.

(vi) Malleability. A material which can be hammered out into thin sheets is malleable. Gold, silver, copper and aluminium are the good examples of malleable materials.

(vii) Ductility. It is the property whereby a material can be drawn out into a wire without cracking. Copper, gold, silver and iron are examples of a ductile material.

(viii) Creep and Slip. Under high temperature and stress many materials flow like a viscous liquid which *is* usually called creep. The creep continues as long as load is applied.

The material will go on elongating as long as there is a load and will ultimately break. At ordinary temperature however, after some elongation, the deformation will stop even if the load is acting. This phenomenon is called "slip".

Therefore, difference between slip and creep is by the application of force beyond its elastic limit the slip occurs and a permanent set is established, while creep causes the material to deform continually so long as the load is present.

(ix) Fatigue. When subjected to fluctuating or repeating loads (or stresses), materials tend to develop a characteristic behaviour which is different from that under steady loads. Fatigure is the phenomenon that leads to fracture the material under such conditions. Fracture takes place under repeated or fluctuating stresses whose maximum value is less than the tensile strength of the material under steady load.

Fatigue Resilience. Under repeated loading and unloading conditions, even ductile materials fracture without any slip. During this process metals change and crystallise their structure which is called "fatigue". Therefore, the property to resist fatigue is called the "fatigue resistance of the material" and is an important quality of engineering purposes.

(x) Resilience. The property of a material to absorb energy within the elastic range is known as resilience, which is desirable in the materials for springing action.

3. Thermal Properties

Conductivity. It is the capacity of a metal to conduct heat from the hot end to the cold end. Silver and copper are the good examples of heat conductors. A material that conducts heat very badly is called heat insulator.

4. Electrical Properties

Conductivity. It is the capacity of a metal to conduct electricity from one end to the other. A metal that allows electric current to pass through it with minimum loss is called a good conductor of electricity. Silver and copper are also good examples of electric conductors. Similarly a material which resists the flow of current through it is called electric insulator.

5. Magnetic Properties

A magnet attracts materials like iron, steel, nickel, cobalt, chromium and manganese which is called magnetic materials. Some materials like copper, zinc, lead, antimony and bismuth are repelled by a magnet which are called diamagnetic. Remaining other materials are called non-magnetic.

20.3 FERROUS METALS AND ALLOYS

Most widely used engineering metals and alloys are the ferrous metals and their alloys in the form of wrought iron, cast iron, carbon steels and alloy steels. The principal raw materials for all ferrous products is pig iron, the product of blast furnace.

The principal ferrous materials used in engineering may be classified as below:

1. Pig Iron

All iron and steel products are derived originally from pig iron, which is the raw material obtained from the reduction of iron ore in blast furnace. The process of reduction of iron ore to pig-iron is known as smelting.

During smelting process coking coal is used to provide necessary heat and lime stone to act as a flux. All the three materials lime, coal and iron-ore are fed in successive layers into the furnace. Their proportions are approximately one-sixth, one-third and one-half respectively. This process of charging goes on continuously. The molten metal from furnace is run off into moulds giving pig-iron. This pig-iron is the basic material for all the products of ferrous metals and their alloys.

2. Cast-Iron

Cast iron is obtained by remelting and refining the pig-iron by means of "cupola furnace". The molten cast-iron from the cupola is directly used for casting the engineering articles. Ccast-iron contains about 2 to 4 per cent of carbon and a small percentage of silicon, sulphur, phosphorus and manganese.

Carbon in cast iron can exist in two forms (a) combined, (b) free form. An excess of combined carbon makes cast iron very hard, brittle and unmachinable. Hard iron on breaking shows a shiny white fracture, often with a glassy appearance. Therefore under this condition it is called as "white cast iron".

Carbon in the free form is also called "graphite" or graphitic carbon, which is generally present in the iron in the form of grey crystalline plates. On breaking, because it shows grey crystalline fracture, therefore, it is called "grey cast iron".

An intermediate stage between these two forms show patches of grey in the white structure, this iron is called "mottled iron". This is a hard and machinable grade of the cast iron. The other varieties of the cast iron in common use are:

(*a*) *Malleable Cast Iron.* The ordinary cast iron is hard and brittle. Therefore, it is unsuitable for light and thin castings or which are subjected to shocks and vibrations. To overcome such difficulty the castings of hard, white iron may be converted into malleable castings by embedding them in powdered haematite and keeping them at a bright red heat (950 to 1000°) for several days, and then cooling very slowly. By this process, the surface is rendered tough and then cooling very slowly. By this process, the surface is rendered tough and shock-resisting, while the core remains hard and brittle. Several, types of furnaces, namely cupola, air furnace and electric furnace are employed for malleable cast iron. There are two varieties of malleable cast iron : 1. white heart, and 2. black heart.

(*b*) *Chilled Cast Iron.* When molten cast-iron is cooled quickly the iron so produced is called "chilled iron". Since cast iron has a higher thermal conductivity than sand, the chilled portion of the casting undergoes rapid solidification and cooling and, thereby, producing a hard surface. Sometimes a casting is chilled intentionally and sometimes becomes chilled accidentally to a small depth. Chilling is done where some parts of the castings are required to have the hardness of white cast iron, while others are required to have relatively soft and tough core of grey cast iron.

(*c*) *Alloy Cast Iron.* These have been developed to overcome certain inherent deficiencies in ordinary cast iron and to give qualities more suitable for special purposes. The main alloys of cast iron are:

(*i*) ***Nickel.*** About 2.2% nickel hardens and strengthens the iron and makes it heat-treatable, like steel. An iron with a higher percentage of nickel is first tempered by cooling down slowly from about 700°C to make it machinable. After machining it is heated again to about 800° C and cooled in an air blast to obtain the necessary strength. To make corrosion and wear-resistant a high percentage upto 20% of nickel is used in iron.

(ii) ***Manganese.*** Cast iron alloyed with 10% nickel and 6 per cent manganese become non-magnetic. A special non-magnetic patented cast iron is known under the name of "Nomag".

(iii) ***Chromium.*** Where improvement in the heat resisting and wearing qualities are of importance, chromium additions with the nickel are often found useful. Cast iron alloys in this manner find wide application for pumps of all types in which frictional, as well as errosive, wear has to be considered. White cast iron containing 10 to 30 percent chromium and 1 to 3 percent total carbon, also exhibit a high degree of heat resistant combined with strength at high temperature.

(iv) ***Molybdenum.*** It is added 0.4 percent to 1.5 percent to cast iron. It improves mechanical properties, increases tensile strength and makes shock-proof to cast-iron. It is also added along with nickel which saves manufacturing cost of alloy cast iron.

3. Wrought Iron

It is iron from which most of the impurities have been removed. It contains less than 0.15 percent carbon and remained about 99.9 percent pure iron. It is produced from pig-iron by any of the process such as *(a)* refining, *(b)* puddling, *(c)* shingling, and *(d)* rolling.

It is tough, malleable and ductile. It cannot be melted, but at a white heat, it becomes soft enough to take any shape under the hammer. It can neither be hardened nor tempered like steel. It rusts more quickly than cast iron. It is soft, easily magnetised and welded. It is used for cores of dynamos, lifting chains, bolts, nuts and rivets.

4. Steel

It is fundamentally an alloy of iron and carbon, with the carbon content varying from less than 0.25 to 1.5 percent. The type of steel is known by its carbon content, (a) low carbon steel—upto 0.25 percent carbon, *(b)* medium carbon steel—from 0.25 to 0.75 percent carbon and (c) high carbon steel from 0.75 to 1.5 percent carbon. It can be produced by any of the following methods:

1. Cementation process.
2. Shear steel.
3. Bessemer process.
4. Open-hearth process.
5. Crucible process.
6. Electric process.
7. Basic oxygen (L.D.) process.

Properties and uses of steel. The properties and some uses of the various types of carbon steel are as given in table 20.1:

Name of Steel	Percentage of Carbon	Properties	Uses
(a) Low carbon or Mild Steel.	upto 0.25%	Ductile, stronger, harder and less malleable. Easily machined, welded, forged and stamped to any shape.	Bolts, nuts, rivets, tubes, plates, bridges, crank axles, shafts, gears, valves, hydraulic cylinders, rams, spindles etc.
(b) Medium carbon steel.	0.25% to 0.7%	Stronger, harder, but less malleable and ductile than low carbon steel, can not be easily worked.	Steel mill rolls, gear wheels, rifle barrels, gun parts, shells, wheel tyres, die blocks, gears, mandrels, set screws, shafts, cold chisels, wrenches, jaws for vices, shear blades, hacksaw frames etc.
(c) High carbon steel.	0.7% to 1.5%	Strong and less ductile. Can be heat treated according to the desire of hardness and toughness, can be made very hard without undue brittleness.	Railway rails, circular saws, rock drills, chisels, punches, dies, leaf springs, coil springs, keys, pins, mendrels, taps, razors, knives, files and other hand tools.

Table 20.1 Types of carbon steels & their properties

5. Alloy Steel

It is a combination of iron and carbon with some other element or elements in varied proportions, the carbon content is usually low. The additional elements give the steel specific properties, those cannot be obtained in plain carbon steels. The common alloy steels and their uses are as given under :

(a) Nickel alloy steel. It contains from 2 to 5 percent nickel from 0.1 to 0.5 percent carbon, which contributes great strength, hardness, elasticity, ductility and good resistance to corrosion. Steel with 25 percent nickel offers the greatest resistance to rusting, corrosion and burning at high temperature. It is used in the manufacture of boiler tubes, engine valves, sparking plugs etc.

Steel with 36 percent nickel has nearly zero coefficient of expansion which is also called as "Invar". Therefore, it is very suitable material for manufacturing parts of measuring instruments.

(b) Chromium, alloy steel. It contains from 0.5 to 2 percent chromium and from 0.1 to 1.5 percent carbon, which contributes great strength, ductility, elasticity and resistance to corrosion. It is used for rollers, balls, and races for bearings.

Steel with 3.25 percent nickel, 1.5 percent chromium, and 0.25 percent carbon is more suitable for armour plates, engine crank shafts, axles and gears.

(c) Tungsten alloy steel. Generally it is used in conjunction with other elements. It contains from 3 to 18 percent tungsten and from 0.2 to 1.5 percent carbon. This steel retains its hardness even when highly heated. Therefore, it is used for dies, valves, taps and cutting tools. It is also a suitable material for permanent magnets.

(d) Vanadium alloy steel. A low or medium carbon steel with 0.2 percent of vanadium has good tensile strength and elastic limit without loss of ductility. Steel with 0.5 to 1.5 percent of chromium, 0.15 to 0.3 of vanadium and 0.13 to 1.1 percent carbon has extremely good tensile strength, elastic limit and ductility. It is used for springs, shafts, gears, pins etc.

(e) Manganese alloy steel. It contains about 1.5 percent manganese and from 0.4 to 0.55 percent carbon, which contributes good strength in both the hot-rolled and heat-treated condition. With 11 to 14 percent manganese and 1 to 1.4 percent carbon, possesses extreme hardness and ductility. They are used for gears, axles, shafts etc.

(f) Silicon alloy steel. It contains from 1 to 2 percent silicon and 0.1 to 0.4 percent carbon. It behaves like nickel steel but some other alloying elements are also added to make it more useful for electrical machineries, engine valves, springs and corrosion resisting materials.

(g) Cobalt alloy steel. The addition of cobalt in steel provides red hardness to a high temperature, hardness and strength. Steel with about 35 percent cobalt imparts high magnetic properties, which may have upto eight times field strength than other steels.

6. Special Alloy Steels

These are commercial steels and have their specific use. The special alloy steels in general use are as given under:

(a) Stainless steel. The best composition of this steel contains chromium 18 percent, nickel 12 percent and molybdenum 2 percent. It resists oxidation and corrosive attack from most corrosive media. Therefore, it is used for cutlery, surgical and dental instruments.

(b) Heat resisting steel. It contains chromium from 23 to 30 percent, carbon less than 0.35 percent. It gives good service at temperatures from 815 to 1150°C. Therefore, it is used for furnace parts, annealing boxes and with other alloys for cutting tool bits.

(c) Magnet steel. Steels having composition 15 to 40 percent cobalt, 10 percent tungsten, 1.5 to 9 percent chromium, 0.4 to 1 percent carbon possess good magnetic properties. Therefore, these are used for permanent magnets, magnets for loudspeakers and other electrical instruments and machines.

(d) High-speed steels. Tool made from this steel has ability to resist softening at high temperatures developed due to high speed of work, therefore, it is named so. Steel having composition 18:4:1 means containing 18 percent tungsten, 4 percent chromium, and 1 percent vanadium, 0.7 percent carbon which it considered to be one of the best all purpose tool steel. Some other elements such as cobalt, molybdenum are also added in it to improve its quality.

(e) Shock-resisting steel. It contains 2.25 percent tungsten, 1.5 percent chromium, 0.25 percent vanadium and 0.5 percent carbon. Another composition of this steel having 2 percent silicon, 0.8 percent manganese, 0.3 percent molybdenum and 0.55 percent carbon is also used for coil springs, leaf springs etc.

(f) Sintered carbide. As the carbides of iron and tungsten formed in alloy or high speed steel are not sufficient to retain their hardness above 1000°C temperature. Therefore, to work on hard materials a very hard material is prepared with tungsten carbide. It is prepared by mixing tungsten metal powder with lampblack and heating the mixture to about 1600°C in an atmosphere of hydrogen until the two substances have undergone the chemical reaction to produce tungsten carbide. The powder carbide is mixed with powdered cobalt metal and water, then the mixture is compressed under high pressure in a desired shape. This method of forming of metal is known as "sintering process". Carbide pieces are used to make the cutting tool tips.

20.4 HEAT TREATMENT OF STEEL

Properties of high carbon or hard steels are considerably changed by some definite heating and cooling treatments, which is known as heat treatment of steel. It happens because the internal structure of steel varies by heating and cooling. The main constituents giving different internal structures to steel are as given under:

1. Micro-constituents of Steel

(a) Ferrite. It is the softest and least strong constituent of steel. It is also the toughest and most malleable and ductile constituent, with highest electrical conductivity and magnetic permeability. At room temperature it can hold only about 0.008 percent carbon in solution.

(b) Cementite. It is a carbide of iron, which is found in all steels containing over 0.87 percent carbon after they have been cooled slowly from just above the critical range. It crystallises in the form of thin plates and is the hardest and most important constituent of steel. It is so hard that it easily scratches glass, and is so brittle that it flies into pieces under hammer blows.

(c) Pearlite. It is a mixture of ferrite and cementite in the ratio 85 to 15 respectively by weight. When a steel contains 0.85 percent of carbon, it consists wholly of pearlite. It is a strong metal and may be cut reasonably well with cutting tools.

(d) Austenite. When a carbon steel is heated above critical or 1110°C temperature, this steel is solid solution of carbide of iron and is called "austenite". It is hard, but its cutting hardness is far less remarkable than its abrasive hardness. It has great toughness and high tensile strength. It is non-magnetic.

(e) Martensite. It *is* next in hardness and brittleness to cementite. It is obtained when carbon steel is cooled rapidly from temperature within or above the critical range. It is composed of needle-like crystals above the critical range. It is slightly magnetic and retains its hardness indefinitely unless warmed.

(f) Troostite. It is obtained from martensite by warming it to about 400°C. It is less hard and less brittle than martensite.

(g) Sorbite. It is obtained by rapid cooling of steel heated to above 400°C. It is tougher and stronger than pearlite. It is softer and more ductile than troostite.

2. Iron-carbon Equilibrium Diagram

When steel is heated, its temperature does not increase at a constant rate. The temperature rises, after proceeding steadily, is momentarily checked and after a short interval, the temperature continues to rise as before. During this period of arrest the metal absorbs heat, but instead of raising temperature, the heat brings about certain changes in the metal. The temperatures at which these periods occur are called "critical points".

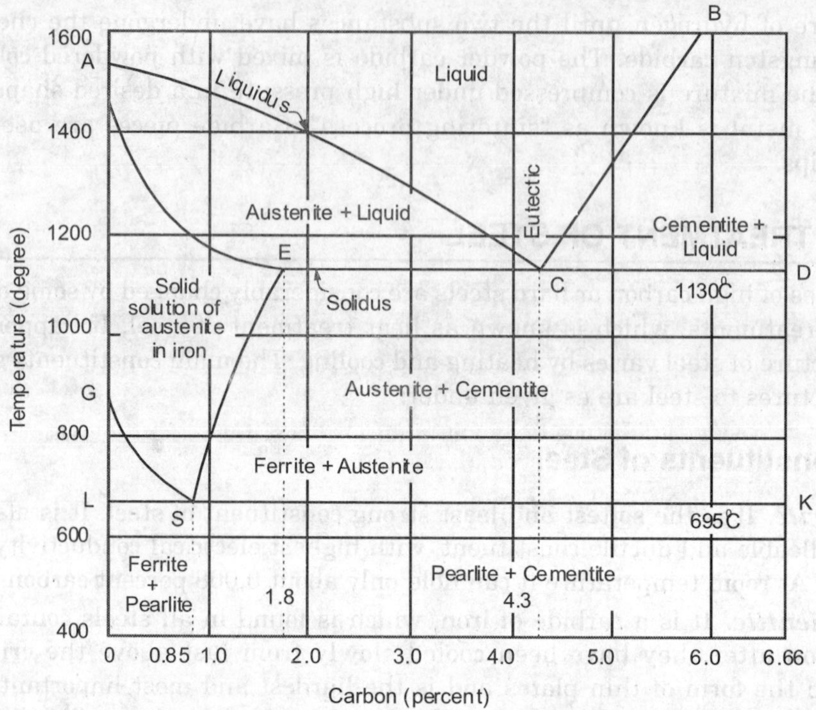

Fig. 20.1 Iron-carbon equilibrium diagram.

Referring Fig. 20.1 the carbon varying from zero to 6.66 percent and temperature ranging from zero to 1600°C. Above the line *ACB* the alloy remains in liquid form, therefore, this line is called the "liquidus". In case of pure iron this stage remains above 1550°C and above 1130°C in case of white cast iron containing 4.3 percent carbon.

The melting temperature decreases as percentage of carbon in the alloy increases, it happens upto a limit of 4.3 percent carbon. Below the line *AECD* the alloy remains in wholly solid form, therefore this line is called the "solidus". Inside the loop *ACEA* the alloy remains in partly liquid and partly solid form. In area *DCESKD* the form is a mixture of austenite and cementite, and in area *AESGA,* the alloy is a solid solution of austenite in iron. In small loop *GLS* it is a mixture of ferrite and austenite. Below line *LSK the* alloy is a mixture of pearlite and cementite. As the *C* is a lowest melting point, therefore, it is called "eutectic point".

3. Atomic Arrangements in Metals

The atoms of solid metals are held together by the electrostatic attraction between the positive ions in the nucleus and the negative electron cloud. In this way the electrostatic attractions form three-dimensional nets or "space lattices" which divide space into equal prisms called "unit cells".

There are different types of space lattices, the structures of most common metals follow one of the three lattice types as given under and their atomic arrangements as shown in Fig.20.2

(*a*) Body-centered cubic structure.

(*b*) Face-centered cubic structure.

(*c*) Close-packed hexagonal structure.

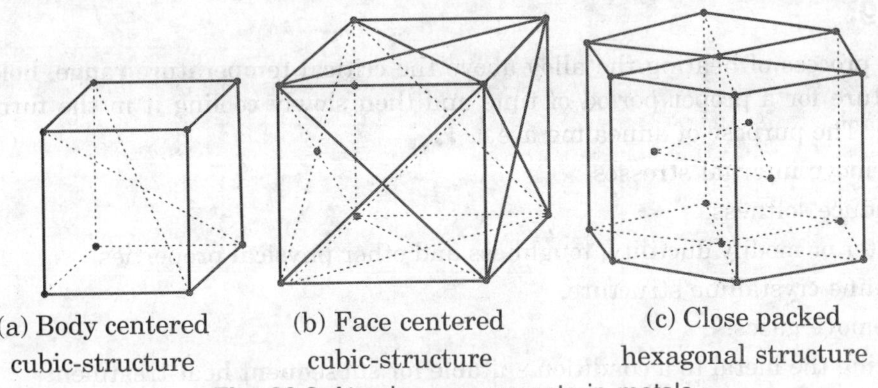

(a) Body centered cubic-structure (b) Face centered cubic-structure (c) Close packed hexagonal structure

Fig. 20.2 Atomic arrangements in metals.

4. Allotropic Forms of Metal

A number of metals exist in one lattice form over a certain range of temperature, but at a somewhat higher temperature the lattice form changes to another type which is stable over that temperature range. Such metals are said to be allotropic.

If we draw a cooling curve for molten pure iron against time, then we find that at four different fixed points the temperature remains steady for a while. These temperatures are called "critical temperatures or arrest points", which generally occur at 1535°C, 1404°C, 900°C and 768°C. Their ranges are as under:

(a) ***Alpha Iron.*** Below 768°C the iron is known as alpha iron, which is also called ferrite, is pure soft variety of iron having high magnetic permeability. Alpha iron has body-centred space-lattice structure, that is, all its atoms are arranged in cubes, with one atom at each of the eight corners of each cube, and a ninth atom at the centre of the cube.

(b) ***Beta Iron.*** Between 768°C and 900°C the iron is called beta, iron. It is just like alpha iron, but is non-magnetic.

(c) ***Gamma Iron.*** Between 900°C and 1404°C the iron is called gamma iron. It is although non-magnetic like beta-iron, has face-centred space-lattice structure, having six atoms at centres of the six faces of the cube in addition to eight at eight corners, *i.e.* 14 atoms in all.

(d) ***Delta Iron.*** Beyond 1404°C and up to melting point, 1535°C, the iron is called delta iron. It is again ferrite with body-centred space lattice structure like alpha iron.

5. Normalising

When steel is cold worked (hammered, rolled, bent etc.) its crystal structure is disturbed, stresses and strains are set up and the metal becomes brittle and weakened. Normalising restores the crystalline structure to a normal condition and relieves the stresses.

The process of normalising consists of heating the metal to temperatures within the normalising ranges usually to about 900°C and cooling in still open air. The minimum period for holding at the desired temperature is 15 minutes and this temperature should not be more than 50°C above the upper critical point of the steel. It produces a homogeneous structure consisting of finely laminated pearlite.

6. Annealing

It is the process of heating the alloy above the critical temperature range, holding it at that temperature for a proper period of time and then slowly cooling it in the furnace or in some medium. The purpose of annealing are :

(*a*) To remove internal stresses.

(*b*) To induce softness.

(*c*) To alter or modify ductility, toughness and other physical properties.

(*d*) To refine crystalline structure.

(*e*) To remove gases.

(*f*) To bring the metal to a condition suitable for subsequent heat-treatment.

If the annealing temperature is high, period is long and rate of cooling is slow, the annealed steel will have coarse grains. If the anneal temperature is only about 60°C above the lower critical, and the cooling starts immediately after reaching this temperature, the grain size will be fine.

7. Hardening

When carbon steel is heated to a temperature, a little above its upper critical point and then cooled by drastic quenching, the normal structural change does not occur. Because the rapid cooling arrests the formation of pearlite, and the austenite is changed into another microstructure called Martensite, which is extremely hard and brittle. If the rate of cooling is not quite so drastic, then the austenite transforms into another structure called Troostite, which is less hard but very tough. An even lower rate of cooling will produce, sorbite, a finely grained structure which is strong and ductile. Therefore, degree of hardness of steel depends upon its rate of cooling and carbon content. Hardening may be considered under the following general heads:

(*a*) *Work hardening.* In which material may receive and retain a desirable or undesirable hardness during working.

(*b*) *Age-hardening.* It is notably manifested in the case of duralumin which, if allowed to age for three or four days after heat-treatments, shows a marked increase in strength and hardness.

(*c*) *Hardening by heating and quenching.* It is employed for iron base.alloys with some minimum carbon content.

(*d*) *Air-hardening.* It is manifested in high speed steels and some of the tungsten alloys which become hardened when slowly cooled in air blast.

8. Tempering

As the hardened steel is brittle, therefore to relieve this brittleness and induce toughness the steel is tempered. Tempering consists of re-heating hardened steel to a temperature below lower critical temperature (700°C), holding it at that temperature for some time and then cooling it usually by quenching. This operation, if carried out below 400°C is called 'drawing', and if performed above 400°C is termed "roughening".

Any grade of steel, when hardened and tempered, becomes much tougher and its resistance to shock is much greater than the same steel when only normalised.

9. Carburising

It is a method of producing hard surface on parts made of low carbon steel. Therefore, it is also called case-hardening. Under this process the part to be case-hardened is embedded in a closed box in some substance rich in carbon and heated to a temperature 900°C to 930°C and maintained at that temperature for 3 to 6 hours, when the gamma-iron formed absorbs carbon. The part is sub sequently hardened and tempered. The thickness of the hardened layer is 1 to 2 mm.

A carburizing mixture consists of 60 percent of wood charcoal and 40 percent of barium carbonate. The active agent in carburizing is carbon monoxide gas, formed by the action of carbon on carbon dioxide discharged from barium carbonate.

10. Cyaniding

It is a process of producing hard surface on low-carbon or medium carbon steels by immersing the steel in a molten salt bath containing cyanide maintained at about 900°C, and then quenching the steel in water or oil. The steel absorbs both carbon and nitrogen. The thickness of hardened skin obtained by this process is no more than 0.8 mm. If only a very thin hard case is desired, powdered cyanide is sprinkled on the heated steel part.

11. Nitriding

This is a process of producing hard surface layer on alloy steels when heated in an atmosphere rich in ammonia gas for periods up to 90 hours. Under this the part is first hardened and tempered in the usual way and machined as required. It is then heated in an electric furnace to 500°C, when liquid ammonia is introduced. The ammonia breaks up into nitrogen and hydrogen, and the chromium and molybdenum, present in the steel, form nitrides, which produce the desired hardening effect.

12. Flame hardening

The process of hardening steel by heating it with the flame of an oxy-acetylene torch and immediately quenching by directing a spray or jet of water to the surface is known as flame hardening. This process is particularly suitable for articles, which cannot be quenched as a whole.

13. Induction hardening

In this process, a high frequency current is passed through a copper inductor block which acts as a primary coil of a transformer. The block is placed around but does not touch the surfaces to be hardened. The heat is obtained due to the induced eddy currents and hysteresis losses in the surface material, and maintained at a temperature of about 750 to 800°C. The heated areas are then quenched immediately by sprays of water delivered through numerous small holes in the inductor block. By this method we can produce hardness on the bearing areas of crankshafts, camshafts, axleshafts etc.

20.5 NON-FERROUS METALS

Non-ferrous metals are those which do not contain iron as base. Their melting points are generally lower than that of ferrous metals. Although ferrous metals and alloys are the cheapest and most common for structural and other engineering purposes, there are fields of use, where non-ferrous alloys are preferred for some reason or other. Some of the properties, which make the use of non-ferrous alloys essential, are given below :

(*a*) Most of the non-ferrous alloys are corrosion-resistant.

(*b*) Some non-ferrous alloys are better heat resistant than ferrous metal and alloys.

(*c*) Non-ferrous alloys like phosphor-bronze and babbit metal are outstandingly suitable as antifriction materials.

(*d*) Non-ferrous alloys of high electrical resistance are required as resistance wires for heating elements, electrical ovens and furnaces.

(*e*) Non-ferrous alloys like constantan and Manganin have negligible temperature coefficient of resistance, these are suitable for use in electrical measuring instruments.

(*f*) Light alloys, like aluminium and magnesium alloys, of high strength are preferred for use in automobile and aircraft industry.

The principal non-ferrous metals used for engineering purposes are as given below:

(1) Copper. It occurs in some minerals, such as copper glance, copper pyrites, malachite and azurite. Extraction may be by the dry or wet process, the former being the most common, carried out in reverberatory or blast furnace proceeded by various stages of ore refinement and followed by various metal refining processes.

It is a soft, malleable and ductile metal of a peculiar red colour. Specific gravity 8.9, melting point 1083°C. It is a good conductor of electricity. It forms very useful alloys with tin, zinc, nickel and aluminium. Copper is largely used in making electric cables and wires, for electric machinery and appliances, in electrotyping and electroplating, in making coins and household utensils, and in making alloys.

(2) Aluminium. It is a white metal produced from the Alumina (oxide) by means of electrical processes. The Alumina is prepared from a clay mineral called bauxite. Bauxite is first purified and then dissolved in fused cryolite. From this solution aluminium is separated by electrolysis at about 900°C.

It is a light metal having specific gravity 2.7 and melting point 658°C. It is a good conductor of heat and electricity. It forms very useful alloys with iron, copper, zinc and other metals. It is used in making utensils, electric wires, and parts of internal combustion engines and aeroplanes. Aluminium powder is used in preparing paints.

(3) Lead. It is extracted from lead ore called galena, which is sulphide of lead. It is a bluish-grey metal. It is so soft that it can be cut with a knife. It marks paper. It is the heaviest of the common metals and having specific gravity 11.36 and melting point 326°C. It is used for making shots and bullets, plates for storage cells, alloys, solders, pipes and sheaths for electric wires, cisterns, water pipes and acid-proof chambers.

(4) Tin. It is obtained from tin stone, an oxide, by a refining process carried out in a reverberatory furnace. It is a brilliant white metal with a yellowish tinge. It is a soft and malleable metal which can be hammered into thin foils. Specific gravity 7.3, melting point 232°C. Tin does not corrode in wet, and dry climates, making it useful as a protective coating for iron and steel. It is also used for tinning copper wire before the latter is made into cables.

(5) Zinc. It occurs in nature as oxide, carbonate and sulphide. The common ores are zincite, calamine and zinc blende. Its method of manufacture consists in heating the oxide ore in an electric furnace, whereby zinc is liberated as a vapour, which is cooled in condensers to obtain metallic zinc.

It is a bluish white, crystalline metal. It is brittle at ordinary temperature, but when heated to about 125°C, it becomes malleable and ductile. Its specific gravity is 6.2 and melting point 419°C. It is used in electric cells, in galvanizing iron and steel, and in making brass and other alloys.

(6) Nickel. It is a silver-white metal. Its specific gravity is 8.85 and melting point 1450°C. It is much used for cooking utensils, and other vessel for heating and boiling. Nickel enters as a constituent into a large number of ferrous and non-ferrous alloys, and frequently finds application as a catalyst in important industrial processes.

(7) Chromium. It is white metal having specific gravity 7.1 and melting temperature 1830°C. It is alloyed with nickel in heat-resisting alloys and with iron and nickel in stainless and heat-resisting steels. It is not easily corroded by atmosphere or chemicals, it is much used as a corrosion-resisting coating for iron and copper, on which it is deposited by electroplating.

(8) Magnesium. It is soft, poor corrosion resistance brittle, lighter than aluminium. It is used extensively in light alloys, and in small quantities as a deoxidant, ignitable charge in incendiary bombs and flares, and also used in photographic flash powder.

(9) Manganese. It is hard, brittle metal having a reddish or greyish tinge. It is used in the preparation of manganese steel which is used where parts are exposed to great wear and tear, *e.g.* railway tracks.

20.6 NON-FERROUS ALLOYS

Non-ferrous alloys having aluminium or magnesium as a base are termed "Light Alloys" while those with other non-ferrous metals *as* base are termed as "Heavy Alloys". The properties and uses of the commonly used non-ferrous alloys are as given below in table 20.2 & 20.3:

(1) Non-ferrous Light Alloys

S.No.	*Name of Alloy Group*	*Constituents approx per cent*		*Properties*	*Uses*
1.	Duralumin	Copper Manganese Magnesium Silicon Aluminium remainder.	4.5 0.7 0.7 0.7	Having light weight, one third that of steel. Strong like mild steel. Can be heat treated and age-hardened.	Aircraft sheets, engines and airframe structural parts, rivets, tubes etc.
2.	Alclad	It is a duralumin sheet but having coating of pure aluminium on each side.		Having all the properties of duralumin with the resistance to corrosion of aluminium.	Aircraft sheets and wing coverings.

3.	Y-alloy	Copper 4 Nickel 2 Magnesium 1.5 Aluminium remainder.	Resistant to corrosion and fatigue. Withstands relatively high temperatures. Can be heat treated and age hardened.	Used for making pistons and cylinder heads.
4.	Hiduminium	Copper 2.5 Nickel 1.5 Iron 1.5 Magnesium 1.2 Silicon 1.0 Aluminium remainder.	Strong as mild steel. Can be heat treated and age hardened at low temperatures	Used for making aircraft structural parts, pistons, cylinder heads etc.
5.	Aluminium bronze	Copper 90 Aluminium 10	It is a copper-aluminium alloy having high strength.	Used for diecasting and pump-rods also in place of steel for non-sparking chisels.

Table 20.2 Non-ferrous light alloys

Non-ferrous Heavy Alloys

S.No.	*Name of Alloy*	*Constituents approx. per cent*		*Properties*	*Uses*
1.	Brass.	Copper Zinc	70 30	Good wearing, antifriction, corrosion resistant.	Used for making pipe fittings, tubing, filter, gauges, bearing bushes etc.
2.	Bronze	Copper Tin	95 5	Good wearing, antifriction, corrosion resistant.	Used for making bearing bushes.
3.	Phosphor-bronze	Copper Tin Phosphorous	89 10 1	Has good tensile strength and resists corrosion by seawater.	Used for bearings and gears.
4.	Solder	Tin Lead	60 40	Low melting point which depends upon the proportion of constituents.	Used for joining the metal sheets.
5.	White metal	Copper Tin Antimony	5 85 10	Tough, antifriction properties, melts at a low temperature when bearing is overheated, and prevents seizure.	Used for making heavy duty bearings.
6.	Gun metal	Copper Tin Zinc	88 10 2	Tough, strong and hard, suitable for sound castings.	Used for making bearings.

7.	Bell metal	Copper 80 Tin 20	Hard and brittle .	Used for making bells
8.	Naval brass	Copper 60 Zinc 39 Tin 1	Has high strength and resists corrosion	Used for marine and engineering castings, condenser tubes, pump parts, motor boat shaftings etc.
9.	Monel metal	Nickel 68 Copper 29 Iron 3 Little quantity of manganese silicon and carbon	High corrosion resistance, excellent mechanical properties.	Used for condenser tubes propellers, pump fittings, turbine blades and chemical and food handling plants.

Table 20.3 Non-ferrous heavy alloys

20.7 CORROSION AND ITS PREVENTION

Corrosion can be defined as the destruction of a metal by gradual chemical or electro-chemical reaction with its environments, such that the metal is converted into an oxide salt or some other compound. The most familiar example of corrosion is rusting of iron when it is exposed to atmospheric conditions. Because of corrosion a metal loses its strength, ductility and other useful mechanical/physical properties.

Although oxygen is the most commonly encountered reacting gas, SO_2 and H_2S from fuels and chlorine and other gases in specific situation cause similar reactions. Oxidation may occur at any temperature, but it is particularly important at high temperatures where a more rapid reaction is likely to occur between metal and air.

Following are the common measures generally adopted to prevent corrosion depending upon the circumstances.

(A) Internal Measures

1. **Purification of metals.** Pure metals have better corrosion resistance than impure metals.
2. **Alloying of metals** with corrosion resistant elements.
3. **Proper heat-treatment of metals.**

(B) External Measures

1. **Design features.** Life of equipment can be prolonged by ‘reducing its corrosion by careful designing, such as avoiding of side by side contact or dissimilar metals or to avoid impingement and turbulence in flowing environment etc.
2. **Application of inhibitors.** Inhibitor may be defined as any substance which when added in a small quantity to the environment reduces the corrosion rate appreciably.
3. **Protective coatings.** These may be either metallic coatings or non-metallic coatings. Metallic coatings may be of zinc, tin, aluminium, nickel, etc. These may be done either

by electroplating, dipping in molten metal, cladding, metal spraying etc. Non-metallic coatings consist of painting, lacquering, plastic coating, enamelling, slushing compound coatings, anodising and other oxide films and chemical dip coatings.

20.8 MISCELLANEOUS ENGINEERING MATERIALS

(1) Adhesives. The various adhesives used in industries are as given under:

(*a*) Animal glues.

(*b*) Albumin and blood glues.

(*c*) Glue made from casein and soybean protein.

(d) Starch glues.

(*e*) Glue made from natural and synthetic resins.

(*f*) Nitrocellulose glues.

The quality of an adhesive is judged by

(*a*) Degree of tack *i.e.* stickness.

(*b*) Strength of bond after setting or drying.

(*c*) Rapidity of bonding.

(*d*) Durability.

(2) Asbestos. It is a mineral substance, fibrous in structure and of a white, grey or brown colour. Chemically it is a silicate of calcium and magnesium. It is an incombustible, fire proof material, which is unaffected by acids and fumes. It is a good insulator of sound, heat and electricity.

It is very widely used in steam-pipe joints, and in manufacturing fire-proof roofing and flooring materials.

(3) Asphalt. It is the solid bituminous deposit, formed in oil-bearing strata, by the removal, through slow, natural evaporation of the volatiles. Natural asphalt is classified in two groups :

(*a*) Rock asphalt.

(*b*) Lake asphalt.

Asphalt is non-inflammable, acid resisting, waterproof, impermeable and resilient. It is used as a damp-proof course in walls and under basements, and in lining tanks, swimming pools, urinals and leak-proof terraced roofs.

(4) Glass. It is a hard, brittle and usually transparent substance It is made by fusing together in a furnace an intimate mixture of one or more of the oxides of silicon, boron or phosphorus with one or more of the basic oxides—soda, potash, lime and magnesia and a quantity of waste glass. The product is cooled rapidly to prevent crystallisation. Its melting point varies from 800° to 950°C.

According to composition and properties, glass may be broadly classed under, (a) crown glass *(b)* flint glass, (c) pyrex glass. According to need, glass can be annealed and given desired shape such as steel glass, plate glass, ribbed glass, wired glass, ground glass etc.

(5) Gypsum. It is crystalline hydrated sulphate of calcium. It occurs massive as alabaster, fibrous as satin spar or as colourless transparent crystals, called selenite. It is used in the manufacture of portland cement for rendering it slow-setting and for making plaster of paris.

(6) Cork. It is the light, porous bark of the cork oak tree. When ground and baked, the natural resin in the cork binds the material into a homogeneous mass, which can be pressed into flexible sheets or boards. Loose cork is used as a non-conducting covering for pipes carrying steam or hot water. Cork sheets and boards may be used to form a non-conducting lining for scientific apparatus etc.

(7) Plaster of Paris. It is dehydrated gypsum usually obtainable as a white powder. When mixed with water it solidifies with evolution of heat, expanding slightly. This expression in setting makes it particularly suitable for making casts and for filling holes and cracks in plastered surfaces.

(8) Rubber. Natural rubber is an elastic material present in the milky juice of rubber trees. The milky juice, which contains about 30-45 per cent of rubber, is treated in two ways to recover the rubber. By the first method, the rubber is coagulated from the milky juice by acids or heat, and then processed. The second method consists of mixing appropriate compounding material with the milky juice, and precipitating the rubber-compound and casting it direct in the-shape to be used.

(9) Plastics. They are natural, synthetic or semisynthetic organic substances, which under heat and pressure become plastics, and can then be shaped or cast in moulds, extruded as rods, tubes etc. or used in the formation of laminated products, paints, lacquers, glues etc. Plastics may be thermoplastic or thermosetting. Thermoplastic plastics are those which become plastic on heating to about 150°C. Thermosetting plastics are those which become progressively hardened when heated and which, after they are moulded, permanently set.

Plastics are mouldable organic resins having light weight, good surface finish, good thermal and electrical resistance, good resistance to corrosion, and can be easily produced in different colours.

(10) Cement. Natural cement is manufactured by burning and crushing to powder stones containing 20 to 40 per cent of clay. The remainder being carbonate of lime, sometimes mixed with carbonate of magnesia. It is brown in colour and sets very quickly when mixed with water. It is not so strong ultimately as artificial cement. Artificial cement may be classified as under:

1. *Portland cement.* It is made by intimately mixing together calcareous and argillaceous stones containing silica, alumina and iron oxide, burning the mixture at *a* clinkering temperature and grinding the resulting clinker. Portland cement may be:

(*a*) Ordinary,

(*b*) Rapid hardening,

(c) Low heat.

2. *Special cements.* Besides ordinary portland cement other varieties of cement are manufactured and used for special purposes. They may be of following types:

(*a*) Quick-setting cement,

(*b*) Calcium chloride cement,

(*c*) High-alumina cement,

(*d*) Blast-furnace slag cement,

(*e*) White and coloured cements,

(*f*) Acid-resisting cements.

(11) Concrete. It consists of a mixture of cement or lime mortar and stone. The mortar used in concrete is called "matrix" and the hard broken material is known as "aggregate".

The aggregate may be any hard stuff readily obtainable—broken stone, broken brick, gravel etc. A porous material, broken into rough angular fragments, to which mortar will easily adhere, is preferred to a material having a smooth, rounded surface. An aggregate of an absorbant nature should be thoroughly soaked in water before mixing. When lime mortar is the matrix, the concrete is called lime concrete; when cement mortar forms the matrix, the concrete is known as cement concrete. The size of the aggregate differs according to the nature of the work.

(12) Heat insulator. A material which insulates the heat is called heat insulator. The more common materials used as heat insulators are cork, mineral wool, asbestos-cement sheeting, fibre-boards, bituminous felt and reflecting paper.

The materials which retard the transmission of heat are known as heat insulating materials. Some of the common materials used as heat insulators are discussed below in brief:

(*a*) **Asbestos.** It is a heat resisting fibrous mineral. These are used in packing and gaskets for high temperature service, brake lining and friction materials for electrical insulation, roofings, etc. Asbestos paint is made by adding finely grounded asbestos to an oil paint or distemper, which can be sprayed on walls and ceilings of rooms and cabins to render them heat proof and sound proof.

(*b*) **Glass wool.** It is an aggregate of fine filaments of glass blown by air or steam into fibrous form. These filaments are finer than silk. These are good insulators of heat, sound and electricity. These are unaffected by various chemicals.

(c) Thermocole. It is a polystyrene foam with snow white colour and is very light having its normal grade density of 18 kg/cu meter. It has a very low conductivity and is highly resistant to moisture. It is resistant to ageing and is odourless, chemically stable and resistant to fungus, water, salt, etc. It possesses good shock protecting properties. It is used as cold insulation and as a shock protecting packaging material in the packaging of delicate electronic equipment, glass bottles of medicines, crockery, cameras, egg trays etc.

(13) Sound insulator. Materials which offer resistance to reflection and transmission of sound are called sound insulators. Usually the materials which are good insulators of heat are insulator of sound as well. Acoustic plasters, cellular concrete, fibre-boards and cobolts quilt are well-known sound insulators. A sound insulator absorbs sound and does not produce echo, much used on inner walls of assembly halls, theatres and cinema halls.

(14) Refractory materials. They are non-metallic materials that can stand high temperatures. They are used in the construction or lining of furnaces, boilers, flues, regenerators converters, crucibles, dryers and pyrometer tubes. A refractory metal to be used for any purpose must:

(*a*) Stand the high temperature to which it is likely to be subjected without cracking, disintegrating or softening.

(*b*) Stand the action of hot gases; slag, flue dust, abrasion and sudden changes of temperature.

They must be free from material like sodium carbonate and iron oxide, both of which act as flux helping the material to melt at comparatively low temperatures.

(15) Abrasive Materials. These are the materials which are employed for something, grinding and polishing the surface of materials, in order to provide them required size, shape or finish. Hardness and brittleness are the two essential characteristics of an abrasive. In a grinding wheel or abrasive wheel, abrasive particles are bonded together by a bonding material. Abrasives may be either natural or artificial abrasives.

Natural abrasives are the abrasives occuring as minerals of rocks of the crust and may be further classified as :

(*a*) Hard materials, having greater range of hardness e.g. diamond, emery, corundum and garnet;

(*b*) Siliceous abrasives composed chiefly of silica like quartz, flint, quartzite, sand stone, etc.;

(*c*) Soft abrasives used for buffing (*i.e.* fine polishing) metal surfaces in the form of rapidly revolving flexible wheels (made of cloth of leather) and coated with soft abrasives.

Rotten stone, vienna lime and metallic oxides are the common forms of soft abrasives.

Artificial abrasives are the abrasives which are not available in nature but are prepared artificially by synthetic means. Some important artificial abrasives are:

(*a*) silicon carbide, SiC; (*b*) boron carbide B_4C ;

(*c*) Fused aluminium oxide Al_2O_3 (*d*) steel shots ; and

(*e*) crushed steel.

20.9 STRESS

When a material or a solid body is subjected to some load or external force, it offers a resistance. This resistance of the material of the body is called the stress.

Stress is a measure of the amount of force exerted per unit area of cross section. Mathematically it is expressed as,

$$\sigma = \frac{P}{A}$$

where, σ = stress,

P = load applied, and

A = area of cross-section.

The SI unit for stress is Pascal (P_a) which is equal to 1 N/m² (*i.e.* force per unit area).

Classification

Stress can be classified as under:

1. Tensile Stress. If a test specimen is subjected to uniform axial tensile load, it results in tensile stress (Refer Fig. 20.3(*a*). Therefore the tensile stress is defined as the ratio of tensile force to the original area of cross-section.

$$\sigma_t = \frac{P_t}{A}$$

where, σ_t = tensile stress,

P_t = applied tensile load, and

A = Area of cross section.

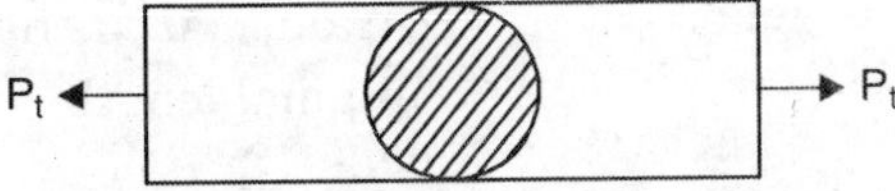

Fig. 20.3 (*a*) Tensile loading

2. Compressive Stress. If a test specimen is subjected to a uniform axial compressive load, as shown in Fig. 20.3(*b*). This results in compressive stress. Therefore, the compressive stress is defined as the ratio of compressive force to the original area of cross-section.

$$\sigma_C = \frac{P_C}{A}$$

Fig. 20.3 (*b*) Compressive loading

where, σ_C = Compressive stress,

P_C = applied compressive load, and

A = Area of cross-section.

3. Shear Stress. In case of an external load P, causing shear, a shear stress will be developed. The shear stress is defined as the ratio of shear force to the shearing area. Fig. 20.4(a) shows a punch, punching a hole of diameter d in a plate of thickness t.

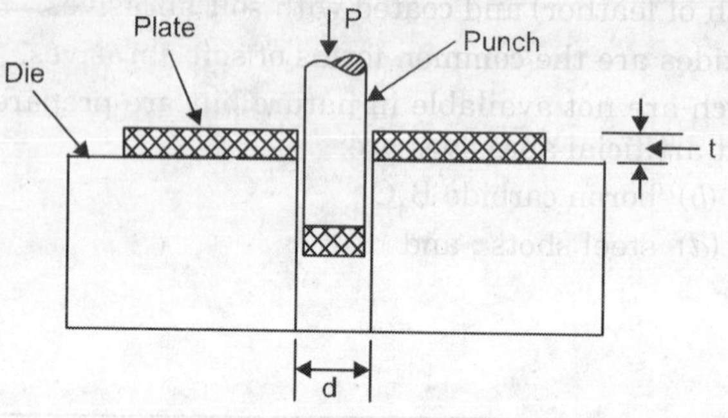

Fig. 20.4 (*a*) Shear loading.

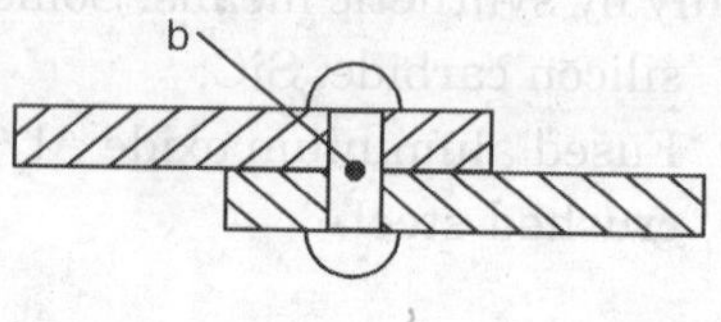

Fig. 20.4 (*b*) Rivet resisting the shear.

It may be noted that in cases of either simple tension or simple compression, the areas which resist the load are perpendicular to the direction of forces. When a member is subjected to simple shear, the resisting area is parallel to the direction of force. Common situations causing shear stresses are shown in Fig. 20.4. (*a*) and (*b*).

In the case shown at Fig. 20.4(*a*), the shear stress,

$$\sigma_s = \frac{P_s}{\pi . d . t} .$$

20.10 STRAIN

Any element in a material subjected to stress is said to be strained. The strain (e) is the deformation produced by stress.

The strain can also be defined as, the deformation caused due to the action of external force per unit length. Mathematically, it is expressed as,

$$e = \frac{\Delta L}{L}$$

where, e = strain, ΔL = change in length, and

L = original length

Classification

According to the manner of loading, the strain may be classified as:

1. Tensile Strain. If a test specimen is subjected to a uniform axial tensile load, the length of the specimen increases from L to $(L + \Delta L)$ as shown in Fig. 20.5. This results in tensile strain. Therefore, the tensile strain is defined as the ratio of increase in length to the original length:

$$e_t = \frac{\Delta L}{L}$$

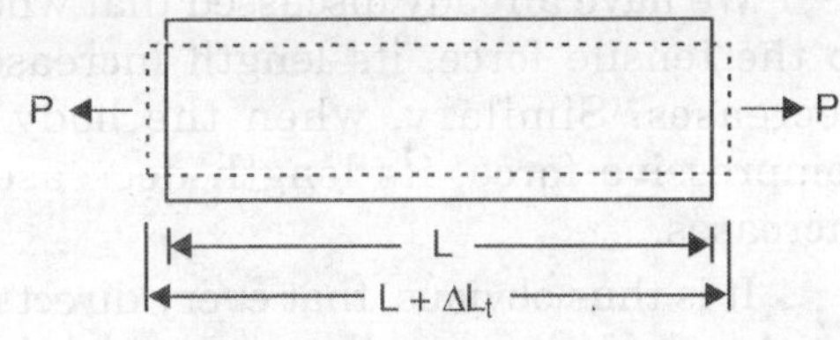

where, e_t = tensile strain,

ΔL_t = increase in length, and

L = original length

Fig. 20.5 Tensile loading.

2. Compressive Strain. If a test specimen is subjected to a uniform axial compressive load, the length of specimen decreases as shown in Fig. 20.6.

This results in compressive strain. Therefore the compressive strain is defined as the ratio of reduction in length to the original length,

i.e.,
$$e_c = \frac{\Delta L_C}{L}$$

where, e_c = compressive strain,

ΔL_c = reduction in length, and

L = original length

Fig. 20.6 Compressive loading.

3. Shear Strain. In case of a shearing load, a shear strain will be produced which is measured by the angle through which the body distorts. In Fig. 20.7 is shown a rectangular block ABCD fixed at one face and subject to force P. After application of force, it distorts through an angle ϕ and occupies new position ABC′ D′. The shear strain (e_s) is given by;

Fig. 20.7 Shear loading

$$e_s = \frac{CC'}{BC}$$

$$= \tan\phi = \phi \text{ rad (as } \phi \text{ is small)}$$

20.11 HOOKE'S LAW

The Hooke's law states when a material is loaded within its elastic limit, the stress is proportional to the strain. mathematically;

$$\text{Stress} \propto \text{Strain}$$

or
$$\sigma \propto e$$

or
$$\frac{\text{Stress}}{\text{Strain}} = \text{a constant} = E.$$

The Hooke's law holds good for tension as well as compression. For tensile and compressive stresses the proportionality constant (E) is known as **modulus of elasticity** or **Young's modulus** (E). For shear stress, it is known as modulus of rigidity (G). It is different for different materials.

20.12 POISSON'S RATIO

We have already discussed that when a body is subjected to the tensile force, its length increases and the diameter decreases. Similary, when the body is subjected to the compressive force, its length decreases and the diameter increases.

It is thus obvious, that every direct stress is accompained by a strain in its own direction which is known as **linear or primary or longitudinal strain** and an opposite kind of strain in every direction, at right angles to it is known as **lateral or secondary strain.**

The ratio of lateral strain to linear strain is known as **Poisson's ratio.** Its value is constant for a particular material but varies for different materials. In most materials its value varies from 0.3 to 0.6.

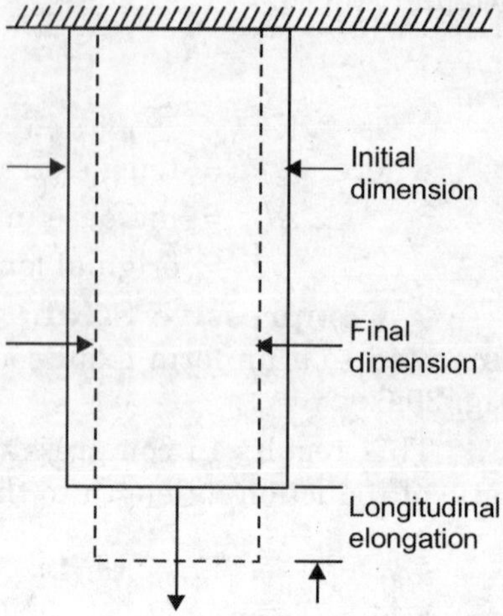

Fig. 20.8 Increase in length due to tensile load

For example, as shown in Fig. 20.8 load is tensile and there is an increase in length, therefore, in this case, linear or primary strain will be tensile; and there will be a corresponding decrease in cross-sectional area of the body, therefore in this case lateral or secondary or transverse strain is compressive.

As, ratio of lateral strain to linear strain is known as Poisson's ratio.

$$\text{Poisson's Ratio, } \mu = \frac{\text{lateral or secondary strain}}{\text{linear or primary strain}}$$

20.13 STRAIN ENERGY AND RESILIENCE

When an elastic body is loaded it undergoes deformation and when it is relieved of the load it regains its original shape. When a bar is acted upon by a tensile or compressive load there is a change in the length and the point of application of the load moves against the resistance offered by the material which gets strained under the action of the load. In doing so, the work is done on the bar. if the bar is loaded within the elastic limit of the material the work done on the bar will be recovered when the load is removed. In other words, the work done on the bar, while loading, is stored in the bar as **strain energy** which can be recovered wholly or partially depending upon whether the bar was loaded upto or beyond its elastic limit. This property is essential for spring materials.

The strain energy stored per unit volume is called the **resilience** of the body. The maximum energy which a body stores upto elastic limit is called **"proof resilience".** Proof resilience is the mechanical property of material and it indicates their **capacity to bear shocks.** Proof resilence per unit volume of piece is called **"Modulus of resilience".** It is an important property of a material which gives capacity of the material to bear impact or shock loads.

20.14 THERMAL STRESSES AND STRAINS

When the temperature of a body is lowered or raised its dimentions will decrease or increase correspondingly. This results in the development of stresses in the body. These stresses are called **temperature (thermal) stresses** and corresponding strains are called **temperature (thermal) strains.**

l = length of bar of uniform cross-section,
t_1 = initial temperature of the bar,
t_2 = final temperature of the bar, and
α = co-efficient of linear expansion

Expansion in the bar due to rise in temperature

$$\propto \alpha\ (t_2 - t_1)\ l$$

If the elongation in the bar is prevented by some external force or by fixing the bar ends, the **temperature strain** thus produced will be given by,

$$\propto \frac{\alpha\ (t_2 - t_1) l}{l} \text{ (Compressive)}$$

$\therefore$ temperature strain developed = $\alpha\ (t_2 - t_1)\ E$ (Compressive stress)

However, if the temperature of the bar is lowered, the temperature strain and stress will be tensile in nature.

20.15 STRESS-STRAIN DIAGRAM (TENSILE TEST)

The tensile test is most widely used test and is carried out on a bar of uniform cross-section. The specimen bar is mounted in the jaws of a testing machine with which gradually increasing load can be applied. The extension (elongation) of the gauge length is recorded continuously and finally a graph is drawn between the stress and strain which is similar to shown in Fig. 20.9.

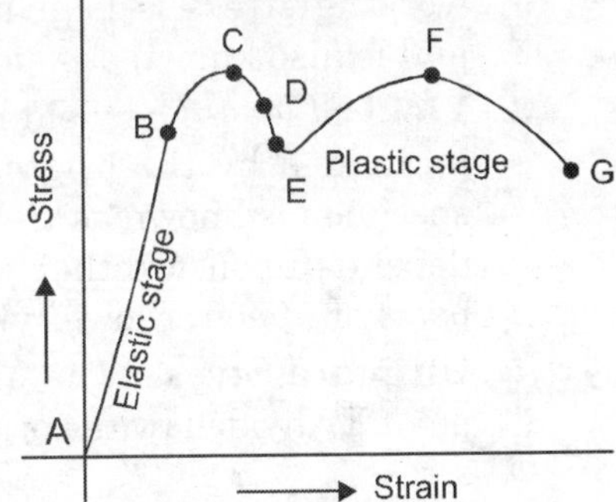

Fig. 20.9 Stress-strain diagram.

This test provides basic design information on the strength of materials. The parameters which are used to describe the stress-strain curve of a metal are the tensile strength, yield strength or yield point, percent elongation and reduction of area.

The stress-strain diagram characterises the behaviour of the material tested. The nature of the curve varies from material to material.

Stress-Strain Diagram for Ductile Materials

Fig. 20.9 illustrates the behaviour of mild steel (a ductile material). It can be seen from the diagram that initially there is a linear relationship between stress and strain from point A to point B. In AB portion the stress is proportional to strain and the limit upto which the curve shows straight line is known as the **proportional limit.** In this region the material follows Hook's law. Beyond this point, by further increasing the load upto C, the deformation completely disappears on removal of load and the material is able to regain its original form. This limit is known as the **elastic limit.** In BC portion Hook's law is not followed.

Beyond the elastic limit (*i.e.*, beyond *C*), any additional loading will result in some permanent change and causes yielding to occur. In the mild steel yielding commences immediately and two points D and E, the **upper and lower yield points** respectively are obtained. On further increasing the load slightly, the strain increases rapidly till F when **neck or waist** is formed. When the point F is reached the deformation or extension continues even with less load and ultimately **fracture occurs** (*i.e.*, material fails or breaks) at G. This is known as **breaking point.**

Stress-Strain Diagram for Brittle materials

Brittle materials has a very low proportional point and do not show yield point. Fig. 20.10 shows a typical stress-strain diagram for cast iron (a brittle material). In brittle materials the failure occurs without necking. Therefore, the ultimate tensile strength and fracture strengths of such materials are very close to each other.

Important terms used in Tensile Text

Refer Fig. 20.9.

1. **Elastic Stress.** The elastic load (corresponding to point C in the Fig.) divided by the original cross-sectional area of the specimen is known as elastic stress.
2. **Working Stress.** It is the stress, which is used for designing the structures or machine components. The working stress is kept much below the elastic stress, and is also known as safe stress.
3. **Yield Stress.** The yield load (corresponding to point D) is divided by the original cross-sectional area of the specimen is known as yield stress. The yield stress value depends upon whether the stress is calculated on the basis of upper or lower yield point load.

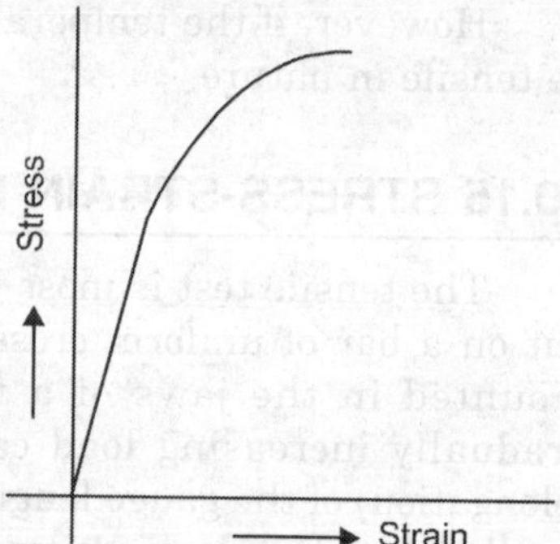

Fig. 20.10 Stress-strain curve for brittle material

4. **Ultimate Stress.** The ultimate load (corresponding to point F) divided by the original cross-sectional area of the specimen is known as ultimate stress.
5. **Factor of Safety.** It is the ratio of ultimate stress to the working stress.
6. **Breaking Stress.** The breaking load (corresponding to point G) divided by the cross-sectional area at the neck of the specimen is known as the breaking stress.

QUESTIONS

20.1 Compare cast iron, mild steel and steel in a tabular form on the following properties: compressive strength, tensile strength, hardness, ductility, forgeability, weldability. How do they stand from the point of view of cost and heat treatment?

20.2 (*a*) State the influence of the following alloying elements on the mechanical properties of cast iron: (*i*) Silicon, (*ii*) Phosphorus, (*in*) Manganese and (*iv*) Nickel.

(*b*) What should be the properties of bearing metals? Name two alloys of copper which are often used in bearings. Which alloys are ideal materials for high speed bearings?

20.3 What is the main difference in the physical properties of cast iron and mild steel? What kinds of steel are generally used for cutting tools and why? How is the required hardness in the cutting tools obtained?

20.4 What are component metals used to produce brass and gun-metal? What are the different engineering uses of brass and gun-metal?

20.5 State the properties and uses of the following :
(*a*) Manganese steel, (*b*) Stainless steel, (*c*) White metals and (*d*) Duralumin.

20.6 Give the composition, properties and uses of the following:
(*a*) Muntz metal (*b*) Invar
(*c*) Phosphor-bronze (*d*) Aluminium-bronze
(*e*) Duralumin (*f*) High speed steel.

20.7 State the materials of which the following are generally made:
(*i*) Razor blades, (*ii*) Water pipes, (*iii*) Railway wheel tyres, (*iv*) Piston rings, (*v*) Commutator segments, (*vi*) Line shafts, (*vii*) Wire drawing dies, (*viii*) Beams and joists, (*ix*) Armature stamping, (*x*) Overhead electric transmission wire.

20.8 Give the composition, properties and uses of the following materials:
(*a*) Phosphor-bronze, (*b*) Gun-metal,
(*c*) Monel-metal, (*d*) Brass.

20.9 (*a*) According to carbon content, steel is classified as high, medium and low carbon steel. In each case state the average percentage of carbon.
(*b*) Write a short note on high speed steel.
(*c*) Compare the properties of cast iron and mild steel.
(*d*) State the composition of two white metal alloys and their uses.

20.10 What is meant by mechanical properties of materials? State their importance in the design of a machine or structure elements.

20.11 Write short notes on:
(*i*) Elasticity, (*ii*) Elastic limit, (*iii*) Yield point, (*iv*) Toughness, (*v*) Hardness, (*vi*) Ductility, (*vii*) Plasticity.

20.12 (*a*) What different kinds of iron are produced?
(*b*) What are constituents of cast iron and how do they vary in grey, white, malleable and nodular cast irons?

20.13 (*a*) How are steels classified?
(*b*) Give a list of various applications of carbon steel.

20.14 What is alloy steel? Classify alloy steels. State the effects of important alloying elements in steel.

20.15 Give a list of common alloy steels and state their uses.

20.16 List and discuss some of the important properties of nonferrous metals and alloys.

20.17 List the properties of aluminium that make it useful engineering metal.

20.18 Define the term "heat treatment". For what reasons steel is required to be heat treated?

20.19 What is the purpose of annealing and how it is done?
Distinguish annealing with normalising.

20.20 Write short notes on:

(*i*) Tempering, (*ii*) Case hardening, (iii) Nitriding, (*iv*) Induction hardening.

20.21 How will you classify engineering materials in order to cater to various engineering requirements?

20.22 What is an alloy? State the important physical and mechanical properties of metals and alloys.

20.23 (*a*) What is the importance of the knowledge of "materials and metallurgy" to an engineer? (*b*) What is the difference between metals and non-metals?

20.24 (*a*) What is the difference between ferrous and non-ferrous metals? (*b*) Define and explain physical properties of materials.

20.25 What do you understand by the term "corrosion"? Suggest some measures for preventing corrosion.

20.26 What are plastics? Write what you know about different types of plastics. Also write about their uses.

20.27 Write short notes on: (*a*) Asbestos, (*b*) Thermocole, (*c*) Glass and Glasswool.

20.28 Give important properties of cork and its industrial applications.

20.29 What are abrasives? Briefly describe various natural and artificial abrasive materials.

20.30 (*a*) Differentiate between ductility and malleability.

(*b*) Differentiate between creep and fatigue.

(*c*) Differentiate between toughness and brittleness.

20.31 Write short notes on:

(*a*) Tensile strain (*b*) Compressive strain.

(*c*) Shear strain.

20.32 What is stress? How these can be classified.

20.33 Draw stress-strain diagram for low carbon steel and mark characteristic points.

20.34 Draw a stress-strain diagram and explain its salient points.

20.35 What is Hooke's law? How does it help in determining the modulus of elasticity?

20.36 Define the following:

(*a*) Proportional limit (*b*) Elastic limit.

(*c*) Necking.

20.37 Draw the stress-strain diagram for mild steel and explain its salient points.

20.38 Define the following:

(*a*) Young's modulus (*b*) Poisson's ratio

(*c*) Hooke's law

20.39 Name the important properties of a material. Why it is necessary for an engineer to know them.

20.40 Define the terms stress and strain. Explain their different types.

20.41 Define resilience. Why this property is important for a material.

MANUFACTURING PROCESSES—MECHANICAL WORKING OF METALS

Chapter **21**

Learning Objectives

After studying this chapter, the students will learn:

- Manufacturing processes and their classification.
- Mechanical working of metals, such as hot working, hot rolling, forging, press forging, extrusion, piercing, drawig or cupping, and cold working.
- Fastenings and riveting.
- Sheet metal working.

21.1 INTRODUCTION

Manufacturing plays an important role in engineering because it changes the form of metal into the final product. The products require a variety of manufacturing processes for the Meterials.

The various processes used in manufacturing are classified into following groups.

1. Primary Shaping Processes: Commonly used primary shaping processes are:

(*i*) Forging	(*ii*) Rolling	(*iii*) Bending
(*iv*) Drawing	(*v*) Extruding	(*vi*) Shearing
(*vii*) Forming	(*viii*) Piercing	(*ix*) Crushing
(*x*) Casting		

2. Mechanical Processes: These processes are performed on casting, rolling and forging so as to obtain the desired accuracy and shape. Following are common machining operations:

(*i*) Turning	(*ii*) Drilling	(*iii*) Grinding
(*iv*) Threading	(*v*) Kurling	(*vi*) Boring
(*vii*) Slothing	(*viii*) Planing	(*ix*) Milling
(*x*) Sawing	(*xi*) Hobing	

3. Surface Finishing Processes: These processes are used to provide a good surface finish to the metal surface of the product. In a surface finishing process, either a negligible amount of metal is removed or a small amout of material is added to the surface of the product. In the surface finishing processes there is appreciable change in dimensions.

Commonly used surace finishing processes:

(*i*) Sand blasting	(*ii*) Buffing	(*iii*) Lapping
(*iv*) Honing	(*v*) Polishing	(*vi*) Metal spraying
(*vii*) Electroplating	(*viii*) Galvanizing	(*ix*) Painting.

4. Joining Processes: These are used for joining two or more pieces of metal parts. Following are the commonly used processes for joining the metal parts.

(*i*) Welding	(*ii*) Soldering	(*iii*) Brazing
(*iv*) Riveting	(*v*) Screwing	(*vi*) Pressing

5. Processes for Changing the Properties: These processes are used to impart certain specific properties to the metal part for specific conditions of use. Commonly used such processes are:

(*i*) Heat treatment (*ii*) Hot working (*iii*) Cold working

Above types of manufacturing processes have been discussed in detail in subsequent chapters.

21.2 MECHANICAL WORKING OF METALS

For the best economy and use of metals, it is necessary to shape the metals into required form. Casting is one of the possible methods of giving desired shape to metals, but it is not feasible in all cases. Mechanical working is another method of shaping metals to desired forms where casting is not desirable. Mechanical working of metals is the shaping of metals in either a hot or a cold state by some mechanical means and does not include the shaping of metals by casting, machining and grinding.

The mechanical working of metal is defined as an intentional deformation of metals plastically under the action of externally applied forces. The mechanial working of metals have been described as hot working and cold working depending open whether the metal is worked above or below the recrystallisation temperature, the temprature at which the new grains are formed in the metal.

Purposes of Mechanical Working

(*i*) To reduce the original block or ingot into desired shapes.

(*ii*) To improve the mechanical properties of the metal through:

(*a*) Refinement of grain size,

(*b*) Directional control of "flow lines,"

(*c*) Break-up and distribution of unavoidable inclusions, particularly in steel.

21.3 HOT WORKING

Mechanical working above the recrystallisation temperature but below the working or burning point is termed as "hot working". The temperature at which new grains are formed is called "recrystallisation temperature", which is 800 C for steel. This is at or near room temperature for lead, tin and zinc.

Advantages of Hot Working

(*i*) By hot working, the grains are broken up and their parts are deformed into small crystals thereby occurring refinement in grains.

(*ii*) Since the strength and hardness decrease at elevated temperatures, therefore low load is required to shape the metal.

(*iii*) The porosity in the metal is largely eliminated.

(*iv*) The original ingot structure is destroyed, slags and other impurities are squeezed into fibres and distributed throughout the mass, and a uniformity is established.

(*v*) Great latitude in shape and size of metal is possible due to reduction of elastic limit.

(*vi*) Directional property resulting from a fibre structure is produced.

(*vii*) Due to refinement of grains in hot working, the mechanical properties of metal is improved.

(*viii*) It is a rapid and economical process.

Disadvantages of Hot Working

(*i*) It produces poor surface finish.
(*ii*) Close tolerance cannot be maintained due to poor surface finish.
(*iii*) It requires expensive tools.
(*iv*) The life of the tools used is less as they have to work at high temperature.

Methods of Hot Working

The principal methods of hot working are as follows :

(*i*) Rolling,
(*ii*) Forging,
(*iii*) Extrusion,
(*iv*) Piercing,
(*v*) Drawing or cupping,
(*vi*) Spinning.

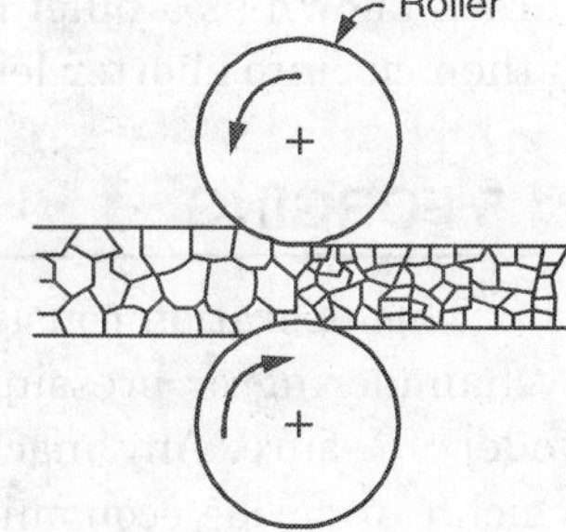

Fig. 21.1 Hot rolling.

21.4 HOT ROLLING

In hot rolling metal is first heated to a plastic stage and then passed between two rolls revolving at the same speed but in opposite direction as shown in Fig. 21.1.

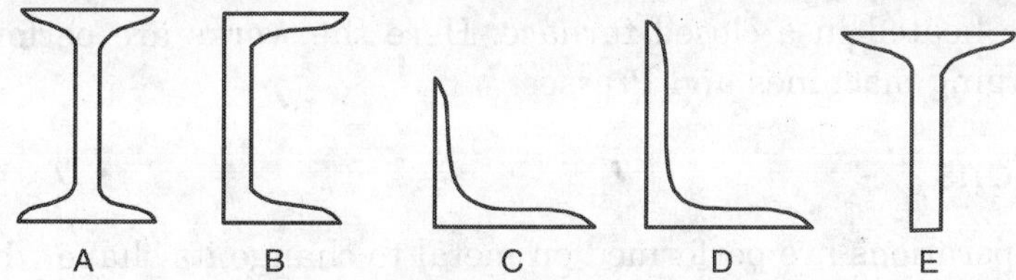

Fig. 21.2 Rolled-steel sections.
(A) I section, (B) Channel section, (C) Equal angle, (D) Unequal angle, (E) Tee-section.

The metal passed through the rolls reduces in thickness and increases in length ; and forming sections such as I-beams, channels, angles, tee-sections, plates, sheets rails etc. as shown in Fig. 21.2.

The metal changes its shape gradually during the period in which it is in contact with two rolls. The processes and sequence of operations through which metal reduces its cross-section from billet to round bar stock are *as* shown in Fig. 21.3.

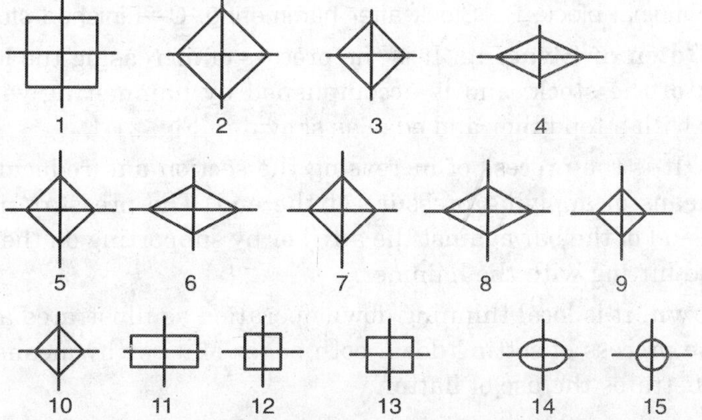

Fig. 21.3 Rolling of a round bar.

Initially rolling starts from an "ingot" which is about 1.5m long, slightly tapering and approximately square in cross-section sizes from 300 mm to 500 mm square.

These ingots are heated in soaking pits to the desired rolling temperature and then rolled on blooming mills between two grooved rolls and formed to "blooms". A bloom is a piece of steel about 3.5 to 5.5 m in length and about 200 to 300 mm square section. When the metal is to be used for bars, blooms are reduced in size to billets by passing through a continuous series of rolls known as a billet mill. The "billet", which has square section about 50 mm to 150 mm, is then cut into shorter lengths which are re-heated for further rolling in the bar mill.

21.5 FORGING

This operation consists essentially of altering the shape and section of a metal specimen by hammering or pressing it at a temperature where metal is entirely plastic and will flow under pressure. Any machine part, tool or utensil which does not originate in a casting and which can not be economically machined from a fabricated section must, with few exceptions, be formed to rough shape, by a forging operation.

The small jobs can be handled by "smithing", where the metal is heated in an open fire or hearth. The shop in which the work is carried out is called "Black Smithy". Here all the operations are performed by hand hammer or small power hammers. But in machine forging the work pieces are heated in a closed furnace. Here the works are performed by means of heavy hammers, forging machines and Presses.

Forging Operations

A number of operations are performed on metal to change its shape when it is bright red hot. The typical forging operations are as given under:

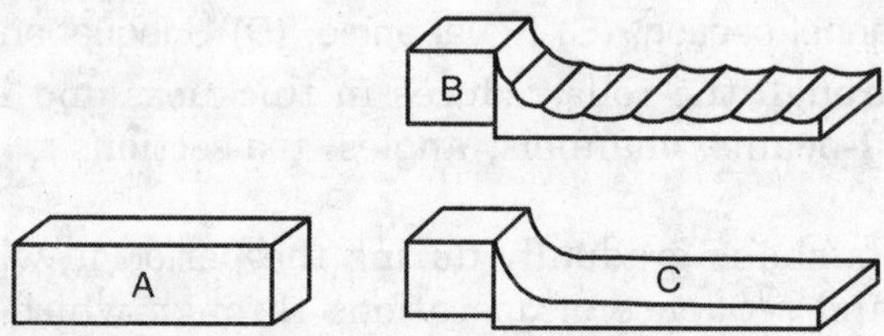

Fig. 21.4 Drawing down or swaging operation.

A—Original block, B—Stock after hammering, C—Finished stock.

(i) **Drawing Down or Swaging.** It is the process of increasing the length and reducing the section of the stock, and is accomplished by hammering with a straight peen hammer or with a top fuller and edge as shown in Fig. 21.4.

(ii) **Upsetting.** It is the process of increasing the section and reducing the length of the stock by means of applying pressure at the end. The pressure may be obtained by driving the end of the bar against the anvil or by supporting on the anvil, swage block, vice etc. and hitting with the hammer.

(iii) **Setting Down.** It is local thinning down operation as illustrated at A, refer Fig. 21.6. B shows the process of setting down both edges of a bar by means of top and bottom fuller. C illustrates the use of flatter.

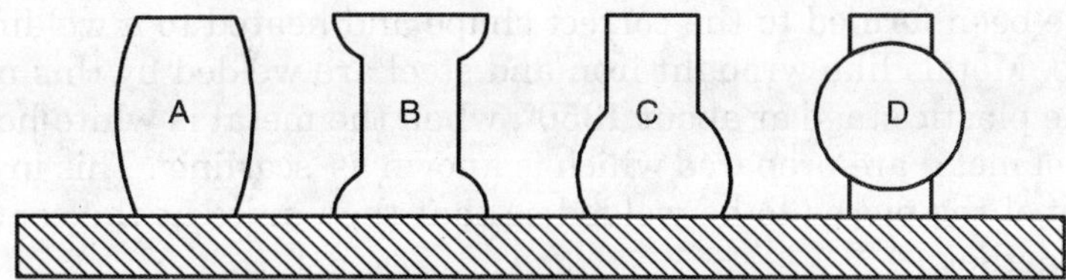

Fig. 21.5 Upset Forging Operation.
A—Effect of heavy hammer blows, B—Effect of light hammer blows, C—Local upsets, D—Upsetting the middle portion of the bar.

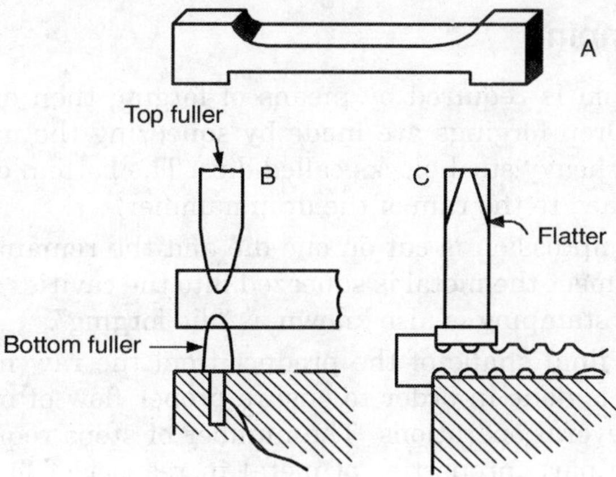

Fig. 21.6 Setting down operation.
A—Finished block, B—Setting down both edges, C—Use of Flatter.

(iv) Punching. It is the process of producing holes, generally cylindrical by using a hot punch over a cylindrical die as shown in Fig. 21.7

(v) Bending. Bending is an important operation of forging which is classified as:

(a) curvilinear, and

(b) angular. This operation is performed by hammering the metal over the edge of the anvil, or over a block of metal as shown in Fig. 21.8.

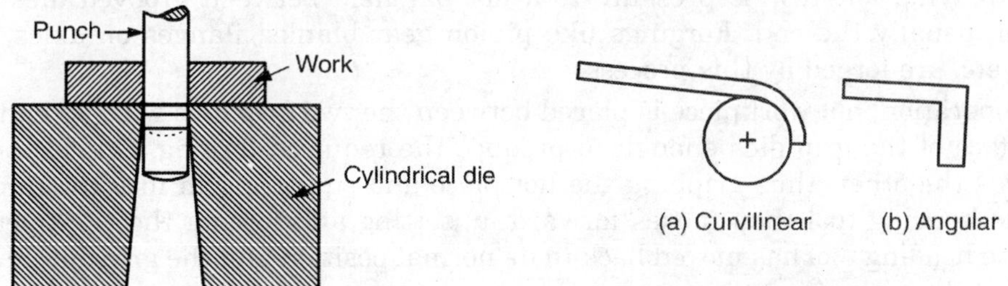

Fig. 21.7 Punching operation. **Fig. 21.8** Bending operation.

While bending the outer and inner surfaces of the metal do not remain same. The inside surface is shortened and the outer surface is stretched, which causes bulging of the side at the inner surface and a radius on the outer surface. If a perfect square bend is required, additional metal must be worked to the place where the bend occurs. When this is bent the additional metal will go to make up the corner.

(*vi*) **Forge Welding.** A forge weld is made by hammering together the ends of two bars which have been formed to the correct shape and heated to a welding temperature in a forge fire. Metals like wrought iron and steel are welded by this method, since they come in the plastic stage at about 1350°, when the metal is white-hot. Before welding, the pieces of metal are prepared which is known as 'scarfing'. This involves the shaping of the ends of the pieces to be welded so that they unite at the centre when they are brought together.

(*vii*) **Cutting.** It is a form of chiselling whereby a long piece of stock is cut into several specified lengths. For hot operation, steel must be heated in a furnace to a light cherry red heat. When cutting with chisels, the hammer blows are directed onto the chisel head.

Drop Forging or Stamping

When mass production is required by means of forging then drop forging is commonly employed to do the job. Drop forgings are made by squeezing the metal at plastic heat into shaped impressions cut in heavy steel blocks called dies. The bottom die is secured to the anvil and the upper die is attached to the ram of the drop hammer.

Generally, half the impression is cut on one die and the remainder in the other, so that when the faces of the dies meet the metal is squeezed into the cavities and a complete stamping is formed. Drop forging or stamping is also known as "die forging".

In drop forging, the final shape of the product from the raw material is achieved in a number of steps and this is done in order to ensure proper flow of metal. The dies generally contain impressions for several operations. The number of steps required varies according to the size and shape of the part, properties of metal in respect of forging and the tolerances required.

Tools, gear blanks, crankshafts, connecting rods, and a great variety of machine parts are produced in this way.

Machine Forging

It is similar to the drop forging but difference is that in machine forging the moveable dies are mechanically actuated by means of cranks and connecting rods, whereas in drop forging the working stroke is produced by falling the ram. Machine forging also called "upset forging" consists of applying lengthwise pressure to a hot bar held between grooved dies to enlarge some section, usually the end. Forgings like pinion gear blanks, flanges on axles, drawbars, valve stems etc. are forged by this process.

In the operation, hot workpiece is placed between the two grip dies with its end extending beyond the face of the grip dies enough to produce the required forging. One of the grip dies slides towards the other, thus gripping the hot piece and squeezing in into the impression in the dies. The heading tool then moves forward, upsetting and forcing the workpiece into the dies. After the heading tool has moved back to its normal position and the grip dies have opened, thus releasing the workpiece.

Press Forging

Generally in drop forging and machine forging a greater amount of energy is absorbed by the machine and foundations. Therefore, these processes are not considered to be economical types. In press forging action is relatively slow squeezing instead of delivering heavy blows and penetrates deeply because it gives the metal, time to flow. Dies may have less draft, and the forgings come nearer to the desired sizes. Therefore press forgings are generally more accurate dimensionally than drop and machine forgings. Presses may be either hydraulic or mechanical power driven. A hydraulic press is more versatile from the stand-point of varying the stroke length and the amount of pressure applied. Whereas the mechanical presses are used for high-speed forging production particularly for forging non-ferrous alloys.

21.5.1 Press

The press is a metal forming machine tool designed to shape or cut metal by applying mechanical force or pressure. The metal is formed to the desired shape without removal of chips. They are used for mass production works of steel metals. They are classified as under:

(*i*) Hand press or fly press or ball press.

(*ii*) Power press.

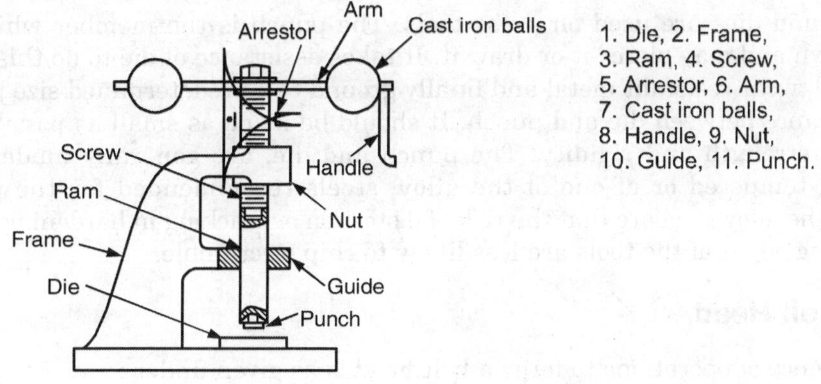

Fig. 21.9. Fly press or Ball press.

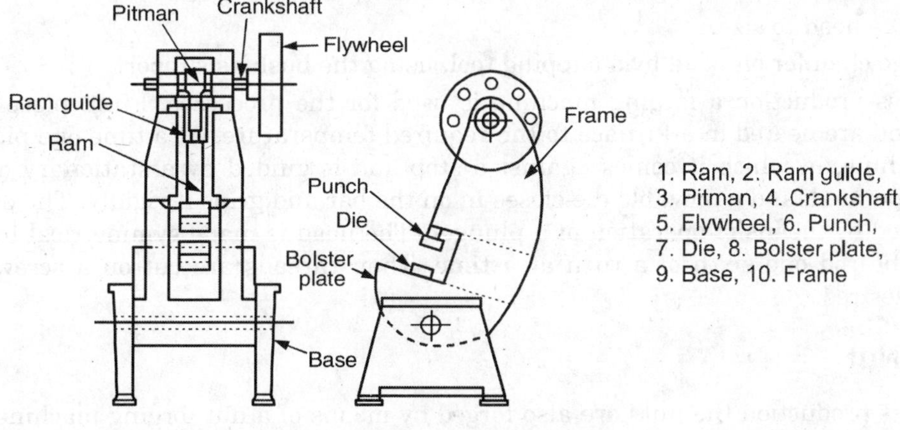

Fig. 21.10 Power press.

***(i)* Fly Press or Ball Press.** It is a very simple press as illuastrated in Fig. 21.9, which is operated by hand. The construction and the function of the different parts are quite clear from this figure, which are self explanatory. The sheet metal to be formed is placed between the punch and the die. The press is operated by a sharp, partial revolutions of the arm (6) by pulling the handle 8 and the kinetic energy is stored up in the two heavy balls (7) mounted on the arm (6). As the ram (3) is forced downwards, the resistance offered by the plate against deformation is overcome by the tremendous trust exerted by the punch (11) on the plate at the expense of the stored up energy and the material is formed to the desired shape.

***(ii)* Power Press.** Referring Fig. 21.10. This type of pressure is driven by power. The rotary motion is obtained from the electric motor, is converted into reciprocating movement of the ram by using different mechanical devices. As illustrated in Fig. 21.10, the punch (6) is fitted on the end of the ram (1) and the die (7) is attached on the bolster plate (8). The flywheel (5) mounted at the end of crank shaft 4 stores up the energy for maintaining a constant downward speed of the ram when the sheet metal is pressed between the punch and the die.

21.5.2 Punch and Die

Punches and dies are used on press tools. The punch is the member which forces the sheet strip down and may pierce it or draw it. It takes assistance of die to do this. It should be made of a hard, wear resistant metal and finally ground to a predetermined size providing just optimum clearance between die and punch. It should be made as small as possible consistent with required strength and rigidity. The punch and die, are generally made of cast steel hardened and tempered or of one of the alloy, steels recommended for the purpose. The advantages of the alloy steel are that the risk of distortion or cracking in hardening is minimised, and the working edges of the tools are less likely to chip or crumble.

Forging a Bolt Head

The sequence of operations to forge a bolt head is as given under :

(*a*) Heat one end of the bar for a sufficient length to make the head.

(*b*) Jump up heated end on an anvil.

(*c*) Flatten head by hammering against the end of a bush, placed directly over the square hole in the anvil, through which the shank will pass.

(*d*) Swage head to size.

(*e*) Forge chamfer on head by a cupping tool, using the bush as support.

For mass production a forging machine is used for the processes. First, pieces are cut from a bar and are heated in a furnace to the required temperature. At a time one piece is fed into the machine, in which it comes against a stop and is guided by a stationary die. Then machine is operated, and a movable die closes in on the bar and grips it rigidly. The end of the bolt is flatened by means of operation of a plunger. The head is made symmetrical hexagonal by turning the rod one sixth of a turn at a time. Then threads are cut on a screw cutting machine as desired.

Forging a Nut

For mass production the nuts are also forged by means of a nut forging machine. A steel bar of rectangular section, thickness of which equals to the thickness of the nut is selected.

The bar of required length is first heated to the desired temperature and then fed into the machine, where a piece of proper length is sheared off by means of a cut-off tool. This piece is then carried into the main dies and pressed against a crowing tool. A hole is punched in the nut with the help of piercing tool. Then nut is automatically forced out of the dies with the help of an ejector.

21.6 EXTRUSION

Extrusion is defined as the process of pushing the heated billet of metal through an orifice provided into a die, thus forming an elongated part of uniform cross section as shown in Fig. 21.11, 21.12 and 21.13. Aluminium and its alloys greatly used for this process. Rods, tubes, moulding trim, structural shapes, brass cartridges, lead covered cables, aircraft parts, flooring strips and many hardware items etc. are produced by this method. The metal can be extruded in three different ways as given under:

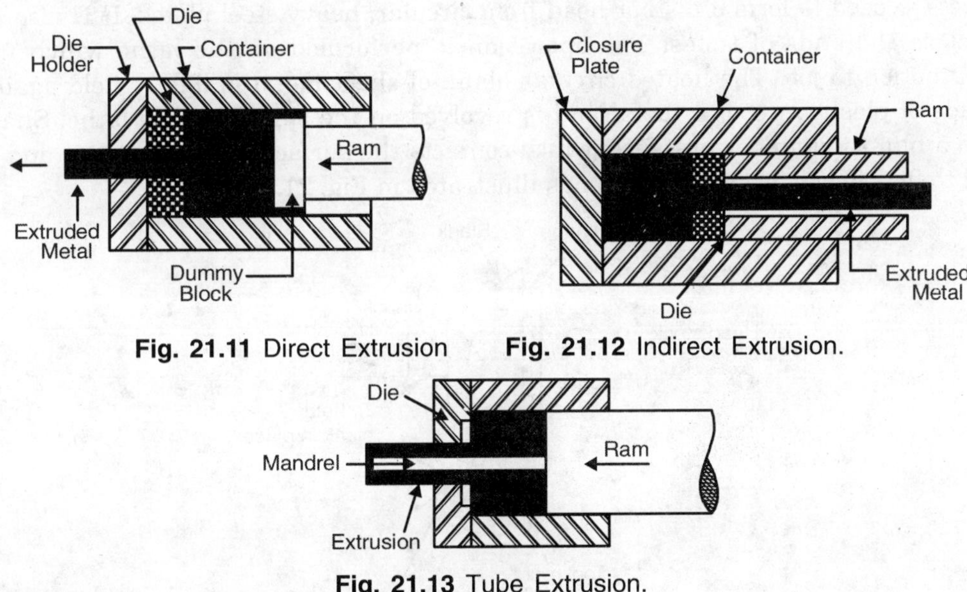

Fig. 21.11 Direct Extrusion **Fig. 21.12** Indirect Extrusion.

Fig. 21.13 Tube Extrusion.

(*i*) Direct and forward extrusion (refer Fig. 21.11).

(*ii*) Indirect or backward extrusion (refer Fig. 21.12).

(*iii*) Tube extrusion (refer Fig. 21.13)

21.7 PIERCING

This process is employed for the production of seamless tubes. It offers the most economical mechanical working process for the manufacture of seamless tubes. It consists of passing the hot rolled billet at 1100°C between two conical rollers and over a mandrel which helps in piercing and controlling the size of bore as the billet is forced over it, as illustrated in Fig. 21.14.

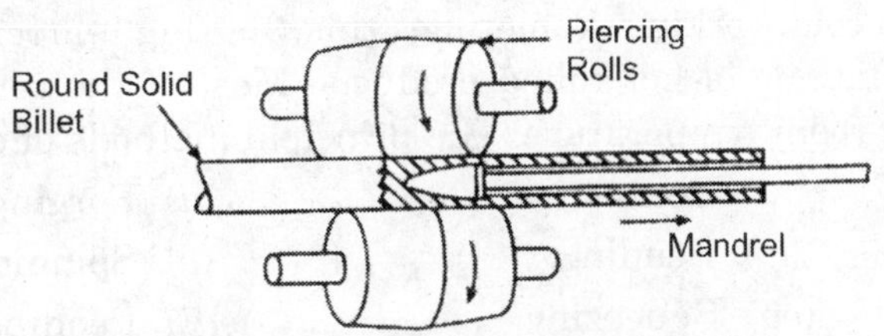

Fig. 21.14 Piercing operation.

21.8 DRAWING OR CUPPING

Drawing is defined as a process of making cup-shaped parts from sheet metal blank by pulling it into dies with the help of punch. Sheet-blanks are heated to working temperatures to provide sufficient plasticity. These are then placed on the die or cavity as shown in Fig. 21.15 and the punch is hammered over the heated blank, which pushes the metal through die to form a cup. So this process is also known as "cuppings".

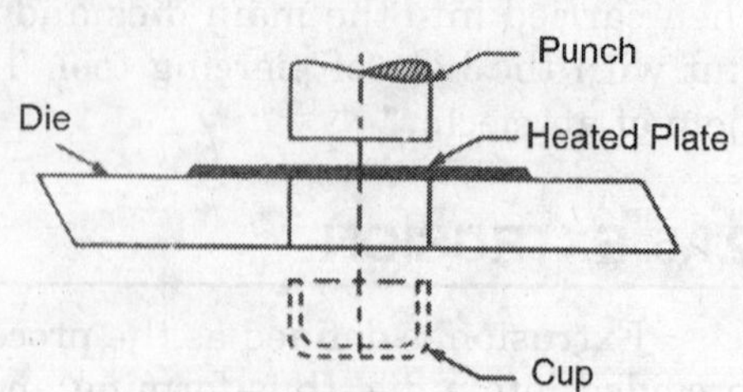

Fig. 21.15 Hot drawing.

21.9 SPINNING

Hot spinning is carried out on thick sheet work to impart circular and desired contours. This process is used to form a dish or head from circular, heavy steel plates. It is also used to neck or close the ends of tubes. The operation is performed on the lathe which imparts rotational motion to job. The heated circular blank of sheet metal is lightly held against the chuck made of plaster, wood or metal and is revolved on the spindle of the lathe. Shaping is done with a blunt tool under pressure, which contacts the surface of rotating part and causes the metal to flow in the desired direction as illustrated in Fig. 21.16.

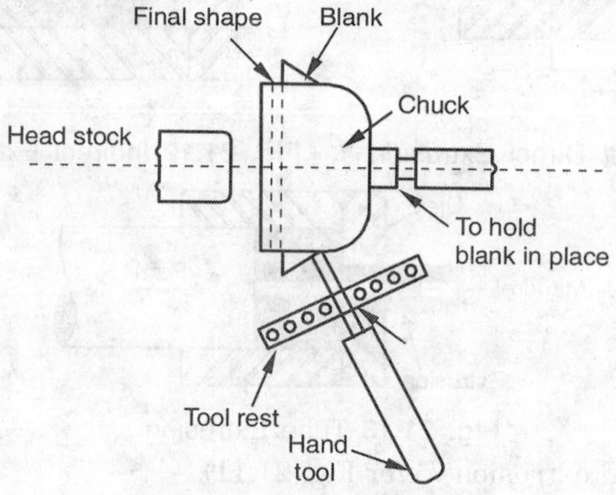

Fig. 21.16 Hot spinning.

21.10 COLD WORKING

The working of metals at temperatures below their re-crystallisation temperature is defined as cold working. Sometimes this working temperature is well above room conditions as in cold finishing of hot rolled products. Most of the cold working processes, however, are performed at room temperature. The principal methods of cold working are as under:

(*i*) Rolling,	(*ii*) Forging,	(*iii*) Drawing,
(*iv*) Bending,	(*v*) Spinning,	(*vi*) Extruding,
(*vii*) Squeezing,	(*viii*) Peening.	

Most of the cold-working processes are performed at room temperature. When material is cold-worked the resulting change in material shape brings about marked changes in the grain structure.

The effect of cold-working on the structure of steel is to distort or elongate the grain in the direction of the flow of metal. Much greater pressures are needed for cold-working than for hot-working. The metal, being in a more rigid state, is not permanently deformed until stress exceeding the elastic limit is passed.

Severe stress, known as residual stress is set up inside metal during cold-working. This effect of cold-working is reduced or minimised by heating the metal into the re-crystallization or annealing range, and finally the metal is brought to its original condition. This sequence of repeated cold-working and annealing is frequently called the *cold-work-anneal cycle.* Cold working is employed chiefly as a finishing operation following the shaping of the metal by hot working.

21.10.1 Cold-Rolling

Items generally subjected to cold rolling are bars, rods, sheets, plates, strips, wires, etc. It is also used to finish the hot rolled components to close tolerances and improve their toughness and hardness. If the object is only to give a clean, smooth finish to the metal only a superficial amount of rolling will be needed. When strength and hardness are to be increased substantially and that the section thickness be appreciably reduced then higher roll pressures and deeper kneading are necessary.

21.10.2 Drawing

Drawing operations involve the forcing of metal through a die by means of tensile force applied to the exit side of the die. Rods, tubes and wires are often given a cold-finishing operation to reduce the size, increase the strength, improve the finish, and provide better accuracy.

21.10.3 Bending

Bars, rods, wires, tubing and structural shapes as well as sheet metal are bent to many shapes in cold conditions through dies. In all metal bending, the metal stressed beyond the elastic limit is in tension on the outside and in compression on the inside of the bend. Stretching of the metal on the outside makes the stock thinner. Bending a flat strip of metal is commonly done by roll forming.

21.10.4 Cold-spinning

Cold-spinning is the operation of shaping very thin metal by pressing it against a form while it is rotating. The method is exactly similar to hot spinning described earlier. In cold-spinning the metal is worked at room temperature. Cup shaped parts having rotational symmetry can be made by the spinning. Stainless steels with thickness ranging from 0.3 mm to 3 mm are readily spun cold and the diameters of parts which are spun cold range from a 50 mm to 200 mm. Since spinning requires considerably more time, spinning is normally used for producing smaller quantities.

21.10.5 Cold-extrusion

The principle of cold extrusion is exactly similar to that of hot extrusion. Of the various processes of extrusion, impact extrusion is essentially a cold-extrusion method. Impact extrusion is limited primarily for making small workpieces from more ductile metals. Disposable tubes of lead, tin and aluminium are used as containers for a wide range of domestic materials such as shaving cream, tooth paste, medicines, shoe polish, etc.

The exactly thin tubes are pressed form slugs, as illustrated in the Fig. 21.17. Heavily constructed mechanical presses are generally employed in the impact extrusion of these collapsible tubes. A small unheated slug of metal is placed in the die cavity, and as the ram descends it drives the punch rapidly into the die cavity, and transmits a very high pressure to the metal, which immediately fills the cavity. The outside diameter of the tube is the same as the diameter of the die, and thickness is controlled by the clearance between the punch and die. As the punch is raised the extrusion is removed by an automatic stripping mechanism. The tubes are then threaded, inspected, trimmed, enamelled and painted.

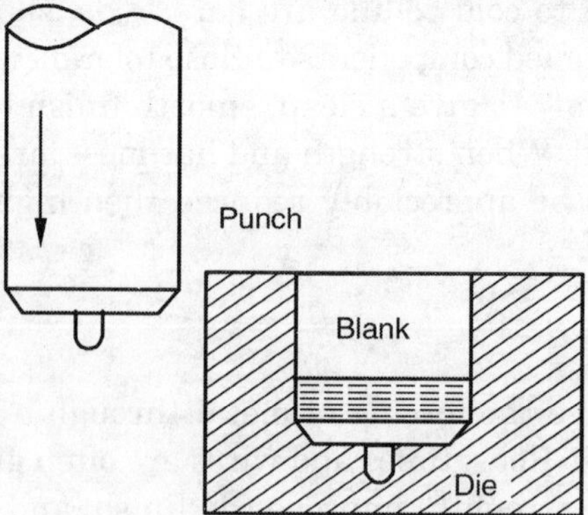

Fig. 21.17 Cold-impact extrusion.

A cold working produces the following effect:

(*a*) Internal stresses are set up which remain in the metal unless they are removed by proper heat-treatment.

(*b*) A distortion of the grain structure is created.

(*c*) Strength and hardness of the metal are increased but ductility is decreased.

(*d*) Smooth surface finish is produced.

(*e*) Accurate dimensions of parts can be maintained.

21.11 FASTENINGS

When two parts of any structure, machine, engine or assembly are joined together, it is called Fastening. The type of fastening being used depends upon the circumstances of joinery and the nature of force acting on the joint. Fastenings are generally divided in two groups as given under:

(*i*) Temporary Fastenings. The fastening under which parts may be unfastened and refastened without destroying the parts and fasteners is called temporary fastening. Joints made with the help of bolts and nuts, screws, couplings, cotters and pins, split pins, keys and splices are the examples of temporary fastenings. These fastenings are generally performed by hands.

***(ii)* Permanent Fastenings.** The fastening under which the main parts and fasteners are destroyed while unfastening and refastening is not easily possible is called permanent fastening. The interference fits, riveting, soldering, brazing, welding, adhesive bonding etc. are the examples of permanent fastenings. They are generally performed by mechanical methods.

21.12 RIVETING

Rivets are used for joining two or more plates to form a permanent joint. Riveting is extensively used on boilerwork, ship-building, structural construction and aircraft work. Rivets may be made from aluminium, duralumin, copper, brass for light works, and for heavy work mild steel and wrought iron rivets are used. Rivets are classified according to the shape of their head as illustrated in Fig. 21.18. They are such as:

(*a*) cup or snap head, (*b*) pan head, (*c*) conical head,

(*d*) countersunk head, and (*e*) bifurcated type.

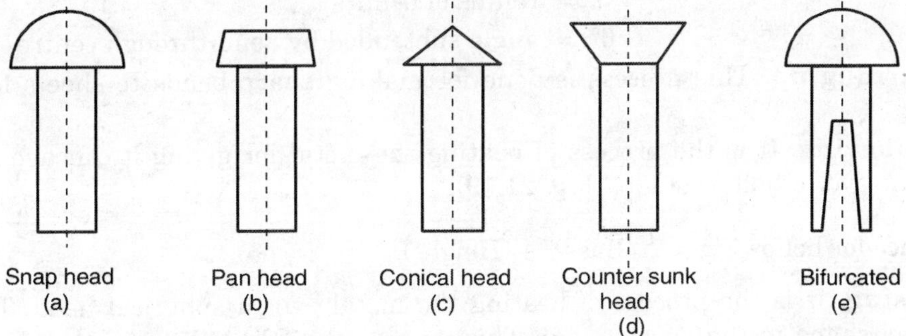

Fig. 21.18 Types of Rivets

Riveting may be either hot riveting or cold riveting. In hot riveting, the rivets are heated to red hot before being hammered up. Hot riveting is generally done on heavy works and rivets are over 10 mm diameter. Cold riveting is generally done on light work and rivets are made of copper, brass, aluminium etc. First correct holes are drilled in the plates to be joined. For proper alignment the drilling of the second plate may be carried out, using the first plate as a jig. Plate holes should be slightly larger than the rivet size. This allows the expansion of the rivet, while hammering. Then holes of the plates are aligned and a rivet is inserted at one end. The head of the rivet is held in a tool called snap or dolly while the tail of the rivet is hammered with a ball pan hammer. A snap heated tool may be used to give the head the proper shape. By this procedure all the rivets are completed.

21.13 SHEET METAL WORKS

Fabrication of components from sheets and plates is done in Sheet Metal Shop. For working in sheet metal shop, an engineer must have a good knowledge regarding plane geometry and mensuration; development of surfaces etc.

Sheet metal process is generally used for preparing bodies of buses, cars, aeroplanes, boiler shells, chimneys, pipes, boxes, toys and furniture making etc.

For preparing a product, first development of the product is marked on the sheet and then sheet is cut along those marked lines.

When large number of products of same size are manufactured (then a template is made and is used for marking on the sheet.

Operations in Sheet Metal Shop

Following are the important operations and processes carried out in Sheet Metal Shop.

***(i)* Breaking out.** In this process, folds or ribbs in sheets, if any, are removed either with the help of beating the sheets by mallet or by passing the sheets through rollers.

***(ii)* Bending.** Bending is carried out for making cylindrical shapes from sheets. The length of sheet required for bending can be calculated as follows:

$$\text{Length for bending} = 2\pi r \times \frac{\theta^\circ}{360}{}^\circ$$

where, r = radius of bend

θ° = angle subtended by bend through centre.

***(iii)* Turning up.** This process is done for making sharp bends to sheets for seaming, hemming etc.

***(iv)* Hollowing.** It is the process of beating the metal for giving it concave shape. This process is done over a hollow stake. (Fig. 21.19).

$$\text{Allowance for hollowing} = \frac{1}{2}\,(\text{Base})^2 + (\text{Height})^2.$$

***(v)* Raising.** It is the process of beating the metal over a spherical head. This process gives a convex shape to the sheet metal. This process should be done on the sheets, having more than 20 gauge.

$$\text{Allowance for raising} = \frac{1}{2}\,(\text{Base})^2 + (\text{Height})^2$$

***(vi)* Planishing.** This is the process, which gives the final finish to the hollow or raised surfaces by removing minor bends. This is carried out by beating the sheet with the help of planishing hammer. Planishing hammer is a short hammer and has high polish.

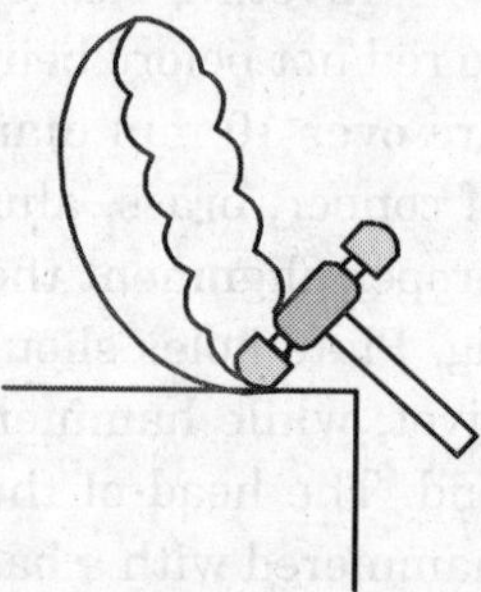

Fig. 21.19 Hollowing

***(vii)* Edge stiffening.** For edge stiffening following are the important ways:

***(a)* Wiring.** In this process, a wire is inserted at the edge of sheet metal articles. This wire adds in the stiffness of edge. Generally, wires used for the black tin plated and G.I. sheets are of mild steel, copper and G.I. respectively.

Allowance for wiring = 2.5 × Dia of wire + 4 × Thickness of sheet.

***(b)* False Wiring.** This process is done like wiring process but in the end wire is taken out so its appearance is just like that as it has been wired, and therefore, known as False Wiring. In this process strength will be less as compared to wiring process.

Allowance = 2.0 × Dia of wire + 4 × Thickness of sheet.

(*c*) **Hemming.** In this process, edge of sheet are folded when folding is done once, as shown in Fig. 21.20, it is called single hemming. The allowance for it is 4-5 times the sheet thickness. When folding is done twice at the edge to give larger strength, as shown in Fig. 21.21, it is known as double hemming. Allowance for it is 10 times the sheet thickness.

Fig. 21.20 Single hemming **Fig. 21.21** Double hemming

(*d*) **Flanging.** In this process, edge of the sheet is folded, at an angle of 90°, to give the shape of flange, as shown in Fig. 21.22.

Sheet Metal Joints

Sheet metal joints are known as seam joints. Various joints in sheet metal are:

(*i*) Lap joint. It is the simplest type of joint. In this, one edge overlap the other and then these are either welded, rivetted or soldered.

(*ii*) Single Seam Joint. This joint is used to join a bottom to a vertical body as shown in Fig. 21.23. For it, allowance is taken as 5 times sheet thickness.

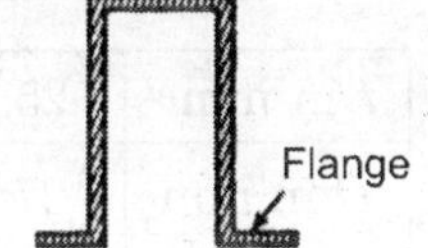

Fig. 21.22 Flanging

(*iii*) Double Seam Joint. It is similar to single seam joint. The only difference is that the formed edge at bottom is bent upwards against the body to give more strength as shown in Fig. 21.24.

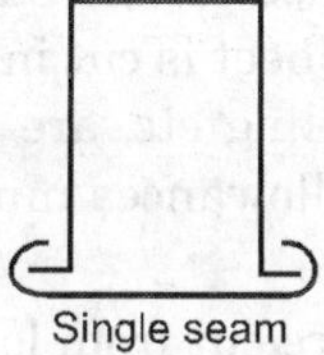

Single seam

Fig. 21.23

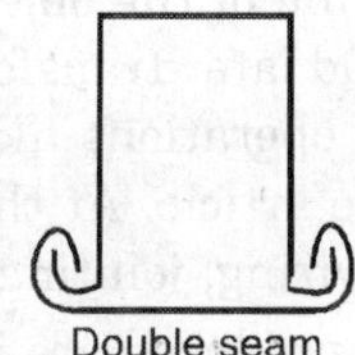

Double seam

Fig. 21.24

Allowance is generally taken as 8 × Sheet thickness.

(*iv*) Grooved Joint. In this folded joints are locked together as shown in Fig. 21.25. This is used where, tension on the sheet is much.

Allowance = 10 × Sheet thickness.

(*v*) Single Plate Butt Joint. In this cover plate of width nearly 10 times the steel thickness is jointed.

(*vi*) Dovetail Joint. In this type of joint, a portion of sheet of length 12 to 20 times, the thickness of the sheet and width 4.5 times the thickness of the sheet is cut to form dovetails as shown in Fig. 21.26.

Fig. 21.25 Fig. 21.26

It is used to joint flat plates with cylindrical shapes.

(*vii*) **Rivetted Joint.** Some of the important formulae required in estimation are:

(*a*) Diameter of rivet, $D = 6\sqrt{t}$ mm.
where t is thickness of sheet in mm.

(*b*) Minimum distance between first rivet centre and sheet edge = 1.5 D

(c) Pitch (p) = 3D. (The distance between centres of adjoining rivets in the same row).

The following table gives an approximate value of D for different values of t:

t in mm	25	22	20	18	16	14	12	11	10	9	8
D in mm	27	26	24	24	22	22	20	20	20	18	18

Blanks Lay-outs

For preparing an article, layout is required to be done on the sheet metal as first step. For this purpose an outline of the object ps either scratched on the sheet of metal directly or first drawn on a paper and later transferred to the sheet. Sheet is cut in accordance with layout and then different other operations like, forming, assembling etc. are performed on it to give the required shape of the article. At the time of layout, allowances must be kept for different operations like raising, wiring, jointing, hemming etc.

When articles to be produced are in lot, then strips are cut from long rolls of sheet. Width of each strip should be slightly more than the blank, so that

(*a*) Material can be handled easily and safely by the operators, specially when working on the machines ;

(*b*) Clean blanks can be cut. The amount by which the strips must be wider than the blanks depends on :

(*i*) Process to be performed:

(*ii*) How they are to be handled, *i.e.* manually or by machine;

(*iii*) Thickness of the sheet;

(*iv*) Accuracy desired. Generally the strip is kept 0.3 cm to 0.5 cm wider than the blank, and the distance between two consecutive blanks (known as bridge) is kept as 0.15 cm.

QUESTIONS

21.1 What is mechanical working and for what purposes it is done ?

21.2 Discuss advantages and limitations of hot and cold working.

21.3 List various methods of hot working. Explain hot rolling in detail.

21.4 What are the forms a steel goes when rolled into structural shape ?

21.5 List different forging operations. Explain any two of them.

21.6 What are the advantages of extrusion over other shaping processes ? State the main applications of hot extrusions.

21.7 Explain briefly with neat sketch the process of wire drawing.

21.8 (*a*) What is piercing ? How is seamless tubing pierced ?

(*b*) What is spinning ? Describe the process of hot and cold spinning stating their advantages and specific uses.

21.9 What is cold working ? List principal methods of cold working.

21.10 Write short notes on:

(*a*) Effect of cold working

(*b*) Rivetting

(*c*) Permanent and temporary fastenings

(*d*) Forging a bolt head.

21.11 Define hot working of metals. What are its advantages and disadvantages?

21.12 Distinguish between hot working and cold working of steel.

21.13 What are the advantages and limitations of cold working processes?

21.14 Describe the various methods of forging.

21.15 Write short notes on:

(i) Drop forge hammering,

(*ii*) Hot extrusion,

(*iii*) Plate bending,

(*iv*) Wire drawing,

(*v*) Hot rolling,

(*vi*) Impact extrusion,

(*vii*) Cold bending.

21.16 Discuss the advantages and limitations of hot working processes.

21.17 Discuss any 4 applications of cold working.

Chapter 22

PATTERN-MAKING, FOUNDRY AND CASTING

Learning Objectives

After studying this chapter, the students will learn:

- Pattern materials, pattern allowances and types of pattern.
- Moulding, moulding processes.
- Melting for casting including crucible furnaces, electric furnaces and cupola furnace.
- Foundry and refractories.
- Castings, casting methods and casting defects.

22.1 INTRODUCTION

Pattern is a model used to form an impression called mould in damp sand or in other suitable material. When this mould is filled with molten metal, and the metal is allowed to solidify, it forms as reproduction of the pattern and is known as casting. The process of making a pattern is known as 'pattern making', while the process of making the castings from moulds prepared by patterns is known as 'foundry engineering'. The entire process of making the castings can be classified in following stages:

(*a*) Pattern-making.
(*b*) Moulding and Core making.
(*c*) Melting.
(*d*) Casting.
(*e*) Fettling.

As discussed above, process of making a pattern is pattern-making. Pattern is the actual model of the desired product including the allowances which are given in addition to the actual dimensions of the product to take care of shrinkage and other changes which might take place during solidification of the molten metal.

22.2 PATTERN MATERIALS

Several materials are commonly used for the construction of patterns. The modern pattern-making employs the use of materials that can be easily shaked and are durable. The type of pattern material chosen depends upon the following factors:

(*i*) The design of casting.
(*ii*) The number of castings to be produced.
(*iii*) The type of casting and moulding process used in the foundry, and
(*iv*) The degree of accuracy and surface finish required.

Following are the commonly used materials for pattern-making.

1. Wood.
2. Metals *e.g.*, brass, cast iron, aluminium alloys, white metal etc.
3. Plaster.
4. Plastic compound.
5. Wax.

1. Wooden Patterns. Wood is the most common material for pattern, the main *reasons* being (*a*) its cheapness, (*b*) ease of availability, (*c*) ease of shaping, (*d*) lightness (*e*) ease with which it can be joined, (*f*) wood can be cut and fabricated into numerous forms by gluing, bending, and carving, (*g*) its surface can be easily made smooth by sanding (*h*) its surface can be preserved by applying a coating of shellac, which is the most commonly used finishing material for wooden pattern.

Wooden patterns are generally used where a small number of castings are to be produced. Wood used for pattern making should be properly dried before it is used. It should be straight-grained, free from knots and free from excessive sapwood. The chief Indian varieties of timber used for making patterns are: Teak, white Pine, Sal, Deodar, Kail, Shisham etc.

2. Metallic Patterns. These are used when a large number of castings are desired from a pattern.

Commonly a metal pattern is itself cast from a wooden pattern called "master pattern". When metal patterns are to be cast from master patterns, double shrinkage must be allowed. For example, if the metal pattern is to be made of aluminium and the castings are to be of cast iron, the shrinkage allowed on the wooden master pattern will have to be 18 mm per metre for aluminium, plus 10 mm per metre for cast iron, making a total of 28 mm per metre.

The metals mainly used for making pattern are: Cast iron, brass, aluminium, white metal, steel etc.

3. Plaster Patterns. Plaster patterns and core boxes are made of gypsum cement. Gypsum cement has a high compressive strength (upto 300 kg/cm^2), and controlled expansion. By means of careful design and use of reinforcements, plaster patterns may be made to have strength upto 1150 kg/cm^2. It can be readily worked with wood working tools. When the plaster is mixed with water, it forms a plastic mass capable of being cast into a mould made earlier by a master pattern or may be swept into desired shape or form by sweep and stricken method. Unlike the metals, gypsum cement expands on solidification and so, by choosing cement of proper expansion coefficient, the shrinkage of casting can be completely offset, and no shrinkage allowance need be given on the pattern.

Owing to these qualities, plaster patterns are very much coming into use for the construction of patterns, and core boxes of small dimensions requiring high precision.

4. Plastic Patterns. Plastics (Phenolic thermo setting resins) are now finding their place as a modern pattern material because they do not absorb moisture, are strong and dimensionally stable, resistant to wear, possess good compressive and flexural strength, better impact strength, good, resistance to chemical attack, have a very smooth and glossy surface, and are light in weight. Due to its glossy surface it can be withdrawn from the mould very easily without damaging the mould. Moreover, the plastic material has a very low solid shrinkage. Phenolic resin plastic best serves the purpose. It is manufactured from a combination of carbolic acid and formaldehyde. Best plastic for patterns is epoxy resin.

For preparing plastic patterns, dies or moulds are first prepared in two halves in metal of plaster of parts. The thermosetting resin is then poured in the required amount into the moulds. When these moulds which carry resin are heated to a specific temperature, the resin solidifies and the patterns are produced.

However, plastic patterns are fragile and thus light section require metallic reinforcing. They do not work well when subjected to severe shocks as in machine moulding viz. jolting.

5. Wax Patterns. These are used in investment casting process. Their use helps in imparting a high degree of surface finish and dimensional accuracy to castings.

Wax pattern is prepared by pouring heated wax into the split moulds or a pair of dies while the latter are kept water-cooled. The dies after having been cooled down are parted off. Now the wax pattern is taken out and used for moulding.

Wax pattern need not be drawn out solid from the mould. After the mould is ready, the wax is poured out by heating the mould and keeping it upside down.

6. Old Casting Used as Patterns. Sometimes in case of emergency when the time is the main factor, old castings are used as patterns.

To increase the thickness and other dimensions so as to provide finish machining and shrinkage etc. surfaces of old castings are lagged with thin strips of wood or leather. The castings to be used as patterns should be well cleaned and made smooth so that they can be drawn out of the mould easily. To provide draw screws, holes can be drilled and tapped in the castings. If necessary, core-prints may be made and attached to the casting.

7. Wooden Patterns with Metallic Coatings. Where metallic patterns cannot be economically employed and wooden patterns are also not found satisfactory due to low strength and lack of high finish, metallic coated patterns are many a time used for short production.

The wooden patterns are sprayed with a thin metallic coating of bismuth. Size of aluminium of about 0.25 mm thickness. The metal is fed in the form of wire into metal spray gun and is heated therein by an oxyacetylene flame. Metal spray is blown in an atomised form through a nozzle under pneumatic pressure over the surface of the wooden pattern.

22.3 PATTERN ALLOWANCES

The patterns are not made of exact size as that of desired casting because such a pattern would produce undersize casting. When the pattern is prepared, certain allowances are given on the sizes specified in the finished component drawing so that the casting produced from that pattern after it has been finished and machined will conform to the specified size.

While designing patterns, the allowance commonly considered are given below:

1. Shrinkage Allowance. All the metals used for casting contract and shrink in size after solidification and cooling in the mould. To compensate for this, the pattern must be made larger than the finished casting by an amount known as "pattern maker's contraction". The pattern maker is provided with a special rule or scale called "Pattern maker's contraction rule". In laying measurements of the pattern, the pattern maker allows for this contraction by using this rule which is slightly longer than the ordinary rule of the same length. There may be different contraction rules for different casting metals.

2. Finishing Allowance or Machining Allowance. The rough surfaces of castings that have to be finished or machined are made to dimensions somewhat over those shown on the finished working drawings. The extra amount of metal which is provided on the surface to be machined or finished is called machine finish allowance. The edges of such surfaces are indicated by a finish symbol or mark *V*, or *f*.

Machining allowance is given due to the following:

(*i*) Castings get oxidized in the mould and during heat treatment. Scales thus formed require to be removed.

(*ii*) For removing surface roughness and other imperfections from the castings.

(*iii*) For obtaining exact dimensions on the casting.

(*iv*) To achieve desired surface finish on the castings.

3. Draft Allowance or Taper Allowance. When a pattern is drawn from a mould, there is always some possibility of damaging the edges of the mould. This possibility is greatly decreased if the vertical surfaces of a pattern are tapered slightly inward. This slight tapper inward on the vertical surfaces of a pattern is known as the 'draft'. Draft is expressed in millimetres per metre on a side, or in degrees. Amount of draft needed depends upon (*i*) the shape of the casting, (*ii*) depth of casting, (*iii*) moulding method, (*iv*) mould material. Generally, a draft of 3 to 6 mm per metre should be provided for precision castings.

The amount of draft also depends on:

(*i*) Method of moulding.

(*ii*) Length of the vertical side of the pattern.

(*iii*) Intricacy of the pattern.

4. Rapping or Shaking Allowance. When the pattern is shaken for easy withdrawal, the cavity in the mould is slightly increased in size. This causes the size of the casting also to increase. In order to compensate for this increase, the pattern should be initially made slightly smaller.

In small and medium sized castings, this allowance can be ignored but for large sized castings or in ones that must fit together without machining or where high precision *is* desired, rapping or shake allowance is provided by marking the pattern slightly smaller to the extent based on experience.

5. Distortion or Camber Allowance. Sometimes the castings because of their size, shape and type of metal, tend to warp or distort during the cooling period. This is due to uneven shrinkage, uneven metal thickness or to one surface being more exposed than another, casting it to cool more rapidly. To overcome this distortion the shape of the pattern is thus bent in the opposite direction. This feature is called distortion or camber allowance.

For example, a casting shaped like the letter U will be distorted with the legs diverging, instead of parallel. For compensating this condition, the pattern is made in such a way that the legs converge but, as the casting cools, the legs straighten and remain parallel. Distortion allowances range from the standard finish allowance upto 20 mm when large castings are considered.

This allowance is considered only for castings having an irregular shape and which have tendency to get distorted *e.g.*, long flat castings with unequal arms.

22.4 TYPES OF PATTERN

The type of pattern selected for *a* particular casting will depend upon several conditions *viz.:*

(*i*) The shape and size of casting.

(*ii*) Number of castings required.

(*iii*) Method of moulding employed.

(*iv*) Anticipated ease or difficulty of the moulding operation.

(*v*) Other factors which may enter the situation because of characteristics peculiar to the casting.

If the pattern is required for producing only a few castings, it should be constructed in the cheapest possible way. Similarly when the casting is of simple and regular shape the corresponding pattern may also be identical and in single-piece. If the casting, on the other hand, is of intricate nature involving changes in section, the pattern may be made in number of pieces involving the use of additional loose pieces, if necessary.

Several of the more common types of pattern are listed and described below:

(*i*) Single-piece pattern
(*ii*) Split pattern
(*iii*) Loose-piece pattern
(*iv*) Cope and drag pattern
(*v*) Match plate pattern
(*vi*) Follow board pattern
(*vii*) Gated pattern
(*viii*) Sweep pattern
(*ix*) Segmental pattern
(*x*) Skeleton pattern
(*xi*) Shell pattern
(*xii*) Built-up pattern
(*xiii*) Boxed-up pattern
(*xiv*) Lagged-up pattern
(*xv*) Left-hand right-hand pattern.

22.5 CORE PRINTS AND CORE BOXES

The holes, recesses etc. of various sizes and shapes in the castings can be obtained without any difficulty by using cores. These cores are made separately by pressing core stand in core boxes.

1. Core Prints. To support the core in the mould cavity, an impression in the form of a recess is made in the mould cavity with the help of a projection suitably placed on the pattern. This projection on the pattern is referred to as 'Core Print'. The core print is thus added projection on the pattern, which forms a seat in the mould for resting the core during the pouring operation. The size and shape of the core print should be adequate so that the weight of the core can be supported during the casting operation.

2. Core Boxes. A core box is essentially a type of pattern made of wood or metal into which sand is rammed or packed to form a core.

The core box is made exactly to the shape and size of the core required. It must conform to the type of the core print used on the pattern. While preparing core boxes, the same accuracy should be maintained as is kept in the construction of the pattern.

22.6 MOULDING

Moulding is a process of preparing a cavity inside a damp sand with the half of pattern, core etc. This cavity is known as the 'mould'. The sand moulds are used only once, as they are destroyed while casting is taken out. Permanent moulds of metals having high melting point like dies are also used.

Moulding Materials

The material used for making moulds is called "Moulding Material". It should be such that mould cavity retains its shape till the molten metal solidifies castings are made in (*i*) Permanent moulds, which are made of ferrous metals and alloys *viz.,* Grey C.I., steel etc., (*ii*) Temporary moulds which are made of refractory sands, resins, wax, plaster of paris, carbon, ceramics, etc.

Permanent moulds are generally used for casting low-melting point materials. They are very costly. Most of the castings are produced using refractory moulding materials mainly refractory sands.

22.6.1 Moulding Sands

The principal material used in moulding is the mould sand.

Moulding sands are classified according to the nature of their origin into two categories:

(*i*) Natural sand

(*ii*) Synthetic sand.

(*i*) Natural Moulding Sand. It is also called green sand. It is taken from river beds or is dug from pits. It contains only the binder and is used as received with water added. It can be used for making moulds as soon as it is received from its sources. It contains binder material from 5 to 20%. It needs additional water 5 to 8%. It may contain considerable organic matter.

This sand has the advantage of maintaining moisture content for a long time. It permits easy patching and finishing of moulds. It is employed for moulds for casting cast iron and non-ferrous metals. As compared to synthetic sand, it is cheaper, needs lesser control. When mixed with bentonite, its properties are improved and it is known as "Semi-synthetic sand".

(*ii*) Synthetic Moulding Sand. It is made in the foundry by mixing a relatively clay-free sand with a selected clay binder *e.g.,* bentonite, water and other materials as required. These are comparatively much better sands for moulding as the properties can be more easily controlled. As compared to natural sand, synthetic sand possesses greater refractoriness (above 1650°C), high permeability, more uniform grain size. It contains no organic matter. It acquires more mouldability. It is easy to control its composition, requires less space for storing. It is more suitable for mass production in mechanised foundries.

22.6.2 Properties of a Good Moulding Sand

(1) Property of Permeability. Molten metal always contains a certain amount of dissolved gases, which are evolved when the metal solidifies. Also the molten metal, when comes in contact with the moist sand, generates steam and water vapour. If these gases and water vapour do not find passage to escape completely through the mould, they will form gas holes and pores in the casting.

Therefore, the sand must be sufficiently porous to allow the gases or moisture present or generated to be removed freely while pouring. This property is known as 'Permeability or 'Porosity'.

Sands which are coarse or have round grains exhibit more permeability. Soft ramming and clay addition in lesser quantities improve permeability. Inadequate permeability causes defects like surface blows, gasholes, mould blasting, etc.

(2) Plasticity or Flowability. This property refers to the ability of the moulding sand to acquire a predetermined shape under pressure and retain the same when the pressure is removed. To obtain a good impression of the pattern in the mould, high plasticity is required of a moulding sand.

Moulding sand should respond to different moulding processes. Flowability of mould sand is its ability to get compacted to a uniform density. It assists moulding sand to flow and pack all around the pattern and take up the required shape.

When energy for compaction during ramming is transmitted through the sand mass (*viz.*, sand slingers), good flowability is very essential. Flowability increases with the increase in clay and moisture content.

(3) Adhesiveness. Moulding sand particles must be capable of adhering to another body. In other words, they should cling to the sides of the moulding boxes. It is due to this property that the sand mass does not fall out of the moulding box and the same can be successfully held in it.

(4) Cohesiveness. The ability of sand particles to stick together is known as 'cohesiveness' or 'strength'. If this ability is not sufficient, it may lead to a collapse in the mould or during con-veyance, turning over, closing or pouring, it may be partially destroyed or damaged.

This property of moulding sand in its moist or green state is known as 'green strength' and in its dried or baked state as 'dry strength'. Sand must have sufficient strength so that it may not be washed away during pouring.

(*a*) **Green Strength.** A mould having adequate *green strength* will not distort or collapse even after removing the pattern from the mould box. Green strength assists in making and handling the moulds. In the absence of green strength, dimensional stability and high accuracy cannot be achieved.

(*b*) **Dry Strength.** If green sand mould is likely to lose its moisture and get dried while waiting for getting poured or when it comes in contact witn molten metal, the mould is *dried* or *skin dried* to withstand pressure and erosive forces of molten metal. Dry sand strength is related to grain size, binder and moisture content. There should be an optimum balance between dry strength and collapsibility of mould sand.

(*c*) **Hot Strength.** It is the strength of sand of mould cavity above 100°C. If hot strength is inadequate, mould is likely to enlarge, break, erode, or get cracked.

(5) Refractoriness. The capability of the moulding sand to withstand the high temperature of the molten metal without fusing is known as 'Refractoriness'. Moulding sands having poor refractoriness burn on to the casting. However, silica sands have high thermal expansion at 575°C.

As compared to casting of low melting point alloys, refractoriness is much more essential in casting of high melting point alloys like steel.

(6) Chemical Resistivity. Moulding sand should not chemically react with the metal as otherwise the shape of the casting will be distorted and its surface spoiled.

Also the sand must not stick to the casting and should strip off easily leaving a clean surface.

(7) Collapsibility. It is that property of the moulding sand, which determines the readiness with which it automatically gets collapsed after solidification of the casting and breaks down in knocking out and cleaning. If the mould or core does not collapse, it may restrict free shrinking of solidifying metal and cause the same to crack or tear.

(8) Fineness. Finer mould sand resists metal penetration and produces smooth casting surface. Fineness and permeability are opposite to each other. Hence, these should be balanced for optimum results. Both these properties can be maintained by using mould coating on highly permeable mould surfaces.

(9) Low Coefficient of Expansion. Mould sand should possess slow coefficient of expansion.

(10) Bench Life. It is the ability of mould sand to retain its properties during storing, handling, or standing *e.g.*, in case of delay.

22.6.3. Types of Moulding Sands

Moulding sands may also be classified, according to their use as under:

(1) Green Sand. By green sand we denote sand in its natural more or less moist state. It is a mixture of silica sand with 18 to 30% clay and water *6* to 8*%*. The clay and water give bonding strength to green sand.

It is fine, soft, light and porous. Being damp, when squeezed in the hand it retains the shape given to it under pressure.

Moulds prepared in this sand are known as 'Green sand moulds'. Cores can also be made out of green sand.

The rammed sand is dense, hence the structure is made porous by venting freely. As the hot metal is poured in it, sharp edges being weak, break and hence sharp edges are avoided in green sand moulding.

Green sand is generally used for casting simple, small or medium sized moulds. Larger output can be obtained from a given floor space as the cost and delay involved in drying the moulds is saved. Coal dust is mixed in green sand to prevent defects in castings.

(2) Dry Sand. The moulds prepared in green sand are dried or baked to remove almost all moisture of the moist sand. The structure in the moulding boxes becomes stronger and compact. Venting is necessary but not *so* much as required in green sand. For larger heavy moulds cowdung, horse manure, horse hair are mixed with the sand of coarser grains.

Depending upon dried mould wall thickness, it may be called a "skin dry mould" or "dry mould". Castings from dry sand moulds possess smoother surface. Dry sand mould possesses greater strength, rigidity and thermal stability.

Dry sand moulds eliminate defects like pinholes etc. Moulds are dried by heating them at 175°C and skin-dried by torch. Skin drying is a localised drying process.

Dry sand moulding *is* employed for large and heavy castings.

(3) Loam Sand. Loam is mixture of clay and sand milled with water to a thin plastic paste from which moulds are built up on a backing of soft bricks. Patterns are not generally used for loam moulding and shape is given to the mould by sweeps.

Loam sand is high in clay, as much as 50% and dries hard. It also contains fire clay, graphite and fibrous reinforcements. It must be sufficiently adhesive to hold on vertical surfaces as it is applied as plaster to the rough structure of the mould. Chopped stray and manure are commonly used to assist in binding. The moisture content is from 18 to 20% to a consistency resembling mortar.

Loam is dried very slowly and completely before it is ready for casting. It is used for casting larger regular shaped castings, *e.g.,* chemical pans and drums etc.

(4) Facing Sand. It forms the face of the mould. It is used directly next to the surface of the pattern and it comes into contact with the molten metal when the mould is poured. It is, therefore, subject to the most severe conditions and hence must possess high strength and refractoriness. It is made of silica sand and clay, without the addition of used sand.

Facing sand may have a thickness of 20-30 mm. Its use reduces the mould material cost.

Different forms of carbon known as facing materials *(e.g.* plumbago powder, ceylon lead or graphite) are used to prevent the metal from burning into the sand. Sometimes they are mixed with 6 to 15 times fine moulding sand to make mould facings.

Facing and layer in a mould usually ranges from 20 to 30 mm. Facing sand is from 10 to 15% of the whole amount of mould sand.

(5) Backing Sand. The accumulated moulding sand from cast up moulds which is used over and over again is known as 'Backing sand' or 'Floor sand'. It is used to fill in the mould at the back of facing layer so that it does not come in contact with the pattern. Old, repeatedly used moulding sand is mainly employed for this purpose. After it is used a number of times, the sharp edges of sand grain become rounded due to high temperature of molten metal and burning of clay content and the sand becomes weak in bonding strength.

The backing sand is sometimes known as 'back sand' because old, repeatedly used moulding sand is black in colour due to the addition of coal dust and burning or coming in contact with molten metal. However before use, backing sand, should be cleaned off the foreign matter like fins, nails etc.

(6) Parting Sand. A mould is generally prepared into two or three boxes. These boxes are separated from adhering to each other by spreading a fine sharp dry sand called 'parting sand'. Parting sand is also used to keep the green sand from sticking to the pattern.

It is clean clay-free silica sand. Some moulders use burnt core sand *i.e.,* core sand cleaned from castings. These are also manufactured parting materials known as 'parting compounds'. They are in powdered form and used by dusting them on parting surfaces.

When the oil has been mixed in moulding sand:

(*a*) lycopodium powder *a* parting compound is used.

(*b*) brick powder is used as parting sand to prevent mould sand to stick to the pattern, the only objection to its use is the high cost.

Parting compounds may be dry or wet. Dry parting substances are fine dried sand, charcoal, ground bone, lime stone, lycopodium powder (a yellow vegetable matter), tripolite a silicate rock, ground nutshells (walnut), talc (magnesium silicate), and calcium phosphate. These are used for wooden patterns.

Liquid parting substances are petrol-cum-jelly mixed with oil, paraffin, and stearic acid. These are used for metal patterns in machine moulding.

(7) Core Sand. Sand which is used for making cores is called 'core sand' or 'oil sand'. It is silica sand mixed with core oil which is composed of linseed oil, rasin, light mineral oil and other binders.

For the sake of economy, pitch or flour and water may be used in large cores. Core sand has been further discussed under core making, section 3.12.

(8) Molasses Sand and Oil Sand. Sand using molasses or oil binders (linseed oil) are called "molasses sands" and "oil sands".

Molasses sand is used small, thin, for producing intricate and accurate castings. It is also suitable for small non-ferrous gears.

22.6.4 Principal Ingredients of Moulding Sands

(*i*) Silica sand grains
(*ii*) Moulding sand binder
(*iii*) Moisture
(*iv*) Additives

1. Silica Sand Grains

They impart refractoriness, chemical resistivity and permeability to the sand. They are of paramount importance in the moulding sand. They are specified according to their average *size* and *shape*.

Grain size and distribution influence sand properties like refractoriness, permeability, plasticity (flow ability), surface fineness, strength etc.

Fine grains will be laying more closely to each other and so permeability will be less. But fine grains will impart more strength to the mould and so tendency of the mould towards deformation will be less. Fine grains give good surface finish.

Coarse and uniformly graded sands impart high permeability high flowability and maximum refractoriness. Normally, moulding sands possesses grain size between 0.1 to 1 mm.

There are three distinct sizes of sand grains: fine, medium and coarse:

(*a*) For small and intricate castings the use of a fine sand is desirable, so that all the details of mould are brought out sharply.

(*b*) Medium sand is used in bench work and light floor work. *viz.*, making machinery castings having sections up to 50 mm.

(*c*) As the size of the castings increases, the sand particles likewise should be coarser so as to permit the gases generated in the mould escape readily.

2. Binder or Clay

Its purpose is to impart necessary bonding strength to the mould sand so that the mould does not lose its shape after ramming. As the quantity of clay increases, the permeability of the mould sand decreases.

Clay may be defined as those sand particles under 20 microns in diameter which fail to settle a distance of 25 mm per minute, when suspended in water. Average clay particles are of colloidal size, clays are essentially aggregates of extremely minute crystalline, usually flake-shaped particles, which can be classified on the basis of their structure and composition into a few groups called 'clay minerals'.

Some of the clays are composed of particles of a single clay mineral, whereas others are mixtures of clay minerals. Some of the clays are composed, entirely of clay minerals whereas others contain mixtures of quartz, organic matter etc.

Clay consists of two ingredients ; fine silt and true clay. Fine silt is a sort of foreign matter or mineral deposit which has no bonding property. True clay imparts the necessary bonding power. When seen under high magnification, true clay is found to consist of extremely minute aggregates of crystalline particles, called 'clay minerals'. Clay minerals are further composed of flake shaped particles about 2 microns in diameter which are seen to lie one on another.

Binders to the moulding sands should be added in minimum required quantity as it reduces permeability. Increasing binder content to a limit increases green compression strength; after which this strength remains practically unchanged with increase in binder content. For bonding mould sands, clay binders are most commonly used.

Clay binders are classified as under:

(*i*) Fire clay

(*ii*) Bentonite

(*a*) Sodium montmorillonite

(*b*) Calcium montmorillonite

(*iii*) Illinite

(*iv*) Limonite

(*v*) Kaolinite

Clays are inorganic binders. Best clay is one which imparts that optimum combination of bonding properties, moisture mould strength, and low cost of producing the casting and produces a thin and adhesive film around the moulding sand grains.

When heated to high temperatures during pouring, clay gets baked and hardened. Clay is added to moulding sand from time to time so as to maintain adequate bonding strength of mould sand.

In natural moulding sand, clay and sand occur mixed in a proportion so that it can directly be used for moulding. But, in synthetic moulding sand, selected clays are added to obtain desired strength and plasticity.

Fire clay. It is a refractory clay usually found in coal mines. Hard black lumps are weathered and then pulverized for use in moulding sand. Particles are about four hundred times as large as bentonite particles. Thus, the same percentage of fire clay imparts lower strength to the mould.

Bentonite. It is the most commonly used clay binder as it produces the strongest bond in mould sand. In India, bentonite deposits are available in Bihar and Rajasthan. Bentonites are weathered product of volcanic ash. They are soft, creamy white powders. They are usually used for synthetic mould sand.

Illite. It is a decomposition product of micaceous material due to weather. It is found in natural moulding sand. Illite particles have thickness and width of 20 and 200 milli microns respectively.

Illite possesses moderate shrinkage due to loss of water. It does not swell like bentonite but gives reasonable strength. It has a softening point of 1370°C. In temperature range of 500 - 550°C, irreversible dehydration takes place.

Kaolinite. It is the residue of weathered granite and basalt. It contains kaolinite 60%, illite 30%, and quartz 10%. Its particles possess thickness and width of 20 and 100-250 milli microns respectively. It is characterized by (*a*) low shrinkage due to less of water, (*b*) very low swelling due to water, and (*c*) non-gel forming. Its softening point is 1650 to 1700°C.

3. Moisture

The purpose of clay will not be served until the requisite amount of moisture is added to it. Clay imparts bonding action and strength in the presence of moisture.

Water, when added to a clay moisture, it penetrates and forms a micro-film coating the surface of each individual flask. The molecules of water which form the micro-film on the flask are not in the original fluid state but they are in a fixed and definite position. On adding more water, thickness of the micro-film increases upto a certain limit after which, the excess water remains in the fluid state. The thickness of this water film varies with the clay mineral. Bonding property of clay depends on the maximum thickness of water micro film it can maintain.

Amount of water added varies from 1.5% to 8%. Water is responsible for the bonding action of clays. It activates the clay in the sand and the clay-sand mixture and develops plasticity and strength. Water added, partly gets absorbed by clay and partly remains free (which is called "free Water").

Absorbed water develops proper bond and green strength. Free water acts as a lubricant, increases improved mouldability but reduces mould strength. Amout of water required to develop optimum properties is determined experimentally.

4. Additives

Materials other than ingredients (viz. sand binder and moisture) are also added to mould sand in small quantities for improving existing properties, developing certain other properties *viz.*, resistance to sand expansion defects, etc.

Additives include:

(*a*) Facing materials and (*b*) Cushion materials

(*a*) Facing Materials. These are used for obtaining smoother and cleaner surface on the casting and help easy peeling off sand from surface of casting while shaking out. Some of the important facing materials are:

(*i*) Sea coal or coal dust. It is finely powdered bituminous coal. Amount of addition varies from 1% for small casting to above 10% for large castings. It is added to mould sand for cast iron to improve surface appearance and easy stripping.

(*ii*) Pitch and Asphalt. Pitch is distilled from soft coal at about 315°C. It is added upto 2%. Asphalt is a by-product of petroleum distillation.

These are added to obtain fine surface finish on ferrous castings and improve hot strength.

(*iii*) Graphite. It may be natural or synthetic. It is added from 0.2 to 2%. It is added to improve surface finish of the castings and mouldability of foundry sand.

(*iv*) Silica Flour. It is pulverized silica which can pass through 200 mesh sieve. It is added upto 35%. It is added to improve surface finish and hot strength, resist metal penetration and minimize sand expansion defects.

(*v*) Coke. It may be added upto 5% and facilitates easy stripping.

(*b*) Cushion Materials. These are added to burn at the time of pouring molten metal, giving rise to space to accommodate the expansion of silica sand of the surface of mould cavity. Thus, they prevent buckling of large felt surfaces of castings due to thermal expansion of silica sand grains.

Following are some of the important cushion materials:

(*i*) Wood flour, Cereal hulls, Cellulose, etc. These are added from 0.5 to 2%. They minimize sand expansion defects, improve flowability and collapsibility of sand.

(*ii*) Pearlite. It is added for the same purpose as wood flour. As it possesses insulating properties, it is used as a riser insulator.

(*iii*) Cereals or Finely Ground-Corn Four. These are added from 0.25 to 2%. They are added to increase green and dry strength, minimize sand expansion defects, and improve collapsibility. They reduce permeability and flowability.

(*iv*) Fuel Oil. It is added from 0.01 to 0.1%. It improves mouldability of sand.

(*v*) Iron Oxide. It is added to improve hot strength.

(*vi*) Molasses and Dextrin. Molasses resists the tendency of the mould to dry out. These are added to increase dry strength and edge hardness of the moulds.

(*vii*) Diethylene glycol. It is added to improve surface finish of the castings. When added to bentonite bonded sand, it improves the tendency of the mould to surface drying.

(*viii*) Ammonium Bifluoride, Boric acid and Sulphur. They are added to moulding sand for magnesium castings so as to prevent metal-mould reactions.

22.7 MOULDING PROCESSES

Moulding Processes are broadly classified as:

1. *Hand Moulding—Used* in piece and small-lot production foundry practice.
2. *Machine Moulding—Used* in large-lot and mass production.

Moulding Processes are often classified according to:

1. Type of Material of Moulding, *viz.*
   - (*i*) Green sand moulds,
   - (*ii*) Dry sand moulds,
   - (*iii*) Loam moulds,
   - (*iv*) Metal moulds
2. Method used, *viz.*
   - (*i*) Bench moulding,
   - (*ii*) Floor moulding
   - (*iii*) Pit moulding,
   - (*iv*) Sweep moulding,
   - (*iv*) Plate moulding.

1. Green Sand Moulding. If the mould is filled with molten metal while the sand is still in the moist state, the method used is known as 'Green sand moulding'. Green sand is generally used for casting small and medium size mould. In this type cost and delay involved in drying moulds are saved. The surface finish of the casting obtained from green sand moulds is not very smooth.

2. Dry Sand Moulding. The method of making dry sand moulds is similar to that of making green sand mould except that (*i*) different sand mixture is used, (*ii*) the mould surface is sprayed with molasses water, and (*iii*) all parts of the mould are dried in an oven (maintained at 150° - 300°C untils the moisture is driven off) before assembling it. Although these are expensive but the castings produced have a better finish.

Dry sand moulds are often used for large work *e.g.*, engine blocks, engine cylinders, rolls of rolling mills, intricate castings and those requiring special smoothness, soundness and accuracy.

3. Skin-dried Moulding. When the moisture is dried from the surface layer of the rammed sand and to a depth of about 25 mm or more by heater or gas torches, the process is known as 'Skin-dried Moulding'.

In a process the advantages of both dry sand and green sand moulding are obtained. Since the time required for drying is less than in case of dry sand moulding the method is less expensive.

4. Loam Moulding. Loam is sand clay milled with water to form a thin plastic mixture to the consistency of mud. This loam mortar is applied as plaster (6 to 12 mm layer) to the rough structure of the mould. The loam sand mould is constructed of porous brick cemented together with loam mortar. Cast iron plates and bars are used to reinforce the brickwork which retains the moulding material. Loam moulds require adequate ventilation so as to open out pores in the otherwise compact, closely knit mass, by artificial means. For this, chopped straw or horse manure is mixed up with the loam sand.

The rough structure of the mould is usually made of brick work and exact shape is given by a rotating sweep around a central spindle, *i.e.*, by using sweep pattern. Loam moulds may also be prepared by using skeleton pattern and strickle made of wood. The surface of loam mould is blackened and dried before assembling it.

Loam moulding is used for large-sized castings where the use of regular moulding method in flasks will be too expensive and inconvenient *e.g.*, large cylinders, round bottomed kettles, chemical pans, large gears etc.

5. **Metal Moulds.** When the impression or the mould *is* made of metal, it *is* known as 'Metal Mould'. Metal moulds are employed for the production of large numbers of identical castings.

The processes which employ metal moulds are:

(*i*) Die-casting process.

(*ii*) Permanent mould process,

(*iii*) Centrifugal process.

6. Bench Moulding. Bench moulding applies chiefly to small sized castings which are light in weight and which can be easily handled.

In bench moulding, the moulds are prepared in moulding boxes or flasks placed at convenient height on a bench. Mould is prepared by the moulder while standing near the bench.

7. Floor Moulding. Very heavy castings or castings having considerable depth or area are moulded in the sand of the foundry floor in much the same way as green sand or dry sand moulding. Thus, the floor itself acts as the drag and this is covered with a cope or the mould is cast open.

8. Pit Moulding. In this method, moulds of large jobs are generally prepared in a pit dug in the foundry floor. This facilitates in lifting the pattern and casting the mould easily. The sand under the pattern is rammed by bedding-in because a pit which functions as a drag cannot be rolled over.

A coke bed is laid on the bottom of the pit. It is covered with straw and then a layer of sand is rammed and levelled. To provide an outlet for the escape of gases generated the coke bed is connected with atmosphere by vertical vent pipes in the corners of the pit. The inside surfaces of the pit are lined with bricks. To complete the mould, generally one box is required for cutting runners, pouring basins, feeders in it.

9. Sweep Moulding. For moulding parts whose shape is that of surface of revolution, sweep moulding is employed.

In this method, a spindle with its base is well placed in the foundry floor. The sand is filled and rammed until the shape and size of the casting is approximately formed by the excavation. Then, a sweep holder is placed in the spindle and the sweep is fastened by bolts and nuts. The mould surface is produced by the sweep profile as the sweep is rotated about the spindle. The spindle is removed after sweeping and the mould is patched at the centre. Then, gate is cut. Now the mould is ready to be poured.

22.8 CORE MAKING

Core is defined *as* that portion of the mould which forms the hollow interior of the casting or hole through the casting. The core means the mass of dry sand which is prepared separately, baked in an oven and then placed in the mould.

Characteristics of Cores and Core Sands, (*i*) The core should have sufficient strength and hardness in both green and dry states so that it may be able to support its own weight and withstand force of molten metal.

(*ii*) The core should be permeable to allow the escape of core gases.

(*iii*) The core should be able to withstand the high temperatures of the molten metal.

(*iv*) The core sand should produce minimum amount of gas when in contact with molten metal. Thus, very high permeability will not be needed. Consequently, greater will be imparted to the core.

(*v*) After the metal solidifies, the core should disintegrate and collapse, otherwise difficulty may be experienced in removing from the casting.

(*vi*) Core sand should be able to give smoothness and finish to the core surface.

Core Material or Ingredients of Core Sands

(*i*) Granular refractories

(*ii*) Binders (self-made and Patented)

(*iii*) Special additives

(*iv*) Moisture.

Granular Refractories. They are silica sand (clean, pure and dry)

Zircon

Olivin

Carbon

Core Binders. For preparing cores, pure sand is used. Core sand has no natural bond. Hence some other materials are added to the sand to act as binders which cement the sand grains into desired shape and impart sufficient strength to cores to prevent breakage, distortion or erosion during core making, moulding and casting.

Various commercial binders available in the market mainly consist of oils, resins, sulphite, liquor, molasses and proteins. Generally, core oils are used as they are very economical and produce better cores. Core oils mainly contain vegetable oil like linseed oil or corn oil. Specially processed mineral oils are sometimes mixed for specific properties.

Resin and pitch are thermoplastic binders. Their action depends on the heat to liquify and disperse them in sand. The powdered binder is mixed with core sand. On heating, the binder liquifies and coats the sand grains. On cooling, the dispersed liquid binds the sand grains together forming a united mass. Rosin which is a form of resin, is obtained by distillation and extraction from pine wood. By-products in petroleum and coal tar industries *viz.,* petroleum and coal tar resins are also used as binders. Use of synthetic resins *is* limited as they are costly. Addition of resin to core oil speeds up drying of core and reduces the volume of core gas. Pitch compounded with dextrine and steam coal is used for large cores.

Phenol and urea-formaldehyde thermosetting plastic core binders are becoming common due to their high strength, low gas formation, collapsibility and resistance to moisture absorption.

Molasses adds hardness to the core but lacks in strength. On the core molasses water may also be sprayed.

Where easily collapsible cores are required, protein binders *viz.,* gelatin, glue, casein etc. are used.

Where high strength, hardness, quick drying and high temperature resistance is required, sulphite liquor obtained from paper and pulp industries in the form of concentrated liquid is used.

Except for very small cores, every core is internally reinforced with core wires or core rods for medium cores, and with cast-iron or fabricated steel devices known as core grids or arborbs for large cores. Reinforcement is required for the following reasons:

(i) To help the core retain its shape while it is damp (green compressive strength of core sand is very low—350 to 1000 gm/cm^2).

(ii) To provide internal hooks or staples by which the core can be lifted safely to a drying plate, and after drying lowered accurately into the mould.

(iii) To strengthen the core so that it withstands the tendency of the fluid metal to break and lift it during casting.

Core Coatings, Core Dressings, Core Washes, or Blackings. These are used like mould coatings. These may be applied by dusting (if in the form of solid powder), by dipping, spraying or brushing (if in the form of liquid).

Types of Cores

The cores used in foundries are classified according to their shape and position in the mould as under:

***(i) Horizontal* Core.** It is the most common type. It is usually in a cylindrical form laid in the mould horizontally. The ends of the core rest in the seats provided by the core prints on the pattern (see Fig. 22.1).

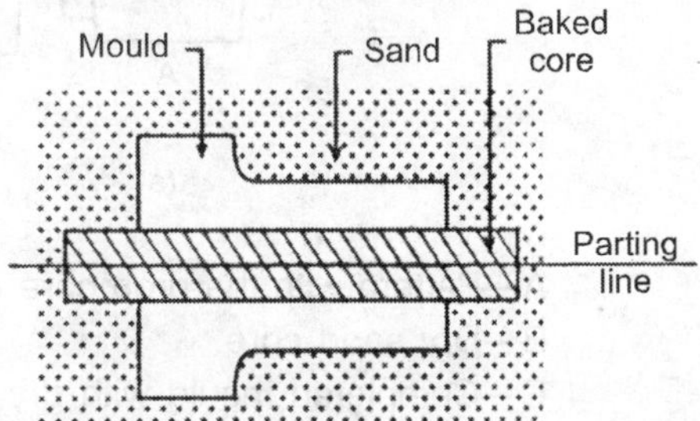

Fig. 22.1 Horizontal or parting line core.

(ii) Vertical Core. The core, when required to be placed along a vertical axis in the mould is known as 'Vertical core'. The upper end of the core is forced in the cope and the lower end in the drag (see Fig. 22.2).

Horizontal and vertical cores are used in foundry work more frequently than cores of any other shape. These cores are, therefore, generally maintained in stock in various diameters and lengths and referred to as 'stock cores'.

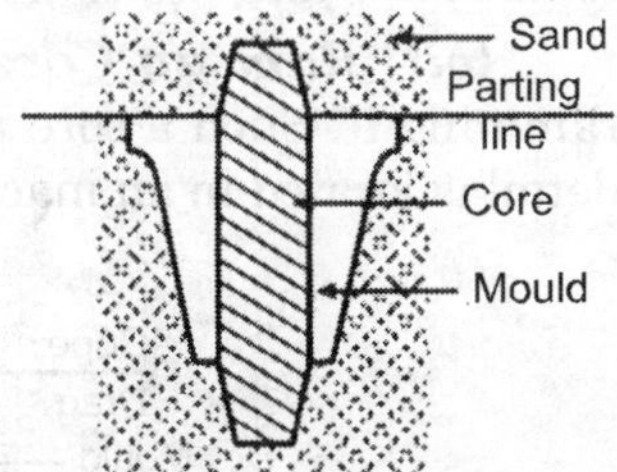

Fig. 22.2 Vertical or Cope and drag core.

(iii) Balanced Core. When the casting has opening only on one side and only one core print is available on the pattern, a 'balanced core' is used (see Fig. 22.3). In such cases, the core print should be sufficiently large to support the weight of the core which extends into the mould cavity and it should be able to withstand the force of buoyancy of the molten metal which surrounds it. For supporting the core in the mould cavity, chaplets are often used.

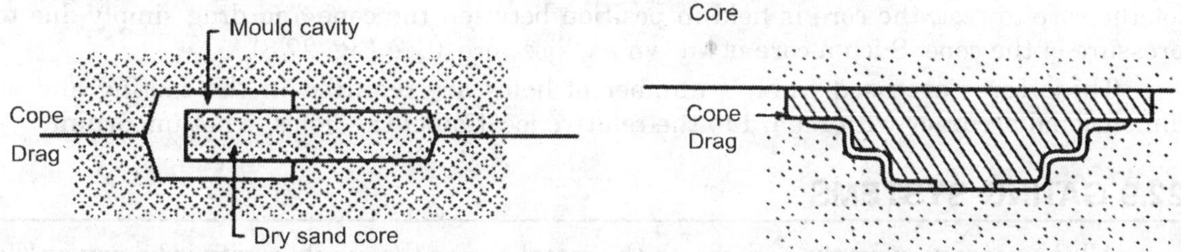

Fig. 22.3 Balanced core.

Fig. 22.4 Corver core.

(iv) Cover Core. When the entire pattern is rammed in the drag and the core is suspended from the top of the mould, a 'cover core' is used (See Fig. 22.4). A 'balanced core' extends horizontally in the mould cavity, whereas a 'cover core' extends vertically downwards.

(v) Hanging Core. If the core hangs from the cope and does not have any support at the bottom in the drag, it is referred to as 'Hanging core' (refer fig 22.5). In such cases, it may be necessary to fasten the core with a wire or rod extending through the cope.

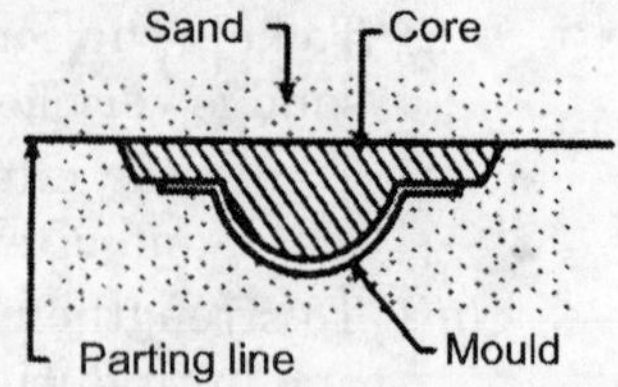

Fig. 22.5 Hanging core.

(vi) Wing Core or Stop off Core. It is used (*a*) When a hole or recess is to be obtained in the casting either above or below the parting line.

(*b*) When it is not possible to place the pattern in the mould such that the recess can be cored directly or with the above mentioned cores.

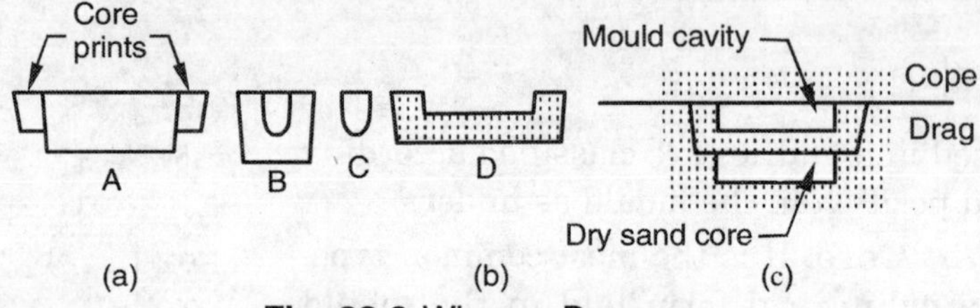

Fig. 22.6 Wing or Drop core

A—Pattern with the drop core prints for a hole above or below the parting surface.

B—Dry sand core.

C—Completed mould with dry sand core.

D—Stop off core.

Fig. 22.6 Wing or Drop core.

The same core at times is also called 'tail core', 'chair core', 'sandle core' etc. according to its shape and position in the mould (See Fig. 22.6).

(vii) Ram up Core. Sometimes the core is set in the mould with the pattern before ramming it. Such a core is known as 'Ram up core' (Refer fig. 22.7). It is used (a) When cored detail is located in an inaccessible position (6) For both interior and exterior portions of a casting.

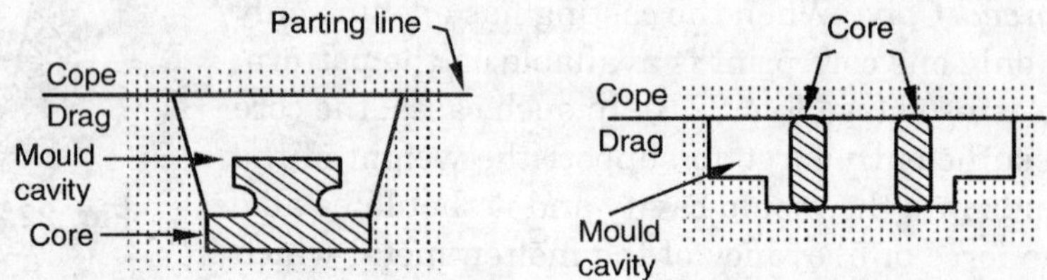

Fig. 22.7 Ram-up core. **Fig. 22.8** Kiss core.

(viii) Kiss Core. When the pattern is not provided with core prints and seat is available for the core to rest, the core is held in position between the cope and drag simply due to the pressure of the cope. Such a core is known as 'kiss core' (See Fig. 22.8).

These cores are used when a number of holes are required in the casting and where dimensional accuracy with regard to the relative location of the holes is not important.

22.9 GATING-SYSTEMS

All the passage ways through which the metal enters the mould cavity is known as 'Gating System'. A pouring basin, sprue, runner, strainer core, gates etc. are included in 'Gating System' (Refer fig 22.9).

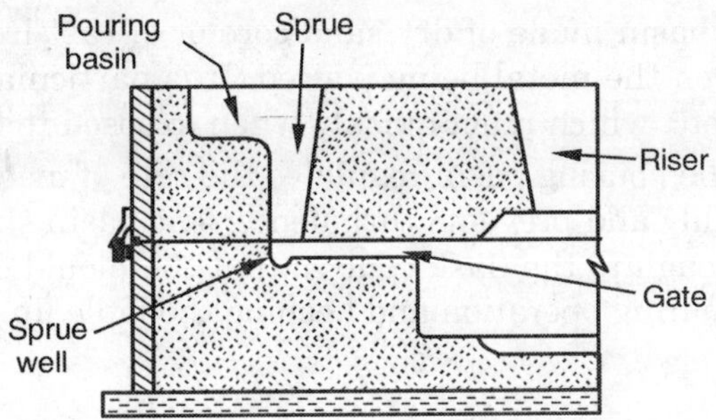

Fig. 22.9 Parts of gating system.

The main requisites of a gating system:

(*i*) The metal should flow through the gating system with minimum turbulence and aspiration of mould gases. Excessive turbulence causes aspiration of air and formation of dross.

(*ii*) The metal should be introduced in the cavity of the mould in such a way that a temperature gradient established on the mould surfaces and within the metal help the directional solidification towards the riser.

(*iii*) The cavity of the mould should be completely filled with molten metal in the shortest possible time.

(*iv*) The casting should be possible with minimum of metal in gates and risers.

(*v*) As the metal flows through the gating system, loose sand, oxides and slag should be prevented from entering the cavity of the mould by providing a proper skimming action.

(*vi*) The design of the gating system should be such that the entering of metal into the cavity of the mould is regulated properly.

(*vii*) The corrosion of the mould walls should be prevented.

Nowadays, the foundries usually design and test their gating system on one or more pilot castings. Then they mount the gates and turners directly on the pattern. To producing castings conforming to the desired specifications, the gating system should be designed carefully for better output and low rejects.

1. Pouring Basin

It is a funnel-shaped opening made at the top of the sprue in the cope. The molten metal is brought from the furnace in the ladle and poured into pouring basin. Different kinds of pouring basins are shown in fig. 22.10.

Fig. 22.10 Different types of pouring basins.

A separate pouring basin made of dry sand core or of cast iron may be placed on the top of the mould, depending on the metal being poured. It is particularly desirable for metals like aluminium, magnesium etc. which react quickly when exposed to air.

A proper design of the pouring basin regulates the rate of metal entry, allows ,the metal to flow into the sprue smoothly and prevents turbulence. The basin size should be large enough. It should be placed close enough to the flask edge so that the mould can be filled quickly, keeping it full during the entire pouring operation and keeping the ladle lip close to the pouring basin.

2. Sprue

It is the vertical passage passing through the cope and connecting the pouring basin to the runner or gate.

The sprue cross-section may be circular, square or rectangular. The size of the sprue usually varies from 10 mm square for small work below 12 kg poured weight to about 20 mm square for heavy castings. The size of the sprue should be chosen with the following considerations:

The sprue should be small enough so that

(*i*) It can be kept full during the entire pouring operation,

(*ii*) The metal may not enter the mould cavity with high velocity causing splattering and turbulence.

Also the sprue should be large enough so that

(*i*) The mould cavity is filled completely without laps, seems or misruns.

(*ii*) A metal head is built up quickly to prevent mould gases from being aspirated into the metal.

If a single sprue is not sufficient for filling the large mould in the required time, two or more sprues and same number of ladles may be employed for pouring.

The base of sprue is usually made one and a half times wider and deeper than the runner. This deeper and widened cavity underneath the sprue is known as 'sprue well'. It serves as a cushion to the falling weight of metal and absorbs its kinetic energy.

3. Runner

In case a mould contains more than one cavity, the common passage-way which supplies the metal to a number of cavities is called a 'runner', and the branches from the runner to the respective mould cavities are known as 'ingates'.

In large castings also, the molten metal is usually carried from the sprue base through a number all round the periphery of the casting and distributed to several gates for feeding to the main mould. The runner may be located either in the drag or in the cope depending on the shape of the casting, but it is generally preferred to have it in the drag.

In order to avoid aspiration and turbulence, the runner should be streamlined. The size of the runner is reduced in cross-section after each ingate by an amount equal to the ingate cross-section so as to obtain a flow of approximately equal metal volumes through each ingate.

4. Gates

The opening from the runner to the mould cavity is called 'gate'. The size and-location of gates is so arranged that (*i*) the moulds may be filled in quickly, (*ii*) cutting of the mould surfaces by the flowing metal is minimum, (*iii*) the cracks do not develop when the metal cools, (*iv*) they can be removed readily without damaging the castings.

5. Risering

A hole cut or moulded in the cope to permit the molten metal to rise above the highest point in the casting is referred to as 'Riser'.

The useful purposes of providing the riser:

(*i*) It enables the pourer to see the metal as it is filled in the mould cavity. If the metal does not appear in the riser, it indicates that either there is some obstruction to the metal flow between the sprue and riser or the metal is insufficient for the mould cavity to be filled.

(*ii*) It helps in easy ejection of the steam, gas and air from the mould cavity while filling the mould with the molten metal.

(*iii*) It also serves as a feeder to feed the molten metal into the main casting cavity to compensate for shrinkage.

(*iv*) Using several risers in case of casting with thin sections imparts greater soundness.

The main requisites of a riser are:

(*i*) The riser must cover completely the sectional thickness which requires feeding.

(*ii*) For effectiveness, the riser must be the last part of the casting to solidify.

(*iii*) The design of the riser should be such that it establishes temperature gradients within the casting so that the casting solidifies directionally towards the riser.

(*iv*) The fluidity of the molten metal should be properly maintained so that the metal may penetrate the portions of the mould cavity which solidify in the end.

22.10 MELTING FOR CASTING

Generally, the metals which are obtained either from the blast furnace, steel making furnaces or other non-ferrous smelting furnaces, are not cast directly into the desired shapes of articles or components due to the following reasons:

(*i*) The metals so obtained are not always in a sufficiently refined state that can be directly cast.

(*ii*) From practical point of view also it is difficult to pour the huge quality of molten metal in the moulds of different shapes and sizes.

The metals which are obtained from the smelting furnaces are first cast into some regular forms like the pigs or ingots. These forms are remelted in foundries for casting the required components.

For remelting purpose, one of the following furnaces may be used:

(*i*) Crucible furnace ;

- (*a*) Pit-type coke-fired furnace
- (*b*) Tilting type or Bale out type gas—or oil-fired-furnace.

(*ii*) Electric furnace:

- (*a*) Direct-arc type
- (*b*) Indirect-arc or Induction type,

(*iii*) Open hearth furnace.

(*iv*) Air furnace or Reverberatory furnace,

(*v*) Cupola furnace.

(*vi*) Converter:

(*a*) Bottom blown or Bessemer Converted

(*b*) Side blown or Tropenas Converter.

22.10.1 Crucible Furnaces

For small foundries, the crucible furnaces are very convenient because the operation is intermittent and various alloys are handled in small quantities.

For melting the metal, it is put in a heated crucible which acts as a melting pot. The crucible is made of clay and graphite by moulding into a standard shape and size. Its size is specified by a number ranging from 1 to 400. The crucible number represents its approximate holding capacity in lb. of aluminium. For other metals, the capacity of a crucible may be determined by multiplying its number with the ratio of their densities (metal: aluminium).

According to the fuel used for heating the charge, that crucible furnaces are classified as under:

(1) Pit-type Coke-fired' Crucible Furnace

This furnace is installed in a pit so is referred to as 'pit type'. It essentially consists of a cylindrical steel shell lined with refractory fire bricks on the inner side. It is closed at the bottom with a grate and covered at the top with a removable lid. The metal to be melted is put in the crucible which is embedded in the burning coke (See Fig. 22.11).

It is most commonly used for melting non-ferrous metals like aluminium, brass, bronze etc., owing mainly to:

(*a*) its low cost of installation,

(*b*) cheap fuel, and

(*c*) easy operation.

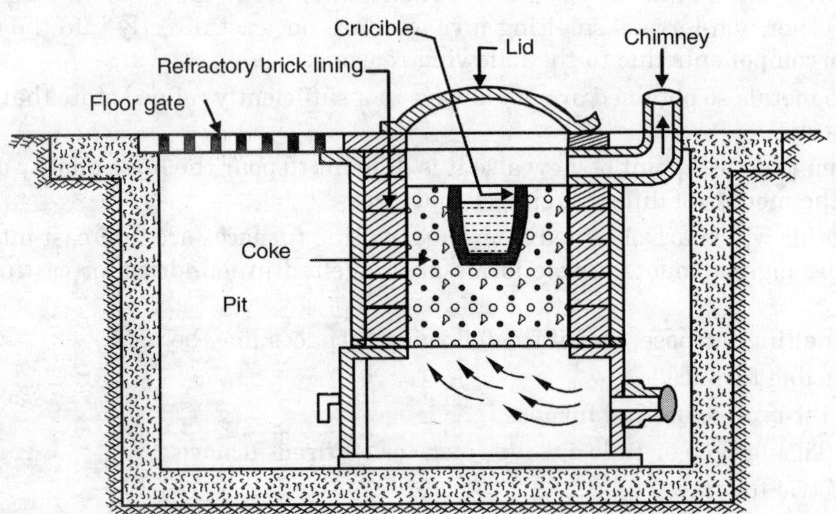

Fig. 22.11 Pit-type cock fired crucible furnace.

Working. A deep bed of coke is kindled and allowed to burn until a state of maximum combustion is attained. Some coke is removed from the top and the crucible is lowered into the furnace. The coke is again added surrounding the crucible on all sides. The metal is then charged in the crucible and lid is replaced to facilitate the chimney draft. Forced draft is generally available from a blower, which is used to help in rapid combustion of the coke. When the metal melts and reaches the desired temperature, the crucible is taken out by means of special long handled crucible-tong and carried away for pouring into the moulds.

(2) Tilting Type Oil or Gas-fired Crucible Furnace

In this furnace, oil or gas is used as a fuel for heating the crucible. The furnace is cylindrical shaped lined inside with fire bricks. The flame produced by the combustion of oil or gas with air is allowed to sweep around the crucible and heat it uniformly (See Fig. 22.12).

Atomised fuel oil or gas is fed through a manifold. It enters the furnace tangentially where it ignites and swirls upwards between the refractory lining and the crucible. The metal is charged in the crucible through the opening in the centre of the head.

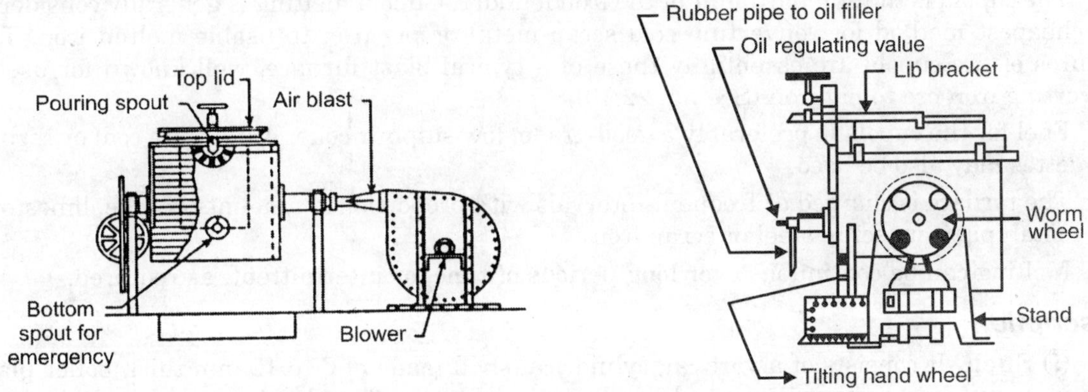

Fig. 22.12 Tilting type oil-fired crucible furnace.

Modern gas and oil fired furnaces are equipped with:

(*a*) automatic proportioning device which produces a neutral flame by regulating the fuel-air ratio.

(*b*) thermostatic device for controlling temperature.

(*c*) regenerative system of heating for economy in gas-fired furnaces.

These furnaces are generally of bale out type or tilting type.

(*a*) The bale out or lift out furnaces are of fixed type but unlike the pit type, are installed on the floor. In order to take out the metal, the crucible is lifted out of the furnace by means of long-handled crucible tongs.

(b) The tilting type furnace are raised above the floor level, mounted on two pedestal trunnions and rotated by means of a geared hand wheel, a hydraulic ram or in case of large sized furnaces rotated by power. The tilting gear is generally designed so that the furnace tilts on a central axis.

22.10.2 Electric Furnaces

Both arc and induction type electric furnaces can be used for remelting purposes. Although adaptable equally well for both ferrous and non-ferrous metals, electric furnaces are mainly used for producing highest quality steels (tool steels, stainless steels), grey irons, brasses due to their high cost of operation.

Usually, these furnaces are worked in conjunction with cupola and Bessemer converter, thus the processes used are known as:

(*i*) Duplexing Process. (*ii*) Triplexing Process.

(1) Duplexing Process. In this process, the iron is more economically melted in a cupola and refined in the electric furnace.

(2) Triplexing Process. In this process, the iron is melted in a cupola, converted into steel in a Bessemer converter and then transferred to electric furnace for altering, adjusting or improving its properties.

In this process, the function of electric furnace is to maintain a constant supply of molten metal at a constant temperature for continuous pouring.

22.10.3 Cupola Furnace

The cupola is the standard unit of the iron foundry. Cupola melting is generally considered the cheapest method for converting cold scrap metal or pig iron to usable molten iron. The features of the cupola are essentially those of a typical blast furnace, well known for use in converting iron ore to pig iron (See fig. 22.19).

Fuel for the cupola is preferably a good-grade, low sulphur coke. Anthracite coal or carbon briquettes may also be used.

The furnace is charged at frequent intervals with proportioned amounts of coke, limestone and metal (pig iron, scrap steel or scrap iron).

Melting can be continuous over long periods of time, or intermittent, as required.

Description

(*i*) Shell. It consists of a vertical, cylindrical shell made of 6 to 12 mm thick boiler plate riveted or welded together and lined inside with acid refractory bricks and clay. Refractory bricks and clay used for cupola lining consist of silica acid (SiO_2) and alumina ($A1_2O_3$). Cupola diameter varies from 1 to 2 metres with a height of from 3 to 5 times the diameter.

(*ii*) Foundation. The shell is mounted either on a brickwork foundation or on steel columns. Most modern cupolas use steel column arrangement; the bottom of the shell being provided with drop bottom door, through which debris consisting of coke, slag, etc. can be discharged at the end of a melt. Their working bottom is built up with moulding sand which covers the drop doors. This bottom slopes (1 - 20) towards the metal tap-hole situated at the lowest point at the front of the cupola.

(*iii*) Tuyeres. Air for combustion of fuel is delivered through the tuyeres which are provided at a height of between 0.6 to 1.2 metres above the working bottom. Total area of the tuyeres should be 1/5 to 1/6 of the cross-sectional area of the cupola inside the lining at tuyere level. Tuyeres have usually a size 50 × 150 mm or 100 × 300 mm. Some cupolas are provided with tuyeres in one row, others have two or more rows.

(*iv*) Wind Belt. The air is delivered to the tuyeres from a wind belt which is a steel plate duct mounted on the outer shell of the cupola.

(*v*) Blower. A high pressure fan or a blower supplies the air at the wind belt through a blast pipe. A valve is provided in the blast pipe to control the supply of air. The blast pressure varies from 265 to 350 kg/m^2 in small, low cupolas to 150 kg/m^2 in large tall cupolas.

(*vi*) Volume Meter. It is sometimes installed to know the volume of air passing. The amount of air required to melt one tonne of iron depends upon the quality of coke and coke-iron ratio.

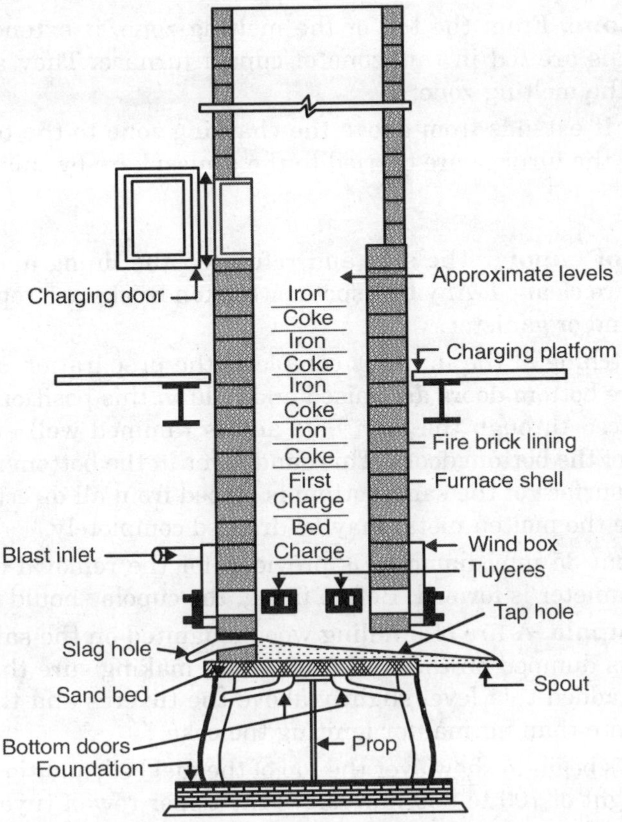

Fig. 22.19 Section through a Cupola Furnace.

In a cupola, assuming that a 10: 1 ratio of iron to coke is used, about 850 cubic metres of air are required to melt one tonne of iron. Higher volumes of air will be required for lower fuel ratios.

***(vii)* Slag-Hole.** It is located at a level about 250 mm below the centres of the tuyeres.

***(viii)* Charging Hole.** It is provided through which metal, coke and flux are fed into the furnace. It is situated 3 to 6 metres above the tuyeres, according to the size of the cupola.

***(ix)* Charging Platform.** It is a large platform or stage usually surrounding the cupola at the level of about 0.3 metre below the bottom of the charging door.

***(x)* Chimney or Stack.** The shell is usually continued for 4.5 to 6 metres above the charging hole to form a chimney.

Zones in a Cupola

The entire shaft of the cupola is divided into four sections:

(i) Crucible zone.
(ii) Melting zone.
(iii) Charging zone.
(iv) Stack zone.

***(i)* Crucible zone.** It is between top of the sand bed and bottom of the tuyeres. The molten metal accumulates here. It is also called the 'well'.

***(ii)* Melting Zone.** It starts from the first layer of iron above the coke bed. It is between 300 to 900 mm above the bed charge. Iron melts in this/zone. Highest temperatures developed in this zone of complete combustion of the coke reach 1600° to 1700°C.

***(iii)* Charging Zone.** From the top of the melting zone, it extends upto the charging door. Charging materials are fed in this zone of cupola furnace. They are preheated in this zone before coming to the melting zone.

***(iv)* Stack Zone.** It extends from above the charging zone to the top of the cupola. The gases generated within the furnace are carried to the atmosphere by this zone.

Cupola Operation

***(i)* Preparation of cupola.** The slag and refuse on the lining and around the tuyeres from the previous run are cleaned. Any bad spots or broken bricks are repaired with a mixture of fire clay and silica sand or ganister.

As soon as the patching of the lining is complete, the preparation of the sand bottom in the cupola is begun. The bottom doors are raised and held in this position by metal props. The bottom sand is introduced through the man hole and is rammed well around the lining and across the intersection of the bottom doors. This sand layer in the bottom is built up to a height of 100 to 200 mm. The surface of the sand bottom is sloped from all directions towards the tap hole so that at any time the molten metal may be drained completely.

An opening of about 35 mm diameter is provided for the removal of the slag and a tap hole of about 20 mm diameter is formed. Before firing, the cupola should be thoroughly dried.

***(ii)* Firing the Cupola.** A fire of kindling wood is ignited on the sand bottom. When the wood burns well, coke is dumped on the well from above making sure that the coke begins to burn too. Coke is then added to a level slightly above the tuyeres and the air blast is turned on at a lower blowing rate than normal for igniting the coke.

As soon as red spots begin to show over the top of the fuel bed, additional coke is introduced into the cupola to a height of 700 to 800 mm above the upper row of tuyeres. For determining the height of the coke bed, a measuring rod prepared to indicate the distance from the sill of the charging door to the top of the coke bed, is used.

Firing is done $2\frac{1}{2}$ to 3 hours before the molten metal is required.

***(iii)* Charging the Cupola.** As soon as the coke bed is built up to the correct height and ignited uniformly throughout, pig iron and scrap may be charged. The proportion of this scrap is ordinarily from 25 to 50% of the total weight of the metal poured. 2 to 4% ferro-man-ganese is used as deoxidiser when considerable steel scrap is used along with pig iron.

Coke and pig iron are charged in alternate layers until the cupola is full. The weight of the metal charged should be from 10 to 15% of the hourly output of the cupola. The thickness of the layers should be kept about 150 to 200 mm.

In addition to charging iron and coke, a certain amount of limestone (the flux used to form the cupola slag) is added above the first metal charge. Besides limestone, fluorsper (CaF_2) and soda ash (Na_2CO_3) are also used as fluxing materials. A flux removes the impurities in the iron and protects the iron from oxidation. The reaction between the lime and silica introduced into the cupola reduces the melting point of the slag and increases its fluidity. Limestone added constitutes about 2 to 3% of the metal charge by weight or 25% by weight of the coke charged.

More commonly a ratio of 10: 1 is kept which means that 1 tonne of coke is required to melt 10 tonnes of iron.

***(iv)* Soaking of Iron.** After the furnace is charged fully, it is kept as such for about 45 minutes. The charge gets slowly heated up since the air blast is kept shut during this time. This causes the iron to get soaked.

(*v*) Opening of Air Blast. At the end of the soaking period the air blast is opened. The top hole is kept closed with a plug called 'bot' till the metal melts and sufficient metal is accumulated. As melting proceeds, the contents of the charge move gradually downwards. The rate of charging should be equal to the rate of melting so that the furnace is kept full throughout the heat.

(*vi*) Pouring the Molten Iron. When sufficient metal collects in the hearth above the sand bed, the slag hole is opened and the slag is run off preferably into a bogie for easy removal. Then, the bot from the tap hole is removed. Molten metal coming out of the spout is collected in ladles and is carried to the moulds for pouring. The same prpcedure is repeated until all the metal is melted and the operation is over.

(*vii*) Closing the Cupola. At the end of the operation, the blast is shut off and the prop is knocked down so that the bottom plates swing open. The remains inside the furnace come out and drop on the floor. These remains are then quenched and removed from underneath the cupola.

Generally, cupolas are run continuously as are blast furnaces, but are worked only for such periods as may be required. At most of the foundries, the melting period does not exceed 4 hours. But, cupolas can be operated continuously for 10 hours or more.

Ladles

They are containers used for moving or conveying the molten metal from the furnace to the pouring area. The molten metal from the cupola or furnace is commonly run into a large receiving ladles from which it is distributed to smaller pouring ladles.

Depending upon the method of carrying them, the various types of ladles are:

(*i*) **Hand Type Ladle.** It is the most commonly used ladle in a foundry shop. It resembles a bucket with a removable, long-handle shank. It normally holds 15 to 30 kg of metal.

(*ii*) **Shank Ladle or Bull Ladle.** It is handled by two men. It holds 30 to 150 kg of metal.

(*iii*) **Bottom Pouring Ladle.** Large castings of the floor and pit type are poured with this ladle. It is widely used in steel foundries. It may also be used for pouring cast iron.

(*iv*) **Tea Pot Ladle.** It is used for small and medium-sized mould.

Ladle handler. It is a mechanical device which enables one man to handle and pour with large ladles. It is hung from a monorail making it easy to be removed.

Lining of Ladles. The inside of the ladle bowl is lined generally with about 25 mm thick layer of fire clay. Large ladles require a thicker lining.

Fire clay is mixed by adding the right amount of water to mix it into a so-called "mud". It is then plastered on the inside of the ladle and allowed to dry.

Pouring the Molten Metal

(*i*) It should be carefully controlled.

(*ii*) It should be done continuously and at a uniform rate until the mould, gates and risers are full.

(*iii*) The temperature of the metal must be just right. If it is too hot, the hot gases will produce blow-holes ; if it is too cold, the metal will solidify prematurely and will not fill the entire cavity.

(*iv*) Temperatures are easily checked with an optical pyrometer. After a sufficient period of experience, one can visually judge the temperature of the metal with remarkable accuracy.

(*v*) An assistant should skim back the floating slag with a skimming bar during the actual discharge of the ladle into the mould.

Scum or slag present on the top of molten metal must not be allowed to enter the mould. Otherwise, slag pockets may form in the casting.

22.11 FOUNDRY

A **foundry** is a place where castings are produced. the casting is a process of pouring molten metal into mould and allowing it to solidify. This process gives strength and rigidity to intricate parts. The mould (or cavity) into which the material is poured is made of some heat-resisting material. Sand is widely used for this purpose, as it can be easily packed to shape and resists high temperature.

Cast iron is widely used for casting, as with this metal, it is possible to have easy control on its properties which include fluidity rate of shrinkage, strength and rigidity. The shape of the object is determined by the shape of the mould cavity.

Commonly used tools and equipment in foundry are various hand tools, moulding boxes, moulding machines, melting equipment, and pouring equipment.

22.11.1 Foundry Refractories

Refractories (heat-resistant materials) are the backbone of the foundry. They form the crucibles and furnace bottoms which contain metal during melting. They comprise of the furnace walls and roof which reduce heat losses and contain gases. Molten metal is carried in refractory ladles and poured into refractory moulds. Refractories are the common requirement of all processes carried out at elevated temperatures.

Foundry refractories are usually, but not always, oxides of elements (MgO, SiO_2, $A1_2O_3$ and mixtures of these and others). Research has resulted in tremendous improvement in the service-ability of traditional refractory materials. A wide range of entirely new refractory material has been created *e.g.* , magnesia from sea water and zirconia.

Classification of Refractories

Refractories for foundry use classified on a chemical basis are:

(*i*) Acid Refractories
(*ii*) Basic Refractories
(*iii*) Neutral Refractories

Acid refractories are oxides of metalloids, *e.g.,* SiO_2 and P_2O_5, and basic refractories are metal oxides *e.g.,* MgO and CaO. In foundry, the best-known acid refractory material is silica (SiO_2) and basic one is magnesia (MgO).

Acid refractories are not used in contact with basic refractories to avoid erosion and weakening of a refractory structure. For the same reasons, basic slags are commonly used only with basic refractories, and acid slags with acid refractories. In case iron and steel foundries, acid slags are used when sulphur and phosphorus elimination from the molten metal is not necessary, but, when these impurities must be removed, basic slags are essential.

Refractory Raw Materials

Over half of all refractories used are obtained from ores of essentially pure silica (quartzite), or from ores that are mainly compounds of silica and alumina (fire clays and similar clays).

Formerly, magnesia (MgO) was obtained by calcining (heating) magnetise, $MgCO_2$. At present, large quantities of MgO are also obtained by chemical treatment of magnesium bearing brine.

Chrome (Cr_2O_3) is obtained from chrome ore of widely varying purity and used in the impure form.

22.12 CASTINGS

Casting is a process of forming metallic products by melting the metal, pouring it into a cavity known as the mould and allowing it to solidify. Commonly used casting methods are:

1. Casting in Metallic Moulds:
   (*a*) Gravity or permanent-mould casting.
   (*b*) Slush casting.
   (*c*) Pressed or corthias casting.
   (*d*) Die casting.
      (*i*) Hot-chamber machines. (*ii*) Cold-chamber machines.
2. Casting in Non-metallic Moulds:
   (a) Centrifugal castings
      (*i*) True centrifugal, (*ii*) Semi-centrifugal,
      (*iii*) Centrifugal.
   (*b*) Precision castings:
      (*i*) 'Lost wax' method, (*ii*) Plaster moulds,
      (*iii*) Shell Moulding.
   (*c*) CO_2 mould-hardening process.
   (*d*) Moulds of wood, rubber, paper etc.
3. Continuous Casting:
   (*a*) Reciprocating moulds.
   (*b*) Draw casting.
   (c) Stationary moulds.
   (*d*) Direct sheet casting.

Most common casting methods are being discussed hereunder.

22.12.1 Casting in Metallic Moulds

Permanent moulds are made of metals capable of withstanding high temperatures. They are recommended only when many castings are to be produced because of their high cost. For large castings and alloys of high melting temperatures, permanent moulds are impractical. But, for small and medium sized non-ferrous castings manufactured in large quantities they can be used advantageously.

1. Gravity or Permanent Mould Casting

This method consists of filling a metal mould as in sand casting. No pressure is used except that obtained from the head of metal in the mould. Therefore, it is also known as 'Gravity Die Castings' (Ssee fig. 22.20).

The process is used successfully for both ferrous and nonferrous castings, although the latter do not present so many problems because of their lower pouring temperatures.

Mould Construction and Operation

Permanent moulds are generally made of dense, high duty cast iron *viz.*, Mechanize or steel. Moulds are made in two parts which can open or close in alignment in a horizontal plane. The parting surface is usually in a vertical plane.

The hollows or cavities in the interior are formed with grey iron or steel cores. Two types of cores used in permanent moulds are:

(*i*) **Fixed Cores.** These cores can be withdrawn with the mould and are fastened to the mould body.

(*ii*) **Movable Cores.** These cores are separate from the mould. They are drawn before parting the mould.

These cores are withdrawn as soon as the metal starts to solidify by means of:

(*a*) Rack and pinion mechanism.
(*b*) Screw mechanism.
(*c*) Hydraulic mechanism.

The moulds being metallic, the heat of the molten metal is removed very rapidly and the metal shrinks very fast, leaving an air space between the mould surface and casting. This air space protects mould material from the effect of heat.

The two halves of the mould may be mounted on a sliding carriage where one half is fixed on one side and other half is movable along a horizontal bed.

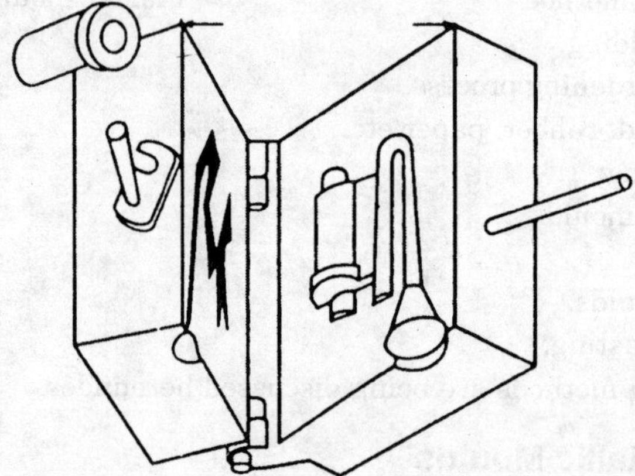

Fig. 22.20 Permanent Mould Casting Process. (Solidified casting with gate and metal core shown in left half of the mould with completed casting in the front).

The simplest type of permanent mould hinges at one end of the mould with provision for clamping the halves together at the other.

In large-scale production, a number of moulds are placed on a rotating table where various operations are performed automatically as the moulds pass through different stations.

The cycle of events consists of pouring, cooling and ejecting the casting, blowing out the moulds, coating them, and, in some cases, setting the cores.

After solidification, the casting tends to cling to one of the mould halves and may not come out on its own accord. Spring-loaded ejector pins operated manually or automatically are used to eject the casting from the mould. Ejector pins are provided in one of the two halves in which the casting tends to cling.

Gating, risering and venting problems are same as in sand casting. Depending on where the junction of sprue and mould cavity is desired, the gates may be top, bottom or side type. Shallow channels about 0.15 mm. deep are made in the parting surfaces for venting purposes.

Thickness of mould walls usually varies from 20 to 25 mm. Ribs and flanges are provided on the outer surfaces for strengthening the mould. Moulds are usually coated with a refractory wash and then lamp black, which reduces the chilling effect on the metal and facilitates the removal of the casting. Castings are frequently removed before they are completely solidified to prevent cracks which may develop as a result of shrinkage strains.

Applications. Permanent mould casting is especially adopted to the quantity production of small and medium-sized castings *e.g.,* aluminium pistons, cooking utensils, refrigerator parts, electric irons small gear blanks etc.

2. Slush Casting

It is a method of producing hollow castings in metal moulds without the use of cores. The mould is filled with molten metal and held stationary until a thin skin of solid metal freezes against the mould walls. The mould is then inverted and the unfrozen metal run out from the casting. A thin-walled casting results, the thickness depending on the chilling effect from the mould and the time of the operation. The casting is removed by opening the halves of the mould.

Precautions. (*i*) The mould should not be inverted too sharply or air will be excluded as molten metal chokes the cavity ; a partial vacuum will be formed, and the still weak solid skin of the casting will collapse at its weakest point.

(*ii*) The casting must be made of a relatively pure metal, as most alloys do not form a strong solid skin.

Applications. For some castings *viz.,* statutory work, only the external features of the casting are important.

Since such castings are not meant for engineering use, uniformity of wall thickness is not an important consideration.

Owing to its obvious drawbacks, this method is used only on a limited scale for low melting point alloys like zinc or lead. It can be used for objects, in which dimensional accuracy is not important *e.g.,* ornamental objects, statues, toys and other novelties, *i.e.,* only for a limited/ amount of art and decorative work.

Slush casting of metals corresponds to slip casting techniques used by ceramists to make ceramic bodies. Parts cast in this fashion are either painted or finished to represent bronze, silver, or other more expensive metals.

3. Pressed or Corthias Casting

This method of casting developed in France by Corthias resembles both the gravity and the slush processes but differs some what in the manner in which the operation is performed.

A definite amount of metal is poured into an open-ended mould and a close-fitting core is forced into the cavity, causing the metal to be forced into the mould cavities with some pressure. The core is removed as soon as the metal sets, leaving a hollow thin-walled casting.

Applications. This process is limited in the use mainly to ornamental casting of open design.

4. Die Casting (Pressure Die Casting)

Die casting refers to the forcing, by pressure of molten metal into a metal die or mould. The term 'die' used in this process implies a metallic mould which is filled under pressure. The pressures, which range from 6 to 3000 kg/cm^2 are maintained until solidification is completed.

Die Casting Dies

The dies resemble the common type of permanent mould, which consists of two halves arranged to open and close with a vertical parting. On a die casting machine, one die half, called the cover die is stationary. The other die half, which moves for the opening and closing is known as the 'ejector die'.

Characteristics of Die Metal

Die casting dies are usually made of an alloy steel having the following characteristics:

(*i*) It should be dimensionally stable.

(*ii*) It should have high resistance to heat.

(*iii*) It should not get soldered to the cast alloy.

(*iv*) It should resist erosion.

(*v*) It should have high wear resistance and toughness.

Die cavities are made with great accuracy usually by machining process. Dies for both the hot and cold-chamber machines are similar in construction because there is little difference in the method of holding and operating them. The life of these dies depends on the metal cost and may range from 10,000 fillings, if brass castings are made, to several millions if zinc is used.

Vents and small overflow wells are provided on one side of a die (*i*) to facilitate the escape of air and (*ii*) to catch surplus metal that has passed through the die cavity. In spite of this provision, there is a certain amount of flash metal which is trimmed off in the finishing operation *e.g.,* grinding or blanking (press shearing).

For large or complex castings, a single-cavity die is used. If the quantity of casting to be produced is large and they are relatively small in size, a multiple-cavity die can be used. A combination die is one that has two or more cavities, each of which is different. Most dies are provided with channels for water cooling to keep the die at correct temperature for rapid production.

Dies. (*i*) for tin and lead alloy casting, dies are made of carbon steels without heat treatment,

(*ii*) for zinc, aluminium and magnesium, dies are of heat treated low-alloy steel,

(*iii*) for copper-base castings, heat treated special alloy steel dies are used.

Advantage of Die Casting

(*i*) Very high rates of production are possible *viz.,* in cold chamber type 75 to 150 per hour and in hot chamber type 300 to 350 per hour.

(*ii*) Close dimensional control can be maintained. In zinc, a tolerance of ± 0.075 mm upto 25 mm size and ± 0.025 mm for each additional 25 mm can be easily achieved,

(*iii*) Thin sections can be cast. Under favourable conditions minimum thickness cast are about 0.5 mm for zinc and 0.875 mm for aluminium.

(*iv*) Surface finish of 1250 micro-mm R.M.S. can be obtained. Hence, appearance is improved and finishing operations are minimised.

(*v*) Details can be reproduced successfully with a high degree of precision.

(*vi*) Die casting dies retain their accuracy and usefulness for a long time. Dies for zinc alloys may produce as many as 5 million castings.

(*vii*) For mass production die casting is very economical due to elimination of machining and mould making resulting in saving of labour and time.

(*viii*) Because of the uniformity in wall thickness less material is required in die casting than in sand casting.

(*ix*) There are no possibilities for sand inclusions and a strong dense metal structure is obtained.

Disadvantages of Die Casting

(*i*) High cost of the equipment and dies used requires sufficiently large quantities to compete economically with other processes. For die casting, minimum economic quantity is considered to be about 20,000.

(*ii*) There is a rapid decrease in the life of the dies as the metal temperature increases.

(*iii*) In some cases, there is an undesirable chilling effect on the metal.

(*iv*) Metals having a high coefficient of contraction must be removed from the mould as soon as possible because of the inability of the mould to contract with the casting.

(*v*) There are certain limitations in the shape of die castings and the process is not adapted to the production of large castings. So far, the maximum size cast is 100 kg in zinc and 30 kg in aluminium.

(*vi*) Die castings usually contain some porosity due to the entrapping of air.

(*vii*) Die casting has, to a large extent, been limited to low-melting non-ferrous alloys.

With a gradual improvement of heat-resisting metals for dies, this process can now be used for numerous alloys.

Die-Casting Alloys

(*i*) Zinc Base Alloys. Over 75% of die castings produced are the zinc base type. The purest grades of commercial zinc, 99.90%, known as Special High grade, should be used, since impurities like lead, cadmium, tin cause serious casting and aging defects.

Zinc alloys are widely used in the automotive industry and for other high-production markets *e.g.,* washing machines, oil burners, refrigerators, radios, phonographs, television, business machines, parking meters, small machine tools, etc.

(*ii*) Aluminium Base Alloys. Many die castings are made of aluminium alloys because of their lightness in weight, resistance to corrosion, high electrical conductivity, ease of applying surface finishes and good machinability. Compared to zinc alloy however, they are slightly lower in physical properties and more difficult to die cast. Their melting temperature is around 1185°F.

Generally, the cold-chamber process is used in casting because molten alloys of aluminium attach steel if kept in continuous contact with it. Although more expensive to operate than the air-injection type, it has advantage of producing sounder castings.

Silicon increases hardness and corrosion resistance.

Copper improves mechanical properties slightly.

Magnesium increases lightness and impact resistance.

***(iii)* Copper Base Alloys.** Die castings of brass and bronze present more problems in pressure casting because of their high casting temperatures (1600 to 1900°F). Heat-resisting alloy steel dies are used to reduce their rapid deterioration. Cost of brass die castings is, therefore, comparatively higher.

Copper-base alloys due to their high temperatures are melted in an auxiliary furnace and ladled to the plunger-type cold-chamber machine.

Copper-base alloys are excessively used in miscellaneous hardware, electric machinery parts, small gears, marine aircraft and automotive fittings, chemical apparatus etc, where high strength, corrosion resistance are important.

***(iv)* Magnesium Base Alloys.** Magnesium is alloyed principally with aluminium but may contain small amounts of Si, Mn, Zn, Cu and Ni. Its alloys are the lightest in weight of all the die-cast metals, being about two-third the weight of aluminium alloys. Their corrosion resistance is inferior to that of the other die-casting alloys.

The principal die easting alloy of magnesium having good casting characteristics (casting temperature 1200 - 1300°F), mechanical properties and excellent machinability contains Al 1%, Zn 0.5%, Mn 0.13%, Si 0.3% max. Cu 0.3%, Ni 0.03%, Mg the remainder. To minimise corrosion Cu and Ni should be kept very low. Melting crucible is hooded, keeping the metal covered by a non-oxidising atmosphere.

Magnesium-base alloys are extensively used for aircraft, motor and instrument parts, portable tools, textile machinery, household appliances etc.

***(v)* Lead Base Alloys.** Pure lead melts at 621°F. When alloyed with about 16% antimony, it melts at about 470 °F, Antimony is the principal alloying element; its percentage ranges from 9.25 to 16.

Principal die-casting alloy of lead containing Ph 80%, Sb 15%, Sn 5%, Cu 0.5% is used mainly as a bearing material for moderate loads ; other applications being weights, battery parts, X-ray shields, non-corrosive metal parts etc.

Presence of tin increases fluidity, hardness, strength and thus improves its bearing characteristics. Use of lead compounds for die casting is limited due to their toxic effect upon workmen.

***(vi)* Tin Base Alloys.** As far as mechanical properties are concerned, they are in about the same category as the lead alloys, but they are high in price. Tin alloys are high in corrosion resistance well suited for use in contact with food and beverages. Also, they have excellent bearing properties, and can be within remarkable close dimensional tolerances.

In most tin alloys Sb 4 to 16% and Pb upto 19% is common. Where die casting remains in contact with food or beverages, use of lead is eliminated. Tin alloys can also be used for low-cost jewellery, pewter etc. Their low melting point and low solidification shrinkage promotes long die life.

Use for tin-alloy die castings is very limited because die casters cannot afford the risk of contaminating zinc alloys with tin.

22.12.2 Casting in Non-metallic Moulds

Each mould material has its own temperature limitations as to the kind of metal for which it is suitable. In many cases, non-metallic moulds are used in precision casting.

Centrifugal casting is included under this heading although castings are made by this process in both metal and non-metallic moulds.

Centrifugal Casting

'It is the process of rotating a mould while the metal solidifies, so as to utilize centrifugal force for positioning the metal in the mould. The metal is forced against the walls of the mould with much greater pressure than that obtained by static pressure in ordinary sand casting. Greater detail on the surface of the casting is obtained, and the dense metal structure has superior physical properties. This method is particularly suited to castings of symmetrical shape.

Methods of Centrifugal Casting are classified as:

(*i*) True centrifugal Casting (*ii*) Semi-centrifugal Casting

(*iii*) Centrifuging.

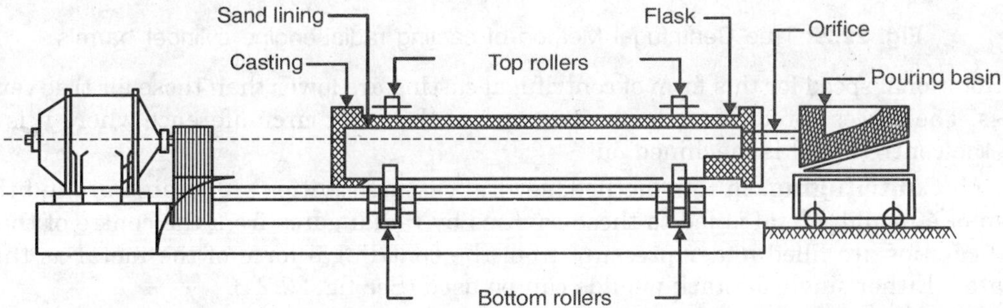

Fig. 22.21 Centrifugal Casting Machine forcasting cast-iron or Steel Pipes.

(*i*) True Centrifugal Casting is used for pipe, liners and symmetrical objects which are cast by rotating the mould about its horizontal or vertical axis. Metal is held against the wall of the mould by centrifugal force and no core is required to form a cylindrical cavity on the inside. The method is illustrated by the casting machine shown in Fig. 22.21 designed for the production of steel or cast iron pipe. The amount of the metal poured into the mould controls the wall thickness of the pipe produced.

Fig. 22.22 shows another example of true centrifugal casting. It illustrates two methods which may be used for casting radial engine cylinder barrels. Horizontal method of casting is similar to the process followed in casting pipe lengths ; the inside diameter being a true cylinder requiring minimum machining.

In vertical casting, the inside cavity takes the form of a paraboloid: the slope of the sides of the paraboloid depending on the speed of rotation-paraboloid 'A' represents a higher rotational speed than paraboloid 'B'. For reducing the inside diameter difference between the top and bottom of the cylinder, spinning speeds for vertical casting are higher than that for horizontal.

(*ii*) Semi-centrifugal Casting is rotating the mould about its vertical axis. In this method, the centre of the casting is usually solid. The centre cavity is machined out later or is formed by the metal passing down around the outside of a core.

This method often used in stack moulding, is illustrated in Fig. 22.23 where two track wheels are cast solid in one mould.

Number of casting made in a mould depends on:

(*a*) the size of the casting.

(*b*) convenience in handling and assembling the moulds.

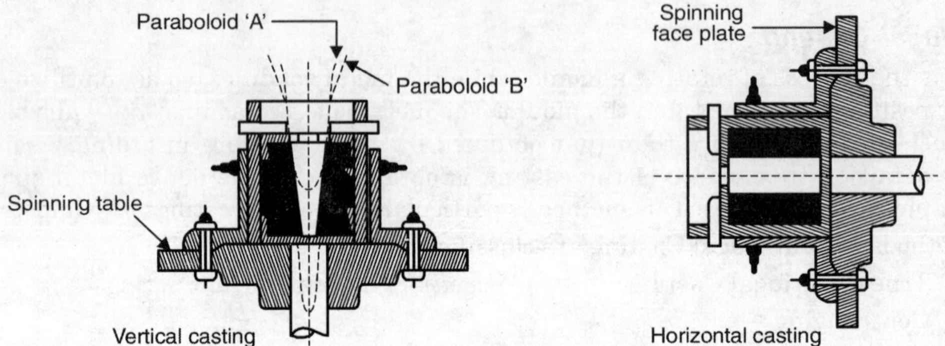

Fig. 22.22 True Centrifugal Method of casting radial-engine cylinder barrels.

Rotational speed for this form of centrifugal casting are lower than those for true centrifugal process. The process produces a dense structure at the outer circumference where it is needed, while the centre metal is machined out.

***(iii)* Centrifuging.** In this method, several casting cavities are located around the outer portion of a mould. Metal is fed to these cavities by radial gates from the centre of the mould. Mould cavities are filled under pressure from the centrifugal force of the metal as the mould is rotated. Either single or stack moulds can be used (See fig. 22.24).

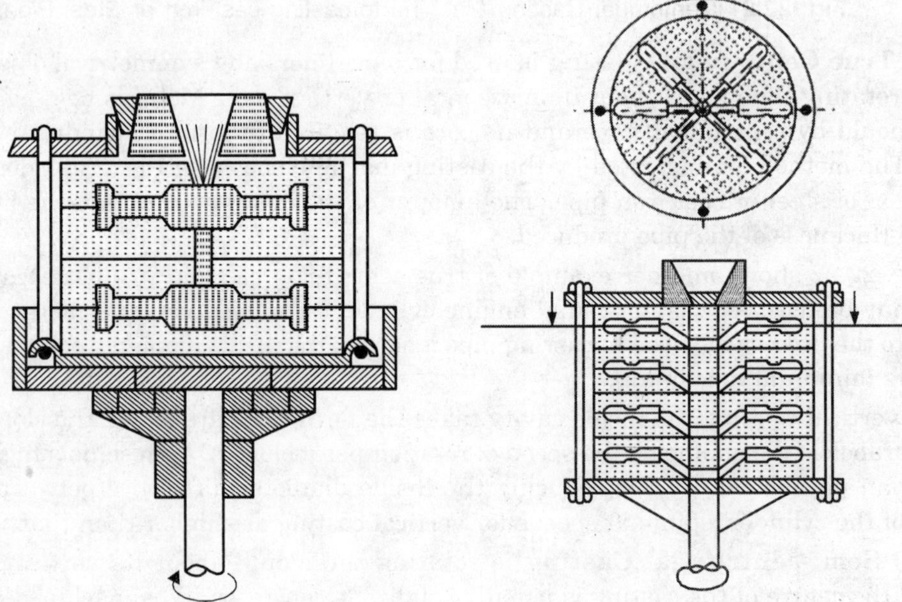

Fig. 22.23 Semi-Centrifuging Casting Process

Fig. 22.24 Centrifuging Process.

This method is not limited to symmetrical objects. It can produce castings of irregular shape *e.g.,* bearing caps, small brackets, etc. Dental industry has used this process for casting gold inlays for many years.

For miscellaneous centrifugal castings, average rotational speed is 200 surface metres per minute. For cast iron pipe, the speeds may be as high as 400 surface metres per minute. Rotational speed depends on:

(*a*) the kind of mould used (whether sand or metal),

(*b*) the manner of rotation (whether horizontal or vertical),

(*c*) the size of casting,

(*d*) the kind of metal being cast.

Too great a speed may result in surface cracks caused by high stresses set up in the mould.

Advantages of Centrifugal Casting

(*i*) It is more economical than other methods.

(*ii*) Cores in cylindrical shapes and risers or feedheads are eliminated.

(*iii*) Castings have a dense metal structure with all impurities forced back to the centre where frequently they can be machined out.

(*iv*) Because of the pressure exerted on the metal, thinner sections can be cast than would be possible in static casting.

(*v*) Any metal can be cast by this process.

22.13 CASTING DEFECTS

Casting defects can be classified into the following categories:

1. Surface imperfections.
2. Defects resulting from incomplete feeding.
3. Gas porosity.
4. External hot tears, cold cracks and warpage.
5. Unfused chills and chaplets.
6. Moulding defects
7. Defects caused by moulding, core-making and gating.
8. Defects caused by mould drying and core baking.
9. Defects caused during closing and pouring the moulds.
10. Defects caused by molten metal.
11. Defects caused by wrong fettling.
12. Defects caused by faulty heat-treatment.

22.14. FETTLING OR CLEANING OF CASTINGS

Whatever type of mould is used, castings will require cleaning and trimming before passing them on for machining, or other operations. Amount of work to be done depends on:

(*i*) The size of castings.

(*ii*) Types of alloys used.

(*iii*) The care and degree of accuracy given to the production of the mould and core.

The noise and dusty atmosphere associated with fettling shops gives the casual observer an impression that this phase in production requires little or no skill. This is entirely wrong because it is the finish produced by the fettler which either definitely enhances or entirely mars the saleability of a foundry's products.

In many foundries making repetition light castings *e.g.,* switch gear etc., the work of the fettler is limited to removing runners and risers and cleaning off any adhering sand. In heavier and more intricate castings, more work is necessary to remove cores, flash metal at joints and surface imperfections.

As temperature rises progressively from white metal alloys to cast steel, the refractoriness of the mould must be improved. Generally, the higher the metal melting points, the greater the surface roughness of the castings. Thus, the heavier castings in grey and malleable cast iron and cast steel call for the most work in fettling.

Fettling is not a precision operation like machining. The main function of fettling department is –

(*i*) speed in removing all unnecessary metal.

(*ii*) the production of a reasonably smooth finish.

Removal of Sand

After *a* casting arrives at the fettling department, the first job is to remove any sand which adheres from the mould and cores.

Method of removing sand depends on:

(*i*) Size of the casting,

(*ii*) Intricacy of the casting.

If design is fairly simple, it will be sufficient to rape with a hammer followed by wire-brushing or shot blasting. If design is more intricate, it takes longer to remove cores and their grid or wire reinforcements. Although a wide range of castings are made in green sand moulds, cores are more usually dried. Dried sand if artificially bonded, does not break easily. At the same time dry sand moulds and cores usually strip well, *i.e.,* once the sand is removed, casting surface requires little cleaning.

On heavy intricate castings, if only picks and bars are employed, cleaning becomes very laborious. Pneumatic picks are, therefore, used to some extent. These have the disadvantages:

(*i*) they are weighty,

(*ii*) considerable dust is created by their exhaust.

However, water sprays in conjunction with pneumatic pick are used to overcome dust problem.

Modern automatic units for smaller castings rotate or tumble the castings under the grit blast. Fig. 22.25 illustrates one such unit where abrasive is mechanically blasted into castings which are being tumbled on an apron conveyor.

Three types of blast cleaning in common use are:

(*i*) **Air Blasting.** A high pressure stream of air propels and (sand blasting) or shot (shot blasting) against the casting surface.

(*ii*) **Water Blast Cleaning.** Water is used instead of air, to propel sand.

(*iii*) **Mechanical Blasting.** Rapidly rotating paddles mechanically-hurl shot or sand on the surface of the casting (Fig. 22.25).

Removal of Gates and Risers

After removal of adhering sand or refractory by one or another of the preceding techniques, gates and risers are removed from the casting. A wide variety of methods are available for accomplishing this, including chipping, flogging, shearing, sawing, abrasive wheel cutting and flame cutting.

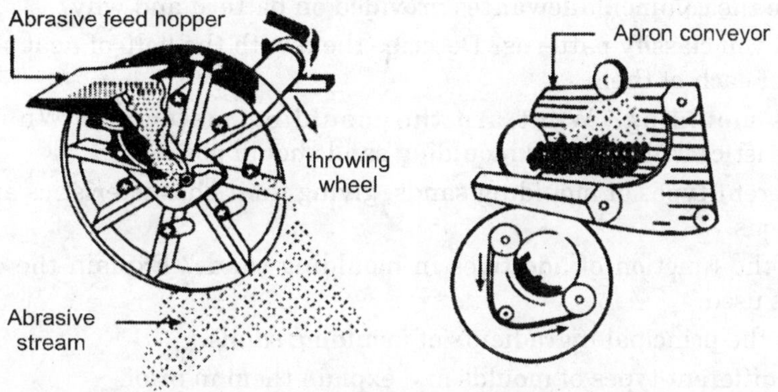

Fig. 22.25 Mechanical Blast Cleaning (Wheelabrator system)

Finish Cleaning

After removing gates and risers, castings are given one or more finishing operations. These are performed:

(*i*) to smooth the gate and riser areas of the casting ;

(*ii*) to remove any excess metal remaining on the casting;

(*iii*) to improve surface finish or appearance.

Preliminary metal removal is obtained by grinding. An abrasive grit of about 25 to 80 mesh is bonded into the shape of a grinding wheel or onto an abrasive belt of canvas. Wheels and belts are available in a variety of different shapes, sizes, grits, grit sizes and binders. The right combination for a given application is chosen.

Grinding machines are either portable or stationary. Portable machines are usually used only for large castings. In softer metals, rotary files *(i.e.,* hardened steels grinding wheels) are sometimes used instead of grinding.

If a smooth surface finish is desired ; castings may be polished, brushed or buffed after grinding:

(i) **Polishing** is usually performed with an abrasive belt machine. It is similar to abrasive belt grinding except that finer grit (80 to 400 mesh) is used.

(ii) **Bushing** employs rotary wire of fibre brushes that remove burns and grinding marks.

(iii) **Buffing** is used to obtain an exceptionally high lustre on cast surfaces. It employs wheels of 'buff' (usually discs of muslin sewn together) called 'mops'. A very fine abrasive mixed with a grease binder is rubbed on the wheel face.

QUESTIONS

22.1 What is "pattern" and "pattern-making"? List common materials used for pattern-making. Discuss their relative merits and demerits.

22.2 What are the factors which govern the selection of a proper material for pattern-making?

22.3 What are the common allowances provided on pattern and why?

22.4 How will you classify patterns? Describe them with the help of neat sketches and state the use of each of them.

22.5 What is moulding? What are the moulding materials? What are the main characteristics which a good moulding sand should possess?

22.6 List different types of moulding sands, giving their characteristics and recommended applications.

22.7 What is the function of additives in moulding sands? Explain the effects of various additives used.

22.8 Describe the principal ingradients of moulding sands.

22.9 Classify different types of moulds and explain them in brief.

22.10 Draw a sketch of a typical mould and name its principal parts.

22.11 Distinguish between green sand moulding and dry sand moulding.

22.12 Write short notes on:

Core box, floor moulding, bench moulding, pit moulding, sweep moulding and machine moulding.

22.13 Write short notes on:

(*a*) Characteristics of cores and core sands

(*b*) Core binders

(c) Skin dried and air dried moulds

(*d*) Gating systems in a mould

(*e*) Risers

22.14 What do you understand by "gating system"? What are the different types of gating systems? Why these are used?

22.15 What do you understand by "foundry"? Sketch a cupola and name its essential parts.

22.16 State the methods of cleaning the castings. What are the common defects of casting? State their causes and remedies.

22.17 What are the main advantages and disadvantages of die-casting? How does a cold chamber die-casting machine differ from *a* hot chamber machine? Describe various alloys commonly cast in die-casting processes.

Chapter 23

WELDING ENGINEERING

Learning Objectives

After studying this chapter, the students will learn:

- Welding and types of welding, namely fusion and pressure welding.
- Gas welding i.e. oxyacetylene welding and oxy-hydrogen welding and air acetylene welding.
- Arc welding types and methods.
- Thermit welding, forge welding and resistance welding.
- Cutting processes namely oxygen cutting, carbon arc cutting and metal arc cutting.
- Bronze welding, soldering, brazing and hard facing.

23.1 INTRODUCTION

Welding is a process of joining similar metals by the application of heat with or without application of pressure and addition of filler material. Welded joints between two parts produce continuous and homogeneous material. Each metal has different weldability. The term weldability means the capacity of being welded into in separable joints. Good weldability for a metal means, it can be welded readily so as to perform its object satisfactorily.

Welding consists of fusing metals together while they are in the plastic or molten state. It is a permanant method of fastening the metals. Heat for welding is generally obtained by gas flame, electric arc or electric resistance.

Brazing differs from welding where metals to be joined are not melted and deeply fused with filler metal. Instead, brazing employs a filler metal which melts at a temperature below the melting point of the metals to be joined and is drawn into the close fitting joint by capillary action. In this chapter we are discussing the important methods of welding, soldering and brazing.

23.2. TYPES OF WELDING

Welding can be classified into following two categories:

(*a*) Fusion or Non-Pressure welding

(*b*) Plastic or Pressure welding.

(A) Fusion or Non-Pressure Welding

This type of welding involves heating of the joining surfaces above their melting point of the metals to be joined, so that they fuse and join each other without pressure. Welding processes falling in this category are:

1. Gas Welding:

   (*i*) Oxyacetylene (*ii*) Oxy-hydrogen

   (*iii*) Air-acetylene

2. Arc Welding:
   (*i*) Carbon arc
   (*ii*) Metal arc
   (*iii*) Gas metal arc (MIG)
   (*iv*) Gas Tungsten arc (TIG)
   (*v*) Atomic-hydrogen arc
   (*vi*) Plasma arc
   (*vii*) Submerged arc
   (*viii*) Flux-cored arc
   (*ix*) Electric slag
3. Thermit welding

(B) Plastic or Pressure Welding

This involves heating of joining surfaces to a temperature range, at which they become plastic and are joined together by the application of pressure. Welding processes of this category are:

1. Forge welding.
2. Resistance welding.
   (*i*) Butt
   (*ii*) Spot
   (*iii*) Seam
   (*iv*) Projection
   (*v*) Percussion.

23.3 GAS WELDING

In gas welding, gas flame is used for heating and melting the joining surfaces and the filler rod. Gas welding is done by burning a combustible gas with air or oxygen in a concentrated flame of high temperature.

Most important combustible gas which gives highest temperature is acetylene. Hydrogen is also used in gas welding of low melting point metals such as aluminium. Temperature of oxyacetylene flame is about 3200°C and is sufficient to weld the steel. Acetylene is a compound of carbon and hydrogen (C_2H_2) containing 92.3% carbon and 7.7% hydrogen by weight: It is a colourless and has a sweet and distinguished odour. Whereas oxygen is *a colourless,* odourless and tasteless gas. Use of pure oxygen speeds up reactions and increases flame temperatures. The reaction of oxygen with organic substances produce large quantities of heat.

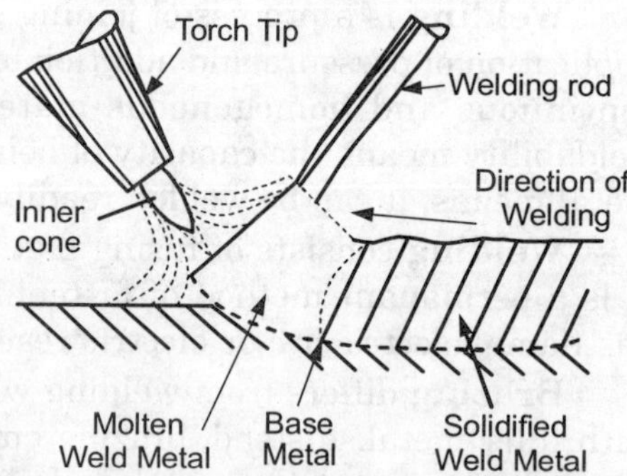

Fig. 23.1 Oxygen-acetylene welding.

23.3.1 Oxyacetylene Welding

Oxyacetylene welding is a fusion-welding process ; the coalescence of metals is produced by an oxygen-acetylene flame. Extreme *heat* is *concentrated on the edges or on the edge and surface of the pieces of metal being joined until the molten metal flows together*. Whether a filler metal should be used to complete the weldment is determined by the type of joint design.

Filler metal is added by inserting it into the molten puddle of the base metals. The puddle then solidifies making the weld bead (Fig. 23.1). The extremely high heat depends on the mixture of two types of gaseous substances, oxygen and acetylene. Oxygen supports higher combustion; acetylene being the fuel for combustion.

Gas Flame

Correct adjustment of the flame is necessary for proper welding oxygen and acetylene can be ejected from torch tip in three possible gas mixtures. Different combinations of oxyacetylene gas flame is shown in fig. 23.2.

(*a*) An equal mixture, for neutral flame.

(*b*) An excess of acetylene for carburising flame.

(*c*) An excess of oxygen, for oxidising flame.

A neutral flame has two definite zones, namely (*i*) a short sharp brilliant cone extending from the tip of the torch, and (*ii*) a faintly luminous and of a bluish colour outer zone. First of these zones develops heat, while the second protects the molten metal from oxidation, since the oxygen in the surrounding is consumed by the gases from the flame. The neutral flame is widely used for welding steel, cast iron, copper, aluminium etc.

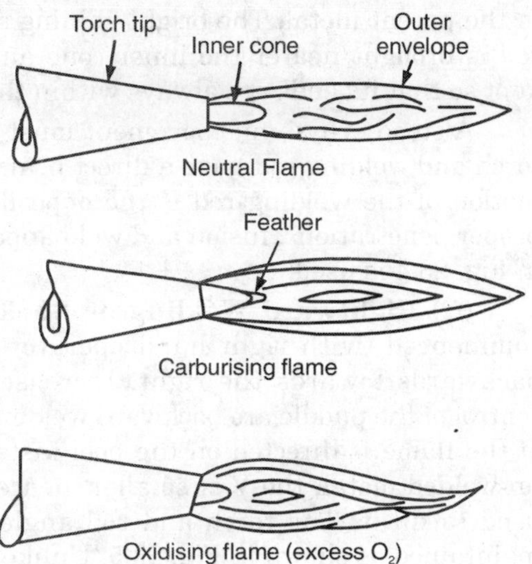

Fig. 23.2 Oxyacetylene gas flame.

A carburising flame has an excess of acetylene, and has three zones, namely (*i*) short sharp inner cone, (*ii*) a whitish colour intermediate zone, and (*iii*) a bluish colour outer cone. The length of the intermediate cone indicates the proportion of excess acetylene in the flame. This type of flame produces a hard and brittle weld while welding steel.

An oxidizing flame has an excess of oxygen, and has two zones, namely (*i*) a short purpish tinge inner cone, and (*ii*) an outer cone. This flame is helpful for welding brass.

Oxy-Acetylene Welding Techniques

The choice of welding technique is dictated largely by metal thickness. Oxy-acetylene welding may be performed with the torch tip pointed forward in the same direction as the weld progresses. This method is called leftward welding or forehand or forward welding. This is applicable for welding material upto 3 mm, thick. The tip also can be pointed back towards the metal which has been deposited and this method is called rightward welding or backhand or backward welding, this is used for materials having thickness 3 mm.

(*i*) Leftward welding or forward welding. This method is suitable for mild steel, cast iron and the non ferrous metals, such as aluminium, brass, bronze and copper. It proceeds from right to left. The welding wire is held in front of the flame and the weld is progressively formed behind the flame as the welding pool is progressed forward as shown in Fig. 23.2. The torch should be held with its nozzle in line with the joint, not angled to one side and in the vertical plane. The welding rod and torch should be held at the given angles as shown. This type of welding results in a better control of the small weld puddle resulting in a smoother weld both at the top and-bottom. The puddle of molten metal is small and easily controlled. A great deal of pipe welding is done using the leftward technique even in wall thickness in excess of 3 mm.

For thin material there should be no side to side movement of the torch, but a swinging movement should gradually increase with increasing plate thickness to ensure that both sides of the V are equally in a state of fusion. For mild steel the inner cone of the flame should be about 3 mm from the molten metal. It should never dip in the metal otherwise hard spots will be caused due to carburising. A greater distance then 3 mm will unduly spread the heat over the job, cause difficulty in melting the parent metal and from an unmanageably large welding pool. Welding should never commence until a clear pool of fully molten metal has been formed in the parent metal. The bright shining surface of molten is readily recognizable. The welding rod is brought nearer the inner cone and drops of molten metal are added one by one. It is kept so that its ends are always within the outer cnvelope of the flame to minimise oxidisation.

A steady rhythmic movement must be attained for sound even welds. The motions of the torch and welding rod have a direct bearing on the quality of the weld. It will be seen that the motion of the welding rod is the opposite of the torch's as shown in Fig. 23.3. This ensures proper penetrations, fusion and weld appearance. For fillet welds, a motion similar to as shown in Fig. 23.4 is used.

***(ii)* Rightward Welding or Backward Welding.** In this type of welding the weld is commenced (with right hand operators) on the left-hand edge of the work and progress backwards towards the right. Increased speeds and better control of the puddle are backward welding technique. The heat of the flame is directed on the pool instead of forwards on to un-welded metal, the *V is* smaller, hence less welding wire is used to fill it. The torch and rod angle should be carefully maintained as shown in Fig. 23.5. Unlike leftward welding the rod is not added drop by drop but is stirred round in the weld, care being taken to take in the top edges on both sides and the root gap in the stirring motion to ensure full build up and root penetration. The motion of the welding rod is the opposite of the torch's as shown in Fig. 23.6. This ensures proper penetration, fusion and weld appearance. When welding plate over 5 mm in thickness by the rightward technique, the flame is moved deeper into the groove and advanced without weaving. The puddling action is supplied by the end of the welding rod.

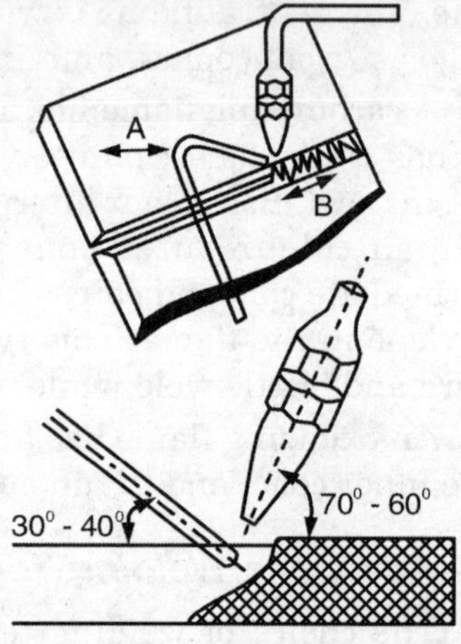

Fig. 23.3 Leftward of Forward welding.

Filler Metals

Gas welding may be carried out with or without filler metal. When the filler metal is used, it may affect the mechanical properties of a gas weld. So the filler metal should fulfil the following requirements.

Fig. 23.4 *(a)* Motion of welding rod is opposite of the torch.

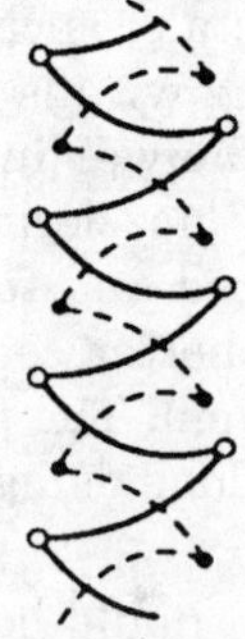

Fig. 23.4 *(b)* Motion used for fillet welds.

1. It should be correct diameter for the thickness of the metal to be welded.
2. It should be free from scale, rust, grease, paint etc.
3. It should not be able to cause boiling or sputter in the weld pool.
4. Should produce a deposit free from blow holes and non-metallic inclusion.

The major elements present in the filler metal or welding rod have their effects on the weld metal. Some of them improve its mechanical properties, others increase the evolution of gases or formation of refractory slag etc.

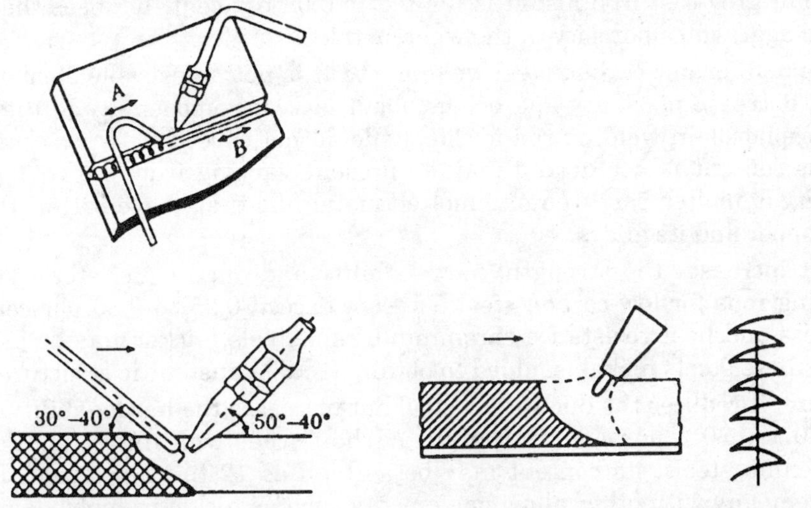

Fig. 23.5 Rightward or Backward Welding.

Fig. 23.6 Motion of the welding rod opposite that of torch in rightward welding.

The following are some of the important elements and their effects:

Carbon. It reduces the weldability of steel when present in excess of 0.3 per cent. With an increased carbon content, so much gas is evolved in the puddle that the molten metal begins to boil and the grains grow in size. The hardness is increased and cracks develop in the weld metal and the heat-affected zone. Carbon reduces the thermal conductivity of steel and the heat-affected area becomes wider. This then produces locked-up stresses in the workpiece. An increase in the carbon content of the welding rod raises the ultimate strength of the weld metal but reduces its ductility. In welding rods for grey cast iron, the carbon content may be as high as 3.5 to 4 per cent.

Manganese. It is an efficient deoxidizer. 1.1 per cent of manganese present in the welding rod raises the ultimate strength of the weld metal to a maximum. An increased content of manganese along with an increased carbon content in a welding rod gives rise to a hardened weld metal and develops cooling cracks. If it is present in relatively large quantities, the manganese and sulphur in the weld metal form manganese sulphide which has a high melting point and dissolves little in the molten iron. For this reason it may remain in'the weld metal as scattered slag inclusions. The proportions of manganese in welding rods for low carbon steel should be kept at 30 per cent of sulphur present. In welding rods for grey cast iron the manganese content is usually limited to 0.5 or 0.6 per cent as it has a chilling effect on cast iron.

Silicon. It is an undesirable impurity in low-carbon welding rods for steel. When present in increased quantities silicates of iron and manganese are formed and forms a pasty blanket of slag. This slag prevents the gases from escaping when the puddle solidifies. As a result the weld metal will contain a large number of gas pockets and non-metallic inclusion. For this reason its contents in low-carbon welding rods is limited to 0.03 per cent. In welding rods for grey cast iron the contents of silicon should be as high as 3 to 3.5 per cent as silicon promotes graphitisation. This is essential if good grey cast iron is to be obtained in the weld metal.

Sulphur. The sulphur content in filler metal is undesirable. It readily combines with iron to form iron sulphide which has a tendency to segregate and form brittle networks at the grain boundaries. The sulphur content in filler metal should not exceed 0.4 per cent for low-carbon steels. For grey cast iron also it is limited to 0.08 per cent. It raises the melting point of cast iron and aggravates porosity in the weld metal.

Phosphorus. For low carbon steel welding rod it is not a desirable element. It produces cold shortness, decrease plasticity and brings about lack of homogeneity in the metal. So the contents of phosphorus in welding rod for steel should not exceed 0.04 per cent. In cast iron, the phosphorus content is useful and may be present varying from 0.5 to 1 per cent. This improves fluidity of molten cast iron and molten metal fills the groove better. It is also useful in welding of copper and its alloys.

Nickel. It increases the strength, hardenability and ductility of steel. The contents of nickel in welding rods for low-carbon steel does not exceed 0.25 to 0.30 per cent. In welding rods for stainless and heat resistance chromium-nickel steels, nickel may be as much as 8 to 20 per cent. In these steels nickel is added to obtain a stable austenitic structure.

Chromium. It reduces the ductility of steel but increases the hardenability. So its content is limited to 0.1 to 0.2 per cent for plain carbon-steel welding rods. For high alloy chromemolybdenum steels, its content may be as high as 12 to 27 per cent. It is generally introduced in company with other alloying elements such as nickel or molybdenum.

Molybdenum. It affects the mechanical properties of steel even when present in small quantities. It raises the hardness and strength of steels specially at high temperatures. Welding rods for chrome-molybdenum steels may contain from 0.15 to 0.6 per cent and more upto 3 per cent for special purpose steel containing chromium nickel and tungsten.

Modern oxy-acetylene welding rods vary greatly in chemical composition, according to the composition of the material to be welded and the strength of welds desired. Rods having a higher carbon content than former low carbon grades and containing higher percentages of silicon and manganese have been developed. In a typical welding rod the composition of various constituents could be 3 to 3.6 per cent of carbon, 3 to 3.5 per cent of silicon, 0.5 to 0.8 per cent of manganese, 0.2 to 0.5 per cent of phosphorus and not over 0.08 per cent of sulphur. Other types of welding rods are alloyed with nickel, chromium, vanadium, molybdenum and other metals. For welding non-ferrous metals, alloy rods are used that produce a relatively fine grain weld free from oxides. A 60 per cent copper and 40 per cent zinc composition modified with manganese and silicon is widely used for welding copper or for braze welding of steel.

Fluxes

In case of high alloy steel, cast iron and non-ferrous metals and alloys a film of oxide is always formed, especially at higher temperatures, on their surface which it carried over into the weld metal, will prevent union between the parent and filler metal. Apart from dissolving oxides they serve to release trapped gases and slag and to remove non-metallic inclusions. Also the fluxes form a blanket of molten slag over the molten pool thereby preventing the metal from further oxidation.

In the case of copper, nickel and iron, no oxides are practically formed in the weld as the carbon-monoxide and hydrogen of the intermediate zone easily reduce them. This is not the case with magnesium, aluminium and zinc and hence requires cleansing agents known as fluxes. Fluxes, while in a sense are cleansing agents but should never replace proper cleaning of base metal. They should also be easy to remove upon completion of welding and should cause no corrosive action on the finished weld.

The requirements of a good flux are:

1. A flux should easily melt and have a melting point lower than the parent and filler metal.
2. The molten flux should readily spread over the molten metal.
3. The flux should readily react with metallic oxides so that the oxides are completely dissolved by the time the molten pool solidifies. The blanket of slag should reliably protect the metal from atmospheric oxidation and come-off easily upon solidification.
4. The flux should have no injurious effect on the metal. Its specific gravity should be lower than that of the metal so that the slag would easily float to the surface of the bath.

The fluxes used for welding are of two types—Basic and acid. Acid fluxes are usually employed in welding non-ferrous metals specially with copper base such as copper, brass, bronze, etc. Basic fluxes are chiefly used in cast iron welding. Some of acid fluxes are borax ($Na_2B_4O_7$, 10 H_2O) and boric acid (H_3BO_3).

Fluxes for use in oxyacetylene welding may be obtained in powder or liquid forms. When using powdered flux, the rod is usually heated and then dipped into the powder. Sometimes the powder may also be sprinkled on the base metal if desired. A liquid flux may be painted on the rod and base metal by use of a brush.

The basic fluxes used for welding of cast-iron consists of equal parts of carbonate of soda and bicarbonates of soda.

Aluminium fluxes may contain portions of alkaline halides. Since chemical characteristics and melting points of metals vary over wide ranges, no single flux can be universally used for all metals. The oxides of most alloyed metals generally have higher melting points than the metals themselves. These oxides, therefore, remain as solids when the filler metal is in the fluid states. As solids they interfere with metal deposition. Appropriate fluxes will form a fusible slag with the oxides. This slag will tend to flow away from the immediate weld area. The slag will also form a coating over the molten metal protecting it from atmospheric oxidation.

For cast iron welding the commonly used fluxes are borax, sodium carbonate, potassium carbonate, sodium nitrate and sodium bicarbonate. Fluxes commonly used for aluminium welding are composed of sodium chloride, potassium chloride, lithium chloride, barium chloride, sodium fluoride, potassium fluoride, flourspar and sodium bisulphate. Fluxes for aluminium welding are hygroscopic and must, therefore, be stored in hermetically sealed cans.

It is important to remove the surplus flux after welding from and near the weld. This is removed by a weak (2 per cent) solution of nitric acid and rinse the weld with water. Fluxes are also used in case of flame hard-facing.

23.3.2 Oxy-Hydrogen Welding

Oxy-hydrogen welding is a gas-welding process where in coalescene is produced by heating with a gas flame or flames obtained from the combustion of hydrogen with oxygen, without the application of pressure and with or without the use of filler metal. Although acetylene is the most commonly used fuel gas ; cheaper fuel gases may be used. The use of cheaper fuel gases instead of acetylene cuts down the cost of the flame metal-working process, simplifies the techniques and procedures and enhances safely to operating personnel.

Oxy-hydrogen welding is a relatively low-flame temperature operation and so is used for the welding of low-melting point metals. It is used for steel on light-gauge section and on small parts. It is also used to a limited extent for braze-welding and brazing operation. The temperatures are low, heat effects will not be as great as with other welding processes and warping and residual stresses will be less. The quality of welds produced by this process is as good as that by any other process of welding. The flame adjustment in this process is difficult. Hence in practice the volumetric ratios of hydrogen to oxygen of $2\frac{1}{2}$ to 1. This ratio can be to a maximum of 6 to 1. The operator usually adjusts the oxygen-hydrogen ratio so as to obtain the most heat.

Hydrogen is normally a clear gas without colour or odour. The oxygen content in a commercial hydrogen should not exceed 0.5 per cent by volume. It is shipped at a pressure of 120 kg/sq cm. into steel cylinders having a capacity of 5 cubic metres. The cylinder is similar in appearance as an oxygen cylinder but is painted with a different colour. The cylinder valve threads are also different. Hydrogen can be produced by (1) the electrolysis of water (2) the steam water gas process (3) the thermal decomposition of methane or natural gas (4) the fractionation of coke-hydrogen gas through refrigeration. It can also be obtained by the reaction of water on a powdered mixture of ferosilicon, caustic soda and lime. Special care should be taken to keep the gas lines and equipment as gas tight as possible. Hydrogen is capable of looking through the tiniest cracks and thus forming an explosion in the surroundings.

Oxy-hydrogen welding is very similar in most respects to oxy-acetylene welding and hence most of the equipment used in oxy-acetylene welding also suitable for this type of welding. Same torches, tips and pipes are used for both. A regulator specifically designed for use with hydrogen must be employed with the hydrogen cylinder. It should be noted that for equal heating effects, oxy-hydrogen welding in general will require a larger size tip that would be necessary for oxy-acetylene welding. It is used mostly for welding aluminium and its alloys and that too for thinner sections metal is used. The welding of lead is also done by this process, thickness of lead upto 10 mm can be done easily by this process. For higher thicknesses the oxy-acetylene welding process is used. Magnesium and its alloys are also welded by oxy-hydrogen welding. Brazing using silver brazing alloys is also done by this process.

23.3.3. Air Acetylene Welding

It is a gas-welding process wherein coalescence is produced by heat with a gas flame obtained from the combustion of acetylene with air, without the application of pressure and with or without the use of filler metal. As the air contains about 4/5th by volume of nitrogen which does not support combustion, the acetylene when burned with air will give a lower flame temperature. The heat available is also low. So air-acetylene flame is used for welding light sections of lead. It is also used in light brazing operations and for soft soldering. The quality of weld is satisfactory. Lead upto a thickness of 7 mm can be welded by air acetylene welding. Brazing of silver brazing alloys also employ air-acetylene flame. Soft-soldering copper pipe joints and electrical connections are done by air-acetylene flame.

The torch used for air-acetylene welding is a special torch where the acetylene flows under pressure and aspirates the proper amount of air required for combustion. An acetylene pressure reducing regulator is used for large operations. For small operations the acetylene is supported from small cylinders fitted with pressure reducing valves. Flame adjustment is obtained by adjusting the amount of air admitted to the burner.

23.4 ARC WELDING

Arc welding is widely used method of joining metal parts. In this, source of heat is an electric arc. An arc is produced between an anode (positive pole of direct current power supply) and the cathode (negative pole). Arc welding is done by producing an arc between the work to be welded and the tip of the electrode. This type of welding has the advantages of (*i*) less heat loss and (*ii*) less oxidation as compared to oxyacetylene flame. An electric arc is an electric discharge in gases, accompanied by high heat and a bright glow.

When two conductors of an electric current are brought together make electrical contact, and then separated, an *electric arc* is formed. A considerable resistance to the flow of the current is thus introduced, the electrode tips are raised to high temperature due to air gap between them.

The electrons which emerge from the negative electrode collide with the molecules and atoms of air, break them up into ions and free electrons and cause the air gap to become a conductor of electricity due to ionisation. This maintains the current through the air gap and sustains the arc.

A higher voltage is necessary for an arc to strike than to maintain it because some of the voltage goes for ionizing the air gap.

An electric arc is made up of the following **regions** (See Fig. 23.7):

(*i*) *The* ***cathode space*** where electrons break away from the cathode towards the anode (the surface of the cathode space is called the cathode spot).

(*ii*) *The* ***arc stream*** which is nearly as long as arc and where the atoms and molecules of air are ionised.

(*iii*) *The* ***anode space.*** where positive ions concentrate.

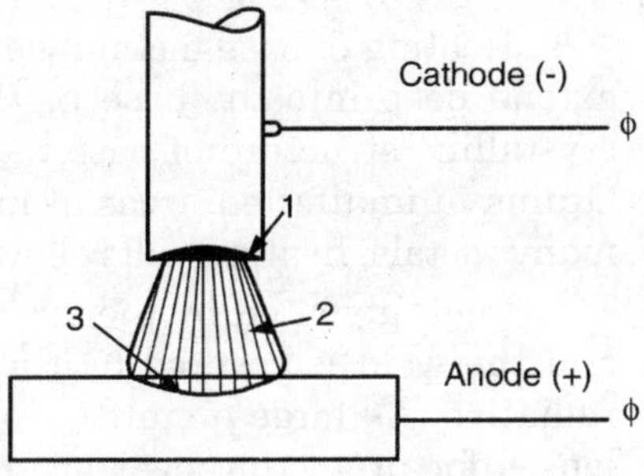

Fig. 23.7 Diagram of an Electric Arc. 1. Cathode space. 2. Arc stream. 3. Anode space.

In both A.C. and D.C. power sources, the arc serves the same purpose. It produces heat for melting metal. If two metal pieces to be joined are placed so that they touch or almost touch one another and the arc from the electrode is directed at this junction, the heat generated by the arc causes the edges to melt. These molten portions alongwith molten electrode flow together. As the arc column is moved, molten puddle solidifies, joining the two pieces.

In the past, most electrodes were bare wire. With bare electrodes, it was difficult to control arc and it cooled too rapidly. Weld bead absorbed oxygen and nitrogen from the atmosphere, turning into oxides and nitrides, and producing a brittle, weak weld.

Most modern electrodes are coated. Function of coating is to form a gaseous shield around the weld, protecting molten metal from atmospheric contaminants. Coating on the electrode burns as the electrode wire melts from intense heat of arc. As electrode wire melts, electrode covering, or flux, provides a gaseous shield around the arc, preventing contamination.

Force of arc column striking the workpiece digs a crater in the base metal. This crater fills with molten metal. As flux melts, part of it mixes with impurities in the molten pool and causes them to float to the top of the weld. When this mixture of impurities and flux cools, slag is formed. Slag improves the weld by protecting the bead from atmosphere and causing it to cool more uniformly ; thus to control grain size. It also helps to design the contour of weld bead itself by acting as an insulator.

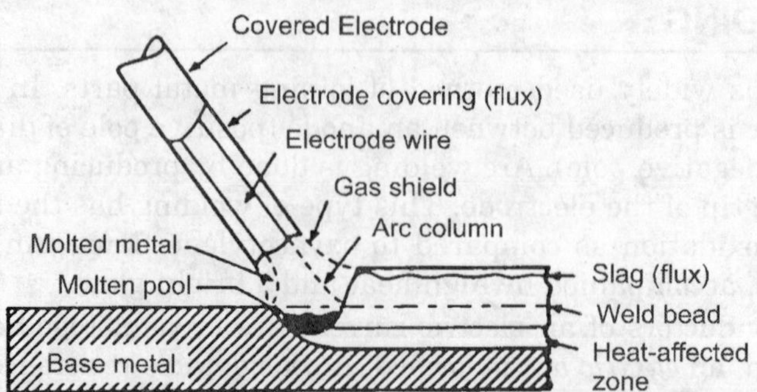

Fig. 23.8 Covered electrode metal deposition.

Arc column reaches temperatures generally of 2750 to 3875°C, which have a harsh effect on the parent metal. The molten pool, which must maintain a temperature of about 1535°C, radiates heat outward and changes the crystals surrounding the weld bead. After welding, the part should be heat-treated to change the grain size in the weld bead and the surrounding area (See Fig. 23.8).

Heating of base metal by arc stream and resulting molten weld puddle of crater generally extend deep into the base metal. Extent of heat-affected zone can be observed by studying crystalline structure of base metal in this zone. It is generally represented by a large grain. Grains in unaffected areas of metal are smaller. It is necessary to treat mild steel. But, with many metals, heating will relieve locked-in stresses through peening or further heat-treatment.

Light given off by the arc column is harmful to eyes ; though major part of it is shielded and the weld is covered by the flux. Head shields are, therefore, used to protect eyes from radiation. As large quantities of infra-red and ultra-violet radiation are given off by the arc, lenses for protecting eyes are extremely important. This radiation can burn the retina and cause burns on the skin. Good welding lenses stop 99.5% of infra-red and ultra-violet rays. Visible light of arc, although, does not contain harmful rays, is uncomfortable to the eyes and will cause blurred vision.

23.4.1. Welding Current Values

The amount of current to be used depends upon many variables, and it is almost impossible to tabulate exact values.

Some typical current values for all position type covered electrodes are given in the table 23.1:

Table 23.1 Current Values for Covered Electrodes

Electrode diameter (mm)	*Flat position (Amps)*	*Vertical and overhead position (Amps)*
2	30-80	30-80
3	80-135	80-120
4	120-175	120-160
4.5	140-250	140-185

These currents usually vary with each brand of shielded arc electrode. Manufacturers of electrodes make recommendations which should, in general, be followed.

Some of the factors determining the amount of current are:

(*i*) Kind of material used
(*ii*) Size and shape of material
(*iii*) Type of joint
(*iv*) Position of weld
(*v*) Skill of operator
(*vi*) Equipment available
(*vii*) Type of electrode
(*viii*) Speed required, etc

Usually, materials considered as good conductors require more welding heat than poor conductors, for they absorb heat more rapidly. For the same reason, those parts which are larger require more current. Some joints, such as lap and tee type require some-what higher currents for the first layer than other types, such as the butt, for the heat is conducted away more rapidly when they are used. Flat welding requires more current than vertical or overhead welding.

At the beginning of a welding job, when the parts are cold, the current should be higher. As the parts become heated, the current may be reduced somewhat. Some operators prefer using a slightly larger electrode to reducing the current. To avoid excessive heating of the work and undue loss of alloying elements by volatilisation or oxidation, electrodes with highly alloyed steel core should be used with currents slightly less than those required by common steel electrodes.

Non-ferrous electrodes need special consideration owing to their different melting point, heat conduction, coefficients of expansion, and other characteristics. No brief instruction for the successful handling of these special electrodes can be given. Also, the amount of current to be used may be limited by the available source of current. Ultimately, the greater the speed of welding, the greater the amount of current.

Arc Welding Equipments:

Commonly used equipments for welding are:

(*i*) B.C. or A.C. Machine

(*ii*) Electrode

(*iii*) Accessories:

(*a*) Cables or leads
(*b*) Electrode holder
(*c*) Lead connectors
(*d*) Ground clamps
(*e*) Wire brush
(*f*) Chipping hammer
(*g*) Face shield
(*h*) Safety goggles
(*i*) Hand Gloves.

23.4.2. Electrodes.

Almost all the shield arc welding done is with covered electrodes due to easy manipulation and contamination free more strong weld deposits. Bare steel rod was used for electrodes for a long time. But, in 1912, experiments with covered electrodes were begun. Coated electrodes were just made commercially available in 1929.

There are two major classifications for coatings on the electrodes; based on the thickness of flux covering.

(*i*) Lightly dusted electrodes,

(*ii*) Semi-coated electrodes.

Materials used commonly for coatings are titanium dioxide, ferromanganese, silica flour, asbestos clay, calcium carbonate, and cellulose, with sodium silicate to hold ingredients together.

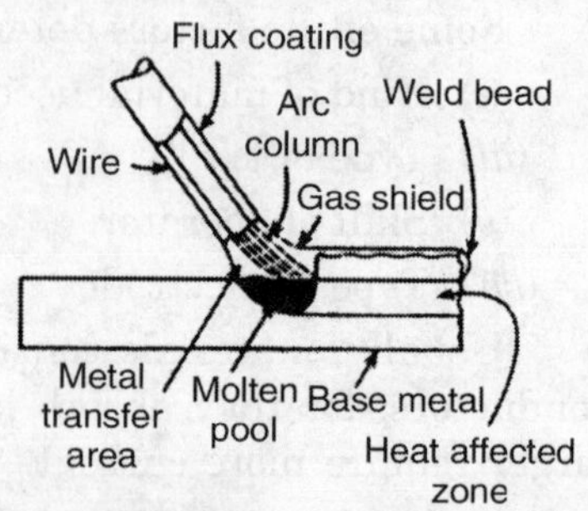

Fig. 23.9 Covered electrode metal deposition.

Coating increase the quality of the weld. Part of the coating burns in the intense heat of the arc, providing a gaseous shield around the arc, which prevents oxygen, nitrogen, and other impurities of air from combining with molten metal to cause a poor quality weld. Its another portion mixes with impurities in the metal floating them to the top in the form of slag. This protects the weld bead from air causing it to cool more uniformly. Slag also helps to control the basic shape of the bead.

The current for manual operation usually ranges from 15 to 500 amps. The voltage ranges from 14 to 24 volts with bare or lightly coated electrodes and from 20 to 40 volts with semi-coated electrodes.

Corrosion-resistant Electrodes. Numbering system for corrosion-resistant electrodes is similar to that used for mild-steel electrodes. Chromium, when subjected to high temperature such as in an arc column, an affinity for carbon is created, *i.e.,* it is attracted to carbon, causing impurities within the weld. Thus, carbon derivatives are not used as fluxing agents.

Manganese and silicon replace the carbon, providing necessary deoxidizers. Titanium oxides are used for promoting arc stability and slag manipulation. Classification of corrosion-resistant electrodes is determined by chemical analysis of the metal to be welded and not by the filler metal. Hence, not all base metals have corresponding electrodes. Most common electrodes are given in the table 23.2.

Table 23.2 Corrosion-resistant Electrodes

Electrode composition (%)	*Base metal*	*Applications*
Cr 18, Ni 8	Similar	Corrosion resistance
Cr 25, Ni 12	Dissimilar	Corrosion resistance
Cr 25, Ni 20	Similar	Surface hardening
Cr 18, Ni 12, Mo 2	Similar and high alloy	High temp, above 600°C
Cr 18, Ni 12, Mo 5	Similar and high alloy	High temp, above 600°C
Cr 15, Ni 35	Similar	High temp, above 1000°C
Cr 15-17	Similar	Corrosion resistance
Crl2	Similar	Abrasion
Cr 19, Ni 9	Similar	Abrasion
Cr 4-6, Mo 0.5	Similar	Abrasion

Generally, corrosion-resistant electrodes deposit a bead containing more than 4% chromium and less than 50% nickel.

Non-ferrous Electrodes. These do not have the same classifications as ferrous electrodes. They depend upon chemical composition of the base metal to be welded. Major non-ferrous electrodes are copper, aluminium, bronze and nickel, as given in the table 23.3.

Table 23.3 Non-ferrous Electrodes

Electrode	*Composition %*	*Base metal*	*Power supply*
Copper	Phosphor-bronze	Electrolytic or deoxidized copper	Ac or dc –
Aluminium	Al 95, Si 5	All alloys (Heat-treatable and non-heat treatable)	Dc$^+$
Aluminium bronze	Al 8, Fe 1, rest Cu (most ductile)	Dissimilar metals or applying hard surface to base metal	Ac or dc$^+$
Aluminium bronze	Al 9—10, Fe 3—5, rest Cu (Harder)	Similar (useful for welding bearings)	Ac or dc +
Aluminium bronze	Al 12—13, Fe 3—5, rest Cu	Dissimilar (weld bead highly resistant to corrosion and extremely hard)	Ac or dc$^+$
Nickel	Ni above 50%, rest Cu	Similar and dissimilar e.g., monel, inconel, nickel-chromium alloy and waste alloy.	Ac or dc +

23.4.3. Equipment for Shield-Arc Welding

(1) D.C. Power Supplies

Power is needed to supply the current, which supports the arc column for fusion welding. Three types of welding power supplies are: (*i*) D.C. motor generators, (*ii*) A.C. transformers, and (*iii*) A.C. transformers with D.C. rectifiers.

Motor generator produces D.C. power. Welding generator is driven by an electric, petrol, or diesel motor. It is designed to operate at 1500, 1800, or 3600 r.p.m. to give optimum current. D.C. power supplies are available in portable, stationary, and multi-station units. Generator capacity is usually designated as 150, 200, or 400 amps.

Anticipated maximum current requirements and the length of time the welder will be used, govern the size of the generator. Duty cycle, based on length of time the welder will be used, is given in percentages based on 10 minutes periods, *i.e.*, a 10% duty cycle means welder will be operated 10% of 10 minutes or 1 minute. Maximum amperage and voltage usually accompany the percentage of duty cycle, giving maximum operating current and voltage values obtainable during the duty cycle. Operating the machine at higher settings for greater lengths of time will deteriorate the insulation within the machine and cause its failure early.

When globules of metal flow across the arc, the arc is short-circuited, causing the arc voltage to drop to zero during welding and may have adverse effects. A large increase in arc current will cause an excessive metal splatter, and a decrease in it will make the arc flutter and go out. Hence, a D.C. welder must be able to remedy fluctuating current due to metal-transfer and maintain an even arc column.

Three major current-voltage characteristics commonly used in D.C. machines to help control fluctuating currents are:

(*i*) Dropping-Arc Voltage (DAV) (*ii*) Constant-Arc Voltage (CAV) (*iii*) Rising-Arc Voltage (RAV)

Machine designed with DAV characteristic provides highest potential voltage when circuit is open and no current flows. As arc column is started, voltage drops to a minimum, allowing amperage to rise rapidly. When length of arc column is increased, voltage rises and amperage decreases. This characteristic is preferred for standard shield-arc welding done manually.

CAV and its modification RAV are preferred for semi-automatic and automatic welding processes, because they maintain a preset voltage irrespective of the amount of current drawn from the machine. These characteristics are sensitive to short-circuiting effect of shield arc mechanism of metal transfer. Therefore, spray arc is used rather than short-circuit arc method of metal transfer.

Spray arc is much like a paint spray gun, working on a principle different from that of short-circuit pinch-off effect of welding with a stick electrode. An advantage of RAV over CAV characteristic is that as amperage requirement is increased, voltage is automatically increased, thus helping to maintain a constant-arc gap even if short-circuiting occurs. This RAV can be adapted for fully automatic processes.

Types of control for d.c. welding machines may be (*i*) a simple rheostat in exciter circuit, (*ii*) a combination of exciter regulators and a series of field taps, (*iii*) a mechanical provision for shifting generator brushes on a self-excited machine to vary the current output, (*iv*) a leakage of magnetic field may be imposed artificially to give desired current output. Rheostat, however, has a maximum number of fine adjustments and is preferred method of controlling amperage, (v) arc welders equipped with remote-control current units, which permit the machine to be located in a remote place of the job site.

Voltmeters and ammeters are sometimes available on welding machines. On some machines, they indicate polarity in addition to efficiency of current and potential values. Voltmeters on d.c. welders range from 0 to + 100 to – 100. Ammeters register both + ve and –ve current values.

Polarity indicates direction of current in the d.c. circuit. Some welding operations demand the flow of current to be changed. For this, a polarity switch or indicator must be on a d.c. welder. When electrode cable is fastened to negative pole of generator and piece to be welded to the positive pole, polarity is negative or more commonly called “Straight polarity”.

(2) A.C. Power Supplies

Transformer welder is usually limited to A.C. power. However, rotation generators, which are engine driven, can be designed to produce A.C. A.C. transformer is usually used in the shop.

A.C. transformer has a primary and a secondary circuit. Primary circuit is independent of secondary circuit. The only connection between the two is by electromagnetic induction. The primary circuit receives the line power and boosts this power, which sets up a perpendicular magnetic field around the primary coil.

These lines of force then induce a like force in the secondary windings, thus causing a further boost in the amperage. Amount of boost or current output can be adjusted either (*i*) by moving the secondary coil closer to primary winding, consolidating the lines of force, or (*ii*) by inserting a reactor coil into the circuit.

If reactor coil is inserted into this circuit, it generally fits on the electrode side of secondary coil. If a reactor is used for adjustment, it may be (*a*) a tapped reactor, (*b*) a reactor with an adjustable coil, or (*iii*) *a* reactor with a rheostat, which controls the amperage output.

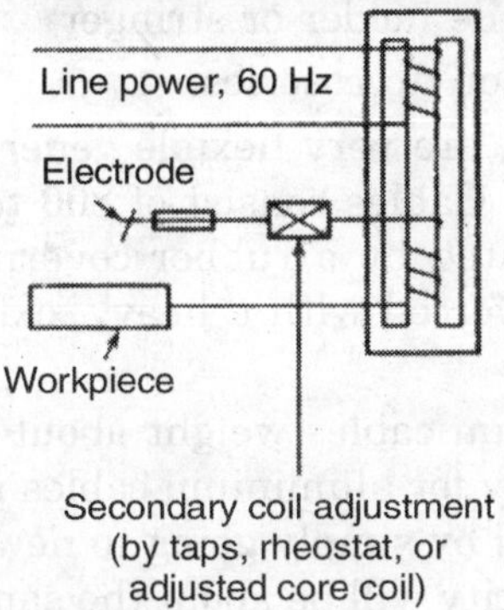

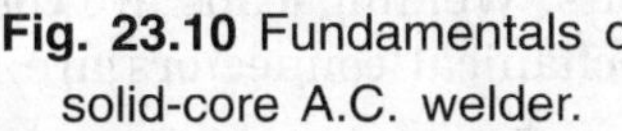

Fig. 23.10 Fundamentals of solid-core A.C. welder.

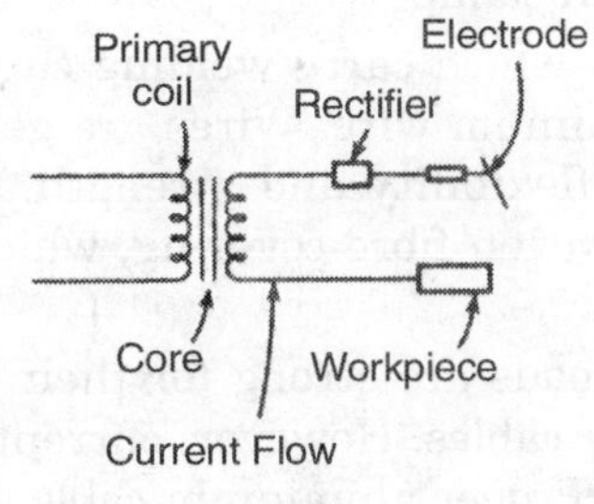

Fig. 23.11 A.C. transformer rectified to D.C.

Rectified Power Supplies. A rectified power supply is an A.C. transformer having a one-way current valve installed on the electrode side of the secondary coil. This valve allows current to flow through it in only one direction. A rectified arc welder has no moving parts, giving it long life. The only moving part is a fan, which cools the transformer, but it is not a basic part of its electrical system.

Rectified transformer welder can supply A.C. straight polarity or D.C. reversed polarity easily, making it most adaptable for shield arc welding. It generally plugs into single-phase or three-phase line voltage like standard A.C. transformer welder. Single-phase power gives a power cycle to the welder every half-cycle. Three-phase power gives three times the power output because three single-power phases enter the power cycle at three different times.

There are two kinds of rectifiers:

(*i*) One uses a silicon diode, which is more efficient and is hermetically sealed, thus impervious to time. Silicon diodes are almost ageless in that they maintain rectifying characteristics indefinitely. It is a preferred type of rectifying unit.

(*ii*) Larger, less-efficient rectifier has selenium plates, which are damaged more easily and attacked by moisture or fumes, causing them to age rapidly. Output characteristic of selenium rectifier change after a few years.

Of the three basic types of welders *viz.*, D.C. welder, A.C. welder and A.C. rectified welder, each has its own advantages:

(*i*) **D.C. welder** permits portable operation. It is better suited for welding thin metal. It efficiently uses a larger variety of stick electrodes.

(*ii*) **A.C. welder** has a lower initial cost, as well as low operation and maintenance cost. It has only one moving part, the cooling fan. It can use common single phase current and operates quietly (see Fig. 23.10).

(*iii*) Rectified A.C. transformer has a simple design and is compact (see Fig. 23.11). It has a quiet operation and high efficiency of D.C. welder, coupled with efficiency of A.C. transformer.

23.4.4 Equipment/Accessories for Arc Welding Machine

These are:

(*i*) Cables or leads
(*ii*) Electrode holder or stringer
(*iii*) Ground clamp
(*iv*) Operator Accessories.

1. Leads, which carry welding current to the work, are very flexible generally made of copper or aluminium wire. Wires are generally very fine. Cables consist of 800 to 2500 wires for maximum flexibility and strength. Wires are insulated by a rubber covering, which is reinforced by woven fibre covering, which in turn is reinforced with a heavy, exterior rubber coating.

Welding leads are strong for their weight. Aluminium cables weight about one-third as much as copper cables. However, current-carrying capacity for aluminium cables is about 60% that of copper. Proper aluminium cable size is determined by simply going to new higher step for similar copper welding lead and current-carrying capacity will be about the same.

Voltage carried by leads varies between 14 and 80 volts. Welding leads are connected by mechanical connectors, soldering, welding or brazing. Mechanical connectors are most used as they can be more easily assembled or disassembled.

2. Electrode holders come in variety of sizes ranging from 150 to 500 amps. They are generally matched to size of lead, which in turn is matched to amperage output to arc welder. They are a means of securing electrode to a mechanical holder, which is easy to use.

Most electrode holders have grooves cut into jaws enabling electrode to be held at different angles for easy manipulation. Larger electrode holders for high amperage provide a heat shield to protect operator's hand from excessive heat and radiation liberated during welding.

3. Ground clamp, which completes the circuit between electrode and welding machine is generally fastened to metal being welded, either with a clamp, a bolt, or some other means, depending on size of metal. Generally, connection must be clean and easy to instal.

4. Operator Accessories. These fall into two groups:

(*i*) Weld bead cleaning accessories
(*ii*) Safety accessories

Main implements for cleaning weld bead arc chipping hammer and wire brush. Power wire wheels, when available, may be used in place of wire brush.

Infrared and ultraviolet rays are harmful to the skin and eyes. A face protection device stops the radiation, which is prevalent in arc welding. Helmet face shield and hand-held face shield are commonly used.

Gloves should be worn at all times during welding process to protect hands from minor burns and radiation by arc column. They should never be used to pick up hot metal or be held in arc column. Clothes worn by welder consist mainly of long sleeve cotton shirt and long pant without cuff. Shoes should be high topped. Some mechanical means to remove fumes produced by arc welding when welding indoors, should be provided like fan or air duct.

23.5 METHODS OF ARC WELDING

Important arc welding methods are:

1. Carbon-arc,
2. Metal-arc,
3. Metal Inert Gas arc (MIG),
4. Tungsten Inert Gas arc (TIG),
5. Submerged-arc,

1. Carbon-Arc Welding

It is a process, in which fusion of metal is accomplished by the heat of an electric arc. Generally, no shielding atmosphere is used. No pressure is used. Filler rod is used only when necessary. This process differs from shield arc welding in that a carbon or graphite rod is used instead of consumable flux-coated electrode.

Function of arc column in carbon arc welding is to furnish heat to the base metal. Heat resulting from the arc between base metal and carbon rod or between two carbon rods is used almost in the same way as the flame of oxyacetylene torch.

Carbon arc welding was commonly used in fabricating aluminium and steel parts before inert-gas welding processes were introduced. It may also be used for welding copper, nickel, monel and other non-ferrous and ferrous metals. Carbon arc can also be used as an excellent source of heat for brazing, braze-welding, soldering and heat treatment.

This process uses either D.C. or A.C. power source. Both these power sources can have manual as well as automatic operation carbon arc electrode holder. Depending upon amperage either of the two may be air-cooled or water-cooled. Generally, electrodes operating at above 200 amps are water-cooled, while those operating at about 200 amps or below are air-cooled.

2. Metal-arc Welding

In this method a metal rod is used as one electrode which the work to be welded is used as another electrode. The temperature produced on the negative and positive electrodes are about 2400°C and 2600°C respectively. The heat of the arc produced melts the metal electrode, which is fused with the base metal, forming a solid union on cooling.

3. Metal Inert Gas (MIG) Arc Welding

This welding process is also known as 'gas-metal-arc (GMA)'. *(MIG welding).* It has characteristics which are similar to those of methods using tungsten or carbon electrodes, but at the same time it extends the use of the process to welding heavier thicknesses and thereby broadens greatly its possible applications (See Fig. 23.12).

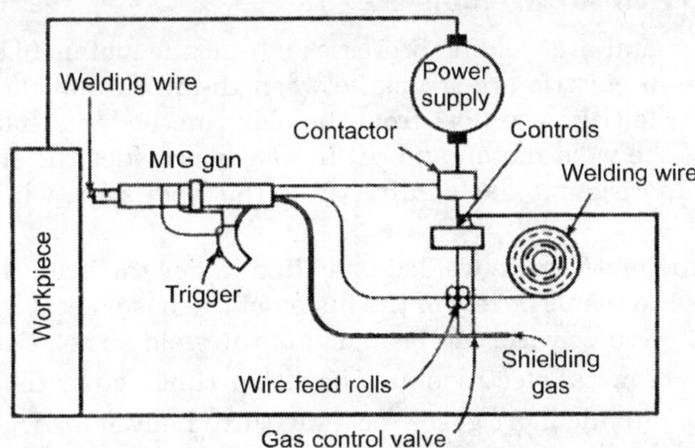

Fig. 23.12 MIG Welding system.

Inert-gas consumable electrode process, or MIG process is a refinement of TIG process. In this process, tungsten electrode has been replaced with consumable electrode. Electrode is driven through same type of collet, which holds a tungsten electrode by a set of drive wheels.

This process consists of continuously feeding a bare or processed filler metal in wire form through a suitable holder. This filler metal, which serves as the electrode, carries the welding current and maintains a welding arc between the wire end and the work.

Wire is continuously fed to the arc at the rate at which it is consumed and transferred to base metal. Arc is shielded by an inert monoatomic gas which flows from the holder nozzle through which the electrode also passes.

In TIG process, tungsten electrode acts as the source of arc column. In MIG process, consumable electrode acts as the source of arc column as well as supply for filler metal (Fig. 23.13)

Arc obtained with this process is non-sensitive since changes in arc length result in smaller changes in arc voltage than are encountered in normal metal-arc welding. Arc consists of a relatively narrow, incandescent cone surrounded by an umbrella-shaped envelope. Outer envelope of the arc provides a cleaning action and thus eliminates the need for welding fluxes.

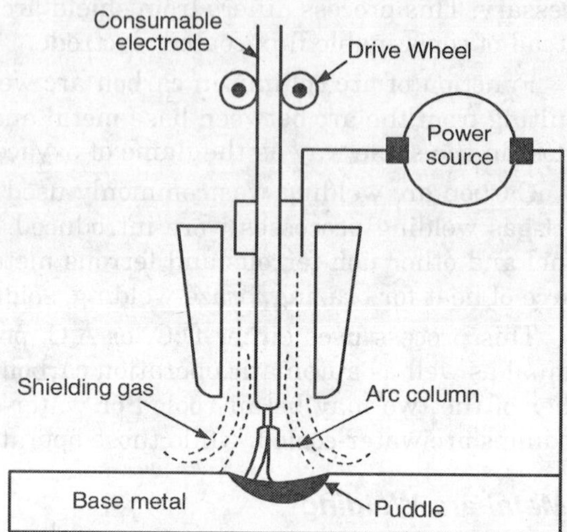

Fig. 23.13 Inert-gas consumable electrode (MIG).

Three basic processes are employed in MIG welding process:

(*i*) Bare-wire electrode process (*ii*) Magnetic-flux process and
(*iii*) Flux-cored electrode process.

All three processes use shielding gases. However, magnetic flux and flux cored MIG processes yield a slightly different weld bead. Weld bead deposited by these two processes has a slag coating as in conventional stick electrode welding. These processes need post weld cleaning.

4. Tungsten Inert Gas (TIG) Arc Welding

It is similar to other metal-arc-welding processes in that the metals to be joined are heated to fusion temperature by an electric arc struck between the work and the welding electrode. However, as indicated by the title, it differs from the older methods in that; (*i*) Electrode does not melt and form part of the weld metal, and (*ii*) Inert gas provides the shield for the molten metal and the electrode. Inert gas is ducted directly to the weld zone, where it surrounds the tungsten.

A localized area of the metal to be welded is melted under the protective blanket of inert gas. As a rule, the weld is the result purely of the fusion of the base metal but, when necessary, filler metal in the form of a rod or wire, can be added to the weld as in oxyacetylene welding.

Non-consuming type of metal electrode may be either tungsten or thoriated tungsten.

Either of the two monatomic inert gases (or a mixture) helium or Argon are used. Gas is fed through a nozzle surrounding the electrode in the head of a suitably designed electrode holder. Flowing inert gas completely envelopes the lower end of the electrode and the work excluding or physically displacing the atmosphere from the molten metal. Total shielding of the puddle from air contamination prevents the formation of oxides, nitrides and other compounds which tend to reduce the strength of the welded joint.

Advantages of TIG Process

(*i*) Main advantage of this process is the elimination of the use of flux. Arc, with suitable current, breaks and lifts oxide film, making the use of flux unnecessary in welding of even rapidly oxidizing, hard-to-weld materials, e.g., magnesium, aluminium etc.

(*ii*) TIG welds are stronger, more ductile, and more corrosion-resistant than welds made with ordinary shield metal arc welding.

(*iii*) As no granular flux is required, it is possible to use a wider variety of joint designs than in conventional shield arc welding or stick electrode welding.

(*iv*) There is little weld metal splatter or weld sparks, which damage bare metal surface as in traditional shield arc welding.

Applications of TIG Process

(*i*) Fusion welds can be made in nearly all commercial metals *e.g.*, aluminium and its alloys, stainless steel, magnesium alloys, nickel-base alloys, copper-base alloys, carbon steel, and low-alloy steels.

(*ii*) TIG welding can also be used for combining dissimilar metals, hardfacing, and surfacing of metals (Fig. 23.14).

TIG arc column is similar to that of standard shield arc welding. (Fig. 23.15)

Amount of plasma generated in the centre of arc column is greater than that generated in standard shield arc stick electrode welding because tungsten electrode has a higher thermionic rating than mild-steel electrode. Thermionic function or emission is the ability of a metal to release its valence electrons.

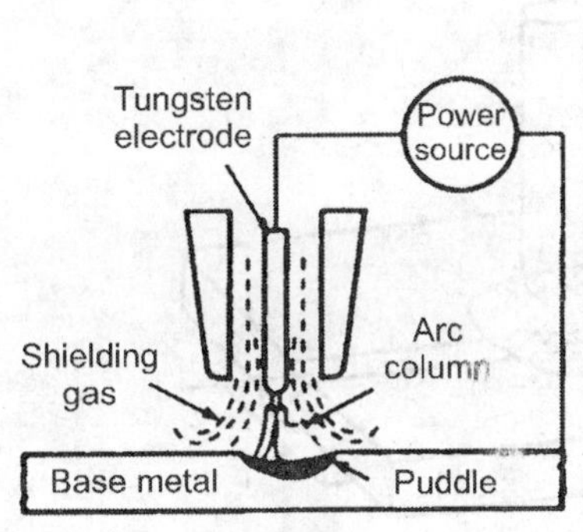

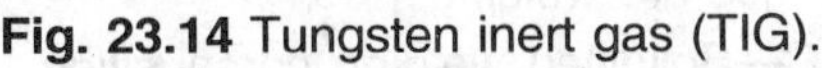
Fig. 23.14 Tungsten inert gas (TIG).

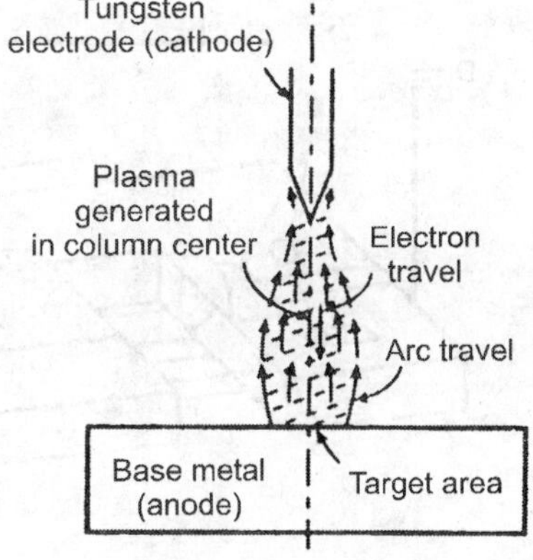

Fig. 23.15 TiG arc column.

5. Submerged Arc Welding

Welding through or under a blanket of flux is being practised by different methods since 1920. Both metallic and carbon electrodes have been used with varying degree of success for hand and automatic processes.

In recent years, perfection of fully automatic and semi-automatic submerged arc welding equipment has increased welding speed considerably; thereby reducing the cost of individual welds.

'Submerged arc welding' is also referred to in industry as 'Hidden Arc Welding' and 'Submerged melt welding'. It is an arc-welding process which utilizes the intense heat generated by an electric arc or arcs between a bare (or, sometimes, highly copper-coated) metal electrode, or electrodes and the work. The welding area is shielded by a blanket of granular, fusible material on the work. The granular material is variously called welding composition, melt, flux etc.

The process may be controlled either mechanically or manually, the former method being used much more widely. The submerged arc process, automatically or manually, creates an arc column between a base metallic electrode and the workpiece. Arc, electrode end, and molten weld pool are submerged in a finely divided granulated powder containing appropriate deoxidizers, cleansers and any other necessary flux.

Fluxing powder is fed from a hopper, which is carried on the welding head. The tube from hopper spreads the powder continuously in front of the electrode along the line of weld. This flux is of sufficient depth to submerge completely the arc column so that there is no smoke or splatter and the weld is shielded from effects of all atmospheric gases. Due to this unique protection, weld beads are exceptionally smooth. (Fig. 23.16).

Flux, adjacent to arc column melts and floats to the surface of the molten pool; then solidifies forming a slag on top of weld metal. Rest of flux is simply an insulator, which can be easily reclaimed. Slag formed by solidification of molten flux is easy to remove.

Generally, the slag cracks off by itself as it cools. Unused flux is removed and placed back into original hopper for reuse. Granulated flux is complex metallic silicate, which can be used over a wide range of metals *eg.*, mild steel, low-alloy and high-alloy steel, chromium-nickel steel etc.

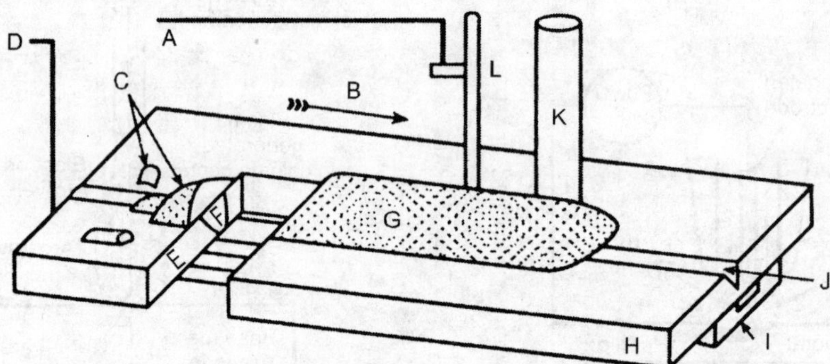

Fig. 23.16 Making a weld by submerged-arc welding.
A—A.C. or D.C. electrode, B—Direction of welding, C—Solidified welding composition, D—Ground, E—Plate metal, F—Weld metal, G—Granulated welding composition, H—Workpiece, I—Weld backing if required, J—Vee if required, K—Welding composition or Flux tube, L—Welding electrode.

Application. It may be applied to the welding of all weights of metals, from light gauges to heavy thicknesses, and welds can be made in one pass in thick sections, which might require several passes by other welding processes. It is also capable of joining medium-carbon steel; heat-resistant steels, corrosion resistant steels, and many high strength steels. Also, it is adaptable to Nickel, Monel, Everdur, and many other non-ferrous metals.

Submerged arc welding process has many industrial applications *e.g.,* fabricating pipes, boiler pressure vessels, railroad tank cars, structural shapes. Many applications of this process have created three general types of equipment:

(*i*) Stationary electrode with work moving, .

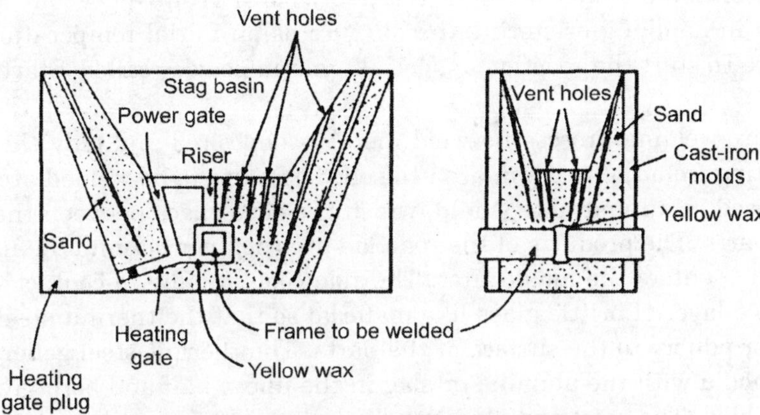

Fig. 23.17 Non-pressure thermit welding.

(*ii*) Electrode mounted on a carriage with work stationary, and

(*iii*) Submerged arc mounted on a self-propelled tractor with work stationary. This, of course, is manual process, in which special flux-hopper electrode holder is hand held.

23.6. THERMIT WELDING

It is the only welding process employing an exothermal chemical reaction for the purpose of developing a high temperature. It is based on the fact that aluminium has a great affinity for oxygen and can be used a reducing agent for many oxides.

Thermit welding is primarily a fusion welding process in which the weld is affected by pouring superheated liquid thermit steel around the parts to be united. In case of thermit pressure welding, only the heat of the 'thermit' reaction is utilised to bring the surfaces of metal be welded in a plastic state and mechanical pressure is then applied to complete the weld (see Fig. 23.17).

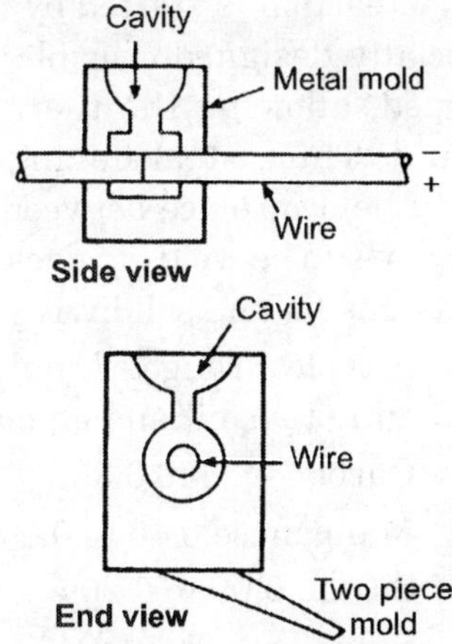

Fig. 23.18 Alumino thermit welding.

It comprises of a group of welding processes wherein coalescence is produced by heating with superheated liquid metal and slag resulting from a chemical reaction between a metal oxide and aluminium, with or without the application of pressure (see Fig. 23.18). Filler metal when used, is obtained from the liquid metal. Basic thermit is a mechanical mixture of finely divided aluminium and iron oxide in form of magnetic iron scale. The proportions are roughly, three parts of iron scale to one of aluminium. This mixture reacts according to the chemical formula.

$$8Al + 3Fe_3O_4 = 9Fe + 4Al_2O_3$$

The temperature resulting from the reaction is computed to be more than 2800°C. Due to the chilling effect of the crucible, however, the temperature of the liquid steel as poured into the mould for welding is slightly lower and has been measured at approximately 2500°C. The reaction is non-explosive and relatively slow, requiring about 30 seconds. No hazard is incurred in handling or storing the mixture as an initial temperature of 1200°C is needed for ignition. To start the reaction, a special ignition powder called 'Starting Thermit' is employed.

In the pressure-welding process no weld metal is deposited and only the heat of metal and the slag resulting from the reaction are utilised. The parts to be welded arc butted tightly together and enclosed in a removable mould, which provides space between the inner surface of mould and the parts. The products of the reaction are then poured into the mould in such a manner that the slag enters the mould first. The slag cools rapidly in contact with the metal parts and provides a layer of brittle glass-like material so that the thermit steel which follows does to fuse with or adhere to the surface of the parts. The thermit steel generally lies in the lower half of the mould with the alumina or slag in the upper half and both give their heat to the parts being welded. The steel and slag are allowed to remain until the ends of the pieces reach a welding heat, where upon the pieces are forced together by means of clamps to make the pressure butt weld. The mould is then removed and the thermit steel then knocked off from around the weld.

In making a weld by this process, the first step is the lining up of the parts and the cutting of a parallel-sided gap at a point where the weld is to be made the width of the gap depending on the size of the section. Around the break or joint, a wax pattern is formed and a refractory sand mould is built, which provides an annular space at the weld. The part is then preheated by means of kerosene and air blown into the mould through an opening. Preheating, which continues until the parts reach red heat, serves to burn out the wax pattern and to dry out the mould.

Reaction is started by means of a small starting thermit placed on top of the thermit in a specially designed crucible (see Fig. 23.19). When the reaction is computed, the crucible is tapped, allowing the thermit steel to run into the mould. Thus, steel having great superheated and being held in place between and around the ends of the parts to be welded, gives up its superheat to these parts, fusing and solidifying with them upon cooling.

A typical range of analysis of the steel produced by forging thermit and employed for welding is:

Carbon 0.25 to 0.35%

Manganese 0.40 to 0.60%

Silicon 0.09 to 0.20%

Sulphur 0.03 to 0.04%

Phosphorus 0.04 to 0.05%

Aluminium 0.07 to 0.18%

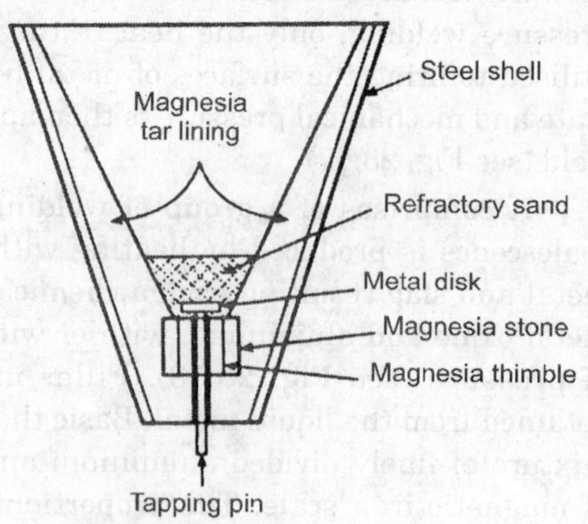

Fig. 23.19 Thermit crucible.

A thermit weld of this composition has an average tensile strength of about 4800 kg/cm^2 with a yield strength of about 2400 kg/cm^2. In fact, although thermit weld metal resembles a cast steel it actually has physical properties closely approaching those of forged steel.

Most **commonly used thermits for welding ferrous metals are:**

(*i*) **Plain Thermit,** a mixture of finely divided aluminium and iron oxide which is the basis of all thermit nurtures.

(*ii*) **Forging Thermit.** It is plain thermit with the addition of ferromanganese and mild steel punching and is used in welding steel.

(*iii*) **Cast-iron Thermit.** It consists of plain thermit vdth addition of ferro-silicon and mild steel punching, used for welding cast iron.

(*iv*) **Wabbler Thermit.** It consists of plain thermit with manganese and carbon additions and designed to produce a hard wear-resistant machinable steel for building up worn wabbler ends of rolls and pinions in steel mills.

In *Thermit pressure welding,* the ends of joints to be welded are pressed together, and a mould is built around the joint (see Fig. 23.20). After the reaction starts, the slag is poured in first forming a glass like surface around the joint, and then the superheated metal added. When heat from slag and metal brings the joint upto welding temperature, the weld is made by pressure, similar to the procedure followed in resistance butt welding.

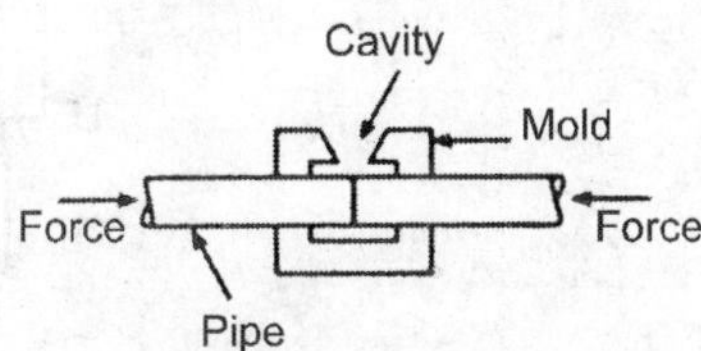

Fig. 23.20 Pressure thermit.

There is no limit to the size of weld, which can be made by thermit welding. It is employed primarily for repairing large parts.

23.7 FORGE-WELDING

It consists of a group of process wherein parts to be welded are brought to suitable temperature by means of external heating generally a furnace, and the weld is completed by pressure or blows. Pressure is applied after the parts have been properly heated. Materials most commonly joined by forge welding are wrought iron and low carbon steels (carbon less than 0.2%).

23.8 RESISTANCE WELDING

Resistance welding is a process where two or more parts are welded together by the combined use of heat and pressure; the heat being generated by a relatively short-time flow of low-voltage high-density electric current across the intended joint location, and the pressure being supplied by contacting electrodes. These electrodes also carry the current to the work pieces. Both electric current and pressure are closely regulated and controlled at all times.

In case of resistance welding processes the required heat, at the joints to be welded, is generated by the resistance offered through the work parts to the relatively short-time flow of low-voltage, high density electric current. Force is always applied before, during and after the application of current to assure a continuous electrical circuit and to forge the heated parts together. The maximum temperature achieved is ordinarily above the melting point of the base metal.

When the current passes through the metal, the greatest resistance is at the point of contact. Hence, the greatest heating effect is at the point where the weld is to be made. Alternating current is generally used, coming to the machine with usual commercial voltages.

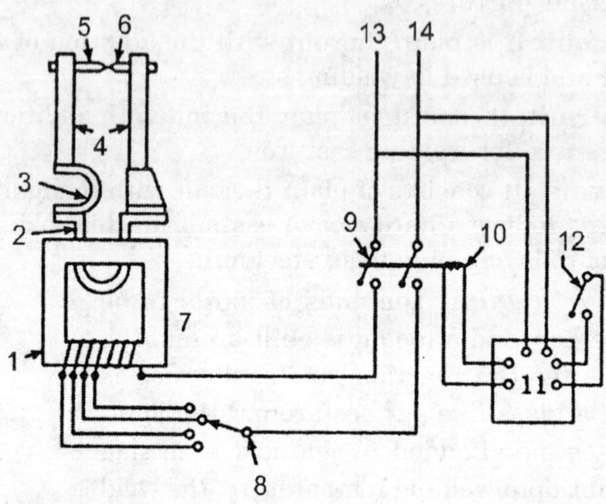

1. Transfomer core.
2. Secondary.
3. Flexible band.
4. Conductors (Horns).
5. Welding electrode.
6. Electrode holder.
7. Primary welding.
8. Heat Regulator
9. Contactor.
10. Contactor coil.
11. Timer.
12. Initiating switch

13., 14. Leads.

Fig. 23.21 Typical circuit of a Resistance Welding Machine.

A transformer in the machine reduces the voltage 4 to 12 volts and raises the amperage sufficiently to produce a good heating current. Amount of current necessary is 5 to 7 kV per square cm of area to be united, based on a time of about 10 seconds. For other time intervals, power varies inversely with time. Necessary pressure to effect the weld varies from 300 to 600 kg/cm^2.

Resistance welding is essentially a production process used for joining light gauge metals, which can be lapped. Usually, equipment is suitable for only one type of job, and work must be moved to the machine. The process is especially adapted to quantity production and includes a large amount of welding done at the present time.

It is the only process, which permits a pressure action at the weld, while allowing an accurately regulated heat application. Also, the operation is extremely rapid. Weldability of a given metal depends to some extent on its melting point. Practically, all metals can be welded by resistance welding, although some few, *e.g.* tin, zinc and lead can be welded only with great difficulty.

In all resistance welding, three factors requiring consideration are expressed in the formula:

$$\text{Heat} = I^2RT$$

where I = Welding current in amperes,

R = Resistance of metal being welded,

T = Time.

Amperage of secondary or secondary current is determined by the transformer.

To provide possible variation of secondary current, the transformer is equipped with a regulator on the primary side to vary the number of turns on the primary coil. This may be seen in the Fig. 23.21 at 1.

For good welds, these three variables, *viz.,* current, resistance, and time must be carefully controlled and determined by the factors, *viz.,* material thickness, kind of material and electrode size.

Timing of welding current is very important. After the pressure has been applied, there should be an adjustable delay until the weld starts. Current is then turned on by the timer and held a sufficient for the weld. It is then stopped, but the pressure remains until the weld cools, thus any tendency for the electrodes to arc is eliminated and also the weld is protected from discoloration.

The pressure on the weld is obtained manually, by mechanical means, by air pressure, by springs, or by hydraulic means. Its application is controlled and co-ordinated with the application of welding current.

Resistance welding may be sub-divided as:

(*i*) Spot welding,
(*ii*) Seam welding,
(*iii*) Projection welding,
(*iv*) Butt welding,
(*v*) Flash welding,
(*vi*) Percussion welding.

1. Spot Welding. In spot welding, coalescence is produced by the heat obtained from resistance to the flow of an electric current through the workpieces pressed together by pointed or domed electrodes. The electrodes are brought to and removed from the workpieces at predetermined times and rate, and a clamping force is applied through the electrodes by some suitable means.

2. Seam Welding. It is similar to spot welding in principle except that circular electrodes, or one circular electrode and a platen are used. The resulting weld is a series of overlapping spot welds made progressively along a joint by rotating the electrodes.

3. Roll spot Welding. In roll-spot welding separated spot welds are made in a row with a seam-welding machine by suitably adjusting the speed of the circular electrodes and interrupting the current at given intervals. The electrodes are not retracted during the welding operations, and the electrode force is not removed between spot welds. The circular electrode provides in effect, the same physical action as a domed or spherical spot welding electrode.

In roll-spot welding, the desired weld spacing is obtained by proper adjustment of electrode speed and current-off time. Roll-spot welds are made with a shorter hold time then spot welds made in the usual manner, and tend to have a some what elongated weld nugget, due to the electrode travel. Single and multiple roll-spot welds may be made as discussed in the section on spot welding.

4. Projection Welding. In projection welding, coalescence is produced by an application of heat and pressure similar to that used for spot and seam welding. Welds are made by localizing the welding current, welding pressure, and heating at one or more predetermined points by the configuration or design of one or more of the workpieces, instead of using pointed domed, or circular electrodes. Localization is usually accomplished by projections, embossments, or intersections.

5. Flash Welding. In flash welding, two parts are placed in an electric circuit adjacent to each other, and an electric potential is applied. A current of sufficient magnitude to cause a flashing action flows as the abutting surfaces are brought together. When sufficient flashing has occurred to heat the metal to welding temperature, the weld is consummated by the application of upset pressure. Metal is expelled from the joint during flashing and upsetting.

Flashing and Upsetting

(*i*) The flashing action, which follows the light contacting of the two parts but precedes any upsetting, provides the necessary welding heat.

(*ii*) Soon after the fusion temperature has been reached, the upset or forging pressure is applied suddenly and the welding current is cut-off. Sufficient plasticity must be obtained in the part to permit proper upsetting action.

(*iii*) The current flowing through the workpieces during the upsetting operation must be sufficient to prevent the material from cooling too rapidly.

(*iv*) Shielding or protective atmosphere e.g., hydrogen, inert, gases etc. may sometimes be used to improve the quality of flash welded joints.

Metal Allowances for Flashing and Upsetting

(*i*) In flash welding, it is necessary to make an allowance in stock length for the metal lost in flashing (or burn-off) and upsetting (or push-up). This allowance usually ranges from one to six times the stock thickness and decreases as the stock thickness increases.

(*ii*) Metal allowance for upset varies in direct proportion with the flashing voltage used but is usually one-fourth to one-third of the total allowance.

(*iii*) No special preparation of the weld surface is necessary if the irregularities are burned off during the flashing period and foreign matter is expelled with the flash.

Clamping of Work. The workpiece to be flash welded are backed up by means of fixed stops on the platens and the upset pressures is transmitted and maintained through these stops.

Flash Removal. Appearance and mechanical performance of a flash welded work often require that both the flash and the extruded material be removed after welding. This material may be removed by some suitable mechanical or oxygen-cutting processes.

Upset Welding. In upset-welding process, two parts are brought into solid contact with one another, and coalescence is produced simultaneously over the entire area of abutting surfaces, or progressively along a joint, by the heat obtained from resistance to the flow of electric current through the area of contact of the abutting surfaces. Pressure is applied before any heat is generated and is maintained throughout the heating period. The workpieces are actually forge-welded together.

Difference between upset welding and flash welding

Upset welding differs from flash welding in that:

(*i*) No flashing from the abutting surfaces occurs at any time.

(*ii*) Heat is developed solely by the resistance between the two workpieces.

Precautions

(*i*) Best results are normally obtained when the parts to be upset welded have equal cross-sectional area and resistance.

(*ii*) Many of the precautions for upset welding are similar to those for flash welding. Work preparation prior to welding, is however, more critical for upset welding.

(*iii*) For good results in upset welding, it is usually necessary to machine the weld surfaces on rolled stock, foreign, and castings, stamping are usually trimmed in a die but sometimes, they are machined.

(*iv*) Surfaces to be upset-welded must be parallel and comparatively smooth.

(6) Percussion Welding. It is a process in which coalescence is produced, simultaneously over the entire area of abutting surfaces, by the heat obtained from an arc produced by a rapid discharge of stored electrical energy, with pressure percussively applied during or immediately following the electrical discharge.

This process is classified as an electrostatic stored-energy system because it has been applied commercially only with the use of electrostatic capacitors. This process does not make use of a welding transformer; instead a condenser discharge is applied directly to the joint.

23.9 CUTTING PROCESSES

Workpieces can be cut from steel plates or other sections by application of fusion welding. The heat required for melting may be provided by a flame, an electric arc or a chemical heat.

1. Oxygen Cutting

In 1887, Thomas Fletcher discovered that oxygen could be used for cutting of ferrous metals. He heated ordinary wrought iron until it turned to a bright cherry red, which happens between 870° to 980°C. He then directed a stream of oxygen toward the heated portion causing a rapid oxidation process to take place, which is now called "Oxyacetylene cutting". Around the turn of the century, first combination preheating tip and pure-oxygen jet-type apparatus was designed.

Before oxyacetylene cutting torch, the only way to separate plate steel was by shearing it, a method limited to cutting low-carbon steels upto a maximum of 40 mm thick and even thinner for alloy steels. Also, mobility of large steel plates created a problem. Oxygen cutting torch eliminated all these problems for steel industries. It was the lightest and most mobile shearing apparatus.

Thickness limit for oxygen cutting is about 1600 mm. Oxygenfuel cutting of ferrous metals is a process of preheating the material to be cut to its kindling or ignition temperature and rapidly oxidizing it by means of a closely regulated jet or stream of oxygen issued from a special tool called a 'cutting blow pipe' or 'cutting torch'. Thus, it is primarily a chemical process based on the remarkable chemical affinity of oxygen for ferrous metals when raised to or above the kindling temperature. Acetylene is most commonly used fuel, gal for oxygen cutting.

When commercially pure oxygen is brought into contact with most steels or iron, at kindling point, a very chemical reaction between them results. In case of plain carbon steel, metal is preheated to a bright-red colour in daylight, reached between 770 and 870°C approx. Gases supplied and rate of progression of cutting should be closely regulated for accurate control.

Only the metal within direct path is acted upon, to form a narrow race having uniformly smooth and parallel walls. Under skilled workmanship, with mechanically guided cutting torches, cutting tolerances as to squareness and longitudinal alignment of the cut may be held within close limits.

Oxygen-cutting Equipment. It consists ordinarily of the cutting torch with separable cutting tips, oxygen and fuel-gas hose with connections, oxygen and fuel-gas pressure regulators, and gas cylinders.

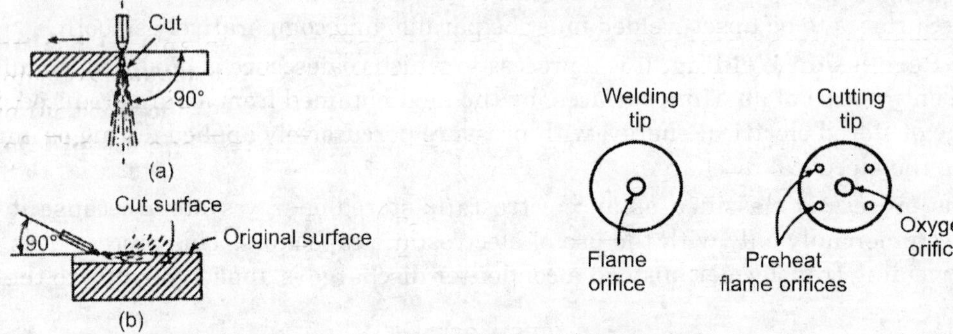

Fig. 23.22 (*a*) Changing position of cutting flame opens a new field of application.

23.22 (*b*) Oxyacetylene torch tips.

Oxygen cutting torch is different from welding torch in two ways:

(*i*) Cutting torch has a high-pressure valve, which is activated by a lever, trigger or button. In addition to oxygen and acetylene adjustment valves, on high-pressure oxygen valve, when pressed, releases a jet of oxygen through the tip.

(*ii*) Welding tip has one orifice, whereas cutting tip has multiple orifices. Cutting tipes have preheat orifices and one central and one central orifice, which directs the flow of oxygen for cutting operation.

Oxyacetylene cutting torch uses the same basic components as a welding torch. A cutting head attached to the torch body is designed to have preheat orifices in the cutting tip as well as a passage for high-pressure pure oxygen. Cutting apparatus may be in a torch designed specifically for cutting, or proper cutting apparatus may be attached to a standard torch body (See Fig. 23.22(a) & (b).

Cutting torches may be of (*i*) premixed or (*ii*) head-mixed type.

(*i*) In *premixed type,* oxygen and fuel are mixed in the body of the torch.

(*ii*) In *head-mixed type,* the most widely used type, oxygen and fuel are mixed just before entering the tip.

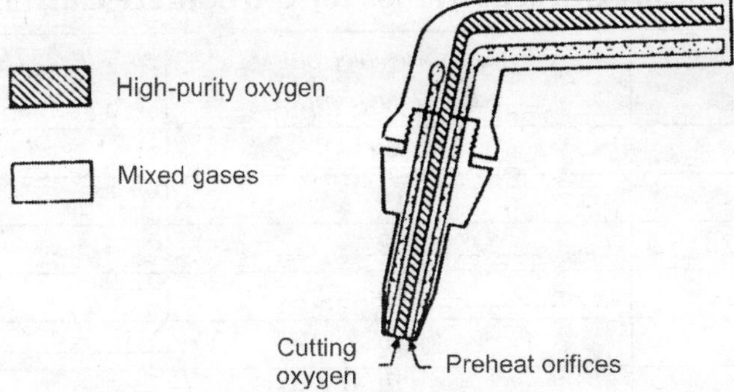

Fig. 23.23 Cutting head.

Torches are further classified as:

(*i*) *Medium-pressure torches,* in which acetylene is used in the range 0.07 to 1 kg/cm^2.

(*ii*) *Low-pressure torches,* in which acetylene is 0.07 kg/cm^2 or less.

The above two classifications are the same as those for welding torches.

Pressure regulators are usually equipped with cylinder and working pressure gauges. Cutting-oxygen pressure regulators are specially designed to deliver oxygen in large volumes and at higher pressures (30 kg/cm^2).

In case of machine oxygen cutting, the torch is held and operated mechanically, but its essential features remain unchanged. Manufacturer's instructions for setting up and operating the equipment should be followed closely to assure to efficient results.

2. Carbon-Arc Cutting

In carbon-arc cutting, an arc is drawn between a carbon or graphite electrode and sections to be cut. Carbon electrodes are most frequently used, but they do not stand up very long under high heat required in arc cutting. Graphite electrodes are preferred at times because they set up lower resistance to the flow of current, thus use of higher currents is possible (see Fig. 23.24).

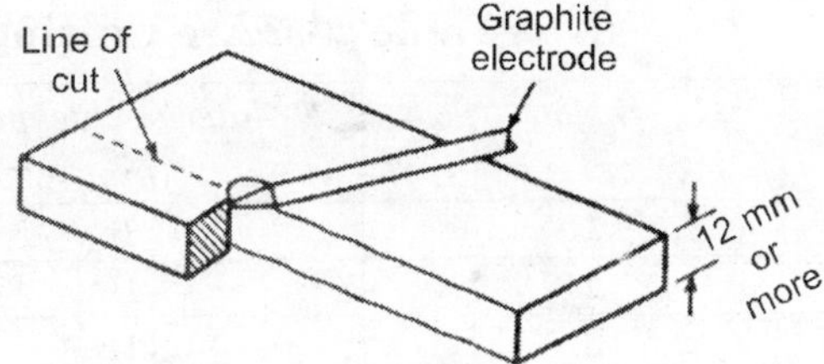

Fig. 23.24 Carbon-arc cutting of heavy plates.

D.C. is generally used in carbon-arc cutting, especially if large electrodes are used. These special holders usually have a long shank as well as a shield to protect the operator's hand from the intense heat.

Application:

(*i*) Carbon arc may be used for cutting both light and heavy plates where rough work is satisfactory.

(*ii*) In cutting plates upto 12 mm thickness, arc is simply advanced along the section to be cut.

(*iii*) Rate of cutting is equal to the rate at which the metal is melted away.

Table 23.4 Current Values for Carbon-arc Cutting

S.No.	*Graphite electrode diameter (mm)*	*Current (Amp.)*
1	6	upto 200
2	10	200-400
3	12	300-600
4	15	400-700
5	19	600-800
6	22	700-1000
7	25	800-1200

(*iv*) Cutting may be done in any position, but best results are obtained when work is in horizontal plane.

(*v*) Angularity of electrode to work depends on work position:

(*a*) For vertical cutting electrode should be pointed downward.

(*b*) With work in a horizontal position, electrode may be, held either vertical or slightly forward.

(*vi*) For plates over 12 mm thickness, cutting procedure should be such that the bottom of the cut is advanced slightly ahead of the cut. Resultant undercut permits molten metal to run very easily: (*a*) If work is in a vertical position, cut should be started at the bottom, (*b*) With work in a horizontal position, electrode should be slanted backward, and cut should be started at bottom surface of the plate. Then, with electrode pointing forward, cut may be made into the plate.

(*vii*) Width of cut depends on (*a*) material thickness, (*b*) operator's skill in cutting a straight line. Of course, the thicker the material, the wider the cut.

Table 23.5 Arc Cutting Speeds with Carbon Electrodes

S.No.	*Material thickness (mm)*	*Current (Amp)*	*Cutting speed (mm./minute)*
1	6	400	375
2	10	400	300
3	12	400	250
4	15	400	175
5	19	400	100
6	25	600	80
7	38.	600	60
8	50	600	45
9	75	600	27.5
10	100	600	17.5
11	150	800	10
12	200	800	7.5
13	250	800	5
14	300	800	3.75
15	350	1000	2.5
16	400	1000	1.5

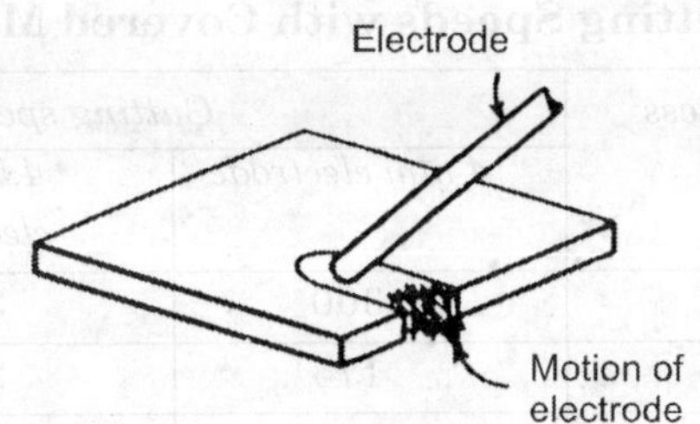

Fig. 23.25 Cutting with covered electrodes.

3. Metal Arc Cutting

In metal arc cutting, an arc is struck between a covered electrode and the work to be cut (see Fig. 23.25).

Due to high cost of electrodes, cutting with covered electrodes is not very widely employed. But, this process may be used to advantage for cutting in many cases, especially when gas torch and carbon electrode are not available.

Due to insulating effect of covering, covered electrodes can be used in tight spots where neithger carbon-arc nor oxygen-cutting process would be practical.

Ordinarily, gas and slag-producing electrodes are used for arc-cutting, with the latter being employed as the same permit the use of higher currents.

Application

(*i*) Cutting of light-gauge metals may be done with covered electrode but this method is not employed very often.

(*ii*) Plates, 10 mm in thickness or higher, can be cut by simply moving the electrode along the line of cut.

(*iii*) For heavier plates, it is necessary to work from the underside of the plate, so that the molten metal and slag may run off or drop.

(*iv*) Cut may be started at the edge of the plate, or a hole may be burned through the plate, and thus the cut may be started at any desired point.

(*v*) For piercing operations, high currents must be used.

(*vi*) To keep metal and slag flowing out of the cut, a short arc should be held and electrode should be kept pushing closely up in the slot, with the electrode tip being pushed right into the molten crater and manipulated with an up-and down motion. Upward motion should be very fast and downward motion should supply a 'pushing' effect, thus, helping to force out the slag and molten metal.

(*vii*) Width of cut does not vary much for electrodes upto 6 mm diameter.

(*viii*) Common sizes of covered electrodes for cutting purposes are 4, 4.5, or 6 mm. Electrodes smaller than 4 mm diameter are not recommended, because under high currents used they heat up quickly and melt away.

(*ix*) Speed of arc-cutting with covered electrodes depends on (*a*) Electrode diameter, and (*b*) Current used.

Table 23.6 shows arc cutting speed with covered metal electrodes for different thickness of plate.

Table 23.6 Arc cutting Speeds with Covered Metal Electrodes

S.No.	*Plate thickness (mm)*	*Cutting speed (mm./min.)*		
		4 mm electrode	*4.5 mm electrode*	*6 mm electrode*
1	6	300	375	525
2	12	175	225	275
3	29	105	140	165
4	25	60	90	115
5	30	40	60	80
6	38	25	40	55
7	45	15	30	45
8	50	M10	25	35

23.10 BRONZE WELDING

Bronze welding is also known as braze welding, and is a process which is intermediate between true welding and true brazing. In this the edges or surfaces are not welded, but a low melting alloy is introduced between them and a joint is produced by adhesion.

In *braze-uelding,* filler material is generally composed of copper and zinc and sometimes tin is added. Filler material must have a melting point below that of the base metal but above 425°C. A cleansing flux is generally necessary to use. Usually, a slightly oxidizing oxyacetylene flame is used. Main limitation of braze-welding is that weld loses its strength as temperature increases, hence can be used at low temperatures only.

It is similar to fusion welding, except that the filler material is non-ferrous and often used with a fluxing agent. This filler material must melt above 405°C but below the melting point of the base metal.

Following are the most important factors while braze welding:

1. Cleanliness of the surfaces which are to be braze welded.
2. Tinning of the base metal with non-ferrous filler material.
3. Complete fusion of additional weld material with non-ferrous filler material previously deposited such as in case of multiple passes.

Basic joint designs which can be used in braze welding are same as used in fusion welding but these joints cannot withstand that much amount of heat, which fusion welds can.

Advantages

(*i*) Main advantage of braze welding is that dissimilar metals can be jointed e.g., high-carbon steel to low-carbon steel, copper to steel, copper to cast iron etc.

(*ii*) Oxygen-fuel consumption for a given job is less than with fusion welding.

(*iii*) Locked-in stresses are not as great as temperatures in braze welding are lower than those in fusion welding.

(*iv*) Braze welding is quite strong upto 260°C.

(*v*) Special heat treatments may not be necessary after braze welding because it has a small heat effect of its own.

Thickness limitation of braze welding. It should be slightly less than that of fusion welding. Lap or butt joints can be brazed 4.5 mm or less thick. Material thicker than 4.5 mm. should be grooved or bevelled or some mechanical joint should be prepared.

The distinguishing feature of the braze-welding process is that the distribution of the filler metal does not depend upon capillary attraction. Also, groove, fillet, plug or slot welds may be made with this process.

Oxygen-fuel braze welding of ferrous metals is often used to joint dissimilar metals or for repair work. All good welding procedures and techniques, which apply to fusion welding, apply also to braze welding of ferrous metals. Braze welding differs from fusion welding in that the base metal, puddle is not in a molten state and the filler material is not ferrous.

Applications. Braze welding may be applied for such fabricating application as for metal furniture, bicycles, automobiles, refrigerators and household appliances. However, its principal use is for repair and maintenance work, particularly where ferrous castings are involved. The most common ferrous metals braze welded are steel and its alloys, galvanized iron, cast iron and chromium-nickel cast iron.

Bronze Welding Filler Metal. It is commonly referred to as bronze welding rod. It is usually a brass alloy containing approximately 60% copper and 40% zinc. Copper silicon alloys and nickel-silver alloys may also be used as filler-metals. Tin, silicon and manganese (about 1%) are added in case of brass alloys to improve the flow characteristics as well as to decrease fumes and add hardness.

23.11 SOLDERING

Soldering differs from brazing in that lower temperature filler metal (below 450°C) are used in the joint. Soft solders are metals or alloys, having a melting range of 150 to 350°C, used for joining most common metals, which melt at temperatures below the melting point of the base metal and in all cases below 450°C. The soldered joint depends for its strength on alloying with the base metal and upon mechanical bonding between the parts.

Soft solders are made up of tin, tin-lead or lead with or without the addition of antimony, silver, arsenic or bismuth to impart special properties, These metals or alloys behave differently than the so-called 'Hard Solders (Silver Brazing Materials).' The strength of the joint is determined by the adhesive quantities and the strength of these alloys.

(*i*) The joint strength obtainable is lower and may deteriorate rapidly if exposed to high temperatures or to corrosive conditions.

(*ii*) Continued stress reversals which may be produced by vibration or a constant load which may cause creep, also results in a very rapid drop in the joint strength.

Soft solders as obtainable commercially in the form of wire (or string), bars, powders or ingots.

Electric connections, wire terminals and similar small parts are typical of the joints made by soft soldering. A form of soldering known as 'wiping' is used in making lead pipe connections.

Fluxes for Soldering

The functions of a flux are (*i*) to clean the joint area (*ii*) prevent oxidation and (*iii*) lower the surface tension of the solder, thus increasing its wetting properties. Fluxes can be classified as under:

1. Non-corrosive Fluxes. Rosin and rosin alcohol are truly non-corrosive fluxes. This type of flux requires that the parts be chemically or mechanically cleaned, yet not polished brightly since the flux has little or no cleaning power and low wettability.

A small addition of glycerine will increase the wetting properties of the flux. The addition of a small amount of lactic acid will increase the efficiency of the flux on the brass parts. The addition of a small amount of turpentine to the flux will help to reduce this stain caused by rosin.

2. Mild Fluxes. Citric acid in water is a low cost slightly corrosive flux. If this flux is used for dip soldering, the acid should be dissolved in alcohol to reduce spatter. The addition of glycerine or suffocated alcohol will increase the wetting action and further reduce the spatter. Levulinic acid in alcohol is widely used as a flux on terne plate and tin plate in the can industry.

3. Corrosive Fluxes. Zinc chloride and ammonium chloride (sal ammonia) are the best known corrosive fluxes used in various soldering. They may be used singly or combined in various solvents. They are quick acting and help to produce efficient joints quickly. Zinc cut in hydrochloric (muriatic) acid, is most commonly used by tinners and repairmen in their work.

The parts to be soldered should be free from all oxide, scale, should oil and dirt to insure sound joints. A flux is generally used to secure good joints except in some cases on aluminium.

23.12 BRAZING AND BRAZE-WELDING

Brazing processes are those wherein coalescence is produced by heating to suitable temperatures above 428°C and by using a non-ferrous filler metal which has a melting point below that of the base metal. The filler metal is distributed between the closely fitted surfaces of the joint by capillary attraction.

With this process, it is possible to produce, at a low cost, a ductile, high-strength joint resistant to fatigue and corrosion. Steel, Copper, Brass, Bronze, Monel, Nickel, Inconel, Cast iron, Cast Brass, Aluminium and Magnesium can be brazed to themselves. Dissimilar metal combinations, excluding aluminium and magnesium, also can be joined by brazing.

Brazing is differentiated from soldering by the melting temperatures of the filler material. If the filler material melts above 426°C, it is brazing; but if the filler material melts below 426°C, such as in tin-lead alloys, the process is then called soldering.

Advantages of Brazing

(*i*) Completed joint requires little or no finishing.
(*ii*) Relatively low temperature at which the joint is made minimizes distortion.
(*iii*) There is no flash or weld spatter.
(*iv*) Brazing technique does not require as much skill as the technique for fusion welding.
(*v*) Process can be easily mechanized.
(*vi*) Process is economical due to the above advantages.

Disadvantages of Brazing

(*i*) If the joint is to be exposed to corrosive media, the filler metal used must have the required corrosion resistance.
(*ii*) All of the brazing alloys lose strength at elevated temperature.

(*iii*) Colour of brazing alloy, which ranges from silver-white to copper-red, may not match the base metal very closely.

In all instances, the liquids temperature of the brazing filler metal must be lower than the solidus temperature of the base metal. For most commercial work, brazing alloy is selected which melts and flows freely at least 93°C below the melting temperature of the base metal.

A filler metal with a narrow melting range is usually desirable. Such an alloy remains solid until brazing temperature is reached, at which time it melts and quickly flows into the joint. Where the fit of the parts is loose, a brazing alloy with a wide melting range will frequently fill the joints more easily than one with a narrow melting range.

Brazing Filler Metals

These are usually classified according to their chemical compositions and applications. When in the molten state, brazing alloys should not oxidize too freely and those oxides which do form should be easily fluxed:

(*i*) Oxides of copper, silver, iron, nickel, zinc, tin, cadmium and silicon are easily fluxed.

(*ii*) Oxides of aluminium, magnesium, chromium and beryllium are increasingly difficult to flux.

Brazing alloys must not diffuse too freely into the base metal at the brazing temperature and must not dissolve too much of the base metal.

A brazing alloy should be available in a convenient form for the brazing process and the assembly for which it is to be used. Sometimes, in selecting a brazing alloy, additional requirements like electrical conductivity, heat conductivity and colour must also be met.

Brazing Fluxes

Fluxes serve the same purpose for brazing as they do for soft soldering. They are not designed to remove oil, grease, paint and other foreign substances. Surface should, therefore, be cleaned before applying the flux.

Action of Brazing Fluxes. Oxides of ordinary commercial metals and alloys, except steel, have higher melting points than the metals themselves. Even when the metals are fluid at the brazing temperature, the oxides are usually pasty or infusible. It is the action of the flux to combine with these oxides, forming a fusible slag which has a lower melting point than the metal, and to allow the slag to flow away from the place of operation.

Chemical Composition and Selection. Melting point and chemical composition of the various oxides are different, no one flux will be satisfactory for all metals. Therefore, flux must be selected not only for the filler metal used but also for the metals to be joined and for brazing method to be employed.

Fluxes are designed to melt and to be active at certain temperatures. They may break down and become useless if heated too long or at too high a temperature. For these reasons, a brazing flux cannot be substituted indiscriminately.

23.13 HARD FACING

This is a non-joining application of welding processes. In this process a lower strength but tough body is coated with a very hard and sometimes brittle surface layer. This is generally employed for repairing wear resistant parts.

QUESTIONS

23.1 List various methods of welding. Describe them in brief.

23.2 (*a*) What do you understand by gas welding?

(*b*) Describe in brief the equipments required for oxyacetylene welding.

(*c*) How would you obtain neutral, oxidizing and reducing flames in a welding torch?

23.3 What is the principle of operation of electric arc welding?

23.4 What are the functions of coatings on shielded electrodes?

23.5 Write a short note on each of the following:

(*a*) MIG (*b*) TIG

(*c*) Plasma-arc (*d*) Laser-beam welding.

23.6 What are the principles of operation of resistance welding?

23.7 What is ultrasonic welding? What are its advantages?

23.8 Explain the principle of thermit welding. What are its advantages over other types of welding?

23.9 Explain oxygen cutting and arc cutting. Which of these two is preferred and why?

23.10 Write short notes on:

(*a*) Welding joints and edge preparation.

(*b*) Leftward and rightward welding.

(*c*) Hard facing, brazing and soldering.

23.11 What is welding? Describe with a neat sketch the principle of electrical arc welding.

23.12 Describe with a neat sketch the spot welding process. What is the principle of spot welding?

23.13 Name the tools and equipment used in electric arc welding.

23.14 Compare the AC and DC arc welding.

23.15 Explain the following in brief:

(*a*) Carbon dioxide welding (*b*) Submerged arc welding

23.16 (*a*) What arc the essentials of a good flux.

(*b*) Name fluxes used for the following.

(*i*) Cast iron (*ii*) Copper and its alloys

(*iii*) Aluminium.

23.17 Describe the defects that generally occur in welding. How are they detected.

23.18 What do you understand by soldering and brazing? Describe the principle of each.

23.19 Explain the process of brazing. Give the advantages of brazing over soldering.

23.20 Write a note on 'fluxes used in brazing.'

23.21 What are the advantages and disadvantages of welded joints over other joints?

23.22 How the welding is classified? Explain them briefly.

Chapter

24

TOOLS, JIGS, AND FIXTURES

Learning Objectives

After studying this chapter, the students will learn:

- Tools and type of tools.
- Cutting tool materials, their characteristics, types of tool materials and standardization of tools.
- Tie and die casting methods.
- Gauges, jigs and fixtures.
- Tool room and its organisation set-up.

24.1 INTRODUCTION

In the metal-working industry, the various working processes fall into two groups, (*a*) the group of *non-cutting shaping,* e.g. forging, pressing, drawing etc., and (*b*) that of cutting shaping by which finished surface of desired shape and dimension is obtained by separating a layer from the parent workpieces in the form of chips, e.g. turning, drilling, milling etc. The process of metal cutting in which chip is formed is affected by a relative motion between the workpieces and the hard edge of a cutting tool held against the workpiece.

The productivity and economics of machining workpieces in a large quantity is greatly affected with the use of tools and workholding devices like jigs and fixtures. These devices reduce the production cost and ensure interchangeability of machined workpieces. They are defined as under:

1. Tool is a device which removes material from a workpiece under stable conditions and gives it desired shape and size.

These may be of the following types:

(*a*) Single-point tools.

(*b*) Multiple-edge tools.

(*c*) Non-mechanical or noncontact tools.

(*d*) The press which is used to handle these tools itself, is called machine tool.

Chip removal in the metal-cutting process may be performed either by cutting tools having distinct cutting edges or by abrasives used in grinding wheels, abrasive sticks, abrasive cloth, etc. The abrasives have a very large number of hard grains with sharp edges which remove metal from the workpiece surface in such operations as grinding,

2. DIE is a device by which materials are changed into their primary shapes and forms. With the use of dies the parts are manufactured with close tolerances and that they do not need further finishing.

3. The Jigs and the Fixtures are the economical means to produce repetitive type of works by incorporating special work holding and tool guiding devices.

A Jig may be defined as a device which holds and locates a workpiece and guides and controls one or more cutting tools.

A fixture may be defined as a device which holds and locates a workpiece during an inspection or for a manufacturing operation. The fixture does not guide the tool.

24.2 TYPES OF TOOLS

1. Hand tools. These are the tools which are held and used by hands for shaping and sizing a work and they do not have any mechanism chips and small pieces of excess metal are cut using the muscular efforts. These include files chisel, saw etc. These tools are common in carpentry and fitting work.

2. Machine tools. A machine tool is a device provided with a mechanism which enables the cutting tool to operate upon a piece of material in order to produce the desired shape, size and surface quality by cutting away the materials in the form of chips. In machine tools workpieces are held against a cutting tool and the material is removed by relative motion between the workpiece and tool through a deformation process.

3. Cutting tools. A cutting tool is used by holding it through Machine tool, which provide relative motion between workpiece and the cutting tool. These are used either for cutting apart or for removing chips. Relative motion is provided by machine tool by a combination of rotary and translating movements either to the workpiece or to the cutting tool or both.

24.3 CUTTING TOOL MATERIALS

Various types of tool materials that have contributed to increased speeds, feeds and rigidity are high carbon steel, high speed steel, cast alloys, cemented carbides, diamonds and ceramics (cemented oxides), stellite (abrasive grains), boron carbide, etc. These materials are generally used for processes like cutting, shearing, forming, drawing, extruding, rolling, etc. A particular property or a combination of such metallurgical characteristics as hardness, strength, toughness, wear resistance and resistance to heat softening required should be considered before final selection.

1. Characteristics of Ideal Material

(*a*) **Hot Hardness.** Tool material must remain harder than the work material at high operating temperatures.

(*b*) Wear resistance.

(*c*) Toughness.

(*d*) Low frictional coefficient at the chip-tool interface.

(*e*) Cost should be within reasonable limits.

2. Types of Tool Materials

Selection of a tool material depends upon the type of service expected from the tool. Principal cutting tool materials are:

(*a*) Carbon Steel. This have carbon contents between 0.80 and 1.50 percent. This is suitable for temperatures upto 200°C only, and hence used at low cutting speeds, and in hand operated tools.

(*b*) Medium Alloy Steels. These has same carbon contents but in addition have alloy contents of tungsten, molybdenum, chromium and Vanadium upto 5 percent. These alloys improve the performance of carbon steels but lose their required hardness at temperatures from 250° to 350°C. Performance of medium alloy steels can be rated between plain carbon steels and high speed steels.

***(c)* High Speed Steels.** This is a general purpose tool material for low and medium speeds. These operate at cutting speeds 2 to 3 times higher than for carbon steels and retain their hardness upto about 900°C. These are generally either high tungsten or high molybdenum or high cobalt high speed steels. Tungsten provides hot hardness, molybdenum or vanadium provides maintain cutting edge and while cobalt improves hot hardness and wear resistance.

***(d)* Cemented Carbides.** These are composed principally of carbon mixed with other elements, generally tungsten carbide. This is then homogeneously mixed with cobalt and is pressed at very high pressure into suitable blocks and then heated in hydrogen. The blocks are then cut and ground into specific shapes and sintered by high temperature heating in presence of hydrogen. A typical analysis for steel machining is 82 percent tungsten carbide 10 percent titanium carbide and 8 percent cobalt. Carbide tools are made by brazing the formed inserts on the ends of steel holders.

These tools are of very high heat (even upto 1000°C), and wear resistance. They can withstand cutting speed of 6 or more than 6 times higher than the high speed steel tools.

***(e)* Ceramics.** These are made by compacting aluminium oxide powder in a mould at about 280 kg per sq. cm or more and then sintered at 2200°C. These are made in the form of tips and are clamped on tool shanks. These have extremely high compressive strength but are quite brittle, and therefore can not be used for tools operating in interrupted cuts with vibrations and for removing heavy chips. These can withstand temperatures upto 1200°C and can be used at cutting speeds 4 times that of cemented carbides. These are generally used for semi-finish or finish turning of cast iron, plastics etc. and where they are not subjected to impact load. These tools, because of very low heat conductivity are generally used without a coolant.

***(f)* Diamond.** These being hardest known material can be used at cutting speeds of 50 times greater than that for high speed steel tools, and at temperatures upto 1650°C. It has a low co-efficient of friction. These are used for cutting very hard materials like glass, plastics, ceramics and other abrasive materials and for producing fine finishes.

Standardization of Tools

The type, fixed size, quality and measure of the tool are considered under this term. The relationships and values of the tool must be considered according to the scientific procedures. A standardisation committee is constituted among the manufacturing engineers, tool designers, tool engineers, technical persons and suppliers of the tools, which also considers all the standards pertaining to the tools. The main advantages of the standardisation of the cutting tools are given as under:

Advantages

(i) It gives good performance of the tools and maintains good standard of quality of the work.

(ii) It reduces tool inventory, volume of tool and storage space, thereby lowering the cost.

(iii) A bulk purchase of such tools is possible which gives economy of quantity buying.

(iv) As it increases the sources of supply, therefore, tools can be purchased at cheapest possible rates.

(v) The interchangeability of tools reduces down time in changing tools on machines.

(vi) It helps in setting up new facilities in the manufacturing process.

(vii) It allows higher utilization of tools.

(*viii*) It promotes safety and increases operating efficiency.

(*ix*) It provides ease in inspection, grinding and maintenance of the tools.

(*x*) It reduces overall production cost of the work.

24.4 DIE

The device by which materials are changed into their primary shapes and forms is called die. The parts manufactured by ordinary casting, forging, rolling, swaging, drawing etc. need further finishing for final product but with the use of dies the parts are manufactured with close tolerances that they do not need further finishing. If the parts are manufactured with further more close tolerances then the metal powders are placed in a metal mould or die and compacted under great pressure.

Die Casting. Die casting is practised now-a-days, refers to the forcing, by pressure of molten metal into a metal die. Therefore, the term die used in this process implies a metallic mould which is filled under pressure. The two most widely used die casting methods are as under.

(*a*) Hot-chamber method.

(*b*) Cold-chamber method.

Dies for both the methods are similar in construction because there is little difference in the method of holding and operating them. They are made in two sections, to provide means or removing the castings. Metal enters the stationary side when the die is locked in closed position. As the die opens, the ejector plate in the movable half of the die is advanced so that pins project through the die half and force the casting from the cavity of fixed cores. The life of the die depends on the metal to cast and may range from 10,000 to several million fillings.

Press Dies. The tools used in most presses come under the general heading of punches and dies. They have an opening to receive the punch, and the two must be in perfect alignment for proper operation. Punches and die are not interchangeable, but must work together as a single unit.

Classification. Dies may be classified according to either the type of press operation performed or their construction as given under:

(1) *According to construction or method of operation:*

(*a*) Simple (*b*) Compound

(*c*) Progressive (*d*) Transfer

(*e*) Hydraulic (*f*) Rubber

(2) *According to the type of press operation:*

(*a*) *Shearing*—punching, blanking, notching, perforating, trimming, shaving, slitting and lancing.

(*b*) *Bending*—folding, angle bending, curling, forming and seaming.

(*c*) *Drawing*—forming tubes, flanges, embossing, bulging, cupping and reducing.

(*d*) *Squeezing*—coining, sizing, flattening, swaging, cold forging, riveting, upsetting, extruding and hot pressing.

24.5 GAUGES

Mass production requires that all parts be made according to rigid specifications and working drawing. The devices by which the operators and inspectors refer that the standards established by the engineering department are maintained in the shop are called gauges. According to the need several kinds of gauges are used, varying widely in shape and size. They are classified as under.

1. Manufacturing Gauges. Manufacturing or working gauges are those used by the machine operators in the actual production of parts. They are made with very small tolerances, the purpose being to keep the size near the centre of the limits of tolerance.

2. Inspection Gauges. Inspection gauges are those used by inspector in the final acceptance of the product. They are to ensure that the product is made in accordance with the tolerance specifications on the working blueprints.

3. Master gauges. In addition to the above two, a third type of gauge, known as a master gauge, is sometimes used. Such a gauge is merely a reference gauge with which inspection gauges are periodically compared. The equipment used consists of such tools as precision-gauge blocks, measuring machines, optical flats, microscopes and projecting equipment. The classification of the gauges as given under is principally according to the shape or purpose for which is used:

1. Plug
2. Ring
3. Snap
4. Length
5. Form—(*a*) screw thread, (*b*) fillet, (*c*) centre, (*d*) drill point, (*e*) angle, (*f*) gear tooth, (*g*) special contour etc.
6. Thickness—(*a*) Precision gauge block, (*b*) feeler, (*c*) wire.
7. Indicating
8. Air-operated
9. Projecting.

24.6 TOOL ROOM ORGANISATION

The superintendent of production control department is generally made incharge of tool room who has to supervise tool department personnel such as foremen, tool inspectors, clerks, group attendants, labourers and tool crib attendants. The tool department is responsible for manufacturing, storing, issuing and maintenance of tools, jigs, fixtures, dies and gauges. Its organiser should have adequate knowledge and familiarity with tools and manufacturing process.

Tool Storage

Tools are stored in tool cribs which are situated at the centre or nearer to the production department. Tool cribs are specially constructed and consist of racks so that tools may be kept in safe custody and can easily be issued whenever desired.

Location of Tool Cribs. Methods for locating tool cribs are as given under.

(*i*) ***Central tool crib.*** The crib of this system is made very large in size and is located at the central place. The central crib consists of all factory tools which are issued according to the requirements of the other cribs.

(*ii*) ***Sub-tool crib.*** From this crib tools are issued to the different departmental tool cribs and from departmental tool cribs, tools are issued for production purposes. The departmental tool crib-consists of tools required by that department. The tools list may vary from department to department and requisition may be placed to the sub-tool crib whenever additional tools are needed.

(*iii*) ***Movable tool crib.*** This crib contains tools needed by the entire plant. It consists shelves for storing the tools and supplies the tools at site against the requisitions of the departments.

Factors Influencing Location of Tool Cribs. Following are the factors which influence the location of tool cribs:

(*i*) Size and layout of the plant.
(*ii*) Allocation of space to departments
(*iii*) Nature of product.
(*iv*) Methods of operation
(*v*) Grade of operators.

Layout of Tool Crib. A tool crib should have all the abilities that of a good store. The factors and considerations on which crib space depends are as given under:

(*i*) The tool should be stored to an approachable height. In the beginning the tools, should be stored in bottom-most bin then in the.upper side bins of the rack.
(*ii*) The size of the bins should be according to the size of the tools to be stored.
(*iii*) A sufficient aisle space should be left between bin to bin.
(*iv*) The space left between bin to bin should be limited to what is absolutely necessary but should not be so small as to overcrowd them.

Record of Issue of Tools

Following are the advantages of maintaining the records of tools issues:

(*i*) It prevent losses of tools.
(*ii*) It ensures maximum utilization of tools.
(*iii*) It ensures good condition of the tools.
(*iv*) It ensures proper handling and prevents the tools from loss.
(*v*) It maintains the history of the tools and guides for further purchasing.
(*vi*) Broken or damaged tools can be replaced immediately with the proper records.

Method of Issuing Tools

The method of issuing tools are described as under:

(1) *Single-check System.* In this system each worker is provided with a brass check on which his clock number is stamped. When the worker desires to have a tool then he hands over his check and gets a tool or tools in exchange. The crib attendant hangs the check on a hook by the place where the tool is kept. It shows that which worker has taken particular tool. The check is returned when the worker deposits the tool. It is very simple but disadvantage is that particulars of worker and tool are not recorded in this system,

(2) *Double-check System.* In this system first check serves the purpose similar to that of the single check system. The second check showing the number of tools issued is hung against the hook number of the worker on a board. This system has also drawback that it does not show that when or for what job the tool has been issued to the worker.

(3) *Triplicate Tool-slip System.* In this system, when a worker wants to draw a tool he fills out a tool-order or tool-slip in triplicate with one writing by means of colour insets. The tool-slip is printed with all the particulars such as:

(*a*) Tool order
(*b*) Date
(*c*) Department where the tool is to be used

(*d*) Bin number
(*e*) Clock number of the worker
(*f*) Quantity of tool issued
(*g*) Size of tool
(*h*) Description of tool
(*i*) Signature of the worker etc.

The three copies are of three distinctive colours. Generally the original is made white, the duplicate yellow and the triplicate of pink colour. One set of slips is used for one tool order and one item only.

The crib attendant issues the tool and returns the yellow copy of the slip to the worker. The white copy is retained by the crib attendant as a receipt and hung it in employee's compartment under the clock number of that worker. The pink copy is filed in the tool compartment which is provided in lock system. All the white copies of the tool orders are fixed on a wall board from where it is visible at a glance that which worker has issued how many tools.

When a workman returns his tool, he also returns the yellow copy of the tool order. If the tool is in good condition, the crib attendant removes the white copy from the compartment and hands over it to the worker, then he also removes the pink copy from the tool records, it means that the tool has been returned.

When the tool returned by the worker is broken, then he has to submit an "explanatory report" approved by the foreman with that tool. The explanatory report specifies the cause of the damage. If the cause is due to carelessness of the worker then action has to be taken against the worker and tool is sent for repairs. After long uses when the tool becomes unserviceable then report is received from production department and tool is declared as scrap and is removed from the stock.

Advantages of Triplicate Tool-slip System.

(*a*) Number of tools received by a worker can be counted.
(*b*) Total number of tools of particular type can be counted.
(*c*) Time of issue of tool and physical position of the tool can be known.
(*d*) The inventory card for each tool is possible.
(*e*) As the explanatory report comes time to time, therefore, durability of the tools can be determined.
(*f*) Slow moving tools can be detected and removed accordingly.
(*g*) Responsibility of the damaged tool can be fixed easily.
(*h*) With due care and proper handling of the tools, the general efficiency of the plant may be raised.
(*i*) Tools are easily indicated when they should be repaired, replaced and declared as a scrap.
(*j*) The system indicates to the purchase department to purchase new tools.
(*k*) Information regarding tool transactions can be given to the accounts department with this system.

(4) *Electric Machine System.* This system consists of an electric machine by which tool order is printed in duplicate. When a workman draws a tool, he signs the original tool order and submits both the orders to the tool crib attendant. The tool crib attendant then issues the tool to the worker and files the original order in the tool record file under the worker's clock number. The duplicate is placed in the bin from which the tool is withdrawn.

When the tool is returned by the worker in good condition then the signed original copy of the tool order is returned to him. The duplicate copy is retained there and is used for further analysis of works. If the tool is broken then the worker returns the tool with broken tool report approved by the foreman. Under this condition the original copy is returned to the worker and the duplicate together with the broken tool report is filed until the case is decided.

Control of Tool Inventory

This consists of control on procurement, quantities and disbursement of tools to different tool cribs. The main records of the tool inventory system are as given under:

(*i*) Tool requisition
(*ii*) Tool inventory ledger
(*iii*) Bin cards.

24.7 JIGS AND FIXTURES

1. Jig. It is a device which holds and locates a work piece and guides and controls one or more cutting tools. A jig consists of a plate, a structure or a box having provisions for holding the components in identical positions one after other and then guiding the tool in correct position, so as to complete work according to the drawing and specifications. Jigs are made of high carbon steel.

2. Fixture. It is a device which holds and locates a workpiece during an inspection or for a manufacturing operation. The fixture consists of work holding devices, which are clamped on the machine table to hold the work in position when it is being worked upon by tools. Fixtures are generally fabricated by using mild steel and grey cast iron.

Objectives of Jigs and Fixtures:

(*i*) Accurate and cheap mass production of items.
(*ii*) Saving of time in handling workpiece and tool.
(*iii*) Semi-skilled workers can be engaged on the job.
(*iv*) Convenience in handling the material.

Differences between a Fixture and a Jig:

(*i*) Fixture holds and positions the work but does not guide the tool, while a jig holds, locates and also guides the tool.
(*ii*) Fixtures are usually heavier and are bolted rigidly on the machine table, while the jigs are lighter for quick handling.
(*iii*) Fixtures are generally used for holding work on shaper, planner, lathe, milling and grinding machines, while the jigs are used for holding the work and guiding the tools in drilling, tapping, boring and reaming machines.

Advantages of Jigs and Fixtures:

(*i*) They eliminate the marking out, measuring and other setting methods before machining.
(*ii*) They enable production of identical parts which are interchangeable.
(*iii*) They increase the machining accuracy of the workpiece.
(*iv*) They increase the production capacity of the man and machine.
(*v*) They reduce the operator's labour and consequent fatigue to him.
(*vi*) They enable semi-skilled operator to perform the operation, which saves labour cost.

(*vii*) They minimise the expenditure on the quality control of finished products.

(*viii*) They reduce the overall cost of machining by fully or partly automatising the processes.

Principles of Jigs and Fixtures Design

Following are the essential factors which must be considered in designing a jig or a fixture:

(*i*) Study of the workpiece.

(*ii*) Study of the machine tools.

(*iii*) Study of the locating elements.

(*iv*) Loading and unloading system.

(*v*) Clamping arrangement.

(*vi*) Power devices for clamping the workpiece.

(*vii*) Clearance required between the jig and the workpiece.

(*viii*) Indexing system and devices.

(*ix*) Tool setting and guiding elements.

(*x*) Fool-proofing arrangement.

(*xi*) The ejecting systems.

(*xii*) Swarf removal system.

(*xiii*) Rigidity and vibration proofing.

(*xiv*) Table fixing arrangement.

(*xv*) Safety devices.

***(i)* Study of workpiece.** Workpiece is studied for deciding the sequence of operations to be performed and designing the jigs and fixtures. One of the surfaces of work piece is first machined to consider it as datum to measure all, other dimensions.

***(ii)* Study of machine.** Before designing the jigs and fixtures, important particulars of machine like its capacity, table size, maximum length of travel, etc. must be known.

***(iii)* Study of locating elements.** Correct location is necessary for accuracy of the final product. The jigs and the fixtures are designed in such a way that all possible movements of the workpiece are restricted. Locating and clamping points are decided to restrict the movement of workpiece in any direction when set at correct position.

***(iv)* Loading and Unloading.** Jigs and fixtures are designed in such a way that it is easy to fix up the workpiece and then remove it, before and after performing the job without spending much time.

***(v)* Power devices for clamping.** Jig and fixture clamps are operated either by hand or by power small, workpieces clamping require limited pressure and hence usually operated by hand. Pneumatic or hydraulic power is required to operate large clamps where high pressure is required for clamping. These are quick, controllable, reliable and easy to operate.

***(vi)* Clearance in a jig and fixture.** In order to accommodate variable sizes of jobs produced by casting or forging, some clearance is provided between jig body and workpiece.

***(vii)* Indexing System.** In order to perform angular operations, indexing arrangement is incorporated in a jig.

***(viii)* Tool setting and guiding arrangement.** Cutter is required to be set relative to the work in a fixture. The tools are guided in jigs by drill bushes fitted in the jig plates.

***(ix)* Fool proofing arrangement.** Jig and fixture must be designed in such a way that it becomes impossible for an operator to insert the workpiece into the jig or a fixture in any position other than the correct one.

***(x)* Ejecting system.** Jigs and fixtures should be designed in such a way that heavy workpieces are ejected mechanically when unclamped. This is possible by introducing wedges, spring loaded or can actuated plungers, which not only saves time but also labour.

***(xi)* Swarf removal.** Jigs and fixtures must have arrangement for swarf removal, as it effects accuracy of location.

***(xii)* Rigidity and vibration proofing.** These should be made sufficiently rigid to take up the cutting load without any deformation. These should also not have vibration when cutting operation is in process.

***(xiii)* Table fixing arrangement.** These should be designed in such a way that they can be attached to machine table.

***(xiv)* Safety devices.** These are designed to assure full safety to the operator. For example, sharp edges are rounded and revolving parts are guarded.

***(xv)* Clamping.** Clamping is necessary to hold the workpiece securely on the jigs or fixtures against the cutting forces.

Factors considered for efficient clamping:

(i) Cutting pressure to be held by the clamps.

(ii) Clamping pressure and whether this can be withstood by work.

(iii) Movement of the clamp while loading and unloading the work.

(iv) Easy removal of the clamp while unloading the work.

(v) Clamp should be simple and foolproof.

(vi) Clamps should not obstruct the operations.

(vii) It should be case hardened to avoid wear of the clamping faces.

(viii) Clamp should be strong enough.

Types of Clamps. Commonly used clamps are:

(i) Flat clamp,

(ii) Screw clamp,

(iii) Pivoted clamp.

(iv) Latch clamp,

(v) Double acting clamp.

(vi) Equalising clamp.

(vii) Cam clamp,

(viii) Swing plate clamp.

(ix) Wedge clamp.

Types of Jigs

Type of jig depends upon the type of work to be machined and its quality required. Some simple drill jigs are:

***(i)* Templet Jig.** A plate having holes at the desired positions serves as a templat which is fixed on the workpiece to be drilled. The drill is guided through these holes of the templet to drill the required holes on the workpiece.

***(ii)* Plate Jig.** This is the improvement over templet jig. In this bushes are also provided in the holes to guide the drills.

***(iii)* Diameter Jig.** These are used to drill radial holes on a cylindrical or spherical workpieces. The work is clamped on a fixed V-block.

***(iv)* Channel jig.** This jig has a channel like cross-section, and the work is fitted within the channel.

***(v)* Ring jig.** These are employed to drill holes on circular flange parts. The work is clamped on the drill body and then the holes are drilled by guiding the tool through the drill bushes.

***(vi)* Box jig.** This has a box type construction and locates the workpiece in it. These are generally used to drill a number of holes on the workpiece from different angles.

***(vii)* Leaf jig.** This has a plate or leaf hinged on the body. Leaf may be swung open or closed on the work for loading and unloading purposes.

QUESTIONS

24.1 Define the terms, "Jigs" and "Fixtures". Explain the objectives of jigs and fixtures.

24.2 What are the essential factors which should be considered in designing a jig or a fixture?

24.3 Write short notes on: tools, clamps, jigs, fixtures

24.4 (*a*) List different types of jigs and fixtures. (*b*) Explain different types of clamps.

24.5 (*a*) Differentiate between hand tools, machine tools and cutting tools.

(*b*) What characteristics are required from a cutting tool material?

24.6 List principal cutting tool materials. Explain their suitability and application.

24.7 Explain the following:

(*a*) Die (*b*) Die-casting

(*c*) Press Dies

24.8 What are the "gauges"? What are its advantages? Explain different type of gauges giving their place of application.

24.9 Write short notes on:

(*a*) Tool room, (*b*) Tool crib,

(*c*) Tool issue system.

Chapter 25

MACHINE TOOLS

Learning Objectives

After studying this chapter, the students will learn:

- Machine tools, their functions and classifications.
- Operations performed on machine tools.
- Cutting speed, feed and depth of cut.
- Limits, fits and tolerances.
- Interchangeable manufacture.
- Lathe machines, their types, construction, working their operations and tools.
- Capstan and turnet lathes, their construction and operation.
- Drilling and boring machines, their construction, types and operations.
- Shaping and slotting machines, component parts and uses.
- Planing machines, its parts, construction, uses and functions.
- Milling machines, its types and uses.
- Grinding machines and types.
- Chip formation, their types and shapes.

25.1 INTRODUCTION

A machine tool is a device by which jobs are shaped in desired sizes by removing the excess material in the form of chips. The main purposes of a machine tool are:

(*a*) to produce certain form,

(*b*) to produce desired finishing of job surfaces.

A machine tool can be defined as a power driven device in which energy is utilized for metal cutting operations. It is capable of holding and supporting the job to be worked and at the same time direct and drive the cutting tool or job or both to perform a definite metal removing operation in order to produce a desired shape of predetermined size. Every machine tool is expected to remove the metal in the form of chips large or small in size, all the machine tools operate on either a rotary type or reciprocating type principle.

There are numerous machine tools used in industries and numerous items are made in the industry by a machine tool by various operations. A machine tool can perform various operations according to the need. Special purpose machine tools are employed for doing certain special jobs. Few examples of machine tools are Lathe, Shaper, Planer, Drilling machine, Milling machine, and Grinding machine.

25.2 CLASSIFICATION

1. According to the Surface Produced

***(i)* Machine tools for cylindrical work.** In this case, either work or tool is rotated and any of the two is traversed for performing operation. Examples of such machine tools are lathes, turret, drilling machine, boring machine, cylindrical and surface grinders.

***(ii)* Machine tools for flat surface work.** In this case, either work or tool is reciprocated and any of the two is traversed in straight path in perpendicular direction after completing the previous operation. Examples of such machine tools are shaper, planer, sloter, broaching machine etc.

The flat surface can also be produced by multi-edged rotating tools with the translatory motion of the job. Example of such machine tools are milling machine, surface grinders etc.

2. Based on the Purpose for which it is used

***(i)* General purpose machine tools.** These machines include lathes, shapers, drill presses, milling machines, grinders, planers, sloters etc., because they are used for general operations for cutting the metals.

***(ii)* Production machine tools.** These are similar to the general purpose machines but consist of multi-station tooling system and designed at a time for one type of job. These include capstan and turret lathes, semi-automatic lathes, production milling machines, multi-head drilling machines etc.

***(iii)* Special purpose machine tools.** These are single purpose and used for mass production. One machine is capable of producing only one type of job. These include gear generators, camshaft grinders, piston turning lathes, thread rolling machines etc.

25.3 OPERATIONS PERFORMED ON MACHINE TOOLS

The main operations which are performed on various kinds of machine tools are as given under:

1. Turning. It is the operation of producing a cylindrical surface on a rotating job.

2. Taper turning. It is the operation of producing a conical surface on a rotating work piece.

3. Shaping. It is operation of producing flat metallic surface by means of single pointed moving tool and the job is fixed on the table.

4. Planing. It is the operation of producing flat surface by stationary single pointed tool and the job is fixed on a reciprocating table.

5. Slotting. It is an operation of producing vertical slot by vertically moving single pointed tool, while the work is kept stationary on the table.

6. Drilling. It is an operation of producing a round hole by pressing a rotating drill bit against the job.

7. Boring. It is the operation of accurately enlarging a hole already drilled in the job.

8. Reaming. It is the operation of accurately enlarging the hole already drilled, by means of a reamer.

9. Screw cutting. It is the operation of producing external threads on the surface of a cylindrical job.

10. Tapping. It is the operation of producing threads on the interior surface of the job, by means of taps.

11. Counter boring. It is the boring of the part of depth of the existing hole.

12. Counter sinking. It is the operation of enlarging to conical shape the end of the existing hole.

13. Grinding. It is the operation of removing the metal of the job by means of a grinding wheel.

14. Honing. Honing is a grinding process mostly for finishing round holes by means of bounded abrasive stones, called hones.

15. Lapping. It is a process to polish or improve surface quality, geometrical precision and dimensional accuracy by means of abrasive materials.

16. Polishing. It is a surface finishing operation performed by a polishing wheel for removing the scratches, tool marks, pits and other defects from rough surfaces.

17. Milling. It is the operation of removing metal by feeding the work against a rotating cutter having multiple cutting edges.

18. Gear cutting. It is the operation of producing teeth on a wheel.

19. Rack cutting. It is the operation to produce teeth on a wheel of infinite radius, *i.e.*, on a straight line instead of a circular surface.

20. Hobbing. It is a process of generating a gear by means of a cutter, called a hole, that revolves and cuts like a milling cutter.

21. Knurling. It is the process of embossing a diamond shaped pattern on the surface of a workpiece. Purpose of which is to provide an effective gripping surface on it to prevent from slipping when operation by hand.

22. Forming. It is the process of producing a convex, concave or any irregular shape on the surface of a cylindrical job by means of turning operation.

23. Facing. It is the operation performed on lathe to finish the ends of the work, to make the ends flat and smooth and to make the piece of the required length.

24. Parting off. It is the operation of cutting a workpiece after it has been machined to the desired size and shape.

25. Broaching. It is a method of removing metal by pushing or pulling a cutting tool called a broach which cuts in a fixed path. Surfaces finished by broaching may be flat or contoured and may be either internal or external.

26. Punching. It is the process of producing holes, generally cylindrical by using a punch over a cylindrical die.

25.3.1 Cutting Speed, Feed and Depth of Cut

***(i)* Cutting speed.** The cutting speed of a cutting tool may be defined as the speed at which the cutting edge passes over the material. Cutting speed is ordinarily expressed in metre, often referred to as surface speed in metre per minute.

***(ii)* Feed.** The feed of a cutting tool is the distance the tool advances into or along the workpiece each time the tool point passes a certain position in its travel over the surface. In the case of turning on a lathe, the feed is the distance that the tool advances in one revolution of the workpiece.

***(iii)* Depth of cut.** It is the thickness of metal removed in one cut and is measured by the perpendicular distance between the machined and non-machined surface expressed in mm.

25.4 LIMITS, FITS AND TOLERANCES

In mass production, due to certain reasons to manufacture the parts with exact dimensions and absolutely identical is not possible. Therefore, success in mass production completely depends upon the inter-changeability of the produced parts. The main technical terms used in production and assembly of the parts are as given under:

***(i)* Nominal size.** This is the term given to a dimension for a part, which is a close approximate to the standard size, say 50 mm hole although hole may be (50 ± 0.01) mm but its nominal size will be 50 mm.

***(ii)* Basic size.** This is the theoretical size from which all limits for that dimension are made.

***(iii)* Actual size.** This is the actual measurement on a particular part.

***(iv)* Limits.** Limits of size are the extreme permissible dimensions of a part.

***(v)* Limits of size.** By the application of a tolerance to a design size two sizes are calculated between which the actual size may be permitted to vary. These two sizes are called limits of sizes. Such as $50^{+0.01}_{-0.01}$ mm then limiting sizes are 50.01 mm (high limits) and 49.99 mm (low limits).

***(vi)* Tolerance.** Tolerance may be defined as the difference between the maximum and minimum sizes allowed or the difference between the high and low limits. Tolerance is calculated by subtracting the lowest size from the largest permissible size, thus

$$50.00^{+0.01}_{-0.01} = \begin{matrix} 50.01 \text{ mm max} \\ 49.99 \text{ mm min} \end{matrix}$$

and the tolerance will be 50.01 – 49.99 = 0.02 mm.

***(vii)* Allowance.** It is the difference between the hole (female part) and mating shaft (male part) to obtain a particular class of fit.

For example, if a hole is dimensioned $50.00^{+0.01}_{-0.01} = \begin{matrix} 50.01 \text{ mm} \\ 49.99 \text{ mm} \end{matrix}$ and shaft which is to fit dimension 50.00 – 0.02 mm, then the maximum allowance is 50.01 – 49.98 = 0.03 mm.

Positive allowance. Under this allowance, the shaft is smaller than a standard hole, and is used in Running fit.

Negative allowance. Under this allowance, the shaft is larger than the hole; and is used in Force fit.

Maximum allowance. Largest hole size – smallest shaft size.

Minimum allowance. Smallest hole size – largest shaft size.

***(viii)* Tolerance unilateral system.** In this system the tolerance is given in one direction only, from the basic size and is usually positive, such as $50^{+0.01 \text{ mm}}_{-0.00 \text{ mm}}$

It means, H = 50.01 mm and L = 50.00 mm.

***(ix)* Tolerance bilateral system.** In this system the limits are extended on both sides of the basic size, such as $50^{+0.01 \text{ mm}}_{-0.01 \text{ mm}}$

It means, H = 50.01 mm and L = 49.99 mm.

where, H = High limit and L = Low limit.

***(x)* Fit.** This is the relationship existing between the mating parts, which governs the amount of play which will remain in the parts while assembling.

There are three principal types of fit:

(*a*) *Interference* in which the shaft is larger than the hole ;

(*b*) *Transition* where the shaft and hole are approximately of the same size, and

(*c*) *Clearance* where the hole is larger than the shaft.

25.5 STANDARD GAUGES

Production work requires speed as well as accuracy in measuring parts. Measuring instruments can be used for this type of work, but many of them are more elaborate than necessary and require adjustments for each individual reading thus wasting the time.

Therefore, to attain the quick measurements required in production work, a measuring device of fixed shape and size is used which is called "standard gauge". Since its use is limited to one dimension and no adjustments are required the operation of inspecting a part requires a minimum of time.

Classification:

Broadly gauges may be classified such as:

1. **According to the Use**

(*a*) **Manufacturing or working gauges.** They are used by the machine operators in the actual production of parts.

(*b*) **Inspection gauges.** They are used by inspectors in the final acceptance of the product.

(*c*) **Master gauges.** These gauges are reference gauges by those inspection gauges are periodically compared.

2. **According to the shape and purpose** for which they are used, gauges are classified as under:

(*i*) **Plug gauges.** A plain plug gauge is an accurate cylinder used as an internal gauge for comparing the holes. These gauges may be either single or double ended. A double-ended, plain gauge has GO and NO GO members assembled on opposite ends, whereas progressive gauge has both gauging sections combined on one end.

(*ii*) **Ring gauges.** They are like hollow cylinders and used to check the diameter of cylindrical parts. A plain, knurled surface of the gauge indicates a "GO" gauge, whereas a "NO GO" gauge is identified by an annular groove on the periphery.

(*iii*) **Taper gauges.** They are made in both the plug and ring styles and, in general, follow the same standard construction as plug and ring gauges. Taper gauges are not dimensional gauges but rather a means of checking in terms of degrees or slope. If size is involved, "GO" and "NO GO" tolerances are indicated on the ends of the gauge.

(*iv*) **Thread gauges.** They are similar to the standard plug and ring gauges but provided in addition with external and internal threads for inspection purposes.

(*v*) **Snap gauges.** These are also known as GO and NO GO gauges. These are used for checking diameters, length and thickness of the parts. The snap gauge may be adjustable also. The size is stamped on the gauge.

(*vi*) **Gauge blocks.** These are rectangular steel blocks machined and lapped after proper hardening. These are, also called slip gauges. These surfaces are so highly polished and true that when rubbed together these will stick together. This rubbing and sticking is called as wringing. Gauge blocks are used for checking snap gauges, comparators and other precision dimensions.

Gauge blocks are available in sets. The most widely set consists of 81 pieces in inch standard and 103 pieces in metric standard.

25.6 INTERCHANGEABLE MANUFACTURE

This term implies that the parts which go into the assembly of the machine can be selected at random from a large number of parts. In such a system of manufacture, selective fitting is unnecessary except possibly for special close-working parts. To make this possible manufacturing methods must be standardized and limits of accuracy specified in details. But extreme accuracy is not always necessary or desirable since manufacturing costs increase greatly as working limits become closer.

Advantages of Inter-changeability

1. Low production cost.
2. Special purpose machines are used. Therefore, product is of great accuracy and interchangeable also.
3. Interchangeable parts can be assembled easily without any delay.
4. Product ensures desired quality of work.
5. Product can be stocked, therefore, seasonal fluctuation of market does not affect the cost of the parts.
6. Due to uniform rate of production, the system and capacity of the plant is improved.

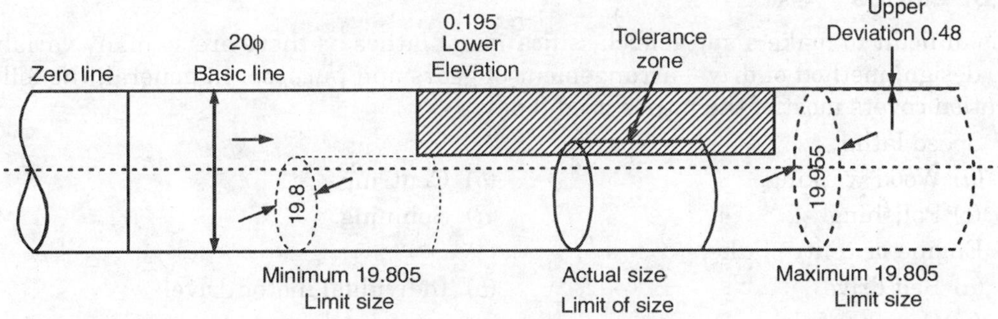

Fig. 25.1 Dimensioning.

7. Ease in performing the repair works since the spare parts remain available always.
8. Increases overall profit to the plant.
9. One spare part is sufficient for variety or number of works.

Terminology (As **per ISO System IS: 919 – 1963)**

(*i*) *Basic Size.* For convenience, a basic size is usually a whole number (*e.g.,* 25,50,60, etc.) and is given to represent the dimension of a part. It is the design size. The allowable maximum and minimum limits of the dimensions of the part are fixed in relation to the basic size. Basic size relates to the position of zero line.

(*ii*) *Upper Deviation.* It is the algebraic difference between the maximum limit and the corresponding basic size.

(*iii*) *Lower Deviation.* It is the algebraic difference between the minimum limit and the corresponding basic size.

(*iv*) *Tolerance.* It is not feasible to make a part precisely to a given dimension due to limitations of cost and inaccuracies of manufacturing methods. It may only be made to lie between two limits, maximum and minimum. Tolerance is the difference between the maximum and minimum limits of sizes. Also, it is equal to the algebraic difference between the upper and lower deviation.

(*v*) *Actual Size.* It is the size of a part as may be found by measurement.

(*vi*) *Zero Line.* It represents the basic size in a graphical representation of limits and fits. Upper and lower deviations are referred to zero line. Positive deviations are shown above and negative deviations below the zero line drawn horizontally.

(*vii*) *Tolerance Zone.* While representing tolerance graphically, zone bounded by the two limits of size of the part (shown by hatched line in the above fig. 25.1) is called the tolerance zone.

25.7 LATHE MACHINE

It is the basic producer of cylindrical jobs. The work is rigidly held in a chuck which is made to revolve while the cutting tool is fed against it. The main movements of the tool are for sliding, surfacing and screw-cutting. "Sliding or Traverse" is meant the longitudinal motion of the cutting tool along the axis of the job. "Surfacing" is meant motion of the cutting tool at right angle to the sliding motion, across the axis of the job.

When a lathe is provided with "Lead Screw" for screw cutting and having power driven both the motions, the sliding as well surfacing, then it is called self-acting lathe.

Types of Lathes

It is difficult to make a suitable classification of lathes as there are so many variables in the size, design, method of drive, arrangement of gears and purpose. In general, the following classification covers most of the lathes used today.

1. Speed lathe.
   (*a*) Wood working. (*b*) Centring.
   (*c*) Polishing. (*d*) Spinning.
2. Engine or centre lathe.
   (*a*) Belt drive (*b*) Individual motor drive.
   (*c*) Gear head drive.
3. Bench lathe.
4. Tool room lathe.
5. Capstan and Turrent lathe.
6. Special purpose lathe.
   (*a*) Wheel lathe. (*b*) Gap bed lathe.
   (*c*) T-lathe. (*d*) Duplicating lathe.
7. Automatic lathe.

***(i)* Speed lathe.** It is a simple power driven machine tool and mostly used for wood turning. Tools are hand operated. Different speeds of the machine are obtained by cone pulley system.

***(ii)* Engine lathe.** It is a general purpose machine tool used in workshop. Tools are mechanically operated. Different spindle speeds are obtained by means of cone pulley as well as by geared head system.

***(iii)* Bench lathe.** It is a small lathe machine mounted on work bench to machine small and light jobs.

***(iv)* Tool room lathe.** This lathe is provided with all sorts of accessories to machine precision jobs.

***(v)* Capstan and turret lathe.** These are production lathe used to manufacture any number of identical pieces in the minimum time. Several different tools may be set on a revolving turret to facilitate in doing several operations on a job.

***(vi)* Special purpose lathe.** They are designed and named after a particular type of job performed on them.

***(vii)* Automatic lathe.** These are high speed, heavy duty, mass production lathes with complete automatic control. Once the tools are set and the machine starts it performs automatically all the operations to finish the job.

Lathe Specifications

A lathe machine is specified as under:

(i) **Length between centres.** It is the maximum length of the job that can be mounted between the lathe centres.

(ii) **Length of bed.** It is overall length of bed including spaces of head and tail stock.

(iii) **Height of centres.** It is the vertical distance of the centres measured above the lathe bed.

(iv) **Swing diameter over bed.** It is the largest diameter of job that can revolve without touching the lathe bed.

(v) **Swing diameter over carriage.** It is the largest diameter of job can revolve over the lathe saddle.

(vi) **Maximum bar diameter.** It is the maximum diameter of the bar stock which can pass through the hole of the head stock spindle.

(vii) **Bed gap.** It is a cut provided in the lathe bed to turn a job of large diameter.

(viii) **Other particulars.** Such as width of bed, spindle nose diameter, centre taper morse number, spindle speeds, number of feeds, metric and English threads, lead screw, power input, floor space etc.

Main Parts of a Lathe

The main parts of a lathe are as described under and shown in fig. 25.2:

1. Bed. It is made of high grade cast iron and this forms the main frame of the machine. The upper surface is plained true with its edges parallel and a parallel slot in the centre. It supports head stock, tail stock and the carriage.

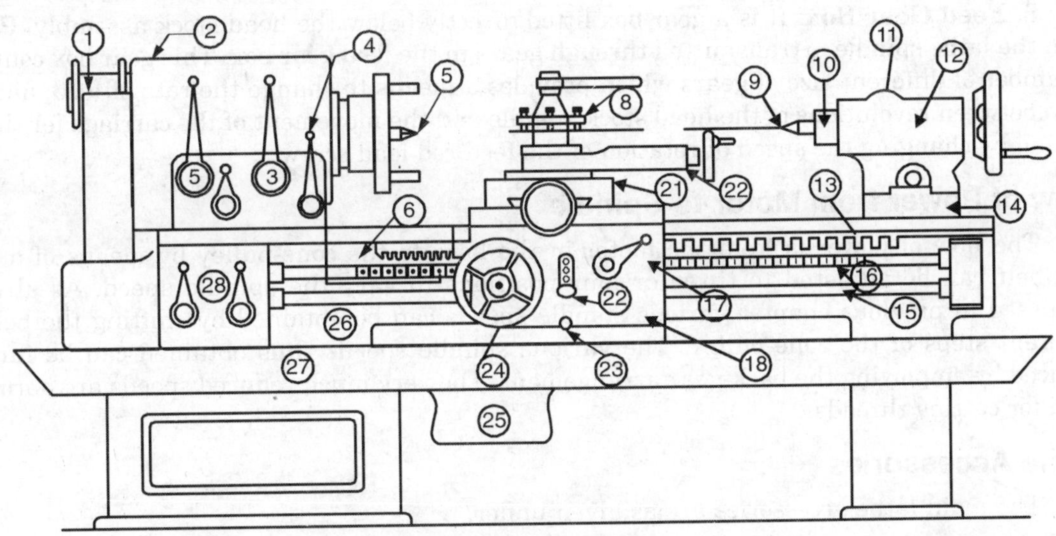

Fig. 25.2 The lathe.

1. Main driving pulley, 2. Head stock, 3. Gear box controls, 4. Driving clutch, 5. Live centre, 6. Gap piece, 7. Thread dial, 8. Tools post, 9. Dead centre, 10. Ram, 11. Ramp clamp, 12. Tailstock, 13. Rack, 14. Tailstock base. 15. Feed rod, 16. Lead screw, 17. Screw cutting engage, 18. Carriage, 19. Cross slide hand wheel, 20. Tools post slide, 21. Cross slide, 22. Reverse for sliding and surfacing, 23. Feed engages, 24. Head traversing wheel, 25. Studs sump, 26. Bed, 27. Tray, 28. Feed box.

2. Head Stock. It is mounted on the left end of the bed and carries the spindle with its driving mechanism. The front end of the spindle is threaded or flanged to attach a chuck. The spindle is made of forged steel and bored taper to suit various lathe centres.

3. Tail Stock. It is a casting in the upper part of which is a screw-operated spindle, the axis of which is at the same height above the bed as the axis of the lathe spindle. It is so fitted that it can be moved along the bed and clamped in any desired position. Its spindle is hollow and the base is tapered to fit lathe centres. It can be offset to turn taper jobs between centres. A friction lock is fitted to lock the spindle in any desired position.

4. Carriage. Devices for controlling the motions of tool are included in this section. The portion which takes on it the cross-slide and compound rest and slides along the bedways, is called saddle and the one which covers the controls for hand and power feed of tool and thread cutting controls, is called apron. The movement of the carriage is controlled by hand wheel and gears carried by the apron.

5. Cross Slide. It is a small table which can be moved across the top of the saddle by means of a screw carried on the upper half, operated by a hand wheel. The automatic feed gears are arranged to operate the cross slide. On the upper side of the cross slide, there is a circular block with a graduated scale in degrees and can be set at any desired angle. On its upper part a tool post is mounted in which the tool can be held by means of a plate and a stud.

6. Lead Screw. It is accurately machined from a special mild steel bar. It is used to guide the carriage at the time of screw cutting. It is geared from the head stock. A half nut is engaged to get the movement.

7. Feed Shaft. It is a shaft with a long keyway, fitted in front of the lathe and controls the automatic feed of the saddle and cross slide.

8. Feed Gear Box. It is a gear box fitted directly below the head stock assembly. Power from the lathe spindle is transmitted through gears to the feed gear box. This gear box contains a number of different size of gears which provides a means to change the rate of feed, and the ratio between revolutions of the head stock spindle and the movement of the carriage for thread cutting by changing the speed of rotation of the feed rod lead screw.

Flow of Power from Motor to Spindle

The flow of power is from the motor in the base to the cone-pulley by means of a belt. The belt can be adjusted to three or four positions to vary the spindle speed. As already discussed in previous chapter various spindle speeds can be obtained by shifting the belt on different steps of the cone-pulley. The various spindle speeds thus obtained can be further reduced by employing the back-gear arrangement. The very much reduced speeds are normally used for cutting threads.

Lathe Accessories

The main lathe accessories are as given under:

(1) Lathe Centres

(*a*) *Ordinary centre.* This is used to support jobs which are centre drilled. It has a morse tapered shank and a 60° cone.

(*b*) *Half centre.* A portion of the cone is removed to permit the tool to turn small diameter.

(*c*) *Square centre.* It is used when centre drill is not available. A small hole is drilled to start with, and a square centre is used afterwards to make a seating.

(*d*) *Cone or revolving centre.* It is used to support hollow jobs, pipes etc. The shank is inserted in the tail stock spindle and the cone revolves with the job.

2. Chucks

(*a*) *Three jawed or self-centring chuck.* The jaws of this chuck hold the piece correctly and is used for holding finished jobs or metals having smooth surfaces.

(*b*) *Four jawed or independent chuck.* This is used for heavy and odd jobs. Each jaw is independent of the other.

(*c*) *Collet chuck.* It is used for holding a piece with a finished surface concentric. A set of collets is provided and the range is limited.

(*d*) *Hydraulic power chuck.* In this chuck, clamping medium is operated by a hydraulic press.

(*e*) *Magnetic power chuck.* In this, the device used is an electromagnet. It is used for holding ferrous metals, mostly for grinding operations.

3. Face Plate

Jobs which cannot be held by the above chucks are held by a face plate with clamps, bolts and nuts. It is necessary to put the balance weights while using this.

4. Driving Plate

It is used for small and precise operations on bars supported between centres, with the help of a lathe carrier.

5. Steadies

(*a*) *Three pointed steady (steady rest).* It is fixed on the bed at the required position and the job is supported by the fingers.

(*b*) *Two pointed steady (Follower rest).* It is fixed on the cross slide and the job is supported by the fingers behind the cutting tool. It travels with the saddle.

6. Change Wheels

A set of 23 gears is provided with a screw cutting lathe to cut different threads. In most of the lathe, the number of gear teeth on the gears ranges from 20 to 120 in steps of 5.

7. Mandrels

It is a hardened piece of round bar used for holding the bored job on a lathe for the purpose of turning the outside surface. It is necessary that the hole in the job must be concentric with the outside surface. Parts like gear blanks, pulleys, bushes etc. are turned with the help of the mandrels.

Methods of Holding the Work on a Lathe

Following are the most important methods of holding a job on the lathe, option of which depends upon the nature of work, nature of operation being performed etc.

(*i*) Holding work in chuck.
(*ii*) Holding work on a face-plate.
(*iii*) Holding between centres.
(*iv*) Holding work on a mandrel.
(*v*) Holding the work on carriage while the tool in chuck.

25.8 LATHE OPERATIONS

Operations which are performed on a lathe either by holding the workpiece between centres or by a chuck are:

1. Straight turning.
2. Shoulder turning.
3. Taper turning.
4. Eccentric turning.
5. Chamfering.
6. Thread cutting.
7. Facing.
8. Knurling.
9. Filing.
10. Polishing.
11. Grooving.
12. Spinning.
13. Spring winding.
14. Forming.

Operations which are performed by holding the work by a chuck or a face-plate or an angle plate are:

1. Drilling
2. Reaming
3. Boring
4. Counter boring
5. Taperboring
6. Internal thread cutting
7. Tapping
8. Undercutting
9. Parting-off.

Operations which are performed by using special attachments are:

1. Grinding
2. Milling.

Turning Long Selender Work

When a long selender workpiece is turned on lathe, it tends to bend at some intermediate point because of its own weight and vibrations caused due to the cutting force that acts on it. To overcome this difficulty when the length of the selender workpiece is 10 to 12 times the diameter rests are used. Two types of rest used in an engine lathe are the steady and the follower rest. When properly set, they provide greatest accuracy in machining and permit heavier depth of cut on the long selender workpiece.

Screw Cutting on a Lathe Machine

It is one of the most important operations on a lathe. It is essential to check that the train of change wheels is correct and that the tool has got the same angle as the form of thread. Tool must be at the centre of the work. A proper lubricant must be used. Speed is to be considered while preparing a lathe for screw cutting. The main formula used for finding the correct number of teeth of the driver and driven gears is as follows

$$\frac{\text{Driver}}{\text{Driven}} = \frac{\text{T.P.I. on the lead screw}}{\text{T.P.I. to be cut}} \quad \text{where T.P.I = Treads per Inch}$$

As it is not possible sometimes to obtain the ratio by means of a simple train of change wheels, a compound train is needed which has two driver wheels and two driven wheels. The formula for obtaining the ratio is still the same.

Screwing in the Lathe by Taps and Dies

Taps and dies are used principally for the manual production of internal and external threads respectively, although with proper mounting they may also be used in machine threading. In a lathe, the work is mounted on a chuck or on face plate and revolved at a very slow speed. A tap or die of required size held on a special fixture is mounted on the tailstock spindle. The axis of the tap or die should coincide exactly with the axis of the work. The tap or die will automatically feed into or upon the work with the help of the special fixture.

Taper Turning

A taper may be defined as a uniform increase or decrease in diameter of a piece of work measured along its length. In a lathe, taper turning means to produce a conical surface by gradual reduction in diameter from a cylindrical workpiece.

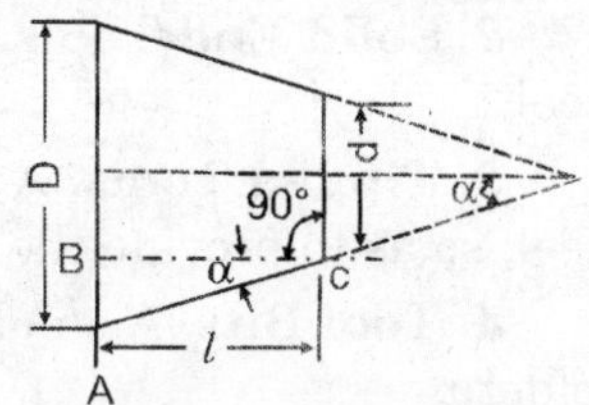

Fig. 25.3 Taper elements.

Referring Fig. 25.3 a taper element having following data

D = large diameter of taper in mm.

d = small diameter of taper in mm.

l = length of tapered part in mm.

2α = full taper angle

α = angle of taper or half taper angle.

The amount of taper in a workpiece is usually specified by the ratio of the difference in diameters of the taper to its length. This is termed as the conicity and denoted by the letter K.

$$K = \frac{D - d}{l}$$

The tangent of the half taper angle can be determined as given under:

$$\tan \alpha = \frac{D - d}{2l} = \frac{K}{2}$$

Taper Turning Methods

A taper may be turned using a lathe by feeding the tool at an angle to the axis of rotation of the workpiece. The angle formed by the path of the tool with the axis of the workpiece should correspond to the half taper angle. A taper may be turned by any one of the following methods:

1. By form tool method.
2. By setting over the tailstock.
3. By swivelling the compound rest.
4. By taper turning attachment.
5. By combination of longitudinal and cross feed method.

25.9 LATHE TOOLS

Single pointed cutting tools are used on a lathe. The tool has a shank which is held in the tool post and a cutting edge which does the turning. The tools may be forged from steel bars or the tips may be welded to low carbon steel shanks. The following are the commonly used lathe tools:

1. Front roughing tool.
2. Finishing tool.
3. Left hand knife tool.
4. Right hand knife tool.

5. Left hand side tool.
6. Right hand side tool.
7. External screw cutting tool.
8. Internal screw cutting tool.
9. External recessing tool.
10. Internal recessing tool.
11. Boring tool.
12. Parting off tool.

According to the construction of shank lathe tools are classified as under:

1. Solid Tools. They are made throughout of one material which is alloy steel.

2. Solid End Tools. End is made of alloy steel and the remainder is made of high tensile steel.

3. Tipped Tools. A cemented carbide tip is brazed at the end of a shank of an ordinary steel, so as to provide the cutting edge.

4. Tool Bits. A small tool bit is provided with cutting edges, which is held in special tool holders.

Tool Angles

The shape of the cutting edge of a lathe tool is formed by the angles show in fig. 25.4, and these vary with the material being cut, rate of feed-and the material of the tool.

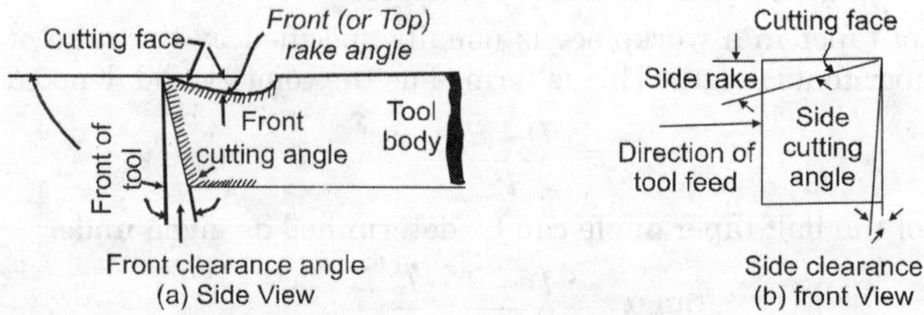

Fig. 25.4 Lathe tool angles.

(*i*) **Front clearance.** It is the angle which the front face of the tool makes with the job.

(*ii*) **Side clearance.** It is the angle which the side of the tool makes with job.

(*iii*) **Top or front rake angle.** It is made between a line at right angle to the axis of the job passing through the cutting point of the tool and a line produced by the top face of the tool towards the shank.

(*iv*) **Side rake angle.** It is made between a line parallel to the axis of the job and a line produced by the top face of the tool towards the side.

Tool Materials

1. High Carbon Steels. Their carbon content ranging from 0.80 to 1.20%, these steels have good hardening ability and, with proper heat treatment, they attain great hardness as high speed alloys. But their drawback is that they lose hardness at high temperature, thus are not suitable for high speeds and heavy duty works.

2. High Speed Steels. They are high in alloy content, have excellent hardenability, and will retain a good cutting edge at high temperature also.

Although there are numerous high-speed steel compositions, they may all be grouped in the following three classes:

(*a*) **18:4:1 High-Speed Steel.** This is an all-purpose tool steel which contains 18% tungsten, 4% chromium and 1% vanadium.

(*b*) **Molybdenum High-Speed Steel.** In above steel molybdenum replaces double quality of tungsten. A molybdenum steel of 6 : 6 : 4 : 2 composition contains 6% tungsten, 6% molybdenum, 4% chromium and 2% vanadium. It has excellent toughness and cutting ability.

(*c*) **Super High-Speed Steel.** It is a high-speed steel having added cobalt from 2 to 15%. One analysis of this steel contains 20% tungsten, 4% chromium, 2% vanadium and 12% cobalt. Tool made of this steel can resist high pressure as well as high temperature.

3. Stellite. It is a non-ferrous alloy composed chiefly of chromium, cobalt, tungsten along with small amount of carbon, silicon and manganese. This metal cannot be forged like high speed steel and so the tips are generally used. This alloy can be deposited by oxyacetylene welding on a high carbon steel shank and after grinding, can serve the purpose of a good tool.

4. Tungsten Carbide. Metal powders of tungsten carbide and cobalt are moulded into the shape of tip and afterwards sintered. Tips may be brazed or clamped on high-tensile steel shanks. Carbide tools permit cutting speeds two to three times that of cast alloy tools. From an economic point of view, carbide tools should always be used if possible.

5. Diamonds. Diamond tip tools are used either for hard materials, difficult to cut with other tool materials or for light, high-speed cuts on softer materials where accuracy and surface finish are important.

6. Ceramic Tools. Aluminium oxide powder is mixed with certain other oxides and a binder and then processed by hot-pressing or by forming and sintering. Heat conductivity of this material is very low, and hence the tools are generally used without a coolant.

25.10 CAPSTAN AND TURRET LATHES

These lathes are development of engine lathes, which are used to manufacture any number of identical pieces in the minimum time. They consist of bed, all geared head stock, saddle on which a four station tool post is mounted to hold four different tools. A tool post fitted at the rear of the carriage holds a parting tool in an inverted position. They do not have tailstock but consist of hexagonal turret mounted on a slide which rests upon the bed. All the six faces of the turret can hold six or more number of different tools; the workpieces are held in collets or in chucks.

These machines are enable to perform a series of operations such as turning, drilling, boring, thread cutting, reaming, necking, chamfering, cutting-off etc. in a regular sequence.

25.10.1 Capstan or Ram Type Lathe

Referring Fig. 25.5. This type carries the hexagonal turret on a ram or a short slide. The ram slides longitudinally on a saddle positioned and clamped on lathe bedways. The saddle, although capable of adjustment, does not move during the operation of the turret. After forward stroke, when the ram is moved backward the turret indexes automatically and the tool mounted on the next face comes into operation. It is used for light duty workpieces.

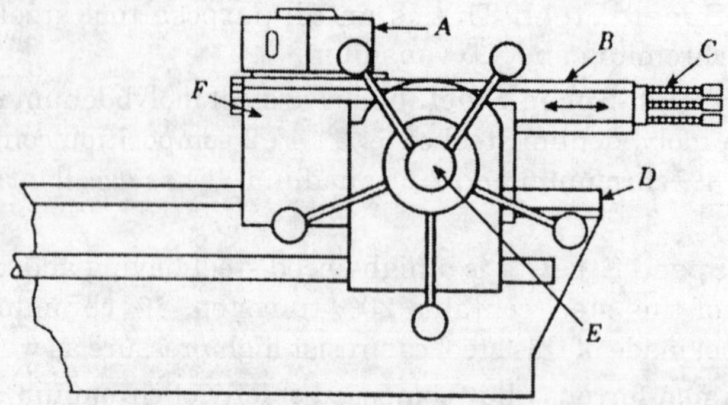

Fig. 25.5 Capstan lathe.
A—Hexagonal turret, B—Auxiliary slide, C—Feed stop rod, D—Lathe bed,
E—Handwheel for auxiliary slide, F—Saddle.

Operations Performed on Capstan Lathe

Main operations performed on capstan lathe are as given under:

1. **Turning.** It is simply producing cylindrical surfaces. Tools are held in cross-slide as well as in main turret.
2. **Forming.** It is producing intricate shapes by the single operation of the tool which is held in cross-slide.
3. **Parting-off.** Cutting away of a completed job from the main bar. Tool is generally held in an inverted position in a tool-post provided at the rear of the cross-slide.
4. **Boring.** It is enlarging of already drilled hole. Tool is held in the main turret.
5. **Screw Cutting.** It is producing the external threads on a cylindrical surface. As the machine is not provided with a lead-screw, therefore, a die-head is used for this purpose.

Advantages and Limitations of the Capstan Lathe

Advantages

1. The movement of the tool can be set with the help of stops, therefore, a semi-skilled man is sufficient to handle the machine.
2. A job can be finished with single setting of workpiece and tools thus the system is economical.

Limitations

1. It takes more setting up time.
2. Needs skilled labour for initial setting up of the job and tools.
3. Production is costlier until unless it is in mass.

25.10.2 Turret or Saddle Type Lathe

Referring Fig. 25.6. In this type the turret is mounted directly on a saddle and the whole unit moves back and forth on the bedways to apply feed. It is heavier in construction and is used for larger diameter barworks. For clamping of job a chuck is provided instead of collet.

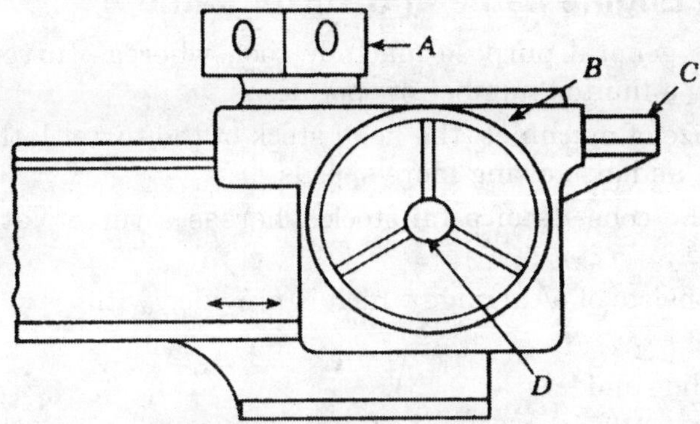

Fig. 25.6 Turret lathe.
A—Hexagonal turret, B—Turret saddle, C—Lathe bed, D—Handwheel for saddle.

Operations Performed on a Turret Lathe

Operations performed on a turret lathe are same that are performed on capstan lathe. But the turret lathe is much more suitable for machining the heavy parts. Following are main operations performed on turret lathe:

1. Turning.
2. Forming.
3. Parting off.
4. Boring.
5. Screw cutting etc.

Advantages and Limitations of the Turret Lathe

Principally this machine is similar to the capstan lathe. It holds the tool more rigidly as the turret head is mounted directly on the bedways. Due to omitting the collet chuck it needs more setting up time. Therefore, to obtain economy it needs *a* production of large number of similar items.

Difference Between Capstan and Turret Lathe

1. In case of capstan lathe the turret is mounted on a ram which slides on a saddle. The saddle itself can be slid and clamped on bedways. Whereas the turret of a turret lathe is mounted on a saddle which slides directly on the bed-ways to the desired length.
2. In case of capstan lathe turret is not mounted directly on the saddle, which is a cause of the overhanging, bending, deflecting and producing vibrations under heavy cutting load. Such troubles are omitted in the turret lathe due to its rigid construction.
3. Capstan lathe is provided with collets of limited sizes thus suits to light bar works. Whereas heavier chucking works can be performed on turret lathe.
4. In case of capstan lathe the length of operation is limited to the slide of the ram, whereas in turret lathe the length of operation can be performed to any limit in single handling.
5. Heavier turret lathes are provided with power chucks like air operated chucks for holding larger sizes of work quickly.

Difference Between Engine Lathe and Turret Lathe

1. Engine lathe is general purpose machine tool, whereas turret (or capstan) lathe is capable of doing a limited number of operations.
2. For the same size of machines, the head stock of the turret lathe is heavier than that of engine lathe, and also owing more speeds.
3. The engine lathe consists of a tailstock whereas a turret lathe is provided with a hexagonal turret.
4. Engine lathe consists of a carriage which is provided with:
   (*i*) cross slide,
   (*ii*) compound-slide and
   (*iii*) tool post, on it. Whereas a turret lathe consists of a cross-slide having a revolving square tool post.
5. Engine lathe is not fitted with adjustable stops to limit the motions of the saddle and cross-slide but the turret lathe is fitted with adjustable stop to limit the motion of the saddle.
6. Rate of production of the turret lathe is more than the rate of production of the engine lathe.
7. Engine lathe is operated by a skilled operator whereas a turret lathe can be operated by a semi-skilled worker.
8. The labour cost required to operate a capstan or turret lathe is less than that required in an engine lathe.
9. Turret and capstan lathes are not usually fitted with lead screws for cutting threads similar to an engine lathe. The threads are cut by die heads and taps. A short length of lead screw called, "chasing screw" are provided for cutting threads by a chaser in a turret lathe.

Circumstances in which Turret (or capstan) Lathes are preferred to Engine Lathes and vice versa

Turret lathes are preferred to engine lathes in the following circumstances:

1. When large number of similar jobs are to be produced quickly at low cost.
2. When operators are not available at low cost.
3. When sufficient capital is available to purchase the turret or capstan lathe.

Engine lathes are preferred to turret or capstan lathes in the following circumstances:

1. When capital is limited and limited number of jobs are to be produced with various operations.
2. When skilled operators are available at cheap rates to operate the engine lathe.
3. When time factor is not the major factor.
4. When jobs are not of repetitive nature.

Automatic Screw-cutting Lathe

Where considerable quantities of similar screws are required, some automatic lathes are preferred. In such machines, once the work is placed in position between the centres, the stops set and the cut started, no further attention from the operator is required until the screw is completed.

The operators of the automatic screw-cutting lathe are similar to those performed manually on an ordinary lathe, and, after the lathe has taken one cut, it follows as under:

1. Withdraw the tool.
2. Return the saddle rapidly to the starting-point.
3. Feed the tool to its original cutting position.
4. Feed the tool forward a few thousandths for the next cut.
5. The saddle traverse must reverse for the cutting stroke.
6. At the final depth the in-feed of the tool must cease.

The lathe is driven from two belts, one for the cutting stroke and the other for the return at high speed. Main advantage of this machine is that a semi-skilled worker can operate several such machines at a time.

Example 25.1 *A truncated cone is to be turned out in lathe from a 80 mm diameter stock. The smallest and the highest diameters are to be 70 mm and 80 mm in a length of 100 mm. What should be the ratio between the velocities of the longitudinal feed and the cross-feed of the cutting tool?*

Solution. Ratio between the velocities of longitudinal and cross feed

$$= \frac{L}{\dfrac{(d_1 - d_2)}{2}} = \frac{100 \text{ mm}}{\dfrac{(80 - 70)}{2} \text{ mm}}$$

$$= \frac{100}{5} = \frac{20}{1}$$

Hence ratio is, **20 : 1 Ans.**

Example 25.2 *A shaft 1000 mm in length is being machined on a lathe. If the spindle executes 500 r.p.m. and the feed is 0.20 mm per revolution, how long will it take the cutter to pass down the entire length of the shaft?*

Solution. Turning time $= \dfrac{\text{Length of cut}}{\text{feed} \times \text{r.p.m.}}$ min

$$= \frac{1000}{0.20 \times 500} = \textbf{10 min.} \quad \textbf{Ans.}$$

Example 25.3 *Find the machining cost of a M.S. bar on a lathe from the following data : R.P.M. of the job = 500. Feed of tool per revolution of job = 0.5 mm. Depth cut = 2 mm. Diameter of raw material = 60 mm. Diameter of finished job = 40 mm. Length of job = 1000 mm. Machining cost = Rs. 3 per hour.*

Solution. Total depth to be machined

$$= \frac{(60 - 40)}{2} = 10 \text{ mm.}$$

Depth of cut = 2 mm.

∴ Number of cuts to be performed over the job from end to end

$$= \frac{10}{2} = 5 \text{ No.}$$

$$\text{Total machining time} = \frac{\text{Length of cut} \times \text{No. of cuts}}{\text{feed} \times \text{r.p.m.}} \text{ min}$$

$$= \frac{1000 \times 5}{0.5 \times 500}$$

$$= 20 \text{ min.} = \frac{20}{60} = \frac{1}{3} \text{ hr.}$$

$$\therefore \text{Cost of machining} = \frac{1}{3} \times 3 = \textbf{1 Rupee. Ans.}$$

Example 25.4 *A steel shaft of 25 mm diameter is turned at a cutting speed of 50 m per min. Find the r.p.m. of the shaft.*

Solution.

$$\text{Cutting speed} = \frac{\pi DN}{1000} \text{ m per min}$$

or

$$50 = \frac{\pi \times 25 \times N}{1000}$$

or

$$N = \frac{50 \times 1000}{\pi \times 25} = \textbf{637 r.p.m. Ans.}$$

Example 25.5 *Find the time required for one complete cut on a piece of work 380 mm long and 50 mm in diameter. The cutting speed is 35 metres per min and the feed is 0.5 mm per revolution.*

Solution.

$$\text{Cutting speed} = \frac{\pi DN}{1000} \text{ m per min}$$

or

$$35 = \frac{\pi \times 50 \times N}{1000}$$

or

$$N = \frac{35 \times 1000}{\pi \times 50} = 222.5 \text{ r.p.m.}$$

$$\text{Machining time} = \frac{\text{Length of cut}}{\text{feed} \times \text{r.p.m.}}$$

$$= \frac{350}{0.5 \times 222.5} = \textbf{3.14 min. Ans.}$$

25.11 DRILLING AND BORING MACHINES

Drilling is the operation of making a hole in a material. The machine tool on which this operation is performed is called 'drill machine'. Other methods of producing a hole are by punching, flame cutting and coring. Sometimes this process is confused with boring, a term which should only be used for the operation of increasing the diameter of a hole previously formed by any of the ways, not necessarily by drilling.

Drills

A drill is a fluted cutting tool used to originate or enlarge a hole in a solid material. Drills are manufactured in a wide variety of types and sizes. In metric system, drills are commonly manufactured from 0.2 to 100 mm. In British system the drills are manufactured in three different sizes as given under:

***(i)* Number sizes.** They are from No. 1 to No. 80. Number 1 is the largest having diameter equal to 0.228″ and the number 80 is the smallest having diameter equal to 0.135″.

***(ii)* Letter sizes.** They are from A to Z, A being the smallest having diameter equal to 0.234″ and Z being the largest having diameter equal to 0.413″.

***(iii)* Fractional sizes.** They are from $\frac{1}{64}''$ to 5″ in diameter.

Types of Drills

The types of the drills commonly used are as under:

***(i)* Flat or spade drill.** It is made from a piece of round tool steel, by forging one of its ends. The cutting angle varies from 90° to 120°.

***(ii)* Straight fluted drill.** It is provided with grooves or flutes parallel to its axis. It is used for drilling in softer materials and sheet metals.

***(iii)* Twist drills.** They are provided with two spiral flutes or grooves that run lengthwise around their body. The point angle of the twist drill varies from 120° to 135°. They may be of:

1. Parallel shank and
2. Taper shank type.

A twist drill may range from 0.2 to 100 mm in diameter.

***(iv)* Core drills.** The cutting edges of these drills are not extended to their centres. Therefore, they cannot originate a hole in solid materials. These drills are intended for enlarging cored, punched or drilled holes.

***(v)* Oil tube drills.** They are provided with oil tubes running lengthwise spirally through the body to carry oil directly to the cutting edges. Cutting fluid is forced through the holes to the cutting point of it to remove the chips, cool the cutting edges and lubricate the machined surface.

***(vi)* Combination centre drill.** It is used for drilling the ends of bars and rods that have to be turned between centres on the lathe. It drills and countersinks in one operation and thus saves considerable time.

25.11.1 Drilling Machines

Drilling machines are made in many different types and sizes. The important types are as given under:

(*i*) Portable drilling machine.
(*ii*) Sensitive drilling machine.
(*iii*) Upright drilling machine.
(*iv*) Radial drilling machine.
(*v*) Multiple drilling machine.
(*vi*) Gang drilling machine.
(*vii*) Deep-hole drilling machine.

1. Portable Drilling Machine. When the work is lying in the field or floor and which cannot be transferred on the table of the drill press, under such circumstances portable drilling machine is used. It is provided with an independent motor which may be pneumatic or electric driven. With this machine hole can be drilled at any angle desired. The maximum drilling capacity of such machines in common use is 18 mm diameter.

2. Sensitive Drilling Machine. Referring Fig. 25.7, it is a hand feed machine which can be mounted on a bench or on the floor. These machines may be direct motor driven or belt driven. It is a high speed machine thus quite suitable for drilling small holes. It is capable for drilling holes from 1.5 to 15.5 mm diameter.

3. Upright Drilling Machine. Referring Fig. 25.8 machines are similar to sensitive drilling machines except that having power feed mechanism and drilling capacity upto 50 mm diameter. There are two types of upright columns:

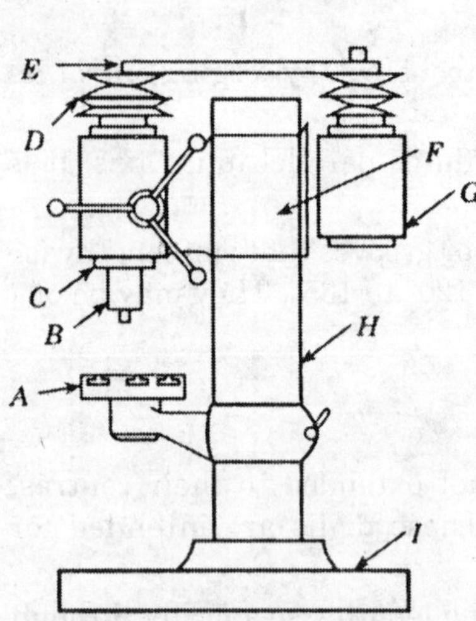

Fig. 25.7 Sensitive drilling machine.
A—Table, B—Vertical Drill Spindle, C—Sleeve, D—Cone pulley, E—V-belt, F—Head, G—Drilling motor, H—Vertical column, I—Base.

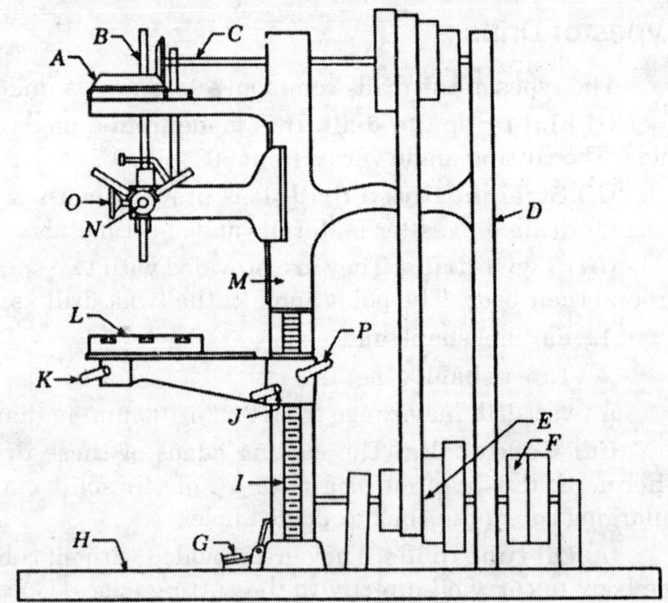

Fig. 25.8 Upright round column drilling machine.
A—Bevel gear drive to spindle, B—Spindle, C—Overhead shaft, D—Back stay, E—Countershaft cone pulley, F—Fast and loose pulley, G—Foot pedal, H—Base, I—Rack on column, J—Table elevating handle, K—Table clamp, L—Table, M—Column, N—Handwheel for quick hand feed, O—Handwheel for sensitive hand feed, P—Table elevating handle.

(*a*) box column and, (*b*) round column.

A box column is better for heavier works.

4. Radial Drilling Machine. This machine is also used for drilling medium to large and heavy jobs. Referring Fig. 25.9 the machine consists of a heavy, round, vertical column mounted on a large base. The column sports a radial arm which can be raised and lowered to different heights. The arm may be swung around the column. Spindle head with feed mechanism is mounted on the radial arm and can be moved horizontally along the arm. The main advantage of this machine is that drill can be moved upon the heavier job to any position desired.

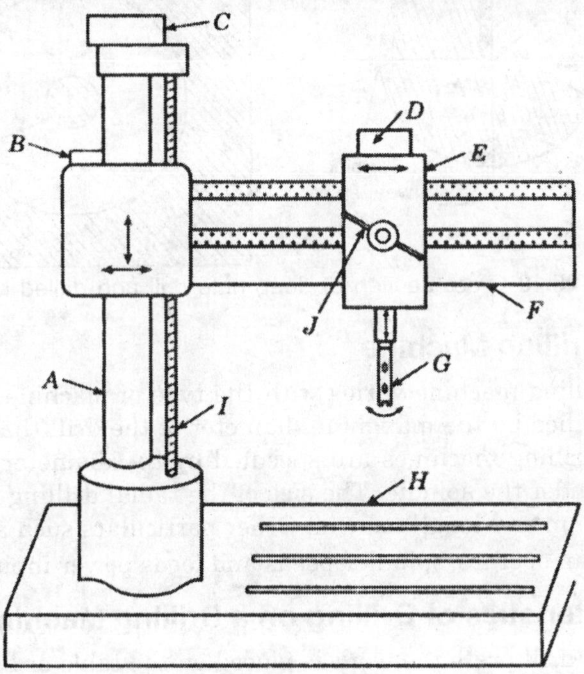

Fig. 25.9 Radial drilling machine.
A—Vertical column, B—Motor for spindle head movement, C—Radial arm elevating motor, D—Spindle motor, E—Spindle head, F—Radial arm. G—Spindle, H—Base, I—Arm elevating screw, J—Spindle feed handle.

5. Multiple Drilling Machine. This machine can drill a number of holes simultaneously. It gives good production and accuracy for interchangeability. This machine consists of several spindles driven by a single motor, and all the spindles holding drills are fed into the work at a time. The distance between the spindles can be adjusted as per desire.

6. Gang Drilling Machine. When several drilling spindles are mounted on a single table, it is known as a gang drilling machine. In some units the spindles are permanently spaced along the table and in some the spindle distance can be adjusted according to the production. This machine has separate motors for each spindle unit. This machine is used in straight line multihole drilling jobs.

7. Deep Hole Drilling Machine. With an ordinary drilling machine the drilling of long holes in rifle barrels, long spindles, connecting rods and certain oil-well, is very difficult. Because, as the hole length increases, it becomes more and more difficult to support the work and the drill properly. For this purpose a special deep hole drilling machine is used, in which work is supported at one end in the head stock and at other end by the carriage. A special drill, end view of which is shown in Fig. 21.10, is supported in drill carriage, and slowly fed into the work. Holes are generally found much accurately drilled with this machine.

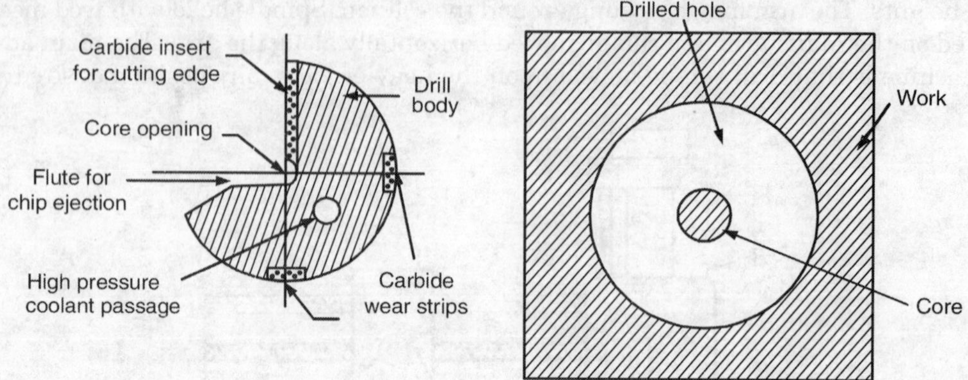

Fig. 25.10 Cross-section of deep hole drill and drilled hole.

Specification of a Drilling Machine

The size of the drilling machine varies with the type of machine being used. A portable drilling machine is specified by the maximum diameter of the drill that can be held in it. The sensitive and upright drilling machines are specified by the diameter of the largest job that can be accommodated under the spindle. The size of the radial drilling machine is specified by the diameter of the column and length of arm. Other particulars such as the maximum size of the drill that machine can operate, spindle speeds and feeds power input, net weight etc.

Operational Characteristics of Drilling on a Drilling Machine and on a Lathe

While drilling on a drill press the work is placed on the table or base of the machine, the drills are held in the spindle by means of a chuck or tapered sockets. In this case work is kept stationary and the drill is rotated. The automatic or hand feed is given to the drill to a required depth.

On the other hand, in drilling on a lathe the drill is kept stationary in tail-stock which is fixed at desired position on the bed of the machine. The job is held in the rotating chuck mounted on the lathe spindle. The feed is given to the drill against the job by means of the tailstock handwheel.

Example 25.6 *How many revolutions per minute must be transmitted to a high speed steel drill of 14 mm in diameter in order to bore into soft cast iron at the rate of 50m/min?*

Solution. Since cutting speed,

$$S = \pi DN \text{ m/min},$$

where,

$$D = \text{Diameter of the drill in m.}$$

$$N = \text{R.P.M. of the machine.}$$

Thus,

$$50 = \pi \times 0.014 \times N$$

$$\therefore \quad N = \frac{50}{\pi \times 0.014} = \mathbf{1136 \text{ r.p.m. Ans.}}$$

Example 25.7 *At what speed a 20 mm drill will run for cutting steel at 25 m per min surface speed?*

Solution. Since,

$$S = \pi DN$$

Thus $$25 = \pi \times 0.020 \times N$$

or $$N = \frac{25}{\pi \times 0.020} = \textbf{398 r.p.m.} \quad \textbf{Ans.}$$

25.11.2 Boring Machines

Boring is the enlarging of a hole that has already been drilled or cored. Principally, it is an operation of truing a hole that has been drilled previously with a single-point or multiple-cutter type tool. Several machines have been developed that are especially adopted to boring work. Most of the common boring machines are the,

(*i*) Horizontal boring machine

(*ii*) Vertical boring machine.

1. Horizontal Boring Machine

Referring Fig. 25.11, in a horizontal boring machine, the work supported on a lathe which is stationary and the tool revolves in a horizontal axis. This machine can perform other operations such as, reaming, turning, threading, facing, milling, grooving, recessing etc. An unsymmetrical and bulky job can also be conveniently machined on it.

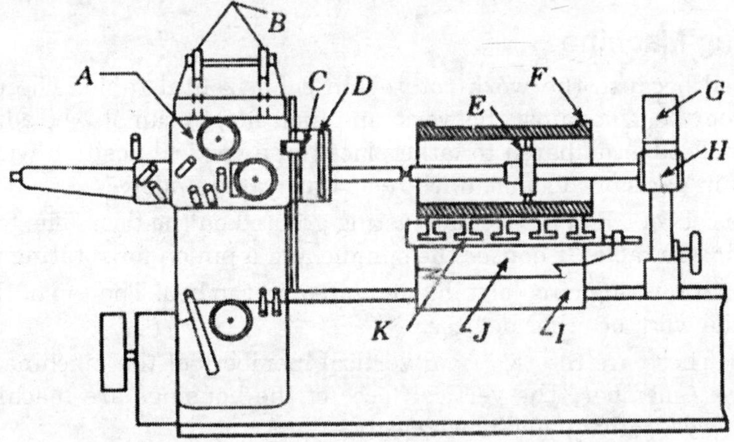

Fig. 25.11 Horizontal Boring Machine.
A—Headstock, B—Pulley for counter balancing weight of headstock, C—Headstock elevating screw, D—Boring head, E—Boring cutter on boring bar, F—Work, G—End supporting column, H—Bearing block, I—Saddle J—Cross-slide, K—Table.

The different parts of a horizontal boring machine are illustrated as under:

1. Bed. It is the foundation member of the machine and supports the saddle and the table over its precision ways.

2. Headstock. It is supported on the column, and may be slid up and down, and can be clamped at any height. It consists of spindle, spindle drive, feed gearing etc. The spindle is used to drive and support the cutting tool and is capable of being fed along its axis to perform boring operation. A machine may have more than one spindle for heavy or light nature of works.

3. Column. It gives support and guide to the head stock and so it should be a very sturdy structure. Counter-weights to balance the head-stock, are also housed in it.

4. Base. It supports the column and contains the driving mechanism for the elements of the column. It may be separate part or an integral part of the bed.

5. End Support Column. Long boring bar held at one end by the spindle, is supported at its end in a bearing which may slide up and down over the end support column ways. Bearing may be adjusted and clamped at any height.

6. Saddle and Table. The table supports the work, and contains T-slots for clamping the work and other devices. The saddle permits the work to be moved longitudinally on the bed. The table may be moved crosswise on the saddle. This can be performed by hand or power.

7. Cross-rail. In a multiple head boring machine, a cross-rail is used to accommodate head stock over it.

8. Floor-plate. Several floor plates may be doweled together to form a rigid base for supporting the job. These plates are permanently set on a sound foundation after due levelling and have the features similar to the clamping table so that the jobs may be supported and held on it.

9. Boring Bar. It supports the cutter for boring operations on jobs having large bore diameters. For short holes the bar may be supported on the headstock spindle end only, whereas for long work the bar is supported on the spindle end and on the column bearing block as shown in Fig. 21.12.

2. Vertical Boring Machine

It is so named because the work rotates on a horizontal table. The cutting tools are stationary which may be fed against the work, and their height can also be adjusted as desired. This machine may also be compared to lathe placed in a vertical position with rotating chuck horizontal. The main parts of this machine are as illustrated under:

1. Bed. It consists of a hollow circular casting grouted on the floor. The top of it is finished having a bearing for the table. It houses the spindle and a pinion for rotating the table.

2. Table. It is a rotating circular table mounted on the bed. The top of the table contains T-slots to mount the work or other devices.

3. Housings. These are the two rigid vertical members of the machine. They are joined at the top by a cross member. The vertical faces of the housings are machined well so that cross-rail may slide vertically up and down easily.

4. Crossrail. It is a rectangular horizontal member mounted on the front faces of the housings. It can be slide up and down. A saddle having tool head can slide lengthwise on it.

5. Tool Head Assembly. It comprises saddle, ram and tool post. The saddle is mounted on the cross-rail and may slide horizontally. The ram holding the tool post may be made to slide up and down in the saddle perpendicular to the table.

Uses of Boring Machines

1. The main use of this machine is to enlarge and finish a hole already drilled.
2. With the boring, a facing operation can also be performed at the same time on this machine.
3. The other operations, such as drilling, reaming, threading and milling can also be performed on this machine.
4. It is a very easy method for machining the large cylinders of engines, pumps, compressors etc.

25.12 SHAPING MACHINES

It is a reciprocating type of machine tool intended primarily to produce flat surfaces. In the system job is fixed on the stationary table and the cutting tool has a reciprocating motion. Tool cuts in forward stroke only except in case of a draw-cut shaper in which the tool cuts in backward stroke of the ram. Due to limited length of its ram stroke, a shaper is conveniently adapted to small jobs.

Types of Shapers

Shapers are classified in a number of ways as given under:

(*i*) According to the type of mechanism,

(*a*) Crank type. (*b*) Geared type.

(*c*) Hydraulic type.

(*ii*) According to the travel of ram,

(*a*) Horizontal type, (*b*) Vertical type.

(*iii*) According to the type of design of the table,

(*a*) Standard shaper. (*b*) Universal shaper.

(*iv*) According to the type of cutting stroke,

(*a*) Push type. (*b*) Draw type.

Size of a Shaper

Its size is usually given by the maximum length of ram stroke. The cross-wise travel of the table in most of the shapers is equal to the length of the stroke. Common sizes vary from 175 to 900 mm.

Main Parts of Shaper

The main parts of a standard shaper are as illustrated in Fig. 25.12.

***(i)* Base.** It is a bed of cast-iron, which is rigidly bolted to the floor. It takes up the entire load of the machine and the forces exerted due to cutting operation.

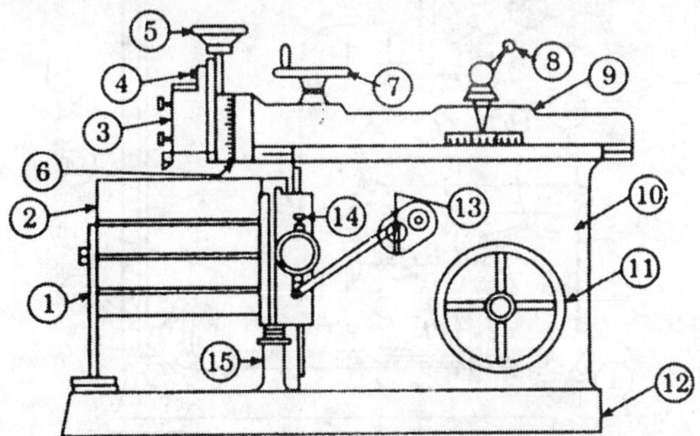

Fig. 25.12 Parts of a standard shaper.

1. Table support, 2. Table, 3. Clapper box, 4. Apron clamping belt, 5. Downfeed hand wheel, 6. Swivel base degree graduations, 7. Position of stroke adjustment Handwheel, 8. Ram block locking handle, 9. Ram, 10. Column, 11. Driving pulley, 12. Base, 13. Feed disc. 14. Pawl mechanism, 15. Elevating screw.

***(ii)* Column.** It is a cast-iron hollow box mounted upon the base. A ram provided with a tool head reciprocates on the top of the column. A cross-rail provided with a table can be guided on its front vertical face. The driving mechanism of shaper is housed within the hollow portion of the column.

***(iii)* Crossrail.** It is an accurately machined casting which can slid vertically with the support of the front face of the column.

***(iv)* Saddle.** It is mounted on the cross-rail which holds the table firmly on its top. The crosswise movement is given to it by means of a screw.

***(v)* Table.** It is a box type casting provided with T-slots both on the top and sides for holding the job. It is bolted to the saddle receives crosswise and vertical movements from the saddle and cross-rail.

***(vi)* Ram.** It is a reciprocating part of the shaper, which is connected to the reciprocating mechanism provided within the column. It slides on the accurately machined guideways on the top of the column. It is provided with the tool-head at the extreme forward end.

***(vii)* Tool-head.** This unit holds the tool rigidly, provides vertical and feed movement of the tool and allows the tool to have an automatic relief during its return stroke.

Uses of Shaper

Following are the general uses of the shaper.

***(i)* Machining flat surfaces.** The small horizontal, vertical or angular surfaces are machined on the shaper.

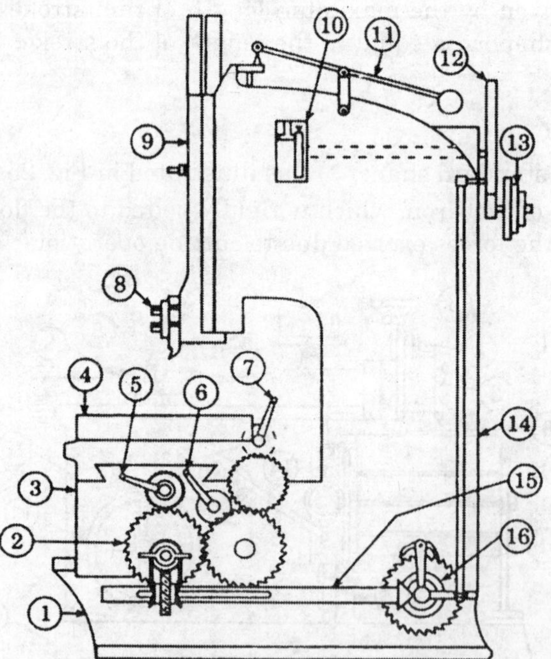

Fig. 25.13 Slotter.

1—Base, 2—Feed gear, 3—Cross slide, 4—Table, 5—Crossfeed handle, 6—Longitudinal feed handle, 7—Circular feed handle, 8—Tool, 9—Ram, 10—Crank disc. 11—Lever for Counterbalance weight, 12—Bull gear, 13—Cone pulley, 14—Column, 15—Feed shaft, 16—Pawl actuating Crank.

(*ii*) **Cutting Key-ways.** A shaft on which a key-way is cut, held horizontally along the stroke of the ram on the table before starting the cutting operation.

(*iii*) **Cutting T-slots.** This operation is completed in three stages. In beginning a plain slot is cut to the required depth. Secondly left side under cutting is done with L-shaped tool. Finally right-side under cut is performed with a similar opposite L-bent tool.

(*iv*) **Gear cutting.** Teeth of big size gears are shaped or generated on this machine by supporting the gear blank on the table against the ram stroke.

25.13 SLOTTING MACHINES (SLOTTER)

The slotting machine also falls under the category of reciprocating type of machine tool like a shaper. The difference between a slotter and a vertical shaper is that a slotting machine has only a vertical movement of its ram whereas a vertical shaper is provided with the extra facility of allowing the adjustment of the ram from its vertical position to nearly 10 degrees from the vertical.

Main Parts of Slotter

As illustrated in Fig. 25.13, a slotter consists of the following main parts:

(*i*) Base. It is a rigid casting which withstands the load and cutting forces exerted on the machine. Its top is machined accurately and is provided with guideways for mounting the saddle.

(*ii*) Column. It is a vertical integral part of the base. The front face of which is machined accurately, and is provided with guideways for sliding the ram; and houses for drilling mechanism.

(*iii*) Saddle. It is machined casting which can slide lengthwise upon the guideways provided on the top of the base. The top face of the saddle is also provided with guideways for giving crosswise movement to the table.

(*iv*) Cross Slide. It is mounted upon the guideways provided on the top of the saddle. It can be moved in the perpendicular direction to the base axis.

(*v*) Rotary Table. It is a circular table having T = slots for holding the workpieces; and is mounted on the top of the cross-slide. A definite circular motion can be given to the table when desired.

(*vi*) Ram and Tool head assembly. Ram is a reciprocating member which can be slid vertically within the guideways of the column. It is provided with a slot for adjusting the position of the stroke. It supports the tool at its bottom end on a tool head.

Size of a Slotter

Like shaper the slotter is also specified by the maximum length of its ram stroke. The size of a general purpose slotter ranges from 80 mm to 900 mm. The other factors of its specification are such as diameter of the table, cross and longitudinal travel of the table, number of speeds and feeds, power of the motor, floor space required etc.

Uses of Slotter

The main uses of a slotter are as given under:

(*i*) For internal machining of blind holes.

(*ii*) For vertical machining of complicated shapes which are difficult to be machined on a horizontal shaper.

(*iii*) For machining dies, keyways, die punches, straight or curved slots etc.

(*iv*) For cutting internal or external teeth on bigger size of gears.

(*v*) For machining vertical faces, T-slots, V-way, and tapered holes.

(*vi*) For making notch on the outside of a workpiece.

Crank and Slotted Link Mechanism

This mechanism is clear from the Fig. 25.14. The pinion 1 receives motion from the motor and transmits to the bull gear 14 on which a slotted link 9 is connected in eccentric position by means of a crank pin 11. The power end of the slotted link is hinged with the base by means of a pivot pin 15, and the upper end is clamped with the sliding ram 2, through ram block 8 and screwed shaft 3 by means of clamping lever 4.

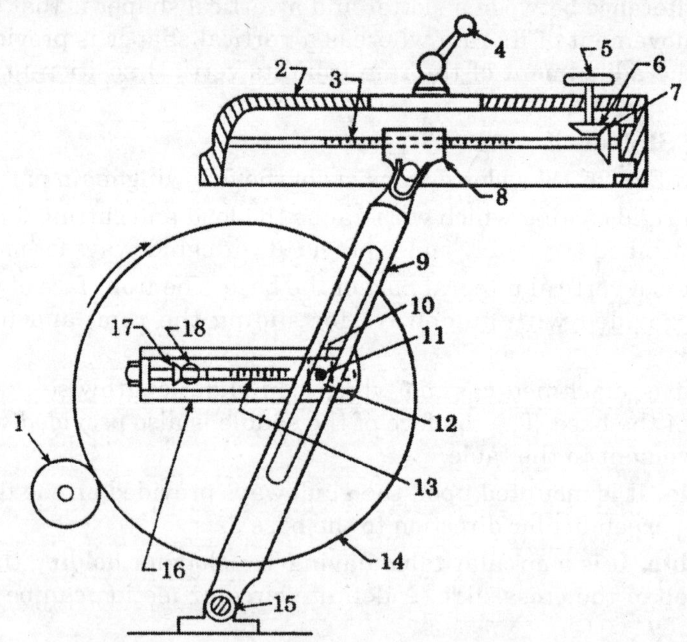

Fig. 25.14 Crank and slotted link mechanism.
1—Driving pinion, 2—Ram, 3—Screwed shaft, 4—Clamping lever
5—Handwheel for adjusting stroke position, 6 and 7—Bevel gears
8—Ram block, 9—Slotted link, 10—Bull gear sliding block, 11—Crank pin
12—Rocker arm sliding block, 13—Lead screw, 14—Bull gear
15—Slotted link pivot, 16—Bull gear slide, 17 and 18—Bevel gears.

For operation, as the bull gear rotates causing the crank pin to rotate, the sliding block 12 fastened to the crank pin 11 will rotate on tho crank pin circle, and at the same time will move up and down the slot in the link 9 given it an oscillating motion which is communicated to ram. Thereby converting the rotary motion of the bull gear into the reciprocating motion of the ram.

(*i*) Adjustment of Stroke Length. For varying the stroke length, the distance between bull gear centre and crank pin 11 has to be changed. Block 12 with its pin 11 is mounted on a screw 13 which if rotated, can move the block 10 away or towards the centre of bull gear as per its direction of rotation. To increase or reduce the stroke length, more or less distance is needed respectively between bull gear centre and pin 11. Rotation of bevel gear 18 will revolve the screw 13 through the bevel gear 17 mounted on it. The clamping lever 4 should be loosened to free the ram while adjusting the stroke length.

***(ii)* Adjustment of Stroke Position.** Again referring Fig. 25.14 to facilitate the positioning of tool travel over any desired length of the job, ram block 8 can be adjusted at different positions with respect to the slotted link. The screw 3 is turned through two bevel gears 6 and 7 which are rotated by turning the hand wheel 5 mounted above the ram. Rotation of the screw 3 will move the block 8 with respect to slotted link 9. For proper adjustment of the stroke position, first set the required stroke length and then with clamping lever 4 tight, move the ram out to its maximum limit. Now loose the lever 4 and rotate the wheel 5 in appropriate direction to adjust the position of ram stroke properly over the job length. After all these adjustments, tight the lever 4 before any use of the machine.

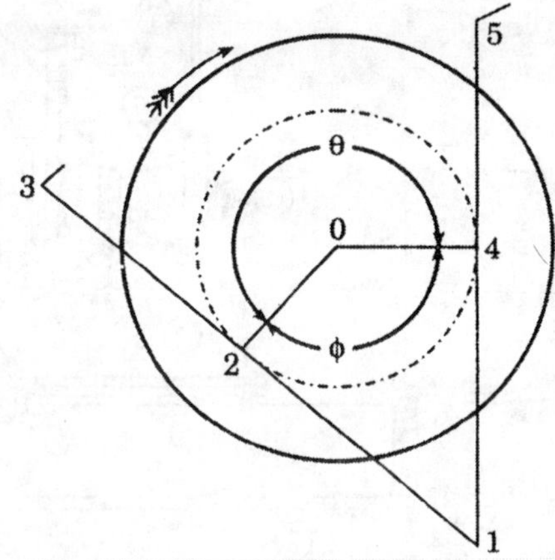

Fig. 25.15 Principle of quick return motion.

Principle of Quick Return

For principle of quick return mechanism refer Fig. 25.15. When the link is in the position 1—3, the ram will be at the extreme backward position of its stroke, and when it is at 1—5, the extreme forward position of the ram 1—3 and 1—5 are drawn tangent to the crank pin circle. The forward cutting stroke, therefore takes place when the crank rotates through the angle θ and the return stroke takes place when the crank rotates through the angle ϕ. It is obvious that the angle θ made by the forward stroke is greater than the angle ϕ described by the return stroke. In this way the return stroke is completed within a shorter time for which it is known as *quick return motion.*

25.14 PLANING MACHINES (PLANER)

A planer is a machine tool designed to remove metal by moving the work in a straight line against a single-edged cutting tool. It is a very large and massive compared to a shaper and capable of machining heavy workpieces which cannot be accommodated on a shaper table. The cuts performed are all plain surfaces, but they may be horizontal, vertical or at an angle.

Main Parts of a Planer

Construction and their details of parts of a double housing planer are illustrated in Fig. 25.16. However they may be described in the following heads:

***(i)* Bed.** It is a strong and heavy structure made out of cast iron and strengthened with cross ribs. The bed length is nearly two times the table length. It consists of V-ways on the top to guide and support the table. It accommodates the table-driving mechanism also.

***(ii)* Table.** Table is also a heavy rectangular casting which reciprocates along the guide ways of the bed. Top of the table consists of T-slots by which job or job holding devices may be bolted upon it. It is provided with a cut on the side by which reversing dogs are clamped at different positions to have varying stroke lengths.

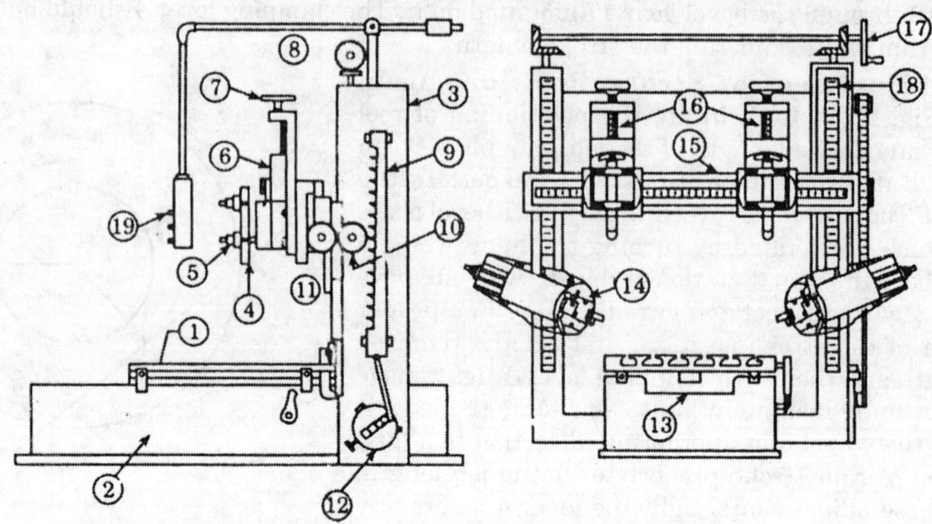

Fig. 25.16 Standard double housing planer.
1—Table, 2—Bed, 3—Column or housing, 4—Claper box, 5—Tool post, 6—Vertical slide, 7—Feed wheel, 8—Feed screw for elevating crossrail, 9—Rack, 10— Feed gear, 11—Saddle, 12—Feed disc, 13—Table rack, 14—Side tool-head, 15—Crossrail, 16—Vertical tool-head, 17—Crossrail elevating wheel, 18—Crossrail elevating screw, 19—Control switch pendant.

***(iii)* Housing.** Housing are the vertical rigid structures bolted on the each side of the bed. The front face of each housing is well machined to provide crossrail for accommodating tool heads, feed and elevating screws.

***(iv)* Cross Rail.** It is a rigid casting which may be raised or lowered on the face of the housing and can be clamped at any desired position. The front face of the crossrail is well machined to provide a guide surface for the tool heads. It consists of screws for vertical and cross feed; and for elevating purposes.

***(v)* Tool Head.** A saddle is fitted on the ways of the cross-rail and its front surface is also provided with finished ways to which the tool head may be fitted. Generally two tool-heads are mounted on the crossrail, which are independently operated. Tool head carries a tool post in which cutting tool is held. This tool post is hinged to the tool head for lifting the tool in its return stroke as found in shaper. Tool head slide is mounted on a swivel plate fastened to the saddle and thus it is easy to swivel the tool head either side of its vertical position.

Size of a Planer

Standard planer size is given by:

(i) Distance between vertical housing.

(ii) Height from top of table to the crossrail when it is at its uppermost position.

(iii) Maximum length of table travel.

Therefore the size of a standard planer is specified by the size of the largest rectangular solid that can reciprocate under the tool. The length of the table is always almost equal to the table travel. Double housing planers range from 750 mm × 750 mm × 2.5 m at the smallest upto 3000 mm × 3000 mm × 18.25 m at the largest size.

Uses of the Planer

A planer is used for the following purposes:

(*i*) To produce large horizontal, vertical or inclined surfaces.
(*ii*) To machine long flat edges of the jobs.
(*iii*) To machine multiple small parts held in line on the table.
(*iv*) To cut keyways on the jobs.
(*v*) For cutting T-slots

Functional Difference and Functional Similarity between Shaping and Planing Machines

The functional differences between a shaping machine and a planing machine are as given under (See table 25.1 also):

(*i*) The planer can accommodate a large work; the shaper can do only small work.
(*ii*) On the planer the work is moved against a stationary tool, on the shaper the tool moves across the work which is stationary.
(*iii*) On planer the tool is fed into the work, on the shaper the work is usually fed across the tool.
(*iv*) The drive on the planer table is generally by gears means. The shaper ram in most of the cases is driven by means of a quick-return link mechanism.
(*v*) Most planers differ from shapers in that they approach constant velocity cuts.
(*vi*) The slides carrying the tools can be swivelled from 0° to 60° in the planer, whereas they can be swivelled upto 45° in the shaper.

The functional similarities between the planing machine and the shaping machine are as given under (See table 25.1 also):

(*i*) Both machine tools are used for machining flat surfaces.
(*ii*) Both machine tools adopt single point cutting tool.
(*iii*) Both machine tools are provided with quick-return mechanism.
(*iv*) Tool-heads can be tilted in both the machines thus produced inclined flat surfaces.
(*v*) In both the machine tools depth of cut is provided by the movement of the tool.

Table 25.1 Difference between Planer, Shaper and Slotter

S.No.	*Planer*	*Shaper*	*Slotter*
1. General Construction	Consists of a bed, a table, columns and crossrail.	Consists of a ram, a table, and a saddle.	Consists of a vertical ram, saddle, cross-slide and a circular table.
2. General Drive mechanism	Table drive is based on rack and pinion system.	Ram drive depends upon the oscillation motion of the slotted link.	Ram drive also depends upon oscillation motion of the slotted link.
3. Operation	(*a*) Work table reciprocates horizontally and provides cutting stroke.	(*a*) Working table does not reciprocate but can be slide cross-wise for providing feed to the work.	(*a*) Work table does not reciprocate but can be rotated and given longitudinal as well as crosswise motion for feed.

4. Cutting stroke	(*b*) The cutting stroke is provided with the horizontal movement of the work table.	(*b*) The cutting stroke is provided with the horizontal movement of the ram.	(*b*) The cutting stroke is provided with the vertical movement of the ram.
5. Feed	(*c*) Tool is fed horizontally or vertically against the reciprocating work.	(*c*) Work is fed against the reciprocating tool.	(*c*) Work mounted on the horizontal table can be fed lengthwise, crosswise and circular against the vertical sliding tool.
6. Uses	Used for machining horizontal large slot surfaces.	Used for machining horizontal flat surface of limited or small sizes.	Used for machining vertical surfaces, slots etc. of limited size and depths.

25.15 MILLING MACHINES

When the metal is removed by means of feeding the work piece against rotating a multipoint cutter as shown in Fig. 25.17, this operation is called *milling*. The work is held on *a* table which controls the feed against the cutter. The device which provides longitudinal, crosswise and vertical movements to the work; and rotary motion to the cutter is called *milling machine*.

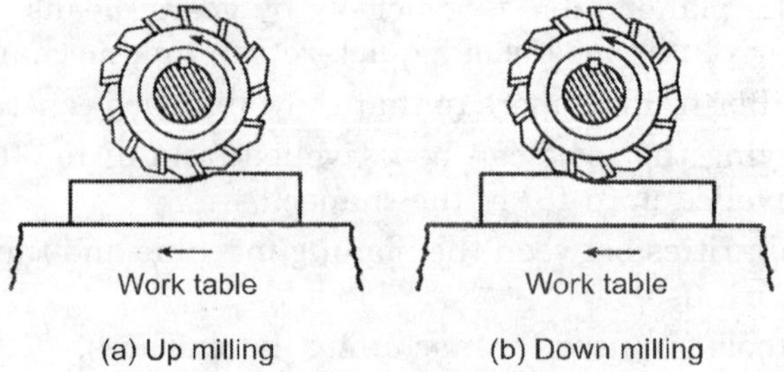

Fig. 25.17 Milling operation.

On a milling machine, flat or formed surfaces can be machined with excellent finish and great accuracy. Therefore most operations performed on shapers, drill presses, gear cutting machines and broaching machines are likewise done on the milling machine.

Types of Milling Machines

There are numerous types of milling machines and to give the exact number of their types is very difficult. But the general purpose milling machines may be broadly classified as given under:

(*i*) Horizontal milling machine. (*ii*) Vertical milling machine.
(*iii*) Universal milling machine.

1. Horizontal Milling Machine

It is most common machine used in general workshop. It consists of a column and a knee as shown in simplified sketch refer Fig. 25.18. The table is mounted on the knee which can slide vertically along the front face of the column, thereby accommodating the works of various heights. A saddle can slide crosswise on the knee and the table lengthwise on the saddle. The cutter mounted in fixed position on the horizontal rotating arbor and the job is fed to this rotating cutter. If desired, several tools can be mounted on the arbor to work at the same time as a gang operation.

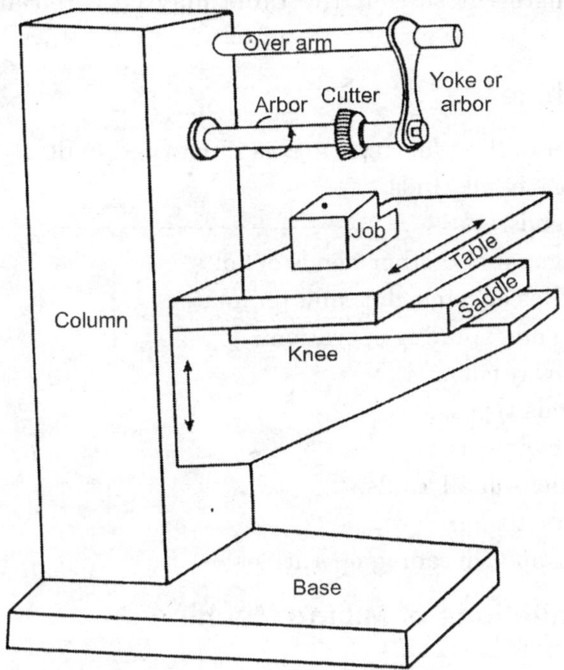

Fig. 25.18 Horizontal milling machine.

2. Vertical Milling Machine

A vertical milling machine as shown in Fig. 21.19 is so called because of the vertical position of the cutter spindle. The table movements are the same as in horizontal milling machines. Ordinary no movement is given to the cutter other than usual rotational motion. However, the spindle head may be swivelled in the vertical plane. Some vertical milling machines are provided with rotary tables to provide grooves on a circular job.

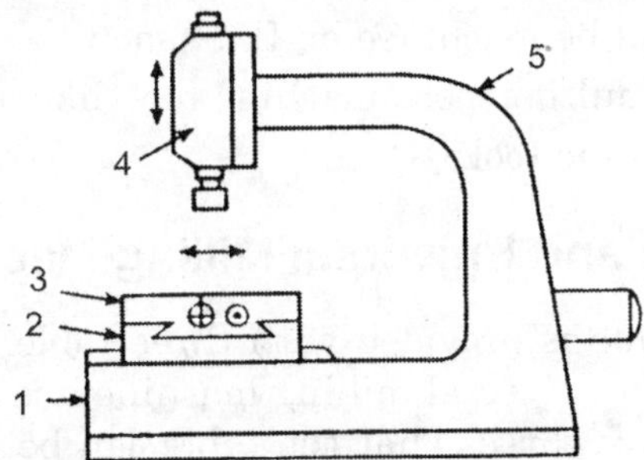

Fig. 25.19 Vertical milling machine.

3. Universal Milling Machine

In appearance it is quite similar to the plain type of milling machine, but differs in that the work table is provided with a fourth movement which permits the table to swivel horizontally. The saddle is made in two parts. The upper one which can be made free over the lower one, can be swivelled to any position horizontally along with table fitted on it.

A modification of the universal machine is called an *Omniversal.* This machine is provided with an additional adjustment so that the table may be tilted and, in that position, fed horizontally.

Uses of Milling Machine

The common uses of milling machine are as illustrated under:

(*i*) Facing operations of all kinds.
(*ii*) Squaring shaft ends.
(*iii*) Making of hexagonal and other heads of bolts.
(*iv*) Forming of profiles of cams, dies and templates.
(*v*) Slotting, slitting and splining operations.
(*vi*) All types of key-way cutting.
(*vii*) Grooves of various types.
(*viii*) Concave and convex surfaces.
(*ix*) Indexing operations of all kinds.
(*x*) All types of gear cutting.
(*xi*) Boring, reaming and threading operations.

Advantages and Limitations of Milling Machine

Advantages:

(*i*) Surface finished by milling is more accurate than that of machined on shaper, planer and slotter.
(*ii*) Work of any shape can be machined more efficiently than on the lathe machine; with the help of form cutters.
(*iii*) Similar and uniform works such as taps, reamers, circular saws, nut faces etc. can be machined more quickly than on other machines.
(*iv*) It is an economical process of machining the part.

Limitations:

(*i*) Cylindrical jobs can not be machined on this machine.
(*ii*) It can not be used as multipurpose machine tool, like a lathe.
(*iii*) It is a very costly machine tool.

Difference Between Plain and Universal Milling Machines

(*i*) A plain milling machine is provided with three table movements longitudinal, cross and vertical, whereas a universal milling machine has a fourth movement of the table in addition to the above three. That the table can be swivelled horizontally and can be fed at an angle to the milling machine spindle.
(*ii*) Universal machine is provided with extra accessories such as dividing head, vertical milling attachment, slotting attachment, rotary table etc. by which extra operations such as spur, spiral, bevel gear cutting; drills, reamers and milling cutter making; drilling, slotting etc. can be performed on it.

(*iii*) The plain milling machine is more rigid and heavier than a universal milling machine of same size.

(*iv*) The plain type milling machine is adopted for manufacturing operations, whereas the universal machine is intended more for tool room work.

Comparison between Slab Milling and Knurling

Plain or slab milling by which a flat surface is produced by a rotating cutter with its axis parallel to the surface being machined.

Knurling is the process of embossing a diamond shaped pattern on the surface of work-piece. It is performed by mounting the knurling tool on the tool post of lathe and the job is rotated between centres or in a chuck. Pressure is exerted by the knurling tool on the job to give impressions of the knurls.

25.16 GRINDING MACHINES

The process by which metal is removed by means of a rotating abrasive wheel is called "grinding". The action of a grinding wheel is very similar to that of a milling cutter. The wheel is made up of many small abrasive grains bonded together, each one acting as a small cutting tool. This process is generally used to smoothen the surfaces by removing the limited quantity of material from the already machined surfaces. Its maximum accuracy is say 0.000025 mm.

Broadly grinding is classified in two groups, say:

1. Rough grinding.
2. Precision grinding.

But according to the type of surface to be ground, it is further classified as under:

(*i*) External cylindrical grinding. (*ii*) Internal cylindrical grinding.

(*iii*) Surface grinding. (*iv*) Form grinding.

Types of Grinding Machines

A grinding machine is designed according to the type of surface to be finished, it may be cylindrical, flat or internal surface. Some special machines are designed as per operations performed on them and are designated accordingly. A general classification of the grinding machines is as given under:

(*i*) Cylindrical grinder.

(*a*) External grinder. (*b*) Internal grinder.

(*ii*) Surface grinder.

(*a*) Reciprocating table type. (*b*) Rotating table type.

(*iii*) Centreless grinding machine.

(*a*) External centreless grinding. (*b*) Internal centreless grinding.

(*iv*) Tool and cutter grinding machines.

(*v*) Special grinding machines.

1. Cylindrical Grinding Machine

A primar idea about a cylindrical centre-type grinder is given by the block diagram referring Fig. 25.20. There are four movements involved in this type of machine:

1. The work must revolve.
2. The wheel must revolve.
3. The work must pass the wheel.
4. The wheel must pass the work.

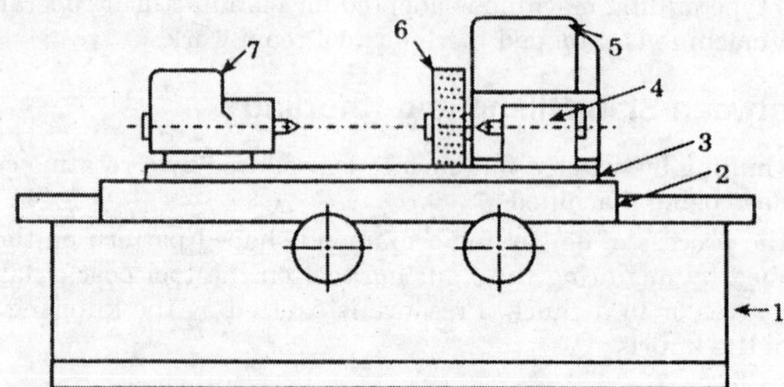

Fig. 25.20 Block diagram of a plain cylindrical centre-type grinder
1—Base, 2—Lower table, 3—Upper table, 4—Tailstock, 5—Wheel head
6—Grinding wheel, 7—Head stock.

This type of machine is essentially a Lathe on which a grinding wheel has been substituted for the single point tool. It consists of the following main parts:

***(i)* Base.** It is the main casting of the machine which supports all the parts and is mounted on the floor, on the top of which guide ways are provided for the table sliding. It also consists of table-drive mechanism.

***(ii)* Tables.** Two tables are provided there, the lower table slides on ways on the bed and provides traverse of the work past the grinding wheel. The upper table that is provided at its centre is mounted on the top of the sliding table; and secures the headstock and tailstock which can be positioned along the table to suit the length of the work.

***(iii)* Headstock.** It supports the workpiece by means of a dead centre and drives it by means of a dog or chuck.

***(iv)* Tailstock.** It can be adjusted and clamped in various positions to accommodate different lengths of workpieces.

***(v)* Wheelhead.** It carries a motor driven grinding wheel and can be moved perpendicularly to the table ways, to feed the wheel to the work.

***(vi)* Cross-feed.** The grinding wheel is fed to the work by the engagement of the cross-feed control lever.

With the cylindrical operations this machine can also be used for grinding contoured-cylinders, tapers, faces and shoulders fillets and even cams and crankshafts.

2. Surface Grinder

To grind flat or plane surfaces is called as "surface grinding". In general the surface grinders are divided into two classes: one class with a reciprocating table and other having a rotating work table. Each type of machine has the possible variation of having the grinding-wheel spindle in either a horizontal or a vertical position. The different four possibilities are clear from the sketches referring Fig. 25.21.

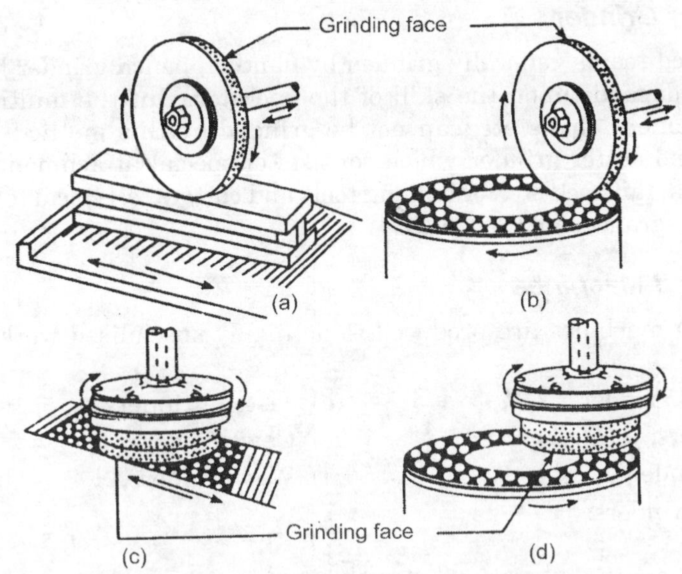

Fig. 25.21 Types of surface grinding machines.
A—Horizontal spindle reciprocating table, B—Horizontal spindle rotary table,
C—Vertical spindle reciprocating table, D—Vertical spindle rotary table.

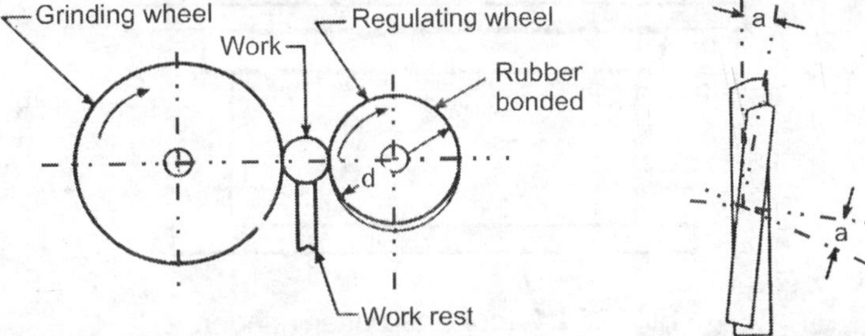

Fig. 25.22 External centreless grinding.

3. Centreless Grinding

In a centreless grinding the work is traversed across the face of the grinding wheel, being supported on a work rest blade and rotation is given between the grinding wheel and a regulating wheel. As illustrated in Fig. 25.22, the large wheel is the grinding wheel, and the smaller one the regulating wheel. Both wheels are rotated in the same direction. The axial movement of the work past the grinding wheel is obtained by tilting the wheel at a slight angle from horizontal.

Internal Centreless Grinding. Referring Fig. 25.23, in this system the work is supported on a supporting roll and is pressed against the regulating wheel by a pressure roll. The direction of rotation of grinding wheel, regulating wheel and the work is clear from the Fig. 25.23. Advantages of centreless grinding include the elimination of work-holding fixtures and the ability of the machine to grind both straight and tapered holes.

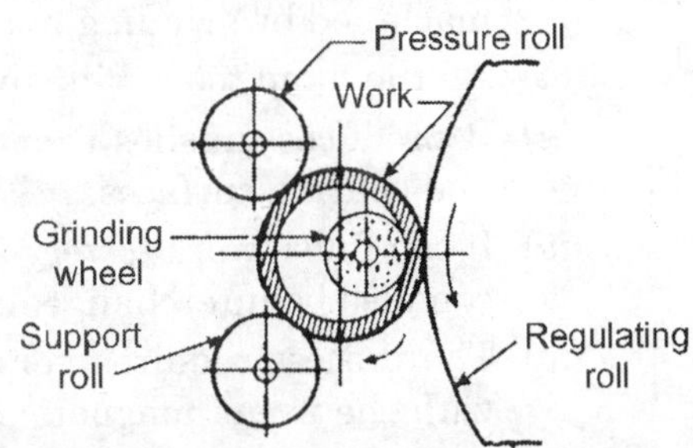

Fig. 25.23 Internal centreless grinding.

4. Tool and Cutter Grinder

A single pointed tool is generally grinded by hand on bench grinder by the operator. The accuracy of angles depends upon the skill of the operator. But the multi-point tools such as milling cutters, reamers, drills etc. can not be grinded by this method. Therefore they are grinded as a tool and cutter grinder which consists of special attachments such as universal head, vise, headstock, tailstock etc. for holding tools and cutters. A typical tool and cutter grinder is shown in block diagram refer Fig. 25.24.

5. Special Grinding Machines

Many grinding machines are produced to do highly specialised work. Some of them are as stated under.

(*i*) Crankshaft grinders,
(*ii*) Piston grinders,
(*iii*) Roll grinders,
(*iv*) Caw grinders,
(*v*) Thread grinders,
(*vi*) Way grinders,
(*vii*) Tool-post grinders.

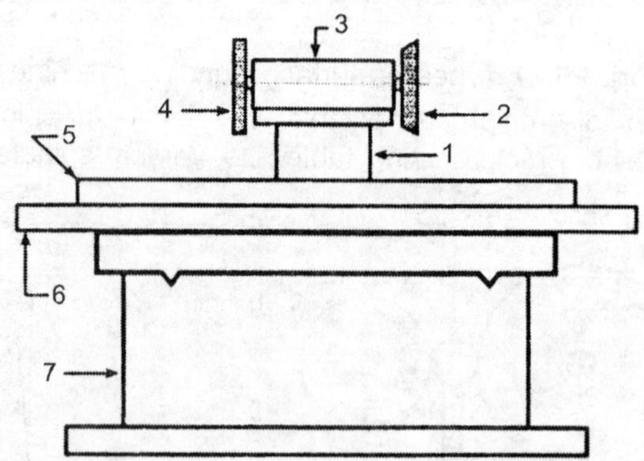

Fig. 25.24 Tool and cutter grinder.
1—Column, 2 and 4—Grinding wheel, 3—Wheelhead, 5—Table, 6—Saddle, 7—Base.

Importance of Grinding

The grinding process is one of extreme importance in production work. It possesses certain advantages that are not found in other cutting process.

(*i*) It is the only method to cut the hardened steel. The important machine parts first machined in an annealed state, then hardened, and finally they are accurately machined by grinding operation.

(*ii*) All the hand tools, bits and cutter etc. are sharpened easily with this process.

(*iii*) It produces finishes that are extremely smooth and, hence, very desirable at contact and bearing surfaces.

(*iv*) It can finish work to very accurate dimensions in a short time, therefore similar product can also be interchangeable.

(*v*) The process requires very little pressure therefore very light work can be permitted with the use of magnetic chucks.

Looking the above advantages some time it is felt that machine shop is incomplete or ill-equipped without the grinding machines.

25.17 CHIP FORMATION

With the advancement in the quality of cutting tool material i.e., with the introduction of ceramic, carbide and diamond tools, a tendency to select high speed and high feed as cutting conditions is gradually gaining ground, which gives rise to a problem of chip control from the stand point of safety of the worker, reduction in floor space, hazards and the tool maintenance. The chip breaker is generally used for this purpose. In the first part of this article the behaviour of chip produced by the tool without a chip breaker is explained and then in the later part effective design of a chip breaker is dealt with.

The first light on chip formation was thrown by Thimme who suggested that the basic mechanism of chip formation is by shear deformation. Since the extensive studies have been made on chip formation by Plispanen, Ernst, Merchant and Okushima.

Principal type of chips obtained in metal cutting (see Fig. 25.25), depending upon the machining conditions are:

1. Discontinous or segmental chips.
2. Continuous chips.
3. Continuous chips with built up edge (B.U.E.).

1. Discontinuous or Segmented Chips

These chips consist of separate plasticially deformed segments which either adhere to each other loosely or are entirely unconnected. The formation of such chip segment was demonstrated many years ago by Thime. Such a chip formation is shown schematically in Fig. 25.25. First of all Krivoukhov used high speed photography to obtain a clear cut picture of discontinuous chip formation at a low cutting speed.

As the tool penetrates the work piece, the stress increases in the layer being cut. This results first in elastic deformation and then in plastic deformation. When the stress exceeds the strength of the metal, the first segment will be sheared along the plane AB (called the shear plane), making an angle ϕ (known as shear angle) with the direction of the tool travel. When the first segment has been sheared away, the tool face compresses the next adjacent layer of metal to form the second segment in the same way, and the process continues and the newly deformed chip segment pushes the previous one along the tool face i.e. away from the work piece.

Such type of chips reduce the tool life, but their handling is easier because of their small size as they can be easily disposed of.

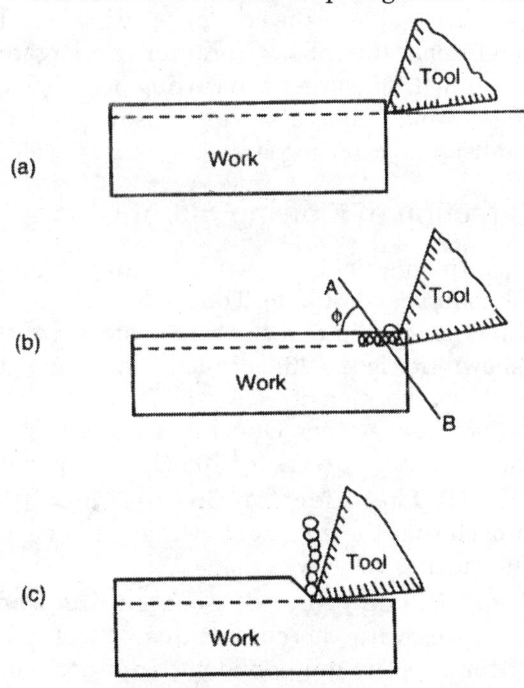

Fig. 25.25 Chip formation.

2. Continuous Chips

This type of chip consists of elements bonded firmly together without being fractured. Under ideal conditions the metal flows by means of plastic deformation, and gives a continuous ribbon of metal. This, even with a microscope does not show any sign of tears or discontinuities. Such type of chips are formed at high speed and with such chips tool life improves and the power consumption is reduced as there is lesser friction at the tool-chip interface.

3. Continuous Chip with B.U.E. (Bailt-up edge)

When a cut is started in ductile metals, a pile of compressed and highly stressed metal forms at the extreme edge of the tool. Owing to the high heat and pressure generated there, this piled up metal is welded to cutting tip and forms a false cutting edge to the tool. This is known as Built Up Edge (B.U.E.). The first correct explanation of the BUE as a "Stagnant Zone" of heavily deformed metal in front of the tool face perhaps was given by Usachev.

The built-up edge grows gradually at cutting edge, and when its growth become sufficiently large, it collapses. A part of it escapes with the chips in the form of very thin flakes adhering underneath the escaping chips. Another part of it gets embedded on the finished surface, while the remaining part remains welded with the tool.

Investigations shows that at low cutting speed of 2 to 5 m per min there is no B.U.E. because a discontinuous chip is produced and the cutting temperature is low. As the cutting speed increases, chip changes to continuous type, flow of the plastically deformed layers is observed and the cutting temperature is such that the formed stagnant zone is not only hardened and arrested on the tool face but is also welded onto it. The cutting speed corresponding to the maximum height of B.U.E. ranges from 10 to 20 m per min. With further increase in the cutting speed, the temperature rises and the stagnant zone, being softend, is reduced in extent.

The value of the cutting speed at which there will be no B.U.E. depends on the machining conditions. Rosenberg Yeramin investigated that the higher the hardness of the steel being machined, the larger the cutting angle of the tool and the thicker are the chips (i.e. the higher the cutting temperature at the same cutting speed.), the lower the cutting speed at which the Built-up Edge disappears.

Direction of Flowing out of Chips

In this field, work has mainly been done by Beiji Okushima and Minato. The angle between the flowing out direction of chip and the centre of axis of the work piece (shown in Fig. 25.26) is called "The angle of flowing out", and can be measured by observing with a universal microscope. As per Okushima the effect of the various factors on the direction of flowing out of chip are:

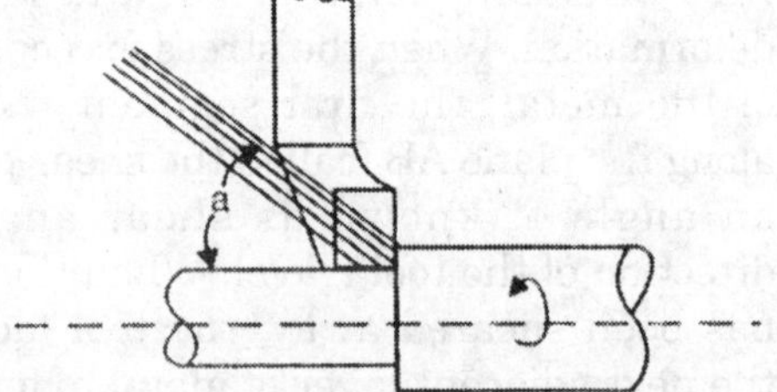

Fig. 25.26 Angle between flowing out chip & centre of axis of work piece

***(i)* The effect of cutting velocity.** It has been proved that the angle of flowing out is hardly affected by the cutting velocity.

***(ii)* The effect of nose radius and depth of cut.** The angle of flowing out of chip increases with the nose radius of tool, and decreases with an increase of depth of cut. The latter effect is quite evident. In other words we can say that, with an increase of the nose radius or with a decrease of depth of cut the chip goes away from the work piece.

***(iii)* The effect of feed.** The effect of feed is small in comparison with other factors, however, with an increase of feed the direction tends to be away from the work piece.

Shape of Chip

The shape of chip changes upon cutting conditions and depends upon the degree of state of chip curl. This field has also been for investigation by Okushima.

According to the chip shape, chips can be classified as:

1. *Straight or curled with a very large radius,* chips are produced from steel at extremely high cutting speeds.
2. *Continuous and irregular helical chip,* is produced, with a gradual decrease in cutting velocity. Here the curl radius of chips decreases gradually but has tendency to increase sometimes.
3. *The flat helical type of chip with equal curl radius* is produced with a lesser cutting velocity.
4. *The equality of curl radius of chip* is disturbed when the depth of cut is less.
5. When the depth of cut is further decreased the chip of disordered block is produced.

The shape of chips changes when an external force acts on it, i.e., the chip is bent by the chip breaker or run against the shank.

The chip which has flowed out of the cutting edge is curled on the horizontal plane and contacts with the side of the tool or work piece. For this reason the chip is bent upwards. If the stress produced in the chip by such deformation is under breaking limit of the mateiral, the chip will be continuous, and if the stress is over the breaking limit, the chip will be broken off. And if the stress is nearly equal to the breaking limit the chip will be broken off with several turns. It has also been noticed that the large feed and the small ratio between the nose radius and depth of cut are effective for chip to break off.

Effect of Cutting Conditions on Chip Shape

Effect of various cutting conditons on chip shape had been a matter of investigation since several years. The various effects as investigated till now are given here in brief.

***(i)* The effect of depth of cut and cutting velocity.** With an increase of cutting velocity and depth of cut, the curl radius of chip increases and the straight type of chip is produced.

***(ii)* The effect of feed.** Above 1.5 mm depth of cut and 0.65 mm/rev, of feed, the broken-off type of one-turn is produced. With an increase of feed, the curl radius also increases, but an excessive increase of feed over certain limit produces the broken off type of chip.

***(iii)* The Effect of Nose Radius.** With an increase of the nose radius, the curl radius usually increases and with its decrease the chip tends to be of the broken off type except in the case of excessively small or large depth of cut.

Chip Breaker

While machining ductile materials like steel and aluminium, a continuous long chip like ribbon is produced. The disposal of such a chip is a problem. This problem becomes serious when the machining is done with carbide and caramic tools (as the cutting speed of these tools is 3 to 5 times to that of HSS tools). At such a high speed of machining the chips leave the cutting edge at a very high velocity which is dangerous to the operator as the chips are hot and sharp. This has led to the development of a device known as chip breaker, and this breaker is located near the cutting edge for obstructing the flow of the chips and make them to curl immediately. The curled chips break either (*i*) because of the tight curling or (*ii*) due to their hitting on the face or shank of the cutting tools.

QUESTIONS

25.1 What is a "machine tool"? What are the main purposes of machine tools? How will you classify the machine tools?

25.2 List various operations performed by machine tools.

25.3 Explain the terms; Cutting speed, Feed and Depth of cut.

25.4 What is Interchangeability? How does it influence production?

25.5 State the importance of limits and fits in large-scale production.

25.6 Explain the terms: tolerance, allowance, clearance, basic size, standard size, nominal size, and limits.

25.7 What is fit? Name the three main types of fits with their uses giving sketches.

25.8 State what is meant by "hole basis" and "shaft basis", which one is preferred and why?

25.9 What is a lathe machine? How would you classify lathe machines?

25.10 How will you specify a lathe machine?

25.11 Draw a neat sketch of lathe machine and indicate main components on it.

25.12 Describe the system of transferring the power from motor to the lathe-spindle.

25.13 List lathe accessories and explain them in brief.

25.14 List different operations which can be performed on a lathe.

25.15 Explain the method used for taper turning on a lathe.

25.16 List various tools used on lathe. Show different angles on the lathe cutting tool.

25.17 Name different tool materials. Specify their suitability.

25.18 Explain in brief about Capstan and Turret lathes.

25.19 Describe the functions of following parts of a lathe:
Head stock, Tail stock, Lead screw, Carriage and Chuck.

25.20 Describe a drill and drilling operation.

25.21 Draw a line sketch of an "ordinary pillar drill" showing important parts on it.

25.22 (*a*) Write a brief note on a "Radial drilling machine".
(*b*) Write a short note on a "Boring machine".

25.23 (*a*) Explain the difference between drilling and boring.
(*b*) Write a note on "multiple drilling machine".

25.24 (*a*) Name the different types of drilling machines.
(*b*) How will you specify a drilling machine?

25.25 Select general purpose machine tools for producing the following:
(*a*) Gear tooth
(*b*) Small hole on a flat plate.

25.26 What is a shaping machine? How will you classify shaping machines?

25.27 Describe a shaping machine with a neat sketch. Mention its uses.

25.28 Show the constructional features of a shaper. Explain quick return motion of a shaper.

25.29 Describe a slotting machine with a neat sketch. What are the uses of a slotting machine?

25.30 Compare shaping machine and planing machine giving functional differences and similarities.

25.31 Describe a planer giving a neat sketch.

25.32 Name the jobs that are generally performed on a planing machine, shaping machine and a slotting machine.

25.33 Write a short note on a milling machine indicating its uses.

25.34 Describe a milling machine with the help of a sketch indicating its advantages and limitations.

25.35 What is meant by milling? Name and describe the principal types of milling machines. State what types of jobs are performed on different types of milling machines? What is the difference between plain milling and universal milling machine?

25.36 Describe with neat sketches the construction, working and functions of a Hobbing machine.

25.37 What are the uses of Grinding machines? Name the different types of grinding machines.

25.38 Describe centreless grinding machine indicating its advantages and limitations. What are its uses?

25.39 Discuss the functions of a precision grinding machine in the production of standard interchangeable parts of machines.

Chapter

AUTOMATION AND ROBOTICS 26

AUTOMATION

Learning Objectives

After studying this chapter, the students will learn:

- Automation, its elements and levels of automation.
- Rationale, advantages, suitability and types of automation.
- Numerical control (NL) machines, their applications, advantages and types.
- Basic components of N.C. machines.
- Servo control systems.
- C.N.C machine control units and their classification.
- Computer aided manufacturing and flexible manufacturing systems.
- Robotics, their applications, components and operations, and robot configuration.

26.1 INTRODUCTION

Automation is the technology by which a process or procedure is accomplished without human assistance.

"Automation" may be defined as, the moving of the procession of materials and parts in and out of machines, and operating of the machines automatically, all without human control. It requires that the machines, tooling and mechanical handling devices between machines be so integrated so as to become one continuous machine.

The term Automation represents a higher degree of mechanization, through which most or all of the operations or activities involved in the manufacturing of a product or products are performed through automatic means, *i.e.,* without direct participation of a human being, except in some aspects like retooling of machine when there is a change of product, readjustments for corrections, etc. The degree of automation may vary from process to process and machine to machine, *i.e.,* it may carry only *partial automation or full automation:* Partial automation means replacement of human activities or involvement by automatic means only partially. In full automation, the human involvement is totally eliminated and the process is entirely carried out and controlled through automatic means along with a proper feedback system. It implies that *full automation* will include all aspects of manufacturing, such as loading, unloading, production, material handling, services, handling of informations, sensing and correction of errors, process control, inspection, parts movement, etc.

Automation is implemented using a programme of instructions combined with a control system that executes the instructions.

An automated system consists of three **basic elements.**

1. *Power* to accomplish the process and operate the system,
2. A *Programme of instructions* to direct the process, and
3. A *Control system* to actuate the instructions.

The relationship amongst these three elements is shown in Fig. 26.1.

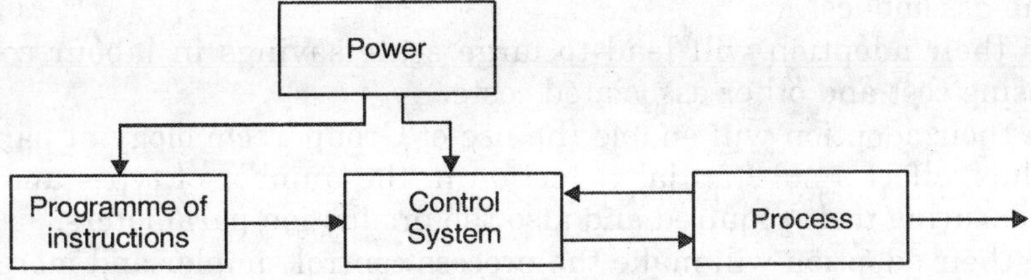

Fig. 26.1 Elements of automated system.

Levels of Automation

Attempts have been made to categorise the levels or degrees of automation based on the human attributes replaced by machine or automatic means. For example, according to *'Amber and Amber's Yardstick for Automation '* all types of automations are divided into 10 levels, from A(0) to A(9), each representing a different level of human attribute replaced by machine in an ascending order. That is, the higher the number of order of automation the higher the level of human attribute replaced by the machine. For example, A(0) level stands for No mechanisation, *i.e.,* no human attribute is replaced by machine, implying that all the activities and operations are performed by human beings only and this includes only hand tools and manually operated machines. The next level A(1) indicates a higher level of automation, *i.e.,* the one in which the *energy* of human muscles is replaced by machine. Similarly, the other orders of automation represent an increasing degree (or level) of human attribute replaced by machine; such as A(2) represents the replacement of *dexterity* (self feeding), A(3) *Diligence,* A(4)—*Judgement,* A(5)—*Evaluation,* A(6)—*Learning,* A(7)—*Reasoning,* A(8)—*Creativeness* and A(9)—*Dominance* (commanding others).

Another method of categorising the levels according to the levels of factory operations. Levels of automation in a production plant are:

1. *Device Level.* This is the lowest level in automation hierarchy. It include the actuators, sensors and other hardware components that comprise machine level.
2. *Machine Level.* Haware at the device level is assembled into individual machines. For example CNC machine tools, industrial robots, power conveyors etc.
3. *System Level.* A manufacturing system is a group of machines or workstations connected and supported by a material handling system, computers and other equipment appropriate to the manufacturing process.
4. *Plant Level.* This is used for complete production system.
5. *Enerprise Level.* This is the highest level, connected with all the functions to manage the company like marketing, sales, accounting, design aggregate planning and master production schedule.

26.2 RATIONALE FOR AUTOMATION

There are several factors which justify adoption of mechanization or automation, of which the production cost is an important one. Mechanization and automation are always justified under the fcllowing situations:

1. Where the environment is highly injurious to human beings, *viz.,* where radioactive substances are to be handled.
2. Where economic feasibility permits it.
3. Where the process is extremely rapid and complex, such that human participation may lead to errors.

4. Where their adoption will facilitate the use of a larger number of standardised parts and sub-assemblies.
5. Where their adoption will lead to large scale savings in labour cost, tooling cost, processing cost and other associated costs.
6. Where their adoption will enable the use of Group Technology in parts manufacture, and, thus, effect a substantial reduction in the number of types and sizes of cutting and measuring tools required and also the production parameters.
7. Where their adoption will make the process control simpler and more effective.
8. Where their adoption will enable maximum utilisation of machines, advanced production methods and tooling in order to maximise production rate and enhance quality standards of the products.
9. When they can be easily adopted without any major alteration in the existing process layout.

26.3 ADVANTAGES AND DISADVANTAGES OF AUTOMATION

1. Advantages

The following are main advantages:

1. Greater output per rupee of labour.
2. More output per rupee invested in facilities.
3. Lesser floor space is required.
4. Increased safety for operators.
5. Automatic production schedules.
6. Increased uniform quality.
7. It transfers skill knowledge and efforts from man to machine, therefore, high rate of production.
8. Less scrap.
9. Less inventories.
10. It requires very less expert staff, rest semi-skilled workers.
11. It is beneficial to the general public, as it raises the income of individuals.
12. Overall production cost is reduced.
13. Human safety is fully ensured.
14. Productivity is increased.
15. Better working conditions are available to the workers.
16. The workplace is tidy and safe.
17. Human fatigue is greatly minimised.
18. After initial setting, there is minimal involvement of the operator in the actual process. As such, a single operator can take care of many machines simultaneously.
19. With the application of Group Technology a considerable saving is affected in terms of design cost, material cost, tooling cost, etc.
20. Less floor area is required.
21. Inventory requirement is minimised due to the use of standardised parts and sub-assemblies.
22. The quality and reliability of the products are greatly improved.
23. The components produced are uniform.
24. Maintenance requirements are minimised.

25. There is an effective control over the production process.
26. Overall profits of the manufacturing concern are increased.

2. Disadvantages

These can be:

1. Greater capital investment.
2. Less manufacturing flexibility.
3. During less demand the management cannot use the automated plant for any other purpose.
4. The immediate effect of automation is increased unemployment.
5. Expensive down time, as the failure of one part may result in the shut down of the whole line where automation is there.

Limitations

Following are the main limitations of automation:

(*i*) High cost of design manufacturing and maintaining automated equipment.

(*ii*) Vulnerability to down time exists because of increasing complexity of automated equipment. However, this is being reduced at an accelerated rate due to improved equipment.

(*iii*) *Loss of flexibility:* It is mainly with fixed automation but the concept of Flexible Manufacturing System (FMS) has largely neglected this restraint.

(*iv*) *More management attention:* Actions in highly automated system occur sometimes at an almost unbelievably high rate and allow little or no time for human decisions. So more and better management personnel is needed as a plant increases its content of automation.

(*v*) *Persistence of automations negative image:* Jobs are eliminated by automation is the negative image but automation also creates new and high skilled jobs.

26.4 SUITABILITY

Automated plants are suitably employed in the following cases:

(*i*) Where a product design is stable for such a long period in which the cost of automation can be written off.

(*ii*) Where the volume for continuous manufacture is adequate and steady for a fairly long time so as to make the installation economical, profitable and to permit the use of single purpose equipment.

(*iii*) Where the manufacturing method demands a high degree of manual work, and adoption of automation presents attractive cost saving potential.

(*iv*) Where the machines are so complex or work at such a high speed, that it is not possible to work with efficiency by manual methods.

(*v*) Where the product is so hazardous that it cannot possibly be controlled by manual methods, *e.g.* radioactive materials.

In short automation is not practically economical if the volume of production is small or where there is a great variety of products manufactured in the plant or where one operator cannot operate several machines efficiently.

26.5 TYPES OF AUTOMATION

Different types of Automation employed in manufacturing are as follows:

1. Fixed type
2. Programmble type
3. Flexible type

1. Fixed Automation

In fixed type, the sequence of process is fixed for one particular product. Such a system can be used for mass production of a product. Its initial cost is high and any changes in product design can be incorporated with difficulty. If, it has to be changed then lot of modifications, new editions etc. are required, which shall be very costly and time consuming.

2. Programmable Automation

In this type of automation, it is possible to accommodate the change in sequence of operations for new product by changing the programme. Such a system is suitable for batch production. For new product, not only programmes have to be changed, but new tools and fixtures may have to be loaded, and machine setting is to be changed.

3. Flexible Automation

In flexible automation systems, no time is lost for changing over the production from one product to a new product. Complete information and programmes for the product desired to be produced are available in computer system and just code for new product has to be fed to computer and changes in all settings, tools etc are done automatically. Such systems are suitable for production of large variety of spare parts for large industrial plants. The spares need not be stocked but the programme for their production can be stored in machine. In this system as an when a spare part is required, machine can be instructed and part can be produced in minimum time without losing time on setting up, loading and arranging tool and fixtures etc.

Thus, flexible automation is an extension of programmable automation. *Features of flexible automation* are:

(*a*) High investment.

(*b*) Continuous production of mixed products.

(*c*) Medium production rates.

(*d*) Flexibility to deal with product design variations.

Flexibility in manufacturing are of following types:

1. Machine flexibility
2. Process flexibility
3. Product flexibility
4. Routing flexibility
5. Volume flexibility
6. Expansion flexibility
7. Operating flexibility
8. Production flexibility

Fixed V/S Flexible Automation

Table 26.1 Describes the difference between fixed and flexible automation.

Table 26.1 Fixed v/s flexible Automation

Fixed Automation	Flexible Automation
1. In this system the sequence of processing (or assembly) operations are fixed by equipment configuration.	In this system the production equipment is designed with the capability to change the sequence of operations, *i.e.* flexible automated system is capable of producing a variety of parts (or products) with virtually no time lost for change over from one part style to the next.
2. Each operation in sequence is usually simple, involving perhaps plain linear or rotational motion or uncomplicated combination of two.	The operation sequence is controlled by a program which is a set of instruction coded so that they can be read and interpreted by the system. New program can be prepared and entered into equipments to produce new products.
3. It is the integration and coordination of many operations which make the system complex and there is some loss of time.	There is no lost production time while reprogramming and altering the physical set up. The system can produce various mixes and schedules of parts or products instead of requiring that they are made in batches.
4. It has high initial investment for custom-engineered equipments.	It also have high investment for custom-engineered equipment.
5. It has high production rate	It has medium production rate but there is continuous production of variable mixture of products.
6. The equipments are relatively inflexible to accommodate product variety	It has flexibility to deal with product design rate.
7. Examples of fixed automation include machining transfer lines and automated assemblv machines.	Example of flexible automation include manufacturing system for performing machining operations.

26.6 NUMERICAL CONTROL (N.C.) MACHINES

As the technology has advanced, *automation* has always been associated with it in order to increase the *productivity,* reduce production costs and improve quality of production. This could be effectively and advantageously introuduced in special purpose and mass production machine tools. However, difficulties were experienced in introducing automation in conventional machine tools, used in batch or small lot production.

Numerical Control (NC) System has made it possible to automate these conventional machine tools too, used in small lot production, and thus, overcome the difficulty specially in respect of the large set-up time.

In a **NC system,** the operating instructions in coded form, such as numbers, letters, symbols, etc., are stored on punched *tapes* or *cards.* The numerical data, containing these instructions, is arranged on the tape in the form of a series of several blocks; each block containing information needed to machine one portion of the component. When the machining of one portion of the component is over, the tape moves forward, by a distance equal to the next block, so that the next portion of the component is machined. Thus, the programmed instructions are automatically implemented in the machine tool in proper sequence without any participation of the operator in it. *Thus, we conclude that* **Numerical Control** *of machine tools is a programmable automation wherein the operating functions of the machine tool are controlled by coded numbers and letters.*

Numerical Control (NC) means *control by numbers.* These numbers are arranged in the form of blocks or series, which carry specific instructions. This is known as the **Prepared programme.** The instructions contained in the programme are read and interpreted by the **NC system** to regulate the different slides of the machine tool, select suitable cutting speeds and feeds and control tool movements, so that the component is machined to the required size and shape.

The Electronics Industries Association (EIA) defines the NC system as follows:

"A system *in which actions are controlled by the direct insertion of numerical data at some point. The system must automatically interpret atleast some portion of this data."*

The computer software is commonly used to facilitate different types of interpolations in NC machines. The main feature of a NC machine tool, apart from the automatic control, is its *feed back machanism* which does away the requirement of a full-time operator of a conventional machine tool.

In NC machine tools one or more of the following functions may be automatic:

1. Starting and stopping of the machine tool spindle.
2. Controlling the spindle speed.
3. Positioning the tool tip at desired locations and guiding it along desired paths by automatic control of the motion slides.
4. Controlling the rate of movement of the tool tip (i.e. feed rate).
5. Changing of tools in the spindle.

26.7 APPLICATION AND SUITABILITY OF NC MACHINES

Applications

The range of applications of NC Machine tools covered is very wide and some common machines using numerical control technology are mentioned here under:

1. NC milling, drilling and boring machines
2. NC lathes and turning centres
3. NC grinders
4. NC punches and presses
5. NC sawing centres
6. NC welding machines
7. NC inspection machines etc.

Suitability

NC machines are quite suitable under the following conditions:

1. Where the components have complex shapes.
2. When the parts are to be made in small batches.
3. Where setups are many and costly.

4. Where precision is desired.
5. When the parts are subjected to design changes.
6. Where the inspection cost is significant.
7. Machining of such components which require 100 percent inspection.
8. Repetitive production of precision parts in small and medium lot sizes.
9. When the amount of metal to be removed is high.

Limitations

Limitations/difficulties encountered in using conventional NC machines are as under:

1. **NC Training Requirements.** NC manufacturing required training of personnel for software as well as hardware. Part programmers are required to be trained to write instructions in desired languages for the machines and also for the manufacturing processes. Similarly machine operators should also be trained for the operation of NC machines.
2. **Higher Investment on Manufacturing Establishment and Running Costs.** The set-up costs of NC machine tools is several times more than for conventional machine tools. Since NC is a complex and sophisticated technology, it also requires higher investment for maintenance in terms of wages of highly skilled personnel and expensive spares. There is a requirement of conditioned environment for operating NC machines, this further adds to the running cost.
3. **Part Programming Mistakes.** There is always a possibility of mistake in programming. Hence three or more passes are required before NC tape is correct. To achieve optimum sequencing of processing steps is also a difficult task.

26.8 ADVANTAGES AND DISADVANTAGES OF USING NC MACHINES

Following are the common advantages and disadvantages in using NC machines:

Advantages

1. *Increased productivity.* Due to reduced setup and lead times.
2. *Reduced scrap.* Due to elimination of human errors and high accuracy of NC systems.
3. *Increased rate of production.* Due to heavy reduction in non-machining times and optimization of machining conditions.
4. *Reduced nonproductive time.* Due to the use of lesser setups, lesser setting time, lesser work handling time, facility of automatic tool changing on some machines, etc.
5. *Reduced lead time for manufacturing.* Due to the requirement of fewer setups and less setup time.
6. *Lesser requirement of jigs and fixtures.* Due to the fact that work and tool positioning is done by NC tape, only simple and cheaper varieties of fixtures may be needed in NC machining.
7. *Reduced tooling cost.* Due to the requirement of simpler and less costly fixtures, the cost incurred on the design and fabrication of fixtures is substantially reduced.
8. *Reduced inspection requirement.* Due to the production of parts of uniform quality.

9. *Higher accuracy.* The degree of accuracy of parts produced through NC machining is so high and quality so uniform that there is no difficulty in interchangeable use of these parts.
10. *Better quality control.* Due to higher accuracy of NC systems, the quality of products is effectively controlled.
11. *Ease of complex machining.* NC facilitates performing of complex machining operations easily, accurately and at much faster rate.
12. *Greater flexibility in manufacturing.* Due to easy adaptability of NC to changes in part design, changes in production schedule, etc., according to requirements.
13. *Reduced inventory.* The firm can do with a smaller inventory due to reduced lead times and requirement of fewer setups.
14. *Greater utilisation of manpower.* Since the operator is not required to attend to a running NC machine constantly, he can be utilised to perform some other functions during that period.
15. *Reduction of human error.* Due to the replacement of all the operator's functions of conventional machining by tape in NC machining, the chances of human errors are reduced to a minimum.
16. *Enhanced safety of operator.* The operator's involvement in NC operations is minimum because the *console,* which operates different NC systems, is located at a distance from the machining area, the operator gets no chance to be exposed to this area.
17. *Better safety of machine tool.* There is practically no chance of any damage to the machine tool due to human error since there is almost no requirement of operator's involvement during the operation.
18. *Better utilization of machine tool.* Due to lesser requirement of setup time and no requirement for operator adjustments and manipulations more time is available for machining, leading to a greater use of machine and, thus, an increased rate of production.
19. *Reduced space requirements.* Lesser storing space is required in NC machining, because a few jigs and fixtures are needed. Also, when a NC *machining centre* is used, it eliminates the requirement of many conventional machine tools, which results in a reduced space requirement for a NC machine shop.
20. *Accurate predictions.* The product cost and rate of production can be accurately predicted because of the programmed machining, which does not allow any deviation of the actual production time from the programmed time.

Disadvantages

1. *Heavy investment.* Which should be justified by full utilisation of NC machines, otherwise the invested amount will be blocked.
2. *Costly and complicated maintenance.* Because a well trained maintenance task force is needed.
3. *Need of well trained and highly skilled manpower.* For part programming.
4. *Surplus Labour.* This happens if NC is made applicable to existing conventional machine tools. In that case, a fairly large workforce becomes redundant and an obvious burden on the organisation.
5. *Costly control systems.*

26.9 TYPES OF N.C. MACHINES

Different types of N.C. nachines operate on different systems of numerical contraol.

The common types of *Numerical Control Systems* used in machine tools are the following:

1. Conventional numerical control (NC)
2. Computer numerical control (CNC)
3. Direct numerical control (DNC).

The details of these systems will now be dealt with in the forthcoming articles.

1. Conventional Numerical Control (NC)

In this system, the programme from which the command signals are derived is stored in numbers, hence it is known as numerical control (see Fig. 26.2).

It is a *hard-wired* NC system, employing *integrated circuits* which are permanently wired and arranged on *printed circuit boards (PCBs).* Also, this NC system is purely a *hardware based system.* It is not easy to change the main features of its *machine control unit.* Some other inherent drawbacks of this system are the possibilities of mistakes in punching of tape, absence of provisions for changing speeds and feeds during machining, unsuitability for very large production, etc. However, it is a simpler and relatively cheaper system than any other type of *numerical control system* and is, therefore, ideal for a beginner.

Till now we have discussed conventional numerical system in the foregoing paragraphs.

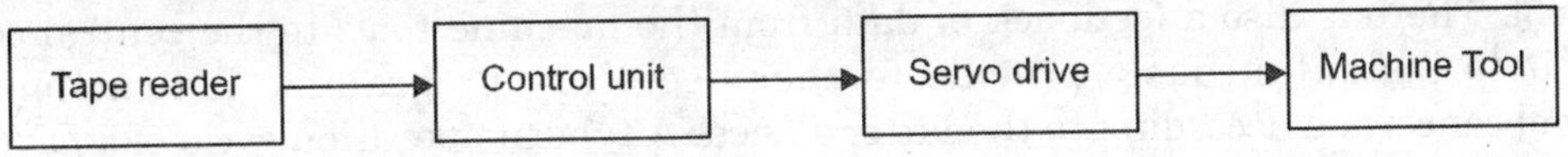

Fig. 26.2 Conventional Numerical Control.

2. Computer Numerical Control (CNC)

It is a *software based* system, in which the *computer* replaces the *control unit* of the conventional NC. It does not carry the *hard-wired logic systems* and all their functions for controlling the machine tool are performed by the *software programme* of the computer. That is why it is known as a *software based system.* Also, since a separate computer is used for each machine tool, with a stored programmable logic, it is known as a *self-contained* NC system for that particular machine tool.

The main objective of using this system is to simplify the hardware of conventional NC and replace it with the software to the maximum possible extent. *The programme* is entered into the computer in its memory, which can be called whenever a part is to be machined. Also, it is easy to edit and modify a programme, if required. Thus, the programmed tape is required only once for feeding the programme into computer memory and the repeated use of tape reader is eliminated, unlike conventional NC. This results in considrable saving of time and cost and increased reliability. An added feature in this system is the *diagnostic software,* which enables easy trouble shooting if the CNC system fails to operate. The problem can be easily detected and rectified through this software. A typical operating cycle of CNC *system* is schematically shown in Fig. 26.3.

The advantages of CNC are related to the 'control system software' which allows a great degree of flexibility not obtained in the case of NC. CNC are being discussed in detail in subsequent paragraphs.

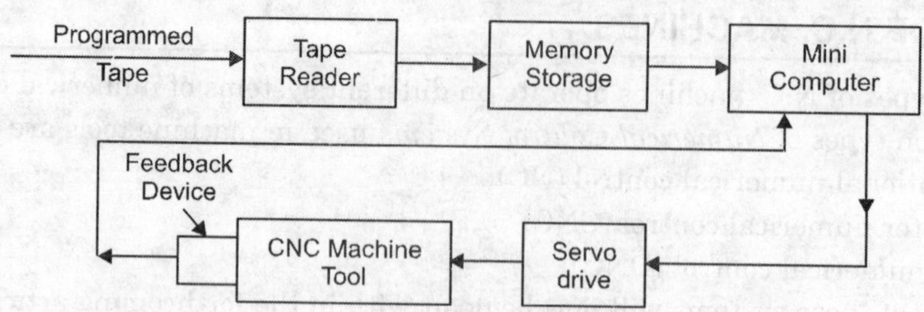

Fig. 26.3 The CNC System.

3. Direct Numerical Control (DNC)

It is also a computerised numerical control system, but it differs from a *CNC system* in that it employs a separately located *Central Computer* and directly controls several machine tools simultaneously, whereas in the DNC each machine tool is separately controlled by its own mini-computer. The central computer, also known as *main-frame computer,* carries a large *memory storage facility* in which the part programmes or machine programmes of all the machine tools, connected to the central computer, are stored in a device like *magnetic disk or drum.*

Whenever an operation is to be performed, the central computer collects the relevant programme instructions from the memory storage and sends these instructions to the concerned machine tool. There is also a feedback of data from the machine tools to the central computer to enable it to instantaneously provide further necessary programme instructions to the machine tool concerned according to the latter's need. A typical operation cycle of a *DNC system* is schematically shown in Fig. 26.4.

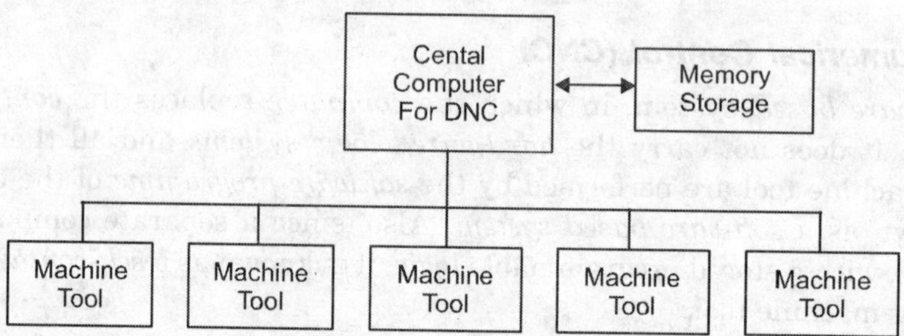

Fig. 26.4 The DNC System.

The most significant and distinct feature of a DNC system is that a single main-frame computer controls a large number of different machine tools simultaneously, providing necessary programme instructions from its memory storage instantly to each individual machine tool linked to it. It implies that the entire manufacturing data is processed and controlled centrally, which facilitates adoption of this sysem in *Computer Integrated Manufacturing (CIM).*

26.10 BASIC COMPONENTS OF NC SYSTEM

A complete **NC machine tool system** consists of the following main *elements* or *units:*

1. A Machine Control Unit (MCU)
2. The machine tool

3. The Drive Units and servo-controls
4. Feedback devices

A schematic diagram of a NC Machine tool system is given in Fig. 26.5.

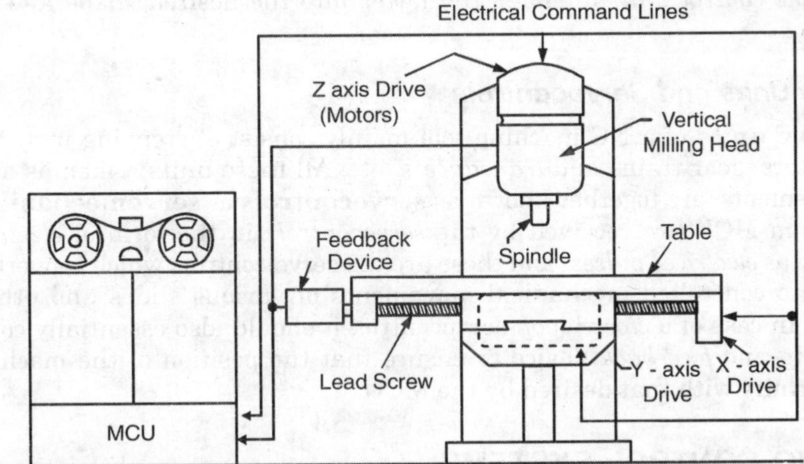

Fig. 26.5 A NC Machine Tool System, showing a NC Knee-type Vertical Milling Machine.

1. Machine Control Unit (MCU)

This is the most important unit of NC machine tool system and consists of many sub-units inside it. The first of these sub-units is a **Tape reader,** which receives the coded data from the punched tape, reads it and passes on the information to the **buffer storage** or **data buffer** via the *decoding circuits.* The function of *buffer storage* is to store the received information, till it is needed, and tansfer it fast to the required area, when needed, to ensure that the machine tool operates continuously. This unit is called the **Data Processing Unit (DPU).** This unit passes on the decoded informatin to the **control unit,** which is another sub-unit of MCU. The *control unit* directs and controls the operations of different drive units of the machine tool through *signal output channels,* which convey the instructions from the control unit to the machine tool. The control unit also receives feedback, through the *feedback devices,* to make sure that the instructions given by it are properly carried out by the machine tool.

From the above, we conclude that a *Data Processing Unit (DPU)* should contain an *input device* (tape reader), a *buffer storage, decoding circuits* and an *interpolator* to convey velocity commands. Similarly, a *Control Unit* should essentially carry *control loops* for controlling motions along different axes, *feed control loops* (if required), *controls for other functions,* viz., change of gears, start and stop of spindle, start and cut-off of coolant supply. etc. The *Machine Control Unit (MCU)* can be separate unit, as shown in Fig. 26.4, as a suspended unit like a pendant, or directly mounted on the machine tool itself. All the principal operative functions of the machine tool, viz. movements of cutting tool, change of tool to different oprations, applications of proper speeds and feeds, etc., and many other functions are controlled by this MCU only.

2. The Machine Tool

It is the principal manufacturing arm of the *NC system.* It receives the raw material and performs different machining operations over it, in accordance with the instructions conveyed by the *machine control unit,* to shape the latter into the desired shape and size of finished article.

3. The Drive Units and Servocontrols

The **drive units** of a NC machine tool mainly consist of stepping motors d.c. motors or hydraulic motors, gear trains and *transducers,* etc. All these units, taken as a group or series of different elements, are together known as **servocontrols** or **servomechanism.** The original commands from MCU are received by the *servocontrols* in the form of *electrical signals,* or more precisely as *electrical pulses,* and these are the servo-controls which convert these electrical commands into controlled mechanical movements of various slides and other parts of the machine tool. In case of a *closed-loop* servocontrol, it should also essentially contain some sort of a *monitoring* and *feed-back* device to ensure that the position of the machine tool slide is truly in conformity with that desired by the MCU.

26.11 SERVO CONTROL SYSTEMS

Servo Controls represent a group or series of electrical, mechanical, hydraulic or pneumatic devices, employed for controlling the positions of slides of a NC machine tool. The following two types of Servo Control Systems are in common use in *NC machine tools:*

1. Open-loop system
2. Closed-loop system.

Open-loop System

It is a simpler and cheaper system than the closed-loop system. Basically there is no change in the sequence of NC system, as it involves feeding *of tape,* interpreting of information by the *tape reader,* storing the information in *Buffer storage,* converting the information into *electrical signals* and sending these signals to the *control unit.* The control unit energises the *servo controls (drive units)* by sending command signals to them. Resultingly, the drive units perform certain motions to move the slides through a desired distance. However, in this system there is no provision to ensure that the slide has actually moved through the desired distance only and that it has actually acquired the desired position as a result of this movement. In other words, what it actually means is that there is no provision for *feed-back* to ensure as to whether the actual slide movement is exactly the same, as was desired, or not. And if not, what error has crept in.

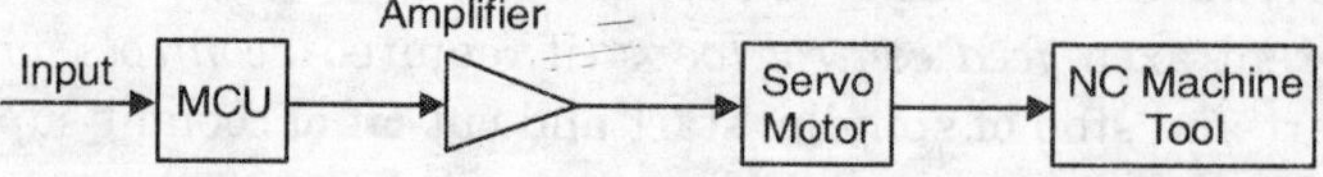

Fig. 26.6 A schematic diagram showing an Open-loop NC system.

This is the main drawback with this system. But, with the use of the **stepping motor** as a *servo-motor* this drawback has been minimised to a great extent. This motor has a specific characteristic that when *elecitric pulses* are fed into it, its output shaft will rotate through a definite angle against each pulse received by it. Thus, the accuracy of this system will entirely depend on the accuracy of the stepping motor, *i.e.,* its ability to correctly rotate its output shaft corresponding to the number of pulses given to it. To further add to the accuracy of the slide movement, *precision lead screws* are used on NC machines. Due to the constraint of limited power available from stepping motors, the use of open-loop system was largely restricted to small NC machine tools. However, with the development of more powerful *electro-hydraulic* stepping motors, this system is now being used on relatively heavier NC machine tools also. It is specially preferred in those cases where a medium size or heavy type conventional machine tool is required to be equipped with a NC system. A typical *open-loop NC system* is schematically shown in Fig. 26.6.

Closed-loop System

This system is almost similar to the open-loop system. But, it carries an additional feature in that a *feed-back device* is incorporated in its electrical circuit. This feed-back device is a *transducer* and is accompanied by a *comparator (comparison unit)*. The command signals, as usual, are sent to the servo motor by the control unit while the transducer feeds back the slide displacement corresponding to these command signals. The comparator compares the actually achieved slide positions with the command signals and the *error,* if any, is fed back to the *control unit,* via an *amplifier*. The control unit sends corrective "commands to the servo motor and this cycle continues unless the signal from feed-back unit and that from the control unit both become equal, *i.e.,* the difference between the feedback signal of actual slide position and the command signal becomes 'zero', indicating that the required adjustments have been made by the servo motor to finish the 'error'.

This system is highly accurate and the 'error', if any, is automatically adjusted. In case of *continuous path control system,* velocity control is also required because the cutting tool has to follow a definite path in accordance with the profile to be machined. A *Techogenerator* is normally used as a transducer for providing the velocity feedback in such cases. A typical *closed-loop NC system* is schematically shown in Fig. 26.7.

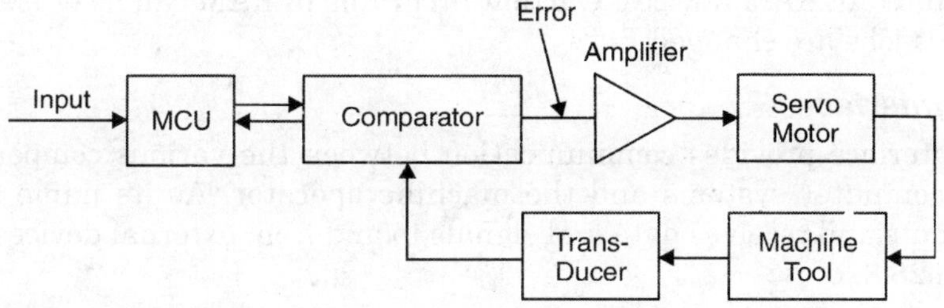

Fig. 26.7 A schematic diagram showing a Closed-loop NC system.

26.12 CNC MACHINE CONTROL UINT (MCU)

The MCU is the hardware that distinguishes CNC from conventional NC. The general configuration of the MCU in a CNC system is illustrated in Fig. 26.8. The MCU consists of the following components and subsystem:

1. Central processing unit,
2. Memory
3. Input/Output interface,
4. Control for machine tool axes and spindle speed, and
5. Sequence controls for other machine tool functions.

These subsystems are interconnected by means of a system bus, as indicated in the figure.

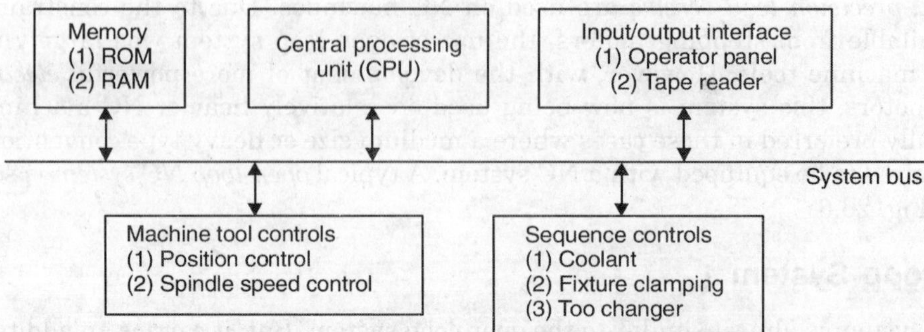

Fig. 26.8 Configuration of CNC machine control unit.

Central Processing Unit

The central processing unit (CPU) is the brain of the MCU. It manages the other components in the MCU based on software contained in main memory. The CPU can be divided into three sections: (1) control section (2) arithmetic-logic unit and (3) immediate access memory.

Memory

The immediate access memory in the CPU is not intended for storing CNC software. A much greater storage capacity is required for the various programs and data needed to operate the CNC system. As with most other computer systems, CNC memory can be divided into two categories. (1) main memory and (2) second memory. Main memory (also known as primary storage) consists of ROM (read-only memory) and RAM (random access memory) devices. Operating system software and machine interface program are generally stored in ROM. These programs are usually installed by the manufacturer of the MCU. Numerical control part programs are stored in RAM devices. Current programs in RAM can be erased and replaced by new programs as jobs are changed.

Input/Output Interface

The I/O interface provides communication between the various components of the CNC system, other computer systems and the machine operator. As its name suggests, the I/O interface transmits and receives data and signals to and from external devices, several of which are indicated in 26.8.

26.13 ADAPTIVE CONTROL

For a machining process, adaptive control is defined as the system that measures certain output variables (such as cutting force, temperature, torque, spindle motor power, tool deflector, machine vibration etc.) and uses these to control speed and/or force. In other words all output process variables can be tried to control input process variables.

Adaptive control is not appropriate for every machining situation. In general the following charactenstic can be used to identify the situation where adaptive control can be beneficially applied.

1. The in process time consumers is a significant portion of machining cycle time. Mathus uses the rule of thumb that adaptive control can best be justified when the cutter is engaged in workpiece more than 40% of the time it is on the machine.

2. There are significant sources of variability in the job for which adaptive control can compensate. In effect, AC adapts feed and /or speed of these variable conditions.
3. The cost of operating machine tool is high. The high operational cost mainly results from high investment in equipment.
4. The typical jobs are ones involving steel, titanium and high strength alloy cast iron and aluminium are also attractive candidates for adaptive control but these materials are generally easier to machine.

The adaptive control is more suitable where there are services of variability. There are a number of sources of variables in machining such as

(*i*) If there is a change in geometry of the cut in the form of changing depth or width of end, their feed rate is usually adjusted to compensate for the variation.

(*ii*) If machinability of work piece varies due to some heat treatment process then speed or feed is adjusted for this variation.

(*iii*) As the machining progresses, the tool begins to dull and the cutting force increases, then adaptive control responds to the tool wearing and reduce feed rate.

The advantage of using adaptive control technique is the more efficient utilisation of machine tool and other resources.

26.14 CLASSIFICATIONS OF CNC MACHINES

(*i*) *Classification based on feedback control:*
- (*a*) Open loop control system.
- (*b*) Closed loop control system.

(*ii*) *Classification based on servocontrol system features*
- (*a*) Point to point control system.
- (*b*) Straight line control system.
- (*c*) Continuous path or contouring control system.

1. Open Loop Control Machine

It controls when there is no provision to compare the actual position of the cutting tool or work piece with the input command value are called open loop system. In this system, electric motor continuously to run until the absence of power from input command signal, indicates that the programmed location is attained.

2. Closed Loop System

In this system, actual output from the system *i.e.*, actual displacement of the machine slide, is compared with the input signal. This system is characterized by the presence of feedback devices in the system. In closed loop system, the displacement can be achieved; to a very high degree of accuracy because measuring or monitoring is used to determine the displacement.

The feedback from the monitoring device is then compared with input signal and the slide position is regulated by the servo system until it agrees with the desired position.

3. Positional (point to point) NC Machine Tools

It is one in which slides are controlled in order to reach a particular fixed coordinate position where machining is to take place. No machining occurs until slide movement ceases.

The movement path is not critical and care must be taken when programming to ensure that the tool does not collide with the component or a projection. The table slide is displaced along both *X* and *Y* axis simultaneously. First at rapid speed and then at a slow rate until the required position is reached.

Positional control is suitable for drilling and boring machines. For applications such as drilling and tapping open loop control system is less expensive.

4. Paraxial (Straight Line) Machine Tools

The above discussed system in which the movement path between one point and the next is not critical satisfy the needs of drilling and boring machines. In other tools where vertical and horizontal milling cutting occurs, while the machine slides are moving, the movement path must follow a controlled route. To produce accurately machined surfaces:

(*i*) The slide movements must follow critically defined instructions.

(*ii*) Set up cutting process.

5. Continuous Path CNC Systems

It is one in which slide motion is more than one axis is controlled continuously and simultaneously. The system enables profiles in 2D or 3D to be contour machined and is used in milling routing = turning and grinding.

26.15 FUNCTIONS OF CNC MACHINES

Functions performed by CNC machines are:

1. Identification Function

In this function, the current value of index of performance of the system is determined, based on measurements collected from the process. Since the environment changes over time hence system performance also changes. Accordingly the identification function must be accomplished continuously over time during system operations.

2. Decision Function

Once the system performance has been determined, the next function is to decide what changes should be made to improve performance. The decision function is implemented by means of adaptive system's programmed algrorithm. Depending on this algorithm, the decision may be the change of one or more input parameters to the process, to alter some of the internal parameters of controller or to make other changes.

3. Modification Function

The third function of adaptive control is to implement the decision. Whereas decision is a logic function, modification is concerned with physicsl changes in the system. It involves hardware rather than software. In modification the system parameter or process inputs are altered using available actuator to drive the system forward a more optimal state.

26.16 COMPUTER AIDED MANUFACTURING (CAM)

In the computer aided manufacturing, a computer is added in the conventional manufacturing system so as to perform most of the activities through computers which were otherwise would have been done manually. In this system requisite informations are fed in a proper language or code in right sequence and entered in the memory of the computer of the manfacturing machine. Whenever required, the operations can also be performed manually by the operation.

Such machines can save a considerable time especially on non-machining operation like setting, adjusting, changing, fitting etc.

Advantages of CAM

In addition to automatic handling of various activities, use of these machines have following advantages:

1. Higher product quality and reliability
2. Reduced scrap
3. Complicated jigs and fixtures not required.
4. Reduced inspection time.
5. Consistency in accuracy
6. Reduction in direct as well as indirect labour cost.
7. Machining on complex shaped work-pieces is also possible.
8. It can store more than one programme in the memory. While performing operations on the work-piece, it is possible to edit data or can do other jobs on the computer.
9. It can perform several other related functions like cost estimating, process planning, process control, process monitoring, computer aided inspection, development of work standards etc.

26.17 COMPUTER AIDED DESIGN (CAD)

In this system, Design Engineer takes the help of computer using different softwares meant for different functions. The designer can test the product under various environmental conditions of temperature, mechanical stress etc. with different cross-section of the product, easily with higher reliability of the result.

Advantages of CAD

1. Less time is required for designing and drafting
2. Negligible chances of error. Error in calculation can be detected through built-in system of checks.
3. Since the designed part can be tested under severest of working conditions and wide variations of the condition, the chance that the product will fail in actual working becomes minimal.
4. CAD has wide range of applications *e.g.* design of parts, products, machine tools, equipment, construction equipment, construction works etc.

26.18 FLEXIBLE MANUFACTURING SYSTEMS (FMS)

Small batches of finished goods or components are expensive to make in traditional manufacturing facilities, because rigid production systems designed for producing high volume, low-variety products are costly and non competitive in the small batch, higher variety market place.

When customers demand greater variety of products and at shorter delivery times, flexible manufacturing systems are appropriate.

Flexible manufacturing system (FMS) can be defined as, a computer-controlled process technology suitable for producing a moderate variety of products in moderate volume. It contains several work stations, each geared to different operations. Workstation machines are automated and programmable. In this system automated material handling equipment are used to move components to the appropriate work-stations, and then onto programmed machines. After completion of one batch, the computer signals for next batch and the machine automatically re-positions and re-tools accordingly, while the just finished batch is then moved to next work station automatically.

An FMS is more flexible than conventional high-volume production systems, and less flexible than a job shop that specializes in one-of-a-kind products. Thus FMS is a 'mid-range' system suitable for moderate variety and modern volume markets.

An FMS is economical to produce a moderate variety of products in modest volume, quickly with high quality at reduced operating costs due to lower direct labour costs. But it requires very high capital investments.

ROBOTICS

26.19 INTRODUCTION

Robotics is a new technology and uses robots. "An industrial robot is programmable multifunctional manipulator designed to move materials, parts, tools or special devices through variable programmed motions for the performance of a variety of tasks". Robots improve productivity by imparting consistency in accuracy, reliability in equality, decreased rejections, reducing cost of operations, reducing errors associated with human operators etc. Recently developed robots can see, think, make facial expression. These robots have body and arm assembly, and a wrist assembly. It can perform various functions like: reach, grap, place, operate etc.

Robot can also be defined as "a device that performs functions ordinarily ascribed to human beings, or operates with what appears to be almost human intelligence." Human-like characteristics are; the robot's capability to respond to sensory inputs, communicate with other machines and make decisions.

Robotics is an automation system *because of the following reasons :*

(*i*) Robots can be substituted for humans in hazardous or uncomfortable work environments.

(*ii*) A robot performs its work cycle with a consistency and repeatability that cannot be attained by humans.

(*iii*) Robots can be reprogrammed.

(*iv*) Robots are controlled by computers and can therefore be connected to other computer system to achieve computer integrated manufacturing.

Advantages from Robotics

Some of the major advantages from robotics are:

1. Robots can do hazardous work which are otherwise very difficult for human being such as handling of heavy items, hot metal slabs handling, working in poisonous, radioactive and chemical atmosphere.
2. Robots have built in capacity for excellent repeatability which improves the accuracy in manufacture.
3. Robots can work uninterrupted round the clock.
4. Robots has consistency in quality of its working.
5. It is more reliable than most of the workers and saves labour cost.
6. Can perform dull or monotonous jobs.
7. Can be used for lifting and moving heavy jobs.

26.20 APPLICATIONS OF ROBOTS

Industrial robots can be used in following fields:

1. *Handling*
   - (*a*) Loading
   - (*b*) Unloading
   - (*c*) Palletizing
   - (*d*) Picking and Placing
2. *Manufacturing*
   - (*a*) Cutting
   - (*b*) Forming
   - (*c*) Coating
   - (*d*) Painting
   - (*e*) Forgoing
   - (*f*) Heat treatment
3. *Assembling*
   - (*a*) Mating
   - (*b*) Gluing
   - (*c*) Welding
   - (*d*) Screw driving
   - (*e*) Rivetting
4. *Testing*
   - (*a*) Gauging
   - (*b*) Varifying
   - (*c*) Inspecting
5. *Other fields*
   - (*a*) Packaging
   - (*b*) Bottling
   - (*c*) Fire fighting
   - (*d*) Pouring hot metals into moulds
   - (*e*) Inserting parts into furnaces
   - (*f*) Forging
   - (*g*) Press working
   - (*h*) Heat-treating
   - (*i*) Laser cutting

Common Applications of Robots

More common applications of robots are discussed hereunder:

1. Material Handling

Material Transfer applications are those in which the robot is used to move workparts from one location to another. In some cases a reorientation of the part may be required in this material handling function. Examples of material transfer robot operations include the following:

(*i*) *Simple pick-and-place operations.* Transfer of workparts from one conveyor to another conveyor.

(*ii*) Pelletizing operations, in which the robot takes parts from a conveyor and loads them onto a pallet in a required pattern and sequence.

(*iii*) *Stacking operations, similar to pelletizing.* Loading parts from a conveyor into cartons or boxes (similar to palletizing)

(*iv*) Depalletizing operations, in which the robot takes parts which are arranged on a pallet and loads them onto a conveyor. Material transfer operations are often among the earliest and most straight forward of robot applications (*e.g.*, pick-and-place, transfer from conveyor to conveyor).

2. Machine Loading

In the typical application, the robot would grasp a raw workpart from a conveyor and load it into the machine. In some cases, the robot holds the part in position during processing. When processing is completed, the robot unloads the part from the machine and places it onto another conveyor.

Production operations in which robots have been successfully applied to perform the machine loading and unloading function include the following:

(*i*) Die casting;

(*ii*) Injection (plastic) moulding;

(*iii*) Transfer (plastic) moulding,

(*iv*) Hot forging,

(*v*) Upsetting or upset forging;

(*vi*) Stamping press operations;

(*vii*) Machine operations such as turning and milling.

In some of these operation (die casting and plastic moulding), the robot only unloads the finished parts. For machining processes, the robot both loads and unloads the machine tool. In upsetting and stamping operations, the robot holds the workpart while it is being processed by the machine.

3. Welding

The welding processes are a very important application area for industrial robots. The applications logically divide into two basic categories, spot welding and arc Welding.

***(i)* Spot Welding.** Because of its widespread use in the automobile industry for car body fabrication, spot welding represents one of the most common applications of industrial robots today. The end effector is the spot welding gun used to pinch the car panels together and perform the resistance welding process. The welding gun used for automobile spot welding is typically heavy. Prior to the application of robots, human workers performed this operation, and the heavy welding tools were difficult for humans to manipulate accurately. As a consequence, there were many instances of missed welds, poorly located welds and other defects, resulting in overall low quality of the finished product. The use of industrial robots in this application has dramatically improved the consistency of the welds.

***(ii)* Arc Welding.** Arc welding is used to provide continuous welds rather than individual spot welds at specific contact points. The resulting arc welded joint is substantially stronger than available in spot welding. Since the weld is continuous, it can be used in airtight pressure vessels and other weldments in which strength and continuity are required.

Industrial robots are used to automate the arc welding process. The cell consists of the robot, the welding apparatus (power unit, controller, welding tool, and wire feed mechanism), and a fixture that positions the components for the robot. The fixture might be mechanized with one or two degrees-of-freedom so that it can present different portions of the work to the robot for welding (the term *positioner* is used for this type of fixture). For greater productivity, a double fixture is often used so that a human helper can unload the completed job and load the components for the next work cycle while the robot is simultaneously welding the present job.

4. Spray Coating

Many large consumer products (*e.g.*, automobiles and appliances) and most industrial products require the application of some form of paint. When human workers apply this paint, the most common method is spray painting. However, the spray painting process poses certain health hazards to the human operator.

For the above and other reasons, specialised industrial robots are being used more and more frequently to perform spray painting and related processes. Spray painting requires a robot capable of executing a smooth motion pattern which will apply the paint or other fluid evenly and avoid runs. To accomplish this, the robot is equipped with continuous path control. The paint spray nozzle becomes the end effector.

5. Assembly

Assembly operations are seen as an area with big potential for robot applications. Batch-type assembly operations seem to offer the most promise for using robots. The reason for this is based on economics and the technological capabilities of the robot. For mass production assembly, the most economical method involves fixed automation, where the equipment is designed specially to produce the particular product. A robot would probably be too slow for mass production.

6. Inspection

Like assembly, inspection is a relatively new area for the application of industrial robots. Traditionally, the inspection function has been a very labour intensive activity. The activity is slow, tedious, and boring, and is usually performed by human beings on a sampling basis rather than by 100% inspection. With ever.'increasing emphasis on quality in manufacturing, there is a trend towards automating the inspection process and towards the use of 100% inspection by machines instead of sampling inspection by human beings. An important role in this area of inspection automation will be played by industrial robots.

26.21 DRIVE SYSTEM FOR ROBOTS

Robot joints are actuated using any of three possible types of drive systems: (1) electric, (2) hydraulic, or (3) pneumatic. Electric drive systems use electric motors as joint actuators (*e.g.*, servomotors or stepping motors, the same types of motors used in NC positioning systems. Hydraulic and pneumatic drive systems use devices such as linear pistons and rotary vane actuators to accomplish the motion of the joint.

Pneumatic drive is typically limited to smaller robots used in simple material transfer applications. Electric drive and hydraulic drive are used on more-sophisticated industrial robots.

Electric drive has become the preferred drive system in commercially available robots, as electric motor technology has advanced in recent years. It is more readily adaptable to computer control, which is the dominant technology used today for robot controllers. Electric drive robots are relatively accurate compared with hydraulically powered robots.

By contrast, the advantages of hydraulic drive include greater speed and strength.

The drive system, position sensors (and speed sensors if used), and feedback control systems for the joints determine the dynamic response characteristics of the manipulator. The speed with which the robot can achieve a programmed position and the stability of its motion are important characteristics of dynamic response in robotics. *Speed* refers to the absolute velocity of the manipulator at its end-of-arm. The maximum speed of a large 'robot is around 2 m/sec (6 ft/sec.). Speed can be programmed into the work cycle so that different portions of the cycle are carried out at different velocities. What is sometimes more important than speed is the robot's capability to accelerate and decelerate in a controlled manner.

In many work cycles, much of the robot's movement is performed in a confined region of the work volume; hence the robot never achieves its top-rated velocity. In these cases, nearly all of the motion cycle is engaged in acceleration and deceleration rather than in constant speed. Other factors that influence speed of motion are the weight (mass) of the object that is being manipulated and the precision with which the object must be located at the end of a given move. A term that takes all of these factors into consideration is *speed* of response, that refers to the time required for the manipulator to move from one point in space to the next. Speed of response is important because it influences the robot's cycle time, that in turn affects the production rate in the *application*.

Stability refers to the amount of overshoot and oscillation that occurs in the robot motion at the end-or-arm as it attempts to move to the next programmed location. More oscillation in the motion is an indication of less stability. The problem is that robots with greater stability are inherently slower in their response, whereas faster robots are generally less stable.

26.22 STRUCTURE (COMPONENTS) AND OPERATION OF ROBOTS

The basic elements (Components) of industrial robots are discussed here under (see Fig 26.9):

1. Power Source

The power supply is the source of energy used to move and regulate the robot's drive mechanisms. The energy comes from three sources: *electric, hydraulic,* and *pneumatic*. Electric drives are clean and quiet with a high degree of accuracy and repeatability. They also offer a wide range of payload capacity, accompanied by an equally wide range of costs.

Hydraulic drives, today's most popular, have high payload capacities, and are relatively easy to maintain, they are, however, rather expensive and not as accurate as either the electric or pneumatic drives.

Pneumatic drives, although limited to smaller payloads, are relatively inexpensive, fast and reliable.

2. Controller

The controller acts like a brain of robot. It performs the function of storing and sequencing data in memory, initiating and slopping the motion of the manipulator, and interacting with the environment.

Controller directs the movements and operations performed by the manipulator. Controller may be either made as an integral part of the manipulator or may be housed in a separate cabinet.

The machine language instructions to the robot, to perform the desired tasks automatically, are input through the keyboard of this unit. The mass storage media devices like floppy drive, magnetic cassete recorder and main frame disk drive can also be used to feed the programmes to the controller.

The controller converts the input programme to suitable signals which actuate the manipulator to perform the desired tasks. It can be equipped with process monitoring capability also.

3. Sensors

As discussed above, the process monitor can 'sense' the information about the desired process parameter, make decisions on the basis of the sensed information, and take suitable action. The current technology relies upon computer and on microprocessor to achieve the control function of the robot control unit (RCU).

Sensors are used to sense the environment, which is essential for intelligent robot.

4. Manipulator

The manipulator is a series of the mechanical linkages and joints capable of movement in various directions to perform the work task.

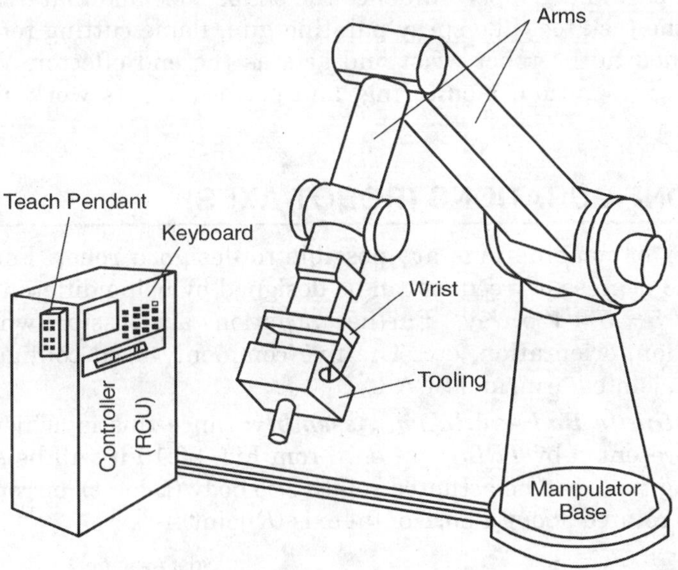

Fig. 26.9 A typical line diagram of a robot.

The manipulator is an assembly of axes capable of motion in various directions. The "wrist" located at the end of the robot arm has from one to three degrees of freedom, depending on the model or make. There is no standard terminology in the industry as yet, but these degrees of freedom may be termed as pitch, yaw, and roll axes or bend, yaw, and swivel axes.

The manipulator is powered by hydraulic or pneumatic cylinders or rotary actuators, and electric or hydraulic motors the manipulator are feedback devices that sense and measure the position and, in some cases, the velocity of each of the various axes of motion and send this information to the control systems for use in coordinating the robot motions (Fig. 26.8).

Main components of manipulator are base, arm, and wrist of the robot joined in resemblance to human arm. On receiving signals from controlling unit, this mechanical unit performs movement functioning desired degrees of freedom. The movement of the manipulator can be described in relation to its coordinate system which may be cartesian, cylindrical, spherical, or articulated. Depending on the controller movement may be point to point motion or a continuous path motion.

The actuation systems which are responsible for converting command signals into desired movements could be closed loop (servo-control) type or open loop type.

5. End Effectors (Tooling)

It is the hand or the gripper of the robot, often called the "end effector".

At the end of the manipulator wrist is a device related to the task that must be accompanied by the robot, called as *end effectors.* End effector are usually either.

(*i*) Gripper for holding a work part, or

(*ii*) A tool for performing some process.

The body and arm of the robot is used to position of end effectors and the robot wrist is used to orient the end effecter.

The design of the end-effector is governed by the nature of the task to be performed by the robot. In part handling applications, nature, geometry, weight and orientation of the part and friction between part and gripper influence the shape, size and material of the gripper. In many applications, the task tool like spray painting gun, flame cutting torch, drilling spindle etc. is directly fastened to the robot wrist and acts as the end effector. Various sensors are provided to help the recognition, monitoring, and performing its work similar to a human operator.

26.23 ROBOT CONFIGURATIONS (ROBOT AXES)

A large number of combinations are possible to design a robot. For example, a robot manipulator having 3 degrees of freedom can be designed by using different combinations of 5 types of joints in 5 × 5 × 5 = 125 ways. Further variations are possible when we consider the sizes, ranges of motion, orientation, etc. The five commonly used configurations are polar, cylindrical, cartesian, jointed-arm, and SCARA.

1. *Polar Coordinate Body-and-Arm Assembly:* Since this is achieved by three joints *T*, *R* and *L*, it is represented by *TRL* notation. From Fig. 26.10 it will be seen that the wrist assembly arm (sliding type) can be actuated relative to body (*L*-joint), or rotated about vertical axis (*T*-joint), or also rotated about a horizontal axis (*R*-joint).

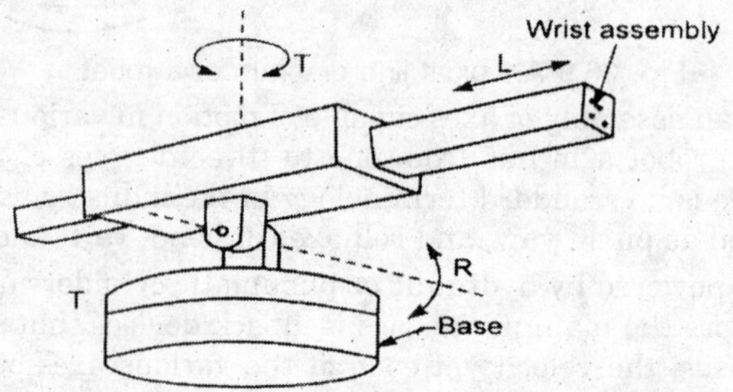

Fig. 26.10 Polar coordinate body-and-arm assembly.

2. *Cylindrical Configuration:* It can be achieved by various combinations like *TLO*, *LVL*. From Fig. 26.11 it will be seen that the arm assembly mounted on a vertical column can be moved up and down (*L*-joint), or rotated about the column (*T*-joint), or the end-or-arm could be moved in or out relative to the axis of the column (*O*-joint).

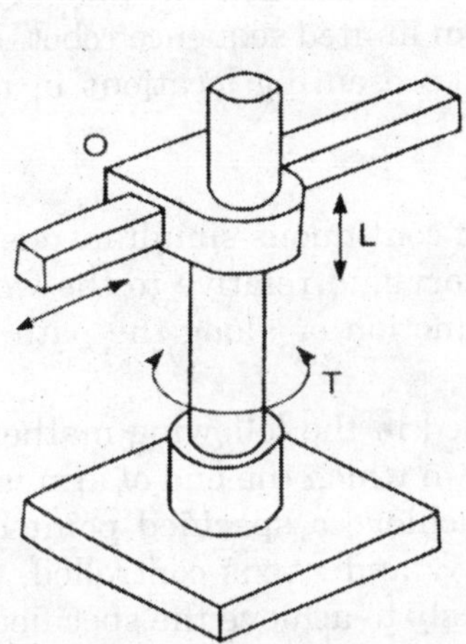

Fig. 26.11 Cylindrical configuration

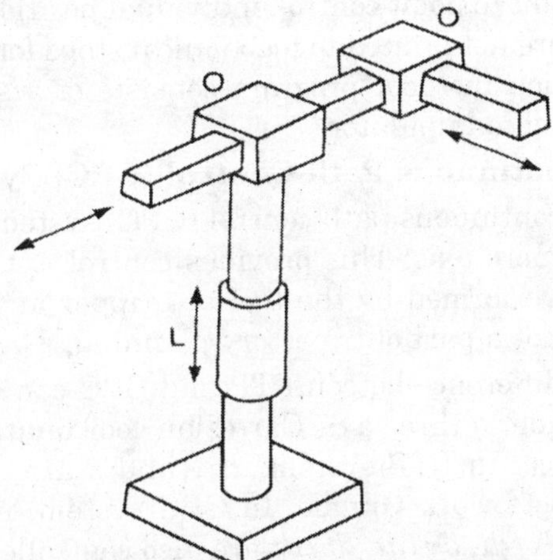

Fig. 26.12 Cartesian coordinate body-and-arm assembly.

3. *Cartesian Coordinate Body-and-Arm Assembly:* (Refer Fig. 26.12). Since all motions in this case are linear type, it is also called rectilinear robot or *X-Y-Z* robot. It incorporates three linear sliding joints, with two being orthogonal and thus represented by *LOO* notation.

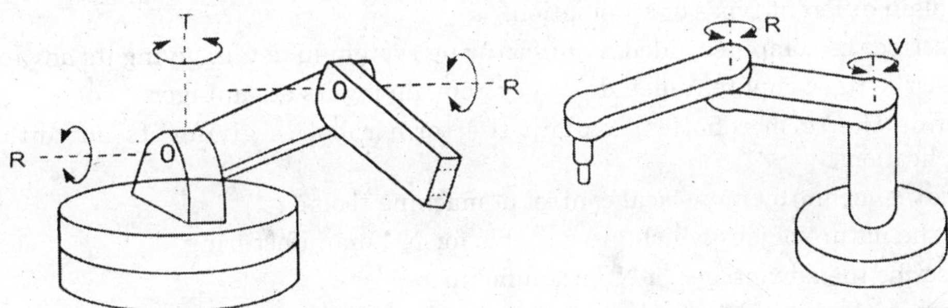

Fig. 26.13 Jointed-arm body-and-arm assembly.

Fig. 26.14 SCARA body-and arm assembly.

4. *Jointed-arm Body-and-Arm Assembly:* (Refer Fig. 26.13). It resembles the configuration of a human arm. Its arm has a shoulder joint and an elbow joint. The arm can be swivelled about the base by combinations of *TRR* or *VVR* joints.

5. *SCARA Body-and-Arm Assembly (Selective Compliance Assembly Robot Arm) can be achieved by VRO joints:* [Refer Fig. 26.14]. This is similar to the jointed arm robot except that the shoulder and elbow rotational axes are vertical. It is very well suited to perform insertion tasks such as for assembly in a vertical direction requiring side to side adjustment to make the two parts properly. Because of minimal orientation requirements, wrist assembly can be avoided.

26.24 MOTION CONTROL

Motion control for robotics can be divided into following two types:

1. Point to Point (PTP) Movement

In point to point control, individual positions of robot arm are recorded into memory. These positions are not limited to mechanical stops for each joint as in limited sequence robot. Instead, each position in robot program consists of a set of values representing locations in range of each joint of manipulator.

2. Continuous Path Control (CPC) System

The continuous path control (CPC) systems is capable of continuous simultaneous control of two or more axis. This provides control of the tool trajectory and relative to the workpart. The task performed by the tool or gripper at a specific TP motion or along the path in CPS motion is not a part of trajectory planning.

The difference beteen PTP and CPC can be distinguished in the following mathematical way. Consider a three axis Cartesian coordinate manipulator in which the end of arm is moved in *xyz* space. In PTP system, x, y and z are controlled to achieve a specified point location within robots work volume. In CPC system, not only are x, y and z axis controlled, but the velocities dx/dt, dy/dt, dz/dt are also controlled simultaneously to achieve the specified linear or curvilinear path. A playback a robot with CPC has the capability for PTP control.

QUESTIONS

26.1 Define the term 'Automation'. How and where it is applicable?

26.2 Discuss the effect of automation on the workers.

26.3 Explain different types of automation.

26.4 Describe the computer aided manufacturing system in detail, giving its advantages.

26.5 Describe the computer aided design system, giving its advantages.

26.6 Define the term robotics, explain the term robotics giving its advantages and applications.

26.7 What is meant by numerical control of machine tools?

26.8 Lit the nature of jobs which are suitable for NC manufacturing.

26.9 Describe the advantages of NC manufacturing.

26.10 Compare the open loop and closed loop types of NC systems.

26.11 Explain the principle of operation of NC feed-back devices.

26.12 What is meant by CNC? Explain.

26.13 How does the operation of a CNC controller differ from that of a hardwired MCU?

26.14 Describe in detail the advantages of CNC over NC.

26.15 What are the advantages of adaptive control (A.C)?

26.16 What is the need of DNC? How can it benefit a CNC shop floor manufacturing?

26.17 Explain various advantages of DNC.

26.18 Describe configuration of a modern CNC system.

26.19 List various applications of robots.

26.20 Describe various functions performed by the controller and manipulator of an industrial robot.

26.21 What is meant by the tooling of a robot? Explain.

26.22 Illustrate with the help of sketches the functions of different types of tooling commonly used in industrial robot.

26.23 How are the axes of robots defined? Show with the help of sketches the working of various types of robots.

26.24 What is an automation? What are the basic element with a block diagram.

26.25 What are the various levels of automation applied to manufacturing?

26.26 List out various benefits of automation. What are the different types of automation employed in manufacturing? Explain briefly.

26.27 How are the CNC machines normally classified?

26.28 Distinguish between fixed and flexible automation.

26.29 What are the advantages and limitations of automation?

26.30 State the advantages and disadvantages of NC system.

26.31 What are the basic components of a NC system?

26.32 Distinguish between NC, CNC and DNC with a block diagram.

26.33 Explain the special features of CNC.

26.34 Discuss briefly with a neat sketch of the circuit diagram, a closed loop feedback control system used in CNC machine.

26.35 Define a robot. What are the different robot configurations?

26.36 What is the status of robot in the family of flexible automation technologies? Is robotics an automation? If yes why?

26.37 Discuss the structure and operation of robots.

26.38 What are the basic components of a robot system?

26.39 What are general characteristics of an industrial work situation that tend to promote the application of robots?

26.40 Explain the common applications of robots.

APPENDIX–I

S.I. Units Names and Symbols

Physical quantity		*Unit*		*Dimensions*
Term	*Symbol*	*Name*	*Symbol*	
Basic units				
Length	l	metre	m	L
Mass	m	kilogramme	kg	M
Time	t	second	s	T
Thermodynamic temperature	T	kelvin	K	θ
Electric current	I	ampere	A	A
Luminous intensity	I_v	candela	cd	
Amount of substance	n	mole	mol	
Supplementary units				
Plane angle	$\alpha, \beta, \theta, \phi$	radian	rad	
Solid angle	Ω, ω	steradian	sr	

*Note.—S.I. system of units is an abbreviation for Systeme Internationale d' Unite's adopted by the General Conference of Weights and Measures in 1960. It is a *coherent* system. In a coherent system all derived unit quantities are formed the product or quotient of other unit quantities.

Velocity = length/time (m/s), volume = length × length × length (m^3)

Force (F) = mg

APPENDIX–II

Important Derived SI Units

S. No.	Physical quantity: Term	Physical quantity: Symbol	Unit: Name	Unit: Symbol	Dimensions
1.	Calorific value, mass	*cv*	joule/kilograme	J/kg.	$L^1 T^{-2}$
2.	Concentration	*x*	kilogramme/kilogramme	kg/kg	—
3.	Conductance, thermal (coefficient of heat transfer)	*h*	watt/metre^2 kelvin	W/m^{2}K (=J/s m^{2}K)	$MT^{-3} \theta^{-1}$
4.	Conductivity, thermal	*k*	watt/meterkelvin	W/mk (=J/s m K)	$MLT^{-3} \theta^{-1}$
5.	Density	ρ	kilogramme/metre^3	kg/m^3	ML^{-3}
6.*	Energy, enthalpy, work	*U, E, H, W*	joule (newton metre)	J(= Nm)	ML^2T^{-2}
7.	Entropy	S	joule/kelvin	J/K	$ML^2T^{-2} \theta^{-1}$
8.*	Frequency	*f*	hertz	HZ	T^{-1}
9.*	Force	*F*	newton	N(= kgm/s^2)	MLT^{-2}
10.	Gas constant	*R*	joule/kilogramme kelvin	J/kg K	$L^2 T^{-2} \theta^{-1}$
11.	Molar mass (molecular weight)	$M(=m/n)$	Kilogramme/mole	kg/mole	
12.*	Moment(vector product)	*M*	newton metre	N m	$ML^2 T^{-2}$
13.	Power	*P*	watt	W (= J/s)	ML^2T^{-3}
14.	Pressure, stress	*p*	pascal (newton/metre^2)	Pa	$ML^{-1} T^{-2}$
15.	Specific heat	*c*	joule/kilogramme kelvin	J/kg K	$L^2T^{-2} \theta^{-1}$
16.	Torque	*T*	newton metre	N m	ML^2T^{-2}
17.	Velocity, linear	*U, V, W, C*	metre/second	m/s	LT^{-1}
18.	Viscosity,dynamic (absolute)	μ	newton second/metre^2	Ns/m^2(kg/ms)	$ML^{-1} T^{-1}$
19.	Vicosity, kinematic	*v*	metre^2/second	m^2/s	$L^2 T^{-1}$
20.	Weight	*W*	newton	N	MLT^{-2}
21.	Specific weight	*w*	newton/metre^3	N/m^3	$ML^{-2} T^{-2}$

*These are derived units with special names and should be avoided wherever possible.

Note.—As joule is a small unit values are generally given in kilo joules (1 kJ = 1000 Joule).

INDEX

A

B

C

D

E

F

G

H

I

J

K

L

M

N

O

P

R

S

W

Z